Processes of Manufacturing

by

R. Thomas Wright
Professor, Industrial Education and Technology
Ball State University
Muncie, Indiana

South Holland, Illinois
THE GOODHEART-WILLCOX COMPANY, INC.
Publishers

Library of Congress Catalog Card Number 86-29563
International Standard Book Number 0-87006-633-1

123456789-87-43210987

Library of Congress Cataloging in Publication Data

Wright, R. Thomas.
Processes of manufacturing.

Includes index.
1. Manufacturing processes. I. Title.
TS183.W745 1987 670 86-29563
ISBN 0-87006-633-1

INTRODUCTION

PROCESSES OF MANUFACTURING is designed for the student seeking a comprehensive introduction to various methods for processing metallic, polymeric, and ceramic materials. The content is organized around the major families of processes: casting and molding, forming, separating, conditioning, assembling, and finishing.

Major elements common to all processes within the family of processes are explored first. Then, a wide range of specific methods to process various industrial materials is presented for each family.

Additional information is presented about manufacturing, industrial materials, automation, process planning, and quality control. This information provides support for a basic understanding of manufacturing material processing.

PROCESSES OF MANUFACTURING is organized and presented in an easy-to-understand format. Ample illustrations are used to support the simple and direct explanation of each concept or process. Study questions at the end of each chapter aid you in evaluating your progress.

R. Thomas Wright

TABLE OF CONTENTS

SAFETY CAUTION

The purpose of this text is to introduce you to the processes of manufacturing. It is not a manual of operation or a set of procedures. Therefore, safety cautions may not be placed in every chapter.

Many of the processes and operations described here will cause injury if performed in an unsafe manner. You are cautioned never to use tools, operate machines, handle materials, or attempt any procedure without making certain that you:

1. Are aware of the dangers represented by the materials, tools, equipment or procedures.
2. Are wearing proper clothing and/or appropriate personal safety equipment.
3. Have received safety instruction from a qualified person or instructor.
4. Have the ability to work safely, following approved procedures.
5. Have followed all applicable federal, state, local, and industrial safety laws, guidelines, regulations, and suggestions.

SECTION 1
INTRODUCTION TO MATERIAL PROCESSING

The manufacturing process comprises three major elements:

- Changing the form of materials to make them suitable for other processing or to make them useful products.
- Providing a structure, such as a company, which efficiently organizes resources and uses them to produce products.
- Raw materials which can be processed into useful products.

Chapter 1

NATURE OF MANUFACTURING

Manufacturing converts raw materials into industrial standard stock which is further processed into finished products. This activity is usually done in one location (a factory). The output (finished goods) is then transported to the point of use.

The manufacturing industry is made up of a number of subindustries. Included in this group would be the automotive industry, the steel industry, the furniture industry, and the home appliance industry.

Each of the individual enterprises which make up the manufacturing industry and its subindustries is a business. They are organized and managed to change material into more useful shapes, compositions, or combinations. They may be in business to change trees into lumber and other standard stock, or lumber into drawer parts, or to assemble numerous parts into furniture. This change is called *form utility*. Each new form the material takes on gives it more use (utility).

Changing the form of materials is done by a series of acts called processes. There are, as seen in Fig. 1-1, two major types of manufacturing processes:

1. Primary processes.
2. Secondary processes.

PRIMARY PROCESSING

Primary processing changes raw materials into standard industrial stock. Raw materials, Fig. 1-2, must first be taken from the earth. Those from inside the earth are *extracted* through mining and drilling activities. Other raw materials may grow to maturity and be *harvested*. Such products are the output of fishing, farming, or forestry activities.

Raw materials can be mineral ores, hydrocarbon liquids, plants, animals, or any of a number of basic material inputs. These raw materials are subjected to various primary processing activities. Metal ores are smelted to separate the metal from impurities. Petroleum is distilled into its various fractions such as gasoline and kerosene. Sea water can be subjected to electrolysis which extracts magnesium and other valuable metals.

The output of primary processing, as seen in Fig. 1-3, is industrial standard stock. Typical of this output are sheets, bars, and rods of steel; boards of lumber; pellets of plastic; lumps of clay; and glass marbles.

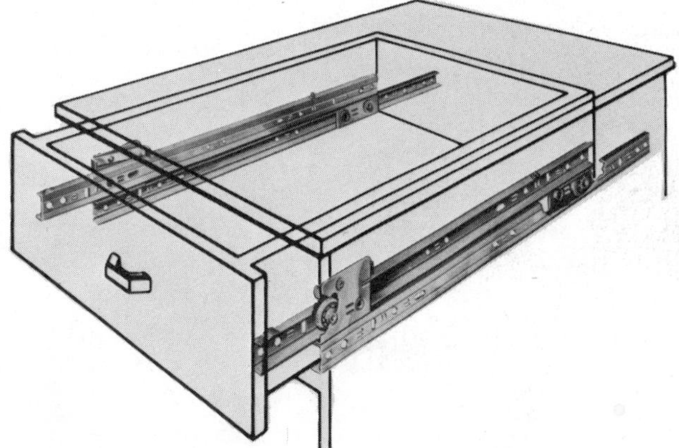

Fig. 1-1. Left. Primary processing produces standard products such as sheet metal. Right. Secondary processing produces products such as drawer guides from sheet metal. (United States Steel, Knape & Vogt)

Fig. 1-2. Raw materials are produced by various activities. A—Mining. B—Harvesting. C—Drilling. (AMAX Inc., Weyerhaeuser Co.)

Fig. 1-3. Primary processing produces material (standard stock) which meets stated specifications. The wire shown is an example of a standard stock. (Alcan Aluminum)

All of these, and the many other outputs of primary processing, need further changes in form to be useful. Plywood must be made into kitchen cabinets, sheets of steel into waste paper baskets, lumps of clay into dinnerware, and ingots of aluminum into siding. Only then is the material truly useful.

SECONDARY PROCESSING

Secondary processing is "form utility" activity. It changes standard stock into useful finished products.

The standard stock may be one of three types of material:

1. Metals (steel, aluminum, copper, etc.).
2. Polymers (wood, plastic, etc.).
3. Ceramics (clay, glass, cement, etc.).

These materials are changed in shape, appearance, and internal properties through six types (or families) of secondary processing:

1. Casting and molding.
2. Forming.
3. Separating.
4. Conditioning.
5. Assembling.
6. Finishing.

These are not pure processes. During the forming activity, material conditioning often occurs. However, the conditioning is not the primary goal of the forming act. Likewise, some finishing activities will remove material. The arrangement of activities by process family does, however, provide a useful way to study material processing. Similarities and differences among related processing techniques are more readily seen when they are studied as a group. Fig. 1-4 shows the relationship between primary processing and secondary processing.

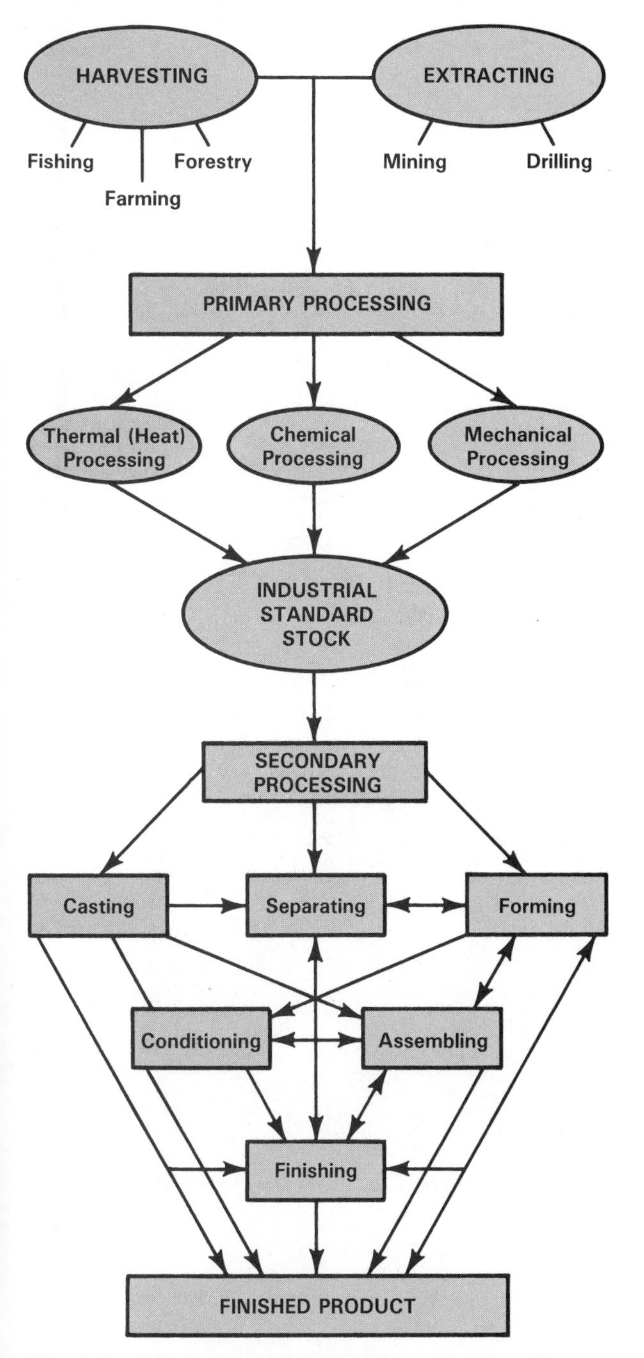

Fig. 1-4. Structure of manufacturing processes. Do you see the relationship between primary and secondary processing?

INTRODUCTION TO SECONDARY PROCESSING

Each of the six secondary process families are different. They have their own common elements. Also, they contribute in their own way to the *form utility* of the material. (Form utility refers to the changes made in materials so they become more useful to the purchaser.)

CASTING AND MOLDING

Casting and molding techniques hold liquid or semiliquid (plastic) materials in a mold cavity. While in the cavity, the material is allowed or caused to harden. The finished product—a casting or molded item—is then removed from the mold.

Casting and molding can produce both simple and complex shapes in a single step. Fig. 1-5 shows the basic principle of casting.

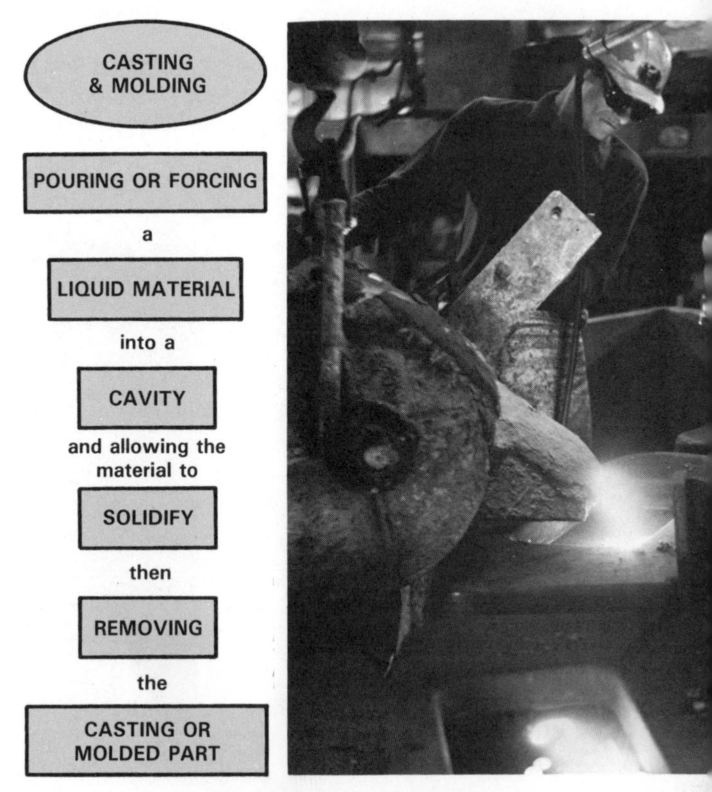

Fig. 1-5. Fundamentals of casting. Simple or complex shapes are produced in single step. (Crouse-Hinds Co.)

Many ceramic, metallic, and polymeric materials may be cast or molded. Castings and molded pieces are produced by either expendable (used one time and destroyed) mold or permanent (long life) mold techniques. Common techniques which will be discussed in Chapters 4 through 8 are shown in Fig. 1-6.

MATERIAL TO BE CAST		TYPE OF CASTING OR MOLDING	
		Expendable Mold Techniques	**Permanent Mold Techniques**
MATERIAL TO BE CAST	**Metal**	Green sand casting Dry sand casting Shell mold casting Full mold casting Investment casting Plaster casting	Permanent mold casting Slush casting Die casting Centrifugal casting
MATERIAL TO BE CAST	**Plastic**		Gravity casting Injection molding Compression molding Transfer molding Dip casting Slush molding Foam molding
MATERIAL TO BE CAST	**Ceramic**		Drain casting Solid casting Centrifugal casting Permanent mold casting

Fig. 1-6. Casting may be done in two different mold types. One is expendable (destroyed in removing the cast part). The other type is permanent (used over and over).

FORMING

Forming techniques, Fig. 1-7, use a *shaping device* and *pressure* to cause material to take on a new size and shape. The material may be placed between or on a die or mold. Pressure is then applied. Under this pressure the material will take on the shape of the die or mold.

Also, materials may be fed between straight or contour rolls. Again, pressure is applied. The material is bent by this pressure as it passes through the rolling machine.

Forming processes must also account for *material temperature*. Typically, forming processes are grouped as either cold forming or hot forming. The material is either worked at a high temperature or at near-room temperatures.

All forming operations change the size and/or shape of the material. They, however, do not change the volume or weight of the material. The same amount of material is present before and after the forming action.

Forming techniques have been developed to process metal, ceramic, and plastic materials. Fig. 1-8 lists the basic processes which are discussed in Chapters 9 through 13.

SEPARATING

Separating removes excess material to produce the desired size, shape, feature (holes, notches, etc.), and surface finish. Separating may be done by cutting

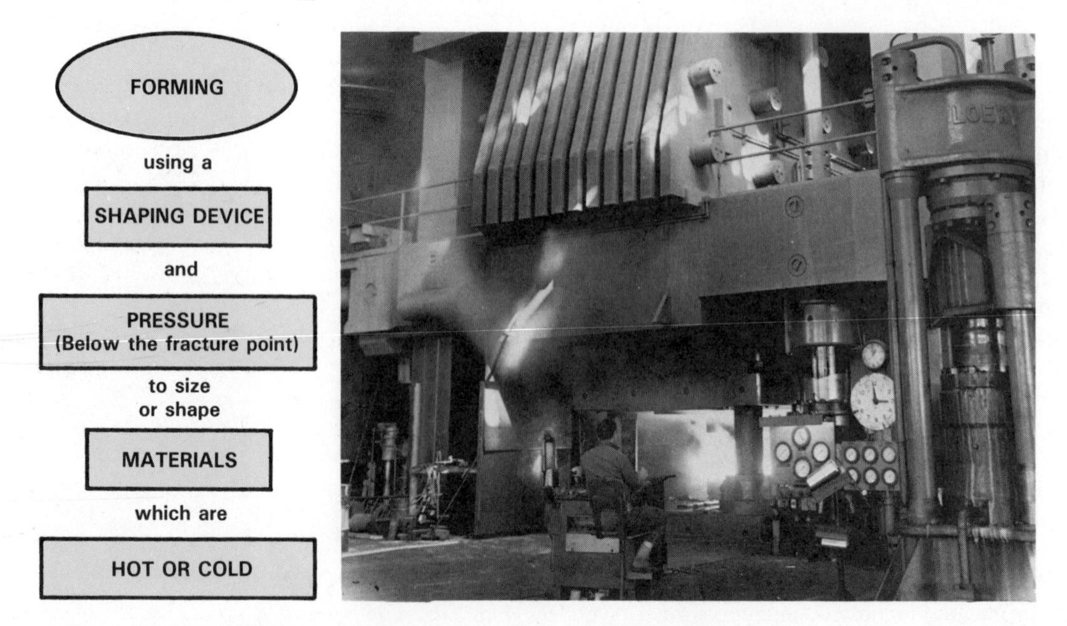

Fig. 1-7. Fundamentals of forming. This press forges parts at extremely high pressures. Forging is a type of forming. (Wyman-Gordon Co.)

MATERIAL	TYPE OF FORMING	
	Cold Forming	**Hot Forming**
Metals	Bend Forming Angle bending Roll bending Seaming Straightening	Forging Extruding Roll forming Piercing Drawing Spinning
Metals	Draw Forming Sheet drawing Bar, wire, and tube drawing Spinning Stretch forming Embossing High velocity forming	
Metals	Squeeze Forming Swaging Sizing Coining Hobbing Staking and riveting Thread rolling Extrusion	
Plastic	Mechanical Forming	Thermoforming Extrusion Blow molding Calendering
Ceramic	Dry pressing Extrusion Jiggering	Pressing Drawing Blow molding

Fig. 1-8. Forming can be done on the workpiece either hot or cold.

(machining) or shearing away the excess material. These processes are shown in Fig. 1-9.

Machining may be done by three common ways. First, the extra material may be removed as chips. This method is sometimes called *traditional* or *conventional* machining.

A second type of machining is called *chipless* or nontraditional machining. This group of techniques uses streams of electrons (high voltage spark), chemicals, or electrochemical action to remove material.

The third type of machining uses *thermal* (heat) cutting to size and shape materials. A flame, a combination of a flame and a high pressure stream of air, or intense light (laser) may be used to cut the material.

Shearing is the second major way to separate material. Shearing fractures (breaks) material along a line between two opposing edges (blades, knives, etc.). Shearing may produce straight-line or curved cuts depending on the shape of the shearing device.

The separating processes may be grouped by the major techniques used. These, which will be discussed in Chapters 15 through 23, are:

1. Turning.
2. Milling.
3. Sawing, filing, and broaching.
4. Shaping and planing.
5. Drilling and related operations.
6. Abrasive machining.
7. Nontraditional machining processes.
8. Thermal cutting.
9. Shearing.

Each of these groups of operations will be viewed from its basic principles including:

1. The *cutting element* (tool, flame, electrical spark, or other agent) used.
2. Cutting and feed *motion* present.
3. *Support* required for the work and the cutting element.

Also, the basic machines and specific operation that are included in each grouping will be discussed.

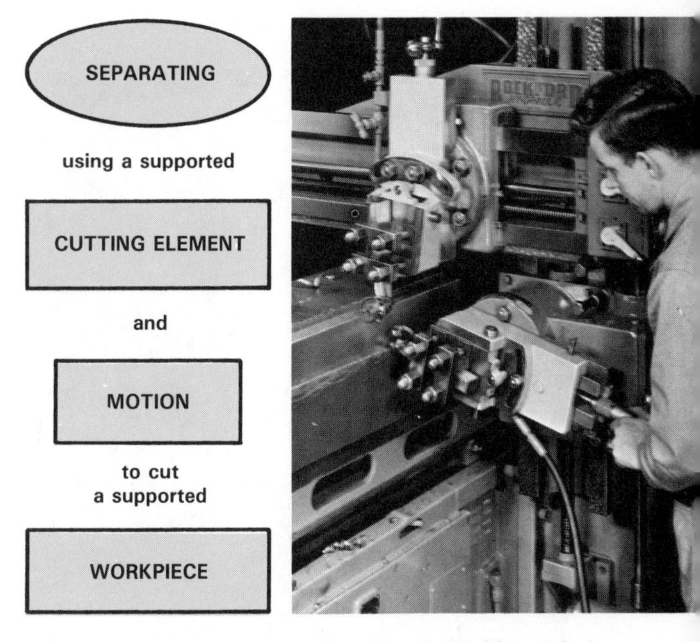

Fig. 1-9. Fundamentals of separating. Material is removed from the workpiece. (Rockford Machine Tool Co.)

CONDITIONING

Conditioning processes, outlined in Fig. 1-10, are designed to change the mechanical properties of the material. The process uses heat, chemical action, or mechanical means to change the hardness, ductility, elasticity, or other properties of the material. These changes usually are not seen but are internal structural changes within the material.

Fig. 1-10. Left. Fundamentals of conditioning. Right. Belt sintering furnaces. (Drever Co.)

The typical conditioning techniques used include:

1. Heat treating.
   a. Hardening.
   b. Tempering.
   c. Normalizing.
   d. Annealing (softening).
   e. Drying.
   f. Firing.
2. Mechanical conditioning.
   a. Cold working.
   b. Shot peening.
3. Chemical conditioning.

These conditioning activities will be discussed in Chapters 24 and 25.

ASSEMBLING

Assembling, Fig. 1-11, is the joining of two or more parts or assemblies. Assembly techniques may be permanent or temporary. They may hold the parts together so that they will not come apart. Or they may be designed for easy assembly and disassembly.

Assembling is basically done by *bonding* the materials together or by using *mechanical force*. The typical assembling processes which will be discussed in Chapters 26 and 29 are classified in Fig. 1-12.

FINISHING

The last secondary processing family is called *finishing*. It is designed to beautify and/or protect the surface of a material. Typical finishing processes, as discussed in Chapters 30, 31, and 32, either coat the material or treat the surface of the material.

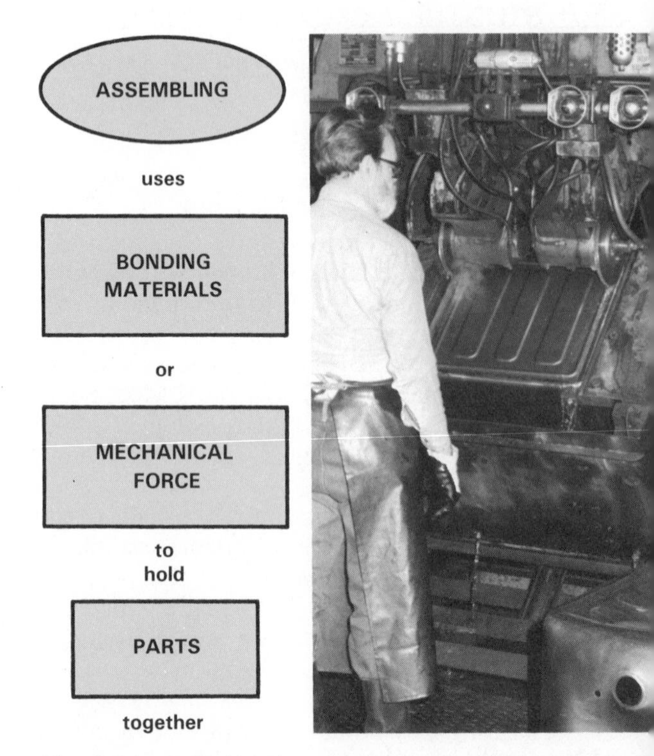

Fig. 1-11. Left. Fundamentals of assembling. Right. Seam welding operation. (Columbus Products Co.)

BONDING	MECHANICAL FORCE
Welding Brazing Soldering Adhesive joining	Mechanical fits Press fits Shrink fits Mechanical fasteners

Fig. 1-12. Assembling processes by classification. Fastening may be permanent or temporary.

Finishing may be done by any one of a large number of processes. These include electroplating, vacuum metalizing, anodizing, painting, galvanizing, and polishing. See Fig. 1-13.

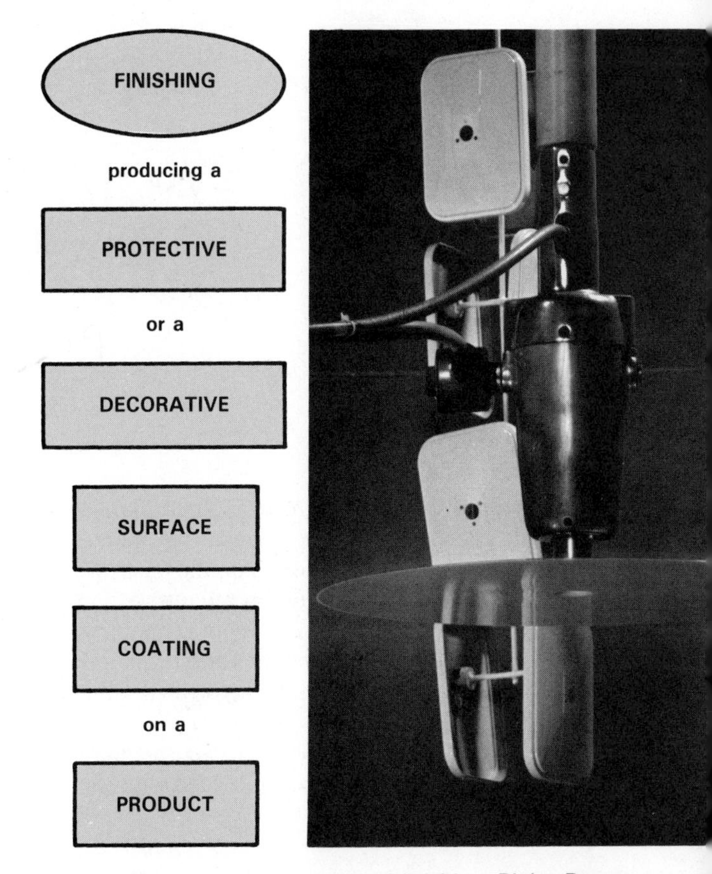

Fig. 1-13. Left. Fundamentals of finishing. Right. Rear-view mirror parts are coated using an electrostatic process. (Ransburg Corp.)

SUMMARY

Manufacturing changes the form of materials through processing acts. The first or primary acts change raw materials into industrial standard stock. This stock is further processed into products. The additional processing is called secondary processing. The major types of secondary processing activities are casting and molding, forming, separating, conditioning, assembling, and finishing. The output of these activities are products which allow each of us to live better and easier.

STUDY QUESTIONS — Chapter 1

1. What is manufacturing?
2. What is the difference between the manufacturing industry and the automotive industry?
3. What is meant by form utility?
4. Describe the difference between primary processes and secondary processes.
5. Name 10 raw materials.
6. What are the outputs of primary processes?
7. Name the three major types of materials.
8. List and define the six types of secondary processes.
9. What is the output of secondary processing?

Chapter 2

ORGANIZATIONAL STRUCTURE OF MANUFACTURING

This chapter will discuss the organizational structure and major operating procedures for manufacturing—a major part of the economic institution. The basic manufacturing activities fall into two main categories. One activity, discussed in Chapter 1, is directly involved with changing the form of materials. This activity as you already know, is called *material processing*.

There are other activities which direct the material processing activities. They are concerned with the efficient and effective operation of the manufacturing enterprise. These activities are called *management*.

MATERIAL PROCESSING

As discussed, material processing changes raw materials into standard stock from which products are made. These activities, Fig. 2-1, are either *primary* or *secondary* processes.

Primary processes change raw materials into usable

Fig. 2-1. Primary and secondary processing. Left. Ore, a raw material, is processed into iron and steel. (Inland Steel Co.) Right. Usable products (dishes) are produced from refined clay. (Syracuse China)

standard materials. Typical standard materials are sheet steel, plywood, lumber, aluminum tubing and rods, sheet cork, plastic films, and refined clay.

Secondary processes change standard stock into products. The processes involve:

1. *Casting, forming* and *separating* materials to create a part of proper size and shape.
2. *Conditioning* materials to produce the desired physical property such as hardness or ductility.
3. *Assembling* parts into assemblies and finally into products.
4. Providing products with protective and/or appearance coating through *finishing* activities.

The combination of primary and secondary processing techniques changes a raw material, like iron ore, which has little value by itself, into hundreds of useful products.

MANAGEMENT

Management is the sum of the practices which see that the enterprise operates efficiently. Carried out by managers, this function guides several important areas of activity. These activities, as seen in Fig. 2-2, move an *idea* for a product through a series of steps. The idea eventually becomes an actual *product* which can be exchanged in the marketplace for *money.*

The major areas of activity, which transform an idea into a product and then sell it, are:

1. Research and Development. This is a managed activity which discovers, develops, and specifies the characteristics of new or improved products and processes.
2. Production. This managed activity engineers the production facility, and produces scheduled products to stated quality standards.
3. Marketing. This managed activity identifies markets for products; then promotes, sells, and distributes the manufactured good.

Two other areas of activity support these three product-centered activity areas:

1. Industrial Relations – the managed activity which develops positive relations between management, its workers and its publics. This activity includes empoyee relations (personnel), union (labor) relations, and public relations programs.
2. Financial Affairs – the managed activity which monitors and controls the company's monetary actions.

Functions or tasks of management are applied within each of the five areas. Goals and courses of action are *planned.* The activities are *organized* into personnel structures and jobs are established. Employees are *directed* and motivated to complete

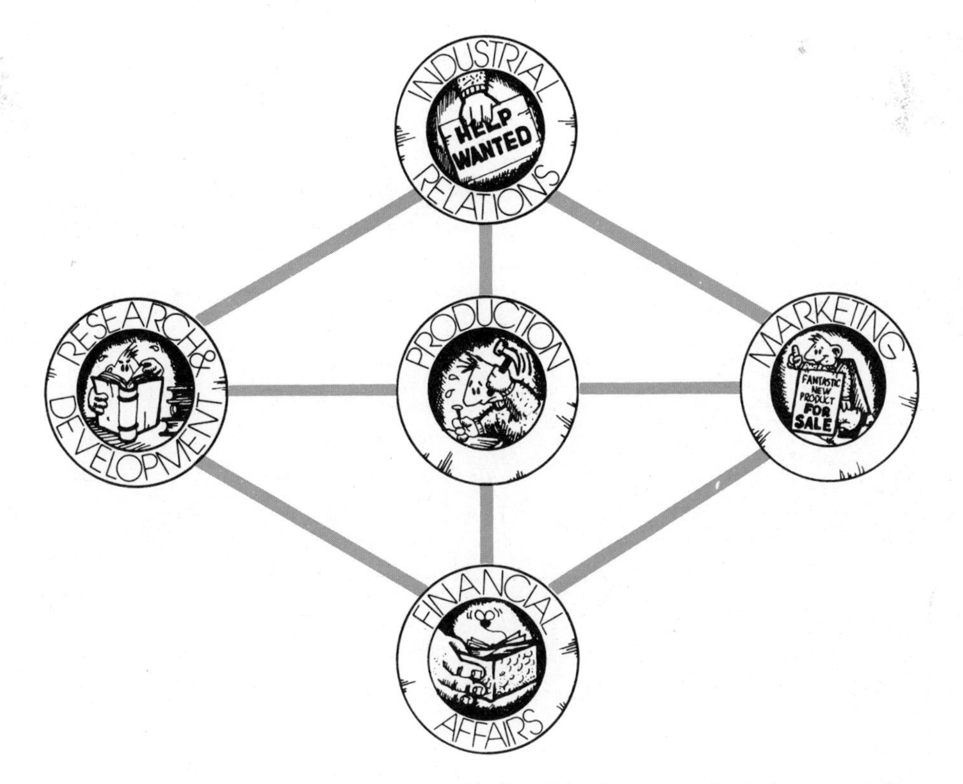

Fig. 2-2. There are five managed areas of activity in a manufacturing enterprise.

their jobs. Finally, the results of the management activity are constantly monitored to *control* the operation, Fig. 2-3.

PRODUCTION

A major activity in every manufacturing enterprise is *production*. The actual making of the product takes place within the managed production activity. The production area, as seen in Fig. 2-4, is responsible for four major activities. These are:

1. Manufacturing engineering – designing and engineering all production facilities and equipment needed to produce approved products.
2. Production Planning and Control – preparing, issuing, and monitoring schedules for the use of human, capital, and material resources within the production area.

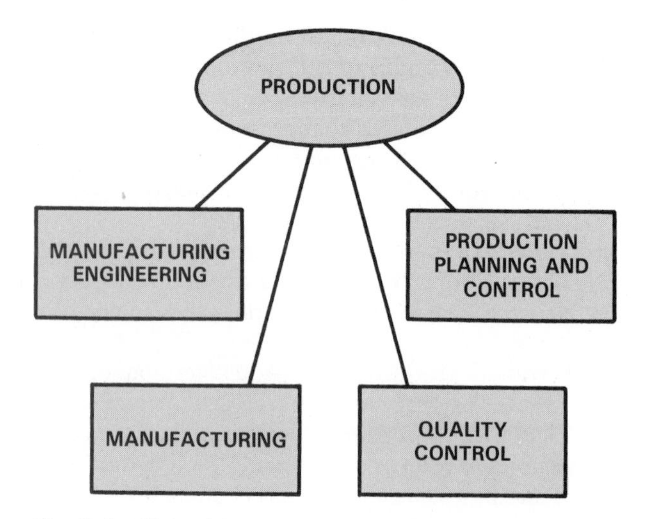

Fig. 2-4. Major functions of the production activity area are divided into four areas of responsibility.

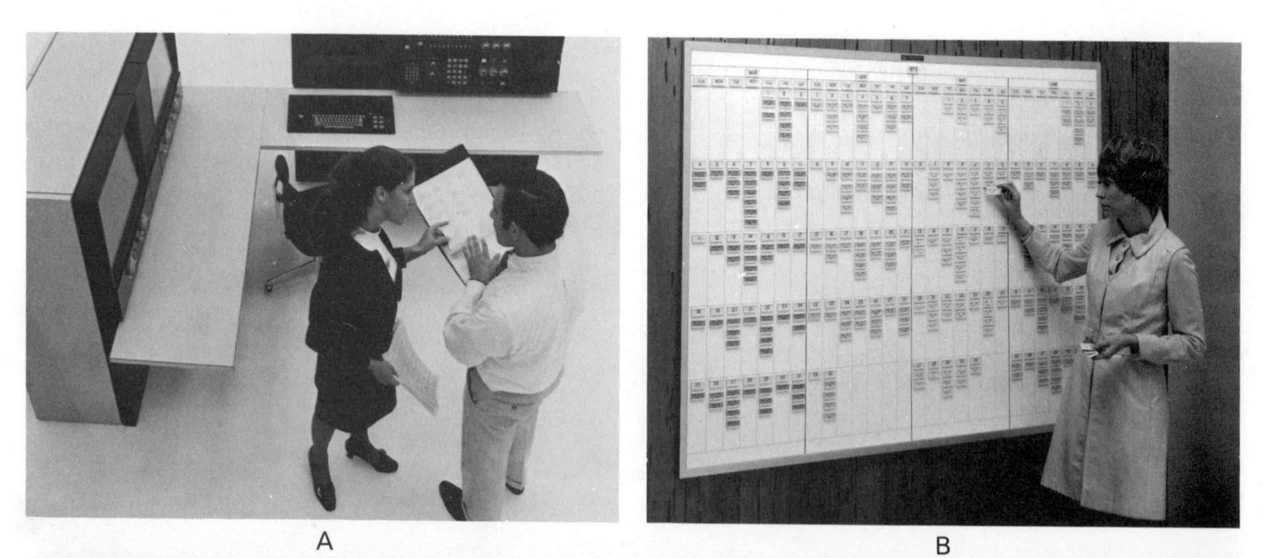

Fig. 2-3. Management plans, organizes, directs, and controls company activities. A – Planning. (IBM) B – Organization and control. (PPC) C – Production activity. (Oldsmobile Div., GMC)

3. Manufacturing—producing approved products by changing the shape, composition, or combination of material, parts, or assemblies.
4. Quality Control—developing and administering motivational and inspection programs to insure products meet stated standards.

All of these activities are designed to change the product idea, as represented by engineering drawings, bills of material, and specification sheets, into an actual product. They change ideas into reality and words and lines into salable three-dimensional objects.

TYPES OF MANUFACTURING

The production of products is carried out by one of three major types of manufacture:

1. Custon manufacture.
2. Intermittent manufacture.
3. Continuous manufacture.

CUSTOM MANUFACTURE

Custom manufacture involves the producing of a limited number of products from a customer's order. These products are usually built to a customer's design and specifications.

Custom manufacture requires highly skilled labor and general-purpose machines. The production rate is slow because drawings and specifications must be constantly read and applied. Also, each operation requires careful setup, performance, and quality checks.

Custom manufacture is an expensive way to produce products. It is used only when cost is not a major concern. Such is the case with custom produced clothing, or when there is need for only a limited number of like items—space vehicles, for example. See Fig. 2-5. Custom manufacture is also used to produce tooling (dies, patterns, and fixtures), special equipment, product prototypes, and models.

Fig. 2-5. Large special-duty equipment, like this 100 ft. overhead crane, is often custom manufactured.

INTERMITTENT MANUFACTURE

Intermittent or job-lot manufacture is used to produce quantities of products that are too small to be continuously manufactured. Typically, intermittent manufacturing plants produce a number of different products (parts, assemblies, or complete products) using the same equipment.

The production may be for the company's use or for outside customers. Some intermittent type operations produce parts for use by other plants owned by the same company. Other intermittent manufacturing plants, called job-shops, bid for production jobs from a number of customers.

In either case, as seen in Fig. 2-6, materials usually move through the plant in lots (trays or baskets of parts). The sequence (order) of operations is determined for each part. The lot is then moved from machine to machine. At each step, the machine is set up as the job dictates.

The scheduled operation is performed on all the pieces in the lot before they are moved to the next work station. The entire lot moves through the plant as a single unit.

Intermittent manufacture, like custom production, requires skilled workers and general-purpose machines. It is, however, a more economical way to produce parts because each machine setup will process a number of parts before being changed. In addition, workers are more apt to be doing the same type of work all day, such as drilling. They, therefore, become more productive. They will also use less of their work time reading and interpreting drawings.

Some intermittent production runs are larger than job lots. There is often a need to run a part or assembly for several days before the equipment is changed to produce a different product. A company may produce enough products to fulfill its needs for a number of weeks or months during a several day run. This type of intermittent production provides some of the benefits of continuous production. An example of this type of production is producing gears of several sizes. Each size may run for one week out of each 10-week period. Another example is the blow

Fig. 2-6. Special products like this 5500 lb. earthmover tire are usually produced using intermittent or job-lot manufacturing practices. (Goodyear Tire and Rubber Co.)

molding of several different plastic bottles. A month's supply of one size may be produced before the molding dies are changed for another size.

CONTINUOUS MANUFACTURE

Continuous manufacture, Fig. 2-7, produces the same product using the same equipment over a long period of time. Quite often the equipment used is special-purpose type machines. The machines are designed and built to produce one specific product or a group of closely related products.

Generally, few workers are required and they may be semiskilled. The typical on-the-line jobs require loading materials and parts into machines, monitoring (observing) the machine during its process cycle, and removing finished parts.

Continuous manufacture produces products at a lower per unit cost than custom or intermittent systems. Product quality is generally higher.

Because of the repetitive operations, continuous manufacture lends itself to automatic controls, use of transfer (automatic part-moving) machines, replacing workers with robots, and other labor-saving equipment.

This method of manufacture is appropriate for the production and assembly of products needed in large numbers. Typical of such products are fasteners (nuts, bolts, and nails), electric motors, automotive parts, and television sets.

Process type industries use continuous manufacture. These industries most often produce standard stock. Plywood, particleboard, and hardboard are typically manufactured in specially designed

Fig. 2-7. Products with high demand, like these toy autos, can be continuously manufactured. (Tonka Toy Co.)

Fig. 2-8. Particleboard is continuously formed by this machine. (American Forest Products Industries)

plants. Also, most chemical and petrochemical plants are continuous process operations. Fig. 2-8 shows a typical continuous process operation.

Continuous manufacture will generally produce the greatest quantity of product at the lowest per unit cost and with the highest quality.

SUMMARY

Manufacturing is responsible for the production of the material goods needed by society. This involves the use of *material processing* activities which are *managed* to insure efficient operation. The managed activities within a manufacturing enterprise are research and development, production, marketing, financial affairs, and industrial relations.

The production activity is responsible for the actual manufacture of the product. The product is produced through a series of primary and secondary processes. These processes are carried on using custom, intermittent or continuous manufacturing systems.

STUDY QUESTIONS – Chapter 2

1. List and describe the two main categories of manufacturing activity.
2. List and describe the four major functions of management.
3. List and describe the three major types of manufacture.
4. The group of techniques which combines two or more parts into a product is called __________.
5. The major managed areas of activity in an enterprise are (select the correct answers):
   a. Financial affairs.
   b. Industrial relations.
   c. Marketing.
   d. Production.
   e. Research and development.
6. Product manufactured to customer's order is more likely to be produced by (custom, intermittent) manufacture.

Chapter 3

INDUSTRIAL MATERIALS

We all live in a world of materials that are used for several different purposes. Materials are used to produce products that make our lives safer, easier, and more enjoyable. Other materials are needed to produce products which transport people and goods, communicate messages and ideas, and provide shelter and comfort. See Fig. 3-1.

Some materials are not changed from the way they appear in nature. Gravel and rock are generally used in their natural state in construction projects. Logs are sawed into lengths but are otherwise unchanged before being burned to provide heat. Most materials, however, need to be refined or converted from raw materials into standard stock. Iron ore, coke, and limestone are used to produce steel. Logs are converted into lumber, plywood, and particleboard. Bauxite is refined into aluminum. Many of the products around us, then, are the result of processed material.

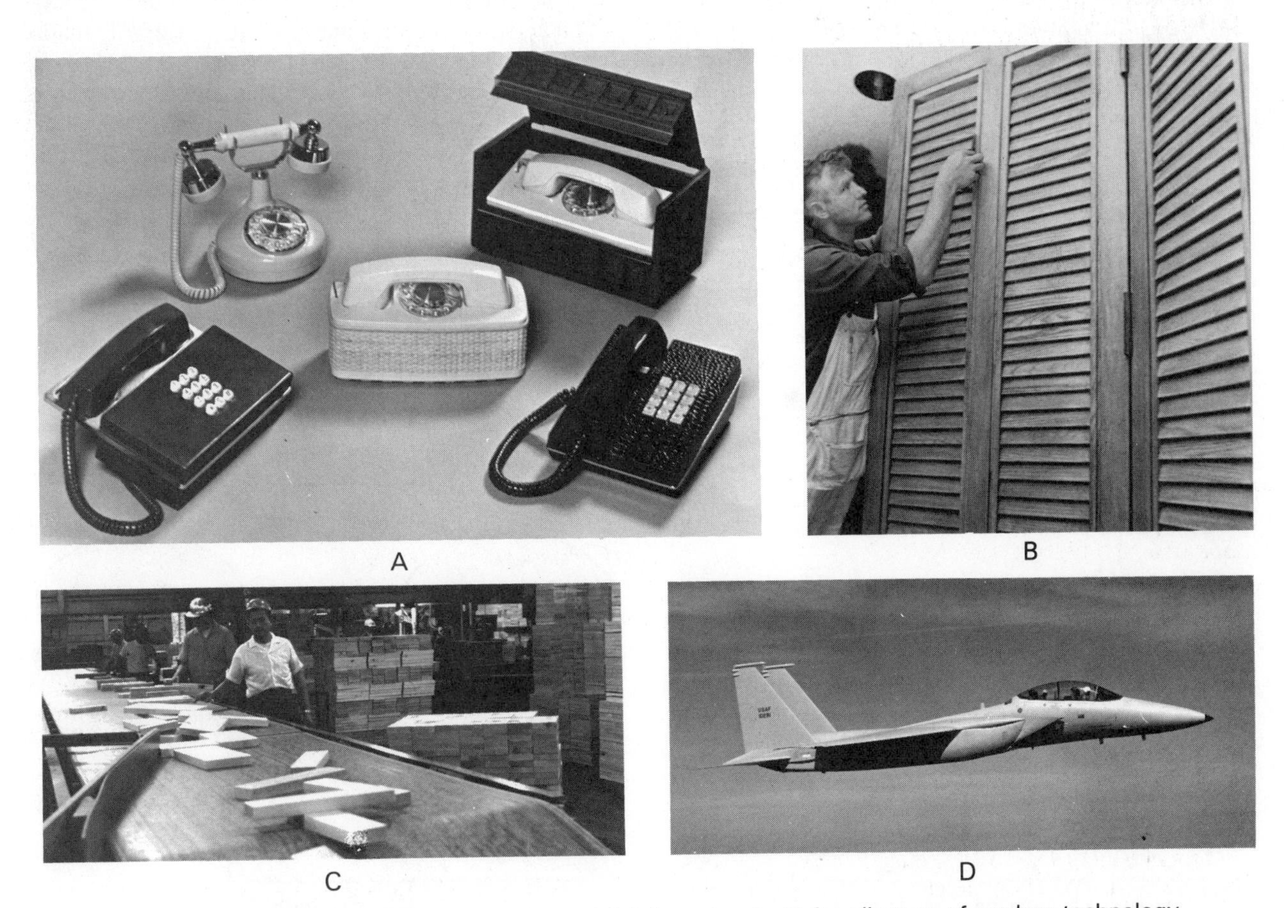

Fig. 3-1. Materials are used in the products fabricated or built for all areas of modern technology. A—Communication. B—Construction. C—Manufacturing. D—Transportation. (Western Electric, Merkel Press, Weyerhaeuser, and McDonnell Douglas)

TYPES OF MATERIALS

Materials may be grouped in several ways. Scientists often classify materials by their state: solid, liquid, or gas. They also separate them into organic (once living) and inorganic (never living) materials.

For industrial purposes, materials are either solid or engineering materials or nonengineering materials. Engineering materials are those used in manufacture and become parts or products. Nonengineering materials are the chemicals, fuels, lubricants, and other materials used in the manufacturing process which do not become part of the product, Fig. 3-2.

This grouping is not exact. It will, however, separate the materials which are processed by the techniques described in this book. Engineering materials, Fig. 3-2, may be further subdivided into:
1. Metals.
2. Polymers.
3. Ceramics.

A fourth type of material sometime listed is called a composite. Materials in this group are made up of two or more materials from the engineering groups. Each of the materials in a composite retains its original characteristics. Examples of composites include wood, concrete, glass reinforced polyester, and graphite polymer advanced composites.

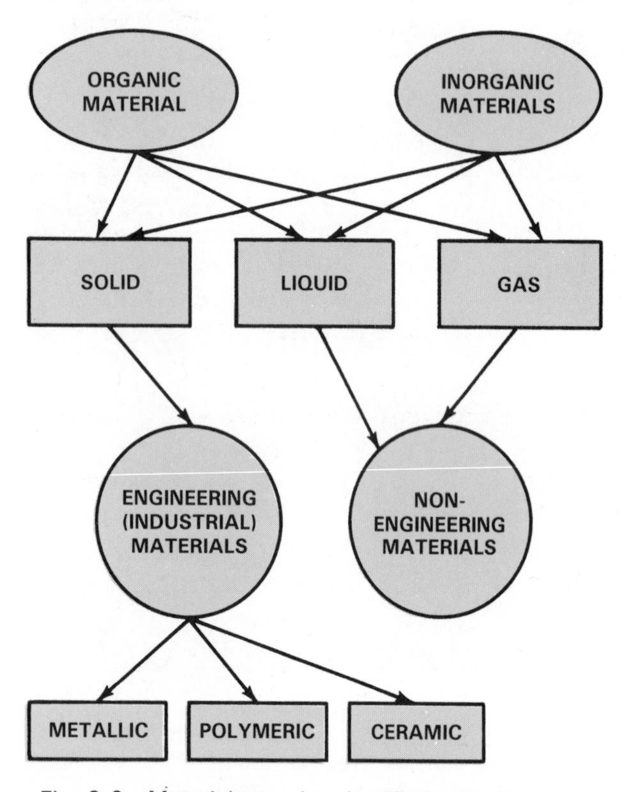

Fig. 3-2. Materials can be classified several ways.

METALS

A metal is an inorganic substance. Metals are made up of one or more metallic elements (copper, iron, aluminum, zinc, tin, etc.). Seldom is a metal used in its pure form. Usually the metallic element is combined with other metals or nonmetallic elements. This combination is called an *alloy*. Steel is an alloy of iron and other elements including carbon, magnesium, vanadium, nickel, and chromium. Brass is an alloy of copper and zinc; bronze is made up primarily of copper and tin. Aluminum may be alloyed or combined with other elements to change its properties. The same is true of gold, silver, lead, and numerous other metals.

All metals are characterized by a close-packed crystal structure. They are conductors of both heat and electricity. Also, metals, as compared to most other materials, are strong, tough, and stable.

POLYMERS

A polymer is an organic noncrystalline (not having a crystal structure) material. It is primarily made up of carbon and hydrogen atoms (hydrocarbons) arranged in long chains. Carbon, a nonmetal, is the base element in these chains.

The term, polymer, covers a broad range of products. As shown in Fig. 3-3, there are two types:
1. Natural polymers: Wood, rubber, cotton, etc.
2. Synthetic (manufactured) polymers: Thermoset and thermoplastic compounds.

The two major polymers discussed in this book are wood and plastics. Wood is an organic composite

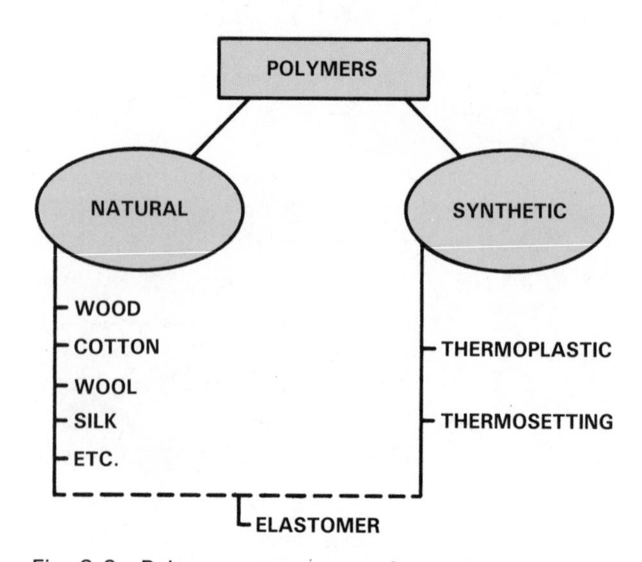

Fig. 3-3. Polymers are a group of organic (once living materials). They comprise two distinct groups.

material made up of cellulose fibers held together by lignin—nature's natural glue. Small quantities of other materials such as minerals, oils, and waxes are present in most woods.

Plastics are synthetic materials primarily produced from petroleum and natural gases. The two basic types, as previously mentioned, are:

1. Thermoplastics: Materials with little or no bonding between the hydrocarbon molecules or chains.
2. Thermosets: Materials with strong bonds between the hydrocarbon molecules.

The lack of intermolecular bonds in thermoplastics allows the material to soften under heat. It will harden again as the material cools. The softening and hardening cycle can be repeated with little or no damage to the material. Thermoplastics can be reshaped as often as necessary.

Thermosets, with their strong molecular bonds, retain their shapes. Once a thermoset material is "set" with heat and pressure, it cannot be reshaped. The bonds within the material are permanent.

Elastomers, either natural or synthetic, are also polymers. These materials may be stretched and will return to their original length. Rubber and neoprene are examples.

CERAMICS

Ceramics are probably the oldest materials used by humans. They were the basic materials of the Stone (a ceramic material) Age and still play an important role in manufacturing.

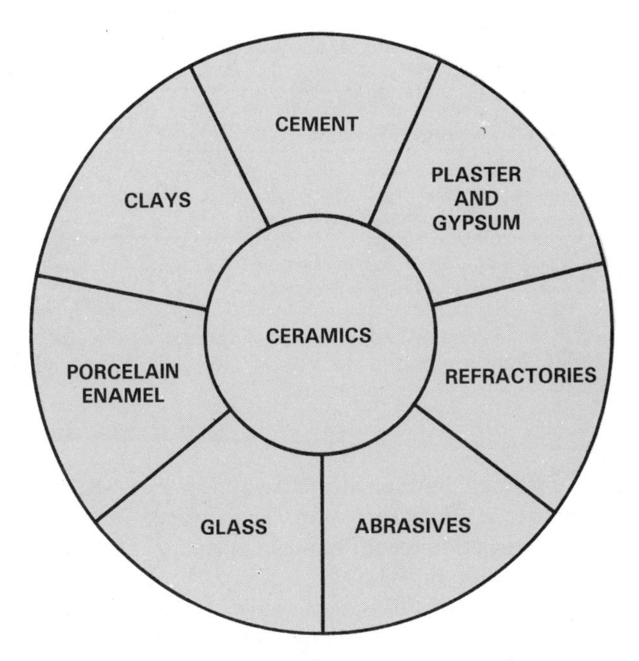

Fig. 3-4. Types of ceramic materials.

A very large family of materials, ceramics are mainly inorganic nonmetallic crystalline compounds.

The list, shown in Fig. 3-4, includes clays used to produce dinnerware, artware, sanitaryware (bathroom fixtures), bricks, and tile; cement, plaster and gypsum, abrasives, glass, porcelain enamel, and refractories (high-temperature-resistant materials).

Ceramics have little or no electrical conductivity. They melt at very high temperatures and are the hardest of the engineering materials. Also, ceramics are very rigid and brittle. They are highly resistant to the corrosive action of a wide variety of chemicals.

Our major emphasis will be upon glass, clays used in dinnerware and sanitaryware, and porcelain enamels.

Every product is manufactured from materials which are carefully selected to meet functional (use) and aesthetic (appearance) requirements. Most products must work well and be reasonably attractive.

Product designers and engineers have thousands of different materials available to them. They choose and specify each material that will be used in the product. Their selection will directly affect the manufacturing processes used and their ability to produce a product of acceptable quality.

The properties of each material must be readily known to designers, engineers, and machine operators. Each material may be identified by a name, grade, and/or identification code. Common metals, plastics, ceramics, and woods should be familiar to all persons using manufacturing processes.

COMMON METALS

Pure metals are seldom used in common industrial products. Pure copper is used in electrical applications, in automotive radiators, and gaskets. Pure aluminum has applications in the chemical and electrical industries. However, most metals are alloys (combinations of two or more elements). There are over 25,000 different iron-carbon alloys (steels) and over 200 standard copper alloys including a number of brasses, bronzes, and nickel silvers. Each of these alloys are identified by a code number.

Three of the most common groups of metallic alloys are:

1. Steels.
2. Copper.
3. Aluminum.

Steel

Steel is an alloy of iron and carbon with other elements added to produce specific properties. The various types of steel can be grouped under two major headings:

1. Carbon steel. A steel in which the main alloying element is carbon. Carbon steels are further divided into three groups.
   a. Low carbon steel. This steel has a carbon content of less than 0.30 percent. It is the most common type and is often called mild steel. It is relatively inexpensive, ductile, soft, and is easily machined and forged. Mild steel cannot be heat treated (hardened) or easily machined. Low carbon steel is a general-purpose steel.
   b. Medium carbon steel. This steel has a carbon content between 0.30 percent and 0.80 percent. Harder and stronger than mild steel, it can be hardened by heat treating. Medium carbon steel is most commonly used for forgings, castings, and machined parts for automobiles, agricultural equipment, machines, and aircraft.
   c. High carbon steel. This type of steel is easily heat treated to produce a strong, tough part. The material has a carbon content above 0.80 percent. It finds wide use in hand tools, cutting tools, springs, and piano wire.
2. High alloy steel. These steels contain significant amounts of other elements in addition to carbon. The common high alloy steels are:
   a. Stainless steel which is produced by using chromium as a significant alloying element along with nickel and other metals. The result is a tough, hard, corrosion resistant metal.
   b. Tool steel which is a special group of high carbon steels produced in small quantities to high quality specifications. Tool steels are used for a wide range of cutting tools and forming dies.
   c. Manganese steel which is an alloy containing 12 percent manganese and one percent carbon. This metal is used in mining, railroad, and construction equipment because of its high tensile strength.

The most common iron-carbon alloys are designated by the AISI (American Iron and Steel Institute) identification system. This system uses the chemical content of the steel to identify it. Each steel has a four digit code. The first two digits divide steels into major groups, Fig. 3-5. For example, 10XX identifies plain carbon steels and 13XX designates high manganese steels. The last two digits indicate the carbon content in "points" where a point is 0.01 percent carbon. A 1040 steel, for example, is a plain carbon steel with about 0.40 percent carbon.

Copper alloys

Copper and its alloys are significant industrial materials. They conduct electrity and heat well, are easily formed, have an attractive appearance, and are highly resistant to corrosion. The Copper Development Association identifies the important coppers and copper alloys, Fig. 3-6.

Copper alloys, like steel, have a coding system which tells the type of alloy being used. The basic categories in this coding system are shown in Fig. 3-7.

CODE	TYPE OF STEEL
1XXX	Carbon steels
2XXX	Nickel steels
3XXX	Nickel-chromium (stainless) steel
4XXX	Molybdenum steels
5XXX	Chromium steels
6XXX	Chromium-vanadium steels
8XXX	Nickel-chromium-molybdenum steels
9XXX	Other steels

Fig. 3-5. Steel coding system is used for identifying alloying elements. The first two digits divide the steels into major groups. The last two numbers indicate the carbon content in percentages.

METAL	DESCRIPTION
Coppers	Metal which has a designated minimum copper content of 99.3%.
High Copper Alloys	Copper used for wrought products which has less than 99.3% copper and does not fall into any other coper alloy group.
Brasses	Copper alloy with zinc as the principle alloying element and can contain iron, aluminum, nickel and silicone. Wrought alloys have three groups of brasses: copper-zinc alloys, copper-lead-zinc alloy (leaded brasses), and copper-zinc-tin alloys (tin brasses).
Bronzes	Copper alloys in which the principle alloying element is not zinc or nickel. Originally "bronze" described alloys with tin as the only or principal alloying element. Today there are phosphorus bronzes (copper-tin-phosphorus alloys), leaded phosphor bronzes (copper-tin-lead-phosphorus alloys), aluminum bronzes (copper-aluminum alloys), and silicon bronzes (copper-silicon alloys).
Copper-Nickels	Copper alloys in which nickel is the only or principal alloying element.
Copper-Nickel-Zincs	Copper alloys commonly known as "Nickel Silvers," in which zinc and nickel are the principal alloying elements.

(From "Standards Handbook—Wrought Mill Products. Alloy Date/2"; Copper Development Association)

Fig. 3-6. Types of copper alloys. A number of different metals can be added to copper to give it special properties.

CODE NUMBERS	ALLOY
C10100-C15500	99.5% or greater purity
C16200-C19500	High-copper alloys (cadmium, beryllium, lead, or chromium)
C20500-C28200	Copper-zinc alloys (Brasses)
C31400-C38600	Copper-zinc-lead alloys (Lead Brasses)
C40500-C48500	Copper-zinc-tin alloys (Tin Brasses)
C50100-C52400	Copper-tin alloys (Phosphor Bronzes)
C53400-C54800	Copper-tin-lead alloys (Leaded Phosphor Bronzes)
C60600-C64200	Copper-aluminum alloys (Aluminum Bronzes)
C64700-C66100	Copper-silicon alloys (Silicon Bronzes)
C66400-C69800	Special brasses
C70100-C72500	Copper-nickel alloys
C73200-C79900	Copper-nickel-zinc alloys (nickel silvers)

Fig. 3-7. There is a system of codes for copper alloys known as the Unified Numbering System (UNS).

Aluminum and aluminum alloys

Pure aluminum and aluminum alloys have many commercial uses. The metals are lightweight, strong for their weight, resistant to corrosion, nontoxic, excellent conductors of heat, good conductors of electricity, and easy to form and fabricate.

Aluminum and its alloys are designated by a number and a letter system such as 1060-H12. The number before the dash designates the composition of the material while the letter and number after the dash indicates the hardness (temper).

The first number in the alloy designation (2XXX) designates the purity and changes caused by alloying. Fig. 3-8 lists the major series of aluminum alloys.

The letter that follows the alloy series number indicate the basic temper (hardness) of the material. These designations are shown in Fig. 3-9.

The designations just given are for sheets, tubing, and bars. Aluminum for casting is designated by another system. However the aluminum-silicon (4xx.x series) alloys are the most common material used in casting followed by aluminum-copper-silver (2xx.x) and aluminum-magnesium-silicon (3xx.x) alloys.

SERIES	DESCRIPTION
1000	Aluminum 99% or higher in purity
2000	Copper as the principal alloying element
3000	Manganese as the principal alloying element
4000	Silicon as the principal alloying element
5000	Magnesium as the principal alloying element
6000	Silicon and magnesium as principal alloying elements
7000	Zinc and the principal alloying element
8000	Other element as principal alloying element

Fig. 3-8. Wrought aluminum alloy designations include both a series of numbers and a letter. The first number designates the purity and changes caused by the alloying.

LETTER CODE	DESCRIPTION
F	As fabricated: no special controls over temperature and strain hardening was employed during manufacturing.
O	Annealed to obtain the lowest strength temper (Hardness).
H	Strain-hardened to increase the strength of the material. This designation is followed by two numbers. The first digit indicates the operations used to harden the material and the second digit the degree of hardness. A zero (0) is soft and an eight (8) is hard.
W	Solution heat treated which is an unstable heat-treatment used for only a few alloys.
T	Thermal treated (other than F, O, and W) to produce a stable temper.

Fig. 3-9. A letter indicates the temper or hardness of the aluminum.

TYPES OF POLYMERS

Both thermoplastic and thermosetting resins are available in a number of different chemical forms. Each resin has application which it best serves. The common resins are listed in Fig. 3-10.

The plastics industry, unlike other materials areas, has not adopted a standard grade designation system.

THERMOPLASTIC	THERMOSETTING
Polyethylene	Phenolics
Polyvinyl chloride	Amines
Polystyrene	Alkyds
Polypropylene	Polyesters
ABS	Allylics
Acrylics	Epoxies
Cellulosics	Silicons
Acetals	Urethanes
Fluorocarbons	
Polyamides	
Polycarbonates	

Fig. 3-10. These are typical thermoplastic and thermosetting materials.

Instead, most engineers designate the plastic by manufacturer's name and number. Presently the ASTM (American Society of Testing and Materials) is developing a system to designate plastic materials under a complex number system called the ASTM D 4000 coding system.

TYPES OF CERAMICS

Ceramic materials are becoming more widely used in modern industrial applications. They may be grouped into the following categories:

1. Clay products. A variety of crystalline structures with a glassy matrix which are used for dinnerware, sanitaryware, bricks, tiles, etc.
   a. Earthenware. Clay products made of kaolin which is fired at a relatively low temperature. Produces a serviceable but weak, porous product used for dinnerware and drain tile.
   b. Stoneware. Clay product similar in composition to earthenware but the material content is more closely controlled and the firing temperature is higher. Stoneware is less porous than earthenware and is used for ovenware and chemical ware.
   c. China. A translucent clay product containing large quantities of quartz, clay, and feldspar crystals, which is fired at high temperatures. Crystals are converted to clear glass during firing. China is widely used for dinnerware.
   d. Porcelain. A clay product similar to china produced without certain fluxes added. Fired at very high temperatures. Porcelain is used for chemical ware and surface finishes.
   e. Structural clays. Clays which contain high percentages of silica, alkali, and iron oxide. The material is used for bricks, wall and floor tile, and drain tile.
2. Refractories. Crystalline body materials noted for their resistance to high temperatures.
   a. Heavy refractories. The most commonly used refractories which are generally produced in straight or curved brick forms. These materials are used widely to line furnaces.
   b. Pure oxide refractories. A special material made from the oxides of selected metals such as aluminum, magnesium, and zirconium. These materials are used for pyrometer tubes, spark plug insulators, wire drawing dies, valve seats, and crucibles.
   c. Nonoxide refractories. A very expensive and little-used refractory because they oxidize readily. Silicon carbide, tungsten carbide, and graphite are some typical examples of nonoxide refractories.
3. Glass. An amorphous (without structure) material which is made of silica with coloring, firing, and other agents added.
   a. Soda lime glass. Common glass made from silica and soda ash which is used for containers, window glass, light bulbs, and inexpensive tableware.
   b. Lead glass. A glass made from the same materials as soda lime glass except lead oxide is added. A high brilliance, clear glass used for crystal ware and eyeglasses.
   c. Borosilicate glass. A glass made from silica and boron oxide which has high resistance to thermal shock and chemicals. Used for ovenware and laboratory glassware.
   d. 96 percent silica glass. A high silica glass which resists chemical action and absorbs little ultraviolet light. Used for chemical ware and sun lamps.
   e. Fused silica glass. A material made from pure silica which is stable over a wide range of temperatures and resists thermal shocks. This glass is used for special applications such as high-voltage electric insulators.
   f. Aluminosilicate glass. A glass made principally of silica and alumina which is very hard and heat resistant. It is used for cookware, combustion tubes, and high-temperature thermometers.

A special group of ceramic material is used for abrasives. These materials are generally crystal bonded oxides and carbides. Three common ceramic abrasives are silicon carbide, aluminum oxide, and synthetic diamonds.

WOOD PRODUCTS

Wood products are important materials especially to cabinet and furniture manufacturing and the construction industry. The major types of materials used are:

1. Hardwood lumber.
2. Softwood lumber.
3. Hardwood plywood.
4. Softwood plywood.
5. Particleboard.
6. Hardboard.

Each of these materials has a separate grading and quality designation system. Also within a specific type of material more than one system is used. For example, redwood is graded under a different system than pine even though both are softwoods. Likewise, Douglas fir has one grading system for appearance boards and another for framing (construction) lumber.

GENERAL STRUCTURE OF MATERIALS

ATOMIC STRUCTURE

All materials are built out of the same components. The atoms of any substance are the smallest unit of a material that will still have the properties of that material. An iron atom will no longer have the properties of iron if it is broken into smaller particles. However, there are smaller particles which make up atoms. All atoms, as seen in Fig. 3-11, have three basic components:

1. Electrons. Negatively charged particles which orbit around the nucleus (center) of the atom.
2. Protons. Positively charged particles found in the nucleus of the atom.
3. Neutrons. Neutral particles found in the nucleus of the atom.

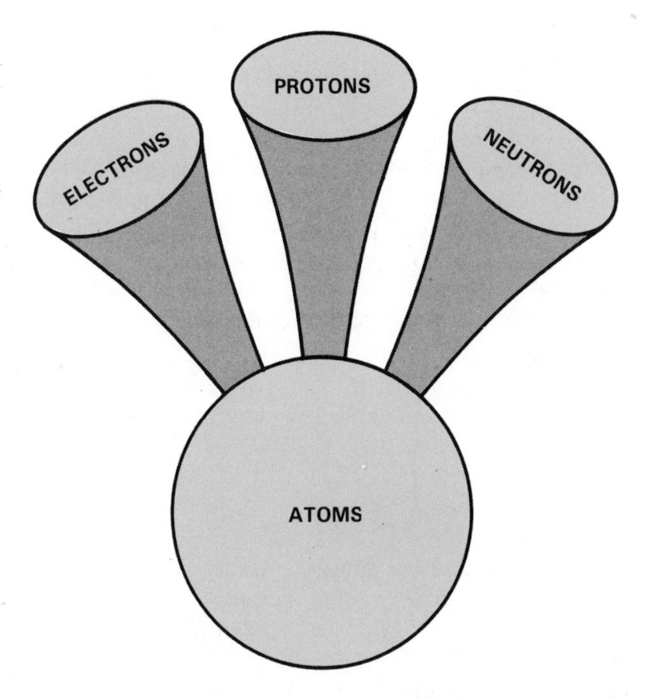

Fig. 3-11. Every atom of any material is made up of these same components.

The electrons, protons, and neutrons of one element are exactly like those found in another element. These three components are the basic building blocks of atoms. The way they are arranged provides us with the 88 different elements or basic materials found in nature. Additional elements have been made by humans but are not found in nature.

Atoms of each element have a basic arrangement of electrons around their nucleus which contains the

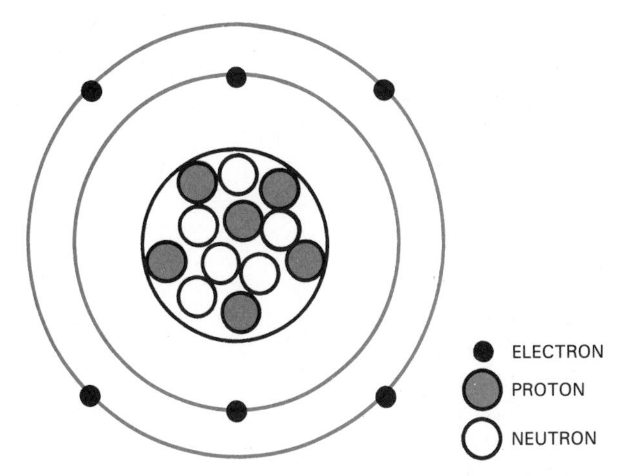

Fig. 3-12. Diagram shows a carbon atom with six electrons (orbiting around the nucleus), six protons, and six neutrons.

protons and neutrons. This arrangement is shown in Fig. 3-12. The electron-proton-neutron relationship determines the chemical properties of the atom and, therefore, the material.

The electrons move in rings or shells around the nucleus much like the planets move around the sun. The first shell is the lowest in energy and can hold one or two electrons. The next shell is a higher energy shell and can have up to eight electrons. As the atom becomes more complex (and therefore, heavier) additional shells of electrons are present. An atom with one electron in its outermost shell will be more active than one with two electrons and so on. If the outer shell is full, the atom will be inactive or inert. At all times a stable atom will have the same number of protons and electrons. Fig. 3-13 shows the electron arrangement for some typical elements.

If, for some reason, the balance between protons and electrons is disturbed an *ion* is formed. If an electron is lost, the ion will have more protons and will be called a positively charged ion. If extra electrons are gained a negatively charged ion is formed. These ions play an important part in forming certain types of materials containing two or more elements.

MOLECULAR STRUCTURE

Some atomic structures (arrangement of electrons, protons, and neutrons) produce atoms with high energy levels which strongly repel similar atoms. This structure produces large (by atomic standards) spaces between atoms within the material. These materials are called gases.

If the energy levels are of lesser strength the material is liquid. As the repelling force or energy level is

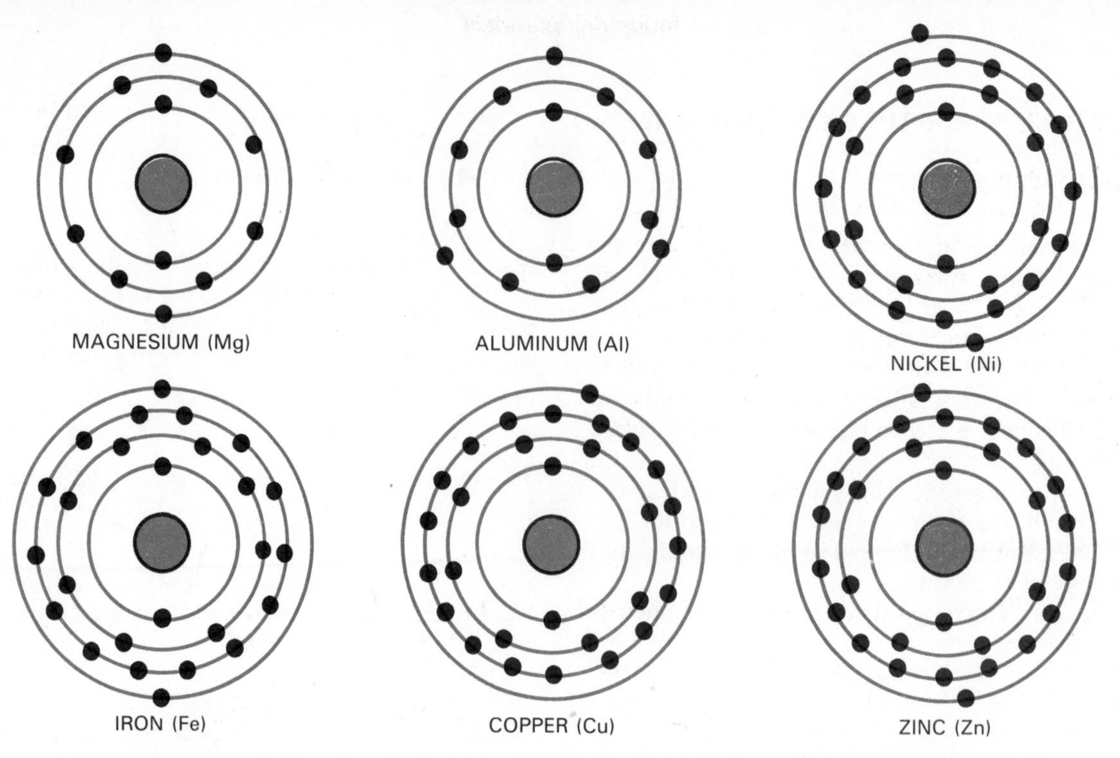

Fig. 3-13. The electron arrangement around the nuclei varies for each different material.

further reduced, solids are formed.

These solid materials may also be called engineering or industrial materials. They are formed by the bonding together of huge numbers of atoms. The atoms may be of the same element or be from different elements. These atoms may be bonded into molecules (two or more atoms) in one of three ways:

1. Covalent bonding.
2. Ionic bonding.
3. Metallic bonding.

Covalent bonding

Covalent bonds are formed when two or more atoms share electrons to fill the outer shells of their structure. Many gases, as seen in Fig. 3-14, are formed through covalent bonds. Also, ALL organic compounds are held together by covalent bonding. As seen in Fig. 3-15, carbon is the basic element in most organic compounds.

Covalent bonds are quite strong because of the sharing of electrons. The diamond is a large molecule

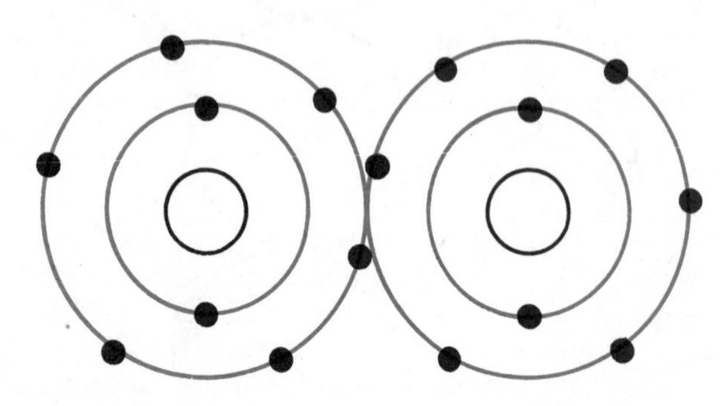

Fig. 3-14. Bonding of oxygen to form O_2 gas. Each atom shares two electrons to reach a stable number of eight in a second ring.

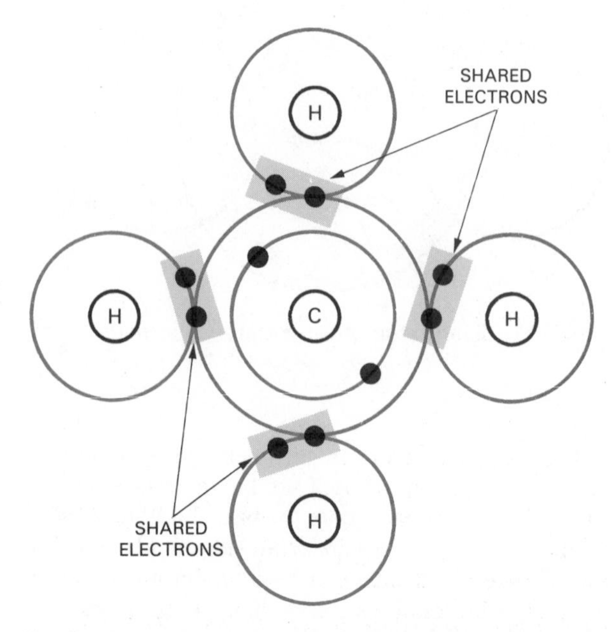

Fig. 3-15. All organic material is held together by covalent bonding. The covalent bonding of carbon and hydrogen (above) creates methane gas (CH_4).

of carbon with each carbon atom sharing electrons with neighboring carbon atoms. Individual plastic molecules are formed in much the same manner and are quite strong. The softness, brittleness, and other "weak" characteristics of many plastics are not due to the covalent bond. These properties are caused by weaker bonds between molecules of the material, not weakness within the molecules.

Plastics and natural polymers are made up of very large numbers of molecules bonded by covalence. Often these molecules are highly complex in their structure.

Ionic bonds

Ionic bonds are formed when one atom gives up an electron to an atom of a different material. The result is that the atom of the first material becomes a positively charged ion and the other atom a negatively charged ion. Since unlike charges attract, the two atoms are attracted to each other.

Common table salt is formed in this manner. Sodium (a metal) has only one electron in its outer ring. This makes sodium very active. Chlorine lacks only one electron to fill its outer electron shell. As seen in Fig. 3-16, the sodium atom gives up its outer electron to a chlorine atom. The resulting positively charged sodium (Na^+) ion is then attracted to the negatively charged chlorine (Cl^-) ion. The result is table salt, NaCl.

Sand and ceramic materials are formed by ionic bonding. They generally are very hard, have high melting points, and resist corrosion.

Metallic bonds

Metallic bonds are formed by metallic atoms giving up the electrons in their outer shell to a common pool of electrons. This pool is called an electron cloud and is evenly distributed throughout the metallic structure. See Fig. 3-17.

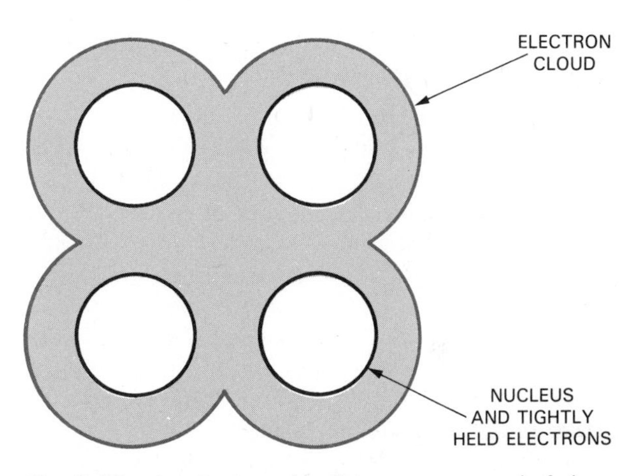

Fig. 3-17. An electron cloud is a common pool of electrons given up by metallic atoms when metallic bonds are formed.

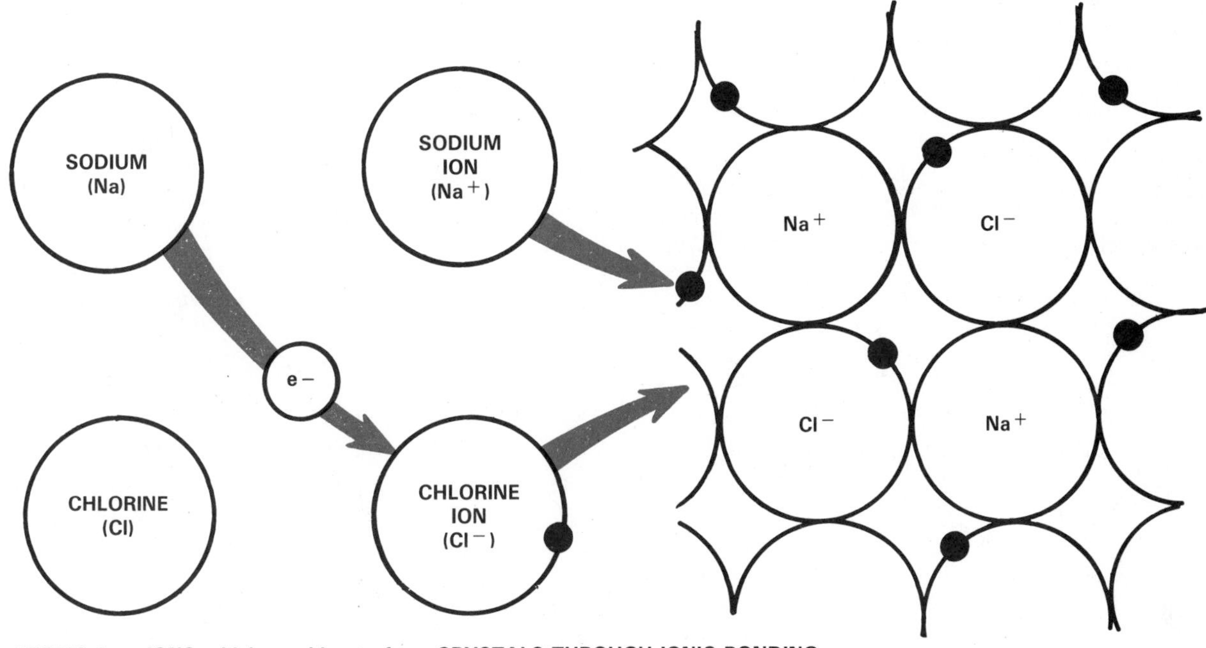

Fig. 3-16. In an ionic bond, one atom gives up an electron to an atom of another material. This is an ionic bonding of sodium and chlorine atoms.

The structure is made up of positively charged metallic ions and the negatively charged electron cloud. The electrons move freely within the structure. The result is a number of positively charged metallic ions repelling each other and a group of free-moving electrons also repelling each other. In addition, the electrons and the metallic ions are attracted to each other. The atoms (ions) space themselves so that all the attracting and repelling forces cancel out. The material comes into equilibrium (balance). At the equilibrium or stable distance, the atoms in a metal resist any force which would change the distance between them. The result is a material which has high compression and tensile strength.

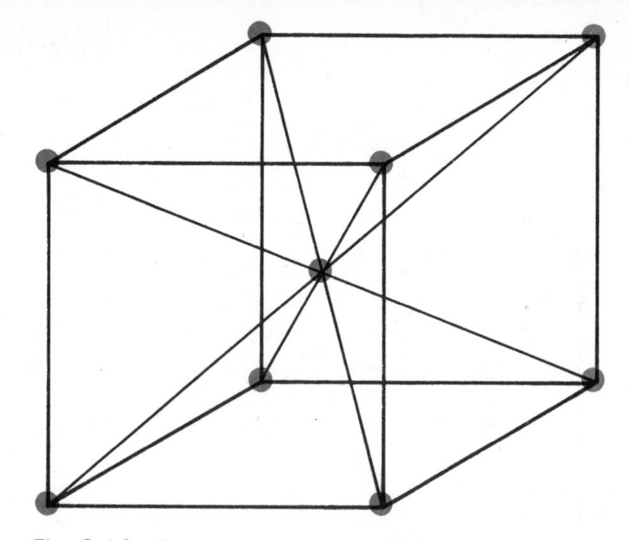

Fig. 3-18. Body-centered cubic lattice has an atom at its center with others forming the outer limits of the cube.

STRUCTURE OF INDUSTRIAL MATERIALS

Each of the three major types of materials have their own unique structure. Metallic, polymeric, and ceramic materials used in manufacturing are almost always a combination of two or more elements. These elements are combined to provide the material with properties and characteristics that suit it to a certain purpose. The way the elements combine produces the material's structure.

STRUCTURE OF METALS

A metal is a material which gives up its outer shell electrons (valence electrons) when it is bonded to another atom. When metallic atoms are bonded together they form a uniform structure. They arrange themselves in a pattern which is repeated throughout the material. Fifty of the chemical elements are metals and act in this manner. Of this group, nearly 40 metals have commercial value. Metals are very common and important manufacturing materials.

The natural structures present in metallic materials are called *space lattices.* They are made up of identical units which are aligned in parallel planes. The result is what is called the crystal or grain structure of the metal. There are 14 types of crystal lattices which are formed by various metals. However, the most common, commercially important metals have one of three lattice structures:

1. Body-centered cubic lattice, Fig. 3-18.
2. Face-centered cubic lattice, Fig. 3-19.
3. Hexagon close-packed lattice, Fig. 3-20.

The black dots in the illustrations represent atoms. The drawings each represent a unit cell. This is the smallest model of the lattice that will exhibit all the properties of the lattice. A single unit cell will not exist alone. To gain stability, it must grow by joining with other unit cells until it becomes a stable mass.

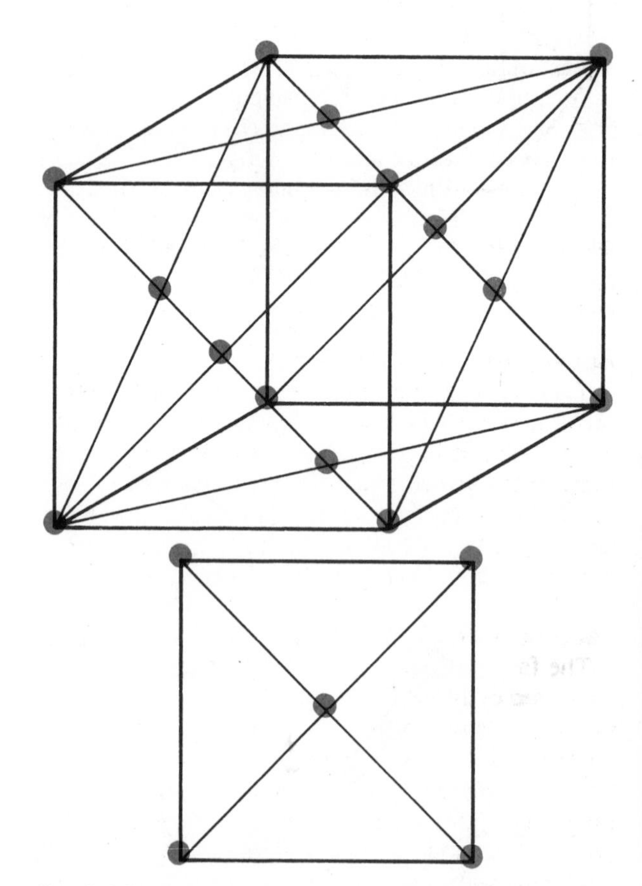

Fig. 3-19. Face-centered cubic lattice. Each face of the cube has an atom in its center.

Each unit cell shares atoms on its outer adjacent surface with other cells. The result is a grain of material made up of millions of tiny unit cells. The illustrations in Figs. 3-18 through 3-20 show only one unit cell so that it can be seen in its simplest structure.

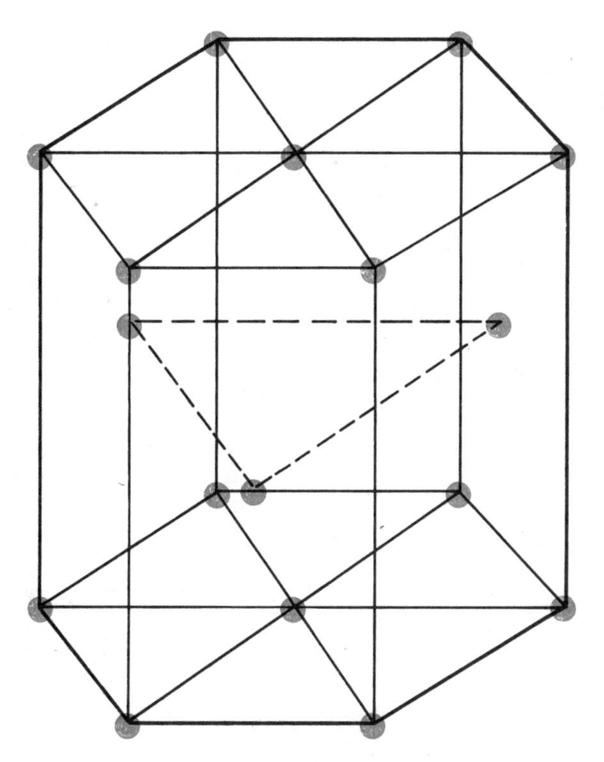

Fig. 3-20. One unit of the hexagonal close-packed lattice has 17 atoms.

Body-centered cubic lattice

The body-centered cubic lattice (BCC) structure is made up of nine atoms. Refer to Fig. 3-18 again. One atom is located at each corner of a cube and the ninth one is in the center. Body-centered cubic type metals generally are brittle and have high melting points. Common BCC metals are chromium, iron (steel), molybdenum, tungsten, and vanadium.

Face-centered cubic lattice

The face-centered cubic lattice (FCC) structure is composed of 14 atoms. There is one atom at each corner of the cube and one in the center of each of the six faces of the cube, Fig. 3-19.

Metals with the face-centered cubic lattice are ductile metals. They are easily formed. Included in this group are aluminum, copper, gold, lead, nickel, platinum, and silver.

Hexagonal close-packed lattice

In the hexagonal close-packed lattice (HCP) structure, Fig. 3-20, there are six atoms at the corners of each of the two hexagonal (six-sided) faces. One atom is located in the center of each face. Three atoms are located at the center of alternating sides of the lattice. Hexagonal close-packed lattice metals are not very ductile. When heated, these metals develop a coarse grain structure, become brittle, and lose toughness. Typical HCP metals are cadmium, magnesium, titanium, and zinc.

Most solid metals have only one space lattice structure. A few, however, can have two different lattice forms while they are in their solid state. These metals are called *allotropics.* The change from one lattice structure to another is called an allotropic change. Iron is the most common allotropic metal. It is a body-centered cubic lattice metal at room temperature. However, when it is heated to 1670° - 2250°F (910° - 1232°C) it changes to the face-centered structure.

Formation of metallic grains

A metal's grain structure is formed as it changes from its molten state to its solid state. When the temperature of molten metal is reduced, atoms in the material become less active. They become attracted to one another and form unit cells. The first cell is a simple unit lattice. This lattice becomes the nucleus (or seed) around which other atoms are arranged. These atoms align themselves around the nucleus lattice to form additional like structures. The lattice structure is formed, as shown in Fig. 3-21, with each unit cell sharing atoms. As cooling continues, the lattice structure grows into a large, three dimensional structure.

Throughout the metal other nucleus lattices are formed and grow during the cooling period. Not all the nucleus cells are in line with one another. They grow until the crystals approach each other. Since the original nuclei were not in line, the contact between the crystals are uneven and unordered.

The growth of the crystals and the resulting grain lines are shown in Fig. 3-22. This drawing shows only a two-dimensional view. Actually, the grains are being formed in three dimensions—thickness, width, and length. These grains are the smallest metal units that can be seen under a common microscope.

The size and number of grains formed will be determined by two factors:

1. The rate at which nucleus lattices are formed (nucleation rate).
2. Speed at which the grains grow.

A high nucleation rate will produce larger numbers of smaller sized grains. High grain growth with lower nucleation rate will produce larger grains.

Grain size has an important influence on the mechanical properties of the material. Coarse grains have lower strength than fine grains. However, coarse-grained material will be easier to machine, while fine grained material produces a better surface finish.

The grain size can be controlled during the solidification process and changed through conditioning (heat treating) techniques. See Fig. 3-23.

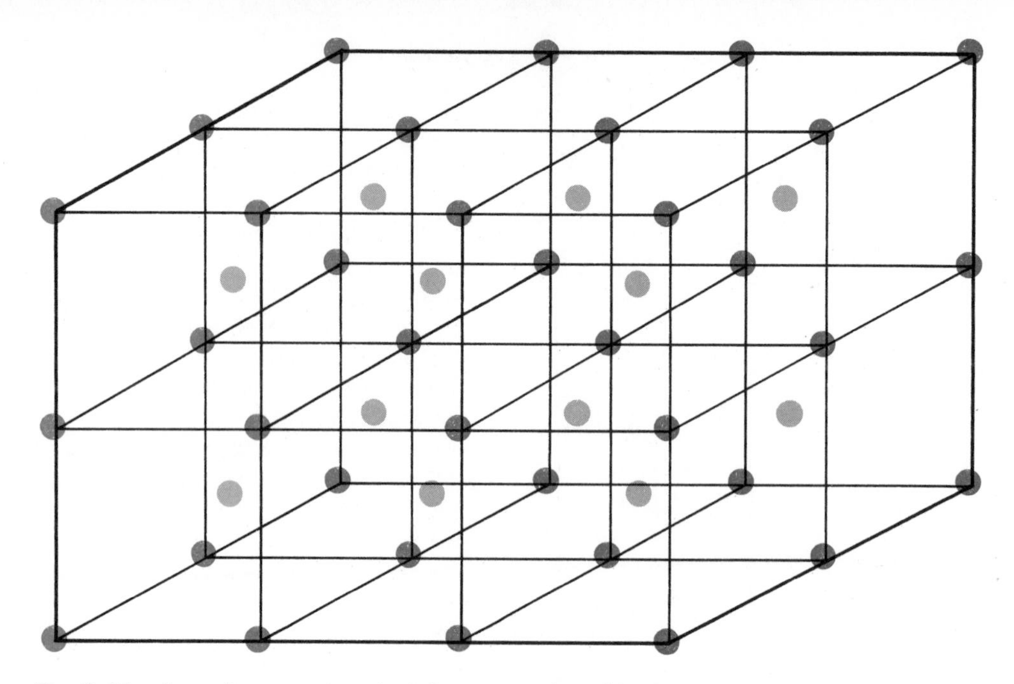

Fig. 3-21. As molten metal cools, it forms a series of lattice units like this. Note that the various lattices share atoms.

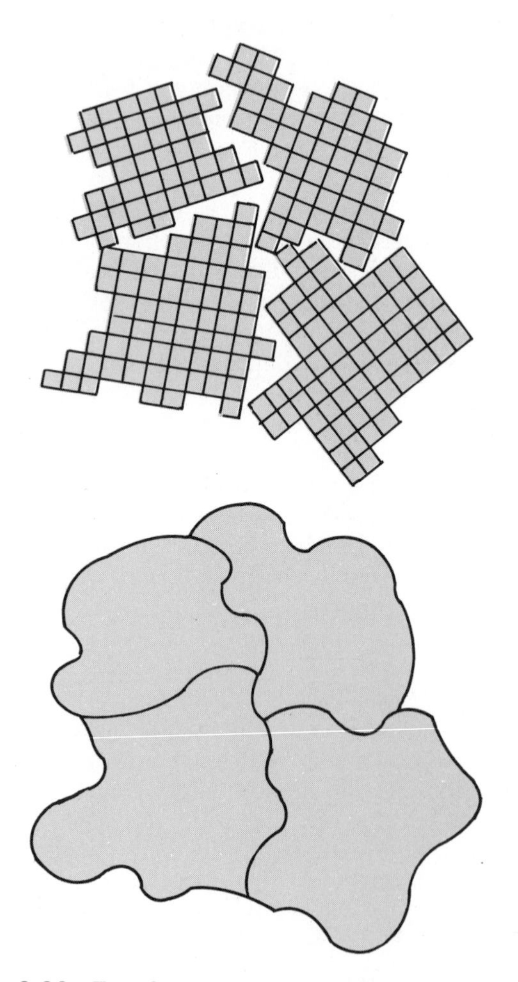

Fig. 3-22. Top. Schematic of lattices forming crystals. Bottom. Grain boundaries or lines between crystals.

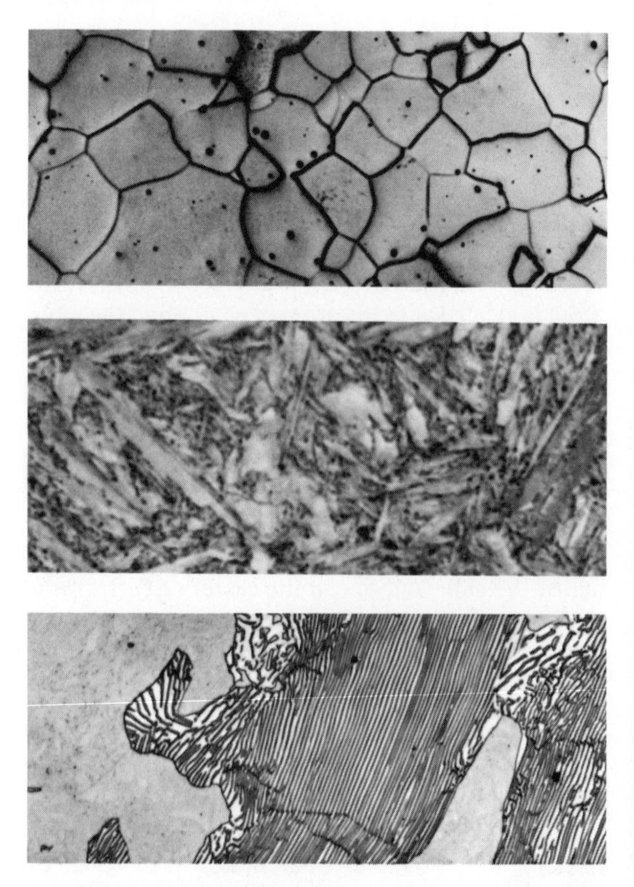

Fig. 3-23. Photomicrographs of metals. Top. Iron magnified 1000 times. Center. Hardened steel showing needlelike martensitic structure magnified 4000 times. Bottom. Mild steel showing pearlite structure. It has been magnified 1500 times.

STRUCTURE OF POLYMERS

Two basic groups of polymers are used in manufacturing. These, discussed earlier, are natural polymers (wood, wool, cotton, etc.) and synthetic polymers (plastics).

Natural polymers

Natural polymers are important raw materials for manufacturing. They may be used either directly or indirectly. Wood can be directly converted into lumber, plywood, shingles, particleboard, and hardboard. Cotton, wool, and other plant fibers may be made directly into fabric. Natural polymers may also be converted into other products. Wood can be chemically converted into rayon, turpentine, wax, and explosives.

The only natural polymer presented in this discussion, wood is the main engineering (solid) material which is a natural polymer. Wood, a vegetable matter, is made up of two major and two minor components. The major components are:

1. Cellulose. This is the most abundant part (about 70 percent) of a tree. It is the basis for most of the useful wood products.
2. Lignin. This natural adhesive holds the structural units (fibers) of wood together. It makes up 20-30 percent of the total volume of wood.

In addition to cellulose and lignin, wood contains other important materials. These components make up a minor part of the total volume of wood but have some major uses. These components are:

1. Extractives – materials contained in wood but which are not part of the structure or functioning of the tree. They provide the wood with such properties as color, odor, taste, and decay resistance. Typical extractives are starch, oils, tannins, coloring agents, fats, and waxes.
2. Minerals – the ash-forming part of wood. These are the materials left when the cellulose and lignin are burned. The minerals are the plant food elements of the tree.

Wood, like all living vegetable matter, is made up of long, thin, hollow cells. They are like very thin tubes or soda straws. The cell walls are composed of cellulose fibers which are held together with lignin. The cellulose fibers are composed of complex organic molecules called cellulose–chain polymers. These polymers are chemically built from simple sugar (glucose) molecules.

The long, thin cellulose tubes are often called wood fibers. They are usually arranged parallel to each other, producing what is called the grain of the wood. The fibers vary widely within a tree and among types (species) of trees. Generally they are about 1/25 in. long in hardwood (broad leaf deciduous) trees. Softwood (conifer) fibers are about 1/8 in. long. Fig. 3-24 shows an enlarged view of a wood sample.

Wood fibers, being hollow, are much different from synthetic polymers which are solid. The density of the cellulose and lignin which make up the fibers are the same. But, the cell wall thickness and the size of the hollow cavity in the fiber may vary widely. These factors cause the density (weight per unit of volume or pounds per cubic food) of various woods to be different. The mechanical properties of wood are almost directly related to the density of the material.

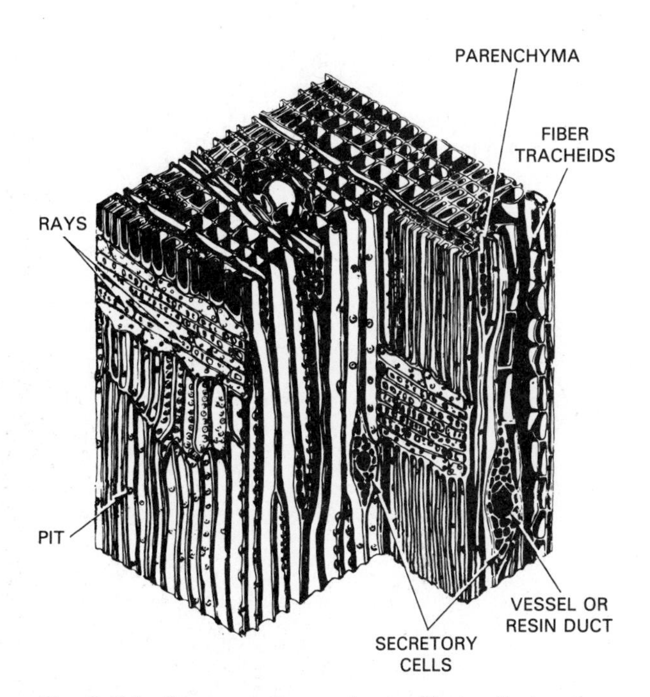

Fig. 3-24. Cross section and magnified cell structure of softwood. The various structures making up wood are held together by lignin. (Forest Products Laboratory, USDA)

Synthetic polymers

Plastics or synthetic polymers are the most widely used polymer in manufacturing. Their use, variety, and acceptance grows yearly. The term, *plastic,* is used in two ways. Originally it meant a material that would flow under pressure. It was any material which could be formed. Using this definition, metals, clays, and synthetic (made by people) polymers are all plastics or have a plastic state – a condition under which they can be formed.

More recently the term plastics is used to describe a group of synthetic organic solids which will flow under heat and/or pressure. These materials are made up of large molecules called polymers. The basic building block of a polymer is a monomer—a simple carbon-based compound. The monomer (meaning single mer) is a very simple organic compound with carbon as its central element. Other elements including hydrogen, oxygen, chlorine, fluorine, and nitrogen are linked to the carbon elements to form the basic organic compound. Fig. 3-25 shows the structure of a simple monomer.

Monomers can be chemically joined by covalent bonding to produce larger, more complex molecules. The action is called *polymerization.* It unites like mers into a polymer (many mers) chain. Fig. 3-26 illustrates polymerization. As can be seen, the polymer has a backbone of carbon atoms with other atoms attached to their ribs. They are held in place by primary bonds.

If two or more unlike mers are united, the action is called *copolymerization.* This process allows for many new plastic materials, Fig. 3-27.

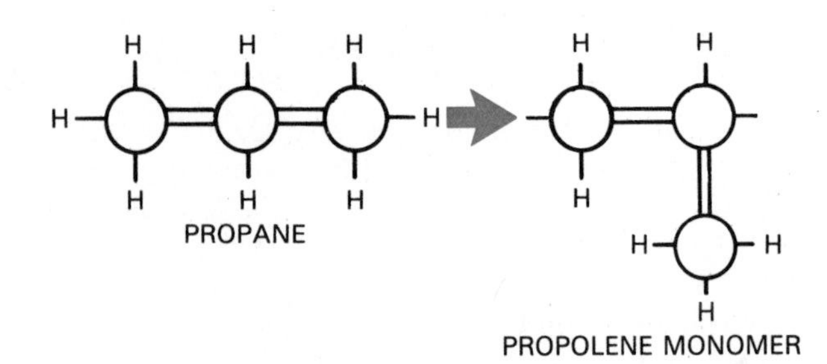

Fig. 3-25. Diagram shows a simple monomer.

The properties of a plastic material are determined by:

1. The elements within the molecule.
2. The structure of the molecule.
3. The strength of the bonds within the molecules.
4. The strength of the bonds between separate polymer chains.

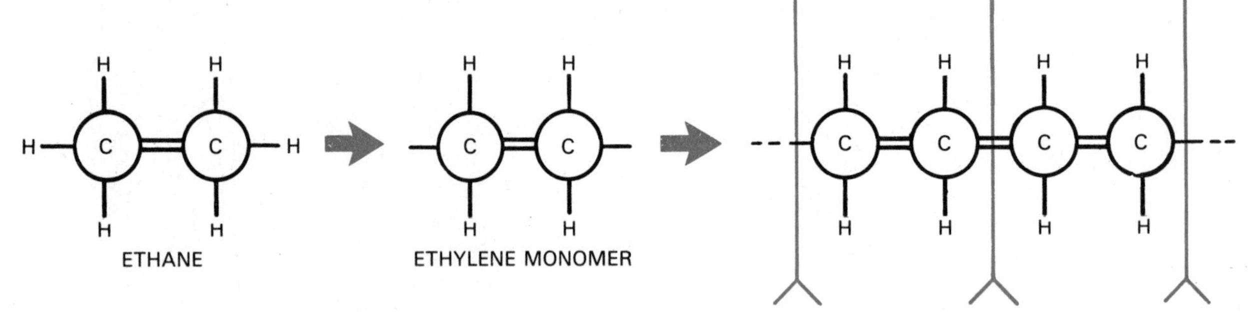

Fig. 3-26. Diagram shows chemical change from ethane to a polymer.

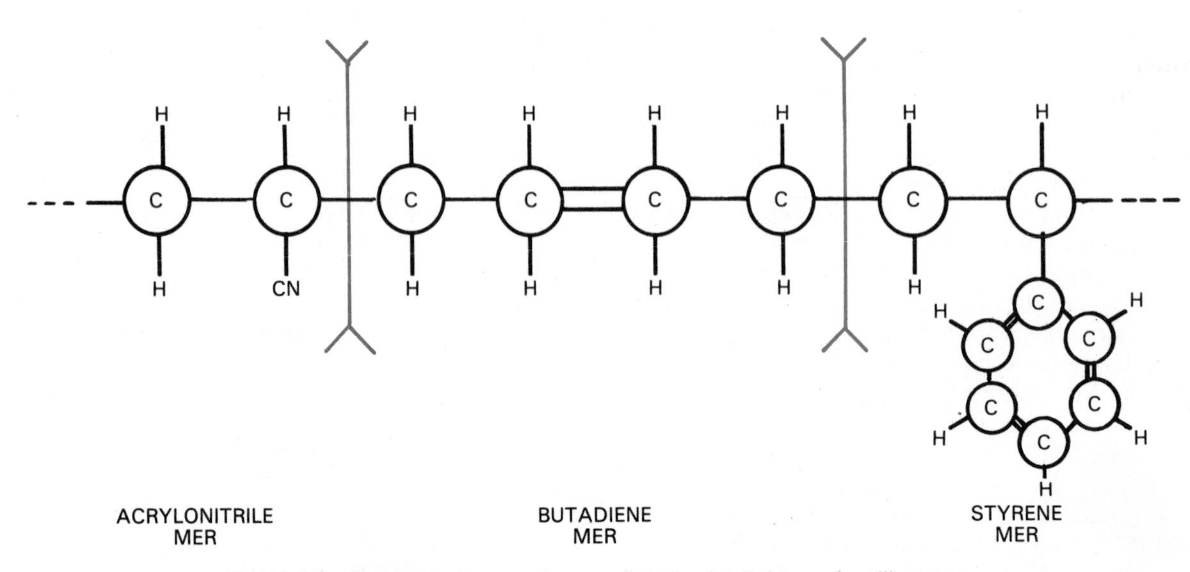

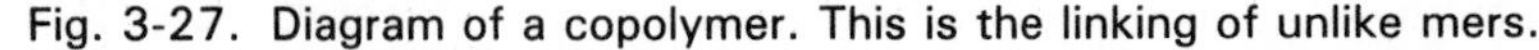

Fig. 3-27. Diagram of a copolymer. This is the linking of unlike mers.

Some plastics have strong bonds between the polymer chains. Separate chains are held together by the same type of bond (primary) that hold the individual polymer chains together. These polymers are called *thermosetting* plastics. The bonds between polymer chains resist breakdown from heat and pressure.

Other polymer chains are held together by much weaker bonds. These polymers are called *thermoplastic* materials. Increased temperatures weaken the bonds while decreased temperatures make them stronger. A thermoplastic material can be formed easily when its temperature is increased and will retain the desired shape if the temperature is reduced. This softening-hardening cycle can be repeated many times without causing damage to the structure of the material.

Elastomers

Elastomers are a special polymer. They may be either natural (rubber) or synthetic (neoprenes, polyurethanes, silicones) polymers. Their structure is very similar to plastic resins. They are generally composed of hydrocarbon (hydrogen-carbon based) molecules which have been polymerized. The main difference between plastics and elastomers is that elastomers have longer, straight polymer chains. They also have weaker bonds between the chains. These structural features allow elastomers, at room temperature, to stretch to at least twice their original length and then quickly return to their original length.

STRUCTURE OF CERAMICS

Ceramics include a wide variety of materials. Particularly important are the clays, the glasses, cement, and concrete.

These materials are composed of complex crystalline compounds. Unlike the simple crystals of most metals, ceramics have unit cells which often contain three or more different atoms. Also, the atoms in the cells are bonded by shared electrons (covalent bonds) or by transferred electrons (ionic bonds). This is much different than the electron cloud present in metallic bonds.

The covalent and ionic bonds in ceramic materials produce a much harder, stiffer material. Ceramics also resist high temperatures and chemical action. The crystalline structure of ceramic materials is greatly different and beyond the scope of this book.

PROPERTIES OF MATERIALS

All materials have their own properties or characteristics. These properties may be arranged into major groups which include:

1. Physical properties.
2. Mechanical properties.
3. Chemical properties.
4. Thermal properties.
5. Electrical and magnetic properties.
6. Optical properties.
7. Acoustical properties.

PHYSICAL PROPERTIES

Physical properties, for this discussion, are restricted to those which describe the basic features of the material. These features are measured or observed without the use of extensive scientific experiments. The common physical properties are size, shape, density, and porosity.

Size is the overall dimensions of the object. These dimensions, for most materials, are given as thickness, width, and length or as diameter and length. An object, as shown in Fig. 3-28, may be said to be 1/2 x 6 x 10 or 1/2D x 10.

Shape is the contour or outline of the object. Contour is given to an object by curved, notched, sloped, or other irregular surfaces, Fig. 3-29.

Density or specific gravity measures the mass of an object. The measurement is by weight for a unit or a certain volume. Typically, density is measured by pounds per cubic foot or kilograms per cubic meter of material. Density allows the mass of one material to be compared with that of other materials.

Porosity is a measure of voids (open pores) in the material. It is generally described as a ratio of open pore volume to total volume of a material. This ratio is expressed as a percentage. Porosity will provide a measure of liquid-holding power, bouyancy, or the ability of air or gas to move through the material.

MECHANICAL PROPERTIES

Mechanical properties mean a material's ability to carry or resist the application of mechanical forces and loads. The material's reaction to these forces is

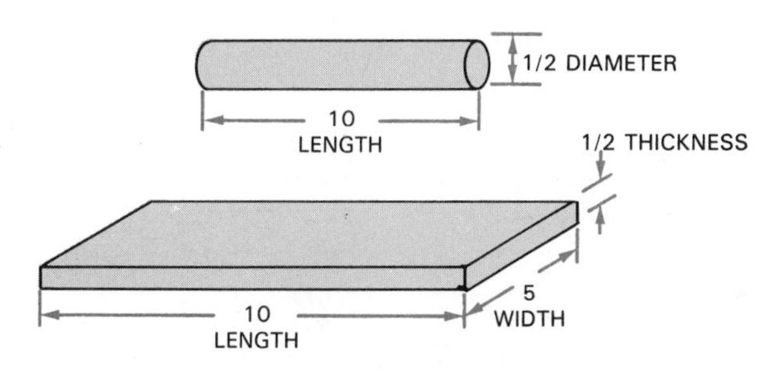

Fig. 3-28. Size specifications are one of the physical properties of engineering materials.

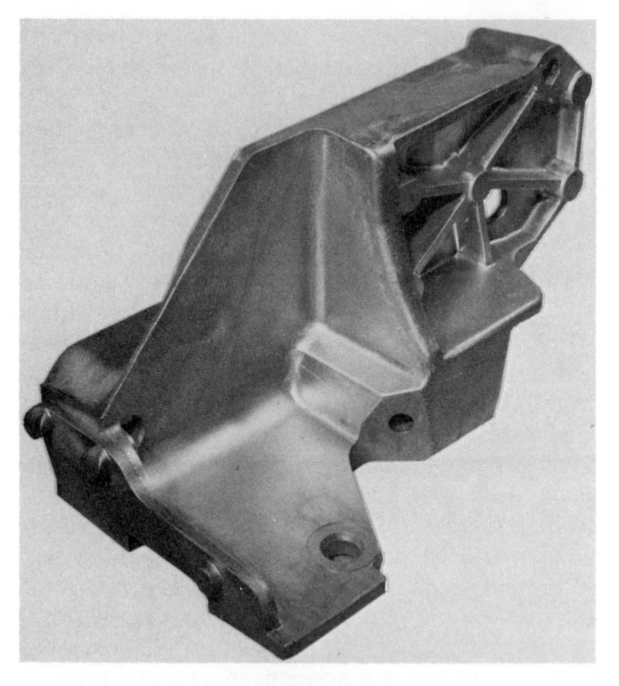

Fig. 3-29. The physical property of shape. This object has a complex shape. (Stahl Specialty Co.)

usually either deformation (shape change) or fracture.

Mechanical properties are probably the most important to manufacturing processing. They determine the extent to which a material may be formed, sheared, or machined.

Typical forces which are applied to a material are shown in Fig. 3-30. These forces—tension, compression, shear, and torsion—are used to form and shape materials. Furthermore, materials must withstand excess amounts of these forces in product applications. Since screws are used to assemble wood parts, they must absorb torsion forces. Rods holding suspended fixtures must withstand excess tension forces. The head of a hammer must absorb compression forces.

Stress-strain

The stress-strain relationship is often used to study many mechanical properties. *Stress* is force applied to material. It is usually measured in either pounds per square inch or kilograms per square centimeter. *Strain* is the change in the length of a material which is under stress. The strain measurements are given in terms of the amount of elongation (increased length) of the material per unit of length. Strain is

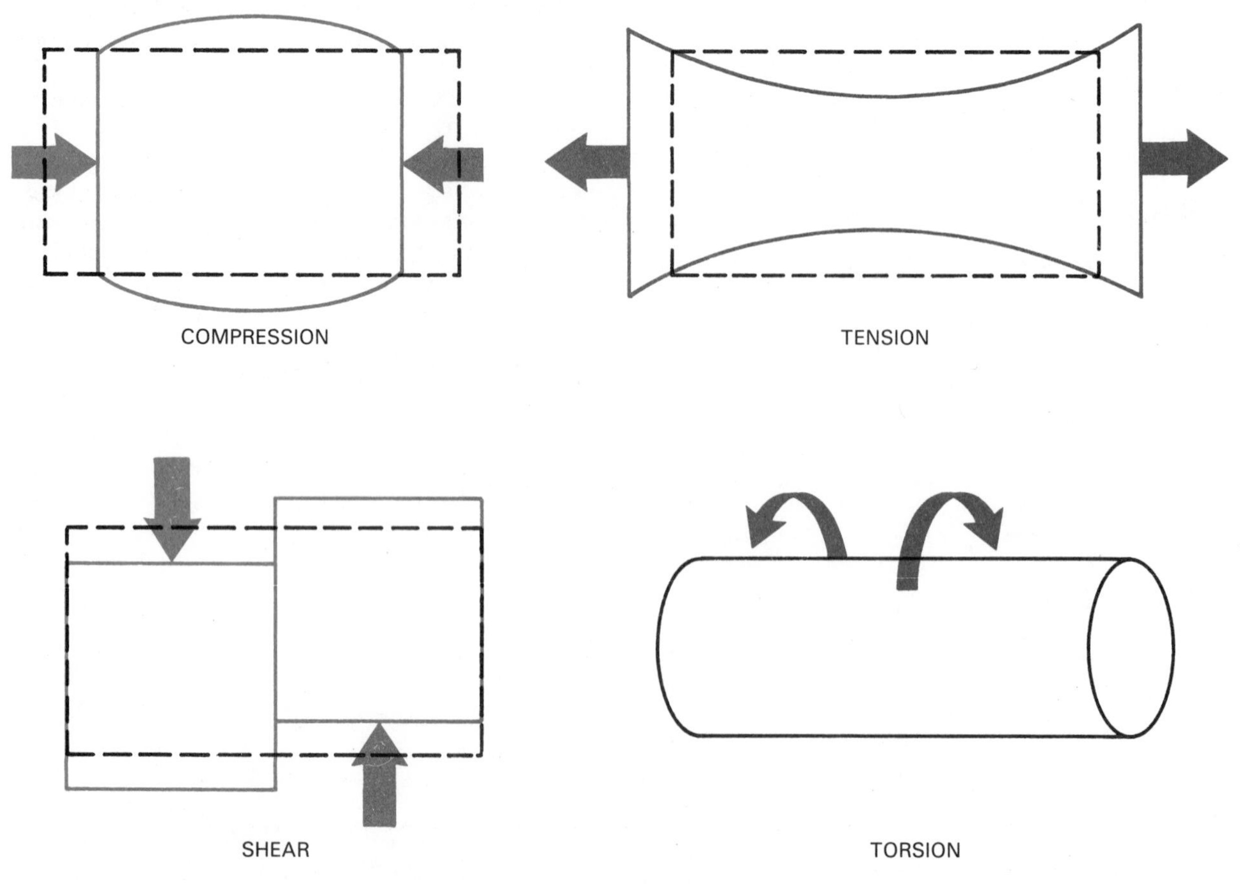

Fig. 3-30. These four types of force may be exerted on materials.

given in thousandths of an inch per inch of material or millimeters (or smaller units) per centimeter of material. For most materials, the elongation of a material under stress is quite small.

The amount of stress and strain is determined using rather simple mathematical formulas. These formulas follow:

STRESS

$$S = \frac{F}{A} \text{ or Stress} = \frac{\text{Load (Force)}}{\text{Cross-sectional area}}$$

Where (in metric):
S = stress in kg/cm^2
F = force in kg
A = area in cm^2
Or (in English measure):
S = stress in psi
F = force in lb.
A = area in sq. in.

Using this formula a 2 in. square bar supporting 2000 lb. would have 500 lb. of stress per sq. in.

$$S = \frac{F}{A} \quad S = \frac{2000}{4} \quad S = 500 \text{ lb./sq. in.}$$

STRAIN

$$e = \frac{\ell}{L} \text{ or Strain} = \frac{\text{Amount of elongation}}{\text{Original length}}$$

Amount of elongation = final length − original length

A stress-strain diagram, like the one shown in Fig. 3-31, is widely used to chart stress-strain relationships. The stress (force per unit area) is plotted on the vertical axis while the strain (elongation of each unit of length) is plotted on the horizontal axis.

As stress is applied, the material first resists permanent deforming. This area is in the materials *elastic range.* This is a range in which the material will return to its original length when the force is released.

Applying additional stress (force) will bring the material to its *yield point.* At this point, additional strain (elongation) occurs without additional force (stress) being applied. Strain above this point is produced with smaller amounts of force. The force also produces permanent changes in the length of the material.

This elongation which is above the material's *elastic limit* (point at which the material will not return to its original length) is called *plastic deformation.* All forming processes, which are discussed in Chapters 9 through 14, use the plastic deformation range of materials.

As stress is increasingly applied above the yield point, additional strain occurs. Finally, a maximum strain is reached and the material begins to fail. Its internal structure begins to come apart. This point is called the material's *ultimate strength* or *tensile strength.* Additional stress may cause a reduction in cross-sectional area (necking) and will finally cause fracture.

Mechanical strengths

A material can be subjected to a number of different types of forces. They may be tension, shear, torsion, compression, or a combination of these forces. Each possible force causes a material to

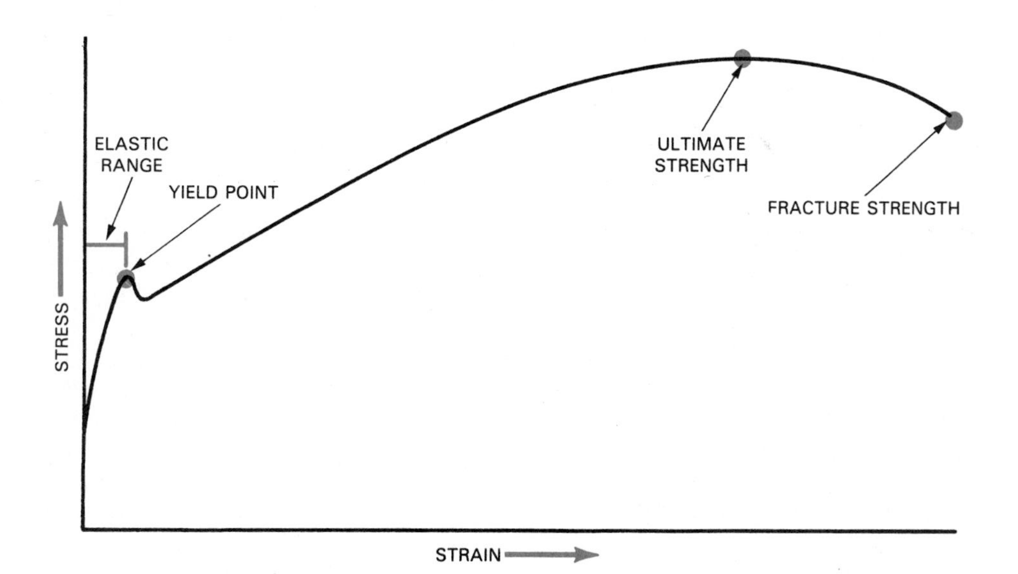

Fig. 3-31. Stress-strain charts are used to show the relationship of forces and how they act on a certain material.

respond in a different way. A material, therefore, has several different mechanical strengths. The strength depends on the force applied.

The most common mechanical strengths are:

1. Tensile strength – the maximum tension load a material can withstand before fracturing. Tensile strength is the easiest strength to measure and, therefore, is widely used.
2. Compression strength – the ability to resist forces which tend to squeeze the material into a new shape. It is basically the opposite of tensile strength. Excessive compression force will cause the material to rupture (buckling and splitting).
3. Shear strength – the ability to resist fracture under shear forces. The shear force is caused by offset forces applied in opposite directions. These forces cause the grains or molecules of the material to slide by one another and eventually fracture.
4. Torsion strength – the ability to resist twisting forces. Forces which exceed the torsion strength (modulus of rupture) will cause the material to rupture.
5. Flexure (bending) strength – the ability of a material to resist the combination of tensile and compression forces. As seen in Fig. 3-32, when a material is bent, the material on the inside of the bend must compress while that on the outside portion must stretch. A material must have flexure strength to undergo bending processes.

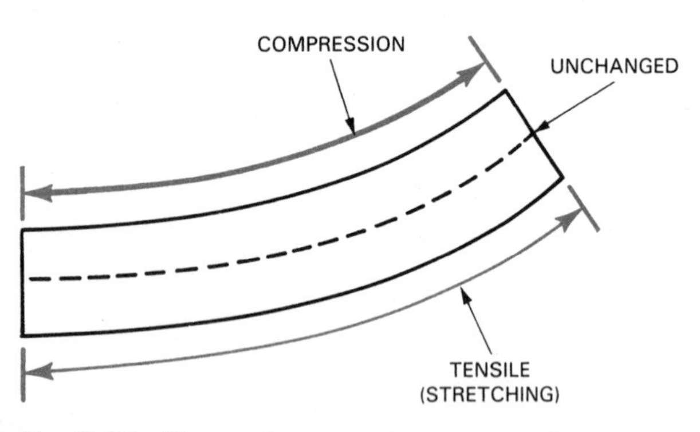

Fig. 3-32. Flexure force creates compression on one surface and stretching on the opposite surface of a material.

6. Fatigue strength – the ability to resist forces which vary in direction and/or magnitude. Typical of forces which cause fatigue are constant bending back and forth (plastic hinge), applying and releasing tension forces (coil spring), or torsion forces (automobile torsion bar).
7. Impact strength – the ability to resist a rapidly applied load. This is a more specific measure of tension or compression strengths. Impact strength determines the ability to absorb a tension or compression load which is quickly applied. Impact strength is often called toughness. The action of a hammer on a nail applies such impacts. Forging dies must have high impact strength.

Plastic flow of materials

In addition to the mechanical properties involving strength, materials have characteristics that govern their behavior during the plastic deformation stage. These properties are usually called ductility and creep.

Ductility is the plastic flow characteristic of a material under normal temperature. The higher the ductility of a material the greater is its ability to be formed without fracturing. Highly ductile materials can be easily bent, drawn into wire, or extruded.

Creep is the movement or plastic flow of material under load over an extended period of time. Creep of a material is affected by the stress level (force applied), temperature of the material, and the length of time over which the stress is applied. Also different types of material have different creep characteristics. Wood creeps very little. Metal experiences slight amounts of creep. Glass and plastic are more likely to experience significant amounts of creep.

Hardness

Hardness is the resistance of a material to penetration or scratching. It accounts for abrasion resistance as well as resistance to denting. Hardness is also directly related to strength. The harder a material the stronger it is. Metallic and ceramic materials are almost always harder and stronger than polymeric materials. A number of different testers have been designed to test the hardness of a variety of materials.

CHEMICAL PROPERTIES

All materials are used in some type of environment. All environments, except a pure vacuum, contain chemicals. These chemicals may be gases (oxygen, hydrogen, chlorine, nitrogen), liquids (water, acids, oils) or solids (other engineering materials). The reactions between the chemicals and a material in an environment is called *corrosion*.

These reactions are complex and require a knowledge of chemistry for a full understanding. In general, there are several types of chemical reactions. These include:

1. Oxidation – corrosion caused by a reaction between the material and oxygen in the air. Iron oxidizes to form iron oxide (rust). Some polymers are weakened and destroyed by a combination of sunlight and oxygen. Rubber is particularly prone

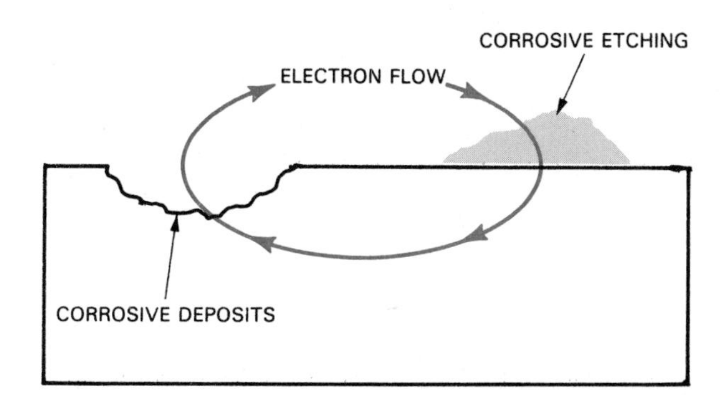

Fig. 3-33. Electrochemical corrosion is caused by an electrical current set up when a material is placed in a moist or liquid environment.

to oxidation which is called aging. As rubber ages it loses its flexibility (elasticity) and strength.

2. Electrochemical corrosion – corrosion caused by electrical currents set up in a material in a liquid or moist environment. Electrical current flows from one point on the material (anodic area) to another part (cathodic). The current removes materials from the anodic area and deposits it at the cathodic area, Fig. 3-33. This action is similar to the operation of the storage battery.
3. Water absorption – the tendency of polymers and ceramic materials to absorb water. This action increases the material's weight and volume. Water absorption will cause many materials to warp and swell, and lose desirable mechanical and electrical properties.

THERMAL PROPERTIES

All materials are subjected to thermal (heat) energy. Heat may cause the material to change its physical form (melt, evaporate), to change in size, or to change in temperature. Heat may also change the internal structure of the material.

The important thermal properties of materials are:

1. Heat resistance, the ability of a material to remain stable with changes in temperature. Ceramic materials have high heat resistance ratings.
2. Thermal conductivity, the ability of a material to transfer heat energy along its length, width, or thickness. The rate of heat transfer for a material is called its coefficient of thermal conductivity. Metals have high coefficients of thermal conductivity. Ceramics and polymers have low conductivity rates. They are therefore good heat insulators.
3. Thermal expansion, the rate of size change per unit of heat applied. Almost all engineering materials expand as they are heated. The rate of expansion is called the coefficient of linear expansion. The rate of expansion varies greatly among materials.

The varying rates of thermal expansion are the basis of the bimetal thermostat. The metal strips of the thermostat have greatly different thermal expansion coefficients. When they are bonded together and heated, the metal with the high expansion rate elongates much more than the other strip. This causes the strips to bend and touch the contact point. This action closes an electric circuit which could start a furnace, turn on a cooling fan, or light a warning light.

ELECTRICAL AND MAGNETIC PROPERTIES

Electrical and magnetic properties are similar. The electrical properties describe the behavior of a material carrying an electrical current. Magnetic properties involve the behavior of materials in an electromagnetic field. These properties include:

1. Electrical conductivity – the ability to conduct an electric current. Engineering materials are classified as conductors (easily carry electrical current) and insulators (resist carrying electrical current). Other materials are called semiconductors and are the foundation of the computer and data processing industry.

   The opposite of electrical conductivity is electrical resistivity (resistance to conducting electrical). If resistivity is an important property, it is often discussed as the dielectric property or strength. The dielectric strength is a specific measure of a material's electrical insulating qualities. Metals generally are considered conductors of electricity while ceramics and polymers are insulators (have high dielectric constants).
2. Magnetic properties – the ability to be magnetized by an external electromagnetic force. Materials with high magnetic properties are said to be permeable or have high permeability. Iron, nickel, and cobalt which are called ferromagnetic materials have high permeability. They are easily magnetized.

OPTICAL PROPERTIES

Optical properties govern the material's reaction to light. These properties include:

1. Opacity – the degree to which a material obstructs visible light. Materials with high opacity are said to be opaque. Low-opacity materials are called transparent.
2. Color – the appearance of the material which depends on the spectrum of light it reflects to the human eye.

3. Transmittance – the measure of the amount of light transmitted by a material. The opposite of transmittance is reflectance. Window glass has high transmittance while polished aluminum has high reflectance.

ACOUSTICAL PROPERTIES

Acoustical properties describe a material's reaction to sound waves. The most important acoustical property is the sound insulating or absorbing quality of the material. This is the ability of the material to absorb sound and is called its sound absorption coefficient. Most ceiling tile has high sound absorption coefficients to reduce echoes in a room.

SUMMARY

Materials are an important part of our daily lives. They are most likely present in some refined or manufactured form. The materials may be metals, polymers, or ceramics. Each of these materials are composed of one or more of the 96 basic elements. The elements are arranged in a specific molecular form called a structure. Metals and ceramics have a crystal structure. Polymers have a structure made up of organic molecular chains.

The structure of the materials provide each material with a set of properties. These properties include physical, mechanical, chemical, thermal, electrical, magnetic, optical, and acoustical. An understanding of these properties is valuable in the selection of specific materials for particular product applications.

STUDY QUESTIONS – Chapter 3

1. List and briefly describe the three types of engineering materials.
2. What is the difference between inorganic and organic materials?
3. What is the difference between thermoplastic and thermosetting plastics?
4. Describe the basic structure and components of an atom.
5. Describe covalent bonding, ionic bonding, and metallic bonding.
6. List and briefly describe the lattice structure found in the commercially important metals.
7. Describe the process by which a metal's grain structure is developed.
8. Describe the basic structure of wood.
9. What is the difference between a monomer and a polymer?
10. What is polymerization?
11. What is an elastomer?
12. What is a ceramic material?
13. List and describe the seven basic properties of materials.

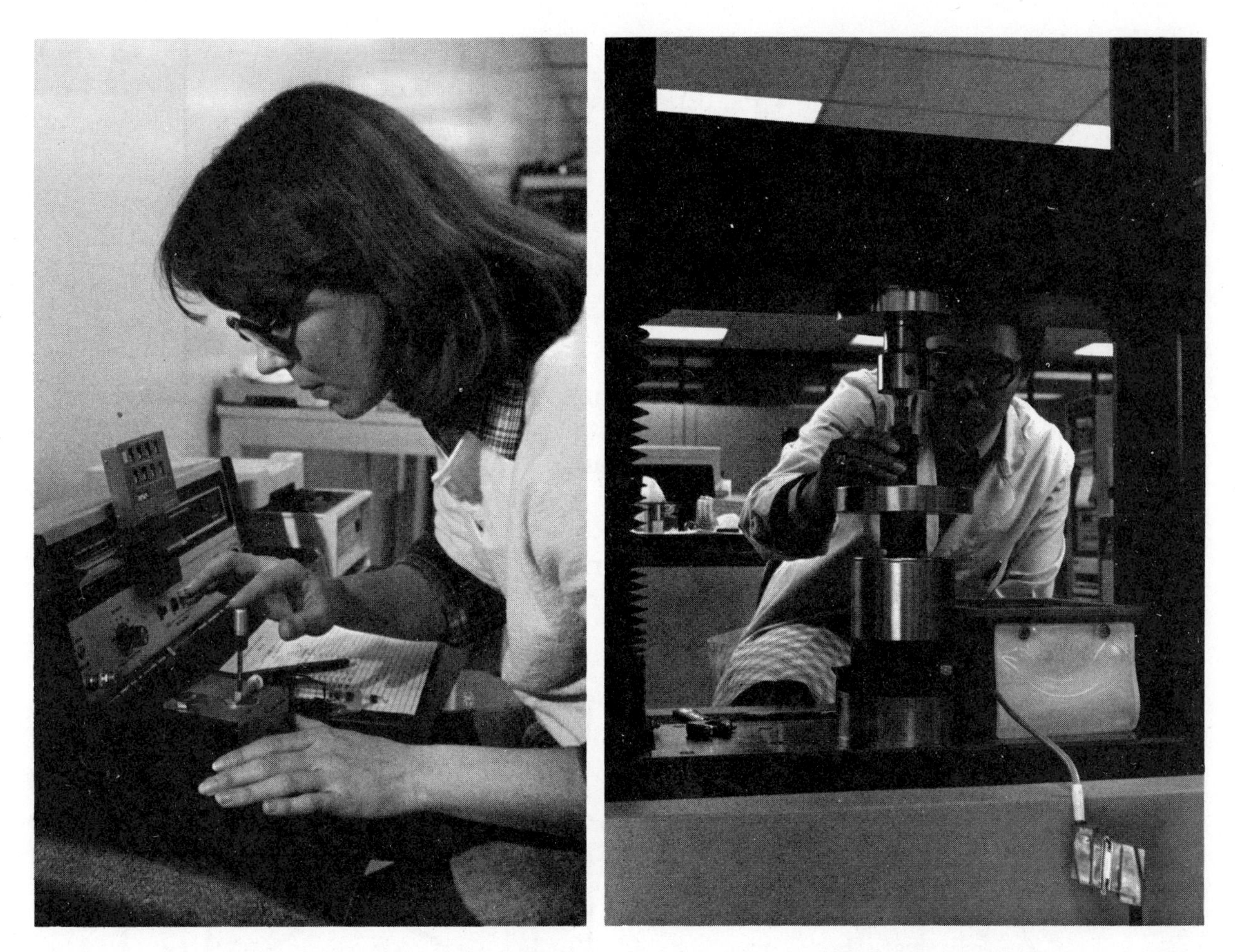

In industry, products undergo testing to meet standards set for strength and other properties. Left. A Beta Backscatter device is used to measure the coating of platinum on an exhaust oxygen sensor. Measurement is made in micro inches. Right. Technician is placing an element-loaded fixture into a burst-tester. The test is part of the quality control procedure for testing automotive parts. (AC Spark Plug—GM)

SECTION 2
CASTING AND MOLDING

Casting and molding are important industrial processes used for converting liquid or plastic raw materials into finished products. Metals, plastics, and ceramics are shaped this way. Major families of molding/casting processes include:

- Metal casting with either expendable or permanent molds.
- Casting and molding of plastic materials.
- Drain casting and solid casting of ceramic materials.

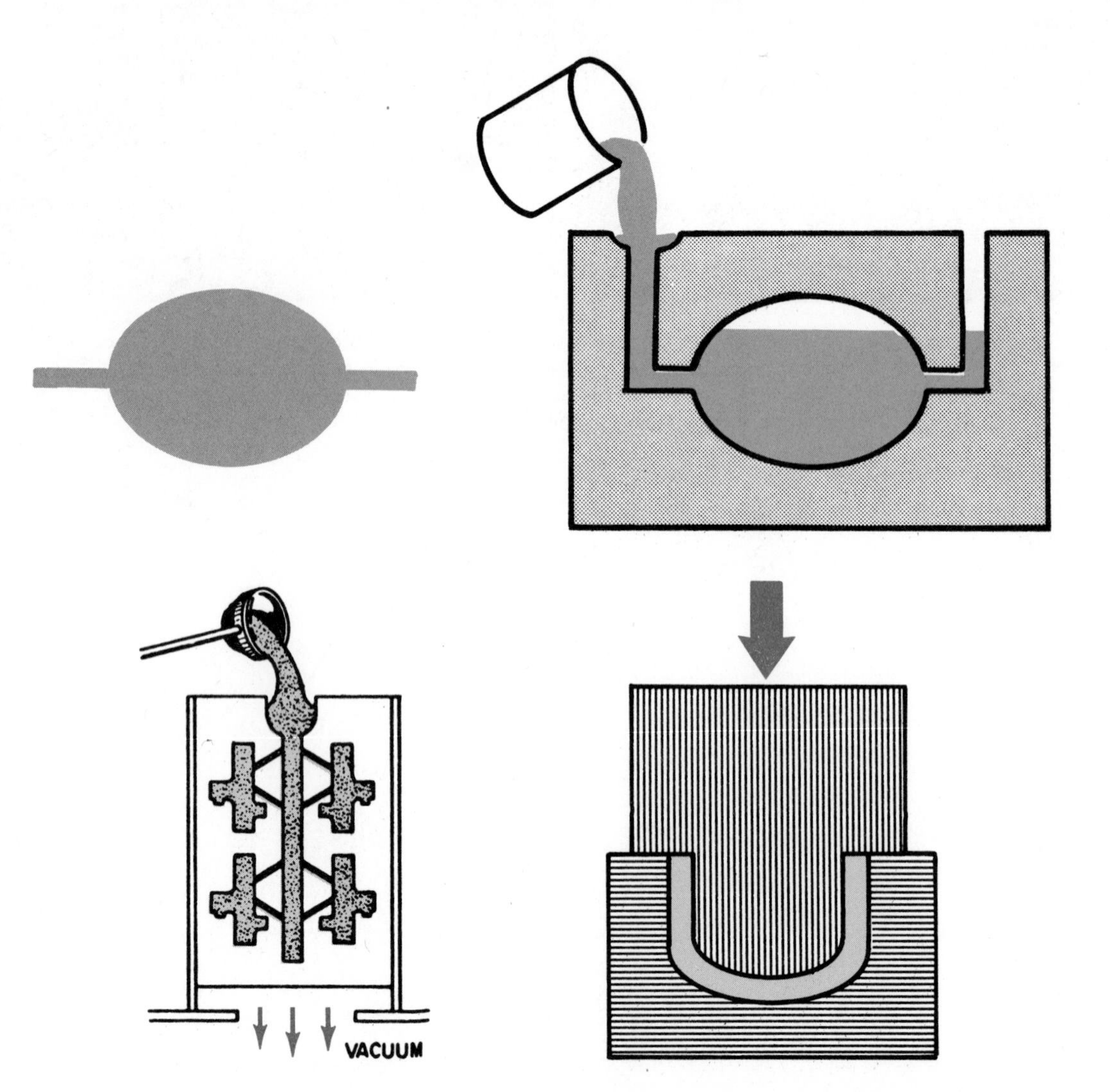

Chapter 4

INTRODUCTION TO CASTING AND MOLDING PROCESSES

Casting and molding is a family of manufacturing processes in which an industrial material is usually changed to its liquid or plastic state. Occasionally the material is already a liquid. It is then poured or injected into a mold cavity and allowed to solidify before being extracted.

If the material is in its liquid state, the process is generally called casting. Molding usually applies to processes that force a material into a mold while it is in a soft or "plastic" state.

HOW CASTING DEVELOPED

Historians do not agree on the actual development of casting as a method of changing the form of materials. Most seem to agree it was the first major step away from chipping material into desired shapes.

The first casting activity probably dates to around 4000 B.C. At that time, inhabitants of the Near East melted copper and copper alloys in small clay-lined furnaces. They cast the molten metal into open clay or sand half-molds. This activity produced crude tools and weapon parts.

Later, pre-Christian artists produced elaborate bronze statues using more complex techniques. The Chinese were making iron castings by 600 B.C. and India made crucible cast steel in 500 B.C. The Indian process was lost to civilization but was rediscovered in England in the mid 1700s.

During the Middle Ages many everyday items were cast from pewter while cannon shot was cast from iron. Later developments allowed humans to cast endless products from zipper parts to room-sized gear blanks using various casting techniques. Also, plastic materials were shaped through molding techniques in ever-increasing quantities.

Casting and molding techniques have grown in acceptance and breadth of application because they provide the most direct route from raw material to finished product as seen in Fig. 4-1. This characteristic makes casting and molding a relatively low-cost process.

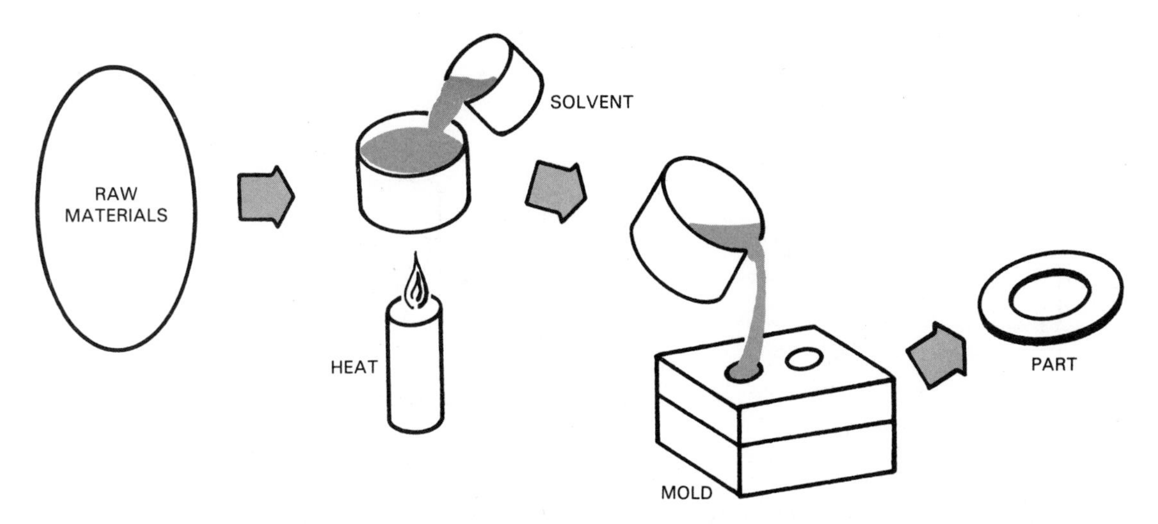

Fig. 4-1. Casting provides a direct route from raw material to a semifinished part.

ADVANTAGES OF CASTING AND MOLDING

Casting and molding have several strong features which account for their success as methods to process materials. These advantages include:

1. Products with complex forms and intricate internal and external shapes may be cast or molded at relatively low costs.
2. Materials, such as certain alloys, difficult to shape using separating and forming techniques, can often be cast or molded.
3. Parts are often simplified. One-piece castings or molded parts may require several pieces when produced by other techniques.
4. Product design changes are often easily incorporated in casting and molding processes.
5. Large, heavy metal parts may be cast. These parts may be economically or technically impossible to produce using other material processing techniques.
6. High volume, low cost production is possible with casting and molding techniques.
7. The number of different casting and molding techniques available permits the production of parts of varying tolerance and complexity out of a variety of metallic, ceramic, and synthetic (plastic) materials.

DISADVANTAGES

However, casting and molding techniques are not without technical and economic disadvantages. Included in these are:

1. Problems with internal porosity (trapped air pockets).
2. Dimensional variations due to shrinkage.
3. Trapped impurities, including solid and gaseous matter.
4. Inability to produce high-tolerance smooth surfaces especially in metals.
5. Inability to produce metal parts with the toughness common to forged parts.
6. Inability to compete with stamping and deep drawing in production of thin, formed parts.
7. Inability to compete with extrusion in producing a uniform cross-section part of varying lengths.

These lists of advantages and disadvantages are only given to represent major considerations. They are general in nature and can only be generally applied. Each casting and molding technique has its own specific economic and technical reason or reasons for use. Thus, an individual selecting manufacturing processes for a particular job must weigh not only the general but also the specific advantages and disadvantages of each possible processing technique. Only then can the best processing technique be selected.

THE CASTING AND MOLDING INDUSTRY

The casting and molding industry may be viewed in several ways. Each view provides its own insight into what might be called the "Casting and Molding Industry" or, more narrowly, the "Foundry (metal casting) Industry." Fig. 4-2 shows two possible ways

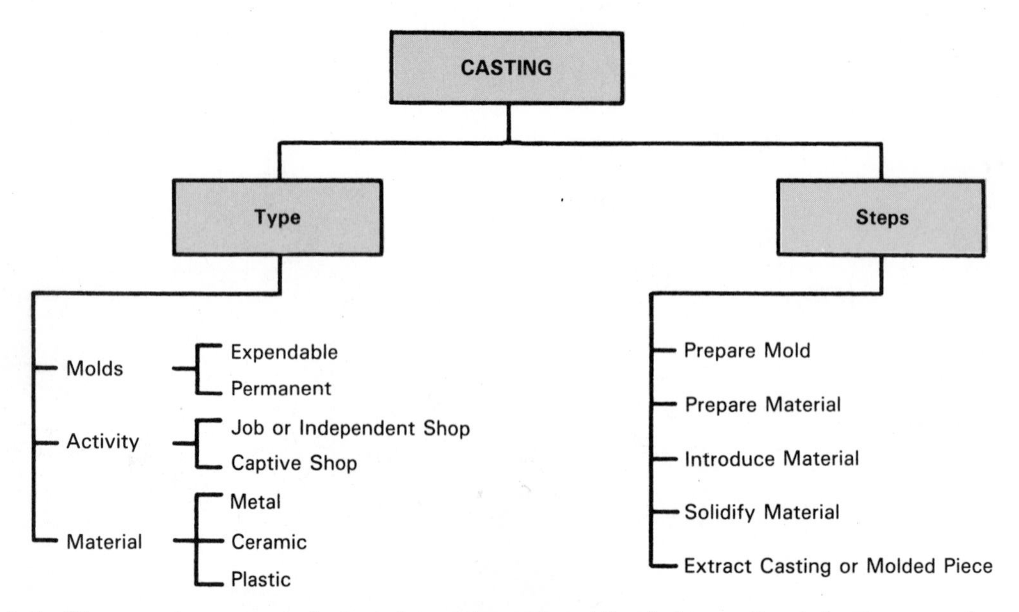

Fig. 4-2. There are two ways to look at the casting and molding industry. One is by the type of casting. The second is through the steps involved in casting and molding.

to gain an understanding of this industry.

One approach views the industry in terms of its "type" (that is, the type of material cast, the type of ownership or business sought, or the type of casting or molding used). The other approach is from the five major steps in producing a casting or a molded part:

1. Preparing the mold.
2. Preparing the material.
3. Introducing the material into the mold.
4. Solidifying the material.
5. Removing (extracting) the piece.

As an aid to general understanding of casting and molding, each view and its elements will be discussed. This will provide a background for later chapters which will present specific casting and molding techniques as they relate to metal, plastic (synthetic), and ceramic materials.

TYPES OF MATERIALS, OPERATIONS, AND TECHNIQUES

The casting and molding industry shapes a wide variety of materials using many different techniques. The output — castings and molded products — serve humans in almost every phase of individual and industrial life, Fig. 4-3. We need only look around us to realize the wide use made of casting and molding. The automobiles or buses we ride in have cast engine blocks, molded plastic dash panels, grills, and trim parts. Cast ceramic parts are found in the radio and the microcomputer fuel control system. Our homes have cast ornamental items and molded trim on appliances, to name a few. Cast and molded items are everywhere.

MATERIALS CAST OR MOLDED

Almost any solid material which can be melted, softened into a plastic state, dissolved or suspended in a liquid, can be cast. If the material is in a liquid state, the process is generally called casting. Molding usually means forcing a material into a mold while it is in a soft or "plastic" state. Typically, the casting industry is divided into metallic casting operations (foundries), plastic molding operations, and ceramic casting areas.

Metal casting operations are usually divided first as ferrous (containing iron) and nonferrous foundries.

Fig. 4-3. Typical cast products include many items readily recognized by the public. (Stahl Specialty Co.)

They are further classified as follows:

FERROUS	NONFERROUS
Gray iron	Aluminum alloys
Malleable iron	Brass-bronze
Ductive iron	Zinc alloys
Steel	Magnesium alloys
	Titanium

Many foundries cast more than one metal. This is particularly true of nonferrous foundries.

The metal to be cast is chosen because of its characteristics. These include:

1. Fluidity (flowing characteristics).
2. Chemical stability when in contact with air and mold materials.
3. Resistance to bonding to mold materials.
4. Its ability to meet the requirements of the end product.

Pure metals are seldom cast because they are too soft and weak for economic use. They are also subject to excessive shrinkage as they change from a liquid to a solid. Therefore, alloys (mixtures of metals) are usually cast.

Plastic molding operations use both thermosetting and thermoplastic materials. The thermosetting plastics take permanent shapes during the molding process. They cannot be reheated and reused. The distributor cap of an automobile is a good example of a molded thermosetting material.

The thermoplastic material, on the other hand, can be reclaimed. Each time it is heated it will soften and can be remolded.

Clay materials also can be cast. They must be suspended in a liquid then introduced in the mold. Until the clay product is fired (hardened) in a kiln, it, like a thermoplastic material, can be reclaimed and reused. After firing, it is like a thermosetting plastic. It is in a nonreversible material state.

TYPES OF CASTING AND MOLDING OPERATIONS

There are two basic types of casting and molding operations. They are classified by their customers and the organizational structure. These two, as shown in Fig. 4-4, are independent shops and captive shops.

INDEPENDENT FOUNDRIES

Independent foundries and molding operations bid and contract to a number of companies for work. They are independent companies selling their ability to produce castings of specified standards at the best available price.

Some independent operations are set up to handle small quantity orders and are called *job shops*. Others, called *production shops* depend upon long runs of a few different castings. This high-volume production allows them to produce products at a lower unit cost.

CAPTIVE SHOPS

Captive shops are separate plants or areas within a larger plant. They produce castings or molded parts for their owners. They are a specialized part of a larger manufacturing company. Their output is mainly used by the parent company. Seldom are their castings sold to other companies.

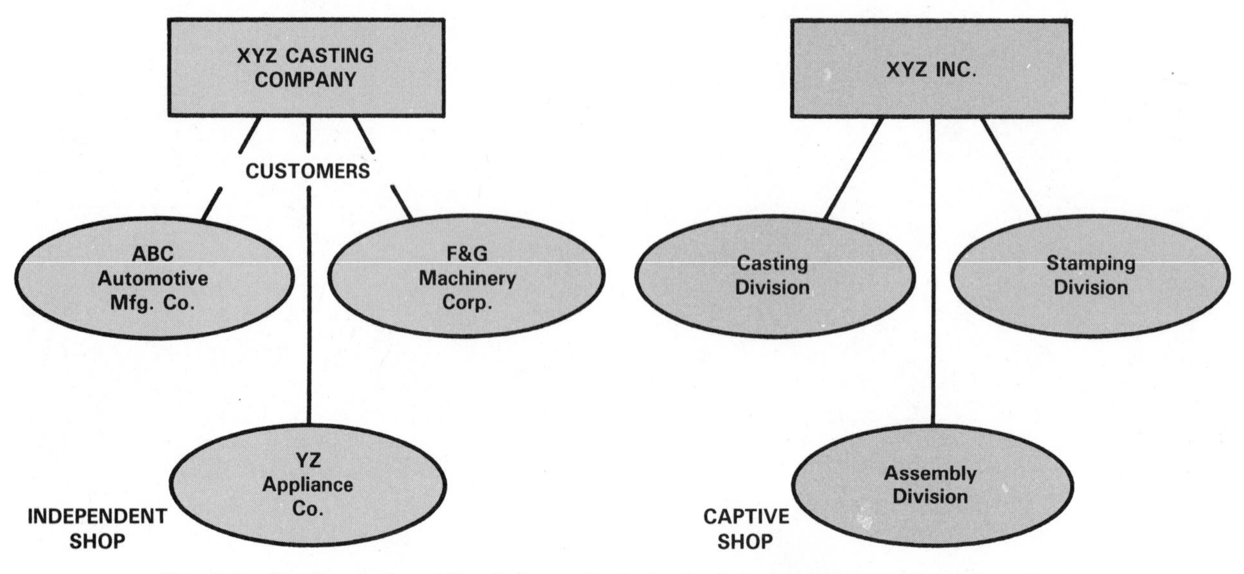

Fig. 4-4. Casting and molding industry includes both independent and captive shops.

CASTING AND MOLDING PROCESSES

Casting and molding processes often take their name from the molds they use. These molds may be separated into two categories: expendable and permanent. Expendable molds are used only once and are almost totally limited to metal casting. Permanent molds may be used for many casting cycles and for casting or molding metallic, ceramic, and synthetic materials. Fig. 4-5 lists the major casting and molding techniques which will be described in more detail in Chapters 5-8.

STEPS IN CASTING AND MOLDING

The five basic steps listed in Fig. 4-2 are common to all casting and molding techniques. They are universal steps no matter what material is being cast or molded.

Casting and molding of metal, plastic, or ceramic material requires that a mold be produced. Then the material must be prepared for the process. It must be made liquid or changed to its plastic state. If the material is already liquid, a hardening agent is often added.

The liquid or softened material is then introduced (poured or forced) into the mold cavity and allowed or forced to harden. The hardened material, called a casting or molded piece, is then removed. Additional manufacturing processes such as machining, conditioning, and finishing may be required to change the casting or molded piece into a finished product.

PREPARING THE MOLD

All casting and molding techniques require a container of the desired shape in which the molten or plastic material may solidify. This container is called a mold. Molds are of two basic types; *expendable* and *permanent*.

Expendable molds

Expendable or one-shot molds are used almost exclusively in metal casting techniques. They are destroyed to remove the casting. Therefore, a new mold must be produced for each casting cycle.

The production of an expendable mold requires two major steps, Fig. 4-6. First a pattern (model) of the part must be produced. The pattern is a duplicate of the part which has been modified or changed. The changes are required by the casting technique chosen and the material to be cast. The major modifications are *shrinkage allowances, machine finish allowances, size tolerances,* and *draft.*

Shrinkage allowance

Almost all materials shrink when they change from a liquid to a solid. An allowance for this shrinkage must be made. If the pattern were the exact size of the desired finished part, the finished casting produced would be too small. Therefore, the pattern is made "oversize" to allow for material shrinkage. Typical

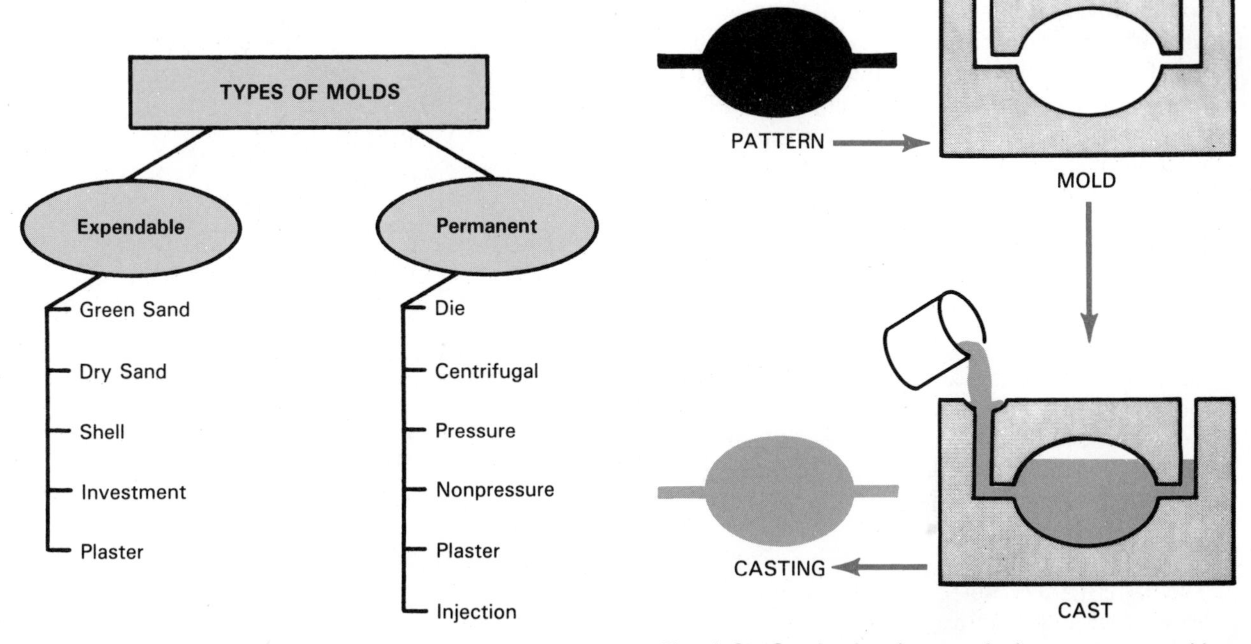

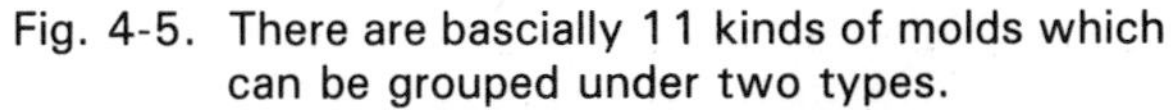
Fig. 4-5. There are bascially 11 kinds of molds which can be grouped under two types.

Fig. 4-6. Casting involves producing a pattern, making a mold from the pattern, and producing a casting.

allowances are as follows:

Cast iron	1/10-1/8 in. per ft.
Steel	1/4-3/16 in. per ft.
Aluminum	1/8-5/32 in. per ft.
Magnesium	1/8-5/32 in. per ft.
Brass	3/16 in. per ft.
Bronze	1/8-1/4 in. per ft.

To aid the patternmaker, special *shrink rules* for various shrinkage allowances are manufactured. On a 1/8 in. shrink rule, the 12 in. mark is actually 12 1/8 in. from the end. A pattern made with this shrink rule will automatically be "oversize" in all dimensions so that the finished casting will be the desired size.

Machine finish allowance

Patterns must also have *machine finish allowances* built in. For example, if you plan to bore a hole to 1/2 in. diameter in a casting, then the hole in the pattern must be made smaller to allow material to be bored out. Likewise, if a surface, like an engine cylinder head, is to be machined, a material allowance is added. The size of the machine allowance depends on the metal used, the shape and size of the part, the tendency of the casting to warp, and the method of setup and machining to be used.

Size tolerance

Additional size tolerances must be built into the pattern. Like the output of all manufacturing processes, no two castings will be exactly the same size. Mold cavities may vary in size due to "shake." (Shake is caused by rapping the pattern to aid in its removal.) Also, distortion of the casting during its cooling causes size differences. Typically, castings can be held to the following tolerances for the materials listed:

Gray cast iron	± 1/16 in.
Malleable iron	± 3/32 in.
Cast steel	± 5/32 in.
Aluminum alloys	± 5/64 in.
Magnesium alloys	± 11/64 in.
Brass	± 3/32 in.
Bronze	± 1/8 in.

Draft

Draft must be built into the pattern. It is a slight slope given to all vertical faces of the pattern. This slope, as seen in Fig. 4-7, permits the pattern to be removed from the molding medium without breaking up the mold walls.

Other considerations

The patternmaker must also select the proper pattern material and pattern type. These decisions are based on length of run and durability of the pattern material.

Most short-run patterns are made of wood because the material is relatively cheap and easily worked. Typically, white pine and mahogany are used because of their straight grain, light weight, and resistance to warping. Patterns for longer runs generally are constructed of a metal alloy.

In some cases, expendable patterns can be used. Common pattern materials are polystyrene, styrofoam, and wax. These patterns remain in the mold. The molten metal vaporizes and burns out the plastic pattern during the pouring step. Wax patterns are allowed to melt and flow out of the mold, leaving the desired cavity.

The pattern is the basis for the second step of expendable mold making, the construction of the mold. In general, expendable (one-shot) molds are made of inexpensive, easily shaped materials. Common among these are sand and plaster. The material is shaped around the pattern which is usually removed leaving a cavity of proper shape and design. The actual making of expendable molds will be discussed in detail in Chapter 5.

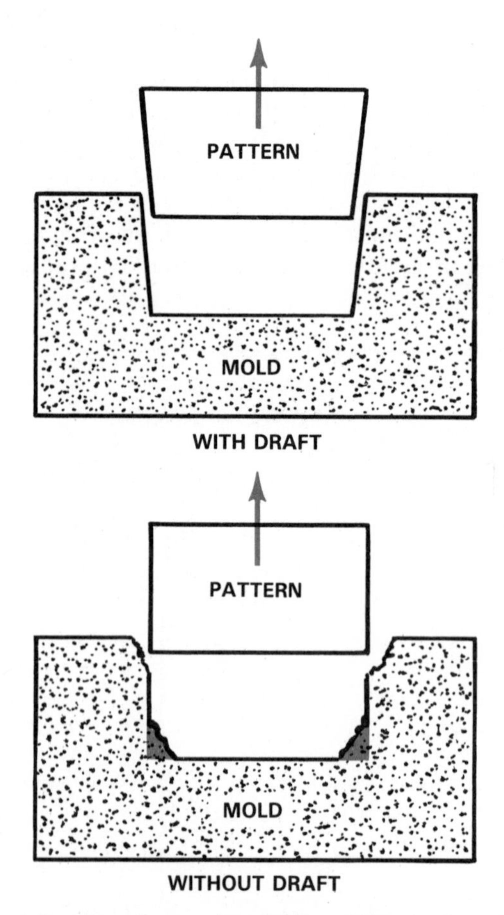

Fig. 4-7. The effects of draft. Top. With draft, a clean cavity is produced. Bottom. Walls are broken when the pattern does not have draft.

Permanent molds

Permanent molds are designed to produce a number of castings before they are discarded or repaired. They are made of a durable material into which the proper cavity has been machined or otherwise formed.

The cavities are produced with the same concern for shrink, draft, machining, and tolerance that was shown for expendable molds. In addition, a method must be developed to allow for opening the mold and removing the casting or molded piece. Also built into the mold is a technique for aligning the mold halves as they are closed for the next cycle.

Permanent molds are widely used for casting and molding metallic, ceramic, and synthetic (plastic) materials. These techniques will be discussed in detail in Chapters 6, 7, and 8.

PREPARING THE MATERIAL

After the type of mold has been selected and proper patterns (if needed) have been made, the method for making the material molten or "plastic" must be determined. Fig. 4-8 shows the major methods of preparing materials.

HEATING

Heat may be used to melt or soften the material. The heating method chosen is directly related to the material to be cast or molded, the economics and size of the melting operation, and the quantity of molten material needed.

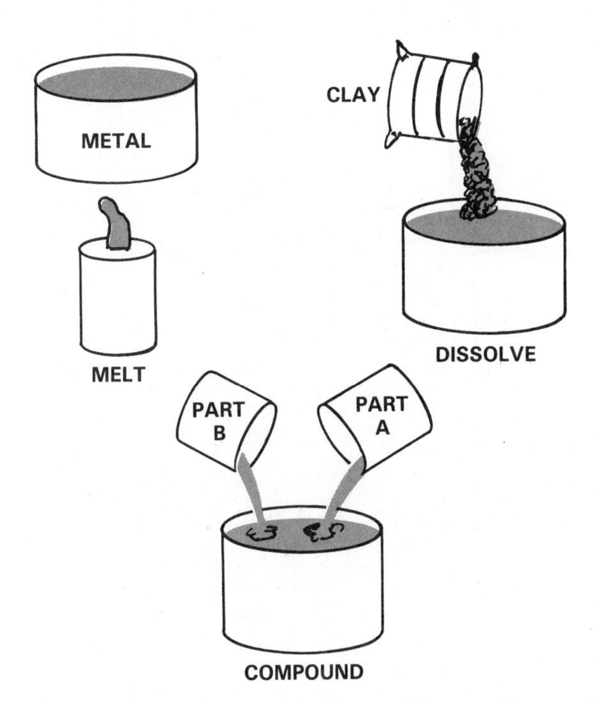

Fig. 4-8. Material can be prepared for casting and molding by melting, dissolving, or compounding (mixing).

When metals are being cast, a furnace is usually used to melt the material. Common melting furnaces are the:

1. Crucible furnace.
2. Cupola furnace.
3. Electric arc furnace.
4. Induction furnace.
5. Air furnace.

Each of these furnaces must meet certain needs. Ideally, they should:

1. Provide adequate melting temperatures.
2. Be capable of holding molten metal at required temperatures without serious changes in the chemical composition.
3. Be economical to use.
4. Hold environmental contamination to a minimum.

Crucible furnaces

The crucible furnace is the simplest and probably the oldest method of melting metals. The basic furnace consists of a fire-brick lined steel shell with a movable cover. Inside is a refractory (a material able to withstand high temperatures) crucible made of silicon carbide bonded with carbon, a clay-graphite mixture, or a metal. The charge (material to be melted) is heated to its molten state by burning oil or gas and passing the resulting hot gases over the crucible.

There are three basic types of crucible furnaces: *lift-out, fixed,* and *tilting*. In the lift-out type, Fig. 4-9,

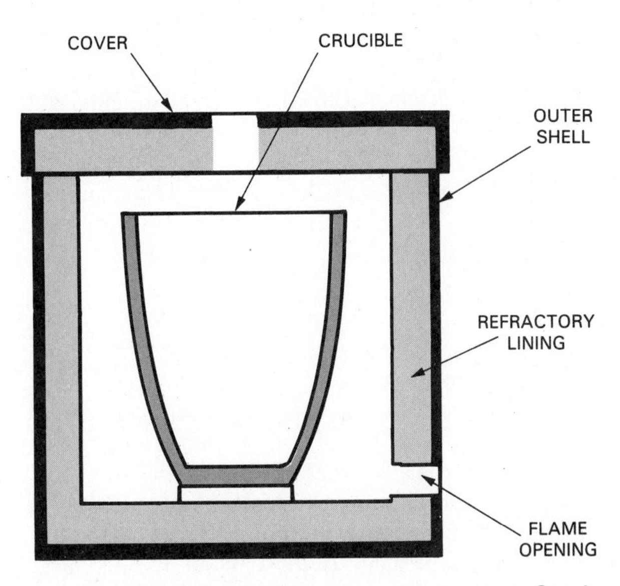

Fig. 4-9. Diagram of a lift-out crucible furnace. Crucible is removable.

the crucible is inside the furnace shell and is lifted out when the metal is ready to pour. The lift-out type crucible furnace is limited to about 100 lb. of molten metal.

The fixed crucible furnace has stationary crucibles which are removed only for repair and replacement. The metal is removed from the furnace by dipping it out with ladles.

The tilting crucible furnace is mounted on a pivot which allows the whole furnace to tip for metal removal. The metal is usually poured into transfer ladles which are used to move the molten metal to the molds for pouring. Capacities of tilting crucible furnaces range up to approximately 1000 lb.

Crucible furnaces provide great flexibility. Changing materials and alloys to be melted is fairly simple. However, capacity is limited, the working conditions are not very good, and a great deal of labor is required. For these reasons, the crucible furnace has limited use. It is most often found in low production operations and in educational and experimental facilities.

Cupola furnaces

The cupola furnace is the oldest and most widely used furnace for melting gray iron, ductile iron, and malleable iron. The cupola provides a simple and inexpensive method of continuously melting pig iron and scrap iron. Basically a steel shell lined with fire brick, it is closed at the bottom and open at the top as shown in Fig. 4-10. A charging (loading) door is located about 10-15 ft. above the bottom of the furnace. Tuyeres (air intakes) are built above the slag and tapping spouts.

The cupola has five zones:

1. The crucible or hearth on which the molten metal will accumulate.
2. The tuyeres which allow blasts of air to enter and support combustion.
3. The melting zone which is the hottest area of the cupola. Here metal, coke, and limestone (flux) are added.
4. The charging zone.
5. The stack.

Operation of the cupola furnace

The operation of a cupola involves several steps. First rags, wood, coal, coke, or other combustible material are placed on the rammed sand floor and lit. An initial charge is placed on top of the fuel.

After the initial charge becomes hot, alternating layers of metal, coke, and limestone are added and allowed to be heated. A blast of air is introduced through the tuyere. The metal will melt and accumulate on the hearth.

The molten metal may be continuously withdrawn through the pouring spout. Clay plugs may be used to stop the flow so that batches of metal may be taken from the furnace.

The cupola furnace is widely used because it is so simple. However, its simplicity also leads to many difficulties. The chemical action within the furnace is very difficult to control. A proper balance must be maintained between metal, coke, and limestone if the output is to have the correct chemical and physical properties.

Electric arc furnace

The electric arc furnace is one of two major electric melting furnaces used to melt metal for the casting industry. The induction furnace is the other.

The electric arc furnace, as shown in Fig. 4-11, is a refractory-lined steel shell. It has a roof which can be rotated to open the furnace for loading. Three adjustable carbon electrodes extend through the roof.

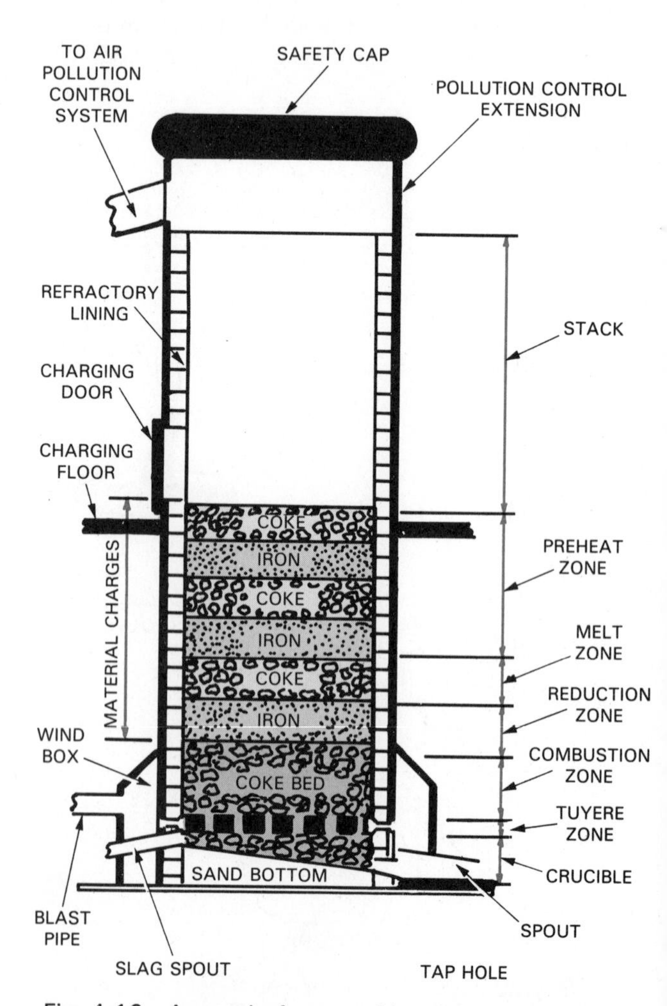

Fig. 4-10. A cupola furnace. Materials are fed in at charge door. Molten metal is removed at crucible spout.

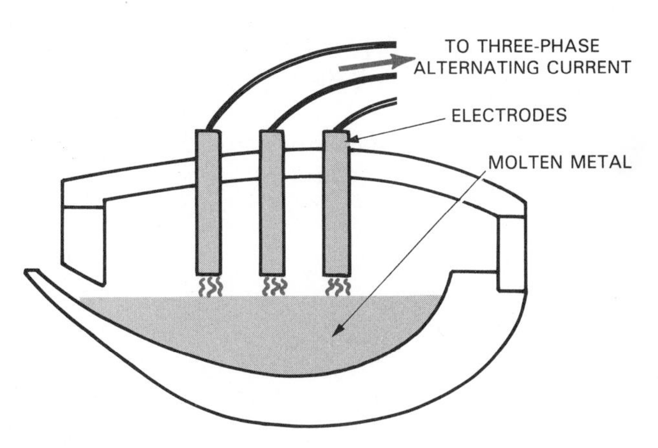

Fig. 4-11. Electric arc furnace. Electrodes extend from the roof and create an arc with the metal charge.

The operation of the furnace involves:
1. Raising the electrodes.
2. Rotating the roof to the "open" position.
3. Placing the charge in the furnace.
4. Rotating the roof to the "closed" position.
5. Lowering the electrodes to their arcing position.
6. Turning on the power. (Fairly low voltage and high current are used.)
7. Arcing between the electrodes and the metal charge creates the necessary heat to melt the metal.

Electric arc furnaces are usually in the less-than-25-ton range but may extend upward to 100 tons. Their use has grown considerably because of rapid melting rates and an ability to hold metal at a desired temperature. Pollution control is also easy to achieve.

Induction furnaces

The use of the induction furnace in foundries has been growing. Its primary advantages, like the electric arc furnace, are rapid melting rates and easy pollution control.

Two basic types of induction furnaces are in use. They are the high frequency crucible type and the low frequency core or channel type, Fig. 4-12. Both types operate by inducing current into the metal charge. However, each has unique design features.

The *crucible type* has a coil of copper tubing wrapped around it. This coil carries a high frequency (up to 30,000 Hz) alternating current. As the alternating current is applied to the coil, an alternating (expanding and contracting) magnetic field is induced around the coil. This magnetic field induces a high alternating current on the surface of the metal inside the crucible. (This action results from the same scientific principle upon which electrical transformers operate.)

The heat generated by this induced alternating current quickly moves toward the center of the metal and causes melting. Water is circulated through the outside copper coil to keep the furnace from overheating.

The *channel type* furnace is the older of the two and provides efficient melting. It consists of an upper unit in which the metal is melted. The coil or power unit is connected to the bottom of the furnace. The power unit contains an air-cooled conductor coil (transformer primary) and a U-shaped channel through which melted metal may pass.

This channel forms the secondary of the transformer. As power is applied to the coil, molten metal placed in the channel is heated by induction. Conduction (rising of warmer, lighter materials) causes the melted metal to rise. Colder nonmelted metal melts as it is contacted by the rising molten metal. A circular action, as shown by the arrows in the figure, continues. The furnace is never shut off or emptied except for repairs. Molten metal is removed as needed

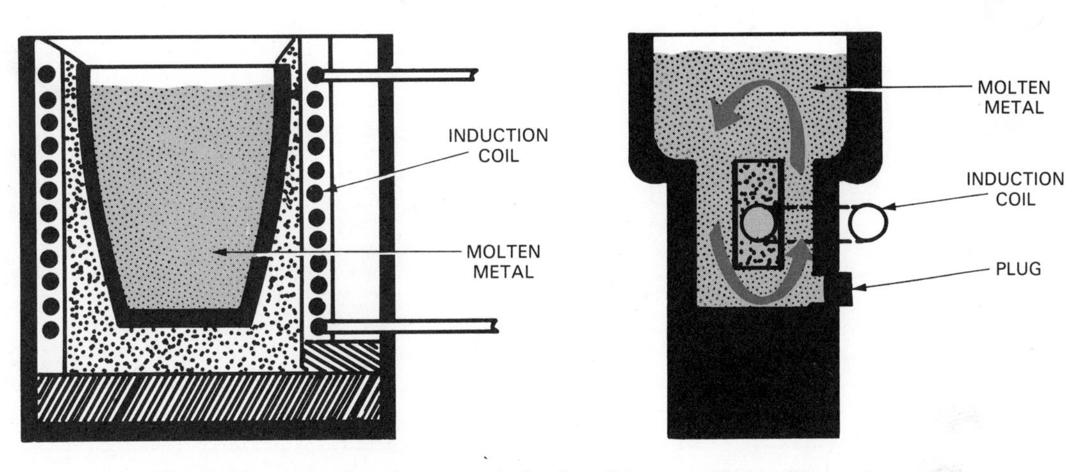

Fig. 4-12. Induction furnaces. Left. Crucible type. Right. Channel type.

and new metal charges are loaded. This furnace requires molten metal to start and, therefore, is not suitable for intermittent or occasional use.

Air furnaces

Air furnaces, as shown in Fig. 4-13, have limited application. They are used only in malleable-iron foundries. The charge is loaded through removable bungs that cover the furnace. Air heated by fuel oil or pulverized coal passes over the charge and causes melting. Slag forming on the top of the melt protects the molten metal from contamination and creates an insulation layer. The metal is removed through side spouts. Other types of heat sources are used for nonmetallic materials. Glass is compounded then melted in a specially designed furnace. Plastic and nonferrous metal materials are often electrically heated directly in the molding machines.

DISSOLVING

Materials are also prepared for casting by dissolving. Clay and other ceramic materials are dissolved or suspended in a liquid (usually water). The liquid carries the material into the mold. This liquid ceramic material suspension is called slip.

COMPOUNDING

A third preparation technique involves a material which, as *compounded,* is already a liquid. A typical material of this type is an acrylic plastic syrup. This material is prepared for casting by adding a chemical catalyst which will start a chemical conditioning reaction. This reaction changes the internal chemistry of the plastic which also changes it from a liquid to a solid. This conditioning action is fully discussed in Chapters 7 and 25.

INTRODUCING THE MATERIAL INTO THE MOLD

The two basic techniques for introducing materials into a mold, as shown in Fig. 4-14, are pouring (gravity) or forcing the material into the cavity.

GRAVITY POURING

Gravity pouring simply involves elevating the material above the mold and allowing it to freely flow into the mold. In the gravity pouring of heated materials, primary attention must be given to maintaining the proper pouring temperature during the movement of the material from the furnace to the mold. Often a ladle is used. Its size and design is determined by the size and number of castings to be made. Fig. 4-15 shows a typical ladle used to pour metals.

The pouring of ceramic slip is often done by pumping the material through a hose which allows the operator, Fig. 4-16, to fill the molds. Slip can also be poured from metal vessels into the casting molds.

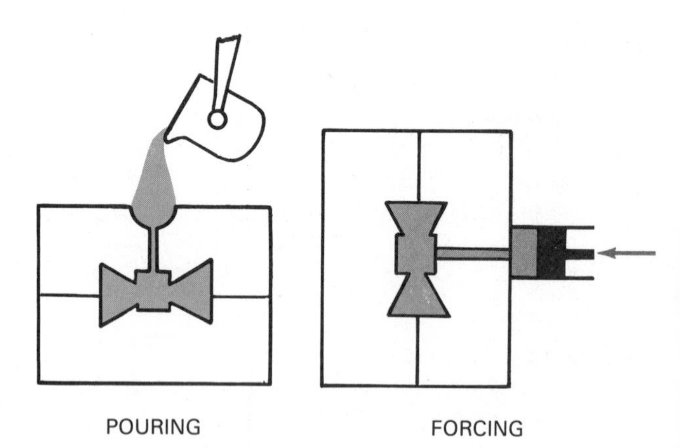

Fig. 4-14. Material is poured or forced into molds.

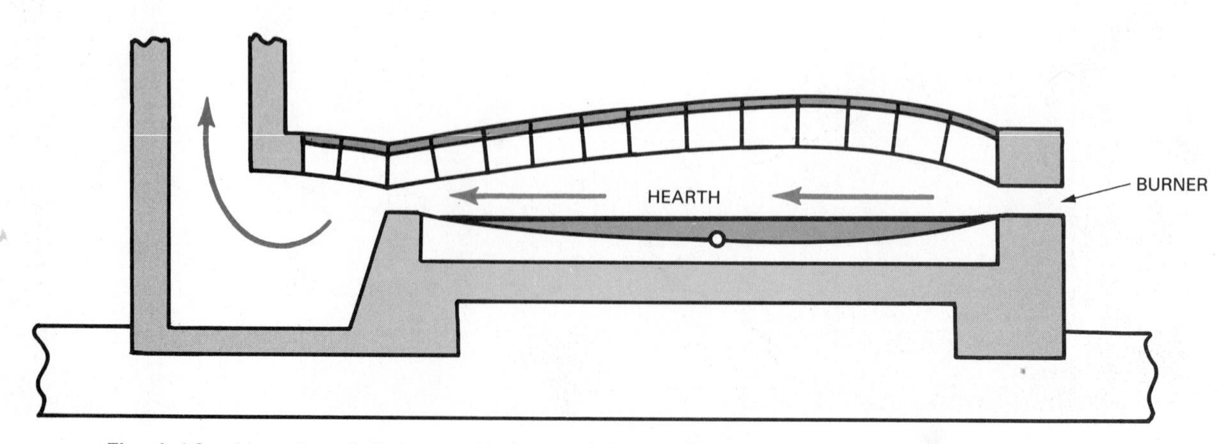

Fig. 4-13. Air or hearth furnace. Air (arrows) heated by the burner is passed over charge.

FORCING

Forcing materials is the second common method used to introduce the material into the mold. Some molding machines ram or auger-feed the material into the mold or die. Others use centrifugal force to "throw" the material into the desired location.

SOLIDIFYING THE MATERIAL

Once the material has been introduced into the mold it must be solidified (made solid). The process is generally the reverse of the technique used to cause the material to be liquid or plastic. If heat was added to soften or melt the material then controlled cooling (removing heat) must take place.

Removing the heat can be done by allowing natural conduction and radiation to occur. The heat may be absorbed by the mold, then radiated into the atmosphere. Cooling liquids may be circulated through the mold to absorb the heat, then radiate it to the atmosphere. This is the same principle used by an automobile engine cooing system.

If a liquid was used to dissolve or suspend the material, drying (removing moisture) must be allowed.

Fig. 4-16. Molds are being filled with liquid slip. (Kohler Co.)

Fig. 4-15. Aluminum is being hand poured into a permanent mold. (Aluminum Company of America)

Fig. 4-17. Left. Solidified cast iron pipe is being stripped from permanent mold. (American Cast Iron Pipe Co.) Right. Expendable molds are being vibrated apart and castings removed. (Crouse-Hinds Co.)

The drying action must be controlled so that warping and cracking will not occur. Often the moisture is absorbed into the mold to start the drying action. The final drying then takes place after the piece is removed from the mold. This technique is widely used in casting ceramic products.

If a hardening agent or catalyst was added then the *chemical action* must be allowed to take place. Adequate control of the hardening action must be maintained. This is done by accurately measuring the hardening agent during the compounding stage and, in some cases, cooling the material during the hardening stage. This will keep material and mold from overheating during the curing (hardening) stage.

REMOVING THE PIECE

The final step in the casting process requires removing the casting or molded piece from the mold. Removal, as shown in Fig. 4-17, is accomplished by destroying (breaking) an expendable mold or manually or automatically opening a permanent mold. Permanent molds often have ejector pins or rods to push the finished piece out of the mold so the next casting or molding cycle can start.

SUMMARY

A variety of materials can be cast or molded using numerous techniques in either independent or captive shops. These techniques all require that:

1. A mold be produced.
2. The material be in its molten or "plastic" state.
3. The material be poured or forced into the mold.
4. The material be allowed or be forced to change to its solid state.
5. The casting or molded piece be removed from the mold.

STUDY QUESTIONS – Chapter 4

1. Define casting and molding.
2. List the advantages of casting and molding.
3. List the disadvantages of casting and molding.
4. What are the five steps for producing a casting or a molded part?
5. List characteristics which make a material suitable for casting.
6. What is the difference between a captive casting shop and an independent foundry?
7. What are expendable molds?
8. What are permanent molds?
9. What is a pattern?
10. What allowances and changes must be made in the pattern to make sure the casting is the proper size and shape?
11. What are the three ways of preparing materials for casting and molding?
12. Name the five types of furnaces used in melting material for casting.
13. What are the two ways of introducing materials into a mold?
14. How are materials solidified in a mold?

Chapter 5

CASTING METALS
EXPENDABLE MOLD TECHNIQUES

Metal casting techniques may be grouped in several ways. They may be classified by the method used to place the molten metal in the mold. Or they may be described as precision (highly accurate) or nonprecision casting techniques. However, for this discussion, the techniques will be grouped by the type of mold they use.

In Chapter 4 the concept of expendable and permanent molds was introduced. Expendable molds, you may recall, are those which must be destroyed to extract the casting. This procedure requires a new mold for each pour.

Most expendable mold casting techniques are used to form metals. The most common of these techniques in use today are:

1. Green-sand mold casting.
2. Dry-sand mold casting.
3. Shell mold casting.
4. Full-mold (cavityless) casting.
5. Investment casting.
6. Plaster mold casting.

GREEN-SAND CASTING

One of the oldest methods of producing shaped metal objects is *sand casting*. The process was first practiced by the metal smelter. After the metal was extracted from its ore through crushing and melting, it was cast in molds. Demand for castings grew and division of labor resulted. The foundry soon became separate from the ore mining and smelting business.

ADVANTAGES OF GREEN-SAND CASTING

Today, green-sand casting accounts for about 80 percent of the more than 20 million product tons of metal castings produced each year. This domination of the foundry business is due to the great flexibility associated with green-sand casting. These castings can be produced in sizes ranging from a fraction of an ounce to over 100 tons. Also, the shapes possible are almost limitless.

Specifically, green-sand molds are so widely used because they:

1. Are strong enough to hold the weight of metals.
2. Have high resistance to fusion by heat.
3. Resist change in size and shape when heated.
4. Allow gases generated during the metal pouring phase to escape through the sand.
5. Generate a minimum of gas.
6. Are easily produced.
7. Resist erosion from flowing metals.
8. Allow the casting to pull away from the mold as it shrinks during its cooling phase

THE GREEN-SAND MOLD PRODUCTION

The making of a green-sand mold involved ramming sand around a pattern inside a container called a flask. This task usually includes eight steps as shown in Fig. 5-1.

A. The bottom half (drag) of the flask is placed upside down on a molding board. The drag half of the pattern is placed inside the flask.
B. Parting compound, which allows the two halves of the mold to be separated and the pattern to be removed, is dusted over the exposed surfaces. A layer of sand is riddled (sifted through a riddle) over the pattern and rammed (compacted). The drag is then filled, rammed, and struck (leveled) off.
C. A bottom board is placed on top of the drag and they are turned over. The molding board is removed and the top (cope) half of the flask is positioned using aligning pins. The cope half of the pattern is inserted.

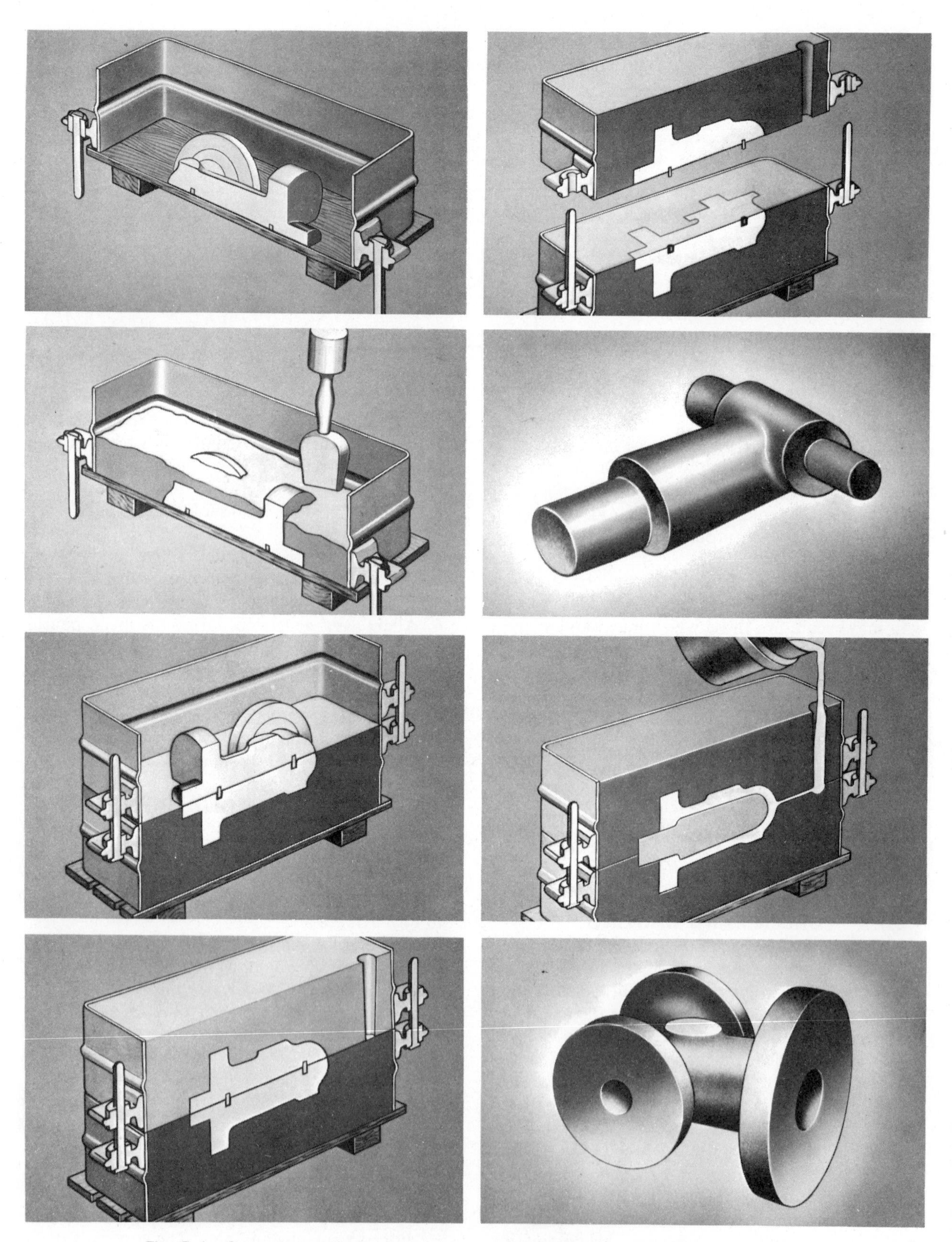

Fig. 5-1. Steps for producing a green-sand mold. Steps are described in the text. (Gray and Ductile Iron Founders' Society)

D. Step B is repeated for the cope part of the flask.
E. A gating system to allow metal to enter the mold cavity is formed. The system generally includes:
   a. A sprue which is an opening (usually vertical) through which the metal enters.
   b. A runner or runners (usually horizontal) which lead the metal into the mold cavity or cavities.
   c. A gate which controls the flow of the metal into the cavity.
   d. Sometimes a riser which is a reservoir connected to the cavity. It provides metal to the casting to offset the shrinkage due to the solidification process.

   Risers and sprues are formed during the ramming of the cope or cut into the cope before the flask is separated. Runners and gates are also either formed during the molding process or after the flask is opened. Venting (piercing the sand with a small diameter (wire) is often done at this point.
F. The flask is separated and the pattern halves are removed. Any cores (sand inserts) needed for added detail are placed in the mold cavities.
G. The mold is closed. Metal is slowly poured into the mold and allowed to solidify.
H. The mold is destroyed to recover the casting.

This process may be further understood by examining:
1. Patterns.
2. Cores.
3. Molding sands.
4. Molding methods.
5. Metal solidification.

Patterns

Patterns are the base upon which sand molding is built. From patterns come molds, Fig. 5-2. Not all patterns are alike, nor are they used in the same way.

Actually, there are three major types of patterns:
1. Loose patterns.
2. Match-plate patterns.
3. Cope and drag pattern plates.

Each type has its own specific uses and is designed with the same consideration for shrink, machining, tolerance, and draft that was discussed in Chapter 4. These considerations are summarized in Fig. 5-3.

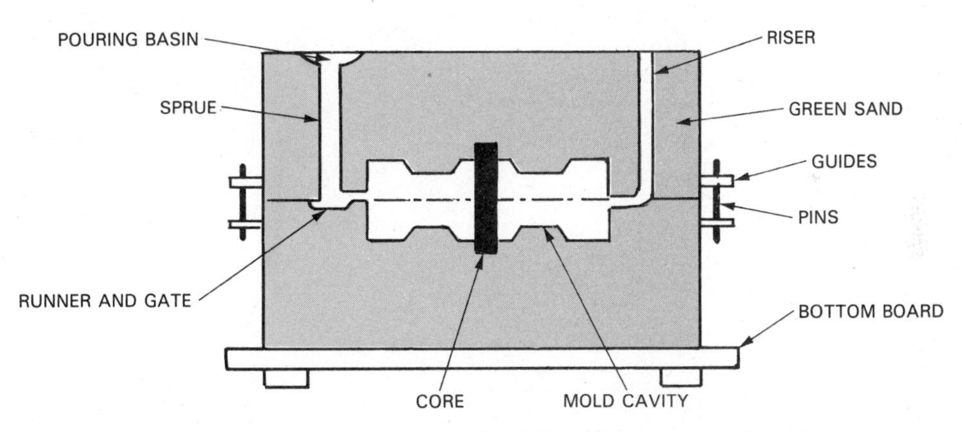

Fig. 5-2. Typical green-sand mold would look like this in cross section. Note names of parts.

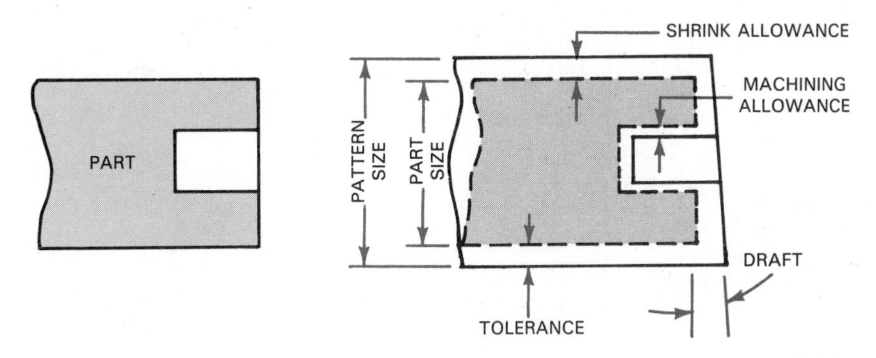

Fig. 5-3. Shrink allowance, machining allowance, draft, and tolerance are all taken into consideration in construction of a pattern.

Loose patterns

Loose patterns are the easiest and most economical to make. They are an exact copy of the part with proper allowances added. Most loose patterns are made of wood but occasionally metal, plaster, wax, or plastic is used. Loose patterns are principally used in hand molding operations and their use is, therefore, limited by the slow and costly nature of the molding process.

Loose patterns are of three major kinds: one-piece, split, and gated. See Fig. 5-4. One-piece patterns usually have one flat side and relatively simple detail appearing on the other side. This allows the pattern to be placed on the molding board which forms the parting line for the mold. One piece patterns, as shown in Figs. 5-5 and 5-6, are used for very short runs or for experimental work.

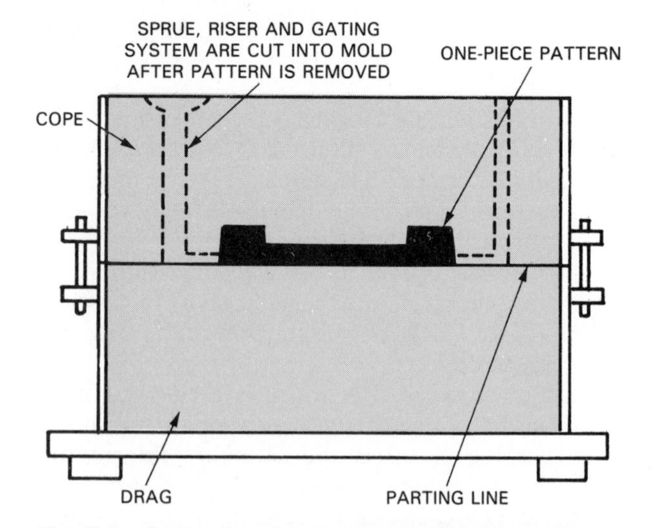

Fig. 5-5. A one-piece pattern has one flat side as shown in this cross section.

Split patterns

Split patterns are used for more complex shapes which do not have one flat face. The pattern is made to split (part) along a plane which corresponds with the parting line of the mold. Draft must be in all directions from this parting line. This allows for a portion of the mold cavity to be produced in the drag (lower) part of the mold and the remainder of the cavity to be in the cope section. Taper pins are used to align the two parts of the pattern and allow for their easy parting. Split patterns, as shown in Fig. 5-7, are used for moderate sized production runs.

Gated patterns are modified one-piece or split patterns which have gates and runners added to the pattern. This addition eliminates the hand cutting of those parts of the gating system.

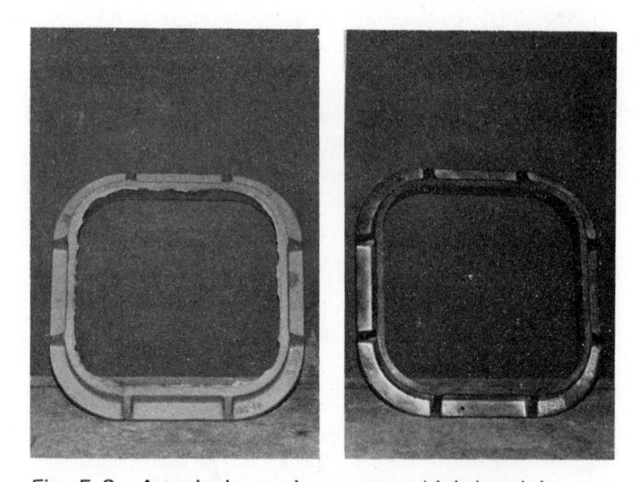

Fig. 5-6. A typical one-piece pattern (right) and the part cast from it. (Pressure Cast Products)

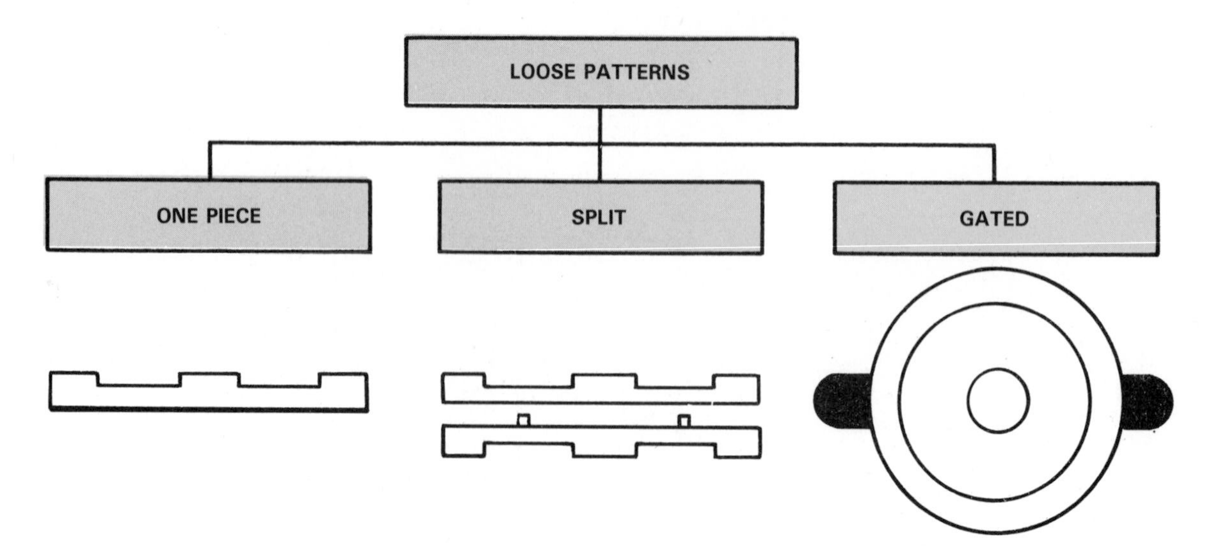

Fig. 5-4. There are three types of loose patterns.

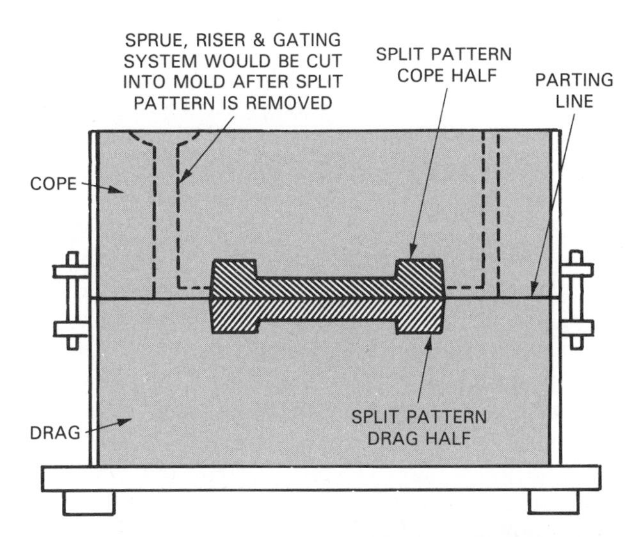

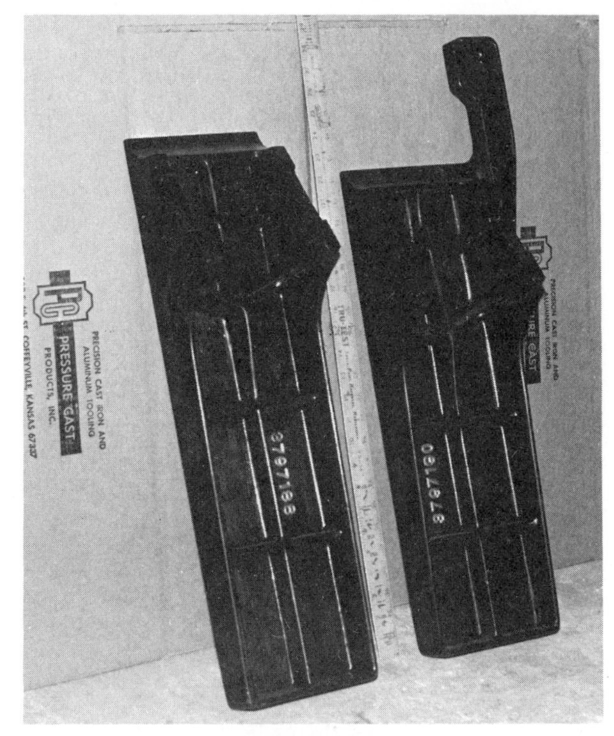

Fig. 5-7. Split patterns are used for casting of more complex shapes. Top. Cross-sectional view of pattern arrangement in the mold. Bottom. Typical part cast from a split pattern. (Pressure Cast Products)

Match-plate patterns

Match-plate patterns were developed for high volume, high speed machine molding of small castings. As seen in Fig. 5-8, a match-plate pattern has the cope part of the pattern on one side of a plate and the drag part on the other side. Runners and gates are almost always included. Match plates may have wood pattern parts on each side but are usually metal cast from a master match plate.

During the molding process, the match plate, which has aligning holes, is placed between the cope and drag halves of the flask. Both sections are rammed and then the match plate is removed by separating the cope and drag.

Match plates cost more than loose patterns. However, increased production rates, greater dimentional accuracy, and the ease of producing duplicate match plates quickly justify the additional expense.

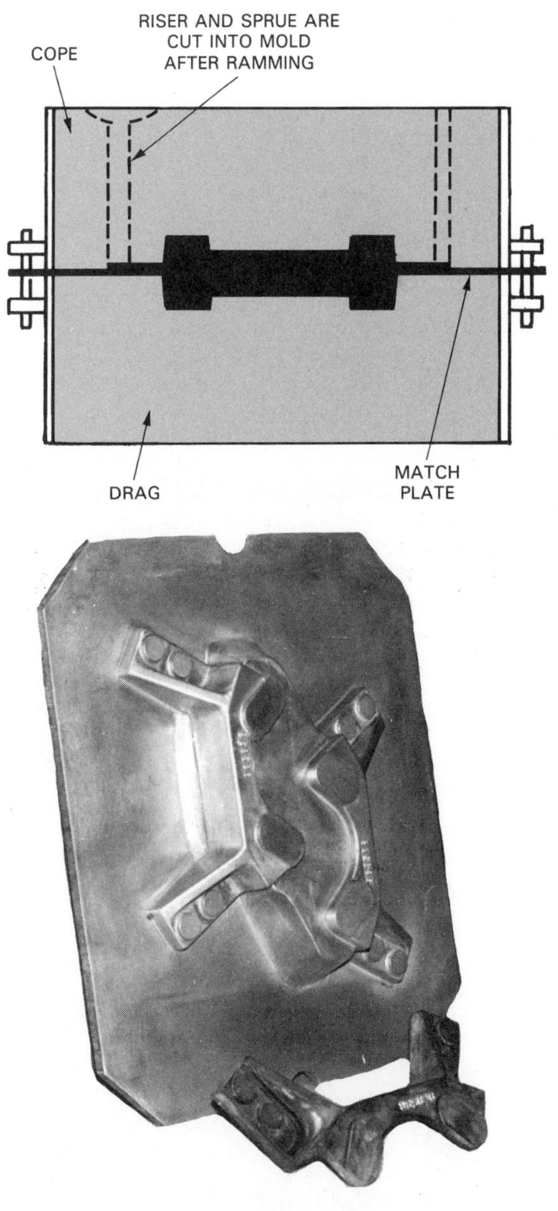

Fig. 5-8. A match-plate pattern. Top. Cross section of mold with match-plate pattern in place. Bottom. Match-plate pattern is shown with actual cast part. (Pressure Cast Products)

Cope-and-drag pattern plates are similar to the match plates except that the cope portion of the pattern is on one plate and the drag section on another. This overcomes the weight problem associated with large match plates. With a separate cope pattern plate and drag pattern plate, each half of the mold may be produced independently. Often this is done by different workers or separate molding machines.

Separate cope-and-drag plates, like the ones pictured in Fig. 5-9, are more expensive to produce than match plates. They, however, provide the ability to produce larger molds or very high quantities of smaller molds.

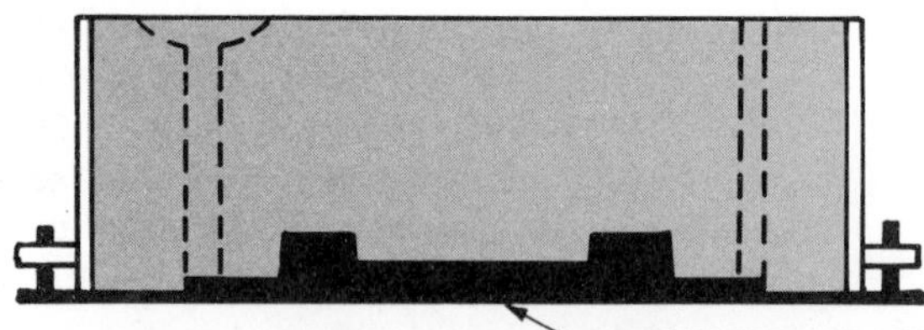

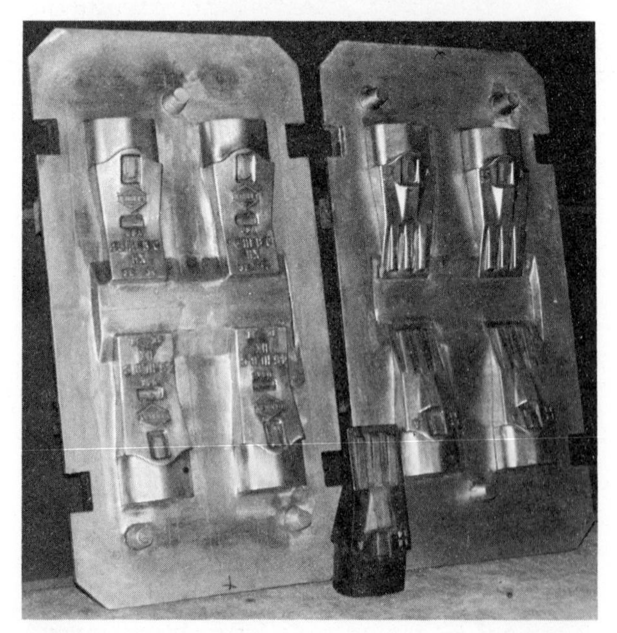

Fig. 5-9. Cope and drag pattern. Top. Unlike match-plate pattern, cope-and-drag pattern has two plates with half the pattern on each. Bottom. Actual cope-and-drag pattern plates and part cast from them. (Pressure Cast Products)

Cores

Often, a pattern cannot produce all needed details in a mold. Detail design features like holes, grooves, undercuts, and deep recesses are difficult and often impossible to produce during the initial molding process. These features are developed by adding cores to the mold cavity. Cores are, by definition, inserts in molds to form design features.

Whenever cores are to be used, *core prints* are included in the original patterns. These prints are voids formed in the mold to provide a "seat" for locating and positioning the core. (See Fig. 5-11.) The core will also be manufactured with the same shape or "print."

Cores may be made from numerous materials including green sand, metal, plaster, and ceramic. However, most cores are made of core sand which is a combination of sand and an organic binder. This mixture provides the three essential characteristics:

1. Green strength for molding.
2. Cured strength for casting.
3. Collapsibility for removal.

Cores may be made by hand, using a corebox. This is a wood or metal box containing a reverse image of the desired shape. Numerous machine techniques, including *coreblowers,* are also used to automatically fill coreboxes. Producing a sand-resin core involves:

1. Shaping the sand.
2. Baking the core at about 650°F (343°C).
3. Finishing the core by cleaning and assembling core parts, and by sometimes coating the core with refractory slurry.

Cores may also be made by the CO_2 process. This involves packing a mixture of dry core sand and sodium silicate in a core box. Carbon dioxide gas (CO_2) is forced through the mixture causing a chemical reaction which hardens the core. Very large cores (6 ft. in length or diameter) are possible with the CO_2 method.

The finished cores are then placed in the mold cavity, Fig. 5-10. They may be supported by their ends, balanced, placed vertically or be fixed by a number of other techniques as shown in Fig. 5-11.

MOLDING SANDS

The sand used in green-sand molding must be carefully selected and controlled. It must meet four basic requirements.

1. Withstand high temperatures (refractoriness).
2. Retain its shape when packed in a mold (cohesiveness).
3. Allow gases to escape through it (permeability).
4. Permit the metal to shrink away from it as the metal solidifies (collapsibility).

Silica sand is well suited for the task. It can withstand high temperatures, has low cost factor, is

Fig. 5-10. Foundry worker is placing cores in green-sand molds. (Crouse-Hinds Co.)

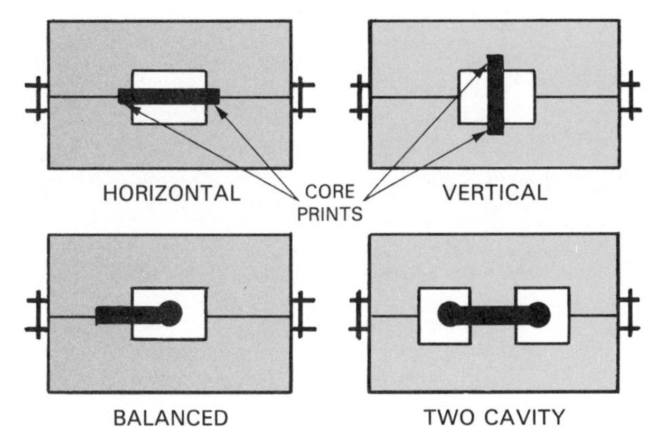

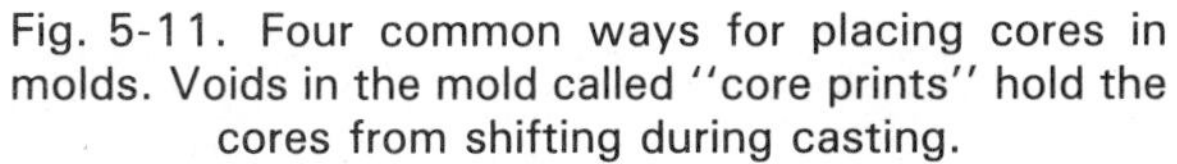

Fig. 5-11. Four common ways for placing cores in molds. Voids in the mold called "core prints" hold the cores from shifting during casting.

available in a wide range of grain sizes and shapes, and has a long life. However, it lacks cohesiveness. To develop this necessary quality, clay is added in amounts that may range up to 30 percent of the total weight of the mixture. The actual amount of clay added will vary greatly with the sand used and the molding application.

Also, the strength of the clay-sand mixture can be increased up to a point by the addition of water. The term "green sand" refers to moisture or a mixture of sand, clay, and moisture. Typically, green sand has 2 to 8 percent water by weight. Other materials may be added to improve certain properties.

Processing of the sand mixture after each molding cycle is generally practiced. This insures that a uniform mixture of sand, clay, moisture, and additives is maintained. Constant laboratory checks indicate the need to add ingredients to maintain this proper balance. The sand mixture is mulled (mixed and worked) to insure that each sand grain in uniformly coated with clay and other additives.

Molding methods

Molding is basically done by hand or by machine. Hand molding is slow, heavy work requiring considerable worker skill.

Several types of molding machines have been developed to increase the speed of the molding process, to improve quality of the mold, and to reduce human fatigue. Five basic machines are commonly in use, Fig. 5-12. These are:

1. Squeeze molders.
2. Jolt molders.
3. Jolt-squeeze molders.
4. Sand slingers.
5. Flaskless molding machines.

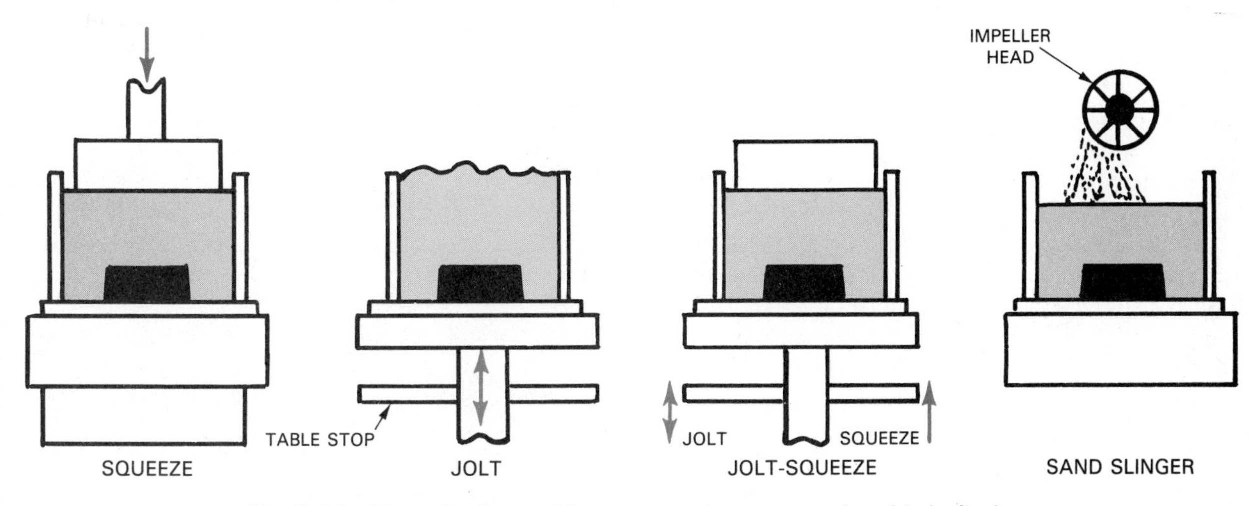

Fig. 5-12. These basic machines are used to ram sand molds in flasks.

Squeeze molders

Squeeze molders use pressure to compact the sand. The flask with its pattern plate is filled with sand. A flat plate or a flexible diaphragm is forced over the sand and air pressure is applied. The pressure compacts the sand around the pattern.

Squeezing compacts the sand more firmly near the squeezing head and less so as the distance from the head increases. The flat plate squeezing machine is least effective while the flexible diaphragm machine provides more uniform compaction around irregularly shaped patterns.

Jolt molding machines

Jolt molding machines are probably the simplest of all molding machines. A flask with a match plate is placed on the machine table. Sand is placed in the flask. The table is then raised a few inches by compressed air and allowed to drop. The resulting "jolt" causes the sand to compact.

Jolt-squeeze machines

The jolt type machine compacts the sand firmly around the pattern but only slightly at the top. For this reason, a simple jolt type machine is rarely used.

Instead a *jolt-squeeze machine,* as shown in Fig. 5-13, is often used. A flask with a match plate in position is placed upside down on the machine table. The top half of the flask (the inverted drag) is filled with sand from an overhead hopper, leveled off and a bottom board is placed on top of the sand. The jolting action of the machine compacts the drag portion of the mold. The assembly is turned right side up and the cope is filled with sand. A pressure board is placed on the sand and the pressure platen is lowered to squeeze the mold.

After the pressure cycle, the pressure platen is swung out of the way. The cope is lifted from the mold and the match plate is removed. The gating system is completed and the mold is reassembled.

Jolt-squeeze-rollover machine

Some machines automatically turn the mold between the jolt phase and the squeeze phase. This machine is called a *jolt-squeeze-rollover molding machine.* Also, mechanical means may be used to separate the mold and remove the match plate. This added feature is called "pattern draw."

Sand slinger machines

Uniform sand packing is difficult when large molds are being produced. A machine called a *sand slinger* uses a rotating impeller head to "throw" sand into the mold. The impelled sand is packed in the mold. Adjusting the speed of the impeller will adjust the degree

Fig. 5-13. Jolt-squeeze molding machines are more common than the simpler jolt type. (Osburn Mfg. Co.)

Fig. 5-14. Flaskless molding machine is capable of producing continuous castings. (Herman Corp.)

of compaction achieved with the sand. If additional compaction is needed, pneumatic rammers are often used.

Because of the high volume of sand that can be impelled and the variable speed of the impeller, highly uniform large molds are produced in a matter of minutes.

Flaskless molding machines

Continuous flaskless molding machines, Fig. 5-14, are basically an automatic squeeze molding machine. These machines, however, form the mold cavities on the outside faces of the green-sand "block." These blocks are then placed face-to-face, producing an endless mold. See Fig. 5-15.

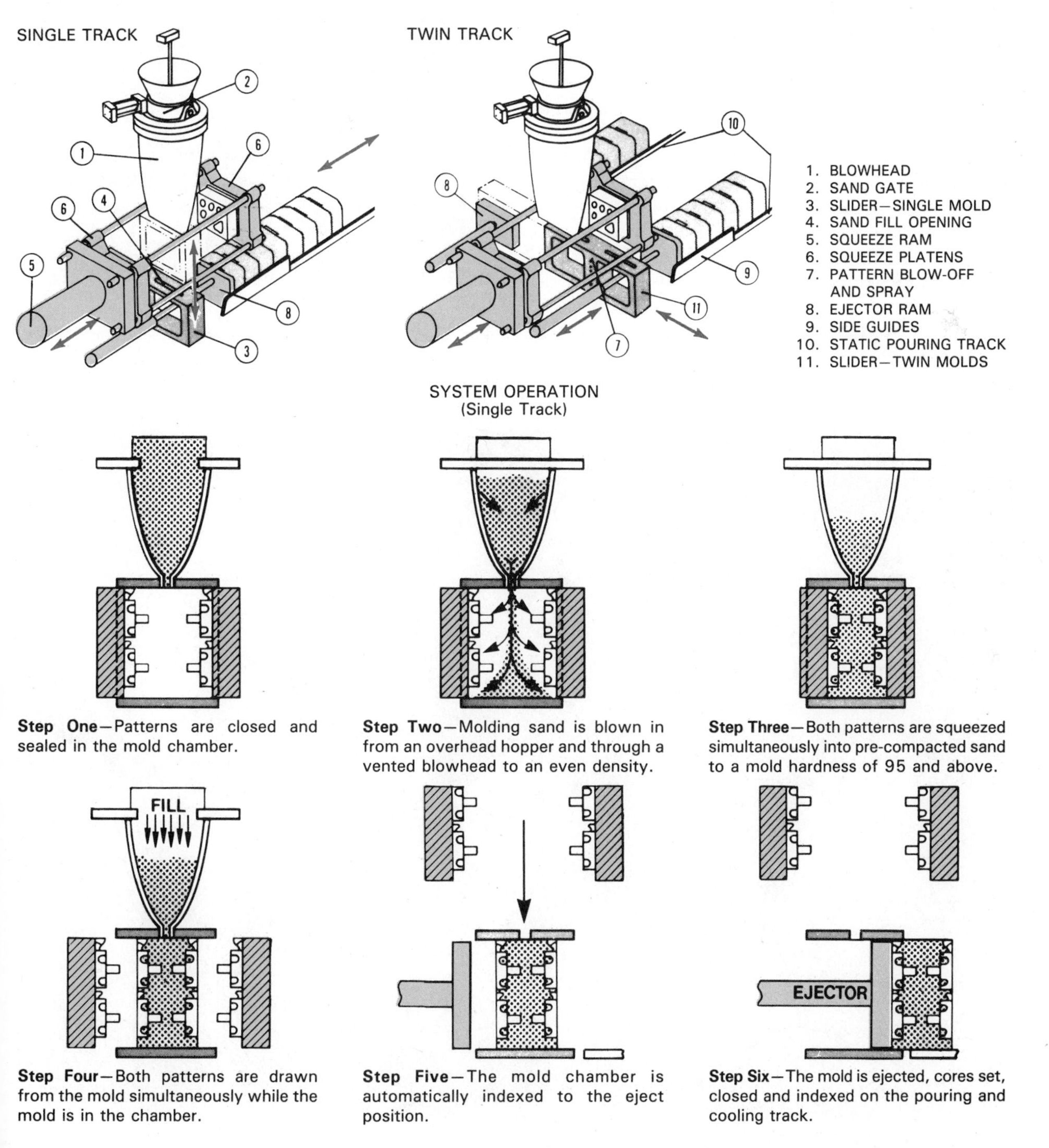

Fig. 5-15. Basic operation of a continuous flaskless molding machine. (The Herman Corp.)

Metal solidification

Once the molds are prepared, molten metal is poured into them, as shown in Fig. 5-16. Almost immediately, cooling starts. The molten metal gives off some of its heat energy. This energy is first lost at the mold walls, causing crystal formation to begin in the cooling metal. As the metal cools further, these crystals form *dendrites* (crystal spears) perpendicular to the mold walls as seen in Fig. 5-17. As the melt continues to cool, more dendrites form until the entire melt is solid.

The first dendrites formed start to shrink into solid cyrstals called *grains* while the inner molten material is still cooling. The slower the cooling the longer the dendrites which, in turn, produce a large grain structure in the metal. Fast cooling creates a fine grain structure. Thus, by controlling the cooling rate of the metal, its grain structure can be controlled.

The solidified metal, called a *casting,* is then extracted by breaking apart the sand mold. This is often done by vibrating the mold. The sand falls through a conveyor which delivers the castings to bins or other containers.

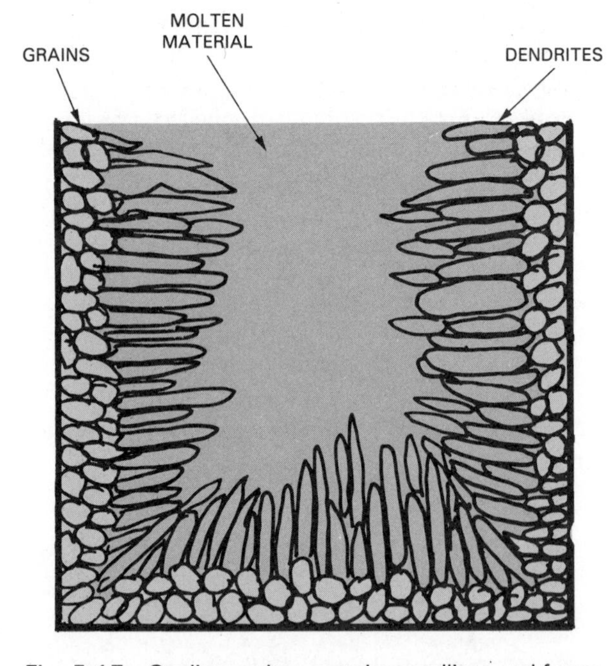

Fig. 5-17. Cooling molten metal crystallizes and forms familiar shapes shown above.

Fig. 5-16. Workers make pour into flaskless molds. (Budd Corp.)

DRY-SAND MOLD CASTING

Dry-sand molds are a second type of expendable molds. A *dry-sand mold* is produced using the same techniques as green-sand molding. However, an oil binder is used instead of clay. The sand and binder need no moisture to develop the required strength. The sand and the oil binder are mulled to coat each grain of sand with the binder. The mold is then produced and baked in an oven. During the baking process, the oil polymerizes and bonds the grains of sand, producing a very strong mold.

A variation of the dry-sand mold is the *skin-dried mold.* In this process, a standard green-sand mold is surface dried to a depth of about 1/2 in. The surface is then coated with a liquid containing a refractory (ability to withstand heat) material. The refractory coating is hardened with heat.

ADVANTAGES OF DRY-SAND MOLDING

In both dry-sand processes, dry-sand cores may be placed in the mold cavities before they are closed. Dry-sand molds are widely used in steel foundries because of their superior strength and ability to withstand higher temperatures. They also produce castings with better surface finish and with closer tolerances. Dry-sand molds cost more to produce than do green-sand molds. Therefore, their advantages and limitations must be weighed against cost factors in choosing between the two mold types.

SHELL MOLD CASTING

A new casting technique was developed during World War II by a German inventor, Johannes Croning. This process was first called the Croning Process but is now widely known as the *shell molding process.*

THE PATTERN

Shell molds are quite simple and use the same sand binder principle as dry-sand molding. The process involves carefully producing a metal pattern for the cope and the drag halves of the mold. These two pattern parts may be placed side by side on the same plate or on two different plates. All sprues, risers and gates are included on the pattern plates. These plates are similar to the cope and drag pattern plates used in green-sand molding.

THE SHELL MOLD PROCESS

A mixture of fine sand and a thermosetting phenolic resin is prepared. About 5 lb. of resin are added to each 100 lb. of sand. This mixture is placed in a dump box. The pattern, Fig. 5-18, is heated to 400-500°F (205-260°C) then placed on top of the dump box. The box is inverted and the sand resin covers the hot pattern. The heat melts the resin building a layer of adhered sand on the pattern.

After a predetermined period of time the dump box is returned to its original position. The sand not

Fig. 5-18. These pattern halves are for engine cylinder castings to be made by the shell molding process. The two parts may be placed side by side on the same plate or they can be on separate plates. Note that sprues and gates are part of the pattern. (Shalco Systems)

adhering to the pattern falls away leaving a hard sand shell. The thickness of this shell varies with the length of time the dump box was inverted and the sand-resin mixture was in contact with the hot pattern plate. The pattern and the newly formed shell are placed in an oven to complete the curing (hardening) process.

The hardened shell is removed (stripped) from the pattern by ejector pins which are built into the pattern. Cope and drag halves of the mold are glued or clamped together to form a complete mold, Fig. 5-19.

The mold may be poured without any additional support or may be placed in a container and backed with shot or sand to support the weight of the material being cast. Fig. 5-20 shows the several steps in the shell molding process.

Fig. 5-19. Bonded shell molds are stored away until they will be used. (Shalco Systems)

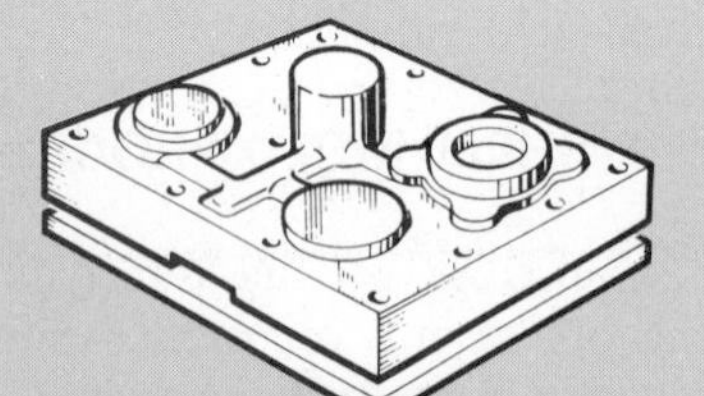

Metal Pattern is required for shell molding. Surfaces must be smoother, dimensions closer, than those used for sand casting.

Sand and Resin mixture is dumped freely onto heated metal pattern as first step in making shells.

Heat Is Applied to cure resin-sand mixture and make a rigid, firm, easily-handled shell. Each shell mold consists of two united halves—cope and drag.

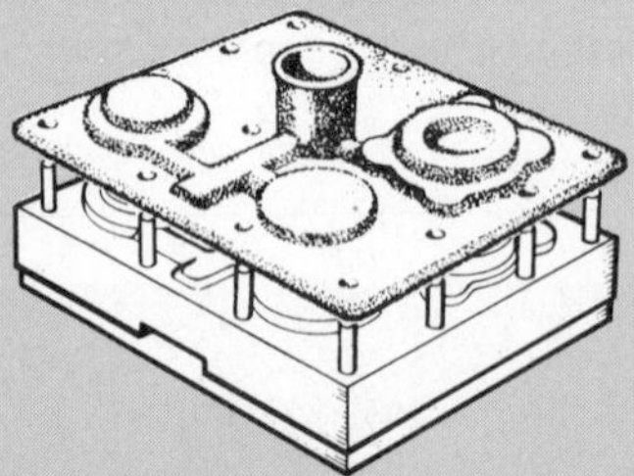

Cured Mold is ejected from pattern plate and is ready to be assembled for casting, or can be stored for long periods without deterioration.

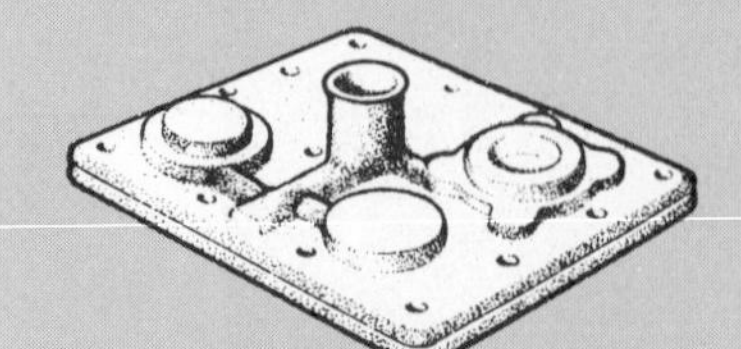

Ready for Casting. The two halves of the mold (cope and drag) are bonded together and the completed molds are ready for pouring. This is accomplished by placing molds in a horizontal position on trays which are suspended from a moving conveyor. Generally there are two or more molds on each tray.

Casting. As molds pass by a pouring station, which also is on a moving belt, they are filled with molten iron.

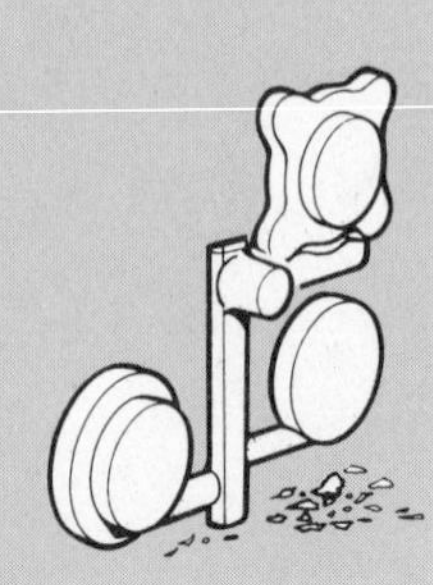

Gate of Castings emerges from the expendable resin-sand shells.

Fig. 5-20. Steps for producing a shell mold casting. (Eaton Corp.)

ADVANTAGES/DISADVANTAGES OF SHELL MOLDING

The shell molding process is easily adapted to automation, maintains close tolerances (.002-.005 in.), provides fine surface finish and requires little casting cleaning. It also uses less sand and requires little operator skill.

However, the process requires precision metal patterns which in turn, require costly equipment and skilled craftspeople. Also, additional costly equipment is needed to produce the shells. Whenever precision is needed or when production quantities are moderately high, shell molding's low labor costs, high productivity, and reduced need for later machining of the casting quickly overcomes the initial cost of patterns and equipment.

FULL MOLD CASTING

Full mold casting is a patented process which uses a pattern made of expanded polystyrene. The pattern is destroyed during casting.

THE PROCESS

The process is simple in concept. It involves embedding the pattern in compacted sand. Molten metal is then poured directly into the foam pattern causing it to vaporize. The vaporized pattern, as shown in Fig. 5-21, leaves a cavity duplicating the original pattern.

ADVANTAGES OF FULL MOLD CASTING

The full mold process produces castings of complex shape without regard for draft and parting lines. The patterns are easily and more cheaply produced than are wood ones. When large quantities of patterns are needed, they may be produced on automatic molding machines.

The process is particularly useful for:

1. Complex shapes which would require many cores or loose piece patterns.
2. Very large parts where pattern weight is of concern.
3. Very short runs or for single castings.

INVESTMENT CASTING

Investment casting is one of the oldest casting techniques, having been used in China and Italy centuries ago. Dentists have also used the process since the turn of the century. It did not, however, come into common industrial use until World War II when high quality jet turbine blades and supercharger buckets for aircraft engines were in great demand.

THE INVESTMENT CASTING PROCESS

Investment casting involves several major steps as shown in Fig. 5-22:

1. A master pattern for the part is produced using some easily worked material—metal, wood, plastic, or clay.
2. A master die is produced from the master pattern. Usually rubber, metal, or plastic are used to make these two piece mated dies.
3. Wax patterns are produced by pouring or injecting molten wax into the master die and allowing the wax to harden. The wax may be made out of beeswax, carnauba wax, paraffin, or other similar materials. Mercury may be used instead of wax

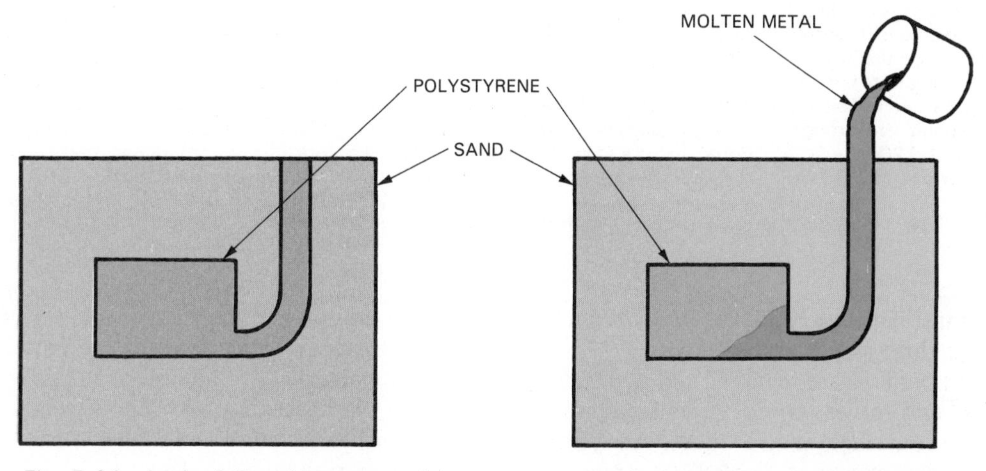

Fig. 5-21. In the full mold process, the pattern remains in the mold and vaporizes as the molten metal enters.

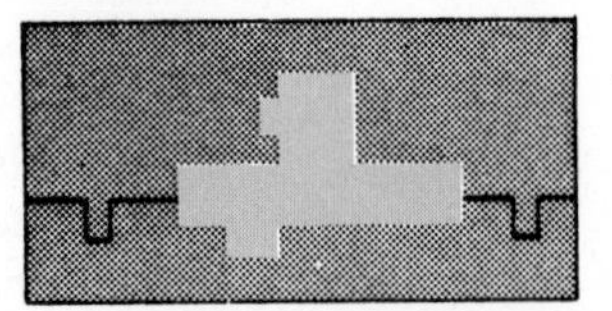

Step 1, MAKING THE TOOL. The tool, known as a wax pattern die, may be made from steel, aluminum or a low fusing alloy. Although relatively inexpensive, it must be made with great skill since the accuracy of the finished part is so dependent upon the initial wax pattern.

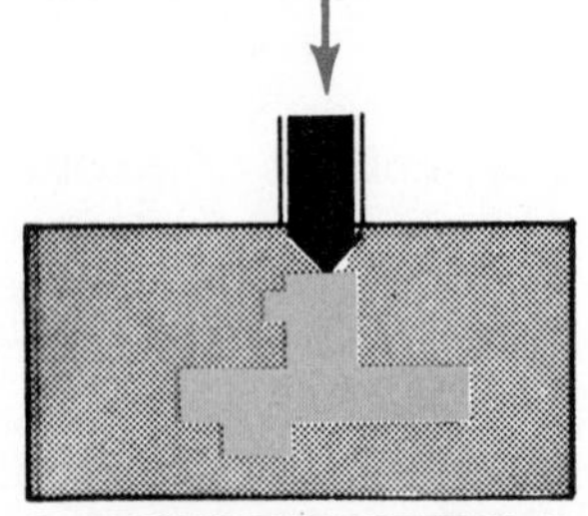

Step 2, MAKING THE PATTERN. Special wax is injected into the tool under heat and pressure to completely fill the cavity detail of the die.

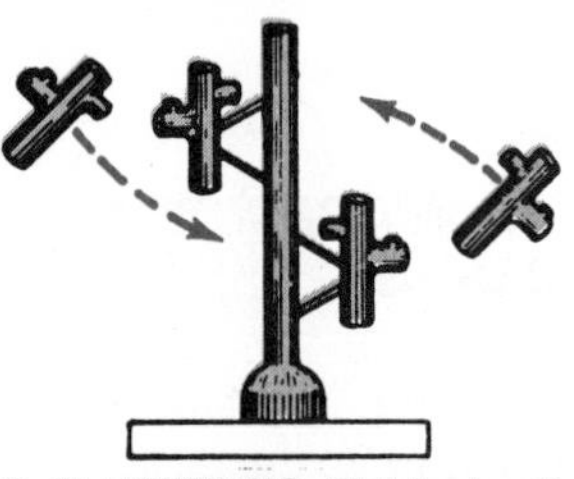

Step 3, CLUSTERING. Finished patterns are mounted on a wax sprue to form a cluster or "tree." The number of patterns will vary depending upon the individual size. There may be a simple pattern or as many as 50 or more.

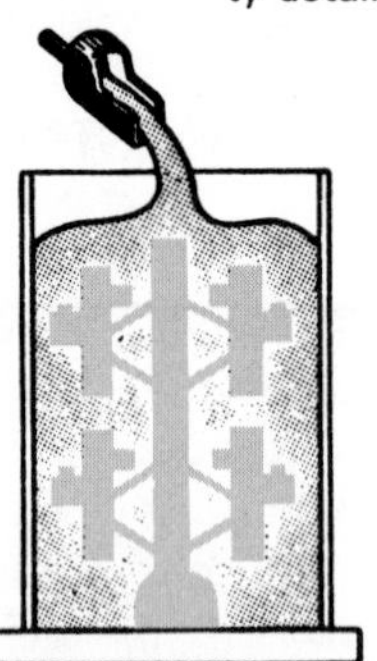

Step 4, INVESTING. Solid mold—The cluster of patterns is dipped into a refractory slurry and after drying is enclosed in a steel flask. A hard-setting molding material is then poured in, completely covering the patterns. This will subsequently solidify in air.

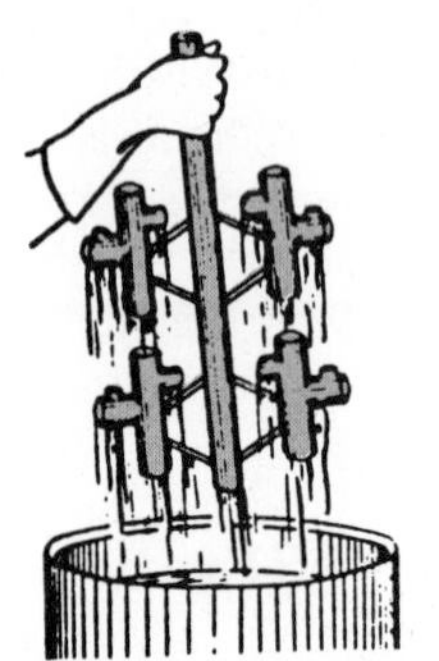

Shell method—The mold, in this case, is a monolithic shell formed by dipping the cluster of patterns into a ceramic slurry and then sprinkling with a refractory grain. This stuccoing process is continued until the desired thickness of shell is obtained.

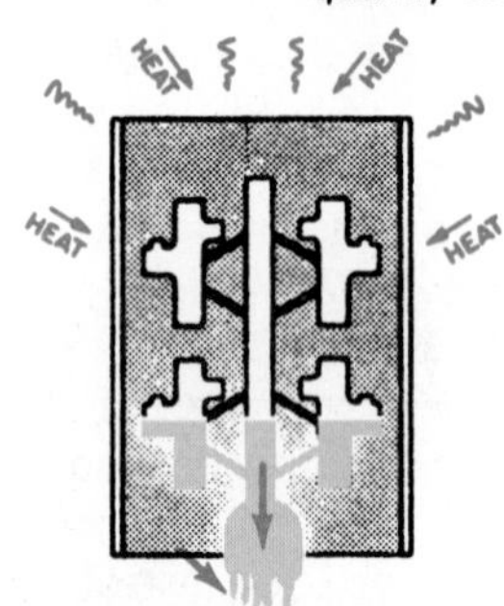

Step 5, PATTERN REMOVAL. The flasks or shells are placed on steam tables, in ovens or autoclaves where, under moderate temperature, the wax is melted out. The flasks are then put into high-temperature furnaces to burn out the remains of any pattern material. Here the expendable molds are brought up to temperatures of 1200 to 3000 °F, in preparation for pouring.

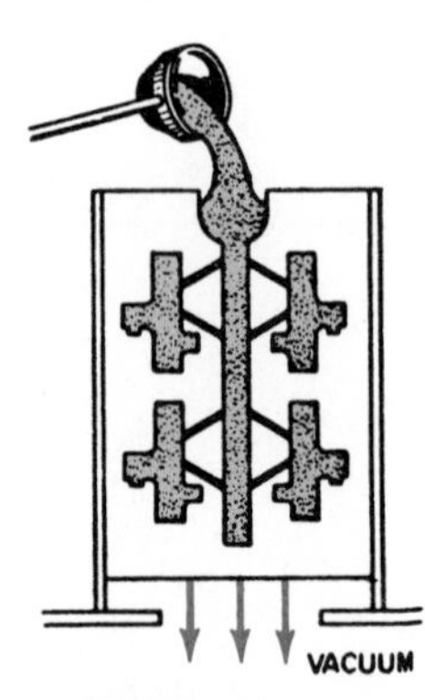

Step 6, CASTING. The heated flask or shell is inverted and the molten metal is poured in, filling the cavity detail to form the castings.

Step 7, CLEANING. After cooling, the castings are broken free of the investment and cut off from the sprue. Following a grinding operation for gate removal and any other trimming work which may be required, the parts are generally sandblasted.

Fig. 5-22. Steps in investment casting. As with full mold process, the patterns are destroyed during the process.

but it must be frozen and remain frozen during the next three steps.

4. The wax patterns are removed and assembled to runners and sprues. Often, several patterns are attached to a single central sprue. Also, several pattern parts may be assembled into a more complex pattern. The assembly is done by wax welding—melting wax to form fusion joints. This clustering of patterns provides for more economical production of the finished castings.
5. The cluster of patterns attached to the common sprue are dipped in a thin slurry of refractory materials and binders and immediately dusted with refractory stucco. This step produces a thin

Fig. 5-23. An investment mold wax pattern is shown being removed from the slurry tank. (Arwood)

layer of investment material which has an extremely smooth surface next to the pattern.

6. The cluster now can be repeatedly dipped in the investment matter, Fig. 5-23, to build up the desired wall thickness or it can be cast into a solid block of investment material in a metal flask. The first technique prepares the mold for ceramic shell molding while the second produces a solid investment mold.
7. The investment is allowed to harden. If a solid investment mold is being produced, it is vibrated to remove air bubbles and to insure that the investment has settled around the cluster.
8. The wax (or frozen mercury) is melted out of the mold. The solid molds are placed upside down in a continuous oven which first melts the wax and allows it to escape, then cures the mold. The process takes approximately 16 hours.

   The wax in the ceramic shell mold may be removed by heating or by dissolving it in a solvent (trichloroethylene).
9. The mold is preheated to 400 to 1800°F (204 to 982°C), depending on the alloy to be cast. This action insures uniform flow to all parts of the mold and allows the metal and the mold to shrink together during the cooling phase.
10. The metal is poured into the mold, Fig. 5-24. The flow of the metal may be aided by air pressure,

Fig. 5-24. Foundry workers pour investment molds shown in Fig. 5-23. (Arwood)

a vacuum drawn in the mold, or by centrifugal action.
11. After the metal is solidified the mold is broken away from the casting by a hydraulic press or by sand blasting.

ADVANTAGES OF INVESTMENT CASTING

Investment casting is usually limited to small parts weighing under one pound but the process has no specific size limitations. The process is complex and expensive and, therefore, is used primarily when:
1. The shape is complex.
2. Accuracy is critical.
3. Material is difficult to machine.
4. Fine surface finish is needed.
5. The part has thin or delicate cross-sections which would break under other molding processes.

PLASTER MOLDS

Plaster molds are similar to solid investment casting molds except that they are two-part molds. Plaster molds use special casting plaster made of gypsum-plaster base.

THE PLASTER MOLD PROCESS

The molding process is similar to green-sand molding in which a drag half is first produced followed by the cope section. In each case, the gypsum-plaster slurry is poured around an aluminum

Crane operator is pouring metal from a ladle into a row of green-sand molds. Operation is entirely machine controlled. (Cleveland Crane & Engineering, Div. of McNeil Corp.)

pattern match plate which contains all sprues, gates, and runners. In some cases wood or epoxy patterns are used. The mold halves are opened and the pattern removed.

The mold halves are dried or dehydrated by heating to a temperature ranging from 400-1400°F for a period determined by the size and shape of the mold.

The plaster mold halves are reassembled with the addition of any needed plaster cores. The metal, usually aluminum or brass alloys, are poured into the molds. Iron and steel cannot be cast in plaster molds. Their melting temperatures will destroy the mold.

After the metal has solidified, the plaster mold is broken away. Sprues and runners are removed and the casting is cleaned.

Plaster molds lend themselves to producing limited number of high-quality nonferrous castings. Size of castings may range from fractions of an ounce to hundreds of pounds.

SUMMARY

The six molding techniques discussed do not include all expendable mold techniques. They do, however, provide a basis for understanding others now in use and techniques not yet developed.

Also, you will note that each has its own advantages and limitations. A trade off between cost, surface fineness, tolerance, production rates, and many other factors is made as one chooses one casting technique over the other.

STUDY QUESTIONS – Chapter 5

1. What is the difference between precision and non-precision casting techniques?
2. Describe the steps in producing a:
   a. Green-sand casting.
   b. Dry-sand casting.
   c. A shell mold casting.
   d. A full mold casting.
   e. An investment casting.
   f. A plaster mold casting.
3. List and describe the major types of patterns.
4. What is "green sand?"
5. What are the three essential characteristics for green sand?
6. What are the major allowances included in all patterns?
7. What are the advantages of:
   a. Green-sand casting?
   b. Dry-sand casting?
   c. Shell mold casting?
   d. Full mold casting?
   e. Investment casting?
8. Describe the main characteristics of the common molding machines.
9. Describe the metal solidification process.

Chapter 6

CASTING METALS PERMANENT-MOLD TECHNIQUES

Casting metal using expendable molds can become costly when large numbers of the same casting are needed. New molds must be continuously produced, used once, and then destroyed. All of this activity uses large amounts of labor and requires new mold materials or constant reconditioning of the old mold materials.

To overcome these limitations of expendable mold casting techniques, engineers and inventors have developed a number of casting techniques with reusable (often called permanent) molds. The molds are not truly permanent but will produce a large number of castings before wear and other damage requires repair or replacement.

Four major casting techniques based on reusable molds are:

1. Permanent-mold casting.
2. Slush casting.
3. Die casting.
4. Centrifugal casting.

PERMANENT-MOLD CASTING

The entire group of casting techniques just listed uses a permanent or reusable mold or die. They are, therefore, called *permanent-mold casting techniques*. However, in the United States, permanent-mold casting also describes that casting technique which *allows gravity or a low-pressure feed system to force molten metal into a metal or graphite mold.*

This process is used primarily in casting nonferrous metals. However, it is used with some iron and steel.

PERMANENT MOLDS

Molds used for permanent-mold casting must be designed to open so that the casting may be removed without damage. These molds, as seen in Fig. 6-1, are made in two or more parts. The mold parts are often cast from steel or alloy cast iron.

A

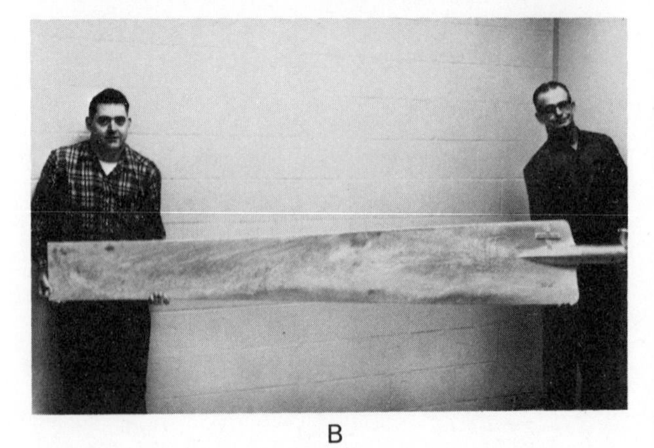

B

Fig. 6-1. Permanent-mold casting machine. A—Part is ready to be removed. B—Molded part. Usually, the size range for this type casting is 1 to 50 lb.

The shape of the mold cavity and the gating system is produced in a rough form in the casting. The final cavity and gating system is then machined to close tolerance. Simple shaped cavities may be produced through machining alone.

Steel cores may be built into the mold. However, provision for withdrawing the cores before the mold is opened must be designed into the mold. Sand cores may also be used but these must be placed in the mold by hand before the mold is closed. Fig. 6-2 shows the parts of a typical permanent mold.

THE PERMANENT-MOLD CASTING PROCESS

The permanent-mold process may be divided into two major types—gravity feed and low-pressure feed, Fig. 6-3. In the gravity feed method the metal is usually poured by hand into the mold. The casting is removed as soon as the metal has solidified. The castings, which are very hot, relatively soft, and easily distorted, are stacked on racks to cool. The mold is then blown clean.

About every four cycles it is coated with graphite to reduce the casting's tendency to stick to the mold. Throughout repeated casting cycles the mold is held at about 500°F. This temperature may be maintained by timing the frequency of pours or by applying additional heating or cooling. The heated mold insures even flow of the metal within the mold cavity.

Gravity fed permanent-mold casting accounts for a very small portion of all types of permanent-mold casting. It is limited to basically nonferrous (aluminum, copper or magnesium) alloys. Some iron and steel is cast using graphite molds.

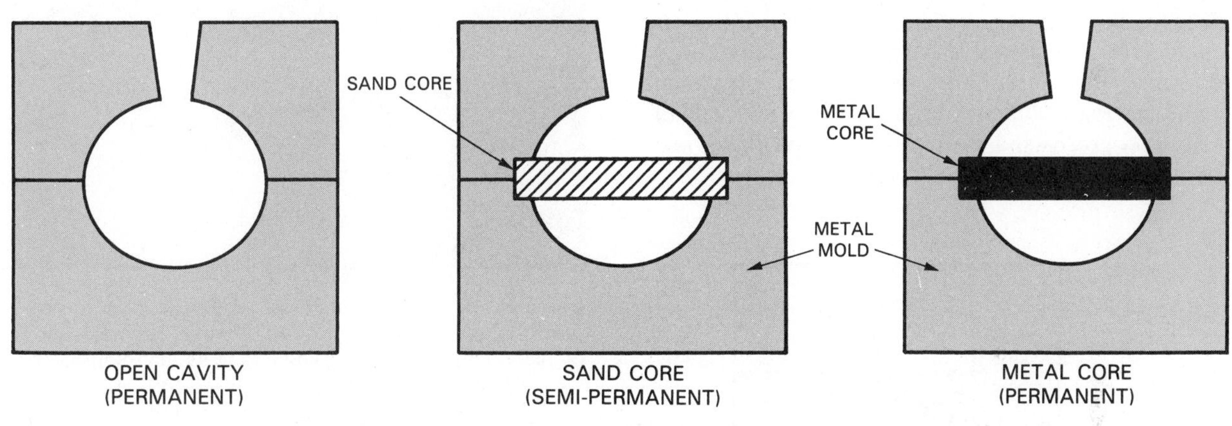

Fig. 6-2. Permanent molds may have a single cavity, sand cores, or metal cores.

Fig. 6-3. Metal may be poured or forced under low pressure into permanent molds. This photograph shows a pour being made under low pressure. (Amsted Industries)

A modification of the original permanent-mold casting technique is the low-pressure feed system. This system is not nearly as widely used as the gravity fed system. It uses low pressure to feed the molten metal into the mold. In this process, the mold, as seen in Fig. 6-4, is located over an induction furnace which is sealed. A pressurized inert gas forces the molten metal up the furnace stalk into the mold. Occasionally, vacuum pumps are used to remove the air from the mold to increase the speed of the operation.

Pressure permanent-mold casting is used in the production of railroad car wheels, Fig. 6-5. This application requires a graphite mold instead of a steel mold.

ADVANTAGES OF PERMANENT-MOLD CASTING

Permanent-mold casting techniques can produce castings to a ± 0.010 tolerance which have a good surface finish. Also, the particular chilling action of the mold produces desirable properties in many alloys. Castings of simple shapes produced by permanent mold casting techniques have a definite cost advantage over green sand castings because of higher production rates and reduced scrap.

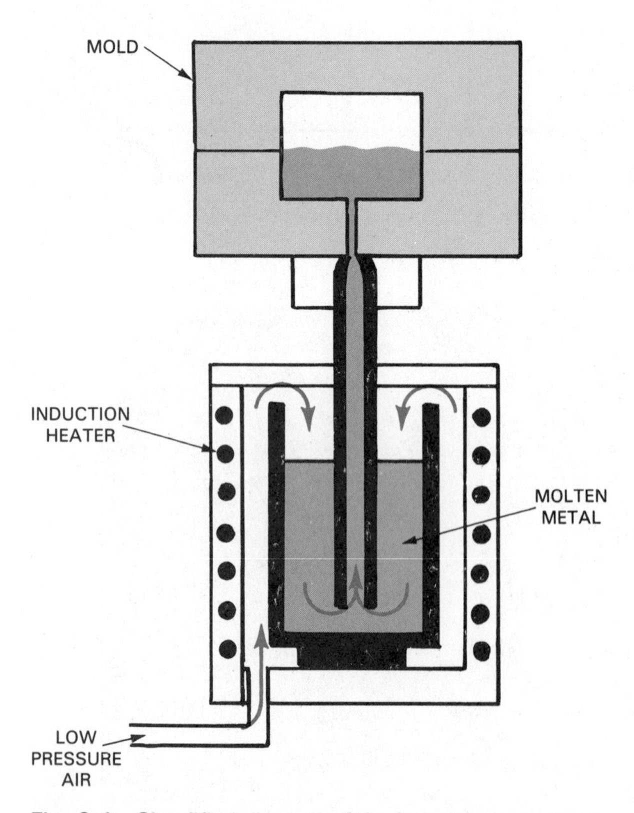

Fig. 6-4. Simplified diagram of the low-pressure permanent-mold process shows air pressure forcing molten material into the mold.

Fig. 6-5. A hot railroad car wheel is lifted from a graphite permanent mold. The wheel is solid enough to be placed on the finishing line five to seven minutes after it has been poured. (Amsted Industries)

SLUSH CASTING

Slush casting is a variation of the permanent-mold technique. It is designed to produce hollow castings without the use of a core.

THE SLUSH CASTING PROCESS

The process, like permanent-mold casting just described, involves pouring a metal into a mold cavity. Typically, tin, lead, or zinc-based alloys are used.

As the metal starts to cool a skin of solidified material begins to form on the mold surfaces. At a predetermined time the two-part mold is inverted allowing the remaining molten metal to pour or "slush" out of the mold cavity leaving the metal skin behind. The thickness of this skin will be determined by the metal and mold temperature and the time allowed for the skin to form. The two-part mold is immediately opened and the casting removed. The casting cycle is then ready to be repeated.

ADVANTAGES OF SLUSH CASTING

Slush casting is used primarily to produce statuary, ornamental works, and toys. The objects have a reasonably fine surface finish but have rough and irregular inside walls. Often the castings are finished to represent gold or silver or are painted to add color.

DIE CASTING

Die casting, like all permanent-mold casting techniques, introduces molten metal into a metal mold or die. Die casting, however, differs in that high pressures are used to force the meal into the dies. These pressures can run as high as 100,000 psi, but typically run in a range under 10,000 psi. Most die casting applications cast zinc, lead, tin, and other low-melting alloys. Using special dies, steel can be die cast but is still seldom used.

DIE CASTING DIES

The dies used in most applications are two-part steel. For special applications, refractory tungsten and molybdenum dies are used. The dies are designed with all mold cavities, cores, gates and runners, and ejector pins included. The dies are generally water cooled to guard against excess heat buildup in the die.

THE DIE CASTING PROCESS

The die casting process is fairly simple in concept. Mated dies are mounted in horizontal or vertical die casting machines, Fig. 6-6. One die is attached to a stationary platen while the other is on a movable platen. The machine closes the die halves. A plunger forces molten metal into the die where the metal cools. The die opens and the casting is ejected (forced out

Fig. 6-6. Vertical and horizontal die casting machines. Left. The horizontal machines are often over 25 ft. long. Right. Vertical models can exceed 20 ft. in height. (Wicks Machine Tool Co.)

of the die). The cycle is then repeated.

Two basic machines are used to carry out this process. These are the *hot-chamber* type and the *cold-chamber* type. The primary difference between these machines is in the method of introducing the molten metal into the die.

Hot-chamber die casting

The hot-chamber machine, as shown in Fig. 6-7, moves the metal directly from the melting chamber through a channel called a gooseneck. The gooseneck may remain in the molten metal or be lowered by cam action into the melt for filling. The metal is then forced into the die by a ram or air pressure.

The steel melting pot and gooseneck will react with metals with high melting points. Therefore, the machine is limited to metals which melt below about 800°F (427°C). Typically the hot-chamber machine is used for zinc die casting but may also cast tin and lead.

Hot-chamber machines operate at high speeds. It is possible for a machine to exceed 500 "shots" per hour on small castings.

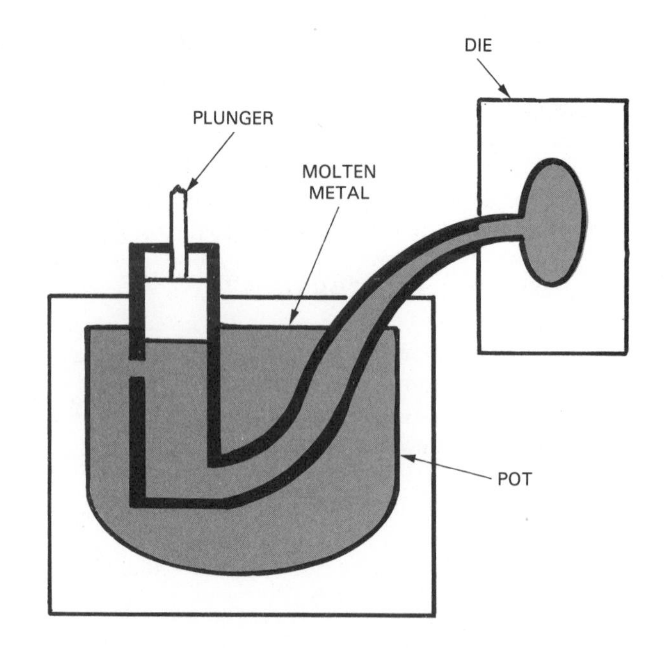

Fig. 6-7. Diagram of a hot-chamber die casting machine. Plunger may generate pressures of 2000-4000 psi.

Cold-chamber die casting

The cold-chamber machine, Fig. 6-8, uses either a hand ladle or a plunger in a transfer tube to move enough metal for one shot from the melting pot into the cold chamber of the die casting machine. The metal is then forced into the closed die by a plunger. Machines using the hand ladle filling system are generally the horizontal type while the machines using the transfer tube are vertical cold chamber machines.

Aluminum, brass, bronze, and magnesium are commonly cast in a cold-chamber machine. These metals, which melt at higher temperatures, can be handled by the cold chamber without excessive reaction between the molten metal and the chamber.

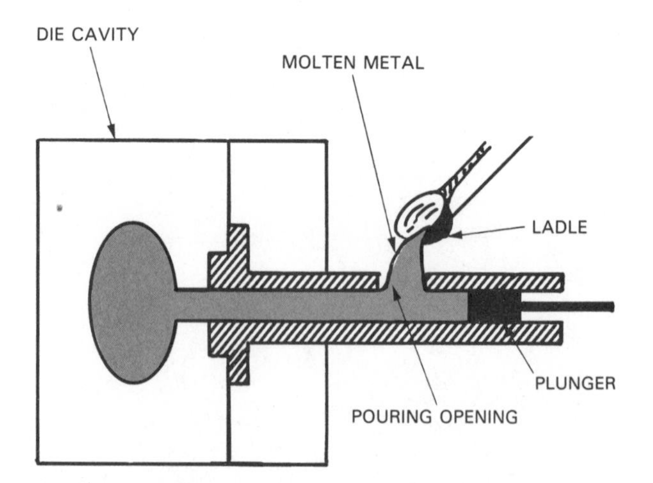

Fig. 6-8. Diagram of a cold chamber hand-fixed die casting machine. It uses either a hand ladle or a plunger to introduce the material into the die.

ADVANTAGES OF DIE CASTING

Die casting produces castings at a relatively high rate of speed. Hot-chamber machines are faster than cold-chamber machines. The production rates usually range up to 250 cycles per hour. Close tolerances of ± 0.001 to ± 0.003 and thin cross sections are possible because of the pressures involved. Die casting also produces a fine surface finish on a part which has higher strength/weight ratio than most other cast products. Little or no machining is required of a die cast part.

These advantages must be weighed against high initial cost of the machines and dies. Also, almost all die castings have flash (thin extensions around the casting) at the parting line of the dies. This flash must be sheared or ground away. However, these disadvantages are quickly overcome by the cost advantages of the high production rates.

CENTRIFUGAL CASTING

Centrifugal casting uses the centrifugal force generated by rapidly rotating the mold around an axis to place the molten metal into the mold. The process refers more to the use of centrifugal force rather than a specific type of mold to be used. In fact, either expendable or permanent molds may be used.

TYPES OF CENTRIFUGAL CASTING

Centrifugal casting techniques are usually grouped into three categories.

1. True centrifugal casting.
2. Semicentrifugal casting.
3. Centrifuge casting.

True centrifugal casting

True centrifugal casting is used primarily to produce cylindrical hollow objects like those shown in Fig. 6-9. It spins the casting around its own axis. No risers or center core are needed.

The centrifugal casting process, as shown in Fig. 6-10, may be done horizontally, at an incline, or vertically. The spinning action of the mold exerts extreme pressures, requiring extremely sturdy machines. The machines may develop forces in excess of 100 G (G equals the force of gravity).

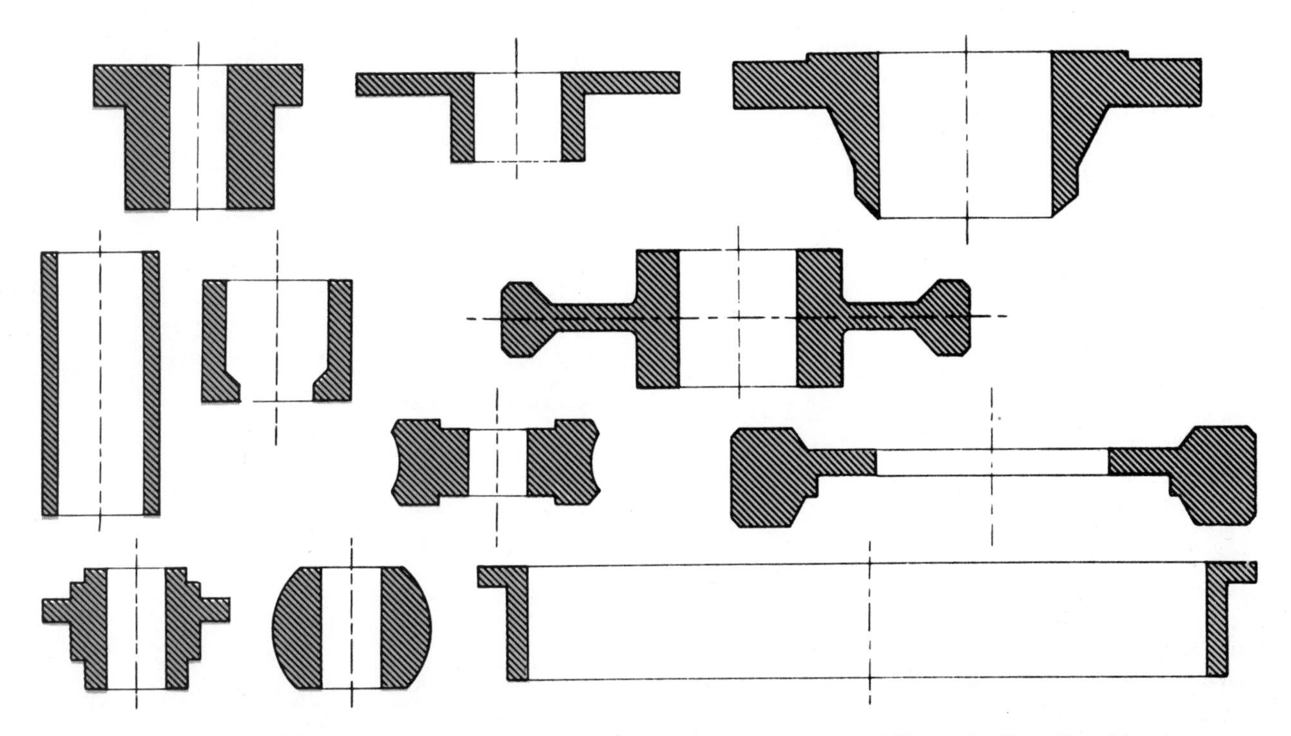

Fig. 6-9. These are axial sections for typical centrifugal castings. (Wisconsin Centrifugal Inc.)

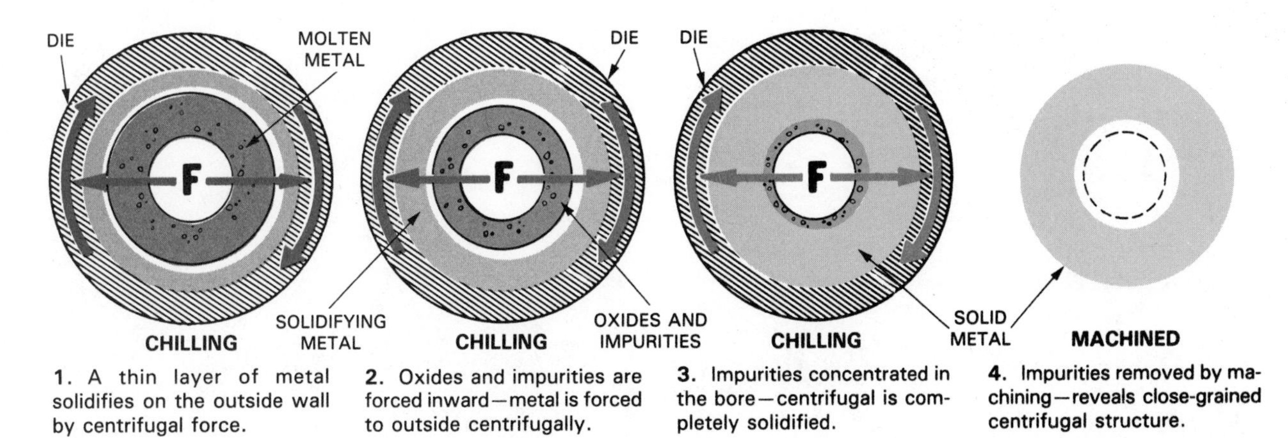

1. A thin layer of metal solidifies on the outside wall by centrifugal force.

2. Oxides and impurities are forced inward—metal is forced to outside centrifugally.

3. Impurities concentrated in the bore—centrifugal is completely solidified.

4. Impurities removed by machining—reveals close-grained centrifugal structure.

Fig. 6-10. Step-by-step process for centrifugal casting. Metal, being heavier, is thrown to outside as die rotates. (Wisconsin Centrifugal Inc.)

The process uses either a sand lined flask or a steel liner which is inserted in the flask. The desired finish on the outside of the object will dictate the mold used. The outside of the object may be any shape (round, hexagonal, etc.) but the inside will be round.

Molten metal is added through the end of the flask or mold, Fig. 6-11. The unit is then spun at 200-1000 rpm. While still molten, the metal is distributed evenly by centrifugal force on the surfaces of the mold. The spinning action continues until the metal solidifies. The flask or mold is then opened to remove the cast product, Fig. 6-12.

The most common products produced are cast-iron pipe, liners for internal combustion engines, sleeve bearings, large steel tubing and other symmetrical parts. Plastic, concrete, and ceramic products may also be produced using the centrifugal technique.

Products cast by the centrifugal process contain more uniform grain structure. They are also free from shrinkage which causes most casting defects.

Semicentrifugal casting

Semicentrifugal casting spins a round casing full of metal around its own axis, Fig. 6-13. The spinning process is used to position the metal in all parts of the mold. Since the mold is full, a central feeding reservoir, risers, and cores are needed. It is essential that the central feeding reservoir contain ample metal to feed the mold as the metal solidifies.

Fig. 6-12. Cast pipe (arrow) is being removed from a centrifugal casting machine. (American Cast Iron Pipe Co.)

Fig. 6-11. Metal is being poured into pipe casting machine. (American Cast Iron Pipe Co.)

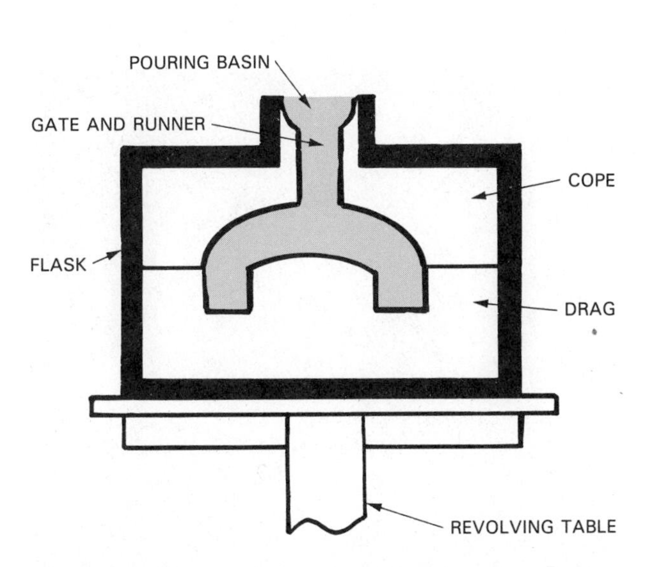

Fig. 6-13. Diagram for semicentrifugal casting. Proper gating is required and, where necessary, cores must be provided.

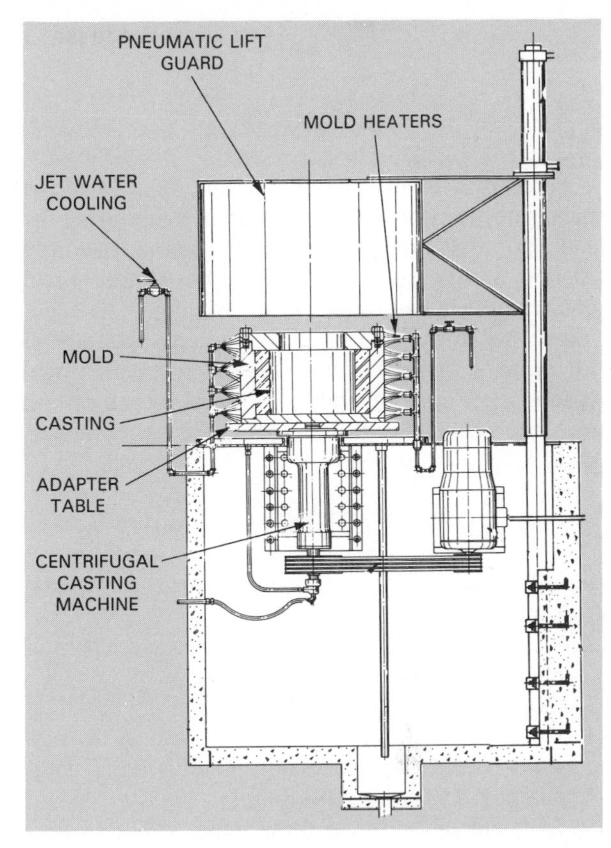

Fig. 6-14. A vertical centrifugal machine which can be used for semicentrifugal casting. Note provisions for heating and cooling the mold.

Typically, semicentrifugal casting machines, as shown in Fig. 6-14, are vertical. They are used to produce solid products such as wheels and gear blanks. The castings will be less dense at the center because of the low centrifugal force close to the axis. Also the center section will often contain air and inclusions. For this reason only casting which will have their centers machined out are cast by the semicentrifugal process.

Centrifuge casting

Centrifuge casting involves attaching a number of molds around a central downgate as shown in Fig. 6-15. The molds are arranged like spokes around the hub of a wheel. The spinning (centrifugal) action is used simply as a technique to fill the molds.

Centrifuging is used to produce relatively small parts like turbine blades, jewelry, valve bodies, pillow blocks, yokes, plugs, and other parts which may have very intricate shapes. Often, centrifuging is used to fill investment molds but other molds made of green or dry sand, steel, or plaster may be used.

The products of centrifuging are of high density. They are free of most impurities and air inclusions.

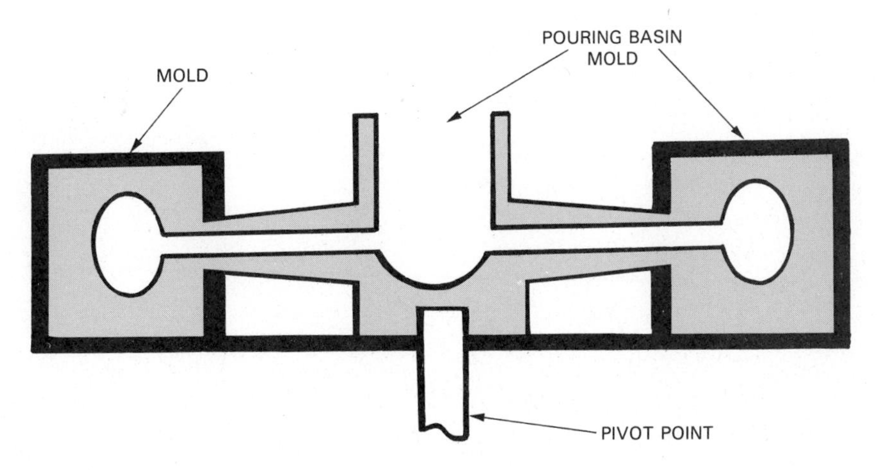

Fig. 6-15. Centrifuge casting is often used for investment casting.

SUMMARY

Permanent-mold casting techniques provide a means of producing castings without a new mold for each pour. The processes include permanent-mold casting, slush casting, die casting (hot-chamber and cold-chamber), and centrifugal casting. The techniques generally produce higher quality products at low unit costs. However, except for centrifugal casting, they do not readily lend themselves to casting iron and steel and are therefore limited to nonferrous metals.

STUDY QUESTIONS — Chapter 6

1. What are the major processes which use permanent molds?
2. Describe the operation of the low-pressure permanent mold process.
3. Describe the slush casting process.
4. Describe the die casting process. What is the major difference between a hot-chamber and a cold-chamber die casting machine?
5. List and describe the three major types of centrifugal casting.
6. List the advantages of:
   a. Permanent-mold casting.
   b. Slush casting.
   c. Die casting.
   d. Centrifugal casting.

Centrifugal casting is capable of producing large ferrous castings. This cast-iron pipe is being pulled from a vertical centrifugal casting machine. (American Cast Iron Pipe Co.)

Chapter 7

CASTING AND MOLDING PLASTIC MATERIALS

Casting and molding provide basic ways of sizing and shaping plastic materials. The materials are powders or granules that are softened to their plastic state or they are a liquid.

Some of the processes are called "casting" while others are called "molding." In either case, the material is shaped while in its liquid or plastic state. This is done inside or around a mold or die. The material is then allowed to solidify.

Basic manufacturing processes used with plastic or synthetic materials which fit the above definition are:

1. Gravity casting.
2. Injection molding.
3. Compression molding.
4. Transfer molding.
5. Rotational molding.
6. Dip casting.
7. Slush molding.
8. Foam molding.

GRAVITY CASTING

Plastics formulated for casting are those synthetic materials which are available in liquid form. Acrylics, nylons, and epoxies, among other materials, can be cast. Casting is probably the simplest method of shaping plastic materials. The process requires no fillers to be added to the material. Also, *no pressure* is needed to force the material into the mold.

THE CASTING PROCESS

As in all casting processes, a mold or die must be made. Several mold-making techniques are possible for casting plastic materials.

The use of an expendable lead mold is probably the simplest. A steel *mandrel* (round, tapered rod) in the shape of the product is first produced. The mandrel is dipped into molten lead and a thin coat of lead forms around it. The mandrel is removed from the container of lead and the lead shell is allowed to cool. The shell, when slipped off the mandrel, is the mold.

The plastic material is then poured into the shell and allowed to harden in an oven at relatively low temperatures (below 200°F). The lead shell is then peeled away. This process will work only when the object is tapered so that the mandrel can be withdrawn from the shell.

More complex molds may be made by casting a latex or silicone rubber material around the pattern. After it has hardened, the mold is rolled off the pattern.

Still more complex molds may be made from tool steel. They resemble the permanent molds discussed in Chapter 6. These molds must be able to contain a low viscosity (thin) liquid at temperatures up to 418°F (200°C). Also, they must allow for the normal shrinkage associated with most casting processes.

The casting operation, Fig. 7-1, usually involves six major steps:

1. Selecting and preparing a plastic material. This could be a powder or other solid dissolved in a liquid solvent (casting acrylic, etc.), a liquid resin, or a monomer melted (such as nylon).
2. Adding a catalyst or activator which causes the chemical action to harden or polymerize the material.
3. Mixing the ingredients.
4. Pouring the mixture into the mold.
5. Curing the casting in an oven or in the atmosphere.
6. Removing the casting.

PLASTIC CAST PRODUCTS

Typical gravity cast products are epoxy molds, dies and bushings, nylon slides, gear ring blanks, gears, and encapsulated or embedded (potted) electrical components. These products may be any size. In theory, there is no limitation to the size of a plastic casting. Products ranging from gears weighing less than an ounce to epoxy insulators weighing in the tons have been cast. The nylons and the epoxies are used most often in the casting process.

INJECTION MOLDING

Injection molding is the most common process used to produce parts and products from thermoplastic materials. These materials can be repeatedly changed from solid to liquid without changing their chemical structure. Also, injection molding is used to a limited extent on thermosetting plastics. These materials, under heat and pressure, complete nonreversible chemical bonds.

THE INJECTION MOLDING PROCESS

Simple injection molding involves four stages.

1. A powder or granular synthetic material is changed into its viscous (plastic) state.
2. The viscous material is forced, under pressure, into a mold of a desired shape.
3. The material is held under pressure in the mold until it solidifies.
4. The mold is opened and the molded object is ejected.

INJECTION MOLDERS

An injection molding machine is very similar to a hot chamber die casting machine used for nonferrous metal casting. The machines range in size from those injecting (shooting) a fraction of an ounce to machines capable of shots in excess of 800 ounces. The machine as shown in Fig. 7-2, has two major units—the *injection unit* and the *clamping unit.*

Injection units

There are four basic types of injection units used in injection molding machines.

1. Standard plunger.
2. Two-stage plunger preplasticator.
3. Screw preplasticator.
4. Reciprocating screw.

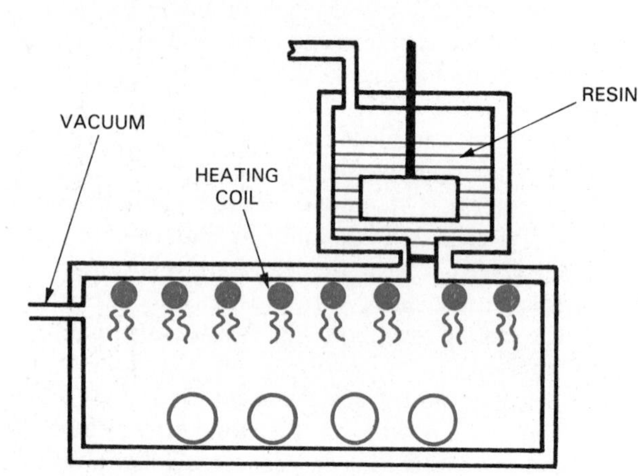

1. Fill resin tank and dry components.

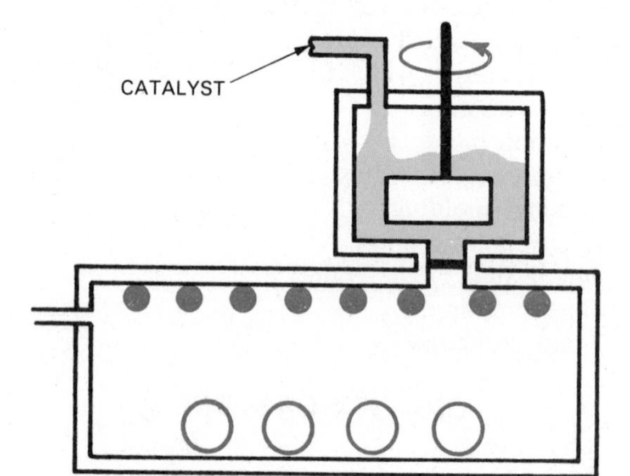

2. Add catalyst to resin and mix.

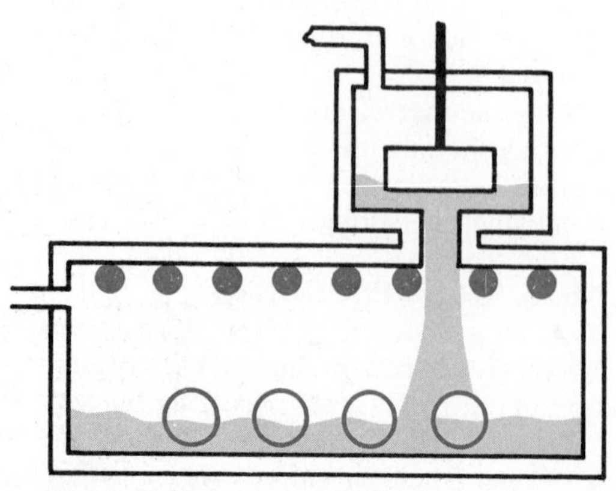

3. Encapsulate (cast) components.

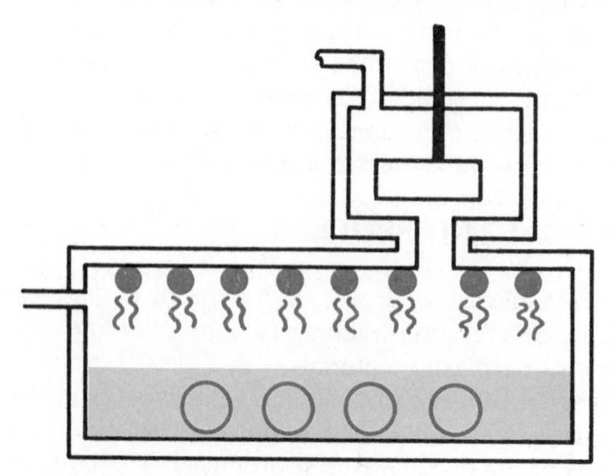

4. Cure casting and remove.

Fig. 7-1. Encapsulating or potting is a casting process done in a vacuum chamber.

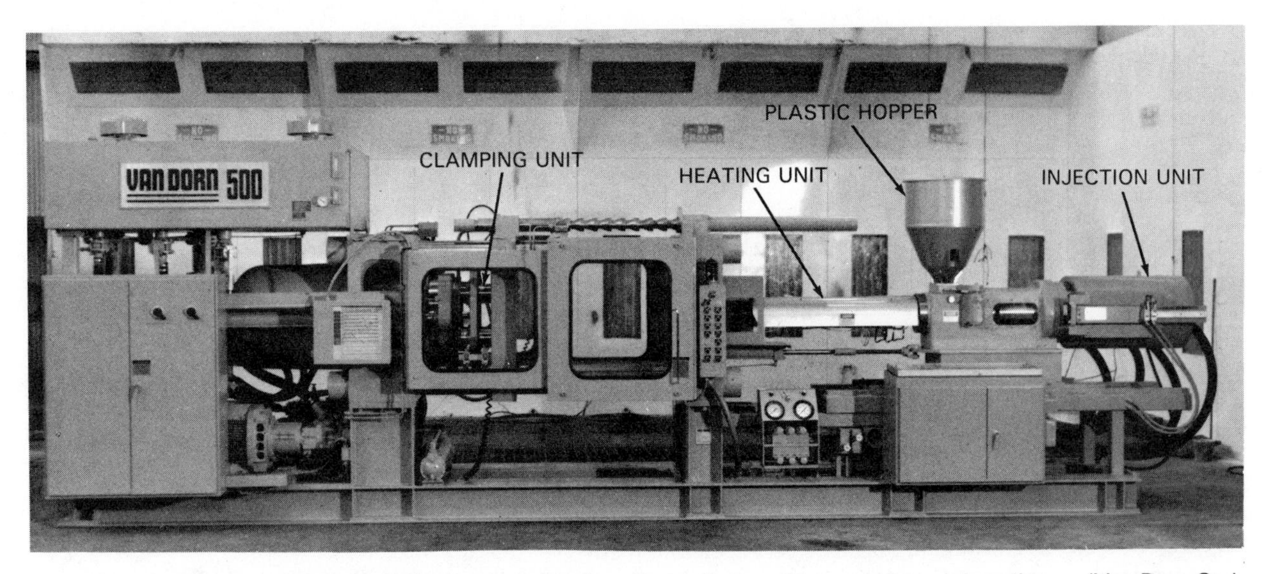

Fig. 7-2. An injection molder is similar to a hot chamber die casting machine used in metal casting. (Van Dorn Co.)

Standard plunger. The simplest injection unit is the standard plunger. This unit, shown in Fig. 7-3, is used primarily in small machines. The unmelted plastic granules or powder feed into the chamber when the ram is in its withdrawn position. When the ram moves into the cylinder, the unmelted plastic is forced through a spreader unit which contains a heated *torpedo* (spreader). There it is melted through a combination of heat and pressure.

When the plunger forces the unmelted plastic around the torpedo, the melted plastic already in the machine is pushed forward. Increased pressure from the plunger movement causes the melted plastic to flow through the nozzle into the mold cavity. Pressure is maintained until the mold cavity is full and the material at the gates of the cavity is solidified. The pressure may then be released. The mold is opened when the part is entirely solid.

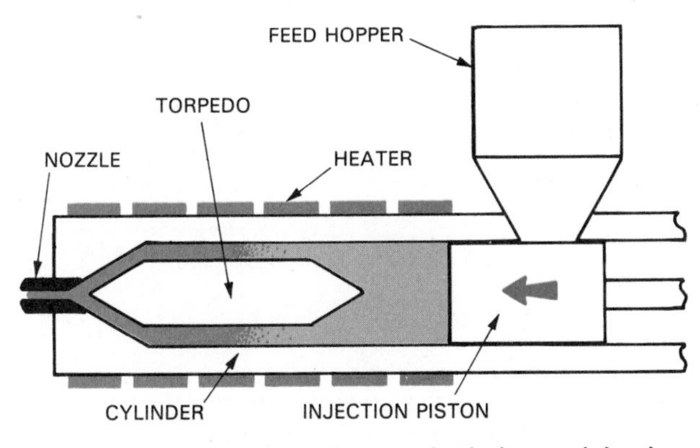

Fig. 7-3. A schematic of a standard plunger injection molding machine.

Two-stage plunger preplasticator. The second type of injection unit involves two separate stages. See Fig. 7-4. The first stage plasticizes (heats) the material and forces it into the shooting chamber. This part of the machine is very similar to the standard plunger injection unit.

The second stage also contains a plunger which forces the material from the shooting chamber into

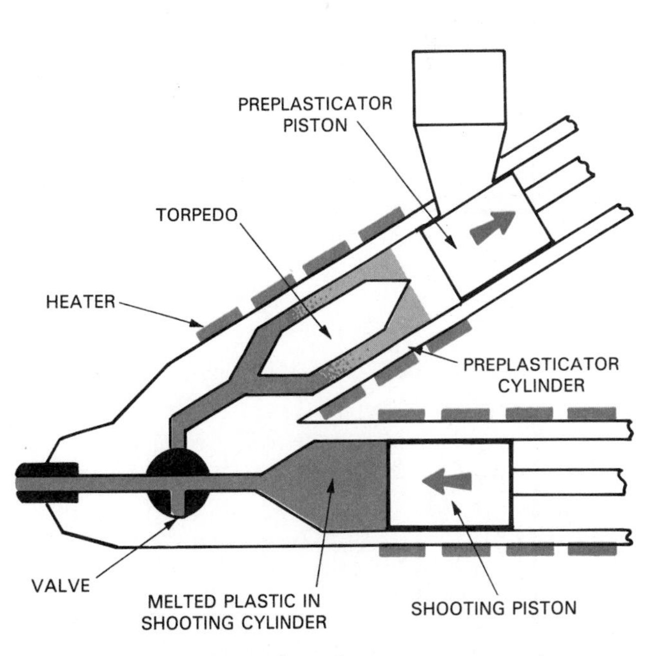

Fig. 7-4. A schematic of a two-stage plunger preplasticator injection molding machine. Plastic is heated in first stage (upper cylinder) while second stage forces melted plastic into mold.

the mold. The use of two stages reduces the pressure drop common to the standard plunger unit. Also, exact shot weights may be obtained by controlling the output of the preplasticizer unit. In addition, larger amounts of material under lower temperatures may be molded with each shot.

Screw preplasticator. The screw preplasticator unit is identical to the two-stage plunger unit except the preplasticator phase uses a screw extruder to prepare the material. This unit delivers a more uniform material to the shooting chamber than the two-stage plunger unit.

Reciprocating screw. The reciprocating screw injection unit is the most common type of unit used on large injection molding machines, Fig. 7-5. It uses the advantage of the screw preplasticator unit without the need of a separate shooting chamber. Also, it reduces the time from melting to shooting which is very critical when using heat sensitive materials.

The reciprocating screw unit has three major sections: feed zone, compression zone, and metering zone. The screw root diameter increases from the feed zone to the metering zone. This, in turn, decreases the thread depth.

As the screw turns, it forces the plastic along and compresses it into a smaller area until it becomes a semifluid mass. As the screw continues to turn, the melted material fills the chamber in front of the screw. The screw is forced back as the chamber continues to fill. At a given point the screw hits a limit switch and stops turning. A nonreturn valve closes so molten plastic cannot be forced back along the screw.

The screw then becomes a plunger forced forward by a hydraulic cylinder. This movement injects the plastic into the mold. Pressure is held until the material in the gates of the mold solidifies.

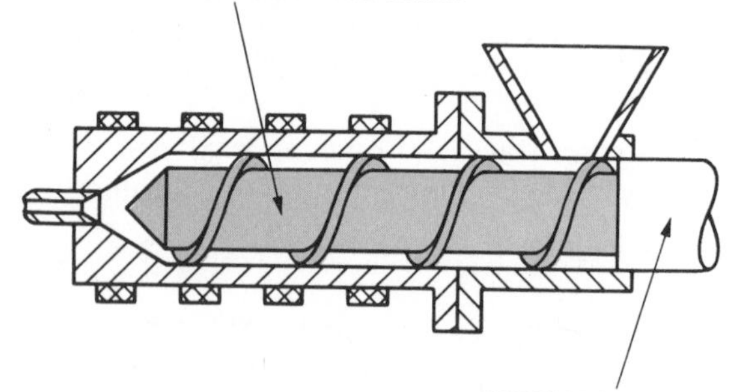

Fig. 7-5. A schematic of a reciprocating screw injection molding machine. Screw also acts as a plunger to force plastic into the mold. (General Motors Corp.)

Clamping units and molds

The molds for injection molding, as shown in Fig. 7-6, have a sprue that carries metal to the runners. Runners carry the metal to the mold cavity or cavities. Gates control the flow of the material into the cavities. Many injection molds also have cooling channels so that water may be constantly circulated through the mold. The water carries away heat so that a mold temperature between 150° and 200°F may be maintained. This temperature range allows free flow of the material but also controls the material's freezing (solidification) period. With water-cooled molds, machines can often make two to six shots per minute.

The molds are held by the second major part of the injection molding machine, the clamping unit. The clamping unit must hold the mold closed during the injection cycle and solidification phases of the process. It must then open so that the molded part may be ejected. Finally, the clamping unit must be adjustable for different mold heights unless the machine is to be used with only one mold.

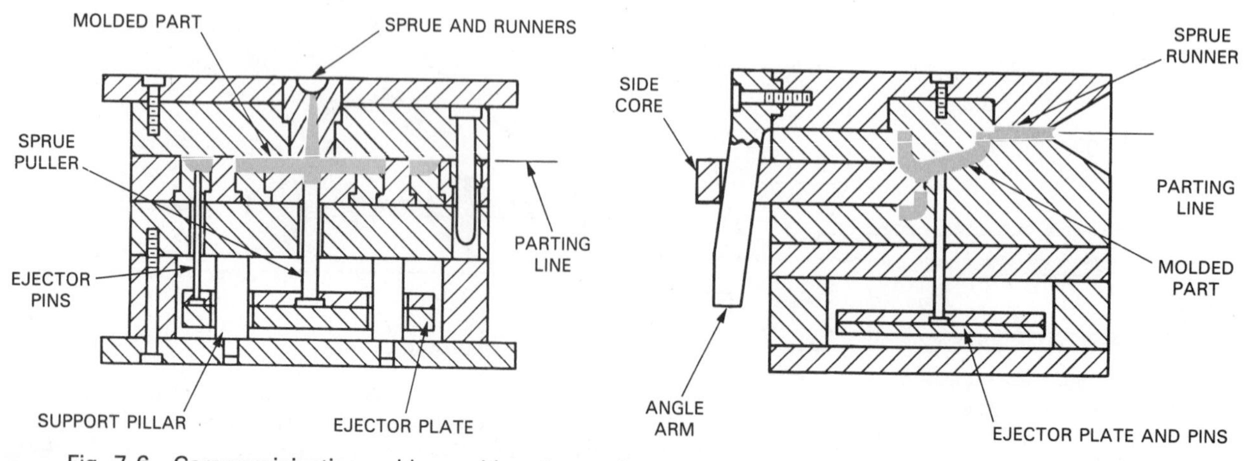

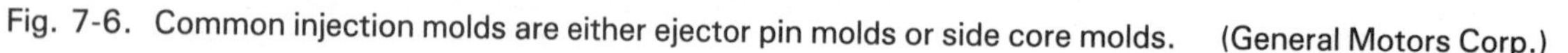
Fig. 7-6. Common injection molds are either ejector pin molds or side core molds. (General Motors Corp.)

The mold clamping units use mechanical (toggle), hydraulic, and hydro-mechanical clamps. The clamps close the movable die half against the stationary half and exert the force necessary to resist the injection pressures. These pressures can range from 1 to 2 tons for small machines to over 3000 tons for large-capacity molders.

ADVANTAGES OF INJECTION MOLDING

The injection molding machine is one of the most economical methods for producing plastic items in large quantities. It heats the material to soften it, molds it under pressure, and cools it to harden it. In most cases, the products of the process come from the mold as a finished part. Only the sprue and runners must be removed. The process is rapid, produces little waste, and can be highly automated.

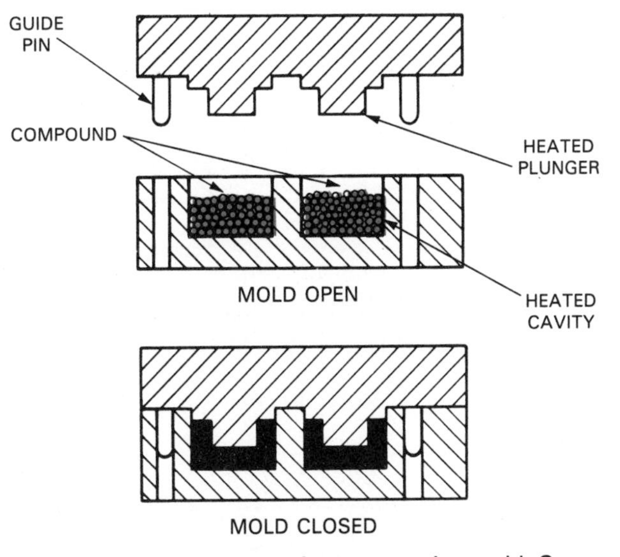

Fig. 7-7. Sectional view of a compression mold. Compound is brought to melt temperature by heated mold half. (General Motors Corp.)

COMPRESSION MOLDING

Compression molding was one of the first processes developed to mold plastic parts. It is used almost entirely for thermosetting plastics.

THE PROCESS

The process, illustrated by the drawings in Fig. 7-7, is very simple. Raw thermosetting materials are placed in the bottom half of a heated open mold. The mold temperatures range from 285°-400°F. The material may be a powder, granules, or a preformed tablet. The plastic is heated in the mold.

The mating half of the mold is lowered into the bottom mold cavity. Pressure from 100 to 8000 psi is applied. The plastic first turns liquid and flows to all parts of the mold. The material, being a thermoset (heat-curing), cures and hardens while in the heated mold. This action changes simple organic molecules into complex interlinked molecules. The process of curing is called *polymerization.*

The mold is opened after the curing cycle. Ejector pins or plugs push the part out of the mold. The cycle is now ready to be repeated.

MOLD TYPES

Three major types of molds are used for compression molding. These molds are diagramed in Fig. 7-8. They are:

1. Flash.
2. Positive.
3. Semipositive molds.

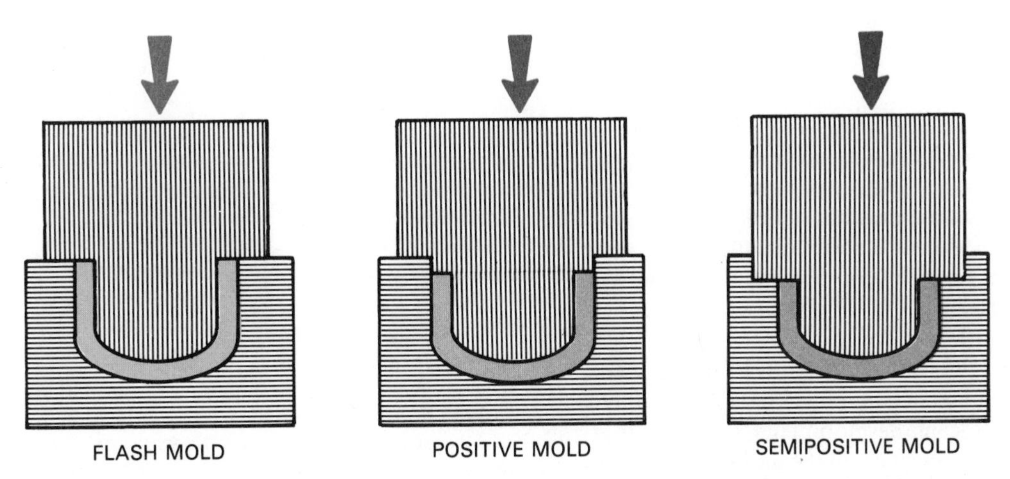

Fig. 7-8. The three types of compression molds have subtle, but important, differences.

Flash molds

Flash molds were among the earliest molds developed. More material than needed is placed in the mold. When the mold closes, the material is compressed until its density equals the force being applied. Extra material then squeezes out between the mold parts. This excess material is called *flash.*

Flash molds are relatively inexpensive to produce. They do not require careful measurement of the plastic material placed in the mold. However, the flash on the molded part must be removed. This adds a machining step to the process. Also, the products vary in dimension and density.

Positive molds

Positive molds are another early design. In a *positive* mold all material is trapped in the mold cavity. The pressure forces the material into its smallest volume. Any variation in the amount of material added will result in variations in thickness or density of the product.

Semipositive molds

Semipositive molds are the most commonly used. They are built with a slight taper to the mating mold half. As the mold closes, the material flows and the excess can escape. Again, the excess becomes flash around the edges of the molded part. During the end of the mold travel, a close fit is produced because of the tapered mold sides. The material in the mold is then trapped and compressed. The full force of the machine is applied during the final movement of the mold half.

The semipositive mold combines the best features of the flash and the positive molds. Material measurement (of the charge) need not be exact because of the flash produced during the first part of the cycle. However, density and thickness are uniform because the exact amount of material is trapped in the mold during its final movement.

ADVANTAGES/DISADVANTAGES OF COMPRESSION MOLDING

Compression molding has definite advantages. The material moves only a short distance into the mold and does not have to flow through runners and gates. Thus, material properties change very little. Mold cost and material losses are low.

However, the time required to heat and then cure the material makes compression molding a rather slow operation. Recent developments in preheating the material before placing it in the mold have partly overcome this disadvantage.

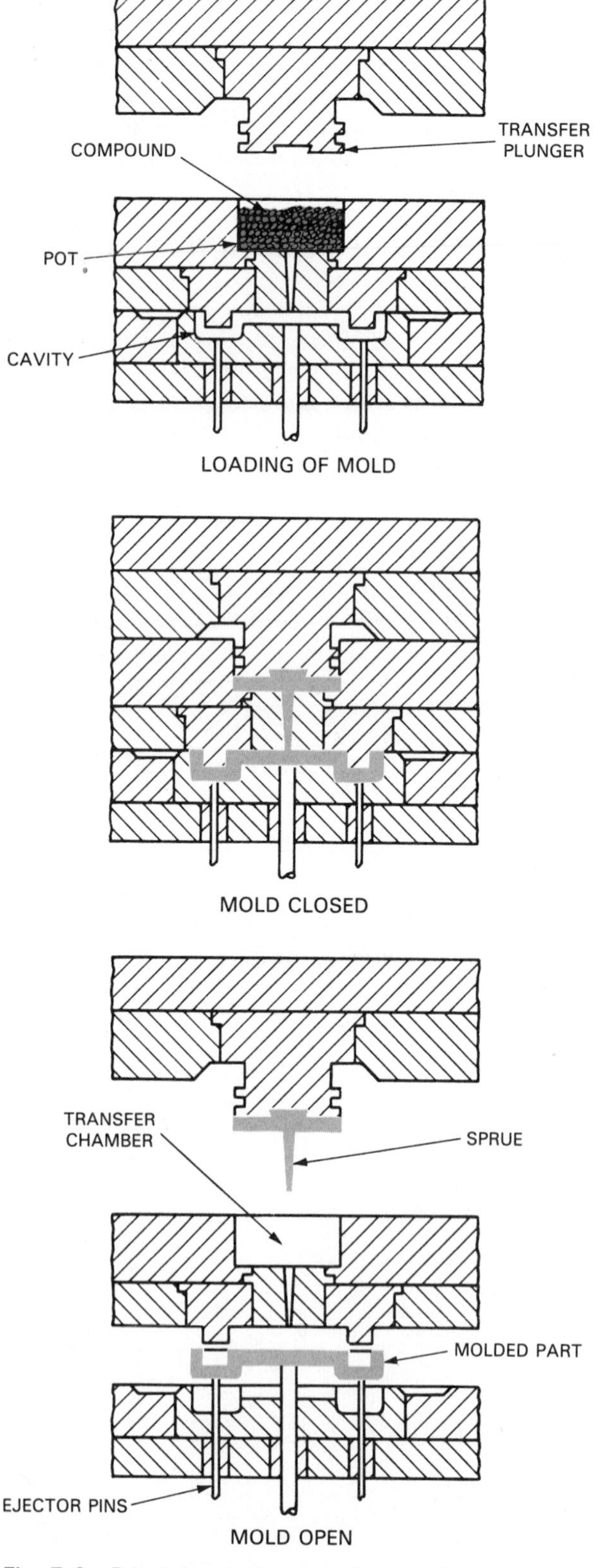

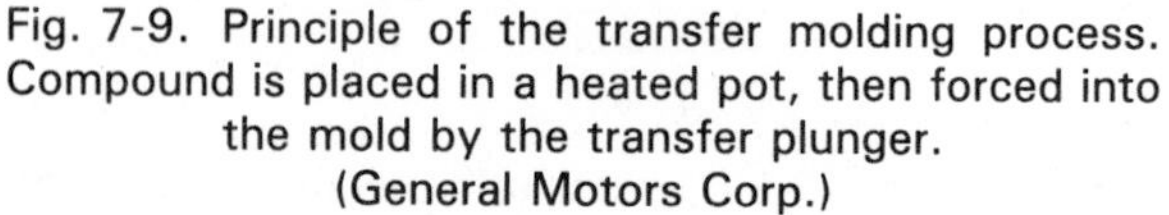

Fig. 7-9. Principle of the transfer molding process. Compound is placed in a heated pot, then forced into the mold by the transfer plunger. (General Motors Corp.)

TRANSFER MOLDING

Transfer molding is a modification of compression molding. It eliminates the turbulence and uneven flow common in high pressure compression molding. The transfer molding process is very similar to cold chamber die casting.

THE TRANSFER MOLDING PROCESS

Like compression molding, transfer molding is used primarily for thermosetting plastics. The process, shown in Fig. 7-9, involves placing the thermoset material in a heated chamber. This chamber, called a *pot,* is in the top on the mold. When the material has reached the proper temperature, a plunger or plug forces the hot, liquid plastic into the mold cavity or cavities. The material flows from the pot to the cavity or cavities through a sprue, runners, and gates.

Still under heat, the material is cured (polymerized) in the mold cavity. After the plastic material has hardened, the mold is opened and the part is removed.

TRANSFER MOLDS

Two major types of molds are used for transfer molding: *integral pot transfer molds* and *plunger transfer molds.* Both operate at relatively low temperatures to prevent material curing in the pot. Typical mold temperatures range from 280° to 380°F.

Integral pot transfer molds

Integral pot transfer molds were the first transfer molds to be used. The pot and the plunger are part of the complete mold. The mold consists of three plates. The top plate, as shown in Fig. 7-10, contains the transfer plug. The middle plate has the pot in its top surface and the top half of the mold cavity in its bottom surface. The bottom plate contains the lower half or the mold cavity.

The mold may be run on a simple flat platen press. When the press closes, it applies pressure on the material by forcing the plunger plate into the pot. Pressure also holds the two lower portions of the mold together during the molding cycle.

The mold clamping pressure and the molding pressure are the same with integral pot molds. This inability to separately control the molding pressure is a disadvantage.

Plunger transfer molds

Plunger transfer molds overcome this disadvantage by using a separate plunger to force the plastic into the mold cavities. A hydraulic cylinder closes the mold. Another hydraulic cylinder moves the plunger. Fig. 7-11 shows the stages in the plunger transfer molding process.

The plunger transfer mold may have high clamping pressure while plunger speed and pressure are controlled. This advantage is particularly important for large molds.

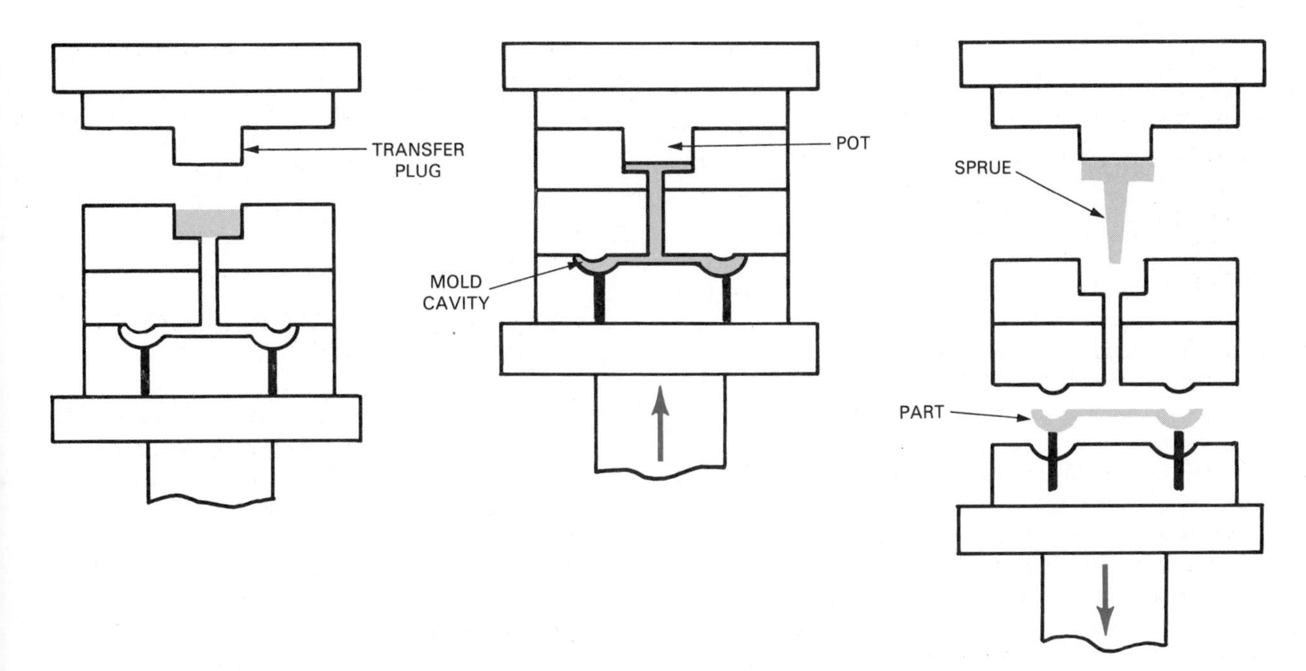

Fig. 7-10. A diagram of the integral pot transfer mold process. Left. Top plate is open and charge is in pot. Middle. Transfer plug forces hot, liquid plastic into mold. Right. Mold is open, part ready for removal.

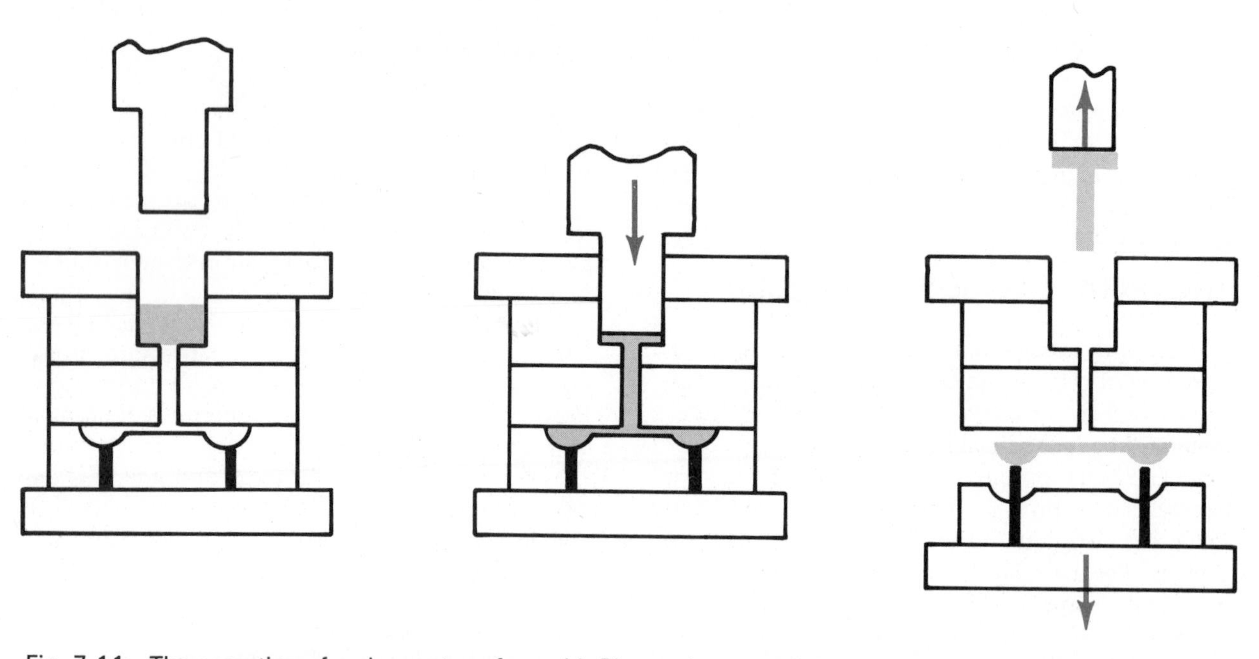

Fig. 7-11. The operation of a plunger transfer mold. Plunger is separate from the mold top. Clamping pressure and molding pressure are different.

ADVANTAGES/DISADVANTAGES OF TRANSFER MOLDING

Compression molding, in general, has several advantages. Metal inserts may be molded directly into the parts because the hot plastic enters the mold cavity slowly and under low pressure. Also, complex parts with large variations in thickness may be accurately molded. These advantages must be weighed against high mold cost and the loss of material in the sprue and runners attached to the part.

ROTATIONAL MOLDING

The rotational molding process was designed to produce hollow objects from powdered plastics, plastisols, and lattices. The mold can be rotated in one or two planes. The process soon found wide use in the toy industry and spread to other applications. As a rule, thermoplastic materials are used; however, on rare occasions, thermosets may be molded.

THE ROTATIONAL MOLDING PROCESS

Rotational molding uses a machine which can rotate a mold on two axes, Fig. 7-12. The axes are perpendicular to each other. A material placed inside the mold will come in contact with all surfaces of the mold because of this twin-axis rotation.

Rotational molding involves several steps.

1. The plastic material, usually a powder though it may be liquid, is placed in the metal mold. The amount of material is carefully weighed to insure proper wall thickness in the product.
2. The mold is placed in an oven and heated to 500°-900°F. The temperature varies according to the material being molded.
3. The mold is rotated while being heated. Temperature and rotation cause the plastic to melt and form a uniform layer on the inside wall of the mold.
4. Once all the material is melted and deposited on the mold walls, the mold is cooled while rotation continues. The cooling action takes place outside the oven. Cooling rate is increased by using moving air, water fog, or water spray.
5. The cooled mold is opened and the finished product is removed.

ADVANTAGES/DISADVANTAGES OF ROTATIONAL MOLDING

Rotational molding produces a product with a fairly uniform wall thickness. A completely closed hollow product can be produced with no waste. In addition, product color can be easily changed. Also, the product, unlike those produced using pressure molding techniques, will be strain free.

Rotational molds are much easier to make and, therefore, cost less. This advantage makes rotational molding well suited for short runs.

However, the high energy consumption from the constant heating and cooling of the mold makes the

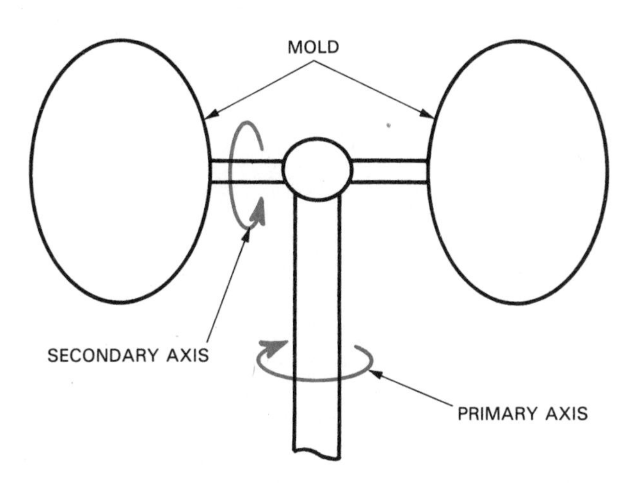

Fig. 7-12. Two-axis rotational molding. The primary axis turns four times slower than the secondary axis.

process less economical for longer runs. Also, time required to load and unload the mold is a definite disadvantage.

New methods for heating the molds and more automatic machines are now being developed. These improvements will lessen some of the serious disadvantages of rotational molding.

DIP CASTING

Dip casting or coating is used to form plastisol materials. These materials are finely divided polyvinyl chloride resins suspended in a liquid plasticizer.

In *dip casting,* Fig. 7-13, a heated metal mold is placed in a container of plastisol material and then slowly withdrawn. The plastisol material forms a layer on the outside of the mold. The thickness of this layer is determined by the temperature of the mold and the length of time the mold is allowed to remain in the plastisol material.

Once withdrawn, the mold is placed in a fusing oven where the material around the mold is cured. The cured product is then cooled and stripped from the mold. Gloves, boots, and tool handle insulators are examples of products produced by dip casting.

Metal products may also be plastic coated using the same process. Instead of a mold, the metal product itself is heated, dipped, and cured. Dishdrainers, baskets, and bobby pins are coated this way.

SLUSH MOLDING

The concept of slush molding, illustrated in Fig. 7-14, is exactly the same as slush casting of metal described in Chapter 6. It is a second way to form a product using a plastisol material. Dipping is the other method.

In slush molding the plastisol material is placed in a mold with a shaped cavity. The mold is heated. The heat from the mold fuses the plastisol material which is in contact with the mold walls. As the heat continues to be applied, the wall thickness of the material increases. When the proper wall thickness is obtained,

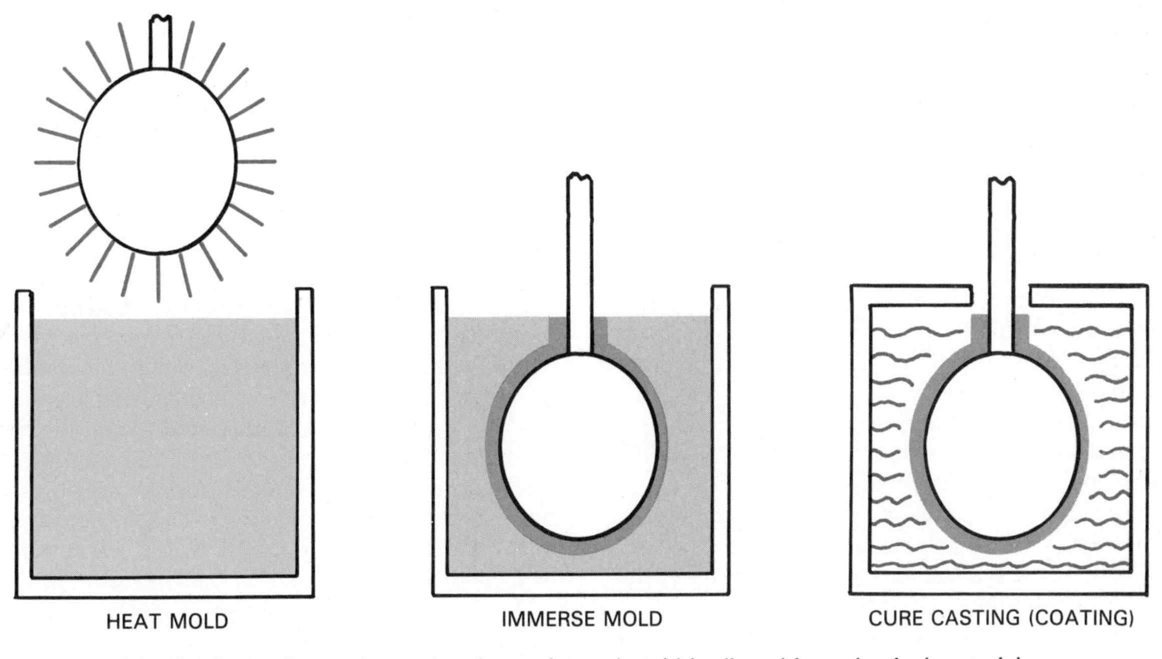

Fig. 7-13. In dip casting or coating, a heated mold is dipped into plastisol materials.

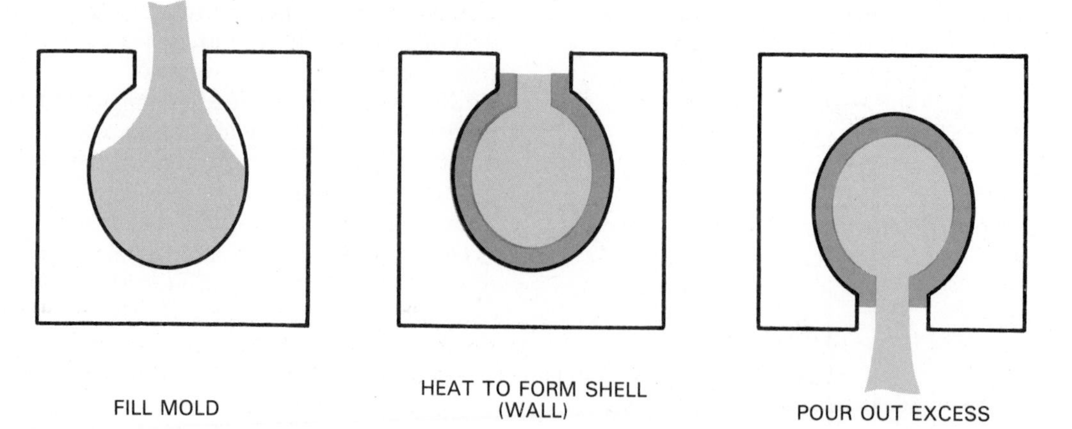

Fig. 7-14. Slush casting is like dip molding except that plastisol is poured into heated mold.

the mold is turned over allowing the remaining liquid material in the center to run out. This leaves a plastic coating on the mold walls. The thickness of this coating is controlled by the temperature of the mold and the length of time the material is left in the mold before dumping.

After dumping, the mold is turned upright and passes through a fusion oven. Here the plastic material cures. The mold cools and the finished, hollow product is removed.

FOAM MOLDING

Cellular or foam plastics are increasing in use and importance. They are used as insulation, in packaging, for furniture and upholstery, and as core materials for construction components. The production and molding of plastic foam material is accomplished through three major techniques:

1. Expandable polystyrene molding.
2. Injection molding structural foam.
3. Extruding thermoplastic foam.

EXPANDABLE POLYSTYRENE MOLDING

Expandable polystyrene (EPS) molding is usually broken down into three major categories: cup molding, block molding, and shape or custom molding. However, the basic process is the same for all three. The difference lies in the size of the bead used and the amount of bonding agent present.

The process basically involves eight steps.

1. The polystyrene beads are pre-expanded. Special machines heat the raw beads with steam or hot air. The lighter, expanded beads rise to the top of the machine and are removed.
2. The beads are allowed to dry, stabilize, and harden. The internal pressure caused by the expansion and cooling is slowly relieved. The stabilization period may take up to 36 hours.
3. After stabilization, the beads are ready to be molded into a final shape. The molding machines are closed and their molds are heated.
4. The pre-expanded and stabilized beads are drawn into the molds from the machine storage hoppers using air or gravity.
5. The mold is clamped shut tightly and low-pressure steam is forced into the mold.
6. The beads expand further under the reaction with the steam until they bond together.
7. The pressure inside individual beads caused by the steam and air is reduced. This is done by cooling the mold with water fog or attaching a vacuum pump to the mold.
8. The mold is opened and the product is removed.

The cycle time varies with the wall thickness and density of the product. Cups can be produced at rates up to 600 cycles per hour while blocks may only run five or six cycles per hour.

INJECTION MOLDING STRUCTURAL FOAM

Injection molding of foamable plastic material can produce, in one step, a product with a solid outer skin and a foam interior. This process, called *reaction injection molding,* uses a material which is a combination of a thermoplastic resin and a blowing agent. The blowing agent produces foam out of the plastic resin. This combination of a resin and a blowing agent is injected into a mold. A solid skin is formed as the material touches the mold cavity. This skin produces a heat resistant barrier which allows the center portion of the material to foam rather than harden, Fig. 7-15.

A second injection molding process for foam is called *sandwich molding.* This process uses a special high-pressure molding machine with two injection units. See Fig. 7-16. One unit injects the skin material. The second unit, using the same sprue injects the core material inside the skin material. As the core material expands and cures, it forces the skin to take the shape of the mold.

This process, also called *co-injection molding,* can produce a two-material sandwich. Hard, expensive materials can be used for the skin while less expensive materials make up the core.

These materials with a solid skin and a foam (cellular) core are called *structural foams.* They have high weight-to-strength ratios.

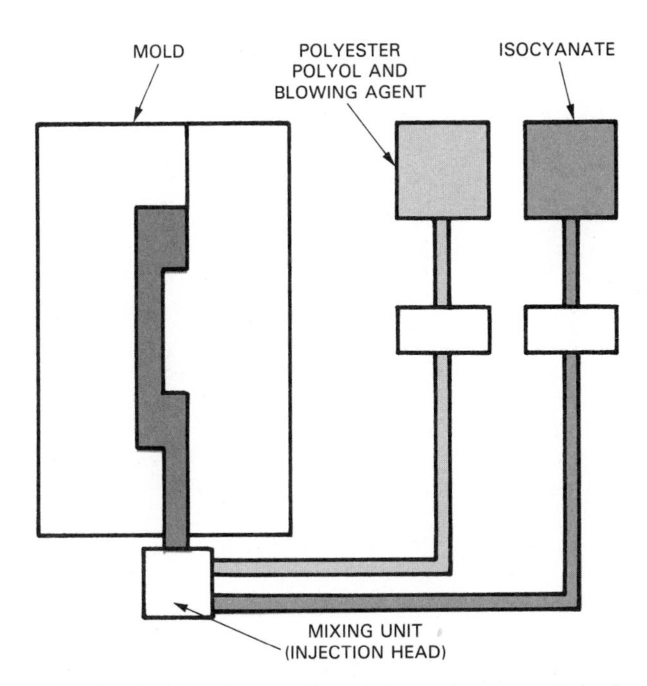

Fig. 7-15. Injection molding of foams is accomplished by using the reaction injection molding system shown here. (RIM)

EXTRUSION OF STRUCTURAL FOAM

Extrusion, which will be discussed in depth in the next section of this text, can be applied to structural foam production. In this process, thermoplastic materials with a blowing agent are heated. They are forced out a cooled shaped die. The die has a torpedo core which causes the plastic to come out as a hollow tube. The tube has a tough skin and a less dense center like products from the injection molding process.

SUMMARY

Casting and molding of plastic material is an important way to produce shaped plastic products. Injection molding of thermoplastic resins and compression and transfer molding of thermoset resins are the most widely used techniques. However, gravity and dip casting and slush foam, and rotational molding produce a wide variety of products. Each technique is chosen because of the materials it can process and the economic and technological factors present with each proposed use.

STUDY QUESTIONS – Chapter 7

1. Describe gravity casting of plastic materials.
2. List and describe some common molds used in the gravity casting process.
3. Describe the injection molding process.
4. Which metal casting process is most like injection molding?
5. What are the major types of injection molding machines?

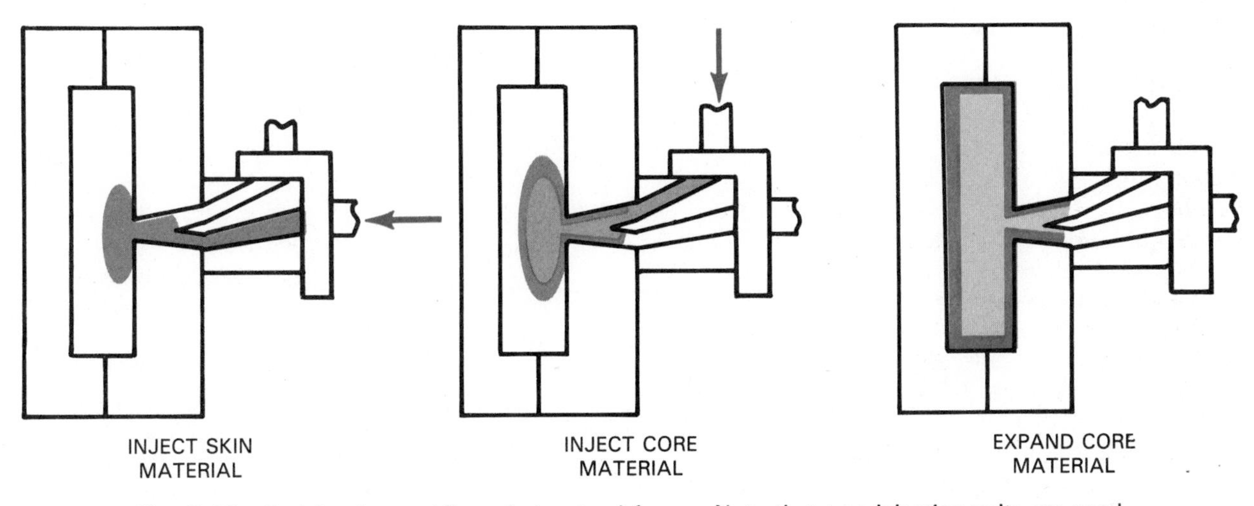

Fig. 7-16. Co-injection molding of structural foams. Note that two injection units are used.

6. Describe the compression molding process.
7. List and describe major compression mold types.
8. Describe the transfer molding process.
9. What are the differences between integral pot and plunger transfer molds?
10. Describe the rotational molding process.
11. Describe the slush casting process.
12. Describe the dip casting process.
13. Describe the methods of foam molding.
14. List the advantages and disadvantages of:
    a. Injection molding.
    b. Compression molding.
    c. Transfer molding.
    d. Rotational molding.

Chapter 8

CASTING CERAMIC MATERIALS

Most ceramic materials are shaped using forming or "plastic shaping" techniques which will be discussed in Chapter 13. However, not all ceramic products can be produced using these methods. The shape of some products makes forming processes impossible to use. Also, not all products can be economically produced using plastic shaping methods.

Two important casting techniques are used to produce many clay products. These are *drain casting* and *solid casting.*

DRAIN CASTING

Drain casting, often called *slip casting* is much like slush casting. It produces a wide variety of hollow products. These products have well defined outside detail and less clear internal shape. Typical of products produced by drain casting are vases, figurines, coffee pots, sinks, toilets, pump impellers and housings for the chemical industry, and porcelain ware for industrial and scientific uses.

DRAIN CASTING PROCESS

The drain casting process involves three major activities.

1. Preparing slip.
2. Making casting molds.
3. Casting the product.

Preparing the slip

Slip is basically a suspension of clay particles and/or nonplastics in water. Chemical agents are added to this suspension to improve its properties.

Nonplastics (material other than clay), such as silica, feldspar, and talc, are added to reduce drying shrinkage, help prevent cracking, improve firing qualities, and reduce heat shock in the product.

Also, deflocculents are added. These agents keep the slip from settling so that a product uniform from top to bottom can be made. In addition, slip with deflocculents drain from the mold more readily. The suspension is mixed to the consistency of heavy cream.

Making molds

The making of casting molds requires two steps: making a model and producing a mold around the model. The model is made the identical shape of the final product. Allowances for shrinkage of the clay during drying and firing are added to these dimensions.

Models are often made from plaster, clay, wood, and plastics using a number of production processes. Fig. 8-1 shows a model being made.

Fig. 8-1. A designer prepares a model of a product which will be slip cast. A master mold will be made from the model. (Lennox China Co.)

The model becomes the pattern which will produce the inside cavity of the slip casting mold. The plaster mold usually has two parts. The mold must:
1. Be the proper shape.
2. Be hard.
3. Have uniform porosity.
4. Allow the clay to shrink during drying.

The model is placed in a "box" which allows one half of it to be above a center divider. The box, divider, and model are soaped. Then plaster is carefully poured into the box and allowed to dry. This process produces the first half of the mold. The plaster mold half and model are removed from the "box." The model is removed from the mold half. The mold is carefully cleaned. The mold half and pattern are reassembled and placed in another "box." Again, the surfaces are soaped and the second half of the mold is poured. After it has dried, the mold is removed, cleaned, and prepared for use. Fig. 8-2 shows typical slip casting molds.

Plaster molds are considered permanent. However, they have a very limited life. A mold with fine detail may produce only 20 casts while a simple small mold will usually exceed 50 casting cycles. Large molds often last through 100 casts.

Fig. 8-2. These slip casting molds will be used to cast lavatories. (Kohler Co.)

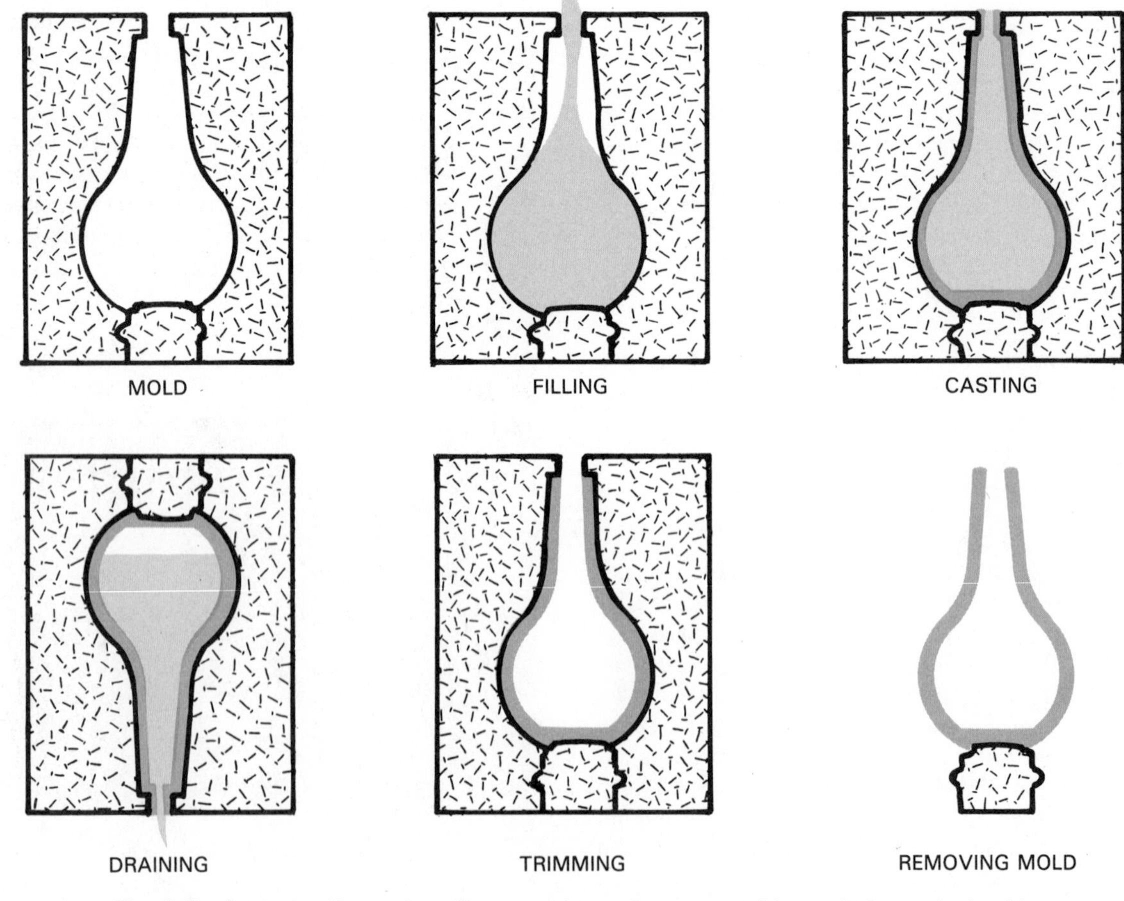

Fig. 8-3. Steps in slip casting. First step is to clean, assemble, and clamp the mold.

Casting the product

The actual casting of a product includes several steps, as seen in Fig. 8-3. These include the following activities:

1. The mold is cleaned, assembled, and clamped together.
2. Slip is poured or pumped into the mold, Fig. 8-4.
3. The plaster mold absorbs moisture from the slip. An even layer of clay builds up on the inside surfaces of the mold.
4. The slip is allowed to remain in the mold until proper wall thickness is achieved. Thin walled products may require 10-15 minutes; an hour or more is required to produce a 1/2 in. thick clay wall.
5. When the proper wall thickness is reached, the slip is poured off.
6. The clay is allowed to dry in the mold until it shrinks from the mold walls.
7. The mold is opened and the damp product is removed, Fig. 8-5.
8. The mold is cleaned and dried for the next cycle.
9. The damp product is prepared for further processing.

Fig. 8-4. Toilet tank slip casting mold are being filled two at a time. (Kohler Co.)

Fig. 8-5. A slip cast product is being removed from its mold. The lavatory is in a ware setter to reduce warping as it dries. (Kohler Co.)

SOLID CASTING

Solid casting is a second technique used to cast ceramic material. It uses slip and a plaster mold like drain casting, Fig. 8-6. The major difference between the two methods is that solid casting allows all the slip to harden in the mold.

Extra slip is often added or is available in the mold "riser." This extra material is needed to make up for the shrinkage as the object solidifies. If extra material is not available, a hollow part will be produced. Fig. 8-6 shows that:

1. The mold is cleaned, assembled, and clamped.
2. Slip is poured in the mold.

Fig. 8-6. A solid casting mold in use. Note the cavities in the right half. (Syracuse China Corp.)

3. Most of the water content of the slip is absorbed into the mold. This leaves a damp, solid object.
4. The mold is opened and the product is removed. Fig. 8-7 shows molds being stripped.

OTHER CERAMIC CASTING TECHNIQUES

Ceramic materials other than clays may be cast. Glass is occasionally cast but it is the most difficult way to form glass products. The largest piece of glass in the world, however, was cast. It is now the 200 in. disc at the Hale Telescope, Mount Palomar, California.

Fig. 8-7. Solid cast cup handles are removed from their plastic molds. (Syracuse China Corp.)

An important casting technique for making glass products is *centrifugal casting*. Typically, this technique is used to produce funnel-shaped glassware. This casting method involves five major steps. Fig. 8-8 shows these steps as they apply to casting television picture tube funnels.

1. Molten glass is dropped into a spinning mold.
2. The spinning mold develops centrifugal forces which cause the molten glass to flow up the sides of the mold.
3. A distributor is often fed into the center of the glob to increase the flow of the glass.
4. A grooving disc trims the funnel to the desired height.
5. The finished part is ejected from the mold.

Concrete is often cast into shapes such as patio blocks, stepping stones, bridge and building beams, and pipe. These products are primarily poured into open permanent molds and allowed to cure.

SUMMARY

Many ceramic products, including small porcelain dishes for scientific laboratories, bases, bathroom fixtures, parts for the chemical industry, as well as large concrete members and products, are cast using primarily permanent mold techniques. These products may be produced from:

1. Clay slips which harden by water removal.
2. Sand.
3. Aggregate.

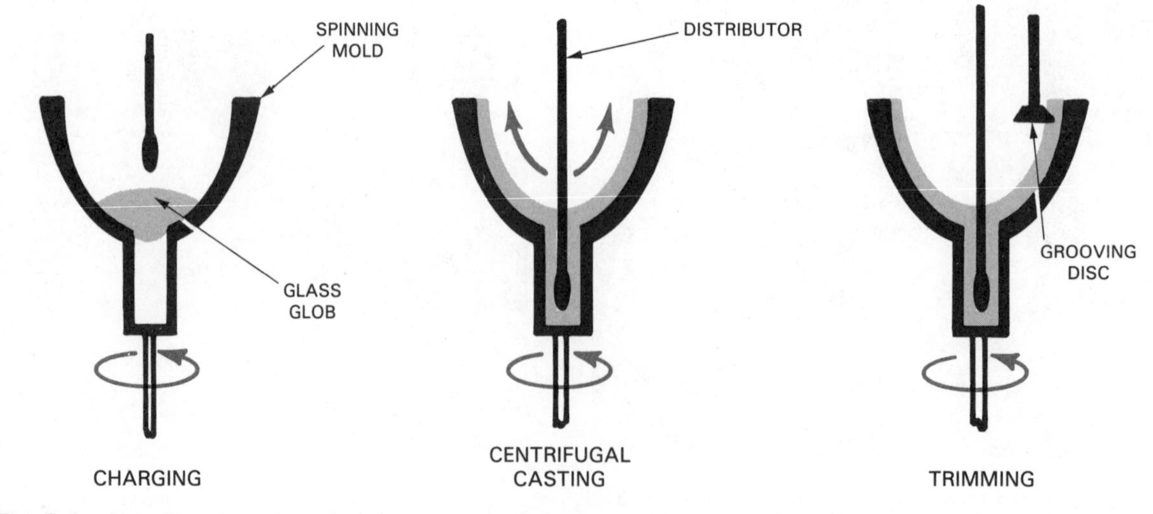

Fig. 8-8. Centrifugal casting of glass products. Spinning mold causes glass glob to take on mold shape.

4. Cement mixtures (concrete) which harden by chemical action.
5. Glass (occasionally) which hardens by cooling.

All provide us with products that are versatile, useful, and durable.

STUDY QUESTIONS – Chapter 8

1. Describe the drain casting process.
2. Which metal and plastic casting processes are like drain casting?
3. ____________ is basically a suspension of clay particles and/or nonplastics in water.
4. How are molds for ceramic casting produced?
5. Why are ceramic casting molds made from plaster?
6. What is the difference between solid casting and drain casting?
7. Describe the centrifugal casting process as it is used to cast glass products.

Worker removes solid cast cup handles from their gating system. (Syracuse China Corp.)

SECTION 3
FORMING

All forming processes have in common the deformation of material to force it into a new shape. This is accomplished by stretching, compressing, or compacting.

Forming processes discussed in this section include:

- Hot forming—forging, extrusion, roll forming, piercing, drawing, and spinning.
- Cold forming—bending, drawing, and squeezing.
- Forming plastics—thermoforming, extrusion, blow molding, calendaring, and mechanical forming.
- Forming ceramics—pressing, extruding, drawing, blow molding, and jiggering.
- Forming powdered metals—extrusion, rolling, slip casting, centrifugal casting, HER compacting, and isostatic forming.

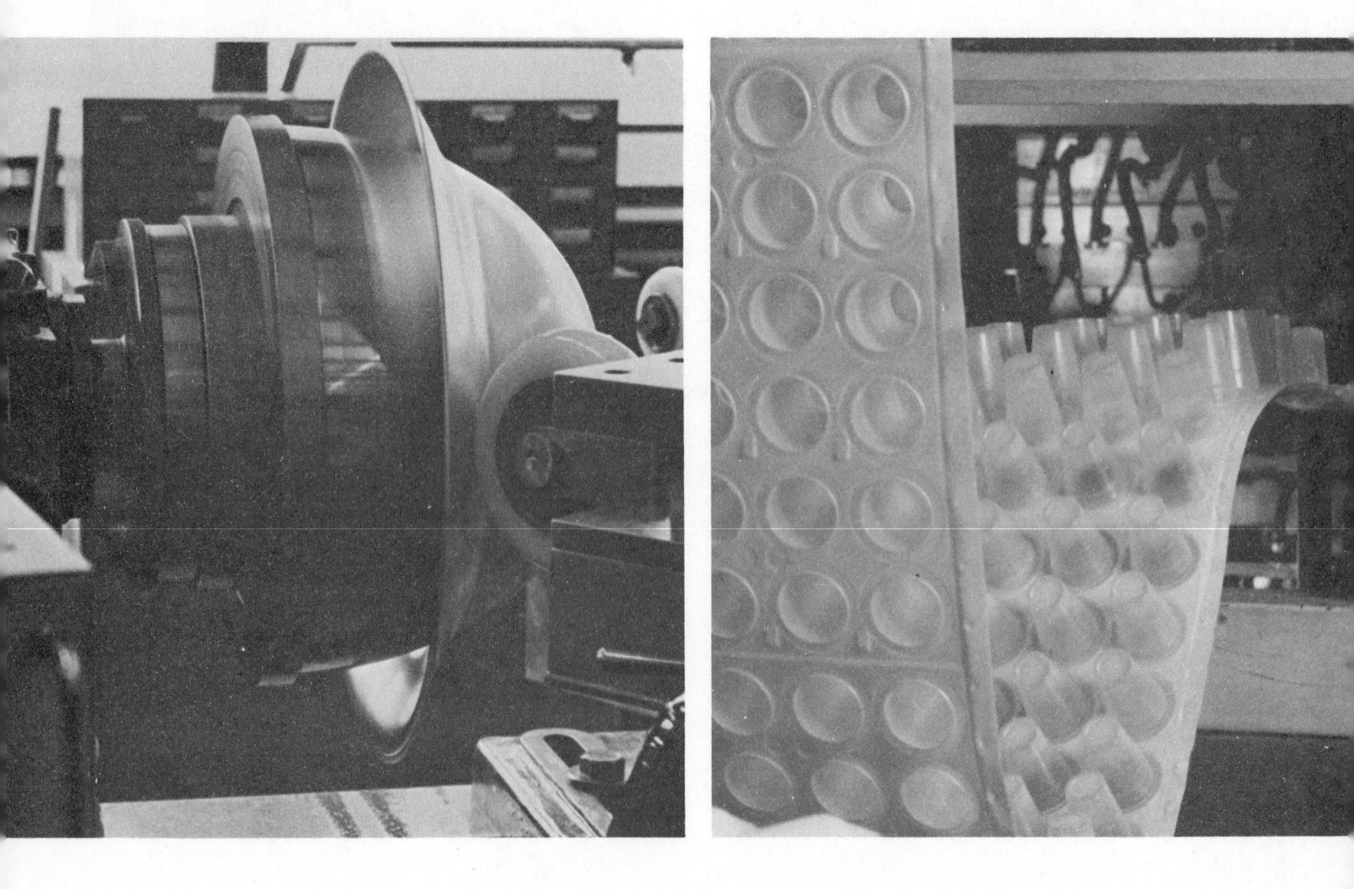

Chapter 9

INTRODUCTION TO FORMING

In terms of tonnage of material processes, the most widely used method of changing the shape of metals, ceramics, and plastics, is *forming*. Forming is a family of manufacturing processes in which an industrial material is shaped through *plastic deformation* (reshaping the material by causing it to stretch, compress, or bend) or compacting of powdered materials.

The change of shape is caused by applying enough force to make the material permanently take on a new shape. The force must not be so great as to cause the material to fracture. Fig. 9-1 shows some typical forming practices.

BASIC ESSENTIALS OF FORMING

Forming includes all of the manufacturing processes which use force to reshape solid or powdered materials without reducing their weight. The material is reshaped but not cut away. In all forming practices three essentials are either present or considered.

1. A shaping device.
2. Material temperature.
3. A method of applying force.

SHAPING DEVICES

All forming practices use shaping devices to create the desired shape to the material being processed. The shaping device is to forming what the mold is to casting. It gives size and shape to the material. In casting, material in its liquid or plastic state is forced into a mold and allowed to harden. Forming uses the shaping device and pressure to cause solid or powdered material to take on new shape.

A B C D

Fig. 9-1. Typical forming operations. A—Press forming of a reinforced plastic part. (Budd Co.) B—Glass jar molding. (Glass Packaging Institute) C—Forging steel parts. (Wyman Gordon Co.) D—Stamping television chassis. (Zenith Radio Corp.)

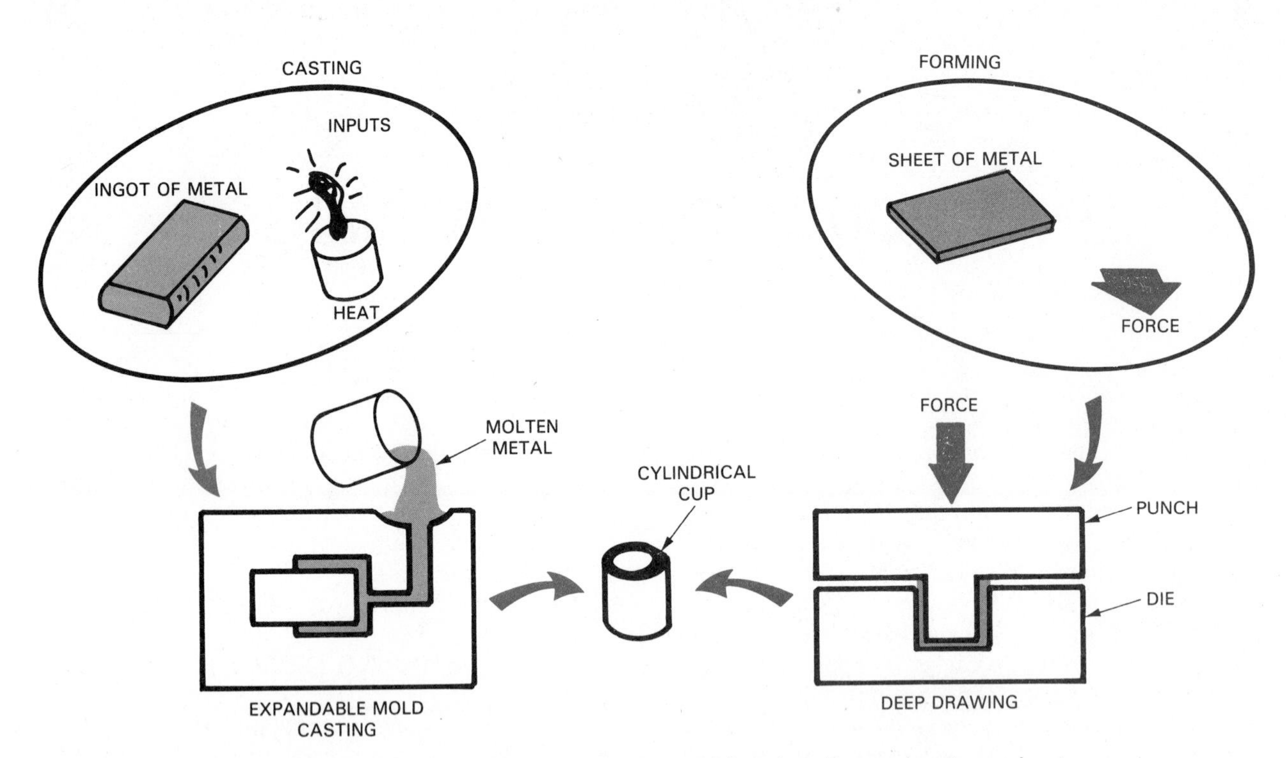

Fig. 9-2. A comparison between casting and forming. Energy inputs are heat for one, force for the other.

Heat is often used to assist in the forming processes. Fig. 9-2 shows these differences between casting and forming.

There are two basic shaping devices used in the many forming techniques practiced by modern industry: dies and rolls. Both of these devices will be discussed, in general. Specific application and unique features of the shaping devices will be presented in Chapters 10-14.

DIES

A *die,* similar to those shown in Fig. 9-3, is used with many forming techniques. A die generally is a shaping device with a cavity or hole machined into it. The holes or cavities shape the material during the forming process. Dies are made of a material harder than the material being formed. Most dies for forming metals are made of tool steel. Steels, epoxies, and other materials are used for dies to form plastic and ceramic material.

Dies may be grouped into three types:

1. Open dies.
2. Mated dies (die sets).
3. One-piece shaped dies.

Open dies

The open die is the simplest. It consists of two flat or simple, curved die parts. One die part is stationary

Fig. 9-3. A set of mated progressive dies. Note that each stage adds additional details to the final shape of the part. (National Machinery Co.)

Fig. 9-4. Open dies are being used in smith forging. (United States Steel Corp.)

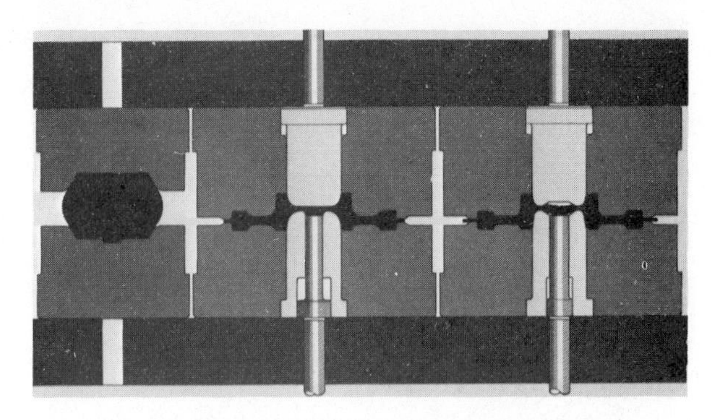

Fig. 9-5. A cross section of a three-stage forging die. Black indicates the hot metal being forged. (National Machinery Co.)

while the other is movable. The workpiece is hammered or squeezed between the stationary and the movable die parts. The hammer and anvil of early blacksmiths were much like open dies.

Fig. 9-4 shows a press with an open die setup. Typically, open dies are used for flat or open-die forging. This process is also called *hand* or *smith* forging and requires high operator skill.

Mated dies

Most dies manufactured are mated or matched dies. These dies are machined in sets. They produce a product of proper size and shape when closed.

Mated dies are of two basic types.

1. Forging dies.
2. Stamping and drawing dies.

Forging dies, as seen in Fig. 9-5, have impressions (cavities) machined in both halves of the die. When a heated metal blank is squeezed between these die halves, it will take on the shape of the cavities. One die set may have several cavities. Each cavity may be one step in the several steps needed to produce the part. Also, several identical parts may be produced at once if each cavity is identical.

Stamping and drawing dies are actually a set containing a punch and a die. The die has a cavity in it. The punch fits into the die and produces the force needed to form the part. Fig. 9-6 is a sketch of a

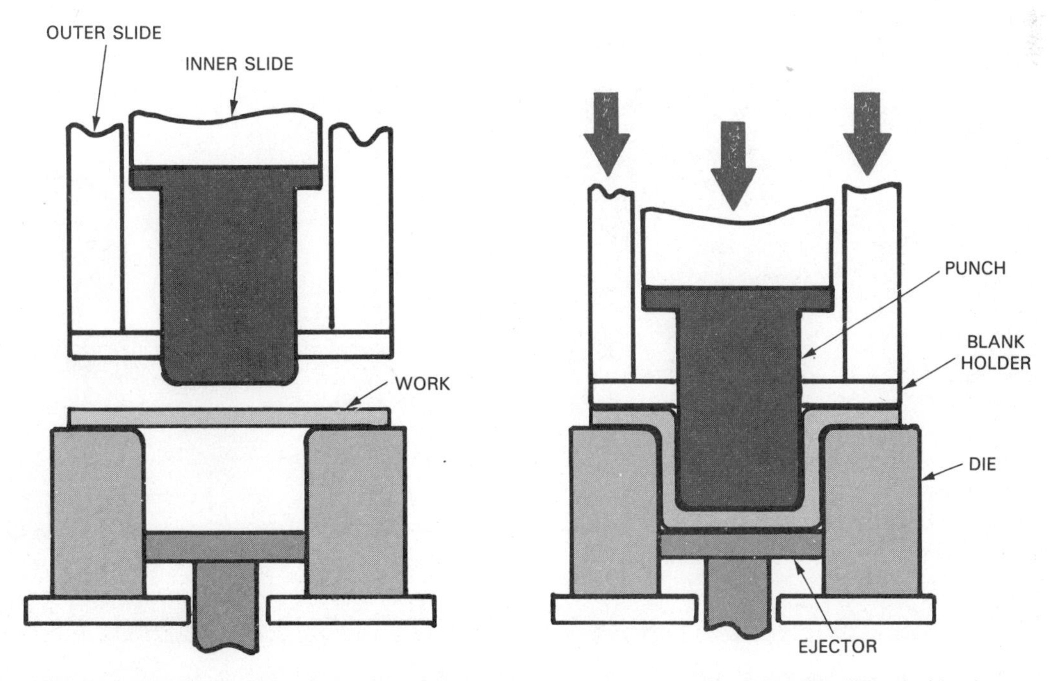

Fig. 9-6. Cross sectional of a conventional double-action drawing die set. A blank holder holds the part during the first action. The punch forms the part in the die during the second action.

typical die set. Often the die set (one punch and one die) is simply called a die without referring to the punch and the die halves of the set.

One-Piece shaped dies

The third type of die is a one-piece shaped die. In the ceramic and plastics industry this type of die is often called a mold.

One-piece shaped dies are used for a number of processes. However, the dies are of three types:

1. External shaped dies.
2. Internal shaped dies.
3. Hollow dies.

External shaped dies are used when the material is forced over the die to create a properly shaped object. Plastic may be drawn around the die (thermoforming), metal may be spun and forced over the die (metal spinning), or the die may be used with several processes where rubber, fluid, or mechanical pressure is used to force material over the die. Fig. 9-7 shows a typical external shaped one-piece die.

Internal shaped dies are similar to the external shaped one-piece dies, Fig. 9-8. The difference is that the material is drawn or forced into a cavity rather than around the die. Fig. 9-9 shows how plastic material may be shaped using internal and external shaped one-piece molds (dies).

Fig. 9-8. A glass parison is automatically positioned in an open mold. When the mold closes, air pressure will press the glass against the walls of the mold to form a jar. (Glass Packaging Institute)

Fig. 9-7. A plaster die for forming a cup. Material is forced around the outside of the die. (Syracuse China Corp.)

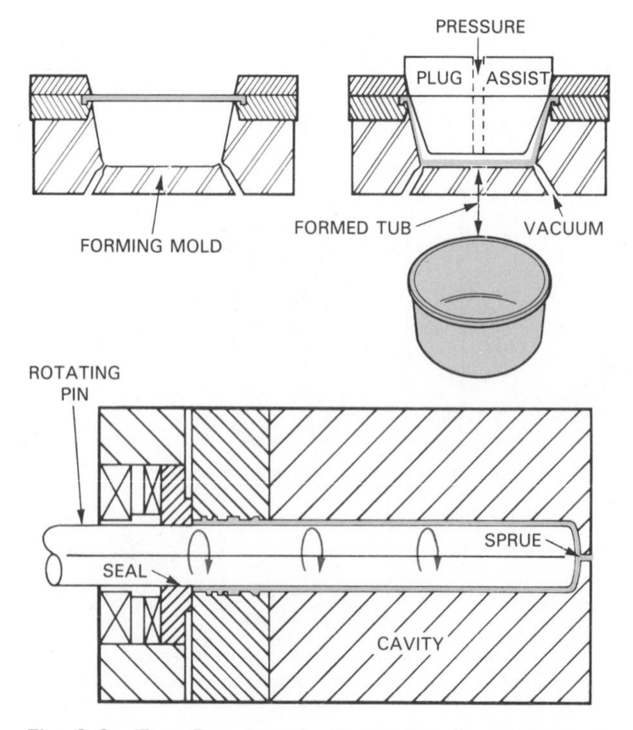

Fig. 9-9. Top. Forming plastic stock using an internal vacuum-forming mold. Bottom. An external shaped rotational mold. (Dow Chemical Co.)

Hollow dies have holes machined completely through them. They are used when a material is shaped by forcing or pulling it through the die. This type of die is used primarily for extruding and drawing metal, plastic, and ceramic materials.

ROLLS

The second major type of shaping device for forming operations are *rolls*. They are used in secondary processing to change the shape of material without changing its thickness. However, in primary processing, rolls are used to reduce the thickness and increase the width and length of ingots, slabs and other basic shapes.

There are two major types of rolls:
1. Smooth.
2. Shaped.

Smooth rolls

Smooth rolls are used primarily to produce curved shapes from metal sheet, bar, and structural shapes. Fig. 9-10 shows an example of smooth rolls being used to form material.

Shaped rolls

Shaped rolls have patterns machined along their length. These patterns may be used to squeeze the material as in *roll forging* and *hot rolling*. This action is seen in Fig. 9-11. Also, rolls are widely used to bend metal into desired shapes. The shapes may include tubing and pipe, architectural trim, and corrugated sheets.

MATERIAL TEMPERATURE

The temperature of the material is always considered during the forming process. Forming techniques may be classified as either *hot-working* processes or *cold-working* processes. Generally, metals and ceramic materials are worked either hot or cold while plastics are almost always hot when formed.

TEMPERATURE AND METAL FORMING

A factor closely asociated with metal forming is the *recrystallization temperature* of metals. Forming is based on plastic deformation of the metallic material. When force is applied to cold material it will deform. The metal is said to flow and the grains of the material will slip. The grains will also elongate in the direction of the deformation if enough force is applied. They will become longer and narrower. This action will produce harder and stronger material due to an action called *work hardening*. However, the metal will also be less ductile. It will tolerate less *plastic*

Fig. 9-10. A heavy three-roll forming machine is being used to bend angle iron. (Buffalo Forge Co.)

Fig. 9-11. Mated rolls are being used to shape wide flange structural beams. (Inland Steel Co.)

deformation (forming) without breaking. An example of the grain structure change caused by cold forming is shown in Fig. 9-12.

The effects of cold-working can be avoided if the material is heated to a point above recrystallization. This temperature is about 50 percent of the absolute melting point of the metal. At this temperature, the internal atoms vibrate more. This action, called *thermal vibration,* gives the grains of the material a greater ability to move.

During hot forming the lattice structure of the metal is broken. New, smaller crystals will form from the broken parts and then grow. If the temperature is held above the recrystallization point, the new crystals will grow to the shape of the original crystals. This is why the process is called *re*crystallization. The effect of hot forming on grain structure is shown in Fig. 9-13.

There is no exact recrystallization temperature. The forming of the new crystals which are like the original crystals depends upon:

1. Temperature. The higher the temperature, the faster the recrystallization action.
2. Time. The longer the metal is held at a temperature where recrystallization will happen, the larger the crystals will grow.
3. Amount of deformation. Recrystallization starts at the location of internal stress caused by the forming process. The greater the amount of deformation the faster the recrystallization.

ADVANTAGES OF HOT FORMING

Both hot forming and cold forming have characteristics which may be seen as advantages. Typically, the choice between hot and cold forming is made by balancing the advantages of each against the particular application.

Hot-working of metals has the following characteristics:

1. The hardness and ductility of the metal is not changed.
2. Porosity is eliminated by forcing small voids out of the material.
3. The structure of the material is improved by reforming smaller, more numerous crystals.
4. Large shape changes are possible without ruptures in the material.
5. Less force is required to reshape the metal.
6. Smaller, faster-acting machines may be used.
7. Impurities are broken up and distributed throughout the material.
8. Material surfaces need not be clean and scale free.

DISADVANTAGES OF HOT FORMING

The advantages of hot-working must be weighed against certain disadvantages. Hot-working of metals produces an oxide scale on the metallic surfaces. The scale creates both a rough surface and a loss of material. Also, close dimensional tolerances are difficult to maintain. Thirdly, carbon is lost from the surface layers of steel during hot working. This loss causes a lower strength outside layer. The weaker layer is subject to fatigue cracks which can then extend throughout the part.

ADVANTAGES OF COLD-WORKING

Cold-working of metals has its own set of advantages. These include:

1. No heating is required.
2. Close dimensional tolerances can be held.
3. Surface finishes are better than hot-formed work.
4. Strength, hardness, and directional properties are improved.

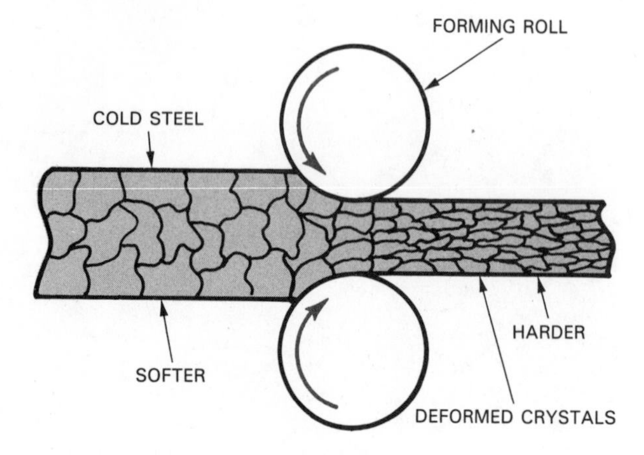

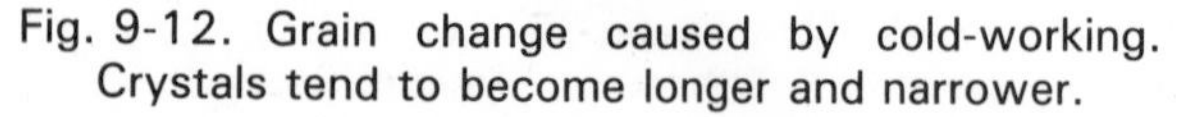
Fig. 9-12. Grain change caused by cold-working. Crystals tend to become longer and narrower.

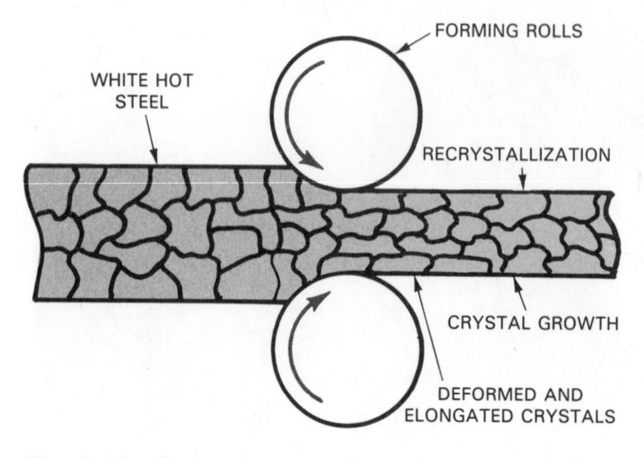

Fig. 9-13. Grain change caused by hot-working. Note similarity of grain shape.

Both hot and cold forming can generally produce parts in large quantities faster and more economically than casting or machining. Often the parts require little additional processing. Investment in heavy-duty equipment and expensive shaping devices is soon recovered by lower per part manufacturing costs.

COLD-WORKING DISADVANTAGES

However, cold-working produces stresses within the material which usually must be removed through heat treating. Also, higher forces and heavier equipment are needed for cold forming. In addition, the metal material must be clean and scale free.

Forming temperatures of plastic and ceramic materials

The temperature required for forming plastic and ceramic materials depends upon the material, itself. Generally the material cannot be worked both hot and cold. Most plastics must be heated so that they can be formed over or poured into molds. Cold plastics will either break or become so elastic that they return to their original shape. Heating allows the polymers to retain their new shape when they cool.

Whiteware ceramic materials are cold formed. The material, being "plastic" in its manufacturing state, requires no additional heat. However, glass must be hot to be formed into new useful shapes.

APPLYING FORCE

As shown in Fig. 9-14, four forces are used to squeeze, stretch, twist, or fracture a piece of material. If applied properly, force causes the material to take on a new shape. This action, called *plastic deformation,* is the basis for all forming processes.

Plastic deformation means that the material has taken on a new permanent shape. When force is applied to them, materials go through three stages. They:

1. Deform elastically. The material becomes longer and narrower when the force is applied. However, the material returns to its original shape when the force is removed. Most materials have a short elastic range. A rubber band has a large elastic deformation range while glass has virtually none.
2. Deform plastically. The material becomes elongated under force. When the force is removed, the material shortens slightly but retains its "plastic" elongation. This action is called "forming" and takes place with stress above the material yield point and below its fracture point.
3. Fracture. As was just stated, force beyond the deformation range causes the material to break.

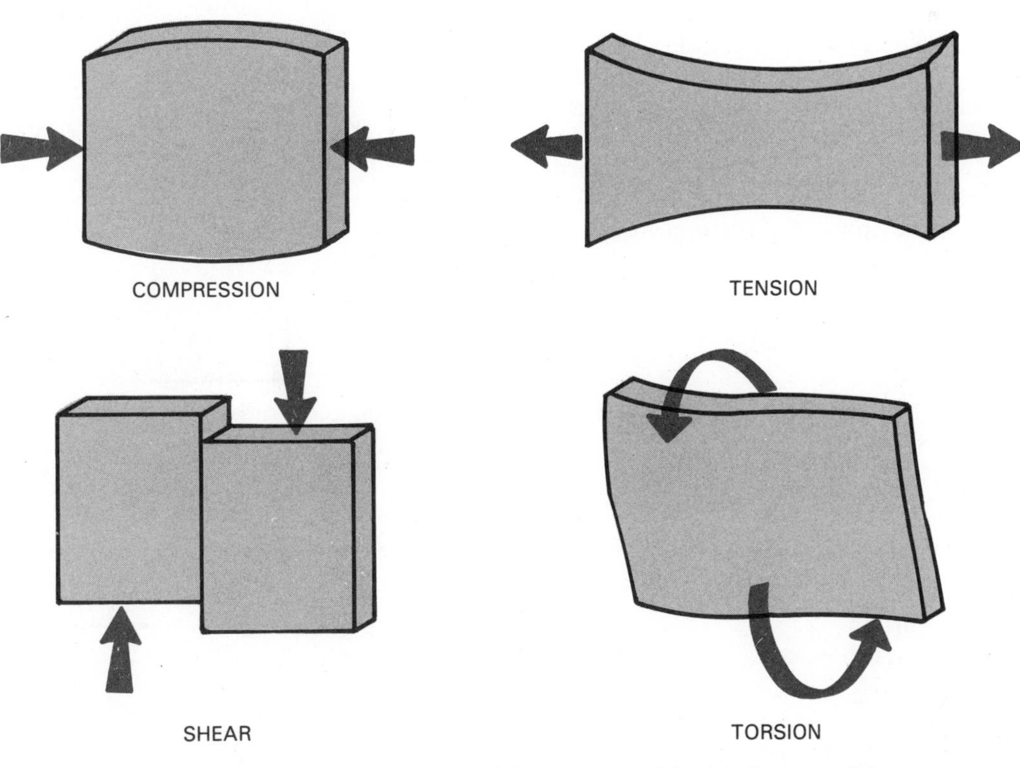

Fig. 9-14. These four types of force are used in shaping material.

This happens because throughout the deforming stages the material work hardens and becomes stronger. However, a hard, strong material is less ductile. It breaks easier.

Finally, as additional force is added, the material passes the balance between strength and strain and becomes very thin at one point along its length (necks out). Any additional load will cause the material to fracture. Fig. 9-15 shows a typical stress-strain curve. Notice the elastic, forming, and fatigue ranges and the fracture point.

The major machines for applying forming forces include:

1. Presses.
2. Hammers.
3. Rolling machines.
4. Draw benches.

Other specialized methods and machines are used for specific forming processes. These unique items will be discussed in the specific chapters on forming that follow this chapter.

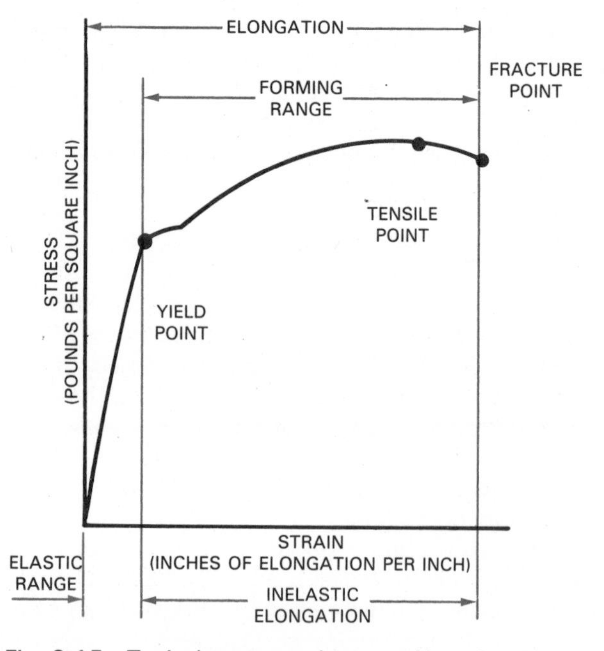

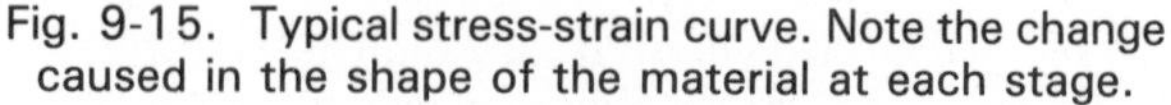

Fig. 9-15. Typical stress-strain curve. Note the change caused in the shape of the material at each stage.

PRESSES

Presses are probably the most widely used forming machines. They are used for a wide variety of operations where compression or squeezing force is needed. These operations include forging, drawing, ironing, sizing, and riveting, to name just a few. Presses are also used to perform a number of sheaving operations including blanking, notching, piercing, and trimming. These operations will be discussed in Chapter 22 which deals with shearing techniques.

A press is designed to provide the power necessary for operating a set of dies while keeping them aligned. To accomplish this, presses generally include four basic systems or components. These elements, as shown in Fig. 9-16, are:

1. A power source.
2. A drive unit.
3. A ram.
4. A frame and bed.

The design of the components depends upon the job for which press is to be used. The designer must consider the force needed, the size of the part to be formed, the desired speed of the operation, and the need for accurate die alignment and close product tolerances.

Press power sources

There are two ways to power a press.

1. Manual power.
2. Generated power.

Manual or human power is used on lightweight presses for forming thin sheet metals. A typical

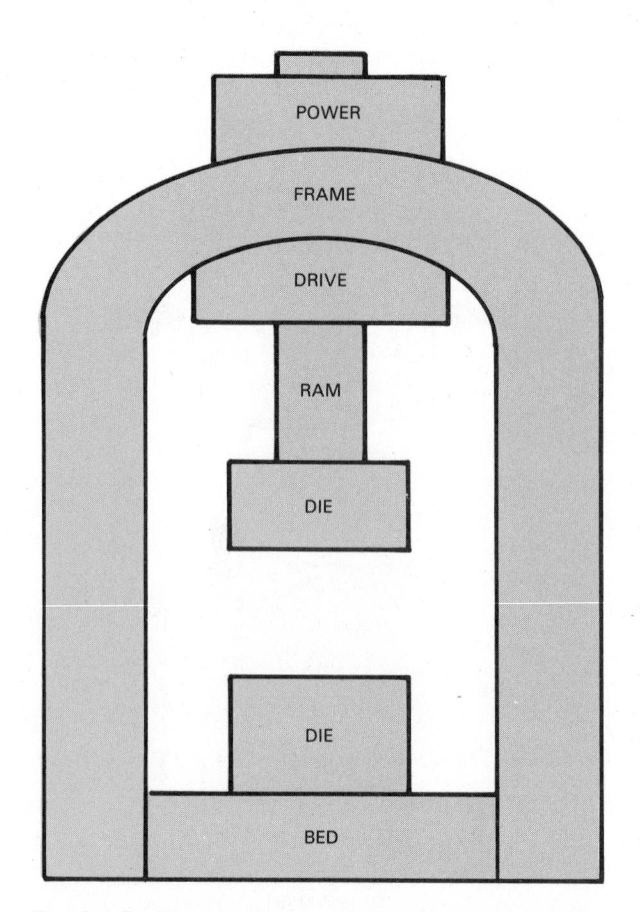

Fig. 9-16. Basic makeup and parts of a press. Every press usually includes these four systems.

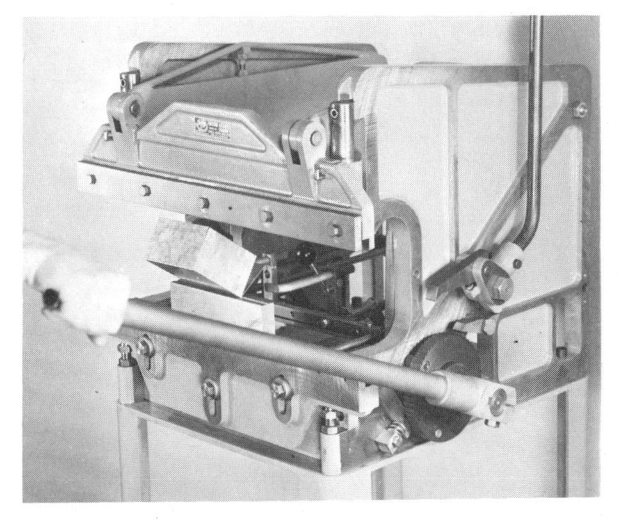

Fig. 9-17. A hand-operated press brake. It is designed to form sheet metal parts. (DiAcro)

Fig. 9-18. A punch press with a flywheel provides uniform delivery of power. (Babcock and Wilcox)

manual press is the press brake shown in Fig. 9-17.

Generated power, such as electrical, mechanical and hydraulic power, is most widely used. These types of power are used on all heavy-duty presses. Often a flywheel is used with electrical/mechanical sources to provide a uniform source of power. Fig. 9-18 shows a press with an electric motor driving a flywheel. These components provide the power to drive the ram.

Press drive units

The press drive unit applies power to the ram. There are a number of different drive units, as shown in Fig. 9-19. These units are:

1. *Crank.* The crank-driven press is the simplest. Its motion resembles a harmonic (uniform U-shaped) curve. The ram accelerates, reaching its maximum speed at the midpoint of the stroke. It then decelerates. Maximum forming action occurs at the midpoint of the stroke. Crank drive is used for basic shearing operations and simple drawing processes. These operations take advantage of the midpoint speed common with crank presses.
2. *Eccentric.* Eccentric drives have short strokes and dwell (period of no movement) at the bottom of the stroke. These features lend themselves for moving blank holding rings for drawing operation and for shearing operations. Cam drives are similar in operation to eccentrics.
3. *Knuckle-joint.* Knuckle-joint drive mechanisms are fast-acting systems. They also provide high force (mechanical advantage) at the bottom of their stroke. At this point, the two arms approach their vertical position. This action provides high force over a short distance. Knuckle-joint presses are often used for coining and sizing where fast action and high load capacity are needed.
4. *Rack and Gear (pinion).* The rack and gear, Fig. 9-20, is not widely used. Its main application is for presses requiring a long ram travel. The stroke is slow and the motion is uniform.
5. *Screw.* A screw drive uses a friction disc which turns a flywheel. The flywheel hub is attached to a screw. As the friction wheel turns the flywheel, the screw moves down. As the flywheel is lowered, it speeds up because it contacts a faster moving (feet per minute) part of the friction wheel. The motion of the screw is ever accelerating, giving it greatest speed at the end of its stroke. The action resembles a drop hammer. A screw-driven press is often called a *percussion* press.
6. *Toggle.* Toggle mechanisms are used mostly to operate blank holding rings for drawing operations.
7. *Hydraulic.* Hydraulic presses can be made in almost any size. For this reason, most large drawing presses have hydraulic drive units. Forming pressures and length of stroke may be easily

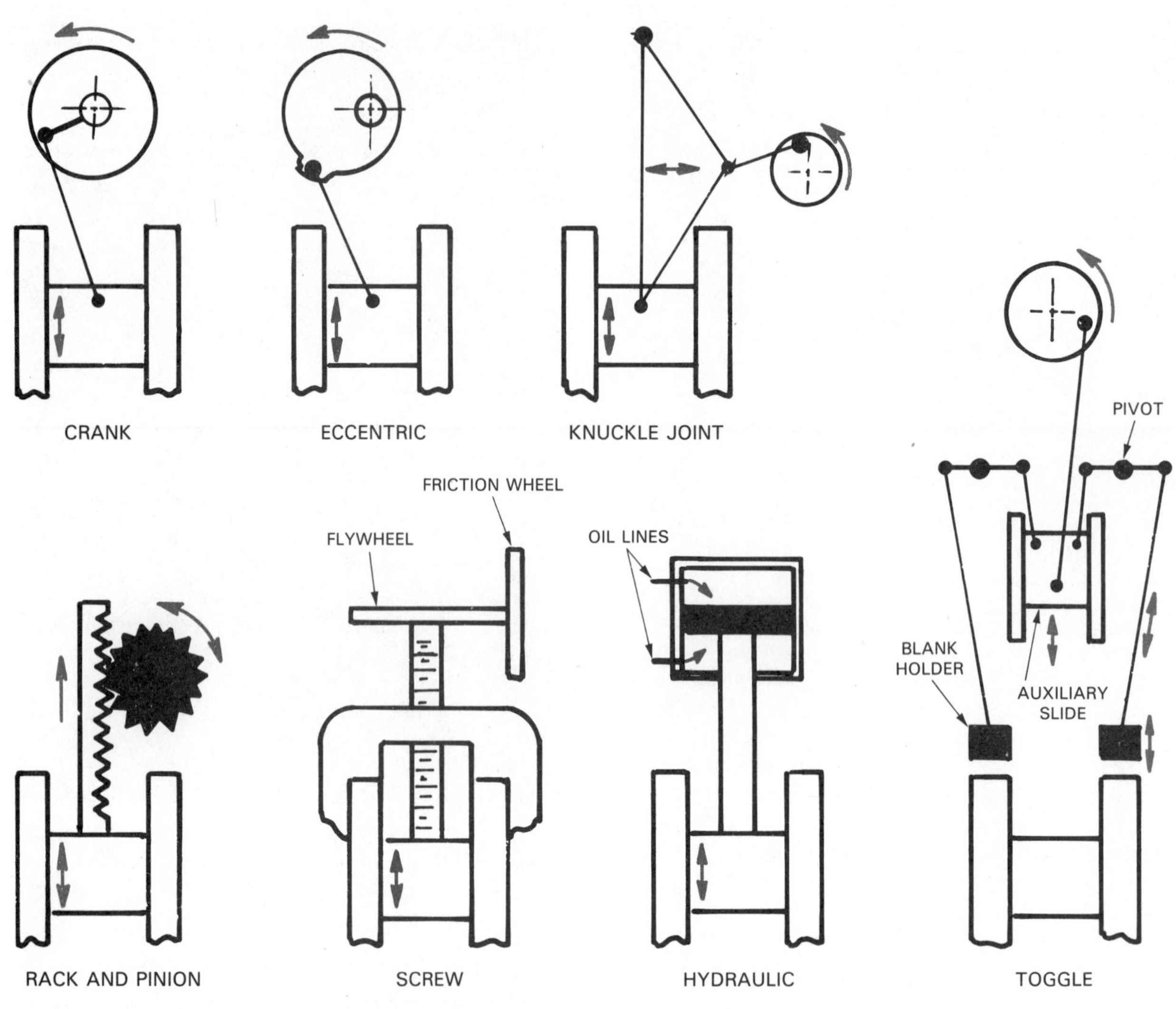

Fig. 9-19. These simplified sketches show the basic types of press drives.

Fig. 9-20. Pressing a ball race into a housing using a bench arbor press. This press is an example of the rack and pinion type. (Dake Corp.)

regulated. Also, hydraulic presses produce their full forming force over the entire length of their stroke. These presses may provide longer strokes than most mechanical presses and also provide constant-speed strokes. Hydraulic presses are usually slow action models but special high-speed (up to 600 strokes per minute) presses have been developed. Fig. 9-21 shows a large hydraulic press in action.

Press rams

There are several important types of ram actions available on forming presses. These include:

1. *Single-acting presses.* Single action presses have a single ram which moves to cause the desired forming action.
2. *Double-acting presses.* Double-acting presses have two actions. Usually the first causes the blank to

Fig. 9-21. A huge hydraulic forging press produces full forming force throughout its entire stroke. (Wyman Gordon)

be held in place. The second action causes the ram to move. This second movement does the forming. Double-action presses are often used for deep drawing of metal.

3. *Four-slide action.* A four-slide press has rams positioned at 90 degrees to each other. These slides operate separately and in a preset order to perform four forming tasks. The four-slide operation, as seen in Fig. 9-22, is used mostly in forming small sheet metal parts.

Bed and frame

Presses are made with two major types of frames. These are C-frame presses and straight-sided presses.

The C-frame construction is the more common. It provides ample access to the dies on three sides. Also, the press can handle wide and long parts.

Typical forms of C-frame presses, Fig. 9-23, are:

1. *Gap (or Gap-Frame) Press.* A rigidly constructed frame with a fixed bed.
2. *Inclinable Press.* A rigidly constructed frame with fixed bed which can be tilted for efficient opera-

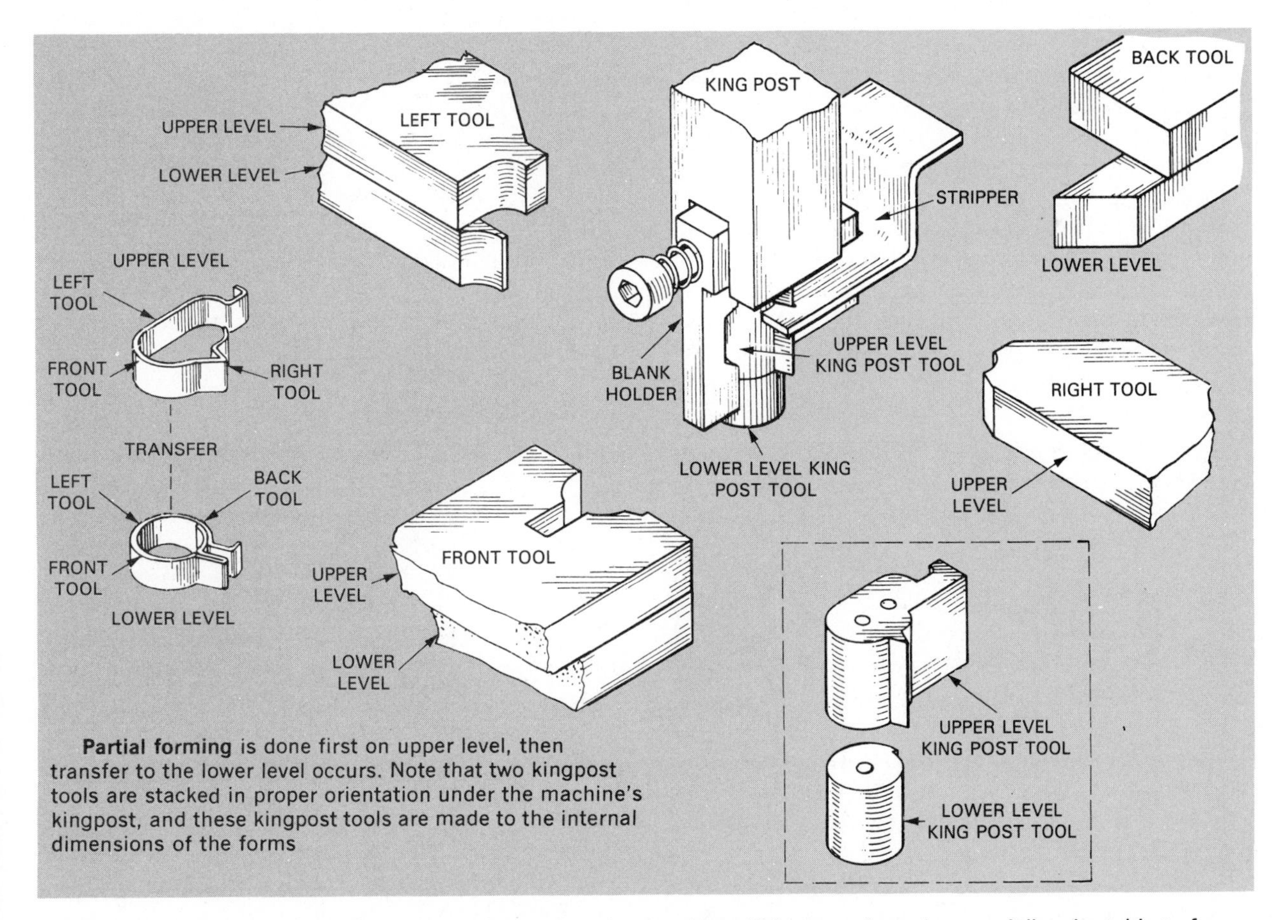

Fig. 9-22. Tooling for each slide of a four-slide forming machine. This type of ram is especially adaptable to forming sheet metal parts. (A.H. Nilson Machine Co.)

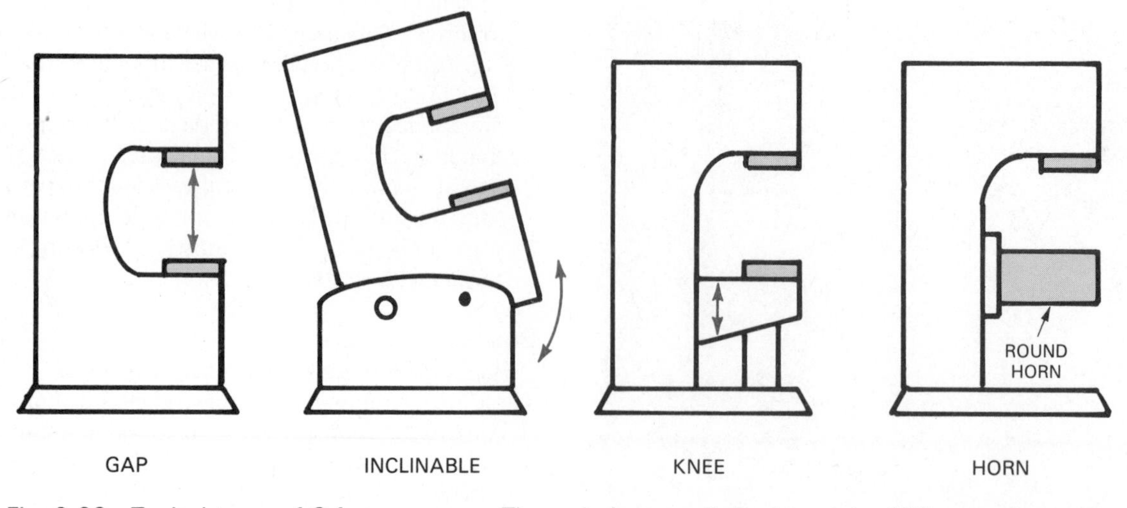

Fig. 9-23. Typical types of C-frame presses. These designs are limited to under 200 tons of pressure.

tion. This press is the most widely used in industry because the incline allows parts and scrap to slide from the press into trays or boxes. Another name for this press is OBI or open-back inclinable press.

3. *Knee (Adjustable Bed) Press.* A press which has a bed that can be adjusted up or down.
4. *Horn Press.* A rigid press which has a cylindrical (round) bed. This press is used for work on round or curved workpieces.

All C-frame presses have a tendency for the top and bottom of the "C" to spring apart under pressure. This causes die misalignment and excessive die wear. Also, close tolerances are more difficult to maintain.

For these reasons presses of larger capacities are usually the *straight-side* type. These presses will exert well over 1000 tons pressure while C-frame presses are usually limited to forces under 200 tons. Most common C-frame presses are in ranges from 1 to 100 ton capacities.

Straight-side presses are of three basic types as seen in Fig. 9-24. These presses are designed to provide large-area beds and long ram strokes. Access to the die areas is more restricted than on C-frame presses.

The solid and arch types provide one-piece frames which insure good die alignment, Fig. 9-25. The tie-rod and pillar types, as seen in Fig. 9-26, have separate beds, columns, and crowns. These units however, are keyed to make sure that the dies align.

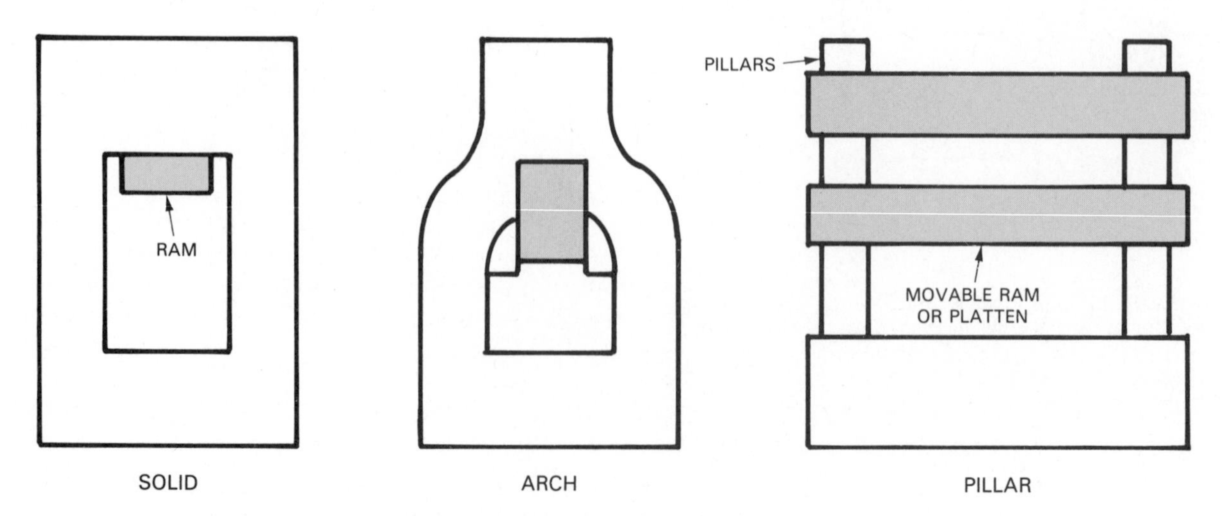

Fig. 9-24. Types of straight-side presses which will exert over 1000 tons pressure.

Fig. 9-25. A—A solid straight-side press. B—An arch drop hammer.

Fig. 9-26. This pillar hydraulic press is a four-column type. (Mesta Machine Co.)

HAMMERS

The construction of a hammer is much like that of a press. However, the hammer is not designed to produce a steady squeezing pressure. A hammer, instead, is an impact device. It produces a sharp blow which creates energy that is absorbed by the workpiece. This energy causes the workpiece to deform.

There are two major types of hammers in use: *driven* hammers and *drop* hammers. A typical driven hammer resembles a gap-bed press. Its action includes the following:

1. The ram is raised by a piston and is driven down on the workpiece.
2. The speed of the upward or downward motion of the piston is controlled by an operator. The operator uses a foot treadle to control the amount of steam allowed to enter above or below the piston.
3. Striking force can be controlled from a light tap to a full-powered hit. Controlling driven hammers requires great operator skill since the falling weights may be as heavy as 25 tons.

Drop hammers look much like an *arch press*. They

Fig. 9-27. A drop forging hammer in operation. It has a one-piece frame. (Wyman-Gordon)

use counter-rotating rolls to raise the ram which is then released and allowed to drop. The force generated by drop hammers is created by a weight (the ram and die) moving freely (by gravity) over a distance.

Drop hammers, like the one shown in Fig. 9-27, may operate at speeds up to 60 strokes per minute. The amount or force applied with each stroke can only be increased or decreased by changing the distance the ram travels. Typical drop hammers range in size from 500 to 6500 lb. This rating is measured by their falling weight.

ROLLS

Rolling machines are the third major type of forming equipment. Rolls are widely used in primary processing to reduce the thickness and increase the width and length of billets. However, this discussion is limited to the use of rolls to form materials without reducing their cross-sectional area. Primarily, there are three types of rolling machines used in accomplishing this task. They are:

1. Three-roll machines.
2. Contour roll machines.
3. Semicylindrical roll machines.

Three-roll machines

Sheet metal, plate stock, and rolled shapes are rolled into desired curves using three-roll machines.

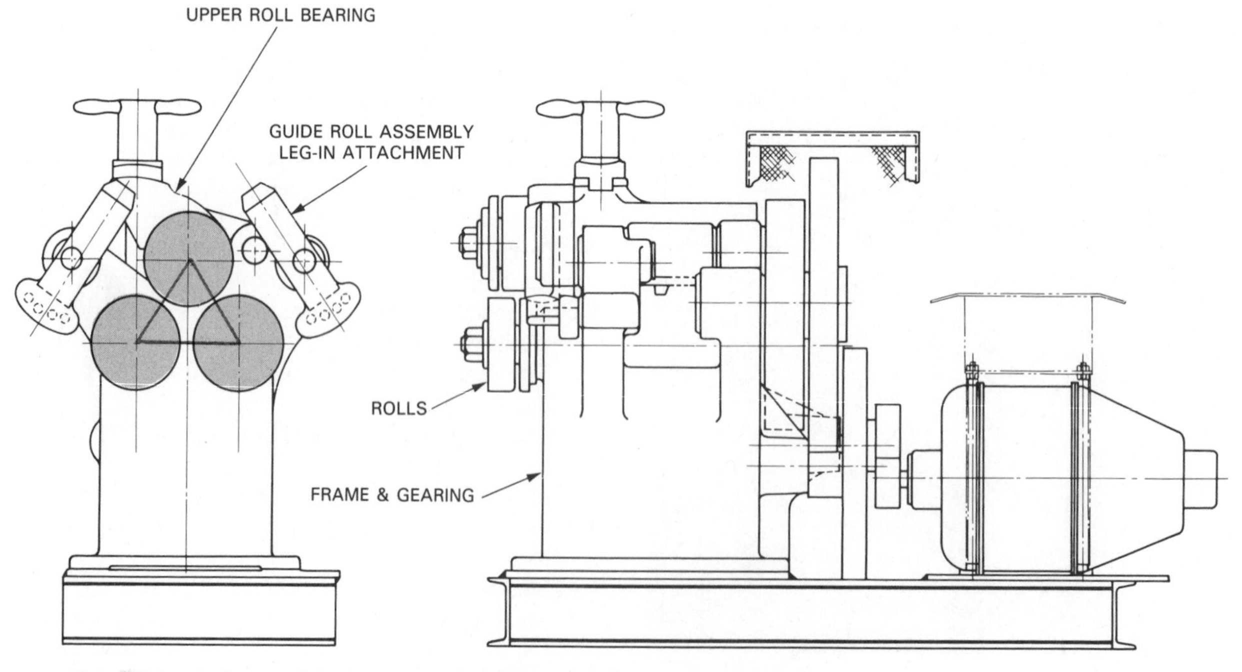

Fig. 9-28. A three-roll forming machine. Note that rollers are placed so that a line drawn from hub to hub would form a triangle. (Buffalo Forge Co.)

Fig. 9-29. A worker is machining a set of mated rolls. (Inland Steel Co.)

The three rolls are arranged in a triangle with all axis being in the same plane. The front rolls grip and feed the stock. The lower, back roll is slowly brought up, causing the material to bend. Fig. 9-28 shows a three-roll machine.

Contour roll machines

Contour roll machines are generally used to perform a number of bending operations in a single pass of the material. They may corrugate sheet stock, form tubing and pipe from sheet material, or bend flat stock into architectural shapes. Fig. 9-29 shows a typical set of shaped rolls.

Semicylindrical rolls

Partly round rolls are used extensively for roll forging. The heated material is placed between the flat portions of the roll and is shaped as the round portions come into contact with the hot metal.

DRAW BENCH

A draw bench, Fig. 9-30, is the fourth major type of forming machine. A typical draw bench can be used for cold reduction (drawing) of bar stock into wire. Also tubing can be reduced in diameter by drawing. A draw bench contains a die holder, a draw head, and a means to pull the draw head.

Fig. 9-30. This bar drawing machine will size hot rolled bars to accurate cold drawn dimensions up to 2 in. in diameter. (Ajax Manufacturing Co.)

A

B

C

These are the most common forming presses. A—OBI or open back inclinable. B—Gap bed. C—Straight side. (Rousselle Corp.)

SUMMARY

The forming process is one of three ways to produce a part from a piece of material. Forming uses plastic deformation to produce a part that weighs exactly the same as the original material. Force, a shaping device, and consideration for material temperature are all elements of the forming act.

Forming produces a solid part from a solid material without weight change. The other two methods of part production do not meet this description. Casting dissolves or melts and pours the material into or around a mold. Separation, which will be discussed later in this text, cuts away material to produce the desired shape. The part will naturally weigh less than the original material because of the weight of the removed (scrap) material.

Forming provides a number of ways to quickly and economically produce large quantities of plastic, ceramic, and metal parts. The forming processes produce the greatest tonnage of industrial materials and must, therefore, be considered as a major group of manufacturing processes.

STUDY QUESTIONS – Chapter 9

1. Define forming.
2. What are the three essential elements of forming?
3. List and describe the two major shaping devices used in forming processes.
4. List and describe the three major types of dies used in forming.
5. How do forging and stamping dies differ?
6. Describe the types of one-piece dies or molds.
7. Describe the two types of forming rolls.
8. Describe recrystallization.
9. Compare the advantages of hot forming to the advantages of cold forming.
10. What are the four forces that can be used to form and shape material?
11. What is "plastic deformation?"
12. List and describe the four major ways to apply forming forces.
13. List the major parts or elements of a press.
14. List and briefly describe the seven types of press drive units.
15. List and briefly describe the four types of C-frame presses.
16. What are the three basic types of straight sided presses?
17. List and describe the three major types of forming roll machines.

Chain hoist positions large die set on bed of forming press. From descriptions of presses given in this chapter, can you tell what type of press it is?

Chapter 10

HOT FORMING METALS

Many metal shapes, such as sheets, bars, and plates, are of little value until they are further processed. A common method for processing them is hot forming. A typical hot forming machine is shown in Fig. 10-1.

HOT FORMING BASICS

Hot forming involves heating the metal to its "plastic" state. Then pressure is applied through a shaping device causing the metal to take on a new size and shape.

Hot forming of metals is always done *above the material's recrystallization temperature*. This point is near room temperature for lead and zinc, between 650 and 900°F (343-482°C) for aluminum and magnesium alloys, 1100 to 1700°F (593 to 927°C) for copper, brass, and bronze, and 1700 to 2500°F (927 to 1371°C) for ferrous (iron-based) materials.

When a metal is heated above its recrystallization temperature, the strength of the material decreases. The grains of the metal are more easily distorted. These grains, however, do not fracture but the structure becomes fragmented.

New and smaller crystals form from these fragments and grow as long as the temperature is maintained. If the temperature is quickly dropped below the recrystallization point, a fine crystal structure forms.

ADVANTAGES

Hot forming techniques have the following advantages:

1. Hardness and ductility of the metal is unchanged.
2. Distorted and strained grains are soon replaced by new strain and distortion-free grains.
3. Tougher metal is produced because:
    a. The grain structure is reformed into a larger number of smaller crystals.
    b. Pores are closed.
    c. Impurities are broken up.
    d. Flow lines in the metal form in the direction of greatest stress.
4. Difficult and extreme shapes may be produced without fracturing the metal.
5. Less force is required than for cold forming; therefore, lighter weight equipment may be used.
6. Impurities are spread through the material giving it a more uniform makeup.

Fig. 10-1. A large forging press is used for hot-working metals. (Wyman-Gordon Co.)

DISADVANTAGES

However, no process is without its disadvantages. For hot forming, these include:

1. The high working temperature needed increases oxidation or scaling on the surface of the metal.
2. Expensive heat resistance shaping devices are required.
3. Close tolerances cannot be held.
4. Maintenance costs are high.

Even with these disadvantages, hot forming is generally more economical than cold forming of materials.

CLASSIFICATION OF HOT FORMING TECHNIQUES

A number of secondary manufacturing processes that are classified as hot forming are important to the industry. These include the six major process groups:

1. Forging.
2. Extrusion.
3. Roll forming.
4. Piercing.
5. Drawing.
6. Spinning.

FORGING

Forging is a forming method which shapes a metal part through controlled plastic deformation. Most forging is done while the metal is hot and sometimes uses heated dies. The forming forces are produced by hammers or presses.

Forging, like most forming processes, produces a part near its final size and shape. Forging, as shown in Fig. 10-2, may:

1. Produce a drawn out part by reducing (compressing) its cross section and, therefore, making it longer.
2. Upset the material by reducing (squeezing) its length to increase its cross section.
3. Create a part by squeezing it to create a cavity.

A principal reason for forging a part is to control the grain of the metal. In rolling operations the direction of greatest tensile strength is the direction the metal was formed. Therefore, parts machined out of solid stock will have straight grain much like a piece of wood.

Controlling grain direction during casting is almost impossible. The grain aligns with the direction of heat flow during the solidification process. However, through forging and proper part design, maximum strength can be developed through controlled grain direction. Fig. 10-3 shows grain direction for a cast,

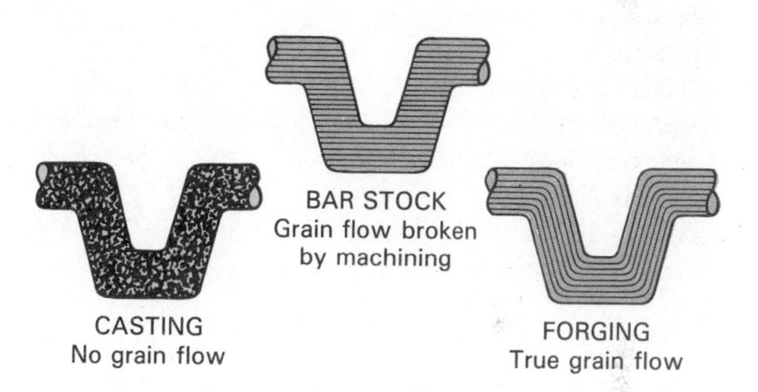

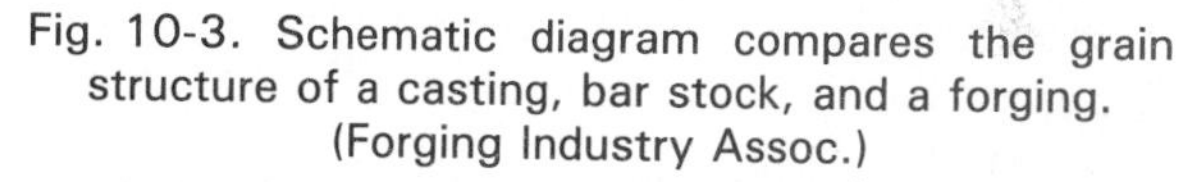
Fig. 10-3. Schematic diagram compares the grain structure of a casting, bar stock, and a forging. (Forging Industry Assoc.)

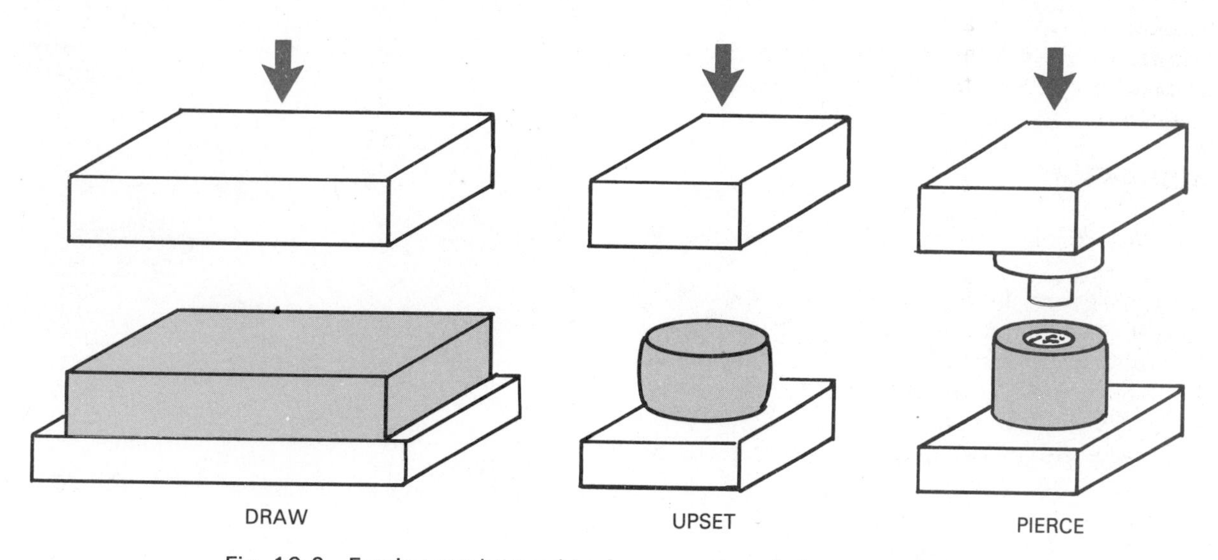

Fig. 10-2. Forging can be used to draw, upset, and pierce a workpiece.

1. INITIAL BREAKING OF CORNERS BEGINS THE CYLINDER-MAKING STAGE.

2. AFTER REPEATED WORKING AND ROTATING, THE STOCK HAS BECOME ROUND.

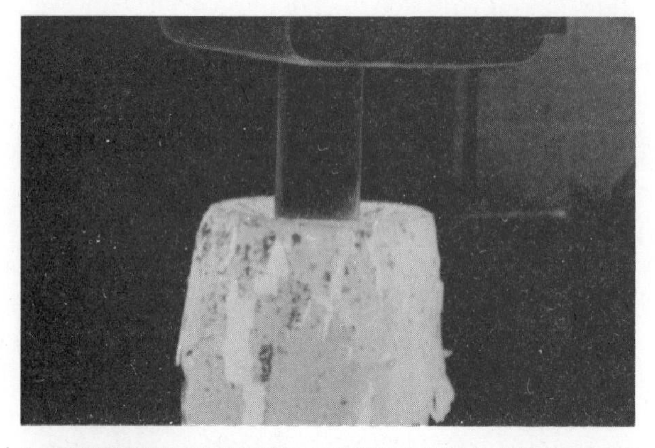

3. PUNCHING REQUIRES A LONG PUNCH CAPABLE OF PENETRATING 75% OF STOCK LENGTH. SCALE CRACKS SHOW HOW THE DISPLACED METALS HAS INCREASED STOCK DIAMETER.

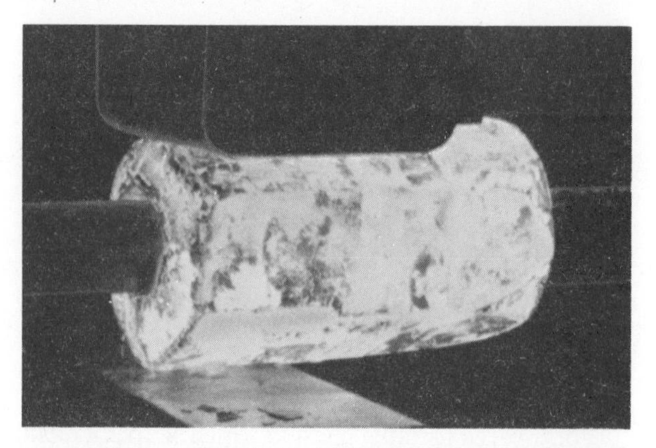

4. AFTER THE MANDREL HAS BEEN INSERTED SOME INITIAL WORKING IS NEEDED TO REGAIN LENGTH LOST DURING PUNCHING.

5. ONCE IN THE SADDLE, REPEATED WORKING OF THE METAL AND ROTATING AGAINST THE MANDREL PRODUCES MAJOR GROWTH OF RING ID AND OD AS THE CYLINDER WALL THINS. LATERAL GROWTH IS MINIMAL.

6. NEARING COMPLETION, THE FORGING PROCESS IS INTERRUPTED TO CHECK FOR FINAL RING ID.

Fig. 10-4. Steps in open die forging. (Forging Industry Assoc.)

a machined, and a forged part. Note the grain flow lines that give the forged part superior strength.

Most common metals and many alloys may be forged into parts varying in size from a few ounces to several tons. Materials may be shaped by one or more of the following common forging techniques.

1. Hammer (smith) forging.
2. Drop forging.
3. Press forging.
4. Upset forging.
5. Roll forging.

HAMMER FORGING

Hammer or open-die forging is the shaping of metal by an impact between flat dies mounted in a forging hammer. This forging technique is basically the same as the method used by early blacksmiths. The main difference is that a mechanical hammer has replaced the smith's hand-held hammer.

In hammer forging, heated metal is hammered to shape by a series of blows. The operator moves the workpiece between blows to shape it, Fig. 10-4.

The forging force may be developed by either a free-falling (gravity) die or a die attached to a steam hammer. Most hammer forging (which is also called open-die forging) uses a double-acting steam hammer. This hammer, shown in Fig. 10-5, uses steam pressure both to raise the upper die half and to power its downward movement.

Production rate and accuracy from hammer forging is low. Therefore, the process is not economical for high-volume production or making complicated parts. What use it enjoys is based on low-volume or specialized forgings. It is used mostly in the early stages of product development and engineering.

DROP FORGING

Drop forging is forming hot metal between aligned impression or cavity dies using a drop hammer. Closed impression dies have the form of the part

Fig. 10-5. A part is being shaped in this 35,000 lb. steam forging hammer. (Wyman-Gordon Co.)

machined or sunken into them.

The dies are mounted on a drop hammer similar to the one shown in Fig. 10-6. The ram is raised and the hot metal stock, or workpiece, is placed in the die cavity. See Fig. 10-7. The ram, along with its upper die half, is allowed to drop on the workpiece one or more times. This hammering action causes the metal blank to flow into all parts of the die cavity. Excess metal (flash) is trimmed off later.

Many drop forging dies have several cavities. This allows the operator to move the part from one cavity to the next. At each cavity the part shape becomes more defined. The product receives its final form in the last cavity. An example of this type of die is shown in Fig. 10-8.

Horizontal impact forging machines using the drop forge principle have been introduced to help automate the forging process. This process automatically passes

Fig. 10-6. A 6000 lb. air-operated power/drop hammer. The blow force and frequency is controlled by the operator through a foot pedal. (Chambersburg Engineering Co.)

Fig. 10-7. Worker is placing a hot workpiece in an open drop forging die. (Forging Industry Assoc.)

Fig. 10-8. A multicavity forge die for producing two engine connecting rods. The sequence of operations is: 1—Material is drawn to length in the far right position. 2—Initial impression is produced in the far left position. 3—Part is given initial shape in the buster position. 4—Part is given additional shape and grain direction in the blocker position. 5—Part is given its final shape in the finishing impression. (Modern Drop Forge Co.)

metal stock from a hopper through an induction furnace. The parts are then forged in a horizontal forging machine like the one shown in Fig. 10-9. The machine has two opposing air-driven impellers (rams). A closed impression die half is attached to each impeller.

With the heated metal positioned between the dies, they move toward each other. Upon contact the dies cause the part to be shaped. The force of the moving dies hitting each other is absorbed by the deformation of the part. The machine gets little or no vibration or shock.

Drop forging and impact forging produce highly accurate parts. The forgings have a grain structure which yields high tensile strength. For this reason, many tools and other small parts are manufactured by these processes.

Fig. 10-9. Horizontal forging allows automation of forging process. Top. Horizontal forging machine. Bottom. Die set for horizontal forging. Product is shown at front center. (KDK Upset Forging Co., and Hill Acme Co.)

PRESS FORGING

Only a portion of the impact produced by either hammer or drop forging is actually delivered to the center of the workpiece. Most of the force is applied to the surface of the work and to the base of the machine itself. Thus, large forgings cannot be produced with the forging hammer. Instead press forging is used.

Press forging applies slow, steady pressure instead of a quick, jolting impact. The energy from the press penetrates the work and causes uniform deforming throughout the part. This is in contrast to the surface penetration of the hammer as seen in Fig. 10-10.

Press forging usually takes only one closure of the dies to complete a forging operation. A hammer may deliver several blows to accomplish the same task.

Press forging uses both mechanical and hydraulic presses. Mechanical presses are normally used in forging parts requiring only moderate pressure ranging from 300 to 16,000 tons. The mechanical press provides its maximum pressure at the end of its stroke. This action causes the metal to complete its flow to all parts of the closed impression die.

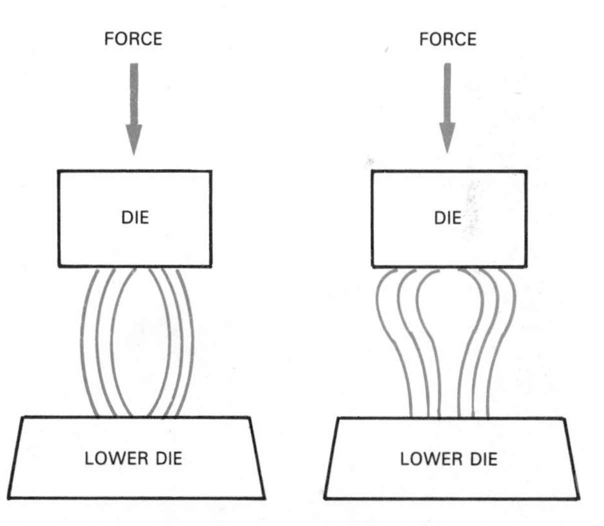

Fig. 10-10. Comparison between the forces in hammer and press forging. Left. In press forging, part is deformed uniformly throughout. Right. Hammer impact causes uneven deformation.

Hydraulic presses are used when greater pressures are needed to forge larger parts and tougher materials. Hydraulic presses are slower acting, often traveling less than 3 in. per second. They can, however, produce more than 100,000,000 lb. (50,000 tons) of pressure. Fig. 10-11 shows some typical press-forged parts. Hydraulic presses provide steady high pressure

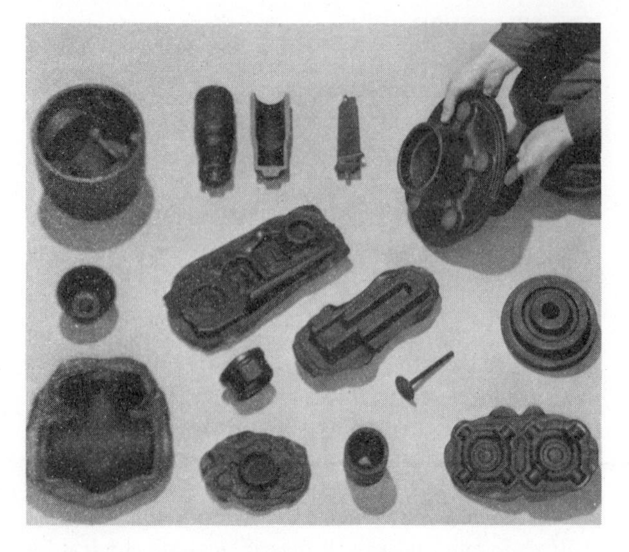

Fig. 10-11. Typical forged parts. Note the flash around some of the parts. (National Machinery Co.)

which is transmitted throughout the workpiece. Fig. 10-12 shows a large hydraulic forging press.

UPSET FORGING

Upset forging, or *machine forging,* accounts for a significant part of total forging production. The technique involves increasing the diameter of the end of a bar by reducing its length. A modification of horizontal impact forging, upset forging may be done

Fig. 10-12. A 50,000 ton hydraulic closed-die forging press. Units like this provide steady, high pressure. (Mesta Machine Co.)

either cold or hot. When the material is worked cold the process is often called *cold heading.*

The technique, as shown in Fig. 10-13, often involves:

1. Placing hot or cold stock in the die similar to the one in Fig. 10-14.
2. The stock is gripped between the die halves.
3. A ram and punch (or punches) enlarges and forms the end of the stock. This may take a single step or several steps, each using a different punch. Typical products from machine forging are gear blanks, valve stems, and flanges on axles. Nails, bolts, and screws are formed by cold heading.

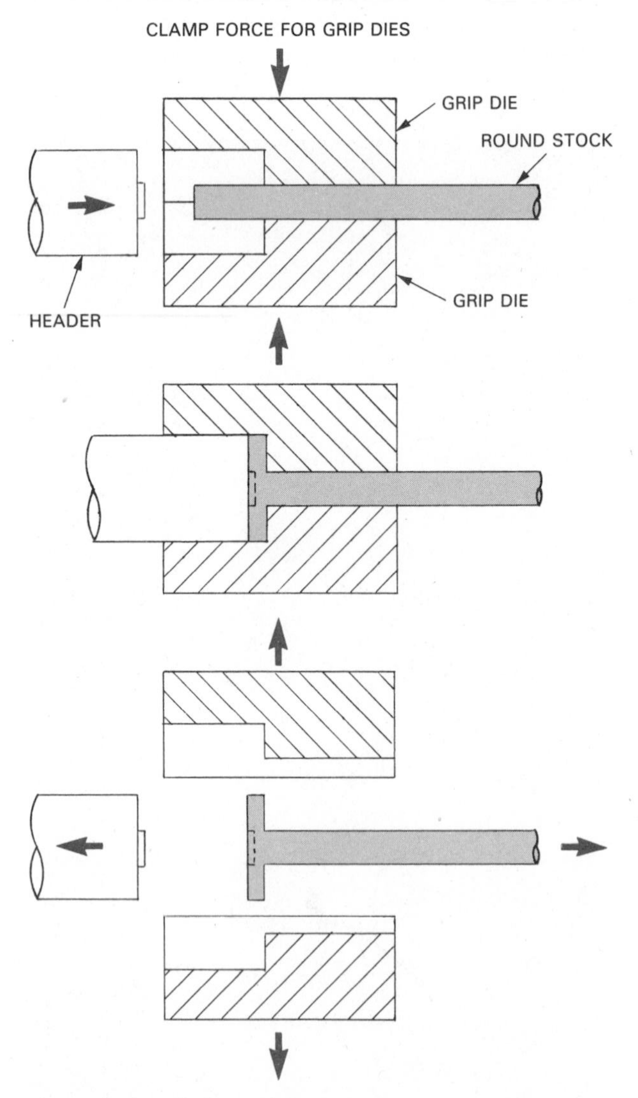

Fig. 10-13. Upset forging sequence. Top. Hot or cold workpiece is clamped between die halves. Middle. Header (ram or punch) is forced against workpiece to enlarge and shape it. Bottom. Header retracts, die halves release workpiece. (Wyman-Gordon Co.)

Fig. 10-14. An upset forging or upsetting die. Note parts held in each station of the die. (Hill Acme Co.)

ROLL FORGING

Roll forging is a very specialized technique which reduces the diameter or tapers short lengths of metal rods or bars. The rolls are not completely round, as shown in Fig. 10-15. A flat segment allows the part to be fed into the rolls. The remaining roll surface is grooved or otherwise shaped to cause one or more forming actions.

Typically, in roll forging, the heated stock is placed between the rolls. They turn, squeezing the material into the desired shape. The stock can then be moved to another set of grooves in the roll and the action is repeated. Several sets of grooves may be machined into a set of rolls. Each set of grooves completes one step in the complete forging action.

Roll forging, being fast, is widely used to make blanks (preformed material) for other forging processes. Blanks are often prepared for such products as connecting rods, levers, and axles. They are later drop or press forged to final shape. Another use of roll forging is the production of parts requiring long, tapered, symmetrical sections.

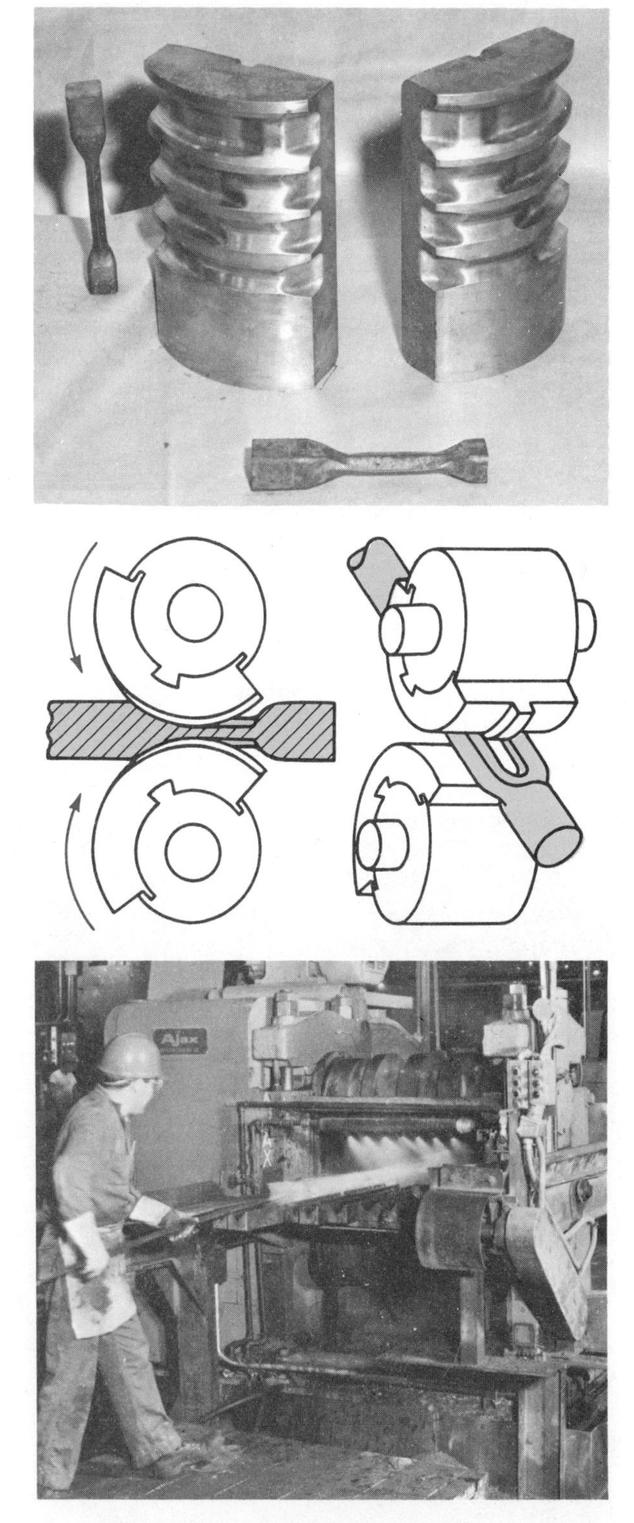

Fig. 10-15. A typical roll forging operation. Top. A set of flat back dies for forging engine connecting rod blanks. Note the pair of blanks. Middle. Schematic of the operation. Part is shown in color. Dies extend around half the roller perimeter. Bottom. Truck I-beam for axle is preformed on a roll forge. (General Motors Corp.)

EXTRUSION

A second major method of forming metals is *extrusion*. This technique forces metal through a hollow, shaped die which is mounted onto a pressure chamber. Extrusion is usually done with hot metal, but cold extrusion is used in some cases.

During extrusion, a metal is compressed in a pressure chamber. When the force exceeds the material's elastic limit, it will flow through an opening called a *die orifice*. The production of an extruded member is shown in Fig. 10-16. Temperatures range from about 700°F (371°C) for magnesium, to 800°F (427°C) for aluminum, to 2100°F (1149°C) for steel.

Pressures required also vary. Magnesium requires about 5,000 psi while steel requires over 100,000 psi. The high temperatures and pressures needed to extrude steel limits use of this process.

Hot extrusion is generally done in a horizontal hydraulic press commonly ranging in size from 250 to over 5000 tons. Specially designed presses have exceeded 25,000 tons.

Hot extrusion is a high-temperature, high-pressure operation. Therefore, extrusion dies must be built of materials which can handle high temperatures and pressures.

The dies must be lubricated to keep the extruded metal from coming in contact with the steel dies. Graphite or oil may be used for nonferrous metals while liquid glass is used for extruding steel.

TYPES OF EXTRUSION

Extrusion is used for a variety of applications from producing brick, hollow tile, macaroni, and a number of plastic and metal shapes. Basically, extrusion can be divided into three major types:

1. Direct extrusion.
2. Indirect extrusion.
3. Impact extrusion.
4. Hydrostatic extrusion.

Direct extrusion

Direct or forward extrusion is shown in the drawings for Fig. 10-17. This technique requires the following steps:

1. A heated round billet is placed in the extrusion die chamber.

Fig. 10-16. Extrusion presses, such as this one, are usually horizontal. The range of sizes is usually from 250 to more than 5000 ton. (Copper Development Assoc.)

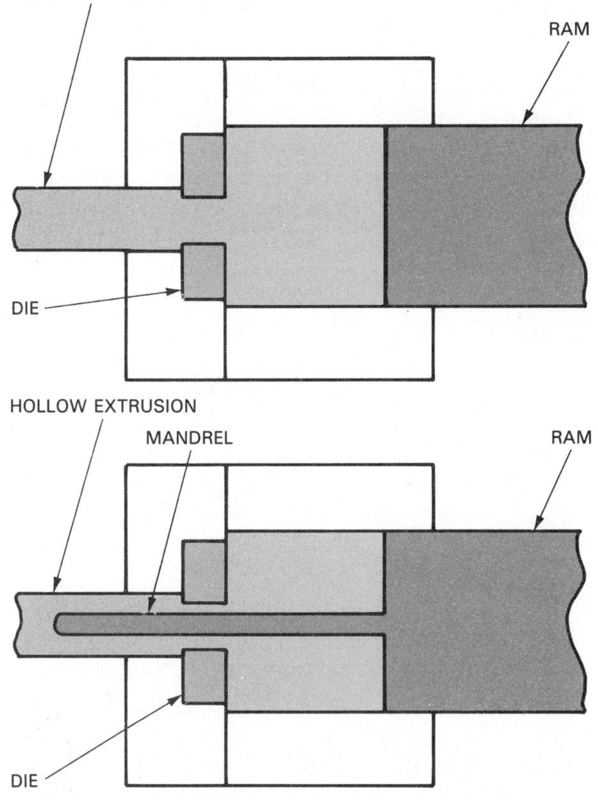

Fig. 10-17. Simplified drawing of solid and hollow shape direct extrusion. Top. Ram forces material through die. Bottom. Mandrel moves through die and extruded material flows around it, taking hollow shape.

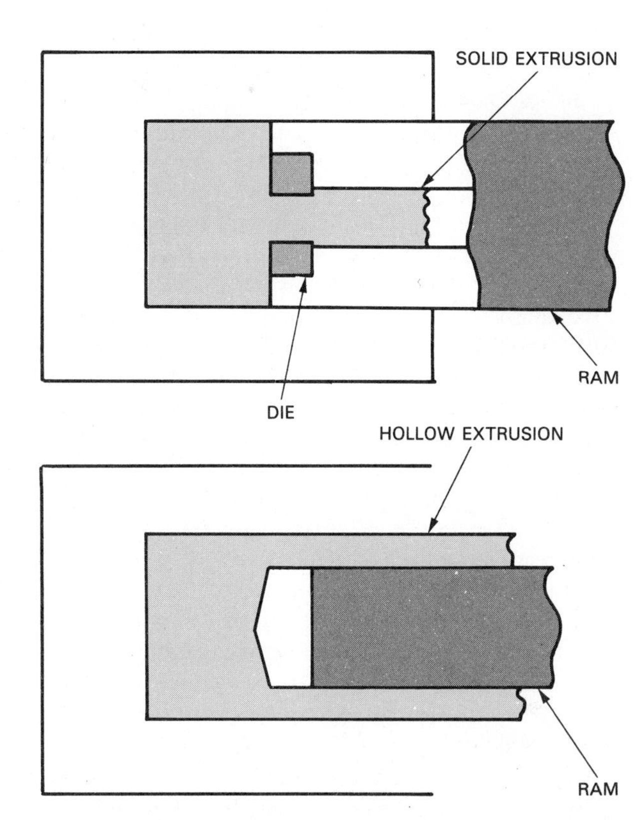

Fig. 10-18. Solid and hollow shape indirect extrusion. Top. Extruded material is forced into the ram stem. Bottom. Extruded material is made to flow around arm.

2. The dummy block (chamber seal) and ram are brought into position.
3. The ram applies force to the hot billet.
4. The metal is extruded through the die opening (orifice) until only a small amount remains.
5. The extruded material is cut off at the die.
6. The remaining material is removed from the extrusion chamber.

A hollow extrusion may be produced by using a mandrel or torpedo. The mandrel may have almost any shape. It may be attached to the ram or remain stationary. The material flowing under pressure around the mandrel and through the die may have various exterior and interior shapes.

Indirect extrusion

Indirect or backward extrusion uses the same principles of heat and pressure as direct extrusion. However, instead of forcing the material through a die orifice, the material is forced back through or along the ram stem, Fig. 10-18.

Indirect extrusion requires less force. However, because it is hollow, the ram is weaker. Also, the extrusion cannot be supported as it can in direct extrusion.

Impact extrusion

Impact extrusion, as seen in Fig. 10-19, is primarily a cold-working process but is also used for hot extrusion. In impact extrusion the punch delivers the impact to the blank. This impact causes the metal to

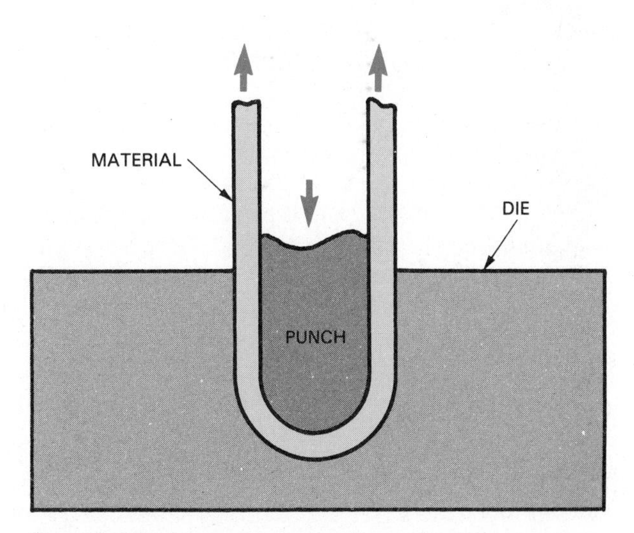

Fig. 10-19. Impact extrusion is used to shape cold or hot materials into lightweight collapsible containers.

flow or "squirt" around the ram or punch. A blast of air removes the part from the ram on its return stroke. The difference between the diameters of the die and the punch establishes the wall thickness of the extrusion.

Impact extrusion is used extensively to produce collapsible tubes from soft aluminum, tin, lead, and zinc and to make zinc flashlight dry cell cans.

Hydrostatic extrusion

In hydrostatic extrusion, pressurized liquid is used for exerting force on the material to be extruded. The pressurized liquid surrounds the billet and acts upon all its surfaces except that touching the die opening. Though usually a hot process, it is sometimes used when the billets are at room temperature or only slightly warmed.

The liquid is pressurized by one of two methods:
1. A ram or plunger is forced into the container. This method is also known as *constant-rate* extrusion.
2. A pump which forces the fluid against the billet. This method is also known as *constant-pressure* extrusion.

Applications and limitations of extrusion

Hot extrusions can produce long, uniform sections of material in almost limitless cross sections, as seen in Fig. 10-20. Producing such shapes is often not possible by any other forming technique. Moreover, extrusion dies are relatively inexpensive to make. They may be set up for use in a matter of minutes. Dimension tolerances are extremely accurate with extrusion techniques.

However, extrusions are limited in length. The common lengths are 20-24 ft. A few extrusions have been produced in the 40-foot range.

ROLL FORMING

Most hot rolling is used to produce primary shapes like sheet, plates, rods and bars. However, roll forming may also be used to convert primary stock into other products. Most of this type of hot roll forming is confined to tube and pipe making. The process involves forming flat stock into round tubular stock and then welding the seam.

The three major pipe-forming processes are:
1. Welding bell technique.
2. Roll forming welded technique.
3. Roll forming resistance welding.

WELDING BELL FORMED PIPE

The first method of forming pipe is not truly a roll forming technique. This system, called the *welding bell process,* is shown in Fig. 10-21. Long, hot strips of thin steel with beveled edges, called *skelp,* are used. This material, which may be up to 40 ft. long, is drawn by a draw bench through a round pipe. Enough pressure is applied at the seam by the bell to cause the edges to fuse (weld). This self fusion is called *forge welding.*

ROLL FORMING PIPE

A second method of pipe forming, as seen in Fig. 10-22, uses a series of rolls to form the pipe. Steel skelp from a roll is passed through a set of forming rolls which bend-form the pipe. As in the welding bell method, the two edges of the skelp are forced into contact by the rolls and a fusion (forge) weld results.

Fig. 10-20. Some typical extruded shapes. In some cases, no other process could produce these items. (Aluminum Assoc.)

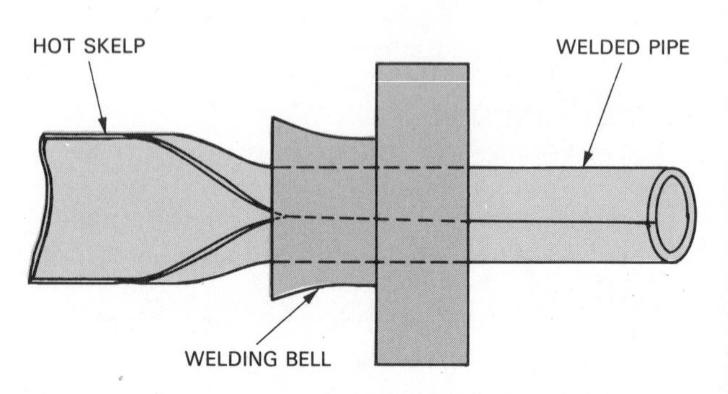

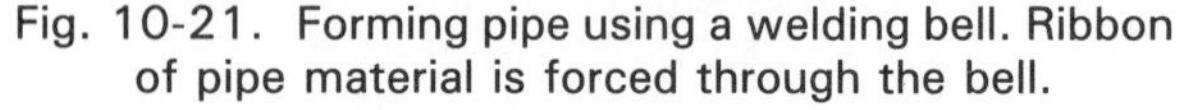

Fig. 10-21. Forming pipe using a welding bell. Ribbon of pipe material is forced through the bell.

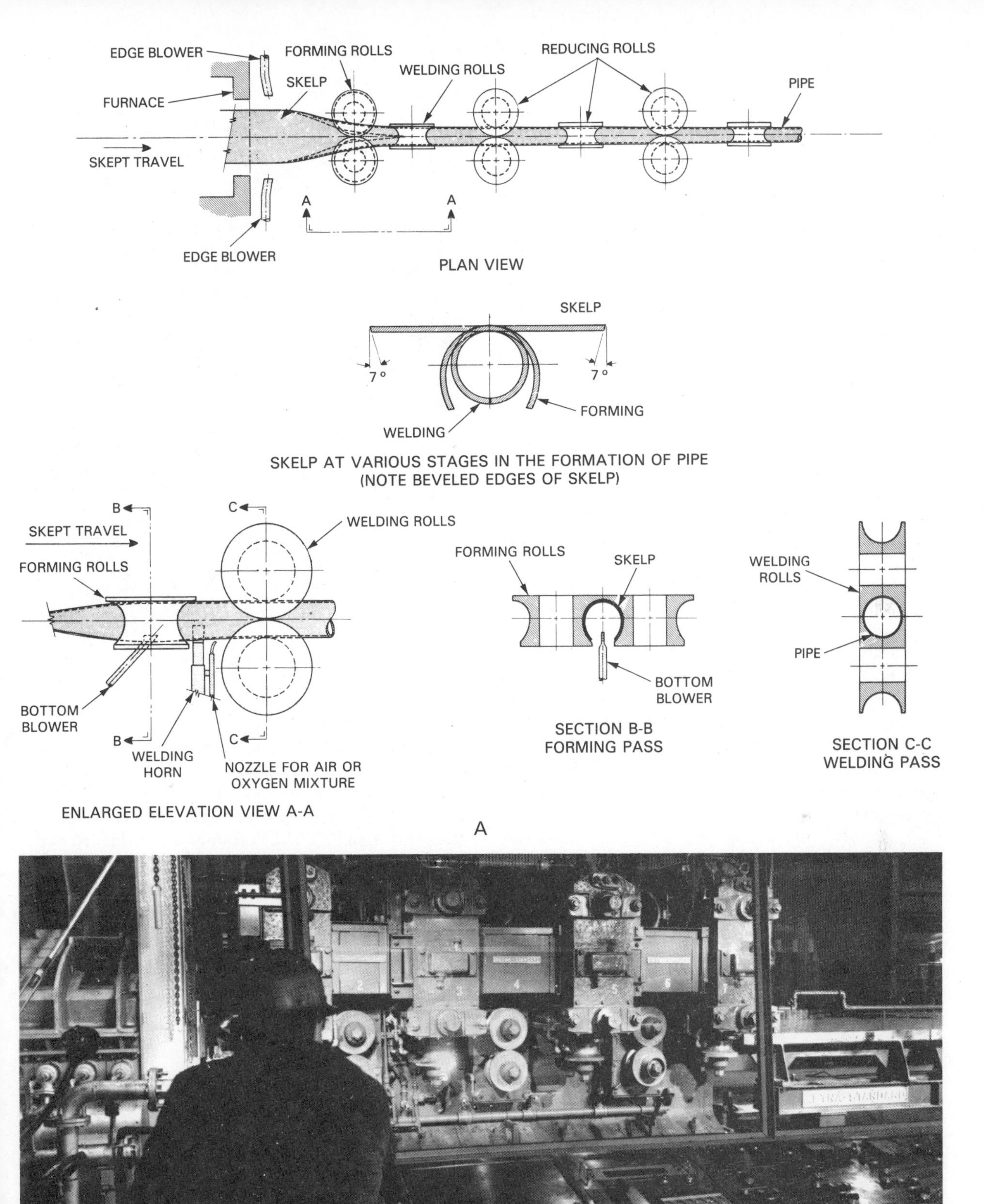

Fig. 10-22. Continuous roll forming and welding of pipe. A—Several views of roller arrangements. B—Forming and welding machine in a butt-weld mill. (United States Steel)

The finished pipe, which may be up to 4 in. in diameter, is sawed into lengths.

RESISTANT WELDED ROLL FORMED PIPE

Larger diameter (up to 16 in.) thinwall steel tubing and pipe is often produced by cold forming a tubular shape whose seam is then resistance welded. The welding takes place by passing electricity through roll electrodes while another set of rolls maintains pressure on the weld area.

PIERCING

The process most often used to produce seamless tubing is called *piercing*. The process, shown in Fig. 10-23, involves several steps. First a heated rod is revolved between two slightly angled rolls. These rolls feed the rod, which has been center drilled, toward a stationary piercing mandrel. The spinning action of the rolls squeeze the rod into an ever-changing ellipse. This constant squeezing action, plus a forward movement created by the rolls, allows the mandrel to produce a hole in the rod. The rod leaves this step as a thick hollow tube.

The pierced rod then passes between grooved rolls and over a plug. This action elongates the tube and gives it the desired wall thickness. Rollers and sizing rolls straighten the tubing and establish the final diameter and wall thickness. Fig. 10-24 shows a tube piercing machine at work.

Fig. 10-24. A tube piercing machine is shown in use. (Timken Co.)

HOT DRAWING

Hot drawing has limited use. It is primarily called upon to produce high pressure tubing and closed end cylinders, from thick plate stock.

Usually, a heated blank is placed between an aligned punch and die. The punch is moved downward forcing the material into the die. This action, shown in Fig. 10-25 produces a cup-shaped part.

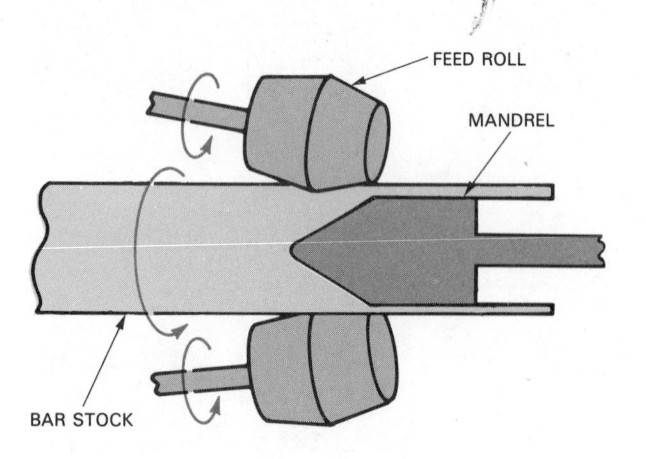

Fig. 10-23. The tube piercing operation. Heated rod, revolving between two rollers, is hollowed by the mandrel.

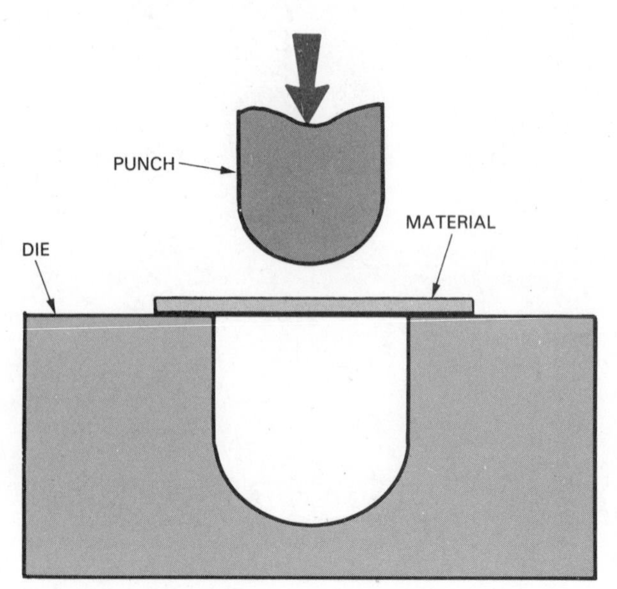

Fig. 10-25. Hot drawing die and punch are similar to those used in impact extrusion.

The process may be repeated drawing the cup wall deeper and thinner. Products made this way include compressed gas cylinders and artillery shells.

HOT SPINNING

Spinning is a technique used to form a disc into a conical shape. The process involves placing the disc in a spinning lathe between a chuck or pattern and a tailstock follower. The pattern and disc are rotated slowly. Pressure is applied at a point on the tailstock side of the disc. A blunt tool or roller is usually used to apply the pressure. The pressure forces the disc to be formed over the pattern.

Most spinning is done with cold materials and a more complete explanation of spinning is included in Chapter 11. Spinning hot metals reduces the pressures required. The process is used primarily for shaping thick hard materials. Steel up to 6 in. thick has been successfully spun. Pressure vessels and tank ends are common products of hot spinning. Hot spinning is also used to shape the ends of thickwalled tubing and cylinders.

SUMMARY

Hot forming is used to shape and size metals while they are above their recrystallization temperature. The resulting products have uniform stress-free grain structure. They are typically stronger and tougher than similar shapes produced by casting or machining operations.

Hot forming may be accomplished by forging, extruding, rolling, piercing, drawing, or spinning. Each process has applications best suited to it. The choice is one of matching the process to the material and its end use.

STUDY QUESTIONS — Chapter 10

1. List the major advantages and disadvantages of hot forming.
2. What are the three major forging actions?
3. How does the grain structure of a forged part differ from bar stock or a casting?
4. List and briefly describe the five major forging techniques.
5. What are some advantages for drop and impact forging?
6. Why is press forging used for large forging?
7. List and describe the four major types of extrusion.
8. List and describe three major ways to roll form pipe.
9. Describe the tube piercing process.

Rod stock is first heated then forged. (KDK Upset Forging Co.)

Chapter 11

COLD FORMING METALS

Many products contain parts which are manufactured from sheet, band, and rod metal. These parts must be sized and shaped to fit their intended use. A common method to accomplish this task is *cold forming.*

Cold forming takes place whenever metal is deformed below its recrystallization temperature. Most cold forming is done at room temperature. However, in some cases, the metal is heated to a point below its recrystallization temperature. This type of forming is sometimes called *warm forming.*

Cold forming techniques have some major advantages over hot forming processes.

1. No energy is used heating the metal.
2. Close tolerances can be held.
3. Higher quality surface finishes are produced.
4. The strength and hardness of the material is increased.

Along with these advantages, cold forming, as compared to hot forming, has some disadvantages.

1. More force is required to produce the same amount of deformation.
2. Greater forming forces require heavier, stronger equipment.
3. Metal surfaces must be clean and free of scale.
4. Cold metals are less ductile.
5. Materials are work hardened during the forming process.
6. The grain structure of the metal is distorted or fragmented.

The sum of the advantages and disadvantages make cold forming particularly useful for large-volume production. High volume allows the manufacturer to spread the relatively high costs of heavy duty equipment and shaping devices over large production quantities. This is not to say that with standardized equipment, some cold forming operations are not economical at moderate production levels. However, the cost advantages of cold forming over other techniques rise as production rates increase.

Cold forming techniques may be grouped into three major categories.

1. Bending.
2. Drawing.
3. Squeezing.

BENDING OPERATIONS

Bending, like all forming operations, involves stressing a material beyond its yield strength and below its fracture point. A bend causes plastic deformation with little or no change in the surface area of the material. There is only slight stretching or compressing of the material. Actually when a bend is made there is a neutral line or axis in the material. This line will be the same length before and after the bend. The material on the inside of the axis will be compressed during the bend. The outside material will be stretched. The compressed side will be slightly shorter and the stretched or tensioned side will be slightly longer than their original dimensions.

There will also be a slight narrowing or thinning of the material at the bend. This is caused by the plastic flow of the material on the stretched portion of the bend.

There are four major types of bending operations.

1. Angle bending.
2. Roll bending.
3. Seaming.
4. Straightening.

ANGLE BENDING

Angle bending involves both simple one-axis bends or multiple bends. Multiple bends produce formed parts. Most angle bends are made on sheet or plate stock. Sheet metal is flat material .006 to .250 in. thick. Plate metal exceeds .250 in.

Simple angle bends on sheet metal may be produced on several machines. The most common are the bar folder, the box and pan brake, and the cornice brake. All these machines produce a bend by gripping the metal then folding the overhanging portion. This action is shown in Fig. 11-1. These machines will generally handle mild steel less than 1/16 in. thick and 8 ft. long.

Heavier gage metals may be angle bent on a press brake, Fig. 11-2. As can be seen, the press brake uses a movable upper forming bar and a die attached to the bed. The upper bar is one-half of the forming die. It is slowly pressed downward into the lower die half, Fig. 11-3. The shapes a press brake can produce are almost limitless.

Angle bending may also be done on a standard punch press using a wide variety of dies, Fig. 11-4. For angle bending, wiping dies, rotary dies, or mated

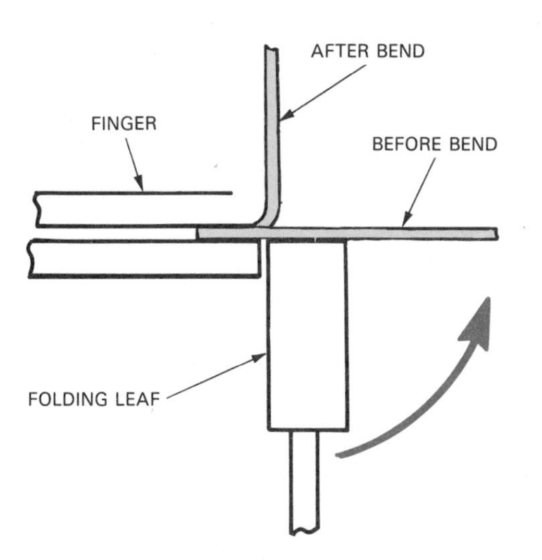

Fig. 11-1. Bar folders, cornice brakes and box and pan brakes use a bending or folding leaf to bend the metal along a hold-down bar or finger.

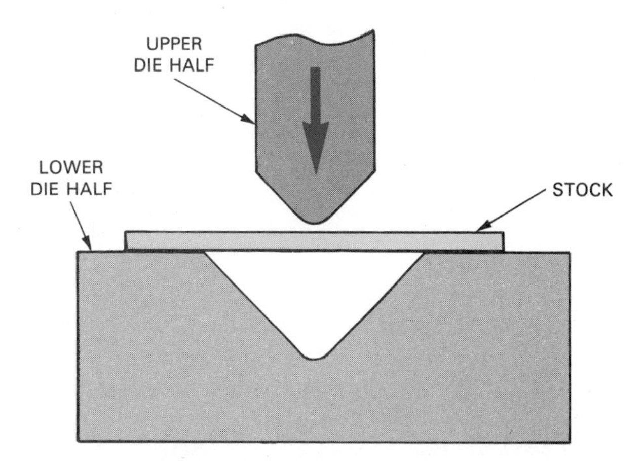

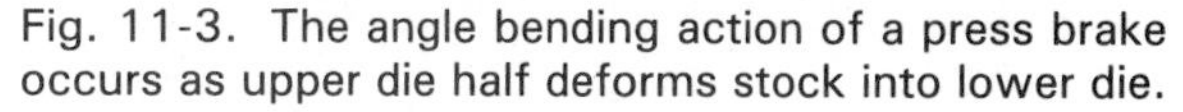
Fig. 11-3. The angle bending action of a press brake occurs as upper die half deforms stock into lower die.

Fig. 11-2. Sheet metal being bent on a press brake. Upper die is a long bar which presses workpiece into vee shaped lower die. (Cincinnati Inc.)

Fig. 11-4. Extruded aluminum shapes are converted into nearly finished bumpers in a single press stroke. (Reynolds Metals Co.)

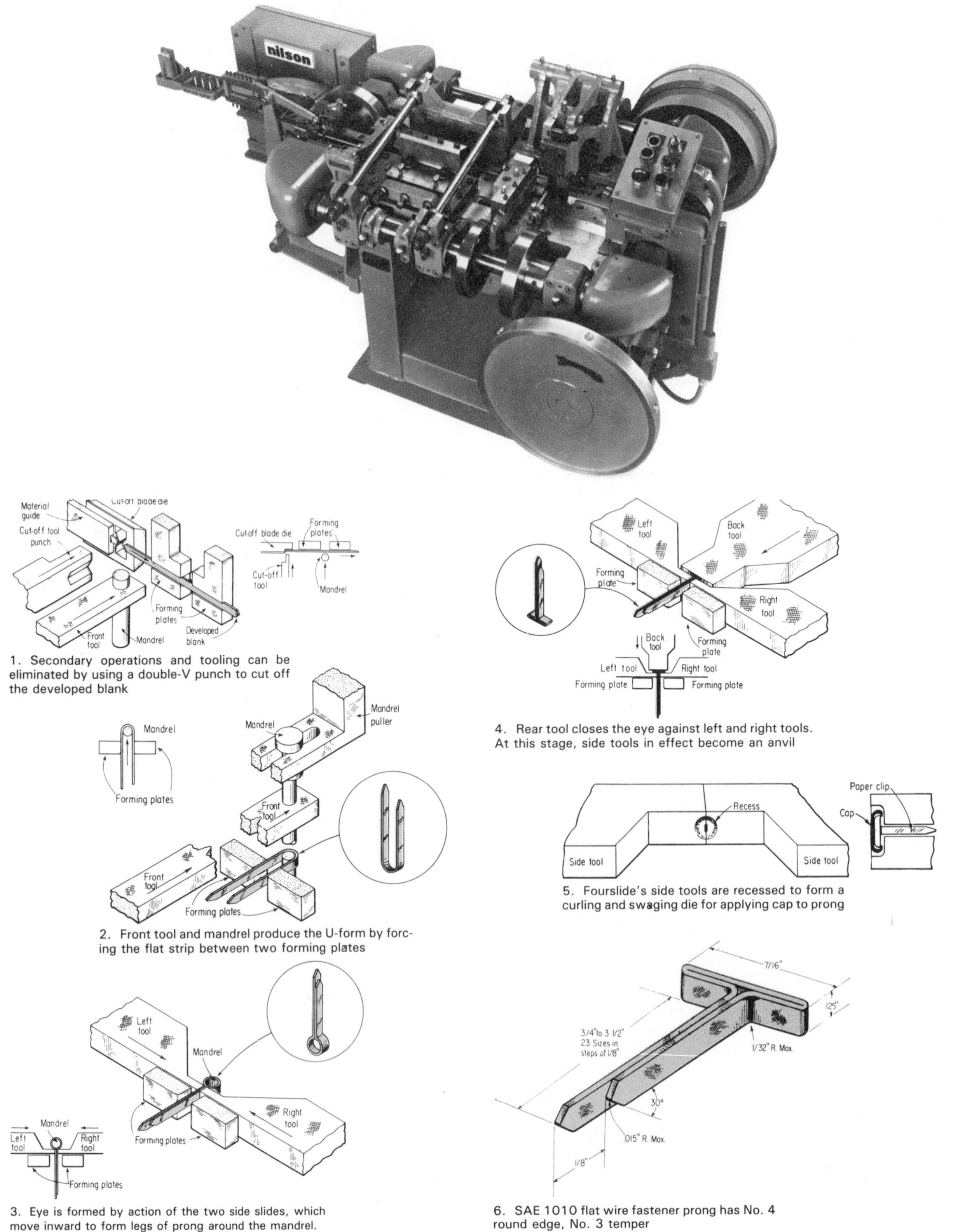

1. Secondary operations and tooling can be eliminated by using a double-V punch to cut off the developed blank

2. Front tool and mandrel produce the U-form by forcing the flat strip between two forming plates

3. Eye is formed by action of the two side slides, which move inward to form legs of prong around the mandrel.

4. Rear tool closes the eye against left and right tools. At this stage, side tools in effect become an anvil

5. Fourslide's side tools are recessed to form a curling and swaging die for applying cap to prong

6. SAE 1010 flat wire fastener prong has No. 4 round edge, No. 3 temper

Fig. 11-5. Complex bends may be made on a four-slide forming machine. Each slide performs a different forming operation. (A.H. Nilson Machine Co.)

dies are usually used. The mated die may produce either single or complex bends, Fig. 11-5. Wiping dies and rotary dies make simple straight-line bends. The action of these dies is shown in Fig. 11-6.

When designing a part for angle bending, these major factors must be considered:

1. The minimum bend radius for the material. Bends sharper than this will cause the material to crack.
2. The springback that can be expected. This factor is the degree to which the material moves back toward its original shape when the forming pressure is removed. While it can be mathematically estimated, actual springback can only be determined through experimentation with the bending process.
3. The size of blank needed to produce a part of proper finished size. Bend allowances must be considered.
4. The minimum length of material which can be bent. The length of the material must be at least the bend radius plus one and one-half to two times the thickness of the metal.
5. The tolerance that can be expected. Tolerances less than 1/32 in. are difficult to hold.

ROLL BENDING

Rolls may be used for producing simple curves in sheets, plates, bars, rods, tubes, and pipes. Matched rolls in a progressive rolling machine may be used to produce complex shapes from sheet stock.

Fig. 11-7. Coils and angle iron being formed on three-roll forming machine. One roll is adjustable to control radius of curve. (Wallace Supplies Mfg. Co.)

Three-roll forming machines

Simple curves are usually produced by three-roll forming machines like those seen in Fig. 11-7. Two basic types are made. One has two front rolls which grip and feed the material. The third roll adjusts upward with each pass of the material to control the curve.

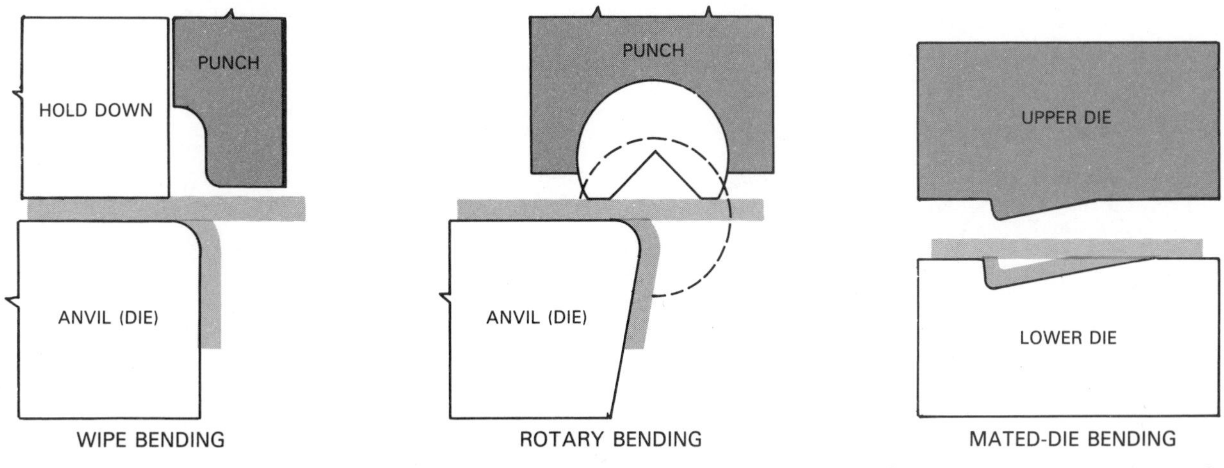

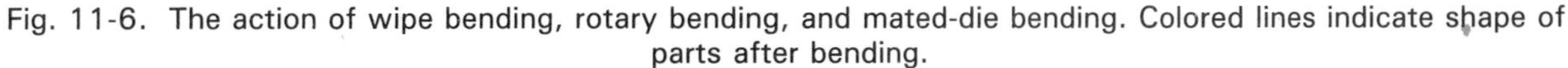

Fig. 11-6. The action of wipe bending, rotary bending, and mated-die bending. Colored lines indicate shape of parts after bending.

The second type has two lower power driven rolls. The upper third roll can be adjusted downward to control the curvature. This type of machine, called the *pyramid* type, cannot be used for thin materials. It is used only to bend heavy bars, rods, and structural shapes. Fig. 11-8 shows the action of the two types of three-roll bending machines.

Continuous rolling machines

Continuous rolling, as shown in Fig. 11-9, is used to produce numerous shapes from sheet metal. Roofing, siding, architectural shapes, tubing, and pipe are but a few of the products of continuous roll forming.

In continuous roll forming a series of mated rolls gradually form the metal into its final shape. The number of rolls required is determined by the complexity of the shape to be formed. Machines can have up to 40 or more forming stations (sets of rolls). Fig. 11-10 shows the steps (sets of rolls) required to produce one architectural shape.

Continuous rolling is a high-volume production technique. It can handle plain, painted, or plated sheet materials at speeds up to 800 ft. per minute. Common forming speeds, however, are about 100 ft. per minute.

Fig. 11-9. A series of mated rolls form corrugated aluminum roofing. (Reynolds Metals Co.)

SEAMING

Seaming is a bending technique used for fabricating sheet metal products. These seams may be produced on special machines or with die sets mounted in press brakes or punch presses. Seaming is an assembly process and is discussed in Chapter 29.

STRAIGHTENING

After processing, metal often needs to be straightened. Straightening or flattening is the opposite of bending.

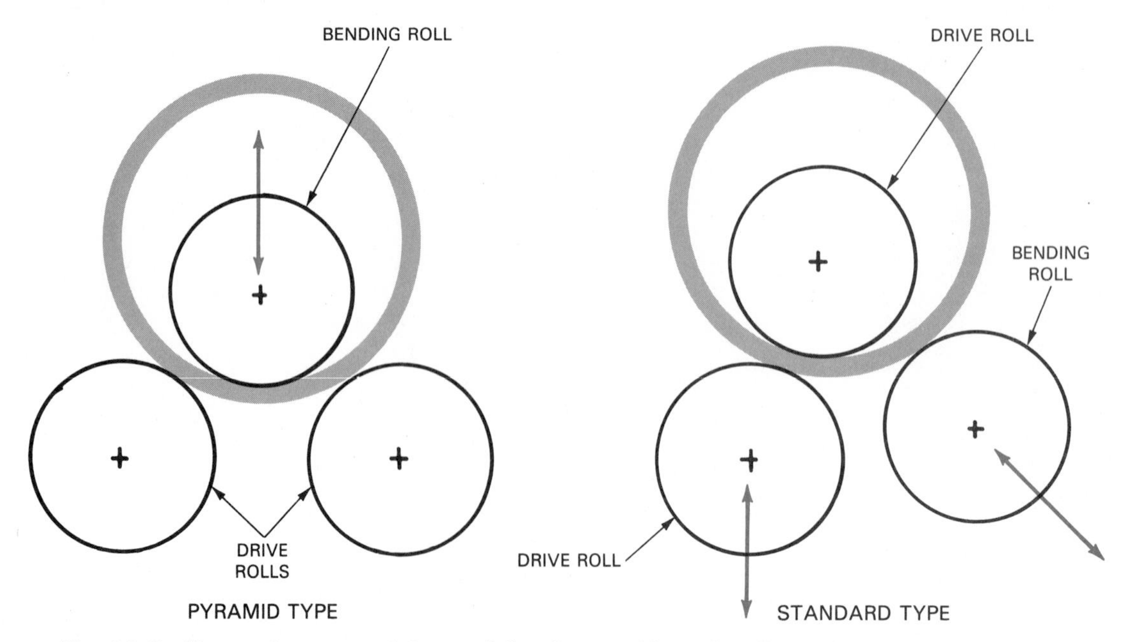

Fig. 11-8. Two major types of three-roll forming machines. Bending rolls, being adjustable, control the bend rate.

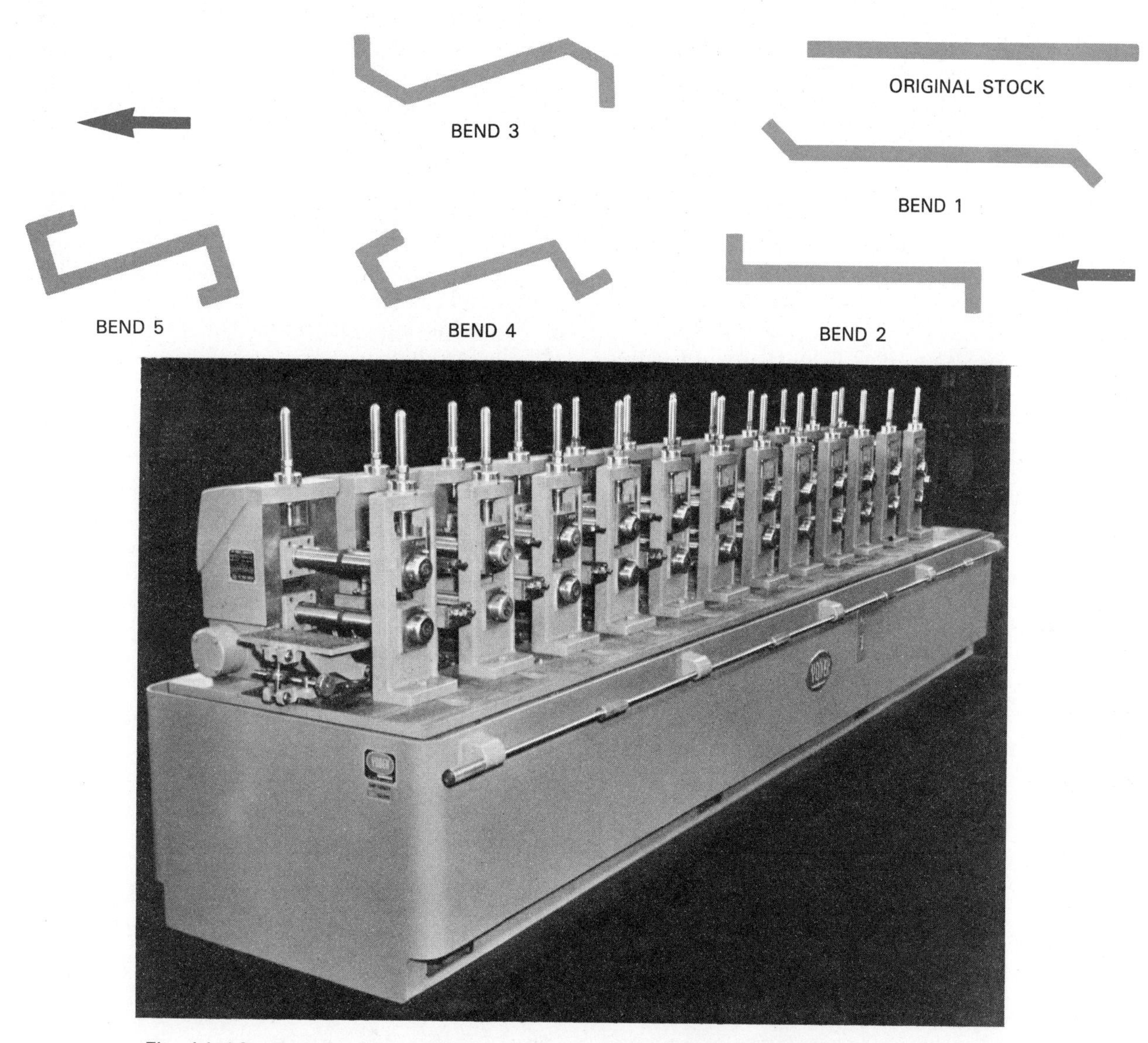

Fig. 11-10. Top. The forming action of a continuous rolling machine produces shapes gradually. Bottom. A twelve-station machine. (Yoder Co.)

Straightening may be done by either of two basic techniques. The first technique *roll straightening,* uses a series of rolls to create a number of reverse bends in the material. Each bend exceeds material's elastic limit and cancels out part of the previous bend. Each set of rolls reduces the bend until the material is finally straight. Fig. 11-11 shows the roll straightening method. The second technique uses stretching forces

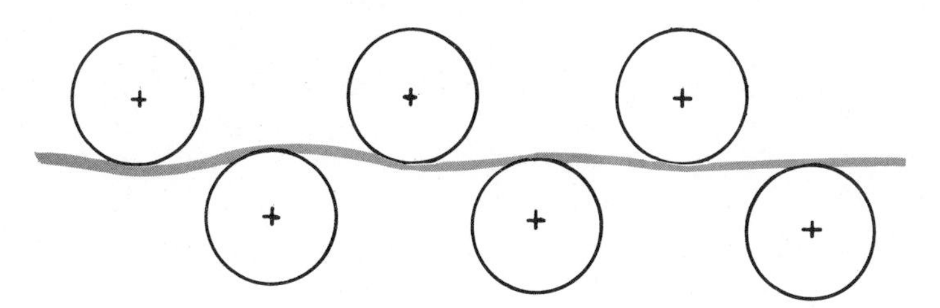

Fig. 11-11. Roll straightening operation removes bends from stock.

(tension) to level the sheet. The material is gripped by its two ends and tension is applied. This method is called *stretcher leveling.*

DRAWING OPERATIONS

Cold drawing involves forcing metal into, through, or around a die to create the desired shape. Drawing may use stretching, compressing, or bending forces. However, stretching and compressing are the common forces used.

Drawing requires a clean material free of major defects and scale. Also, dies used in the drawing operation are usually lubricated to reduce friction and abrasion.

The products of drawing techniques are often smooth and have accurate dimensions. They are usually formed sheet metal products or wire and tubing.

The major cold drawing techniques used to form metals are:
1. Sheet drawing.
2. Bar, wire, and tube drawing.
3. Spinning.
4. Stretch forming.
5. Embossing.
6. High velocity forming (HVF).

SHEET DRAWING

Sheet drawing uses dies to form a wide variety of products from thin, flat (sheet) stock. The process is often called simply *drawing.*

Sheet drawing uses a combination of stretching and compressing forces. These forces deform the material into the desired shape. As in all forming practices, the forces must be in a range above the elastic limit but below the fracture point.

Sheet drawing practices include:
1. Shallow drawing.
2. Deep drawing.
3. Flexible die drawing.

Shallow drawing

Shallow drawing, as seen in Fig. 11-12, forms products in which the depth of the draw is less than one-half the diameter of the blank. It produces products with little thinning of the material. The sheet stock is formed in a press between two mated dies. Fig. 11-13 is a simplfied diagram of the steps. The term, stamping, includes bending and shearing as well as drawing operations. Shallow drawing is used to produce automotive body components, shallow pans, discs, appliance parts, toys, and many other products.

Fig. 11-12. The first step in this progressive forming operation draws a shape from sheet metal ribbon. Note the parts in their various stages of forming (in front of each die station). (E.W. Bliss Machinery Co.)

Deep drawing

Deep drawing is similar to shallow drawing. However, deep drawing forms a product that has a depth greater than its final diameter.

In forming thin material, deep drawing has several problems.

1. As the punch contacts the material, *tension stress* is set up in the blank. The material must stretch around the punch and into the die. The outer edges of the blank, which are not in contact with the punch, must compress. Instead of compressing, they often wrinkle. To counteract this action, a blank holder squeezes the material to stop edge wrinkling and encourage compression. Fig. 11-14 shows an arrangement of the blank holder, punch, and die used for deep drawing and other complex drawing techniques.
2. Deep drawing also requires a large amount of material deformation. This often causes excessive work hardening. Thus, many deep draws cannot be made in one step. Instead, progressive dies are used. Each cavity in the die performs one step in the total forming act. The material is annealed (softened) before each successive step. At each step, the material's depth is increased and its thickness may be reduced. Additional draws, after the first one, are sometimes called *redraws*. The shape formed in the first step is redrawn into deeper shapes at each stage.

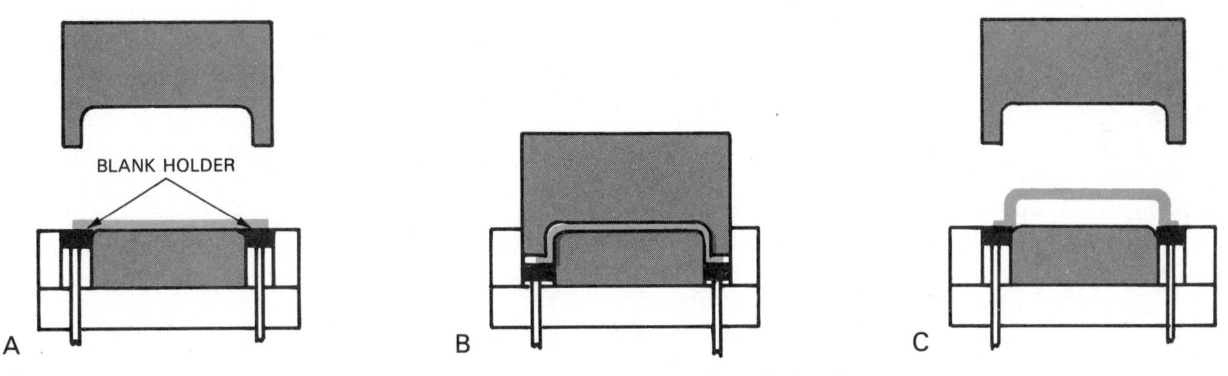

Fig. 11-13. Steps in drawing. A—Blank is placed over the lower die half. B—Upper die descends, squeezing blank against blank holders and forms part. C—Part ejected after forming.

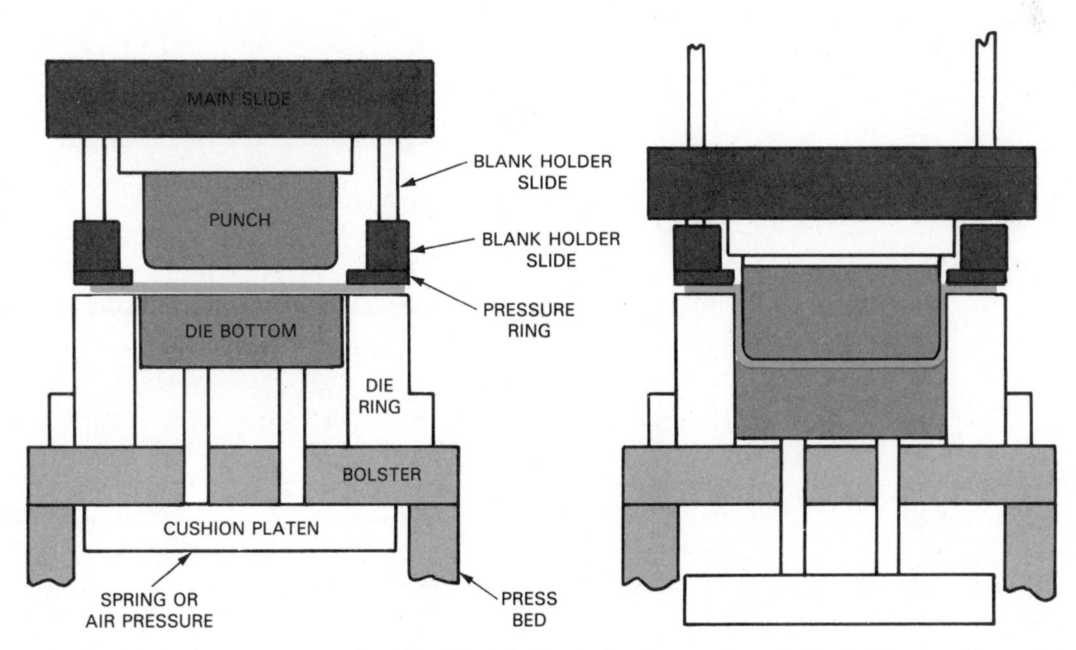

Fig. 11-14. Double action press operation. The blank holder is first lowered to grip blank. The punch then is lowered to draw part to shape.

Often, the final step in a series of drawing steps is called *ironing*. This technique smooths and reduces the thickness of the walls. Fig. 11-15 shows the action of an ironing die.

Deep drawing is often called *shell drawing* because the technique was first used to produce artillery shells. Sinks, deep pans, and equipment cases are products manufactured by deep drawing.

Flexible die drawing

Several drawing methods use a single shaped die. A fluid or rubber is used to provide the forming forces normally created by other die half.

The *Guerin process* uses a solid or laminated rubber forming pad and a *form block* or *punch*. The form block is the shaping device for the process. It may be made out of wood, hardboard, aluminum, or other easily worked material. The process involves placing the blank between the form block and the rubber pad. The back and sides of the rubber pad are surrounded by a steel retaining frame.

When the press is closed the form block presses the blank into the rubber pad. As the rubber compresses, it becomes the mated die half for the form block. The rubber, like confined liquids, applies equal pressure in all directions. The part is therefore evenly forced around the forming block. Fig. 11-16 is a diagram of the Guerin process.

The Guerin process was first used in the aircraft industry. Their need for small quantities of interchangeable parts made the process very useful. The process however cannot be used for high volume production or for deep drawing. It is basically a low-volume production process for products having fairly uniform forming demands.

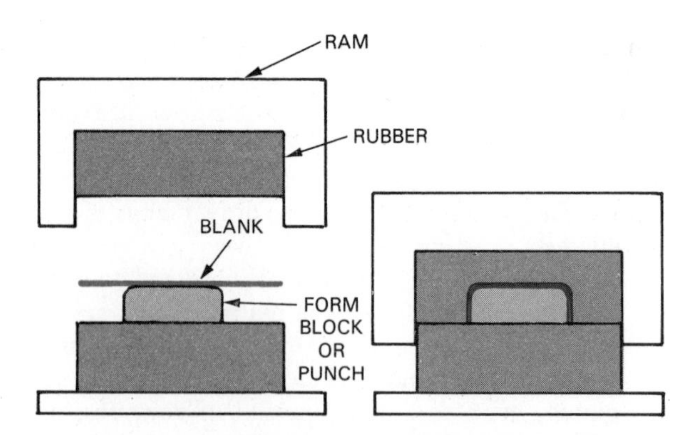

Fig. 11-16. The Guerin process of flexible-die forming. Blank is pressed into rubber pad causing it to take shape of form block.

A refinement of the Guerin process is the *Marform process*. This process uses a blank holder and a downward movement of the flexible die. Products with greater depth and wrinkle-free flanges can be produced using this process.

A second method of flexible-die forming is shown in Fig. 11-17. It is called *hydroforming* or fluid form-

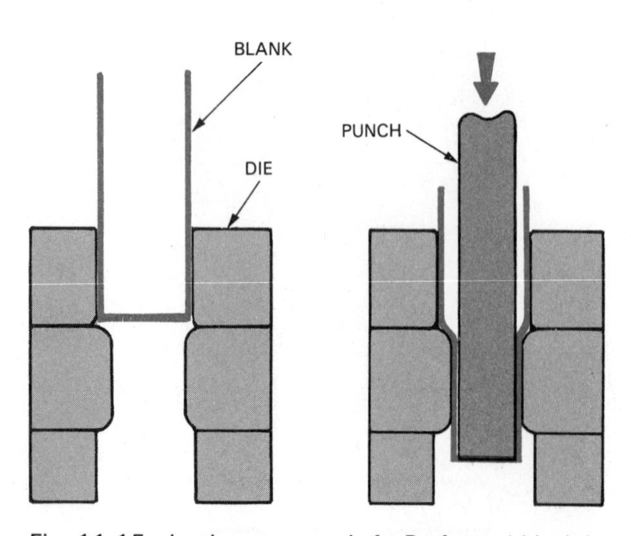

Fig. 11-15. Ironing process. Left. Preformed blank is placed in die. Right. Punch "irons" metal blank into shape of die.

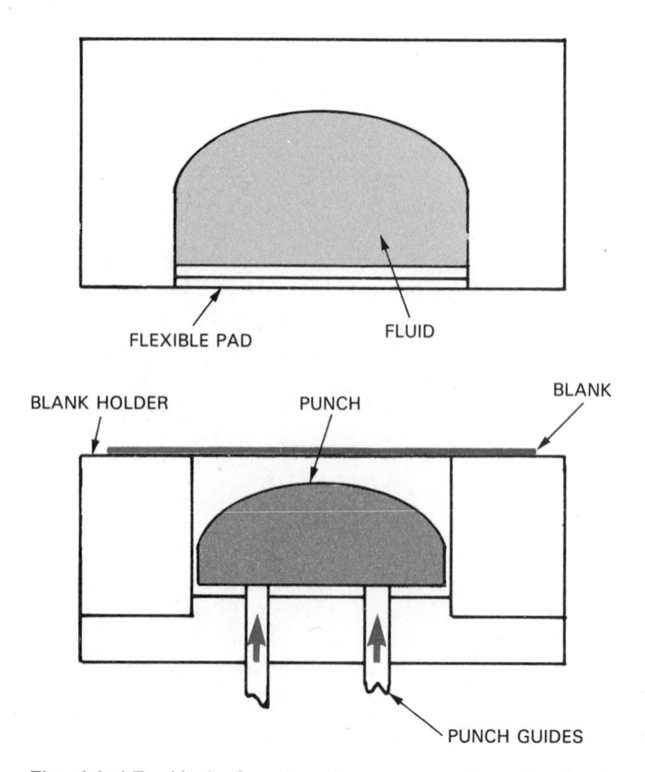

Fig. 11.17. Hydroforming is a second method of flexible-die forming. Flexible element is an oil chamber behind a rubber diaphragm.

ing. This process is very similar to Guerin forming. The basic difference is in the flexible die portion. This portion consists of a chamber filled with oil behind a rubber diaphragm on the forming face.

The blank is placed on the shaping die and forced upward into the rubber diaphragm. The fluid behind the diaphragm exerts equal pressure in all directions, causing the blank to form around the punch.

Hydroforming can be used at high speeds to produce parts with accurate dimensions. Hydroforming tooling is less expensive and can produce complex shapes.

Hydrostatic bulging is a third flexible-die forming technique using fluids or rubber to generate forming pressures. Bulging is causing the walls of drawn shapes, tubing, or other similar parts to expand along a portion of their length.

The part to be formed, as shown in Fig. 11-18, is placed in a die. A rubber or polyurethane punch or oil is placed into the part. Pressure, applied to the oil or the punch forces the oil or punch against the part walls expanding the metal against the die walls. Sometimes small steel balls and grease are used instead of rubber or oil.

Fluid cell process

A variation of the Guerin process, the fluid cell process uses higher pressure than the Guerin and is suited for forming parts that are slightly deeper. A fluid cell or bladder forces a rubber pad against the form block. It exerts nearly uniform pressure at all points.

This uniform pressure on the form block sides permits forming of products with wider flanges than is possible with the Guerin process. Further, shrink flanges, joggles, and beads and ribs can be formed to sharp detail in one operation.

BAR, TUBING AND WIRE DRAWING

Probably the simplest drawing technique is bar, tubing, and wire drawing, Fig. 11-19. This series of operations involves reducing the diameter of a material and thereby increasing its length.

The process involves several steps. The most important of these are:

1. Cleaning and lubricating the material.
2. Pointing the stock.
3. Inserting the pointed end through a die mounted in a die box.
4. Gripping the material with a clamp.
5. Pulling the material through the die.

Dies used for these drawing operations are made of hard materials. Carbide dies are commonly used.

Diamond dies are usually substituted for small-diameter wire drawing. High temperatures are generated by the drawing action. The dies, therefore, are water cooled to insure dimensional accuracy and to extend their life.

Bar and tube drawing

Bar and tubing drawing are similar processes. The material is drawn to size on a machine called a

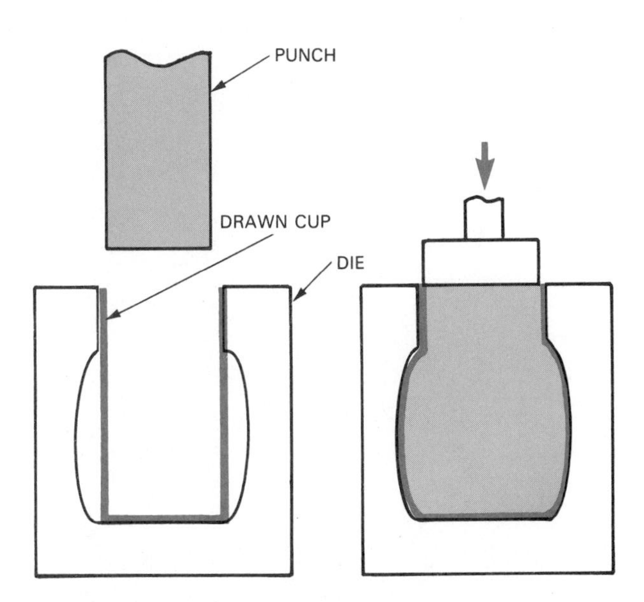

Fig. 11-18. Bulging, a third flexible-die forming technique, expands walls of parts by placing a rubber punch or oil into the part.

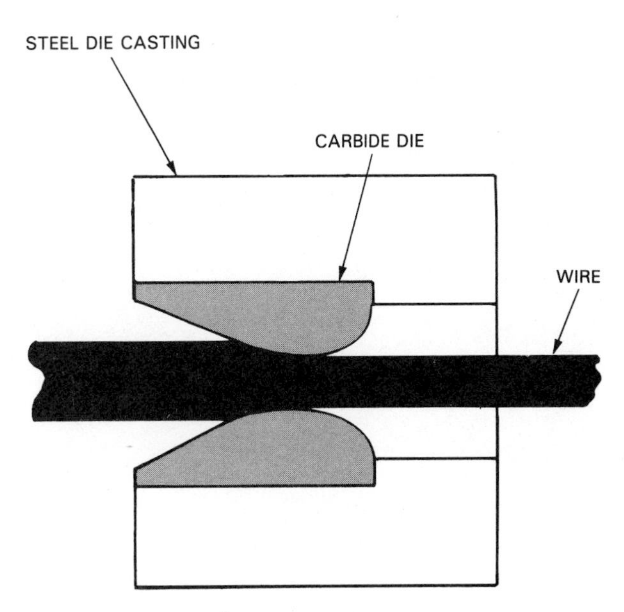

Fig. 11-19. Wire drawing die. Carbide is used for durability.

drawbench. This long horizontal machine consists of a bed, sets of entry rolls, a die box or holder, a carriage with clamping tongs, and a carriage power (pulling) unit. The power unit may be an endless chain or a hydraulic cylinder.

Drawbenches for bar stock are usually not over 40 ft. long and have a pulling force up to 50 tons. Tubing drawbenches can reach 100 ft. in length.

The bar or tube stock is first formed using hot forming techniques. The material receives its final size and physical properties through the cold drawing operations.

Wire drawing

Wire drawing may be done by feeding the stock through one die after another until the proper diameter is reached. The material, fed from a coil, goes through the die, and is wound on a take-up reel. This process is repeated with progressively smaller dies until the final diameter is reached.

However, higher production rates are reached by *continuous drawing.* Stock passes from one die to another in one continuous operation, Fig. 11-20.

SPINNING

Round, cup-shaped parts may be formed by a technique called spinning. The workpiece, a thin metal disc, is shaped while being rotated in a spinning lathe. Shaping takes place as the disc is forced over a wood or steel shaping device called a *chuck.* The force needed to cause the metal to flow over the chuck is generated by a blunt tool. The tool is used to gradually force the metal to take the shape of the chuck.

Manual spinning, as shown in Fig. 11-21, requires great skill. The portion of the disc not in contact with the chuck must always be perpendicular to the axis of the spin. If the operator tries to force the material too rapidly, edge wrinkling and tearing will occur. Automatic spinning machines, Fig. 11-22, produce more uniform results.

Spinning is an old art and at one time was used to produce most deep-drawn metal parts. It is still used

Fig. 11-20. Continuous wire drawing machine in operation. Wire passes through several progressively smaller diameter dies before reaching final diameter. (Ajax Manufacturing Co.)

for low-volume production activities. Typical products of the spinning operation are pressure tank ends and large lighting reflectors.

STRETCH FORMING

Stretch forming, as seen in Fig. 11-23, was developed by the aircraft industry. Two or more sets of jaws grip the metal at its ends. The jaws stretch the material to a point above its elastic limit. The metal is then pulled over and around a single shaping device which also exerts pressure upward. The part is then trimmed to shape. There are considerable trim allowances at the gripping ends and along the sides. These allowances produce a fairly high waste factor.

Stretch forming produces large, shaped pieces such as aircraft wings, tails, or fuselage parts. Other uses include making truck trailer or bus body parts.

EMBOSSING

Embossing is a process using matched dies or rolls to produce a shallow design. This is done without changing the thickness of the metal. The embossed item will have the same design on both sides. That

Fig. 11-21. Manual spinning operation. With great skill, operator forces spinning metal disc into shape of chuck (pattern). (Aluminum Company of America)

Fig. 11-22. Automatic spinning involves several steps. Top. Loading the disc. Middle. Forming the rotating disc with a machine-controlled forming roller. Bottom. Removing the formed workpiece. (Crouse-Hinds Co.)

Fig. 11-23. An aircraft part is stretch formed. (Boeing Co.)

is, on one side the design will be raised while on the other the design is depressed. Fig. 11-24 contains a sketch of a flat embossing operation.

Embossing is used in the manufacture of identification and name tags, jewelry, tableware, and decorative patterns. Rotary embossing can create patterns on thin sheets and foils.

HIGH VELOCITY FORMING

Operations grouped under high velocity forming (HVF) are often called *high energy rate forming* (HERF). They depend upon a very rapidly moving forming force. This motion must be at least 50 fps to be called high velocity.

A number of techniques depend upon high velocity forming forces. The most widely used are:

1. Explosive forming.
2. Electrohydraulic forming.
3. Electromagnetic forming.
4. Pneumatic-mechanical forming.

EXPLOSIVE FORMING

Explosive forming uses the rapidly expanding gases released from an explosive charge to produce the forming force. Typically, the blank is clamped over a shaped die cavity. A vacuum is created within the cavity. An explosive charge is then discharged in a chamber above the blank. The explosion chamber may be air or water-filled. The rapidly expanding gases from the explosion quickly pressurize the upper chamber. This high-velocity force causes the metal blank to collapse into the lower die cavity.

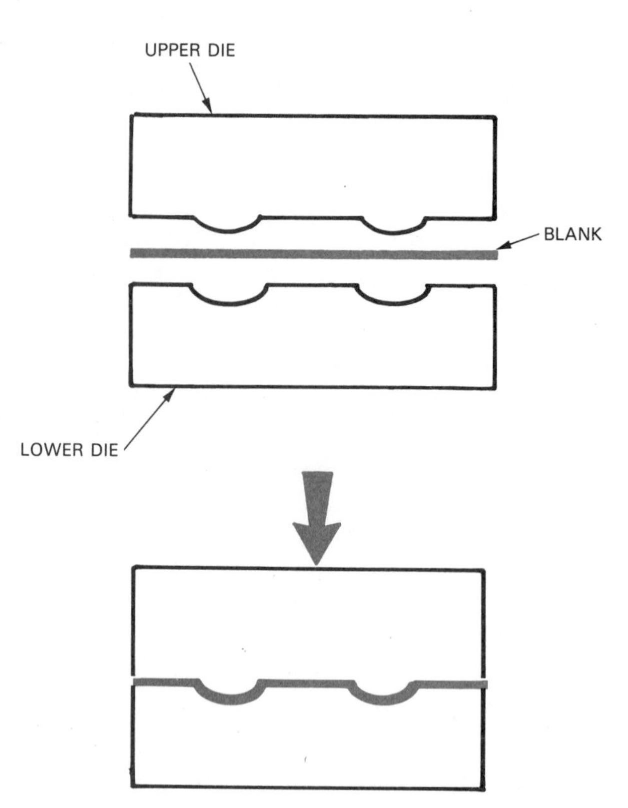

Fig. 11-24. Embossing. The process may use either dies or rolls.

The pressures generated range from up to 100,000 psi (689,000 kPa) for low explosives (cartridge) systems to 2,000,000 psi (13,780,000 kPa) for high explosive systems. This pressure is an intense shock wave and lasts only a fraction of a second.

Fig. 11-25 shows several explosive forming methods. These methods are safe and quiet. They are used for limited-quantity production runs, prototype work, and for large parts which would require very expensive mated dies.

Electrohydraulic forming

Electrohydraulic forming (EHF) rapidly converts electrical energy into mechanical energy to form metallic parts. The mechanical energy is then transmitted to the workpiece by water. A typical EHF system contains two major components:

1. Two-part chamber. The top half can have a vacuum while the bottom half is filled with water.
2. A bank of capacitors connected to a spark gap in the lower half of the chamber.

There are two common methods of electrohydraulic forming. One is called *electrospark forming* or *spark discharge forming*. The other is known as the *exploding bridge wire* method.

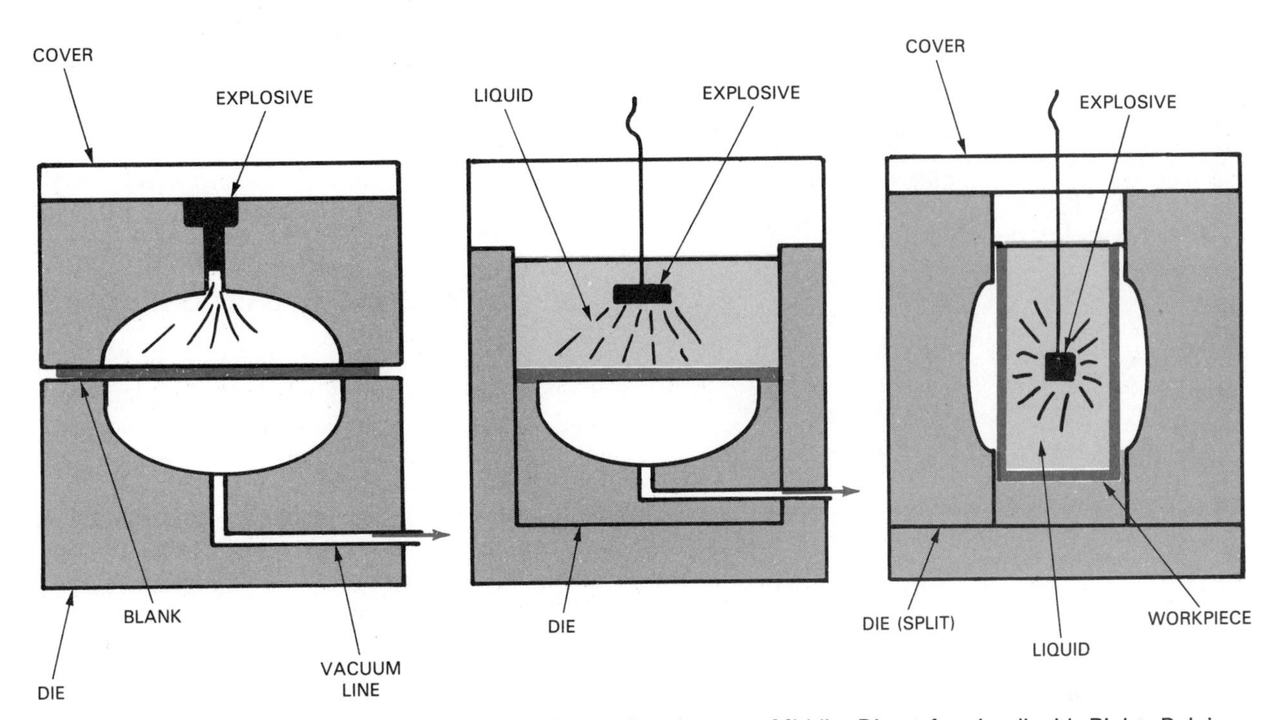

Fig. 11-25. Types of explosive forming. Left. Direct forming-gas. Middle. Direct forming-liquid. Right. Bulging.

Fig. 11-26 shows a typical electrohydraulic forming arrangement. Its operation involves clamping the workpiece in place then charging the capacitors with high voltage. When the switch is closed, the capacitors rapidly discharge sending an electrical spark arcing across the spark gap. The shock waves from the spark are carried by the liquid to the workpiece which absorbs them by deforming against the walls of the upper chamber.

In the exploding bridge wire method a wire filament is stretched between the two electrodes in the lower chamber. As an electric current from the discharging capacitors flows through the wire, the wire vaporizes (explodes). The pressure of the explosion is carried by the medium (liquid or air) and, as in the spark discharge method, deforms the workpiece.

Advantages and limitations

The most important advantage of EHF is its ability to form hollow shapes into complex shapes. It is able to do so with less expense than other fabricating methods. The cost of tooling is usually less than for conventional methods. It is more useful in forming small-to-intermediate sized parts which do not require excessive levels of energy.

The exploding bridge wire method offers more accurate control than the spark discharge method. One disadvantage is that the wire must be replaced after each part is formed. This adds to production time.

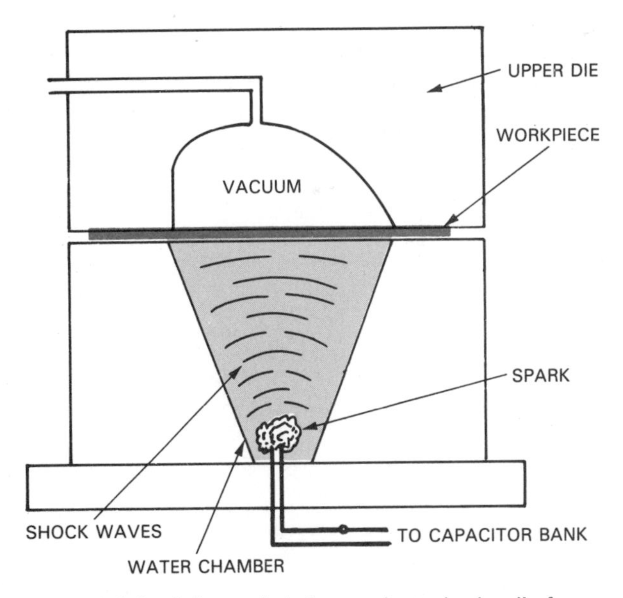

Fig. 11-26. Schematic is for an electrohydraulic forming operation.

Electromagnetic forming

Electromagnetic forming uses the force from an expanding magnetic field to generate high velocity forming forces, Fig. 11-27. This process involves placing the part to be formed inside a coil of wire or placing a coil inside a hollow part. High voltage from a

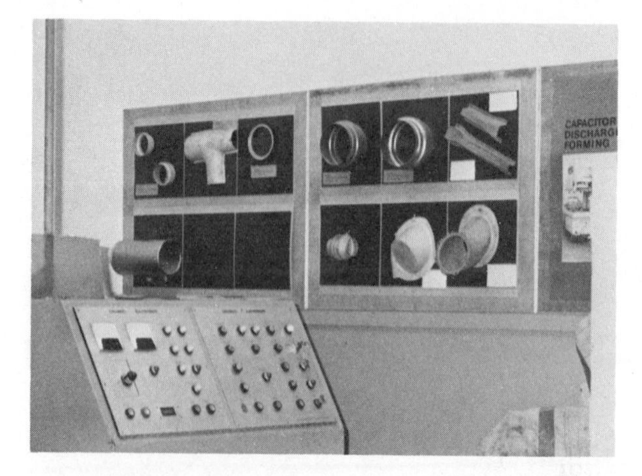

Fig. 11-27. Sample of electromagnet formed parts. Console for controlling the high-voltage process is shown at lower left. (Boeing Co.)

capacitor bank is discharged through the coil. The electrical current from the capacitors induces a rapidly expanding magnetic field around the coil. The changing magnetic field, in turn, induces electrical current in the workpiece. The induced current interacts with the magnetic field and creates the forming force. These forces may be as high as 50,000 psi (7257 kPa) but last only a very short time (several micro seconds). Additional pulses from the capacitor bank may be needed to complete the desired forming action.

Fig. 11-28 shows how electromagnetic forming may be used to expand the workpiece or shrink (swage) the item.

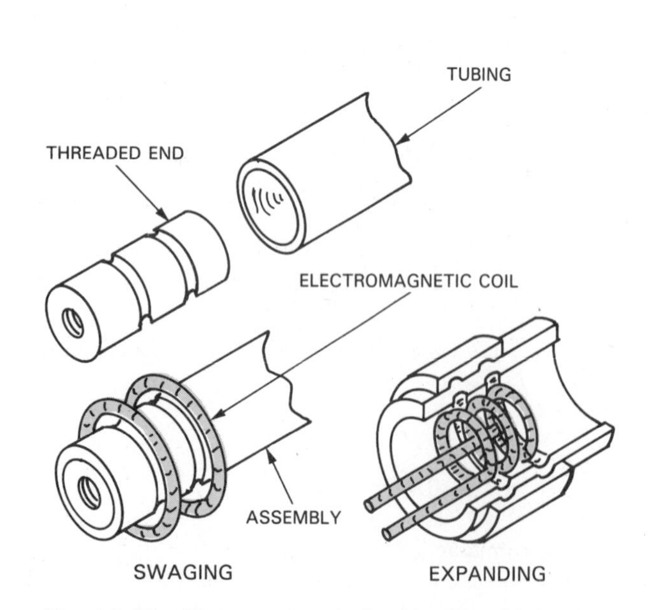

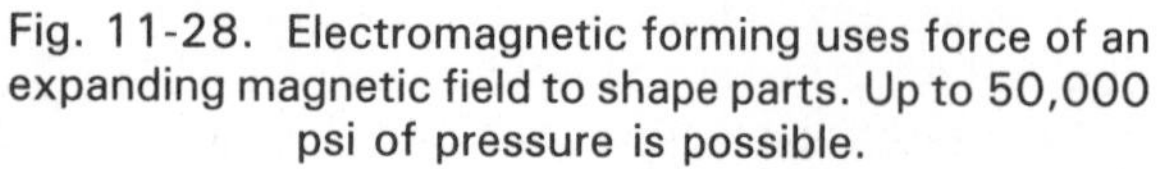

Fig. 11-28. Electromagnetic forming uses force of an expanding magnetic field to shape parts. Up to 50,000 psi of pressure is possible.

Pneumatic-mechanical forming adapts the drop hammer forging principle but increases the velocity of the ram. Compressed gas or a burning fuel-oxidizer mixture power the ram. These techniques can multiply the speed of a common drop hammer ram by two to ten times. Pneumatic-mechanical forming is used in forging extrusion, and metal powder compacting.

SQUEEZING OPERATIONS

Squeezing operations use compression forces to shape materials. Like all other cold forming operations, *cold squeezing* techniques have their hot forming counterparts. Cold forming is used when dimensional accuracy is critical.

Cold forming (cold rolling) may be used in the initial manufacture of metal stock, sheet, bar, and rod stock. Cold forming operations may also be used for secondary manufacturing activities. Typical of squeezing operations are:

1. Swaging.
2. Sizing.
3. Coining.
4. Hobbing.
5. Staking and riveting.
6. Thread rolling.
7. Extrusion.

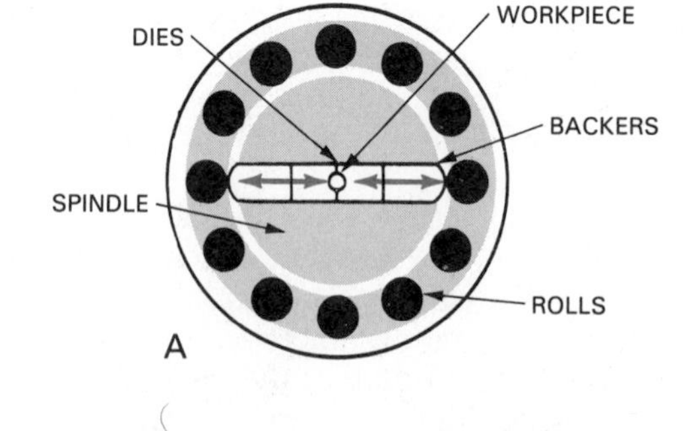

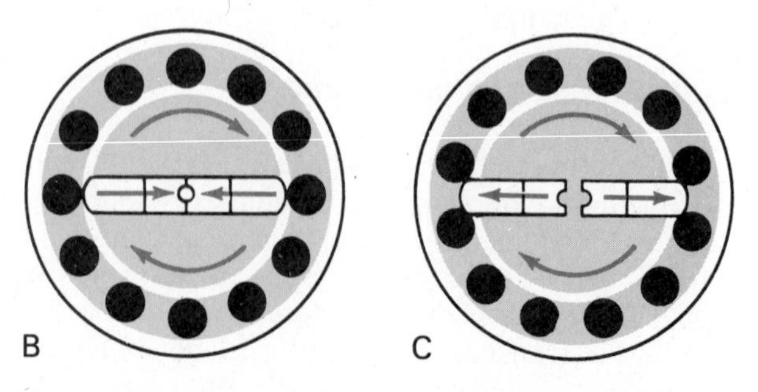

Fig. 11-29. The swaging process. Rotating dies move outward by centrifugal force and are driven inward to strike workpiece when they strike a roller. A—Components. B—Dies compress part. C—Work feeds into dies as they open. (Torrington Co.)

SWAGING

Swaging is basically a forging process. It reduces the diameter or tapers round bars, pipe, or tubing through a hammering action. Cold swaging is often called *rotary swaging*.

The process, as seen in Fig. 11-29, uses a rotating tapered split die and several sets of rollers. As the die set rotates, each pair of rollers closes the die. As the die set closes it gives the round stock in it a quick impact.

The rapidly repeated blows from the die set compress and shape the part, 11-30. The speed of rotation and the number of rollers sets determine the speed of the swaging action.

Depending on the material, size of the machine, and the job, rotary swaging may deliver 1300 to 5000 strokes per minute. The stock may be fed into the die until the proper amount of swaging is done. Fig. 11-31 shows a rotary swaging machine.

A modification of rotary swaging is called the *Intraform process*. The process uses a shaped mandrel which is inserted in tubing or a hollow part. The hammering action causes a hole in the part to take on the shape of the mandrel. The Intraform process is often used to form internal splines and gears, Fig. 11-32.

SIZING

Sizing is an operation designed to give a part or a portion of a part dimensional accuracy. This operation changes the metal thickness and shape through a squeezing action as seen in Fig. 11-33.

Sizing is particularly useful for providing accurate dimension to a portion of a forged or cast part. The base or flats that must mate another part are often

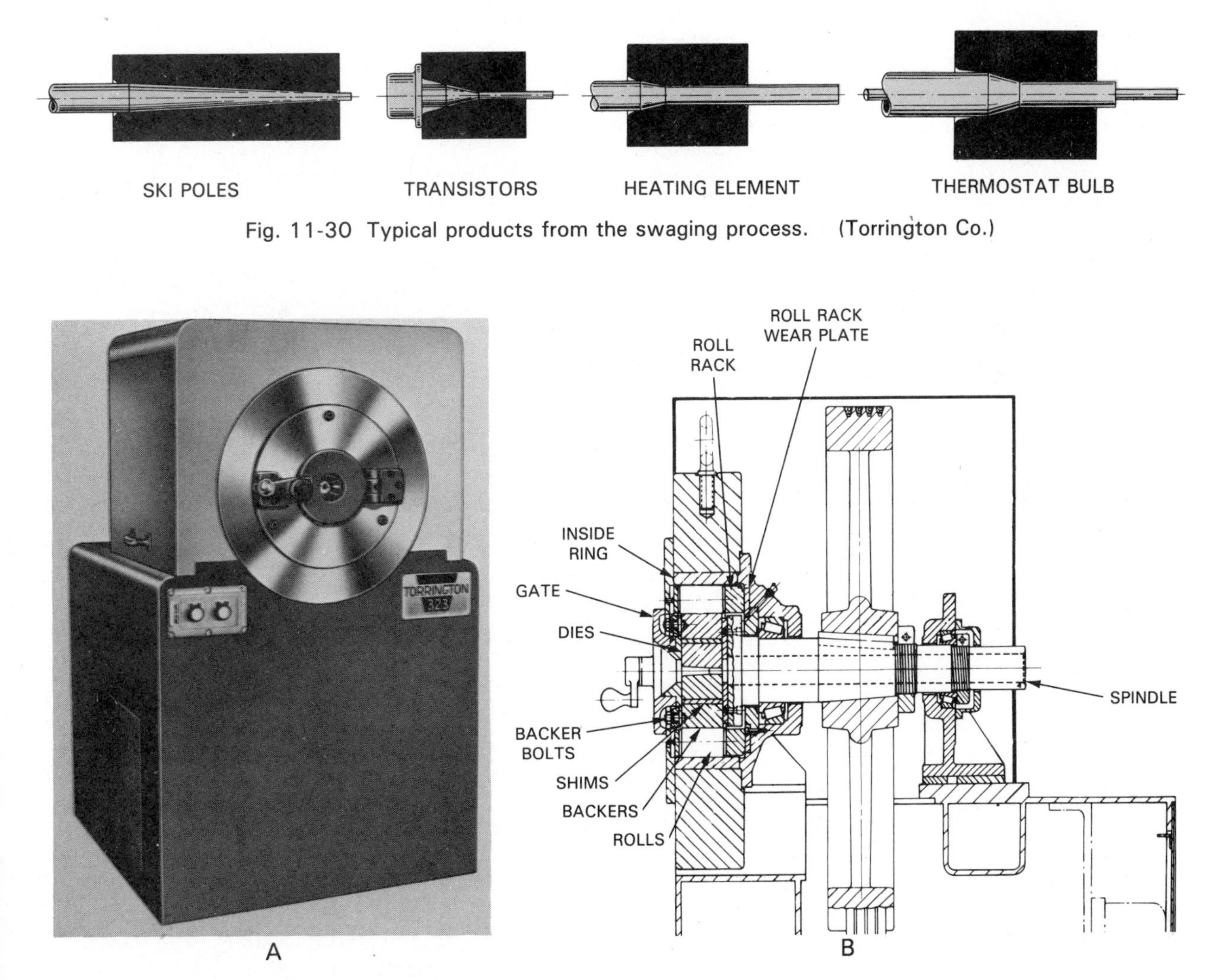

Fig. 11-30 Typical products from the swaging process. (Torrington Co.)

Fig. 11-31. A—Swaging machine. B—Parts of machine. (Torrington Co.)

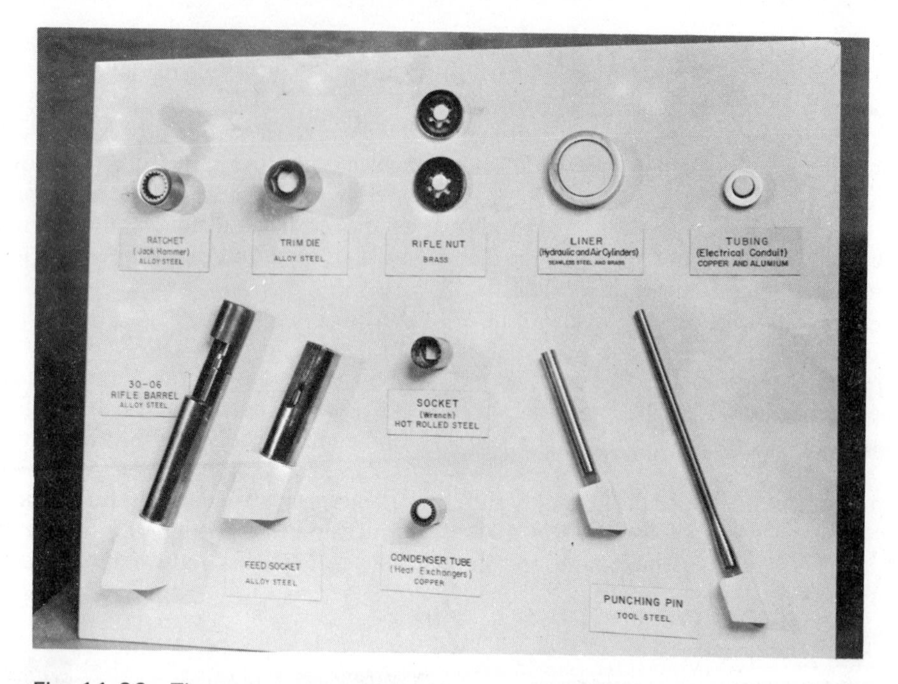

Fig. 11-32. These parts are produced by the Intraform process. Note the complex shapes of the holes. (Cincinnati Milicron Co.)

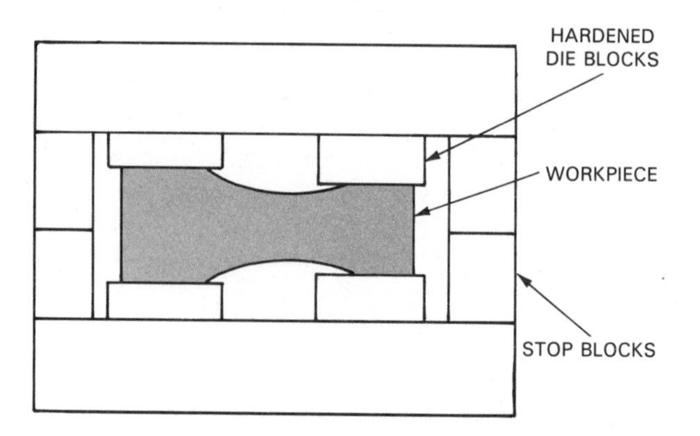

Fig. 11-33. A sizing operation accurately shapes two major dimensions of a workpiece.

sized. Sizing also improves the surface hardness and finish.

COINING

An operation closely related to sizing is coining. In this operation a material is both sized and given surface impressions. This action is accomplished by squeezing the stock under high pressure in a closed die. See Fig. 11-34.

The stock must be accurately sized so it will just fill the die cavities at the end of the pressure cycle. During the squeezing act, the material flows to take

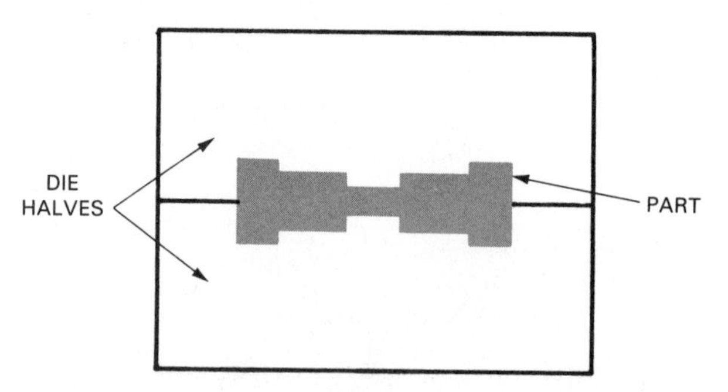

Fig. 11-34. Coining uses closed dies to produce detailed patterns on the faces of the workpiece.

on the shape engraved on the die faces.

The amount of material movement during a coining operation is small. Providing surface detail is the main objective. Coins, medals, and jewelry are common products of the coining operation.

HOBBING

Hobbing or *hubbing* is a cold forming technique used to make molds for the plastics industry. Hobbing involves machining a punch the shape of a desired die cavity. This punch is made from tool steel and is heat treated to develop its hardness. The hardened punch is then carefully polished.

In hobbing, the hardened punch is slowly and

carefully pressed into a soft steel mold block. The block is reinforced against the forming pressures by a heavy steel band. The punch forms the cavity in the steel block.

Often the punch will not be able to form the cavity in one step. The metal deformation in the soft steel block will cause the block to work harden. In such cases, the block is annealed (softened) between the several forming cycles needed to produce the final cavity shape. See Fig. 11-35.

In many cases the punch is used to form several identical cavities in the same mold block. When these cavities are connected with runners and gates several identical parts may be produced in the same mold.

STAKING AND RIVETING

Staking and riveting are assembly operations that will be discussed in more detail in Chapter 29. They are, however, also cold forming practices.

Riveting involves placing a rivet through holes in two or more pieces. The head of the rivet is placed in a shaped cavity in the anvil portion of a press or hammer. A punch upsets (squeezes) and shapes the shank end on the rivet. The final shape is often the same as the original head. This action also draws together the two pieces being assembled. An example of riveting is shown in Fig. 11-36.

Staking is similar to riveting except a separate fastener (a rivet) is not used. Instead a portion of one part extends through a hole in a second part. A punch compresses and spreads the extended portion. This action locks the two parts together. Staking, also shown in Fig. 11-36, is often used to permanently attach rods and shafts to support members.

THREAD ROLLING

Almost all threads made in large quantities are produced by cold forming. The technique used is called thread rolling. The process forms threads by displacement of material rather than cutting away unwanted stock.

There are two basic methods of producing cold formed threads. These methods, as shown in Fig. 11-37 are:

1. Reciprocating die method.
2. Cylindrical die method.

Reciprocating die thread forming uses two flat dies. One of the dies is fixed and the other movable. Each die is grooved along its length. These grooves match the pitch (threads per inch) and shape of the desired threads.

The threads are formed on a rod which is rolled between the stationary and the movable dies. Pressure between the dies causes the metal to deform. Part of the metal is squeezed to form the thread root. Other material is raised to form the crest of the thread.

Cylindrical die thread forming uses two or three grooved roll dies. The two-roll machine uses work rest and two rotating grooved rolls. The work is placed on the rest. The back cylindrical die is stationary and rotates. The front die rotates and is fed into the stock until the proper thread is formed.

The three-roll machine uses three rotating cylindrical dies. The stock is placed between them. All

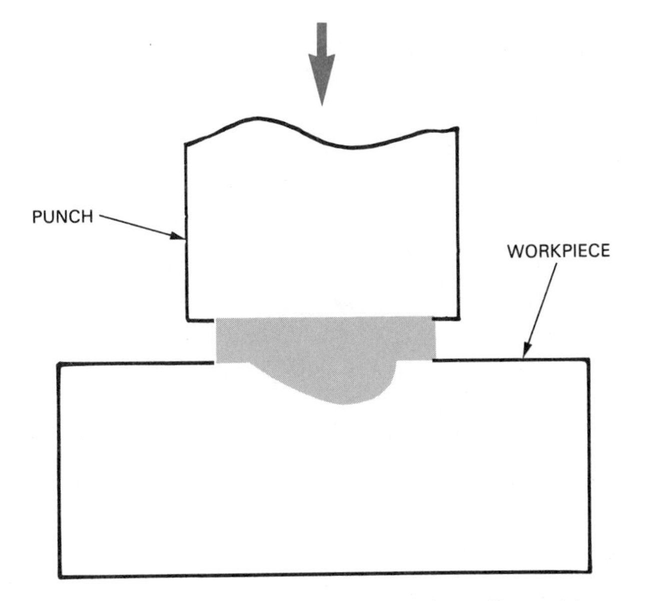

Fig. 11-35. Hobbing is used to produce die cavities.

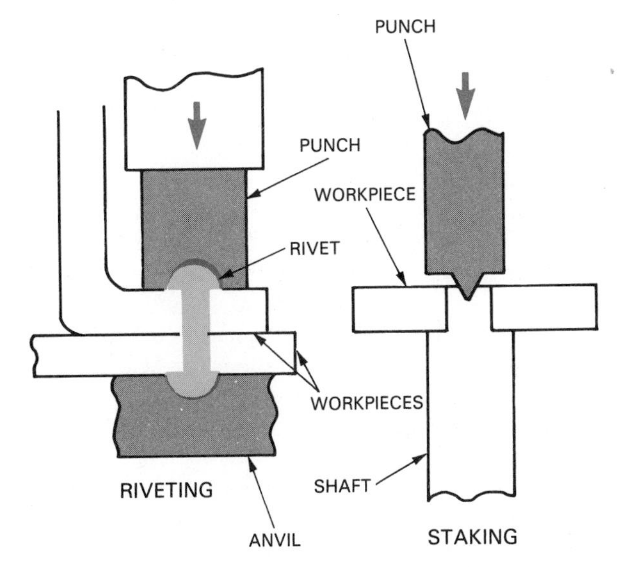

Fig. 11-36. Parts may be assembled using squeezing forces. Left. Riveting. Right. Staking.

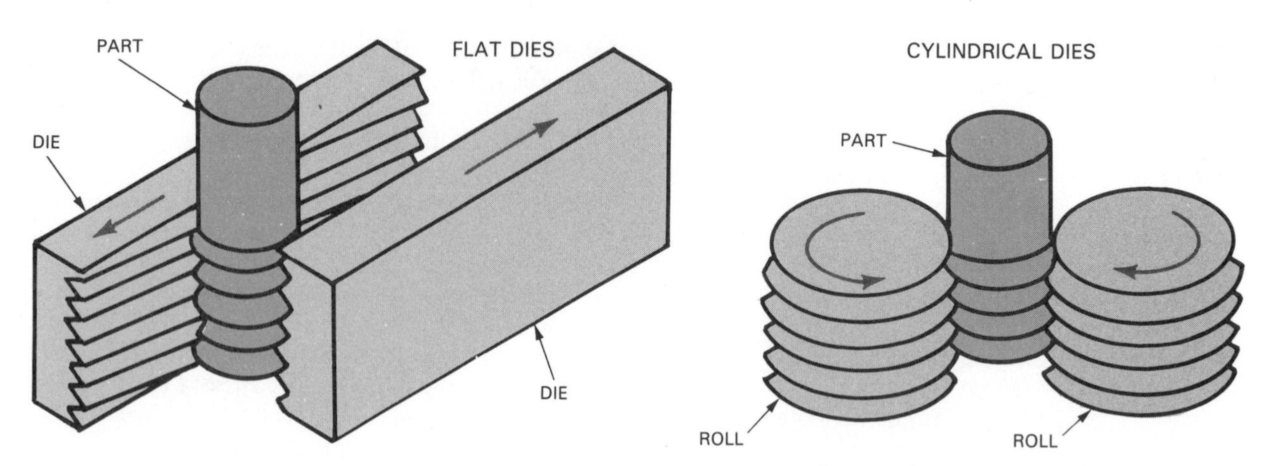

Fig. 11-37. Threads may be rolled on shafts using flat or cylindrical dies.

three rolls feed into the stock to form the thread.

Roll forming of threaded products is fast and economical. The threads produced have good form and excellent strength.

EXTRUSION

The basic concept involved in extrusion was discussed in Chapter 10 under the heading of hot extrusion. Cold extrusion is used mainly on soft ductile metals like tin, lead, and aluminum. In recent years mild steel has been added to the list. Collapsible tubes, flashlight battery cases, and other thin-wall products are cold extruded. The impact extrusion technique is most often used to produce these products.

Cold extrusion is often combined with cold heading to produce a number of parts. Typical of these are motor and compressor rotors and shafts and bolts. The body of the part is shaped through extrusion while the enlarged portion is cold headed.

STAMPING

Often a number of press operations are combined to produce a part from sheet stock. This combination of operations produces a part often called stamping. The term stamping groups together all presswork operations applied to sheet metal. It involves cutting and forming operations performed in a single or a series of operations. The operations can be completed in a single press closing with a mated die or with several press closings and multiple dies or a mated progressive die.

The production of a stamping, Fig. 11-38, may include both forming and cutting (separating) processes. The part may be cut to size and have holes, slots, notches, or features formed in it. These operations remove stock from the original workpiece and are therefore classified as separating activities. These processes are described in detail in Chapter 23. Additionally, the part may be bent, drawn, or otherwise formed. These operations have already been described in this chapter.

The production rate for stampings is high and generally the part needs no further processing. Vast quantities of sheet metal are converted into stampings. They find their way into almost every type of product. Machines, tools, buildings, vehicles, clothing, office machines, household appliances, packages, hardware, electrical devices, sporting goods, and toys all contain stampings. They range from automobile body parts to eyelets in shoes, air-

Fig. 11-38. This part is a stamping. Many cutting and forming operations were used to produce the holes, slots, beads, and bends.
(American Metal Stamping Assoc.)

craft control panels to hinges, Olympic medals to fan blades, dry cell cases to dental tools. The combination of shearing and forming operations to produce stamping is truly an important activity in all manufacturing.

SUMMARY

Cold forming of metal involves deforming metal to create a desired shape. The shape may be created by bending, drawing or squeezing operations. These operations may all be done using hot forming methods. However, cold forming produces a product with greater dimensional accuracy, smoother surfaces, and improved physical characteristics.

STUDY QUESTIONS — Chapter 11

1. List the major advantages and disadvantages of cold forming.
2. What are the three major types of cold forming operations?
3. List and briefly describe the four major types of bending operations.
4. What major machines are used for angle bending?
5. Describe three-roll forming and continuous roll forming.
6. Define drawing.
7. List and briefly describe the major cold drawing processes.
8. Describe shallow drawing, deep drawing, and flexible die drawing.
9. Describe the wire drawing process.
10. Describe the spinning process.
11. List and briefly describe the four major high velocity forming processes.
12. Briefly describe seven common squeezing operations.

Chapter 12

FORMING PLASTICS

Plastics or synthetic materials are becoming a more important manufacturing material. As their importance grows, their processing techniques are refined and new ones are developed.

A number of processes fit the forming definition. Plastics may be forced over, through, or around a shaping device to create the shape. Plastic forming may be done either hot or cold. However, most forming is done with heated materials.

The most common plastic forming techniques are:
1. Thermoforming.
2. Extrusion.
3. Blow molding.
4. Calendering.
5. Mechanical forming.

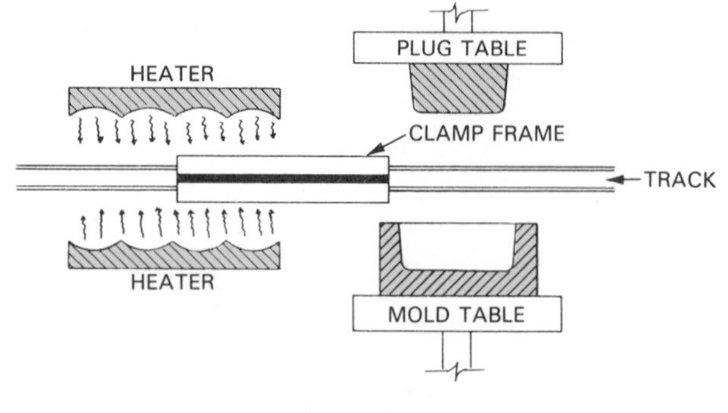

Fig. 12-1. A typical thermoforming process involves heating the material, left, then forming the product, right. (Dow Chemical Co.)

THERMOFORMING

The thermoforming process uses heat, force, and a shaping device to change flat sheet plastic into three-dimensional objects. Most thermoforming activities use thermoplastic sheets. Thermoforming changes flat sheets to three-dimensional objects.

The technique involves two basic steps, Fig. 12-1. First, the material is heated until it softens. Then it is forced into or over a mold using mechanical means or differences in air pressure.

In a few instances, such as the forming of skylights and aircraft windshields, the plastic material is formed without a mold to obtain better optic qualities. In these cases, only vacuum or compressed air and a holding fixture are used.

The material will faithfully reproduce the texture and shape of the mold as it cools. The cooled part is removed and trimmed to final size. Excess material trimmed from the molded part can be recycled through injection molding and extrusion operations.

Thermoforming, as shown in Fig. 12-2, can be used to produce items smaller than a drinking cup and larger than a boat hull. The wall thickness of the product may vary from a few thousandths of an inch up to 1/4 in.

Fig. 12-2. A continuous strip of drinking cups leaves an automatic thermoforming machine. (Brown Machine Co.)

ADVANTAGES OF THERMOFORMING

The thermoforming process technique has a number of advantages including:

1. Size of the product can be widely varied.
2. Low internal stresses and good physical properties in finished parts.
3. High production rates are common.
4. Parts can be predecorated, laminated, or coextruded to get different finishes and properties.
5. Material thickness and color changes are easy to make.
6. Parts can be made light, thin, and strong for packaging and other uses.
7. A number of different plastic materials can be formed with only slight changes in the operation.
8. Machinery and tooling are relatively inexpensive because of low processing pressures.

THERMOFORMING TECHNIQUES

A number of thermoforming techniques have been developed. Each technique is used to manufacture a product with slightly different characteristics. Some common thermoforming techniques are:

1. Standard vacuum forming.
2. Plug-assist vacuum forming.
3. Pressure bubble plug-assist vacuum forming.
4. Drape forming.
5. Matched mold forming.
6. Plug-assist pressure forming.
7. In-line preform forming.

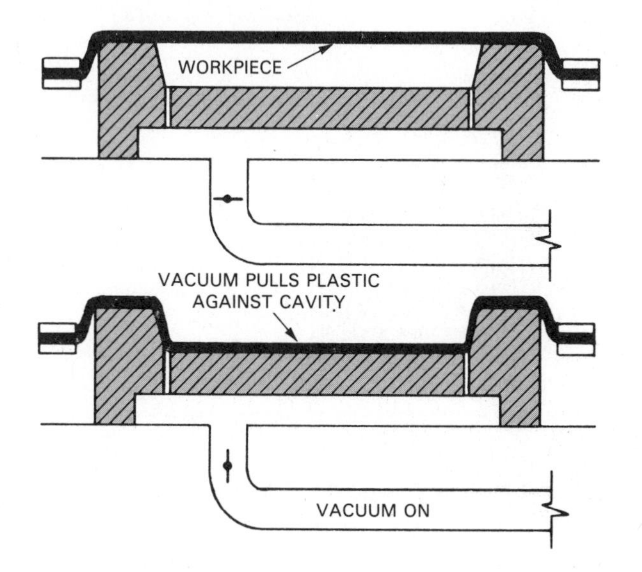

Fig. 12-3. Standard or straight vacuum forming uses no plug to help form plastic. (Dow Chemical Co.)

Standard vacuum forming

Standard vacuum forming, as shown in Fig. 12-3, is the simplest thermoforming method. It uses a mold with a recessed cavity. The cavity has small holes in the bottom through which the air can be evacuated from the mold.

The plastic sheet to be formed is clamped above the mold. Heat is applied to soften the plastic. When the plastic sheet reaches forming temperature, the air in the mold is pumped out, producing a vacuum. Atmospheric pressure above the plastic forces the material into the mold. Upon contact with the mold the plastic cools.

The product produced will vary in thickness. The material touching the mold last will be the thinnest. This is because that material has been stretching the longest and, therefore, has thinned the most.

Plug-assist vacuum forming

Plug-assist vacuum forming allows for deeper draws with more control of wall thickness. This technique involves heating and sealing a sheet of plastic across a mold cavity. A plug which has a shape similar but smaller than the cavity is fed downward. This action pre-stretches the material.

When the plug reaches its maximum downward travel air is pumped from the mold. A vacuum is produced and atmospheric pressure finishes the forming action.

The design of the plug largely determines the wall thickness of the part. The area which touches the plug first will be the thickest. Fig. 12-4 diagrams a typical plug-assist vacuum forming operation.

Pressure-bubble, plug-assist vacuum forming

This technique is very similar to the plug-assist method. The main exception, as seen in Fig. 12-5, is the bubble cycle. The plastic is heated and sealed to the mold. Pressure is then applied to the mold to evenly stretch the plastic upward.

The plug then moves downward in the first shaping stage. A vacuum is next pulled in the mold cavity to bring the part to final shape.

Drape forming

Drape forming draws the heated plastic over a male mold rather than into a mold cavity. The plastic is clamped and heated to the proper forming temperature. It is then lowered over the mold, as seen in Fig. 12-6. In some cases the mold is forced upward into the plastic. As the plastic drapes around the mold, a vacuum is drawn to finish the forming action.

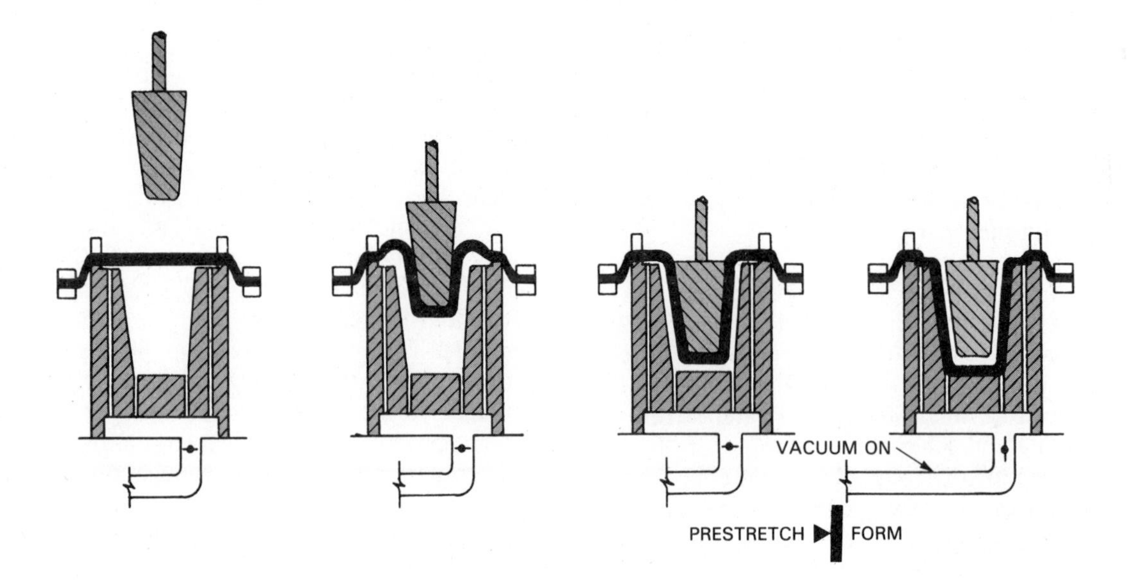

Fig. 12-4. Plug-assist vacuum forming combines action of plug and atmospheric pressure to mold part. (Dow Chemical Co.)

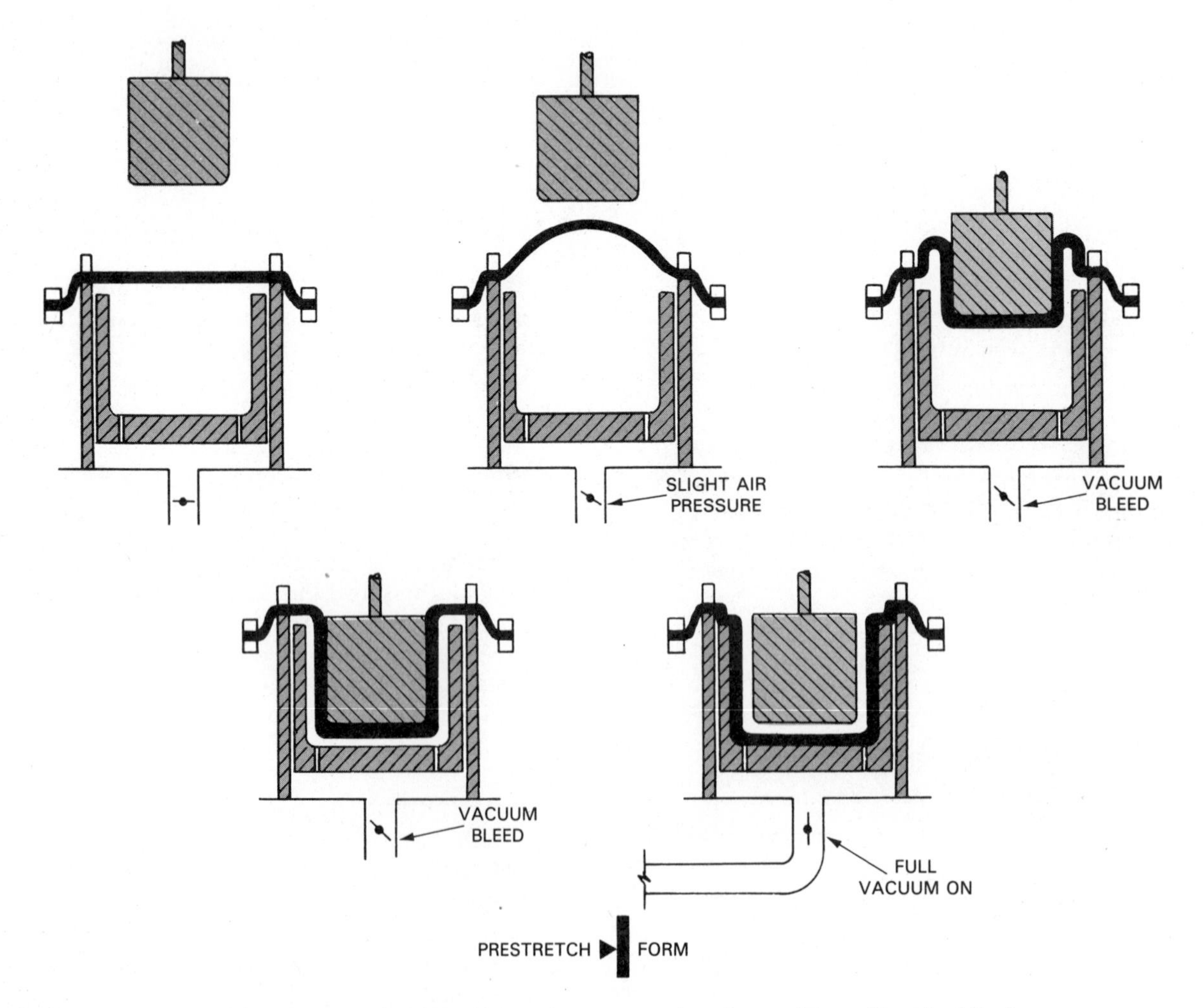

Fig. 12-5. Pressure-bubble plug assist vacuum forming. (Dow Chemical Co.)

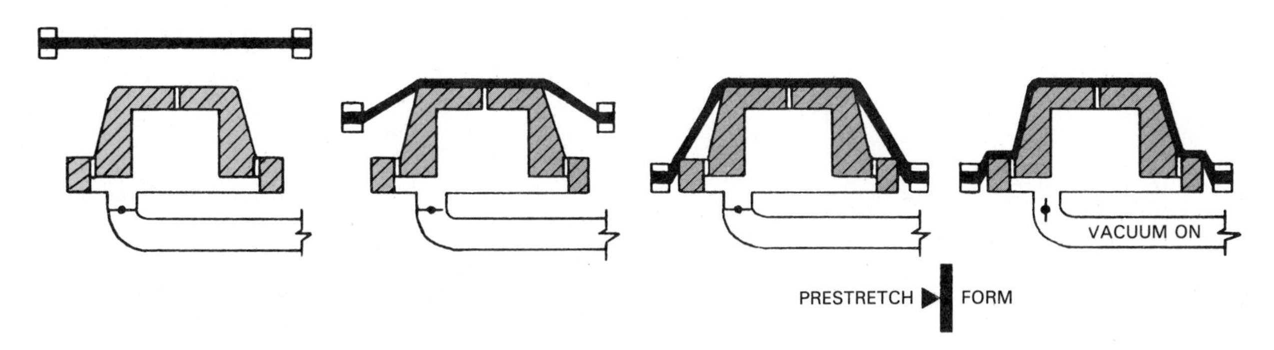

Fig. 12-6. Drape forming stretches plastic over a male mold. (Dow Chemical Co.)

Matched-mold forming

Matched-mold forming is similar to matched-die forming of metals. As shown in Fig. 12-7, the plastic is first heated. Then, matched dies close, forcing the heated sheet into shape.

Excellent dimensional accuracy and surface detail is possible with matched-mold forming. Also, wall thickness can be controlled by the design of the dies.

Plug-assist pressure forming

Plug-assist pressure forming is very similar to plug-assist vacuum forming. The difference lies in the method of causing the material to form against the die. In plug-assist pressure forming, the sheet is heated and a plug moves downward. However, instead of using a vacuum in the mold cavity, air pressure is introduced through the plug. This air pressure forces the sheet plastic out against the mold cavity.

During the pressure cycle, air in the mold cavity is vented to the outside. This action reduces resistance to the pressure molding segment of the operation.

In-line preform forming

In-line preform forming combines extrusion, compression mold, and thermoforming. In this technique, pellets are melted and extruded. Then the extrusion is cut or measured to an exact size preform (shape to be used in another process). This preform or *metered slug* is placed in a compression mold cavity.

The compression mold produces a molten disc with a lip or rim. This molded item, a preform for the next operation, is cooled until the proper temperature for thermoforming is reached.

The compression molded preform advances to the thermoforming stage. Here, using plug-assist, pressure and vacuum thermoforming principles, the final product is produced. Air from the plug side and vacuum from the cavity side insure high quality products.

In-line preform forming is used to produce lipped tub and cup-like containers. Other thermoforming processes which are modifications of the types already described are shown in Fig. 12-8. These include:

1. *Vacuum snap-back.* The heated, sagging sheet is drawn upward around the mold.
2. *Pressure bubble vacuum snap back.* The heated sheet is formed into a bubble; then a plug is inserted. A vacuum draws the sheet to the plug.

Thermoformed products

A wide variety of sheet plastic materials may be changed into useful items. Typically, the work is done

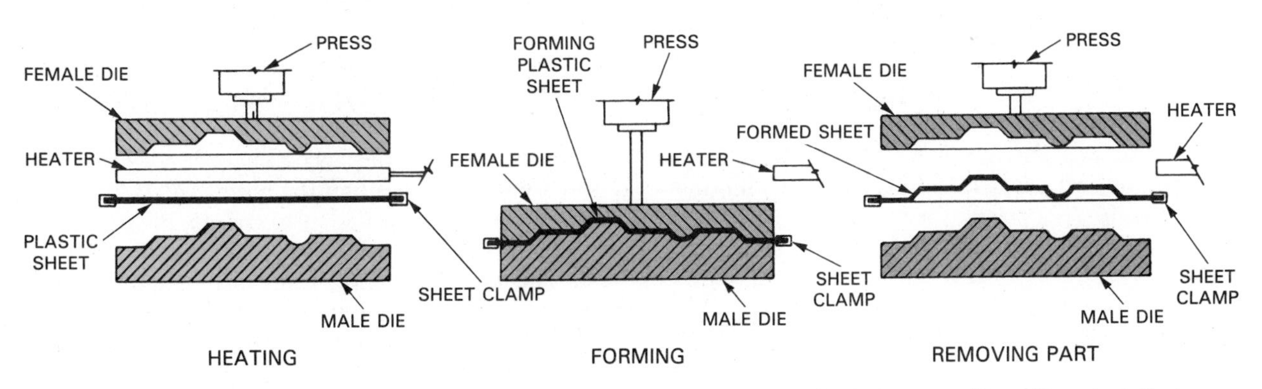

Fig. 12-7. Matched-mold or die forming forces heated sheet to desired shape. (Dow Chemical Co.)

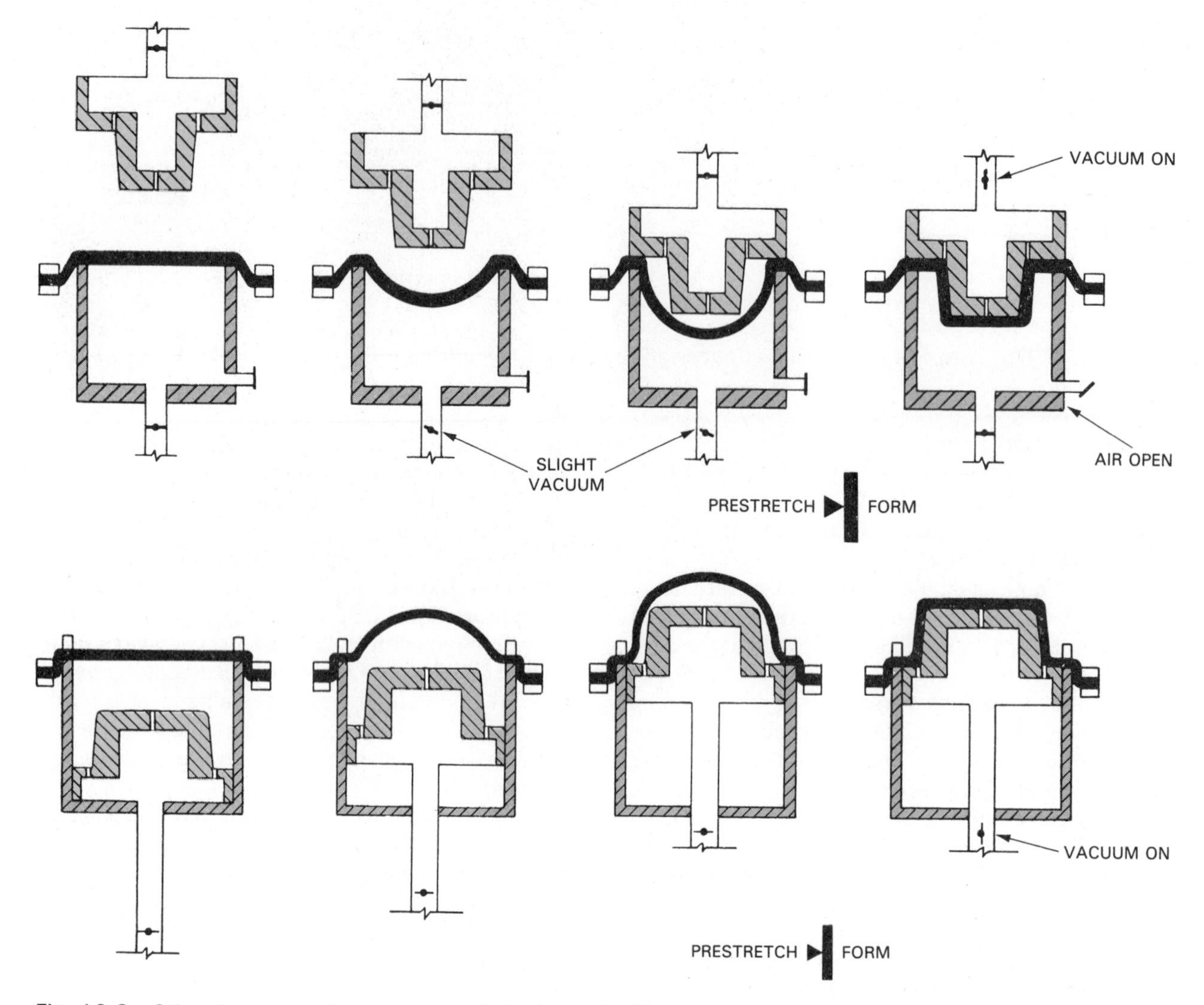

Fig. 12-8. Other thermoforming methods include these. A—Snapback forming. B—Pressure bubble vacuum snap-back or air-slip forming. (Dow Chemical Co.)

on machines like the one pictured in Fig. 12-9.

Among the products produced by these machines are automotive parts, luggage, toys, signs, displays, tote trays, cups, protective covers, housings, guards, picnic ware, and packages.

EXTRUSION

Extrusion was described in metal-forming chapters as using pressure to force a softened material through a shaping die. As a plastic-forming technique, extrusion is used mainly to form thermoplastic materials. See Fig. 12-10.

The process involves:

1. Compounding and pelletizing the plastic. Plastic material is mixed with coloring agents and other additives.
2. Heating the material to the proper plasticity.
3. Forcing the material through the die.
4. Cooling the material.

A typical plastic extruder, as shown in Fig. 12-11, contains a power unit, a hopper to feed raw plastic into the machine, a barrel containing a continuous fed screw, some kind of heating element, and a die holder or adapter.

Plastic is fed into the machine at one end. The screw draws the material along the hardened steel barrel. The barrel may be heated by electricity, oil, or steam. The screw shears the plastic as it moves along the barrel. The shearing action generates enough heat to soften the plastic material. The viscous plastic is delivered as a homogeneous mixture to the die head. The material is forced through the die as additional material is being delivered by the screw.

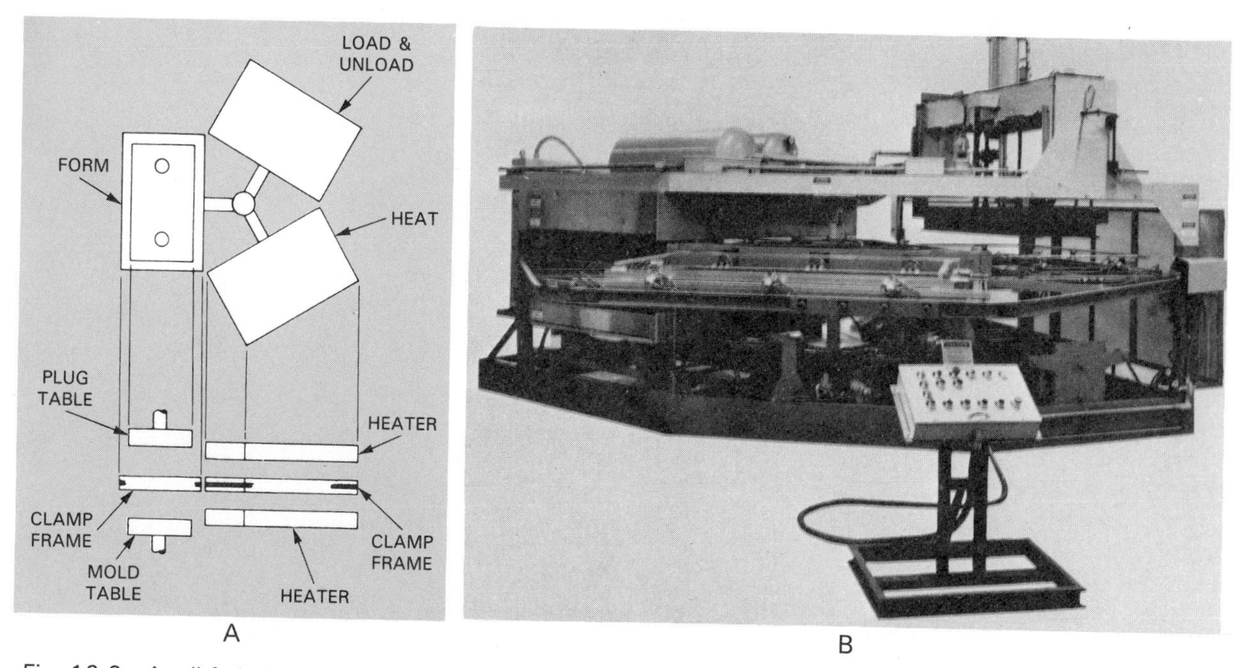

Fig. 12-9. A roll-fed, three-station thermoformer. A—Simple schematic of machine. B—Machine forms large parts from sheet material. (Brown Machine Co.)

Fig. 12-10. Plastic window parts are one product made by extrusion. (Crane Plastics Co.)

The special cross-sectional shapes and lengths possible from extrusion are almost limitless, Fig. 12-12. In addition, sheet material, plastic film, pipe, and tubing may be produced using the extrusion technique. Also, extrusion may be used to coat other materials. Fig. 12-13 shows a technique for extruding plastic insulation materials around electrical wire.

BLOW MOLDING

In blow molding, a softened plastic tube is inflated to fill a cavity. The operation requires only a small opening into which air can be blown. This allows for

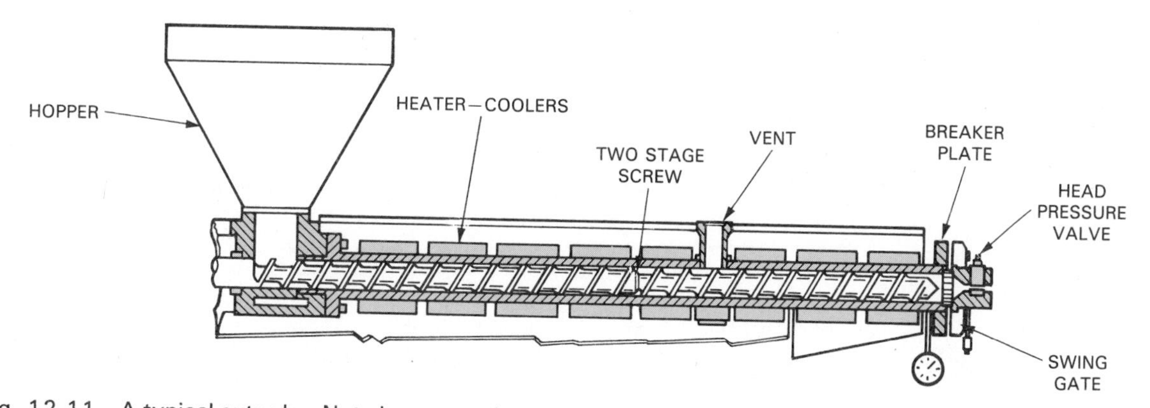

Fig. 12-11. A typical extruder. Note heater-cooler elements. Screw compresses and heats plastics while moving material to extrusion head. (Dow Chemical Co.)

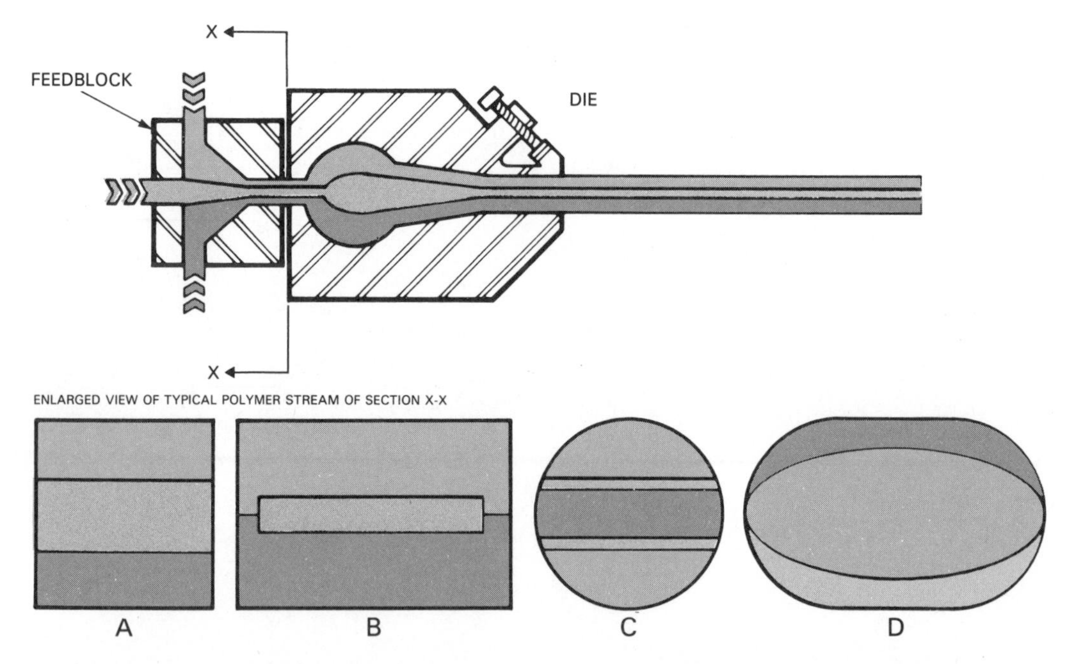

Fig. 12-12. An extrusion system (co-extrusion) to produce a multi-layer material. Top. Schematic arrangement of extrusion machine. Bottom. Cross-sectional views of feedblock. The resins can be delivered through a single feedblock. (Dow Chemical Co.)

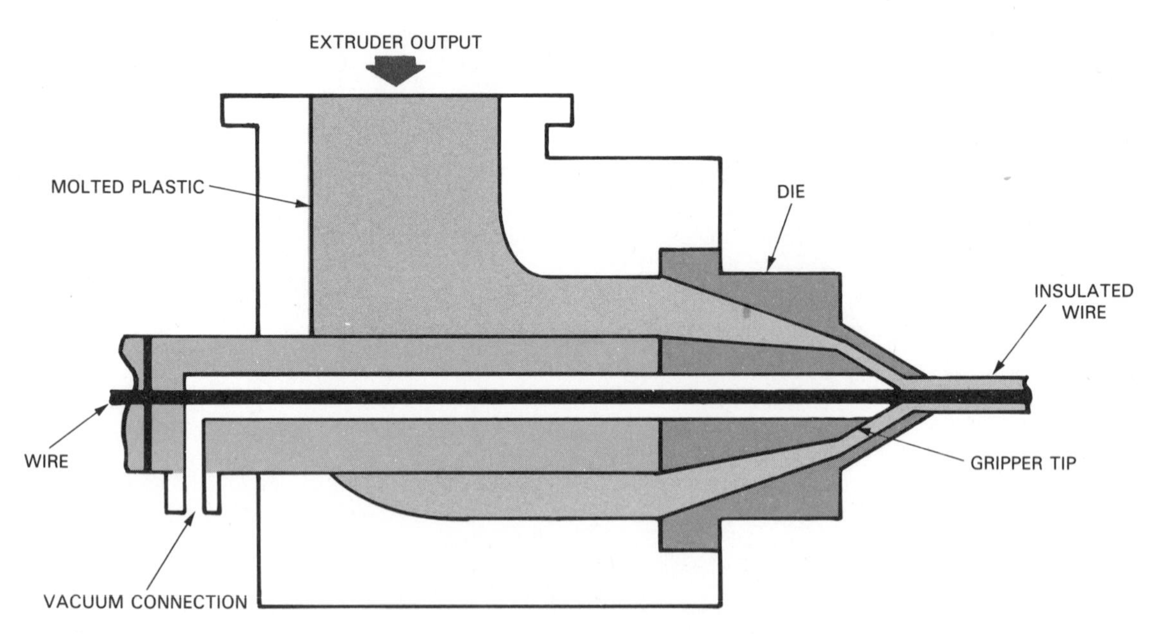

Fig. 12-13. System to extrude insulation on wire. Vacuum draws plastic tight around wire.

the rapid manufacture of products which have an almost solid shell.

Blow molding is used to produce products from thermoplastic materials. The technique is used to manufacture bottles for cosmetics, detergents, pharmaceuticals, beverages, chemicals, and numerous other products. Also automotive parts, hot water bottles, liners for drums and pails, lighting globes, gasoline tanks, and self supporting 55 gallon drums are blow molded.

Blow molding is done using one of two major techniques. These are:

1. Extrusion blow molding.
2. Injection blow molding.

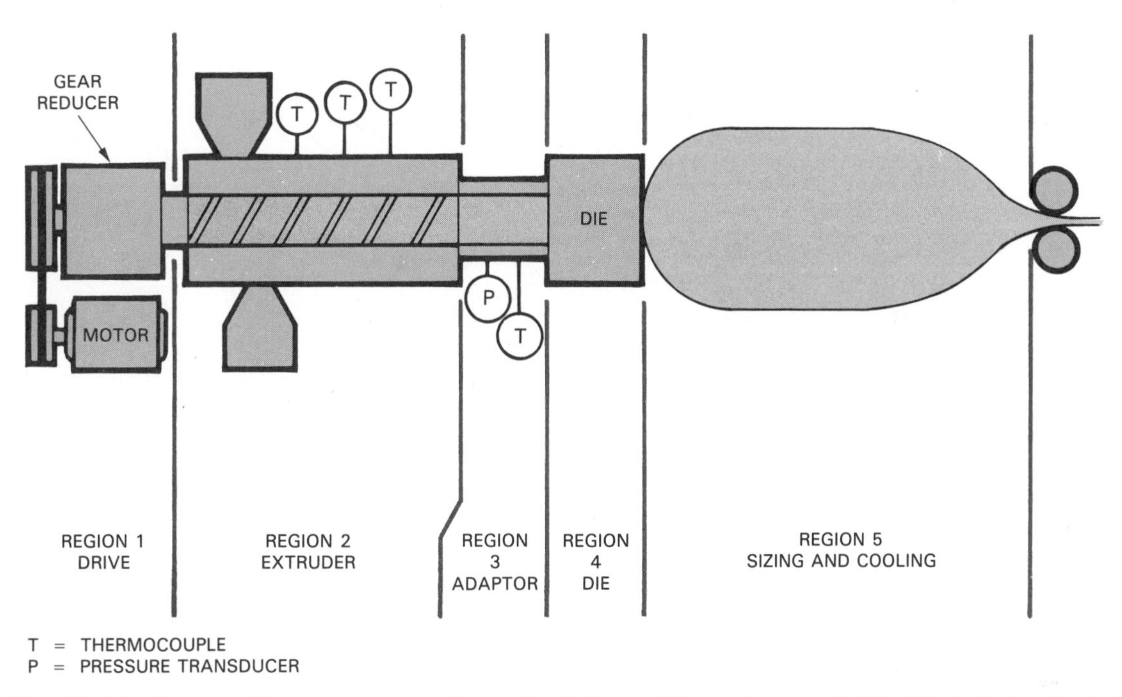

Fig. 12-14. An extrusion system to produce polyethylene tubing or bag stock. The film is first extruded then blown into thin tubing. (Dow Chemical Co.)

EXTRUSION BLOW MOLDING

Extrusion blow molding is the simpler of the two techniques for creating hollow bodied thermoplastic products. The process has five major steps.

1. *A hollow tube of molten plastic is formed.* This tube is called a parison. The parison is formed by the extrusion process which converts solid thermoplastic material into a softened material. The material is then forced through a die to form the hollow tube.
2. *The parison is enclosed by the mold.* After the hollow tube of plastic is formed it must be positioned between the open mold. The mold is then closed and clamped. This action pinches off the parison at the top and bottom. The clamping pressure must be strong enough to hold the parison in position and resist the pressure used during the molding cycle. The clamp also opens and closes the mold for rapid loading and unloading.
3. *The clamped parison is expanded in the mold.* Air is blown into the top of the pinched off parison. The pressure causes the soft plastic material to expand against the mold sides. Air pressure of about 100 psi (689 kPa) is used.
4. *The expanded part is cooled.* The heat contained in the plastic material is transferred to the mold. It is then carried away by circulating water or other liquids. The part is kept in the mold until it is cool enough to retain its shape.
5. *The finished product is removed from the mold.* Extra material which was pinched in the mold is trimmed off. The excess material includes bottom tabs, neck trim, and any flash (material blown between mold parts). The trimmings can be recycled.

Fig. 12-14 is a simplified drawing of an extrusion blow molding machine. Machines of many sizes and with various clamping systems are available.

INJECTION BLOW MOLDING

Injection blow molding, as seen in Fig. 12-15, has three major stages.

1. Parison forming.
2. Product molding.
3. Product ejection.

The first stage is the injection molding process. A parison is formed by injecting plastic around a hollow core pin or rod. Also, the container neck is formed at this stage.

The molded parison is then transferred to the blow molding station by the core pin. The parison is positioned in the product mold. Air is blown through the core pin to expand the parison into its final shape.

The final product is cooled and transferred to the ejection station. There the product is removed from the core pin.

Injection blow molding has several advantages over

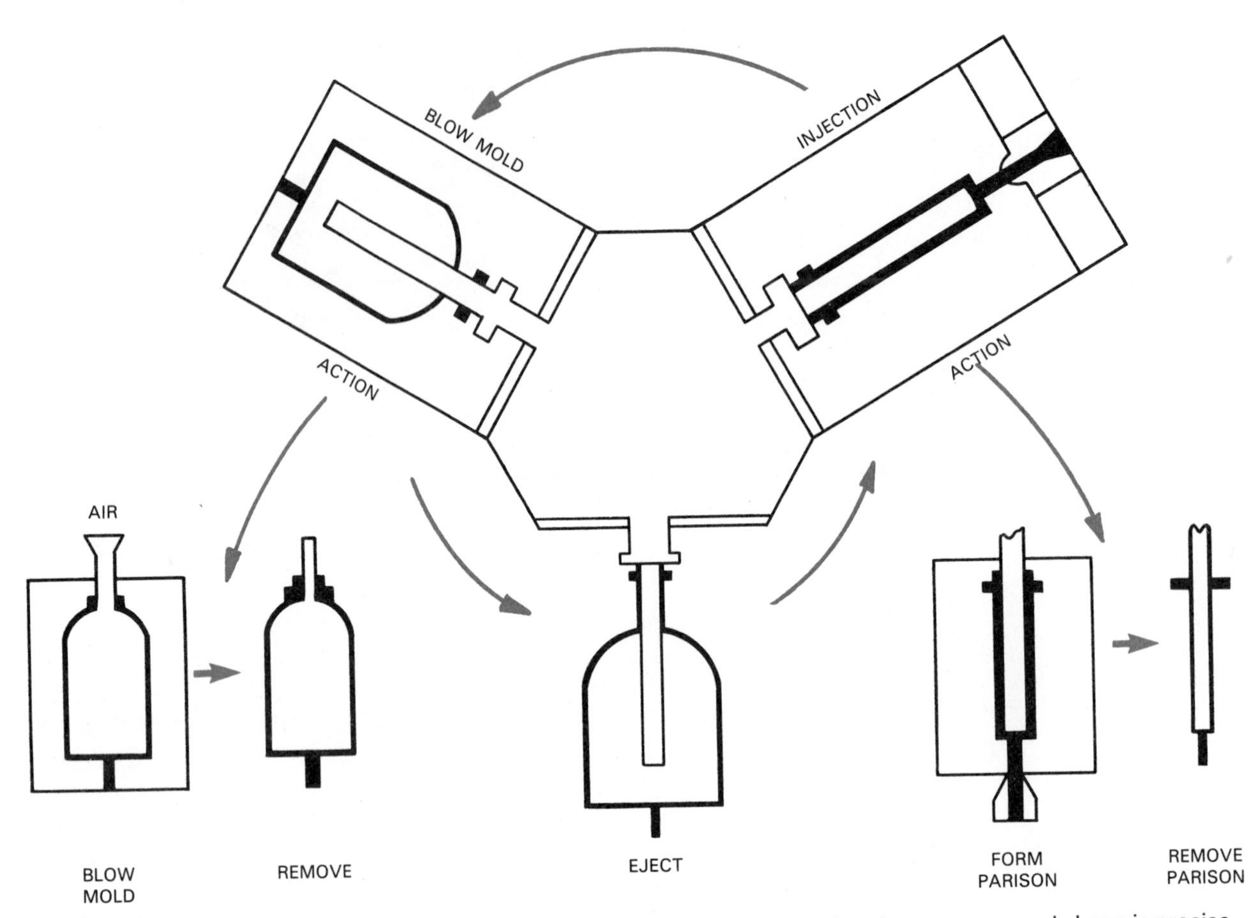

Fig. 12-15. A three-station injection-blow molding operation. Finished product has no scrap and shape is precise.

extrusion blow molding.

1. The product is a scrap-free, complete item when it is ejected.
2. Closer tolerances can be held.
3. Neck shapes are more accurate.
4. Internal and external finishes are superior.
5. Bottle weights are closely controlled.

CALENDERING

Calendering produces thin sheet material or coats a material with plastic. The technique is a continuous forming process somewhat like extrusion. However, calendering uses rolls to form and shape the material. Extrusion uses a die. Fig. 12-16 shows the calendering process applied to forming sheet material.

A widely used application for calendering, as seen in Fig. 12-17, is to coat fabrics and paper with plastic materials. Plastic materials are evenly spread on a roller, then transferred to the fabric or paper stock. Pressure is exerted between the fabric or paper feed roll and the plastic feed rolls. This pressure causes the plastic to penetrate and adhere to the fabric or paper. The thickness of the plastic coating will depend upon the space between the various plastic forming rolls in the calendering machine. The resulting material is widely used for upholstery, tablecloths, draperies, printing press blankets, flooring, and numerous other products.

MECHANICAL FORMING

Several mechanical forming techniques for plastic materials have been adapted from metal-forming operations. Two basic mechanical forming methods that are used for forming plastic materials are:

1. Forging (solid-state forming).
2. Stamping and deep drawing.

FORGING

There has been an increased use of mechanical and hydraulic presses in the forming of thermoplastic materials. The processes strongly resemble press forging of metals. For this reason, the process for plastics is often called forging. However, the process is not

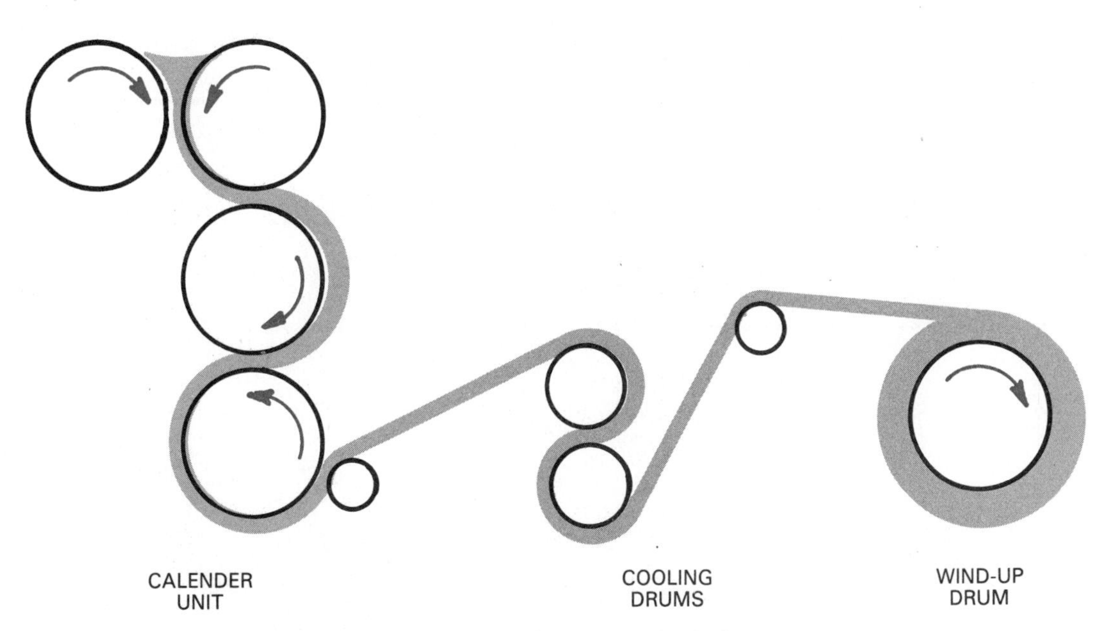

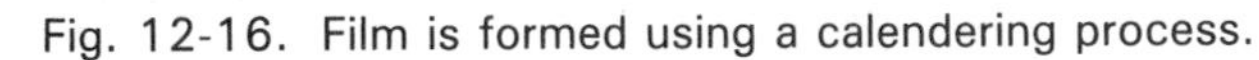
Fig. 12-16. Film is formed using a calendering process.

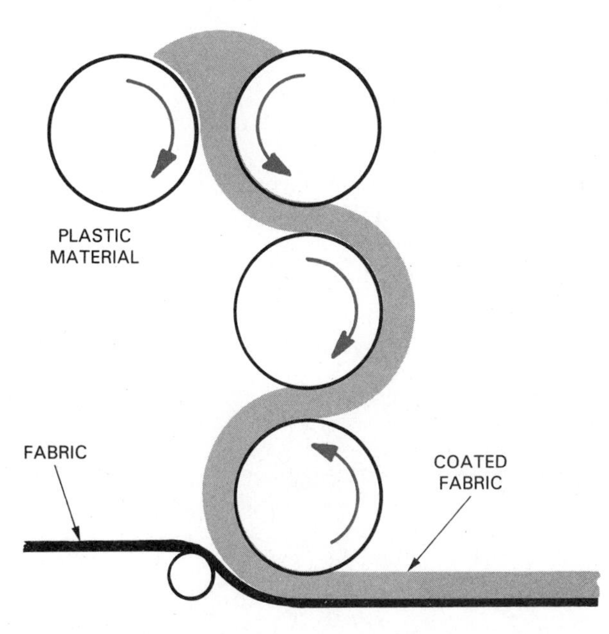

Fig. 12-17. Fabric coating uses a calendering process.

forging but reforming. Therefore, many individuals in the plastics field call the process solid-state forming.

By either name, the process involves preheating a blank of thermoplastic material. The material is placed inside the cavity area of a set of matched dies. The dies are closed by a press. Pressure is applied and the material deforms into the shape of the die cavity. The pressure is maintained after the plastic fills the cavity. This extra pressure time, called *dwell time,* allows the plastic to cool. The dies are then opened and the reformed part is ejected. Fig. 12-18 diagrams the process.

Forging of plastic materials have several important advantages including:

1. Thick-section plastic products can be produced with greater ease.
2. Heavier molecular weight plastics can be formed.
3. Physical and mechanical properties are improved.
4. Tooling and equipment costs are relatively low.
5. Metal forging presses can be used.

Forging of plastic parts has had limited application. The technique is basically used to form high molecular weight polyethylene and polypropylene materials. The products manufactured include gears, pump parts, textile weaving machine parts, and snowmobile parts. In most cases the parts produced must withstand a high degree of wear.

STAMPING AND DEEP DRAWING

Stamping and deep drawing of plastic material is used to produce a product with uniform wall thickness. The process is used to form sheet material through the following steps:

1. The sheet is heated.
2. The heated sheet is placed between matched dies.
3. A press closes the dies.
4. The dies form the part.
5. The part is cooled.
6. The die opens and the part is ejected.

Rubber-pad and hydrostatic forming procedures

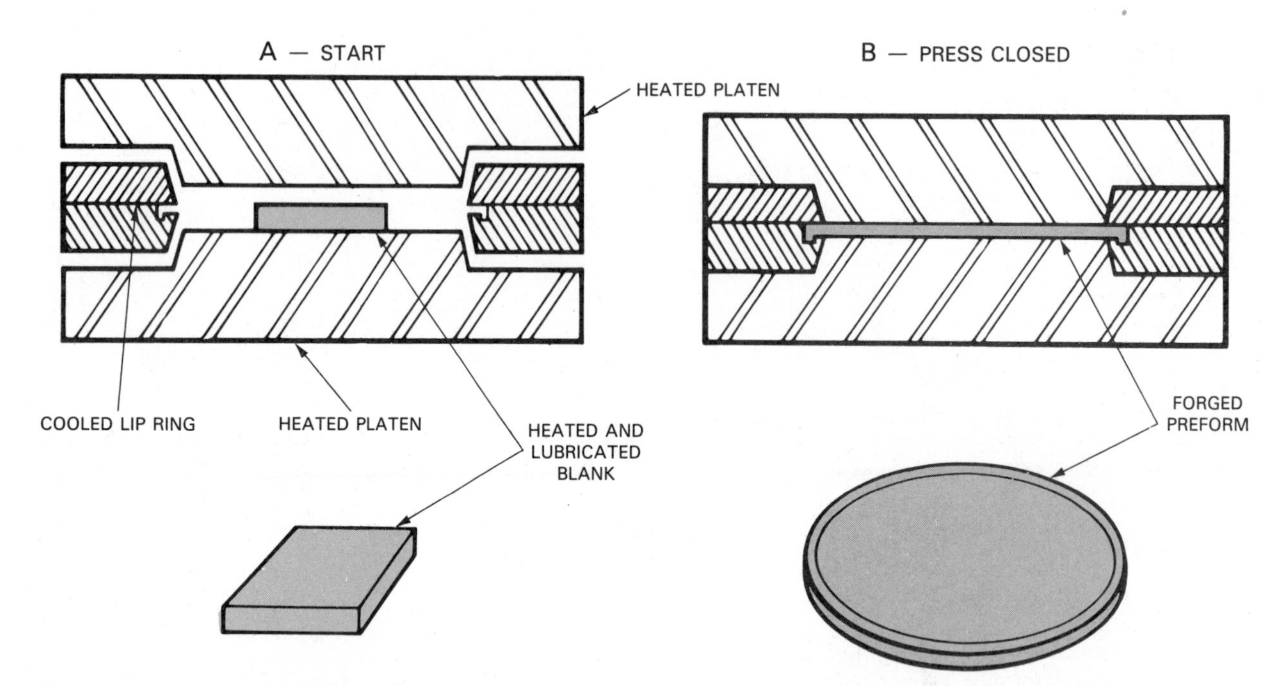

Fig. 12-18. Plastic forging is done on blanks heated to 230° to 280°F (110-138°C). Dow Chemical Co.)

described in Chapter 11 can also be used for plastic stamping.

Stamping of plastic parts is becoming an important process for producing automotive parts. This use usually forms glass reinforced polypropylene materials. These high strength materials are well suited for structural members. They, however, have a poor surface finish which limits their use to unexposed areas. Also deep drawing can be used to form deep tubs and trays for numerous uses.

SUMMARY

Forming plastic materials convert granulated and sheet plastic, extruded tubes, and preformed shapes into a number of products. Common processes used are thermoforming, extrusion, blow molding, calendering, and mechanical forming. Each of the processes uses a shaping device, force, and considers material temperature.

STUDY QUESTIONS — Chapter 12

1. What are the most common plastic forming processes?
2. Describe thermoforming.
3. What are the two major steps in thermoforming?
4. List the advantages of thermoforming.
5. List and briefly describe the major thermoforming techniques.
6. Describe extrusion.
7. What are the major steps in extruding plastic shapes?
8. Describe the major steps in extrusion blow molding and injection blow molding.
9. What is calendering?
10. Describe the major mechanical forming processes.

A high-speed heat-transfer label applicator attaches labels to polyethylene bottles. (Owens-Illinois)

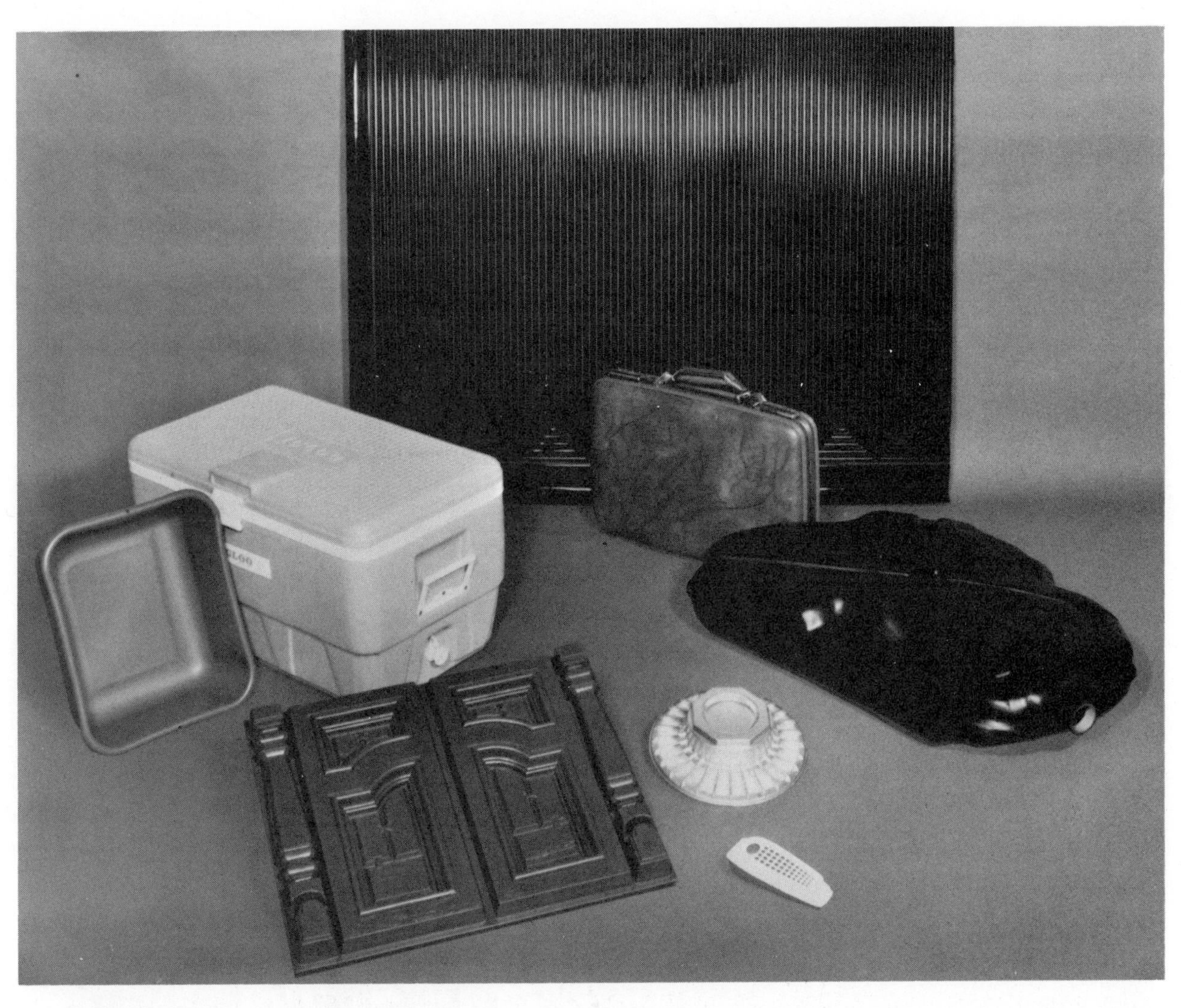

Typical formed plastic products include solar panel, luggage shells, furniture fronts, wheel covers, tubs, and ice chests. (Brown Leesona Corp.).

Chapter 13

FORMING CERAMIC MATERIALS

Many ceramic materials can be formed using pressure and a shaping device. Some of these materials are cold formed while others, like glass, must be hot formed. The typical forming techniques used for the major types of ceramic materials are:
1. Pressing.
2. Extruding.
3. Drawing.
4. Blow molding.
5. Jiggering.

PRESSING

Pressing material between mated dies is a common way to form clays and glass. The process uses the same basic principles as metal and plastic pressure forming techniques. They involve the use of accurately produced dies in a press to cause the material to take on the desired shape.

CERAMIC PRESSING

Pressing ceramic material to shape is widely used in the whiteware (dinnerware, electrical insulators, and art objects) industry and the structural clay products (tile and brick) industry. Typically, three major processes are used:
1. Dry pressing.
2. Ram pressing.
3. Isostatic pressing.

Dry pressing

Dry pressing, as the name applies, forms clay with little use of water. The moisture content of the clay is less than 10 percent. Special binders cause the clay to stick together. Also, release agents are added to keep the material from sticking to the molds.

In dry pressing the clay is placed in matched steel molds. A hydraulic press applies pressure up to 100,000 psi (689,000 kPa) to squeeze and form the material. The material, now a compact, dense, formed part, is ejected from the mold. Glazing and other finishing operations follow.

Dry pressing is used to produce flat, straight-sided objects. The production rates for these items may be increased by using multiple cavity molds and automatic machining. Fig. 13-1 shows a press for producing dry-pressed refractory bricks.

Fig. 13-1. A high speed dry press is used for producing refractory bricks. (Chisholm, Boyd and White Co.)

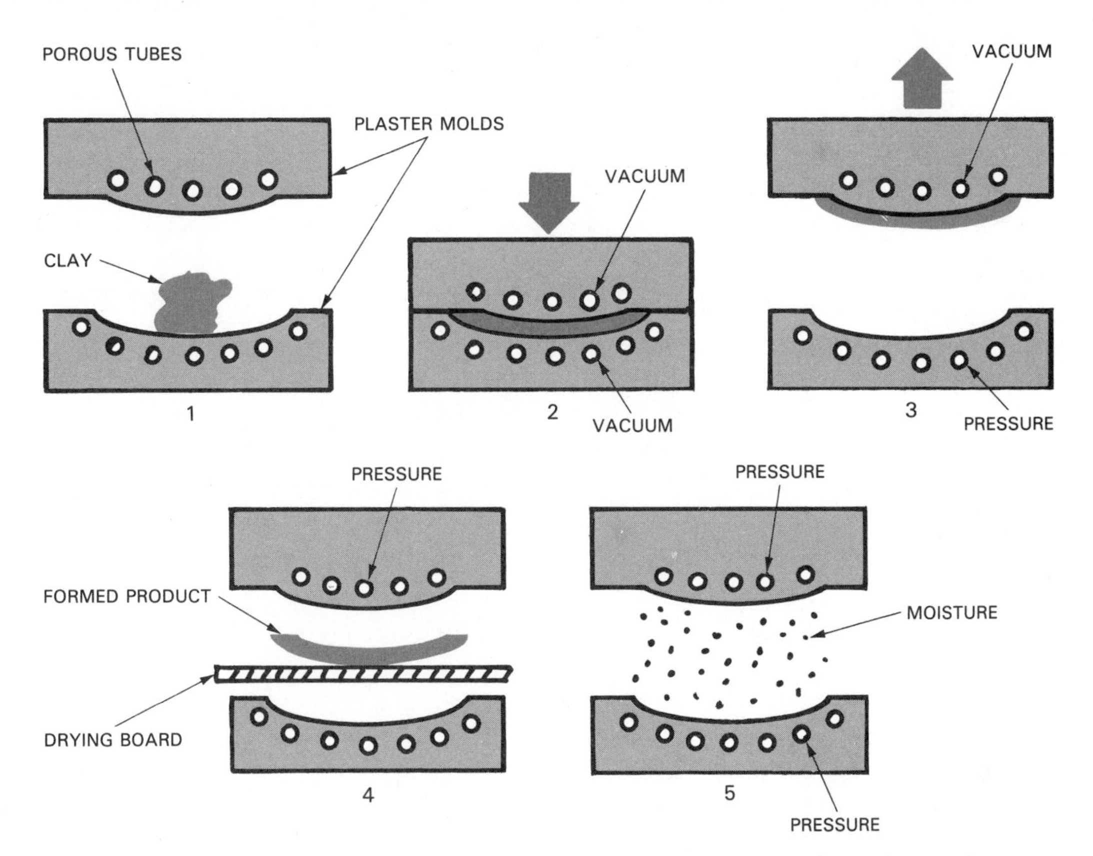

Fig. 13-2. The ram process uses both air pressure and vacuum to form clay products.

Ram pressing

Ram pressing is a technique developed, patented and named by Ram Inc. The process uses two hard plaster matched molds. These molds have porous air lines and passages built into them.

The five basic steps in the Ram process follow. Refer to Fig. 13-2 as steps are described.

1. A slug of soft clay is placed between the open mold halves.
2. A hydraulic press closes the mold squeezing and shaping the clay. Moisture in the clay is vacuumed away through air passages in the mold.
3. The vacuum in the lower mold releases as the mold is opened. The formed piece is held to the top mold by the vacuum in its air lines.
4. The formed piece is ejected from the top mold with a blast of low pressure air.
5. The moisture in the mold is blown out with air pressure through the porous air lines.

The Ram press system is used to produce oval shapes with draft in one direction. Platters, oval serving dishes, and other similarly shaped items are typical products of the Ram process.

Isostatic pressing

In isostatic pressing a high pressure chamber, like the one pictured in Fig. 13-3, molds dry clay materials. This technique uses a rubber mold into which the clay

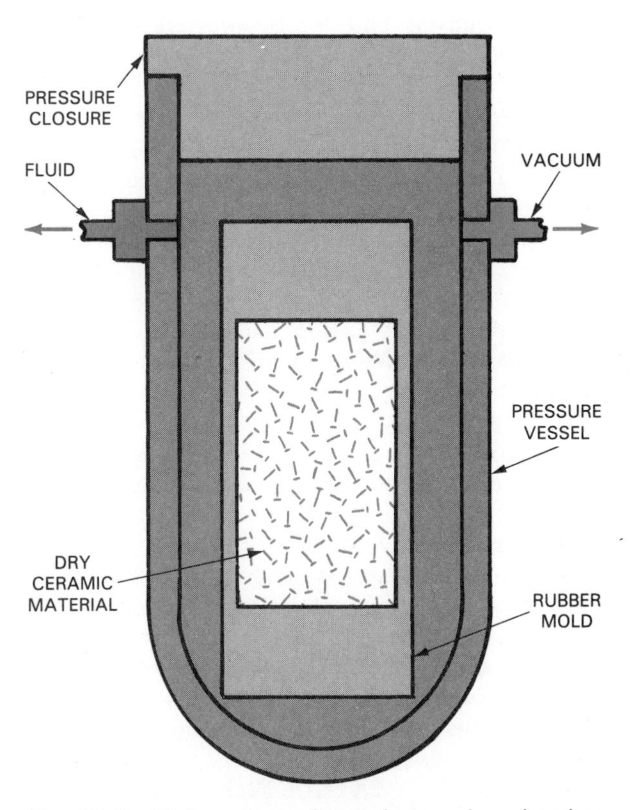

Fig. 13-3. High pressure isostatic pressing chamber combines rubber mold with extremely high pressure hydraulics.

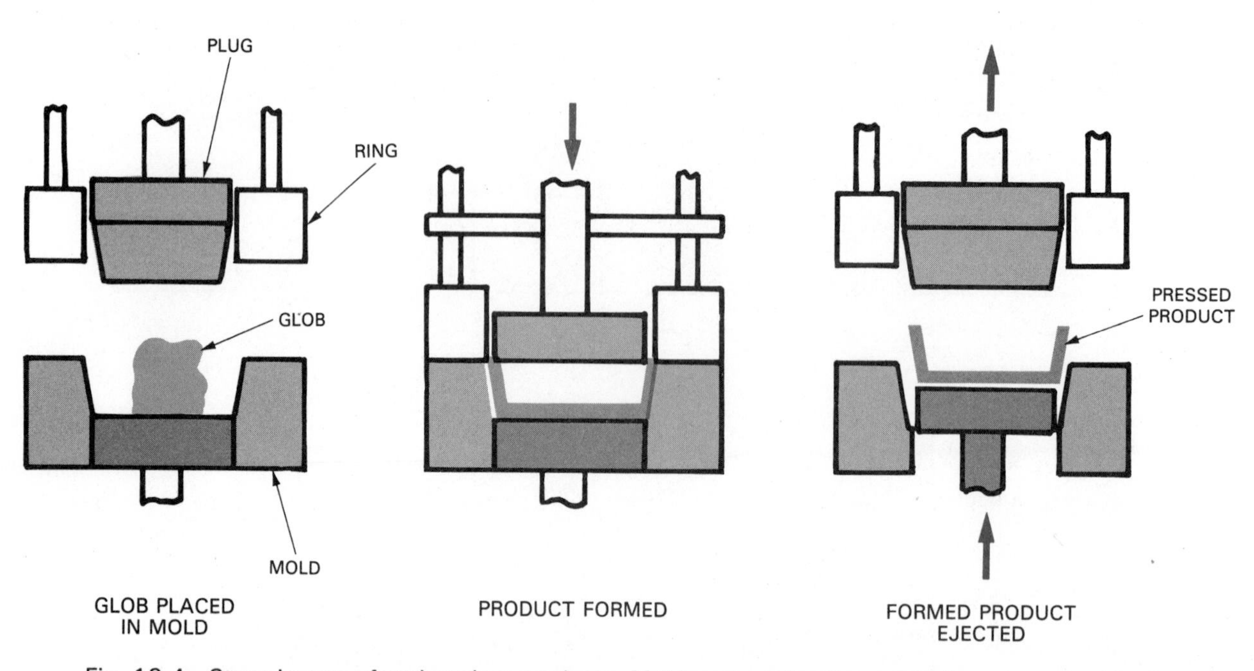

Fig. 13-4. Steps in press forming glass products. Molds are made of metal and are coated with oil.

material is placed. The mold is positioned in the chamber. The chamber is sealed and a liquid is pumped into it under high pressure. Typically, pressures range from 5000 to 20,000 psi (34,450-137,800 kPa).

The pressure compacts the clay material in the mold. After molding, the pressure is released and the chamber is opened. The mold is removed from the chamber and rolled off the formed part.

Isostatic pressing can produce parts with draft in two directions. One important product of this technique is ceramic spark plug shells.

GLASS PRESSING

Glass pressing, Fig. 13-4, uses cast-iron or steel molds coated with oil. The oil forms a carbon layer which helps separate the molded glass item from the mold.

Glass pressing involves placing a gob of molten glass between heated matched mold halves. The mold is closed and pressure is applied. Pressure and mold heat cause the glass to flow, filling the mold cavity. The mold is then opened, allowing the object to cool before removal. Typical products of this process are pie plates, baking dishes, and other flat or shallow forms.

EXTRUDING

Extruding clay products is very much like extruding metal and plastic shapes. The raw materials are placed in an extruding machine and forced through a shaping die. Common extruding machines are either auger fed or piston fed.

AUGER-FED EXTRUSION

Auger-fed extrusion machines, as shown in Fig. 13-5, first knead the clay to improve its workability. The kneaded clay is fed through a shredder into a vacuum chamber.

Called a *de-airing chamber,* it removes air holes and bubbles from the clay. This part of the extruder is sometimes called a *pugging machine.*

An auger in the vacuum chamber picks up the clay and forces it through the forming die. The die may form solid shapes or, with the aid of a mandrel, hollow (cored) shapes.

As the material leaves the die it is supported by a belt conveyor or moving carriage. While on the supporting device, the extruded shape is cut to desired lengths. The cutting is often done with one or more taut wires.

Typical products from auger extrusion are bricks, drain tile, roofing tile, and other structural clay or cement products, Fig. 13-6. All these products leave the extruder with a smooth surface. To complete their manufacture they may require a texturing operation as they leave the die. Bricks are given a face texturing while drain tile may have the manufacturer's name stamped in them.

Auger extruders are high production machines.

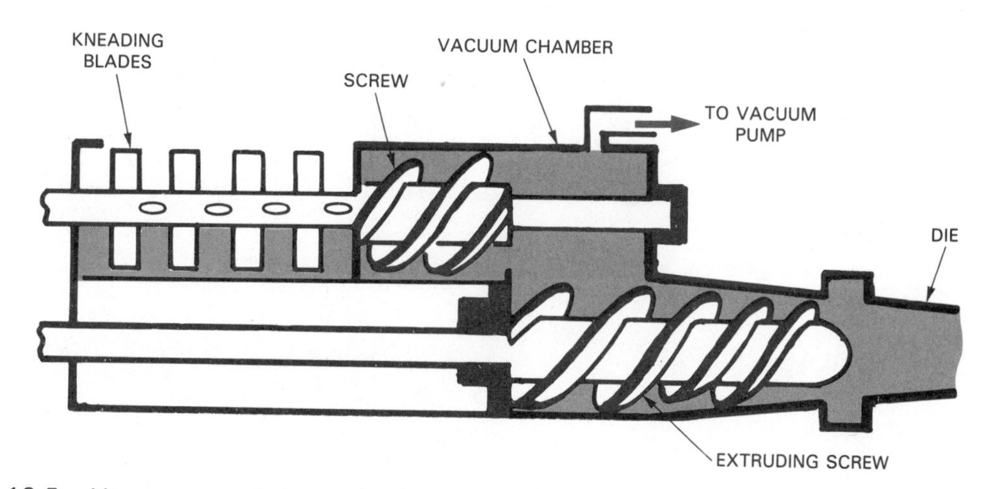

Fig. 13-5. Vacuum auger-fed extruder. Vacuum chamber removes air bubbles from kneaded clay.

Fig. 13-6. Concrete blocks are an example of extruded products.

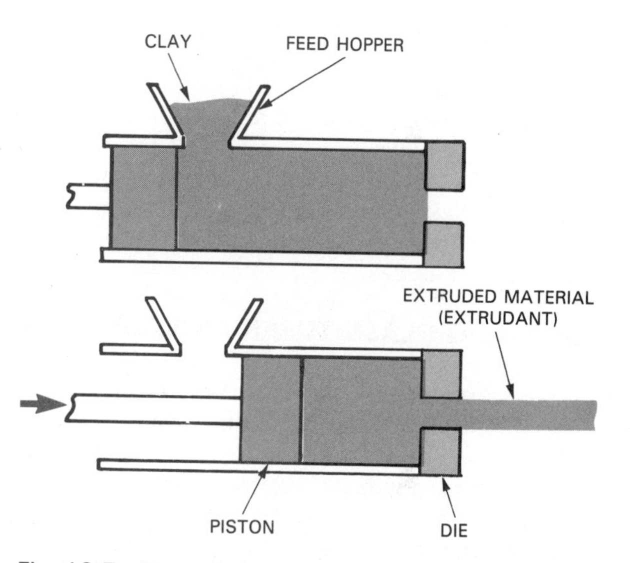

Fig. 13-7. Piston-fed extruder produces products with more uniform density than auger-fed extruder.

Many large models process up to 10 tons of clay per hour. Also, they can produce products up to 3 ft. in diameter.

PISTON-FED EXTRUSION

Piston-fed extrusion uses the reciprocating motion of a piston to produce a continuous extruded column. The machine, Fig. 13-7, uses clay from a pug mill or mixer. The clay is loaded in the chamber. The piston moves through the chamber compressing the clay. The compressed clay is forced through the forming die.

The piston extruder produces a more uniform product than auger-fed machines. The difference in density between the center and the outside of the extrudate (extruded column) is less than with auger extrusion.

Piston extrusion is widely used to form bell end sewer pipe. The pipe section is first extruded then forced into another shaping die. This second die forms the bell end of the pipe. The formed pipe is then trimmed, dried, and fired (heated to harden).

DRAWING

Drawing techniques are widely used in production of glass products. Sheet, rod, tube, and fiber glass are produced by various drawing methods.

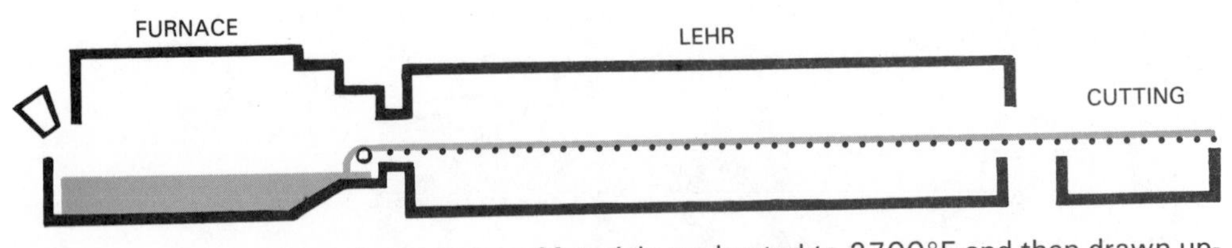

Fig. 13-8. Drawn sheet glass process. Materials are heated to 2700°F and then drawn upward into a sheet and onto horizontal rolls. (Libby-Owens-Ford Co.)

DRAWING SHEET GLASS

Sheet glass is formed in a continuous sheet by drawing molten glass from a glass furnace. As shown in Fig. 13-8, a feeder in the furnace draws the glass up and onto a flattening table. The glass hardens as it moves across the table. Next comes annealing in a *lehr* (annealing oven) to remove internal stress. The continuous sheet leaving the lehr is cut to standard sizes, Fig. 13-9. Sheet glass may also be produced by rolling molten glass, Fig. 13-10, or by floating the glass on a pool of molten tin as in Fig. 13-11. The float process, Fig. 13-12 is actually a casting process. The molten tin bath is the mold for the molten glass while the drawn and roll processes use rollers to size the sheets.

Fig. 13-9. A continuous ribbon of glass moves across delivery rolls. (Libby-Owens-Ford Co.)

DRAWING GLASS TUBES AND RODS

Glass tubes and rods are manufactured on the same machine. Glass is allowed to flow from the molten furnace onto a revolving mandrel. The mandrel is inclined so that the glass collects and flows along its length.

When the glass reaches the end of the mandrel it is hand drawn over a series of supporting pulleys until it reaches an insulated belt. The belt then pulls the rod at a uniform speed. The temperature of the glass and the speed at which the rod is pulled will determine its diameter.

Tubing is manufactured in the same manner except that air is blown through the forming mandrel. The air pressure keeps the glass from flowing together and forming a rod as it leaves the mandrel. The volume of air, the speed of drawing, and the temperature of the glass will establish the tube diameter and wall thickness.

Glass tubing and rods do not require annealing. They are, therefore, cut to length as they leave the drawing belt.

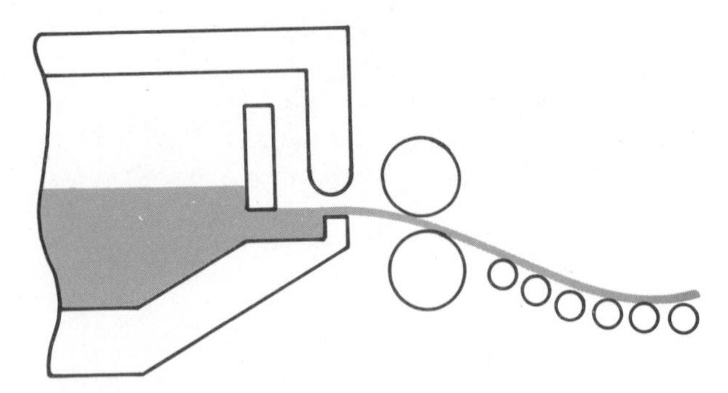

Fig. 13-10. Roll glass process. Large rollers squeeze glass into uniform sheet.

DRAWING GLASS FIBERS

Glass fibers are made from refined glass or glass marbles which are remelted. The marbles are often used because they can be inspected for impurities before the process is started. Also, marbles melt with fewer air bubbles.

Glass fibers are produced by two basic methods. See Fig. 13-13.

1. Continuous fiber method.
2. Discontinuous fiber method.

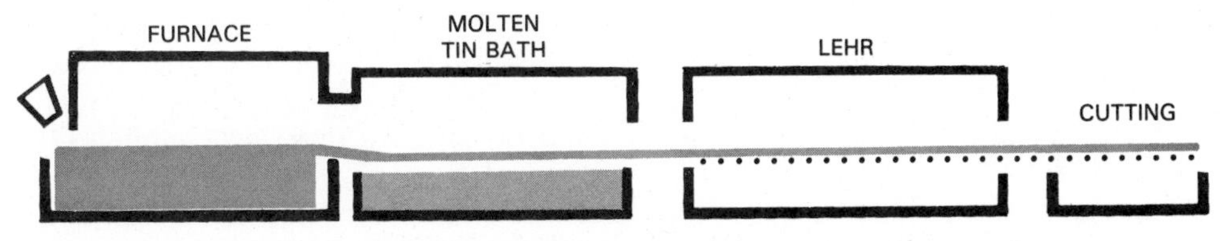

Fig. 13-11. In float glass process, molten glas is formed into a sheet while floating on a bed of molten tin. (Libby-Owens-Ford Co.)

Fig. 13-12. A view of the tin bath area helps illustrate the massive size of the float glass manufacturing equipment. (Libby-Owens-Ford Co.)

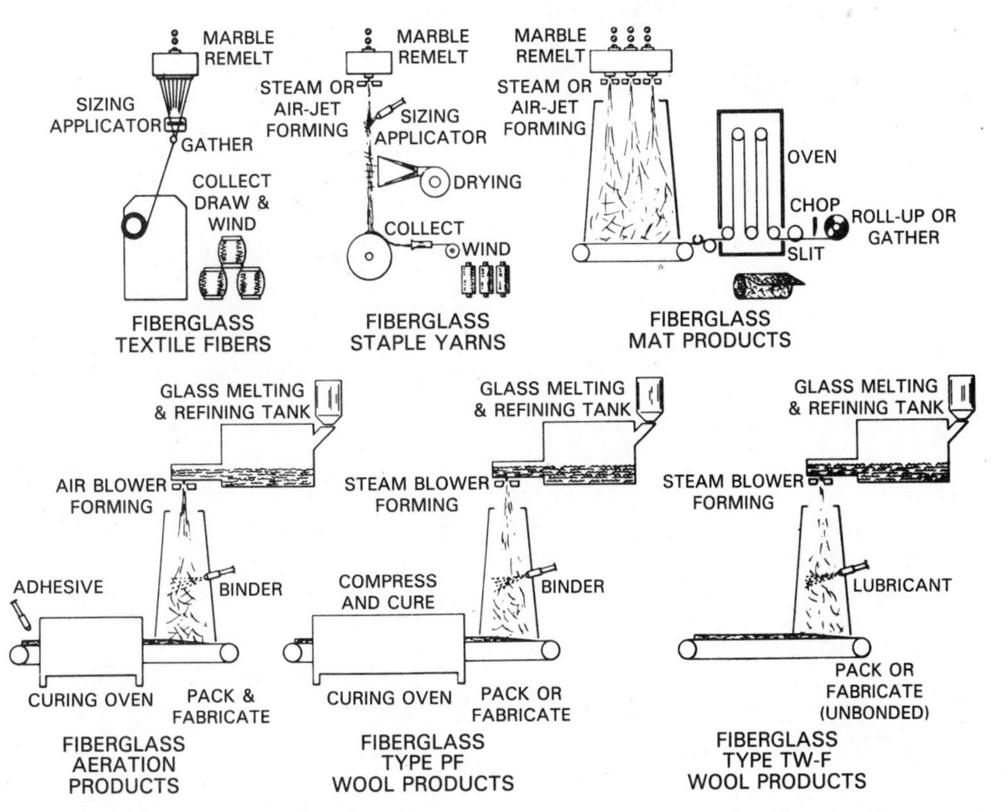

Fig. 13-13. Diagram of the methods of producing fiber products. (Owens-Corning Fiberglas Corp.)

Continuous glass fibers

Continuous glass fibers are produced by continuously melting glass marbles. The glass is melted in small platinum melting units (bushings) using high amounts of electric current.

The melted glass flows from small openings in the bottom of the bushings, Fig. 13-14. As the glass fibers form they are coated with a sizing. The sizing keeps the fiber from becoming scratched and weakened. The sized fibers are wound on rapidly revolving spools, as shown in Fig. 13-15.

Fig. 13-14. Continuous glass fiber forming bushing (top) and formed fiber. Bushing is heated by electric resistance to melt glass. (Owens Corning Fiberglas Corp.)

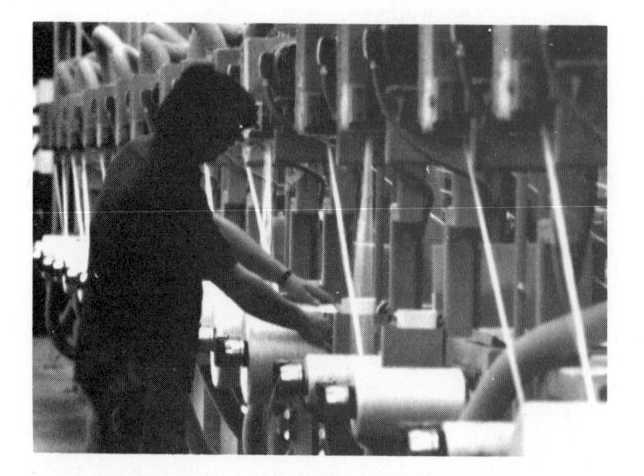

Fig. 13-15. Glass fibers are being wound on spools. (Owens-Corning Fiberglas Corp.)

Continuous glass fibers are used to produce products which must resist chemical corrosion, dampness, heat, and fire. These products include draperies, shower curtains, fireproof clothing, filter bags, and filter cloths.

Also, continuous glass fibers can be reinforced with plastic materials. The fibers are protected from the weakening effect of scratching by the plastic coating. Reinforced glass fibers are used in producing automotive body parts, shower enclosures, boats, fishing poles, hammer handles, golf club shafts, aircraft wings, vaulting poles to name but a new applications.

Discontinuous glass fibers

Discontinuous (short, separated) glass fibers are produced from streams of glass flowing from a melting unit, Fig. 13-16. The streams are hit with a high-speed jet of steam or air. The blast breaks up the streams of glass and forms short fibers. The ends of the fibers form into small lumps or *shots.* These shots usually have sharp tails which irritate the skin of individuals using the fibers.

Discontinuous glass fibers are often called glass wool or rock wool. Glass wool is widely used as insulation for buildings and appliances.

BLOW MOLDING

Blow molding is the basic process for the glass container industry. Millions of bottles, jars, and deep-shaped glass objects, like those shown in Fig. 13-17, are formed each year. These objects are formed

Fig. 13-16. Discontinuous glass fiber building insulation manufacture. Streams of glass are broken up by jet of air or steam. (Owens-Corning Fiberglas Corp.)

Fig. 13-17. Glass jars formed by blow molding are being conveyed from forming machines. (Glass Packaging Institute)

Fig. 13-18. A reheated parison is being placed in an open mold while a formed jar is being automatically removed. (Glass Packaging Institute)

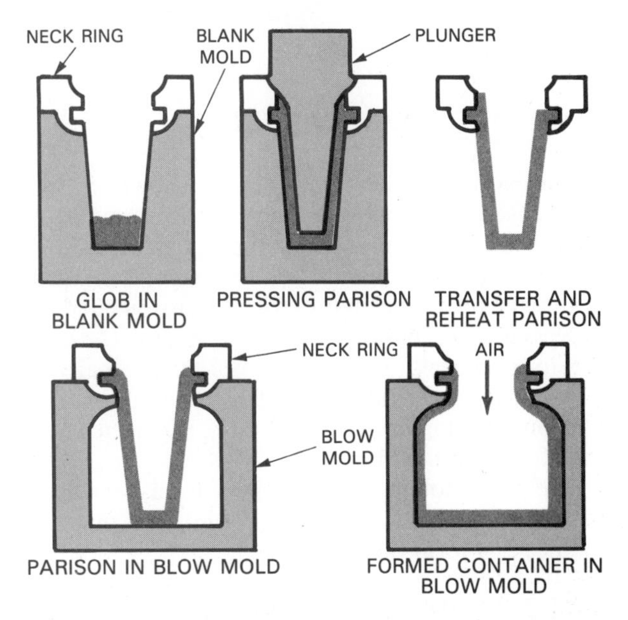

Fig. 13-19. Press and blow glass forming combines two forming processes.

through a basic three-step process using two sets of molds. These steps include:

1. A molten glass gob is formed into a temporary shape. This first shape is called a parison. Usually it is formed in a *blank* or *parison* mold.
2. The parison is reheated to molding temperature. It is then transferred to the *blow mold.*
3. The parison is placed in the mold. Air is blown into the center of the parison. The air forces the parison to expand to match the shape of the mold. The finished container is removed and the cycle repeats, Fig. 13-18.

Molds for glass blowing molding are made of a special cast iron or steel. They are heated to about 400°F (204°C) to improve the molding of the glass. Also, the molds are coated with oil or a mold paste so that the molded glass container can be removed.

The actual molding process is done by four different types of automatic machines.

1. Press and blow machine.
2. Blow and blow machine.
3. Vacuum and blow machine.
4. Ribbon machine.

PRESS AND BLOW MACHINES

The press and blow machines first forms a glass gob into a parison by pressing. The pressing step opens the center of the gob and forms the neck ring.

The partially formed container is transferred to the molding station by the neck ring. The blow mold is closed and air is blown into the parison through the neck opening. The mold is opened and the container is conveyed to an annealing oven (lehr). Fig. 13-19 shows the operation of a press and blow machine.

BLOW AND BLOW MACHINES

The blow and blow machines, as shown in Fig. 13-20, feed a gob into the blank (parison) mold. A blow head is placed over the mold and the gob is blown down into the mold. This action molds the neck of the container. The third step in the parison molding process involves blowing air up into the gob through the formed neck. This step completes the parison forming. Then the blank mold is turned over. The parison, hanging from the neck ring, is reheated and transferred to the blow mold.

The blow molding phase is the same as was used in the press and blow machines. Air is blown through the neck to form the container.

VACUUM AND BLOW MACHINES

In a vacuum and blow machine, Fig. 13-21, the mold is dipped into a container of molten glass. Vacuum draws the glass into the mold cavity. The glass is drawn upward until the neck of the container is formed.

The mold is then raised out of the pool of glass. A knife cuts off the glass and seals the bottom of the blank mold. Next, a blast of compressed air opens the neck of the parison.

The parison is removed and held by a neck ring. Another puff of air through the neck, elongates it in preparation for the next step.

The reheated parison is placed in a blow mold where it receives its final shape. The mold opens and the finished container is conveyed away from the machine.

RIBBON MACHINES

The ribbon blow molding machine uses a thin, continuous glass sheet about 3 in. wide in its operation. As seen in Fig. 13-22, the glass ribbon is formed between two water-cooled rolls. It is then fed under a set of moving blow heads. These blow heads form the parison without the use of a blank mold.

The blow head and molded parison is automatically enclosed in a blowing mold. Additional air is blown into the parison to form the final product. Ribbon molding is used to form incandescent lamp globes, tumblers, and Christmas ornaments at speeds over 1000 pieces per minute.

Of course, the actual molding speed will vary in all the processes. The size of the object being molded and its wall thickness are major factors in molding speeds.

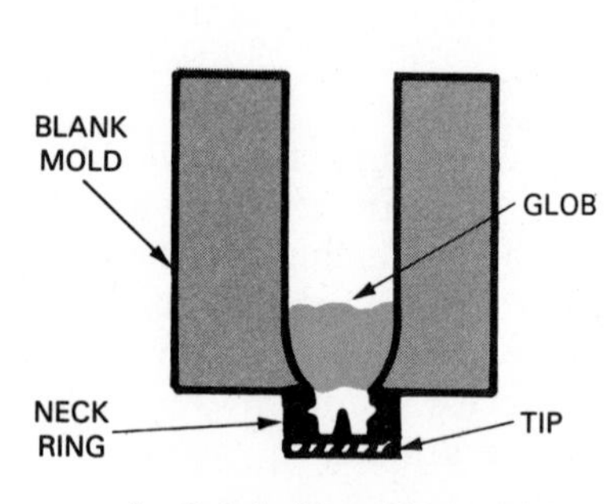

1. Gob fed into blank mold

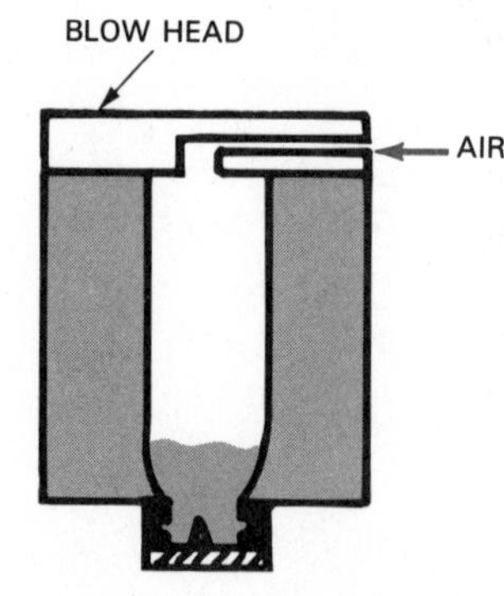

2. Gob blown down into blank mold

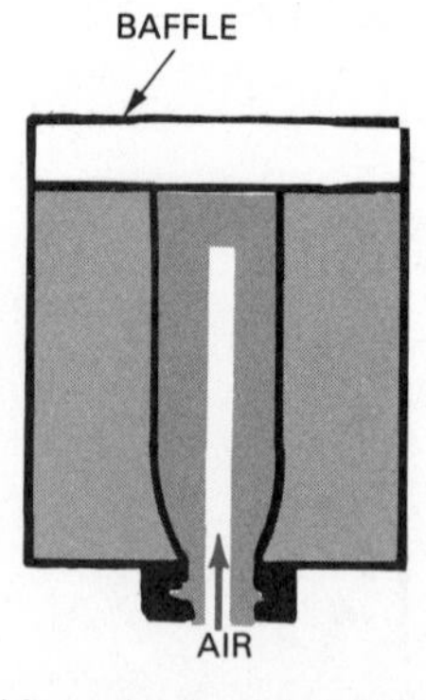

3. Air blown back up into blank mold

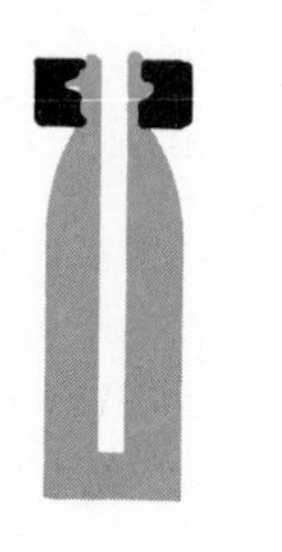

4. Turn over and remove mold

5. Parison in blow mold

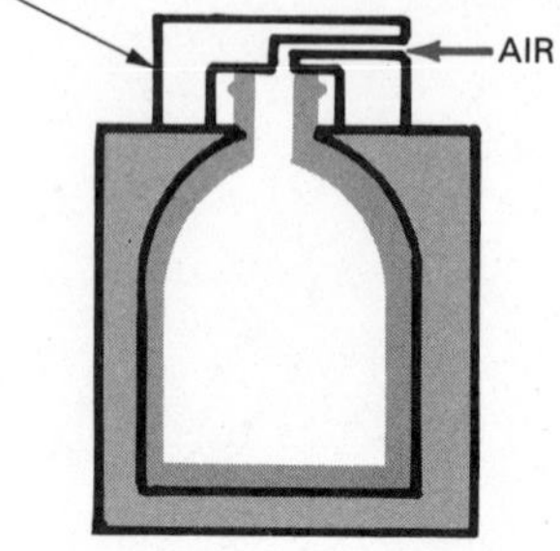

6. Form container in blow mold

Fig. 13-20. The blow and blow glass forming process. Air pressure is used at every step.

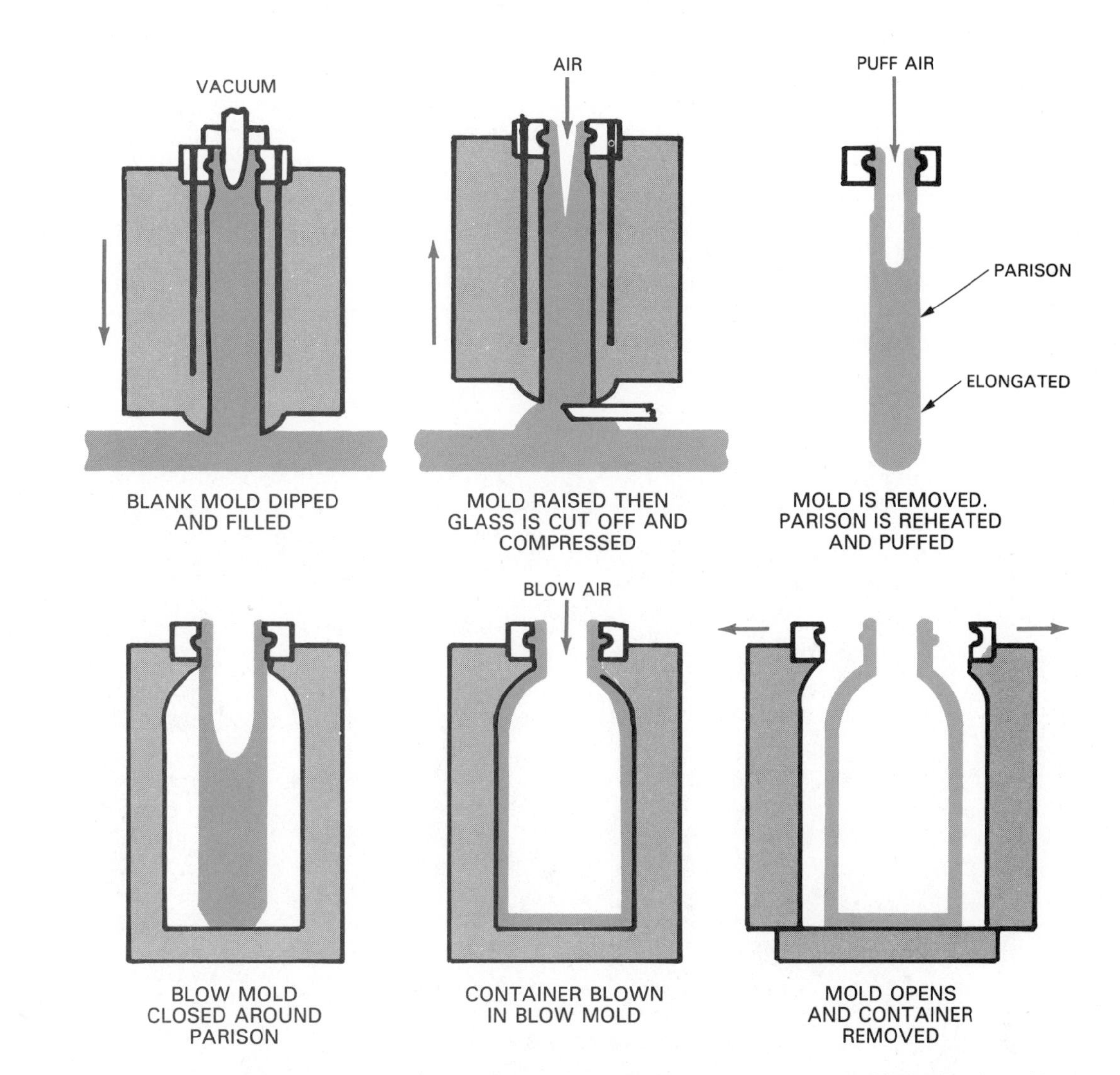

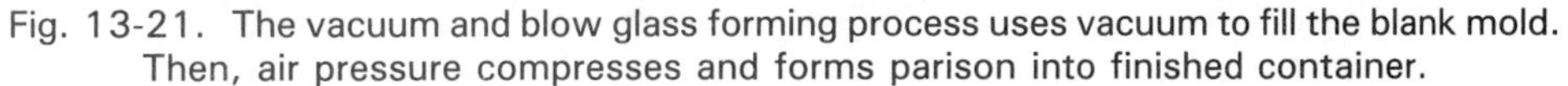
Fig. 13-21. The vacuum and blow glass forming process uses vacuum to fill the blank mold. Then, air pressure compresses and forms parison into finished container.

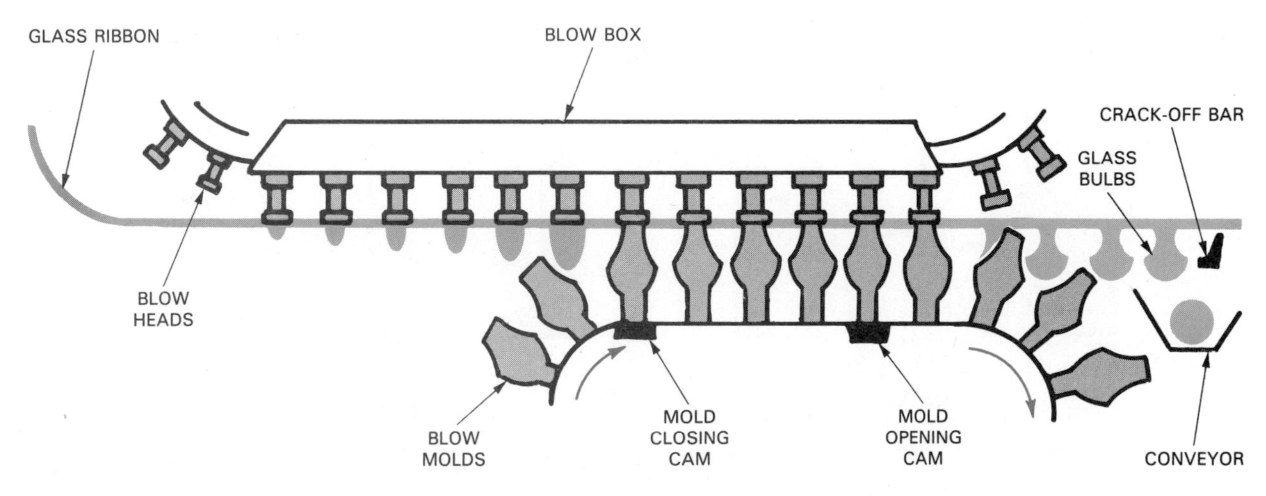

Fig. 13-22. Ribbon glass forming process. Ribbon is formed between cooled rollers. Blow heads form parisons. Light bulbs are formed by this process.

These two factors greatly influence cooling rates for the glass which, in turn, influence molding rates.

JIGGERING

Round, cylindrical clay products may be automatically produced by a technique called jiggering, Fig. 13-23. Round plates, saucers, cups, and bowls are formed using jiggering. Also large electrical insulators are made from several parts which are formed by jiggering.

The technique, as shown in Fig. 13-24, uses a template or roller to force clay against a revolving plaster mold. The template or roller may force the clay against the inside of a mold to produce hollow ware such as cups and bowls. Flatware like plates and saucers are produced by forcing the material against the outside of the mold.

The molded object is allowed to dry for a period of time before being removed from the mold. Additional hand touch-up, drying, finishing, and firing is required before the product is complete.

Fig. 13-23. Automatic roller jiggering machine continuously produces dinnerware. (Syracuse China Co.)

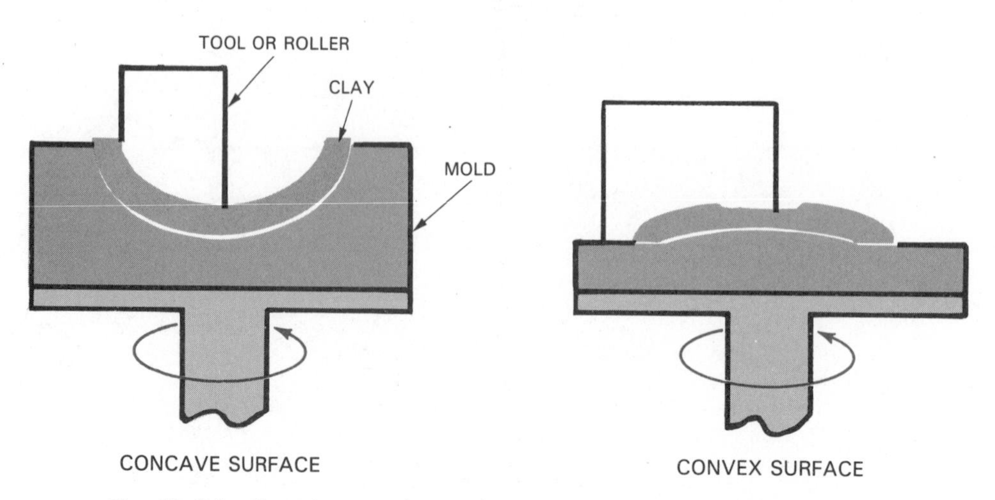

Fig. 13-24. Jiggering may be used to shape inside or outside contours.

SUMMARY

Ceramic materials are formed by using shaping devices and pressure into a wide variety of functional products. Glass and clay material are pressed and extruded into useful shapes. Glass is drawn into tubes, sheets, rods, and fibers to meet many everyday uses. Vast quantities of glass are blow molded into containers, light bulb envelopes, and ornaments. Finally, clay is shaped into dinnerware and other products by the jiggering technique.

STUDY QUESTIONS — Chapter 13

1. List the five forming techniques used for the major types of ceramic materials.
2. What are the three major ceramic press-forming processes?
3. Describe dry pressing, ram pressing, and isostatic pressing.
4. Describe auger-fed extrusion and piston-fed extrusion operations.
5. Describe the major glass drawing processes.
6. Describe the two major methods of producing glass fibers.
7. What are the basic steps in blow molding?
8. Describe the operation of:
   a. press and blow machines.
   b. Blow and blow machines.
   c. Vacuum and blow machines.
   d. Ribbon machines.
9. Describe the jiggering machine.
10. What process would you use to produce:
    a. Bricks?
    b. An oval serving dish?
    c. Spark plug shell?
    d. Glass baking dish?
    e. Sewer pipe?
    f. Sheet glass?
    g. Glass jar?
    h. Christmas tree ornament?
    i. Round dinner plate?

A hot glass parison is being automatically placed in an open mold. (Glass Packaging Institute)

Chapter 14

FORMING POWDERED METALS

Powdered materials are sometimes the basic ingredients for manufacturing processes. Several plastic processing techniques may start with powdered synthetic resins. These processes, discussed in Chapter 12, include rotational molding, compression molding, and transfer molding. Dry pressing of ceramic powders was discussed in Chapter 13. Another important, rapidly growing manufacturing process involves the forming of parts from powdered metals. This process is called *powder metallurgy*.

Powder metallurgy is the manufacture of products from very fine metal powders using pressure and, usually, heat. Powdered metals were used in product manufacture as far back as 3000 B.C. At this time, *sponge iron* was used by the Egyptians. Later, iron powders were used by the Arabs to produce high quality swords. However, not until 1829 was press-bonding of metals successful. This process is the foundation for modern-day powder metallurgy.

Powder metallurgy involves four basic steps.

1. Producing metallic powders.
2. Blending powders and other materials.
3. Compacting the blended mixture.
4. Sintering (heating) the compacted shape.

METALLIC POWDERS

The quality of the metal powders used in the powder metallurgy process largely determines the properties of the end product. Important characteristics of the powder include the size and shape of the particles, their chemical composition, and their purity. The grains should be fine, smooth, and high in purity. These powders are produced by three main processes.

1. Reduction.
2. Electrolytic deposition.
3. Atomization.

REDUCTION

The reduction method uses a gas to reduce a metal oxide to a metal powder. Reduction is widely used to produce iron powder.

One method mixes mill scale (a type of iron oxide) with crushed coke. The materials are fed into a rotating kiln. In the kiln the materials are heated to around 1900°F (1038°C). At this temperature the carbon in coke unites with the oxygen in the iron oxide. The resulting gas is removed through a stack. The remaining material, called sponge iron, is nearly pure.

A second process, shown in Fig. 14-1, mixes iron ore, limestone, and coke. These materials are placed in a tube lined with firebrick and heated in a tunnel kiln. The temperature of the kiln is below the melting point of iron. A chemical reaction occurs leaving porous, sponge-like iron.

The sponge iron is then crushed, screened, and cleaned. The particles are separated by size and packaged for later use. Tungsten, molybdenum, cobalt, and nickel powders may also be produced by reduction processes.

ELECTROLYTIC DEPOSITION

Electrolytic deposition uses the same process that electroplating uses. A plate containing an alloy of the desired metal is placed in a tank. The tank contains an electrolyte (a liquid which will break into charged particles and conduct electricity). A second plate is placed in the tank to gather the metal. Direct current is applied across the plates. The metal leaves the first plate (the anode) and is deposited on the other plate (the cathode).

After a period of time, up to 48 hours, the cathode is removed. The deposited metal is stripped off, crushed, sized, and packaged. Copper and iron powders are common products of the electrolytic deposition process.

ATOMIZATION

Atomization involves melting the metal. It is then poured into a tundish which leads to an atomizing chamber. Within the chamber the molten metal is sprayed with a high-pressure stream of gas or liquid. This stream breaks up the metal into small, solid

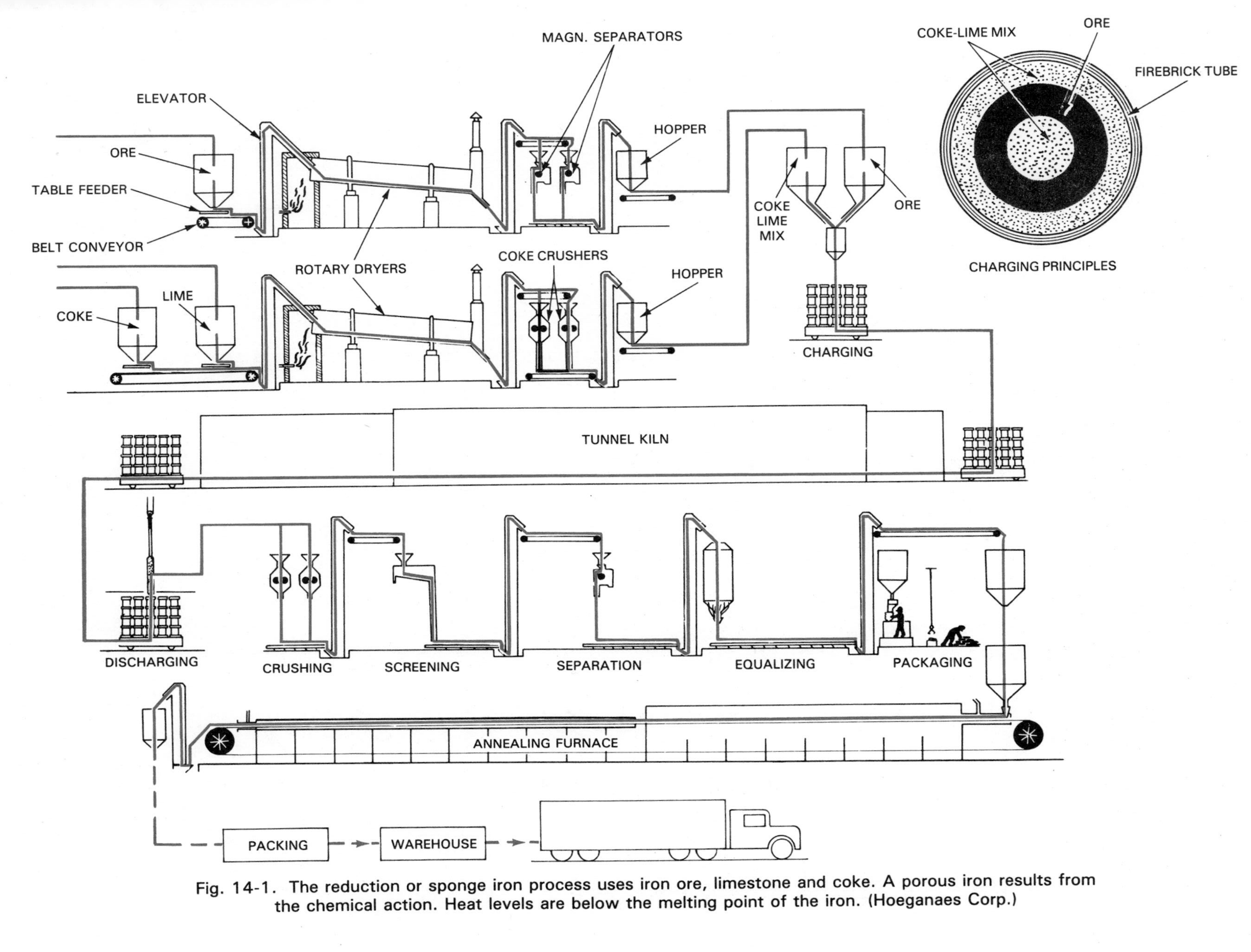

Fig. 14-1. The reduction or sponge iron process uses iron ore, limestone and coke. A porous iron results from the chemical action. Heat levels are below the melting point of the iron. (Hoeganaes Corp.)

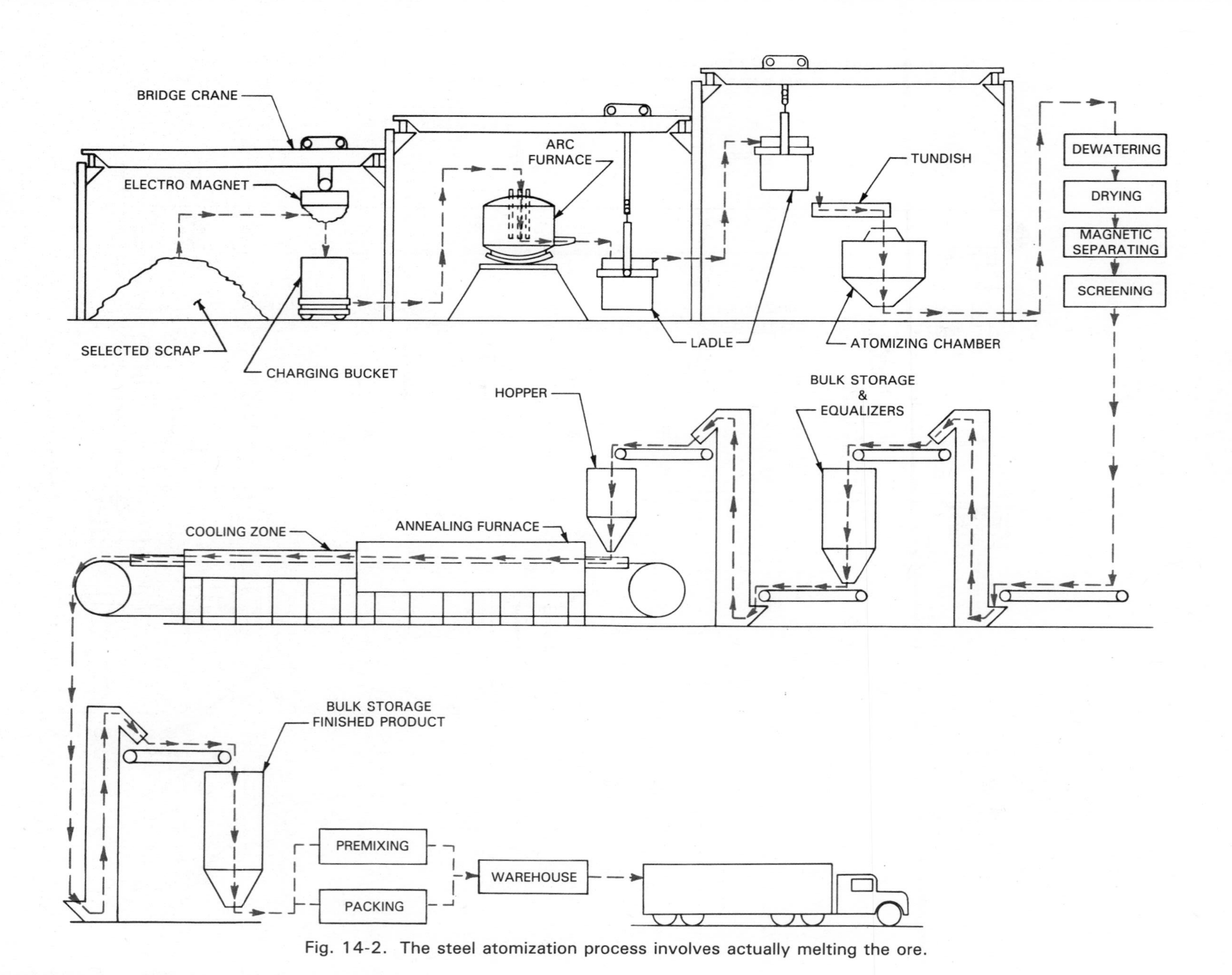

Fig. 14-2. The steel atomization process involves actually melting the ore.

particles. They are then dried and screened for size.

Fig. 14-2 is a diagram of a steel atomizing process. Atomization is also used to produce brass, bronze, aluminum, zinc, tin, and lead powders.

The powders produced by all three methods are quite brittle. They must be annealed to soften them for the blending and compacting stages. Both Fig. 14-1 and 14-2 show annealing furnaces.

BLENDING

The powders produced by reduction, electrolysis, and atomization are ready for blending, Fig. 14-3. Blending is used for any or all of the following reasons:

1. To mix particles of different sizes to produce a uniform particle distribution.
2. To mix powders of different metals.
3. To add lubricants.

The mixing process may be done either wet or dry. Wet mixing reduces dust and the chance of explosion. Also, more uniform mixing is possible when water or a solvent is used.

Fig. 14-3. Blending is the first step in producing a P/M (powder metallurgy) part. This blender can combine over 4000 lb. of powder metal at a time. (Burgess-Norton Mfg. Co.)

Lubricants are often added during mixing. They improve the flow characteristics of the powder in the forming dies. In addition, lubricants reduce die wear during the compacting step. Typical lubricants are powdered graphite, stearic acid, or lithium stearate.

COMPACTING

Forming or compacting of metal powders may be done in several ways. Some of these processes were discussed in earlier chapters. Typical forming practices include pressing, extrusion, rolling, slip casting, centrifugal casting, high energy rate (explosive) compacting, and isostatic (hydro) forming.

PRESSING OR BRIQUETING

Pressing involves placing the blended metal powders in steel dies shaped to the finished parts. The dies are closed using pressures from a few tons to 100 tons psi.

Pressure needed depends upon the material being compacted. The density and hardness of the part increases, up to a point, with increased pressure. Above this, the optimum point, added pressure produces little change in the part's characteristics.

Most dies, as seen in Fig. 14-4, have three parts.

1. A die sleeve or shell which contains the cavity.
2. An upper punch.
3. A lower punch which, with the upper punch, compresses the powder in the cavity.

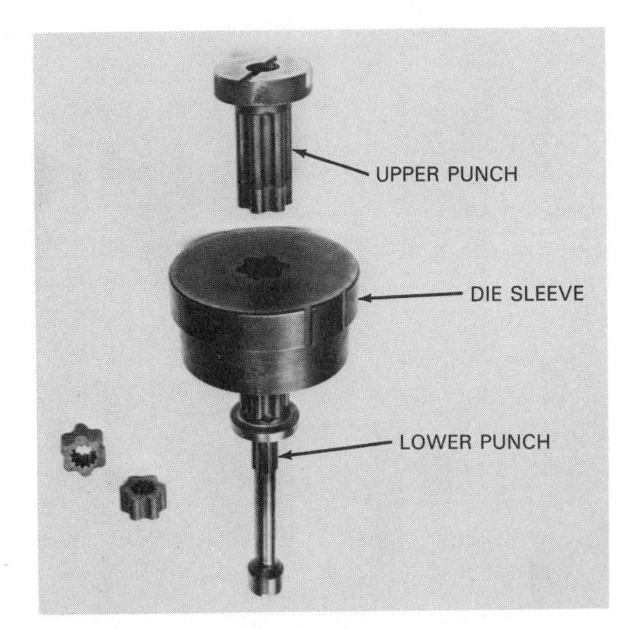

Fig. 14-4. A die for forming P/M parts. Parts produced are shown at left. (Burgess-Norton Mfg. Co.)

A fourth part, Fig. 14-5, is used for parts with holes in them. This part, called the *core,* performs the same function as a core in casting. It allows a hole to be formed in the part, Fig. 14-6.

EXTRUSION

Extrusion is used to produce long shapes from metal powders. Different metals require different extruding techniques. Some materials are mixed with a binder and extruded cold. Others are heated in a controlled atmosphere to form a billet. The billet is then extruded. Still others are sealed in a mild steel tube or can. The can with the powder is forced through the extruding die, Fig. 14-7. The can is removed from around the shaped and compacted powders.

ROLLING

Rolling involves passing powder metals between two rolls. The rolls compress and interlock the powders into a sheet or strip.

Metal powder may be either *cold rolled* or *hot rolled.* In cold rolling, the powder is metered into the roll gap at room temperature. Then the rolled material is sintered to develop a metallurgical bond between the particles of compressed material. This is usually done in a *belt* or *roller hearth furnace.* The hot rolling process is similar in that the powder is fed through

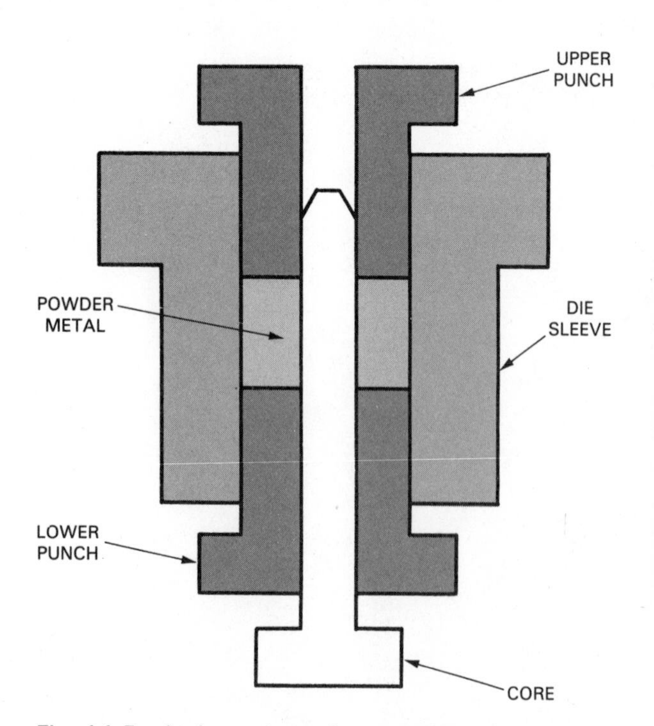

Fig. 14-5. A three-piece die sometimes has a core added for forming P/M parts.

Fig. 14-6. An operator holds a P/M oil pump gear just formed in a die set. (Burgess-Norton Mfg. Co.)

Fig. 14-7. An extrusion press for forming powdered metals such as tungsten carbide. This unit has a capacity of 120 tons psi.

rollers cold and then feeds into a sintering furnace. However, the strip or sheet is fed through a second set of rollers just before it emerges from the furnace.

SLIP CASTING

Slip casting is used to form hollow metal shapes. Metal powders can be mixed with a liquid to form a slurry. The slurry is poured into a plaster-of-paris mold. The mold absorbs the liquid building up a powder metal wall. When the proper wall thickness is reached, the remaining slurry is poured out of the mold. After drying, the mold is opened and the formed part is removed.

CENTRIFUGAL CASTING

Centrifugal casting uses centrifugal force to introduce powders into a mold. The force of up to 400 psi produces a compacted powder shape. Centrifugal casting is limited to heavy powders such as tungsten carbide.

HIGH ENERGY RATE COMPACTING

HER forming uses an explosive charge in a closed chamber to form the powders. This technique generates very high pressures and a high-density product.

ISOSTATIC FORMING

Isostatic forming uses a gas as a forming medium. The same process using a liquid is called *hydrostatic pressing*. The process involves placing the powder in an elastic (rubber) container. The container is placed in a sealed vessel. Gas is pumped into the vessel. The gas exerts equal pressure in all directions. This principle causes the powders to be evenly compacted as the rubber mold is compressed. The resulting part has a highly uniform density.

SINTERING

Sintering heats the green part to 60-80 percent of the melting point of the principal powder. Temperatures above this level may cause "casting" of the part.

Sintering increases the atomic bonds between the atoms in the metal. In short, the solid shapes formed by compacting are firmly bonded by atomic forces.

During sintering, grain boundaries are formed, plasticity, and density are increased, and better mechanical properties are produced.

Sintering is usually done in a continuous furnace, Fig. 14-8. The furnace has a section for loading, a heating zone, and a cooling zone.

Parts are heated to their proper range in the heating zone. Typical temperatures are 1600°F (870°C) for

Fig. 14-8. An operator is placing parts on the belt of a continuous sintering furnace. (Burgess-Norton Mfg. Co.)

copper, 2000°F (1095°C) for iron, and 2700°F (1480°C) for tungsten carbide. The parts are held at these temperatures for 20-40 minutes, then cooled. Fig. 14-9 shows a continuous sintering furnace.

Batch furnaces may also be used. These furnaces process one batch of parts at a time. They produce the same results as a continuous process furnace.

SECONDARY PROCESSES

The sintered part may undergo one or more of several secondary processes. Common secondary processes are:

1. Infiltration. Infiltration is used primarily on iron-based sintered parts. Copper, brass, or copper alloy blanks are placed on or under the part. The part and the blank are reheated. The temperature reaches a point just below the original sintering temperature. The metal blanks will melt at this temperature. They flow into the pores of the part (infiltrate) or are drawn up by capillary action into the part. Infiltration can increase density, hardness, and strength of the part. Strength, alone, can be almost doubled.
2. Impregnation. Impregnation introduces oil or other lubricants into porous bronze or iron bearings. The most common method of impregnation is to place the parts in a tank of hot lubricant. A vacuum tank may be used. In both methods the wax, grease, or oil is drawn into the pores of the bearing.

   Impregnation may also be used to seal pores of parts with solder, other low-melting allows, and plastic resins. This technique produces a smooth surface for plating or other finishes.
3. Sizing and coining. Sizing and coining operations are cold forming press operations. They are used on powder metallurgy shapes to produce parts with more accurate dimensions, greater density, and better surface finish.

Fig. 14-9. A view of a continuous sintering furnace. These furnaces can operate at temperatures up to 2700°F. (Burgess-Norton Mfg. Co.)

Fig. 14-10. Typical powder metallurgy parts. (Burgess-Norton Mfg. Co.)

PRODUCT APPLICATIONS

Powder metallurgy processes are used to produce a variety of products as shown in Fig. 14-10. These include metal composites which add fibers to metal parts; porous parts to control the flow of liquids and gases; cemented carbide tools and dies; magnets, structural parts requiring exact heat-resisting, hardness, and corrosion-resisting properties; metallic fibers for filtering fluids and gases; gears and rotors; motor brushes; ceramic-metal friction parts such as brake linings and friction discs; and electrical contacts.

SUMMARY

Powder metallurgy is a rapidly growing technology used to produce a variety of parts. The basic process involves producing metal powders, blending the powders with other material, compacting or forming the part, and sintering (heating) the part. The part may then be subjected to additional secondary processes. The parts may also be infiltrated with other metals, impregnated with lubricants or sealers, and sized and coined. The product is quickly produced generally without the need for machining.

STUDY QUESTIONS – Chapter 14

1. List the four basic steps used to produce a powder metallurgy (P/M) part.
2. Describe the following processes for producing metal powders:
   a. Reduction.
   b. Electrolytic deposition.
   c. Atomization.
3. List three reasons for blending of metallic powders.
4. Describe seven basic ways of compacting powder metallurgy parts.
5. What is sintering?
6. List and describe three common secondary processes used on sintered parts.
7. Name at least five special products which are produced through the powder metallurgy process.

SECTION 4
SEPARATING

Separating is a family of processes that removes excess material to change the size, shape and/or surface finish of parts or products. The major techniques, which can be groups under machining, shearing, and flame cutting, include:

- Turning operations.
- Milling operations.
- Sawing, broaching, and filing.
- Shaping and planing.
- Drilling, boring, reaming, and tapping.
- Abrasive machining.
- Thermal and chemical machining.
- Shearing operations.

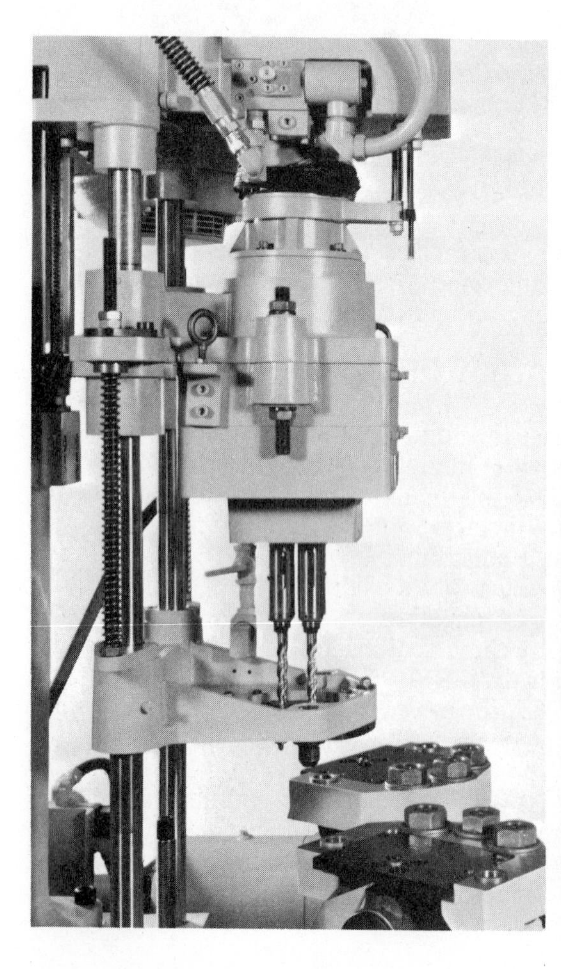

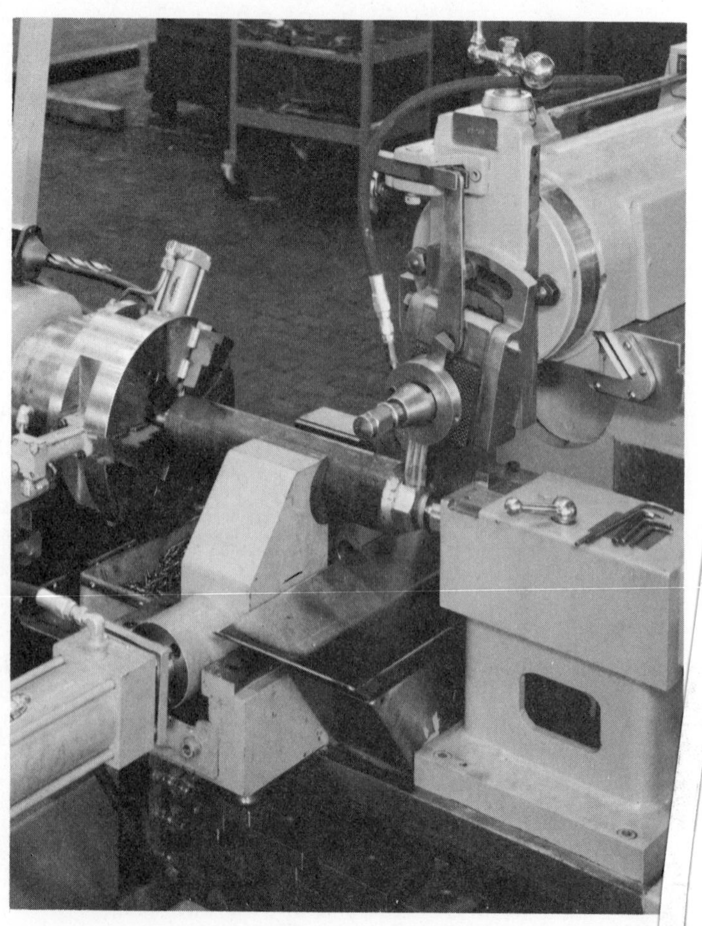

Chapter 15

INTRODUCTION TO SEPARATING

The third method of sizing and shaping materials is separating. Separating is the process that removes excess material to change the size, shape and/or the surface finish of materials. Separating may be done by one of several major techniques. These methods, Fig. 15-1, make up three major groups:

1. Machining—changing the size, shape, or material finish by removing excess material from the workpiece.
2. Shearing—changing the size and shape features by using opposed edges to fracture (break or cut) the excess material from the workpiece.
3. Flame cutting—changing the size and shape by using burning gases to separate excess material from the workpiece.

HOW THE PROCESSES DEVELOPED

Probably the earliest cutting and shaping operations were practiced during the Stone Age when early humans shaped tools and weapons from stone by chipping away excess materials. They created weapons and tools with sharp cutting edges or pointed ends. Using these tools and weapons, they were able to hunt game and harvest vegetation to provide food, clothing, and shelter.

Later, skill in using tools to bore holes was developed. The *bow drill* was, most likely, the first successful design for a tool that could make holes. Until recently, Indians of the southwest pueblos used this type of drill to manufacture handcrafted jewelry.

Among the earliest power-driven, metal-cutting machines was the lathe. It was used before the Middle Ages (about A.D. 500-1400) by early clock makers. Later, more specialized lathes, including the turret lathes, Fig. 15-2, were developed.

Another important machine tool was the milling machine designed by Eli Whitney to help build muskets. Other important separating machines were:

1. Power-driven saws.

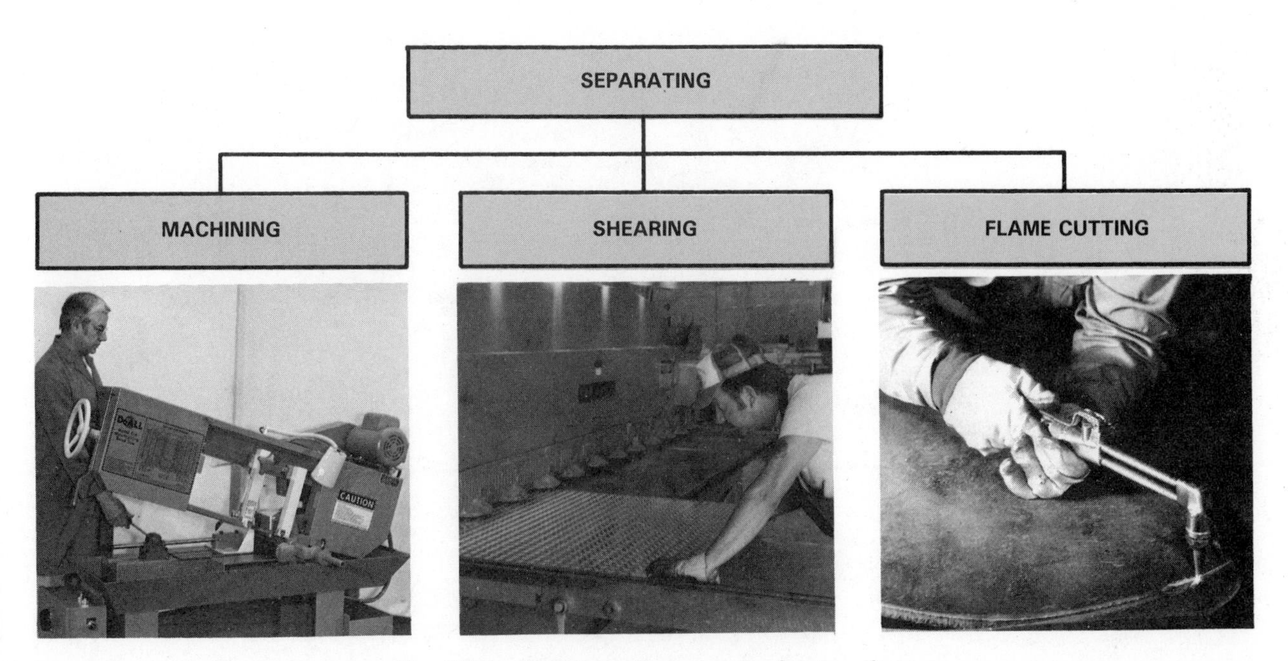

Fig. 15-1. There are three types of separating.

Fig. 15-2. Early turret lathe. Revolving turret holds more than one tool. (Collections of Greenfield Village and the Henry Ford Museum)

2. Metal shapers and planers.
3. Woodworking jointers, planers, shapers, and routers.
4. Grinding machines, Fig. 15-3.
5. Shearing machines.

America has developed the greatest production capacity the world has ever known. The foundation of this capacity is machines for casting and molding, forming, cutting, and machining.

Separating tools that cut and machine materials to shape are perhaps the most important. They help shape and size materials into interchangeable parts. These machines are used to build many of the jigs, fixtures, dies, molds, and patterns used in modern manufacturing. Machine tools are also used to build new machines.

BASIC ESSENTIALS OF SEPARATING

Separating involves all techniques in which excess material is removed to create desired part size, shape, and/or surface finish. These essential elements are present in all separating acts. These are:

1. A tool or cutting element is always used.
2. There is movement between the work and the cutting element.
3. The work, the cutting element, or both are clamped or supported in a desired position.

CUTTING ELEMENTS

A cutting element is the device or agent which causes a material to be separated into its proper size and shape. The cutting element produces a shaped workpiece and scrap (chips, solid waste, and powdered material). There are three major cutting elements plus a group of other agents called nontradi-

Fig. 15-3. Early grinding machine.

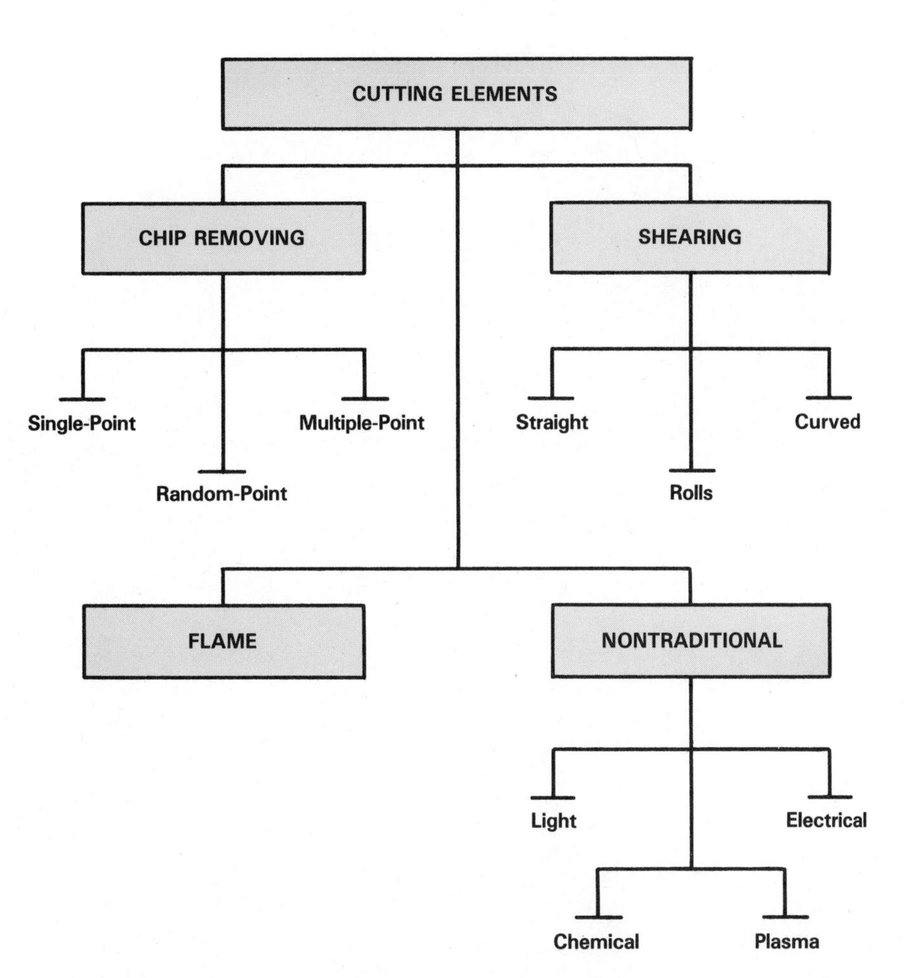

Fig. 15-4. Types of cutting elements and their relationship to one another.

tional. These cutting elements, as seen in Fig. 15-4, are:
1. Chip-removing tools.
2. Shearing tools.
3. Flame-cutting elements.
4. Nontraditional cutting elements.

CHIP REMOVING TOOLS

Chip removing or cutting tools are used for most machining operations. The cutting tool becomes the heart or basic element for hundreds of machining techniques. As listed in Fig. 15-4, there are three basic types of chip removing tools: single point, multiple point, and random point. These tools, Fig. 15-5, are used in various machining operations. Typical examples include:

1. Single-point tools.
   a. Lathe tool bits (metal cutting).
   b. Lathe tools (wood cutting).
   c. Shaper and planer tools (metal cutting).
   d. Single lip router bits.
2. Multipoint tools.
   a. Drills, reamers, and taps.
   b. Broaches.
   c. Milling cutters.
   d. Jointer and planer cutter heads (wood cutting).
   e. Hack, band, and circular saw blades.
   f. Router and shaper cutters (wood cutting).
3. Random-point tools.
   a. Grinding wheels.
   b. Abrasive belts and discs.
   c. Abrasive cut-off discs.

To understand the selection and use of various cutting tools it is essential to know how chips are formed. Also, a knowledge of tool shapes or geometry; cutting speeds, feeds, and depth of cut; and factors affecting tool life is worthwhile. A general discussion of these factors will follow. More specific information is contained in Chapters 16-21.

Chip formation

Basically, chip forming is a physical process common to all machining operations. It is an interaction

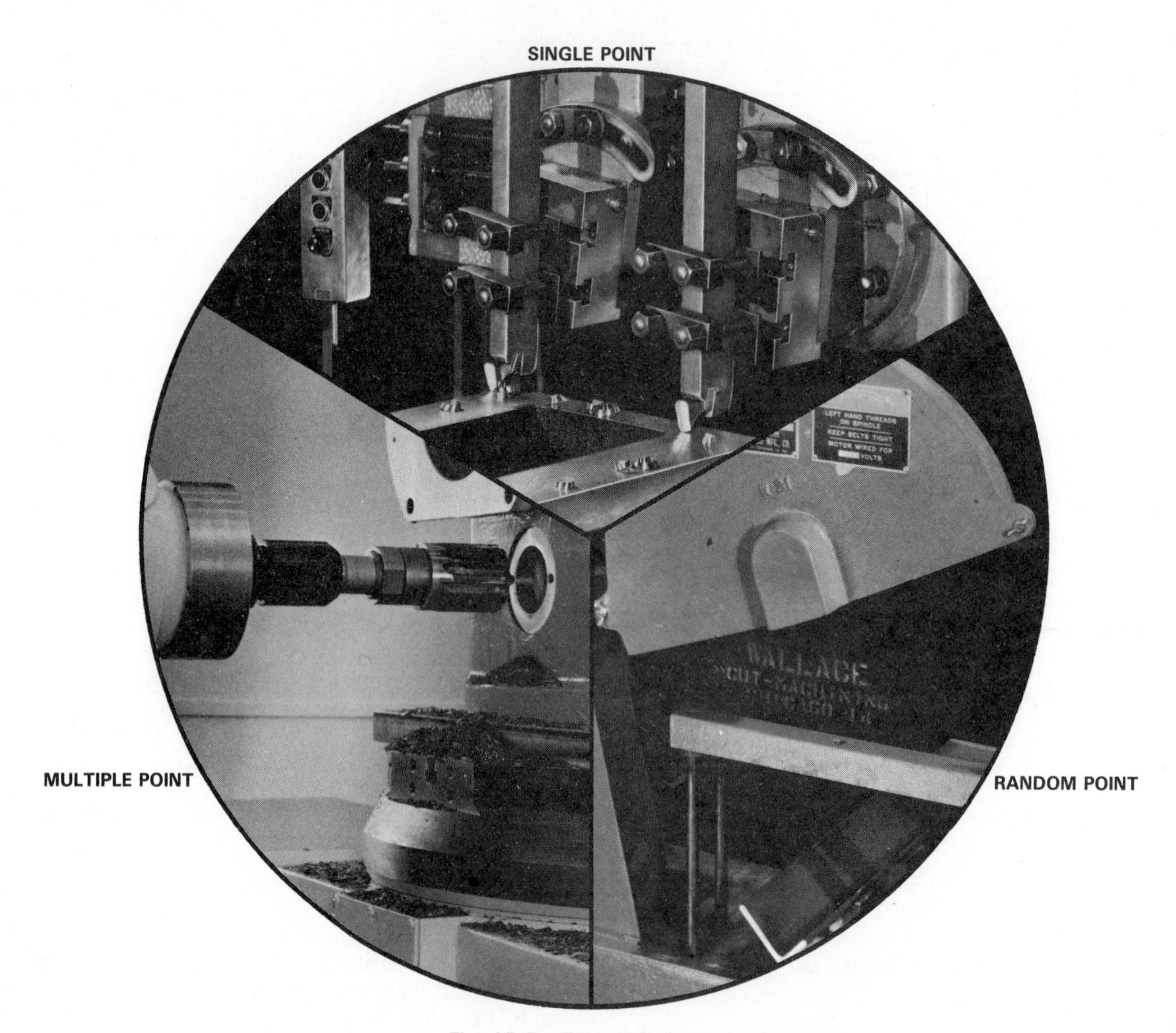

Fig. 15-5. Types of chip removing tools.
(Rockford Machine Tool Co., Wallace Supplies Mfg. Co., and Cincinnati Milacron)

between the cutting tool and the workpiece. The tool must penetrate the work and cause a part of it to be lifted away. Anything else that happens—generating heat or bending the work, for example—either contributes or hampers this action.

The Cutting Action. The cutting action is a very complex physical force. To fully understand the action, a knowledge of physics and advanced mathematics is needed. You can get a basic idea of what is happening by examining the action of a single-point tool being forced across a flat metal workpiece.

Whenever a tool cuts a material it must be driven by a force. This force must be strong enough to overcome two other forces. These forces are the friction between the tool and the work and the atomic structure which bonds the material together.

The friction forces depend upon several factors.

1. The sharpness or keenness of the tool cutting edge.
2. The smoothness of the surface of the tool over which the chip must pass.
3. The shape of the tool.
4. The material from which the tool is made.
5. The material being machined.
6. The coolant (if any) being used.
7. Cutting speed, feed, and depth of cut.
8. Shear angle of the cut.

As the cutting action proceeds, Fig. 15-6, the tool compresses the metal ahead of it. The stress on the compressed material is increased until it reaches the fracture point. The internal grain structure of the material is broken. The separated material, then, flows up the face of the cutting tool. This separating

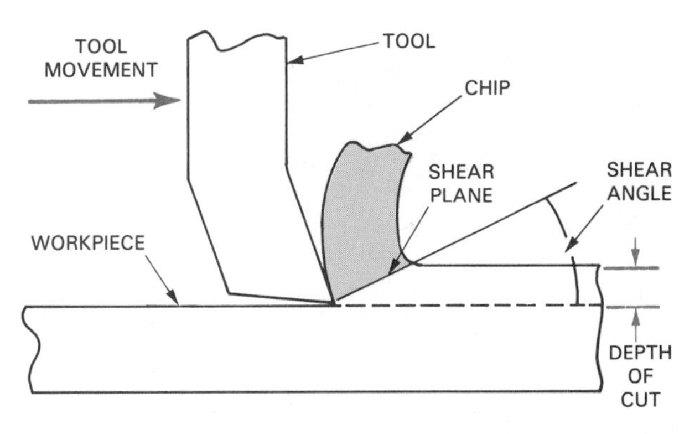

Fig. 15-6. Cutting tool action. As tool cuts through material portion being removed it compresses and finally fractures producing chip.

action is shown in Fig. 15-7. The tool moves forward compressing, shearing, and displacing additional chips. The forward motion of the tool scrapes the cut surface. This action provides a smooth, clean surface.

Multipoint and random-point tools operate on the same cutting principle. The difference between single- and multipoint tools lie in the successive cutting acts performed by each cutting surface as it contacts the material. Instead of a single point being driven across the surface, many cutting edges create the cutting action.

Types of Chips. The cutting action of various tools creates one of three types of chips. There are:

1. Type 1 or segmented chips. Segmented chips are formed when brittle material is cut or when the friction between the tool and the work is high. The material is broken into small pieces along the shear line. Fig. 15-7 shows Type 1 chips being formed. Cast iron and bronze form segmented chips when they are machined.
2. Type 2 or continuous chips. Continuous chips are formed when most metals, plastic materials, and woods are cut. The chip comes off the cutter as a continuous curled chip (or shaving for wood). Fig. 15-8 shows a Type 2 chip being formed.
3. Type 3 or continuous chip with built-up edge (BUE). A built-up edge is formed when heat and pressure cause some of the highly compressed material from the workpiece to adhere to the tool. As the built-up edge grows, it becomes unstable. Parts of it come off the tool and either stick to the chip or the work. The broken off built-up edges cause roughness in the cut. Changing the cutting speed and cutting fluid can eliminate built-up edges. Fig. 15-9 contains photographs of the three types of chips.

Tool geometry (shape)

The shape of cutting tools directly affects the amount of power needed to cut a material. It will also determine, to a large extent, the smoothness of the surface produced. The purpose of tool geometry is to provide a strongly supported, sharp cutting edge that will penetrate the work. To accomplish this purpose, a number of different angles are considered.

1. Relief angles. These are angles which keep the tool

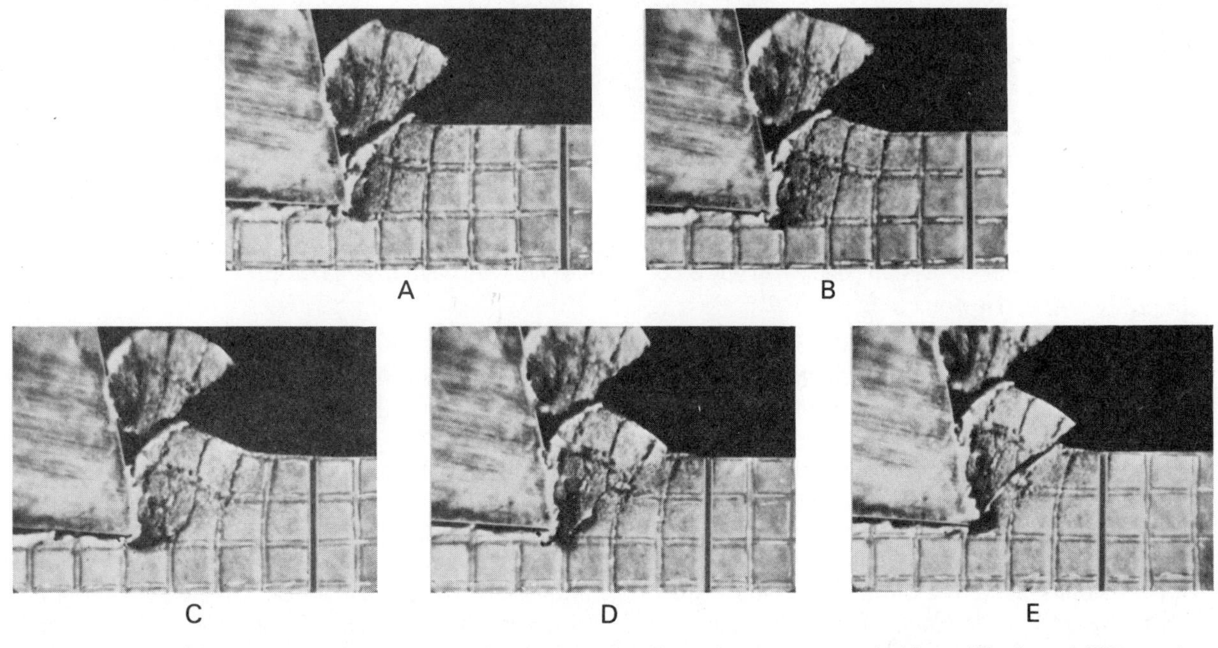

Fig. 15-7. Photomicrograph shows the formation of a Type 1 or segmented chip. (Cincinnati Milacron)

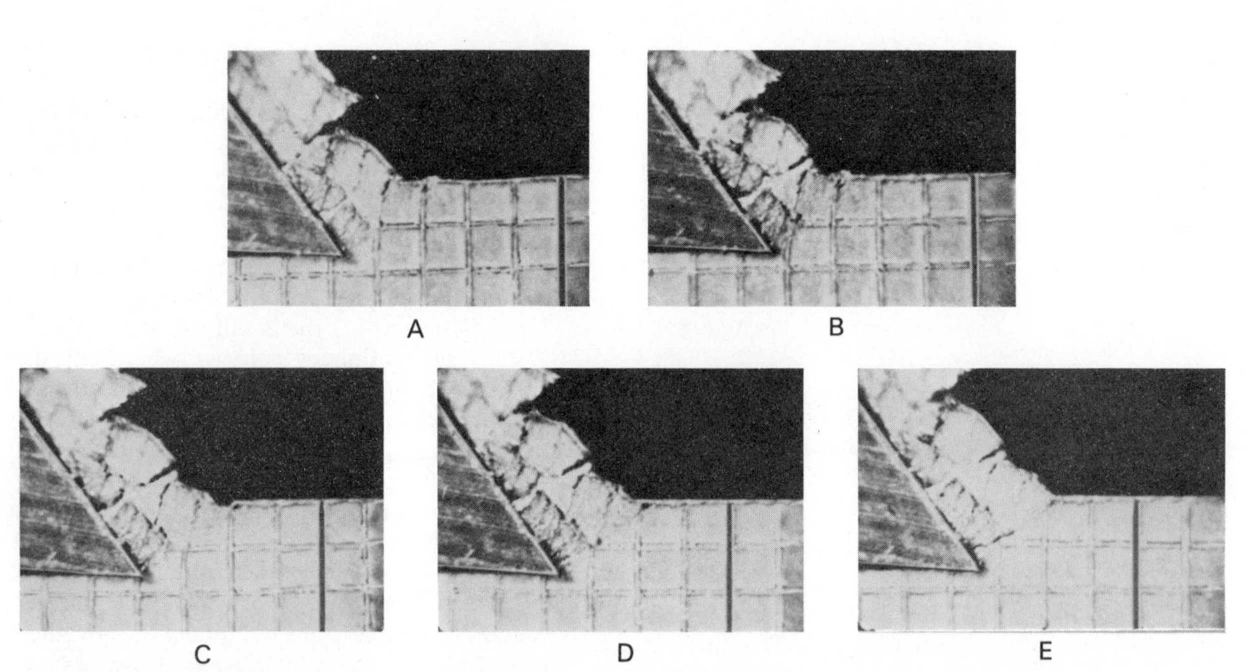

Fig. 15-8. A photomicrograph of the formation of a Type 2 or continuous chip. (Cincinnati Milacron)

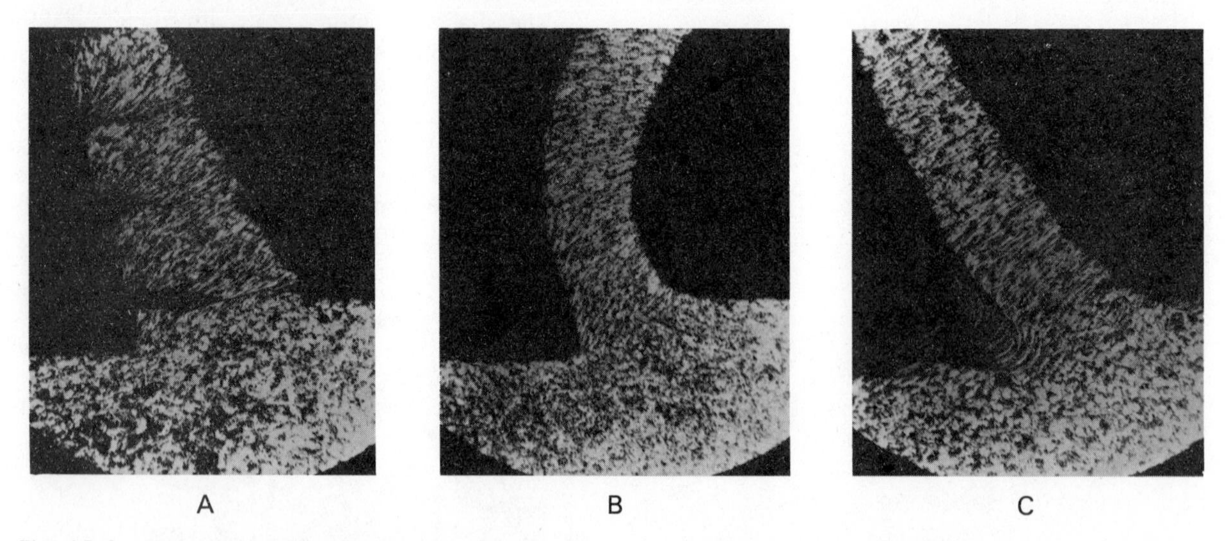

Fig. 15-9. Highly magnified photographs show three types of chips. A—Type 1 discontinuous chip. B—Type 2 continuous chip without built-up edge. C—Type 3 continuous chip with built-up edge (BUE). (Cincinnati Milacron)

from rubbing on the surface of the work.

2. Cutting-edge angles. These angles produce the cutting edge or point.
3. Rake angles. These angles produce the surface over which the chips flow. Rake angles provide the direction for chip flow.

In addition, a nose radius is sometimes considered. The nose radius blends the cutting angles and reduces the chance for tool breakage. The nose radius provides for a smoother cut and greater cutting speed.

Fig. 15-10 shows tool geometry for a normal single-point cutting tool. The actual angles and nose radius will vary with the material being cut and the material from which the tool is made.

Tool materials

Cutting tools are made from a number of different materials. The particular task at hand determines what metals will be selected. No one material will meet all machining requirements. The tool, however, must be

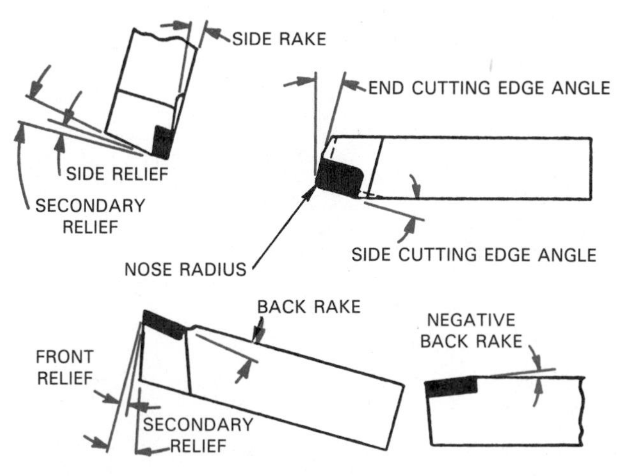

Fig. 15-10 Single-point tool angle nomenclature. (The Norton Co.)

tough enough to penetrate the workpiece and withstand cutting forces.

The main materials used for cutting tools are:

1. Carbon tool steel.
2. High speed steel.
3. Cast nonferrous alloys (cast alloys).
4. Carbides, tungsten and alloyed.
5. Sintered oxides (ceramics).

Carbon Tool Steel. For many years carbon tool steel was the only material available. It has a carbon content of 0.90 to 1.2 percent and often contains other alloying (metals) elements. Carbon tool steel, when properly heat treated, has good strength and hardness. It is a relatively tough material which will hold a keen edge.

Carbon tool steel, however, loses its hardness at temperatures above 400°F (204°C). This factor limits its use to short production runs or low-speed operations where tooling costs must be kept low. Carbon tool steels, though seldom used now for production metal-cutting, are used widely for cutting soft materials like wood and certain plastics. They also are used in cutting nonferrous metals such as brass.

High-speed Steel. High-speed steel (HSS) was first developed for cutting tools at the turn of the century. It rapidly replaced carbon tool steel in most metal machining operations. High-speed steels retain their hardness and cutting ability at temperatures as high as 1100° to 1200°F (593° to 649°C). This ability allows for higher cutting speeds which contributed to the name "high-speed" steel.

High-speed steel is a high alloy of steel in which carbides are held in a lower melting point matrix. The oldest alloy is called 18-4-1 or T1. It contains 18 percent tungsten, 4 percent chromium, and 1 percent vanadium along with about 0.7 percent carbon. T1 high-speed steel is considered an excellent all purpose tool steel. A typical high-speed steel tool is shown in Fig. 15-11.

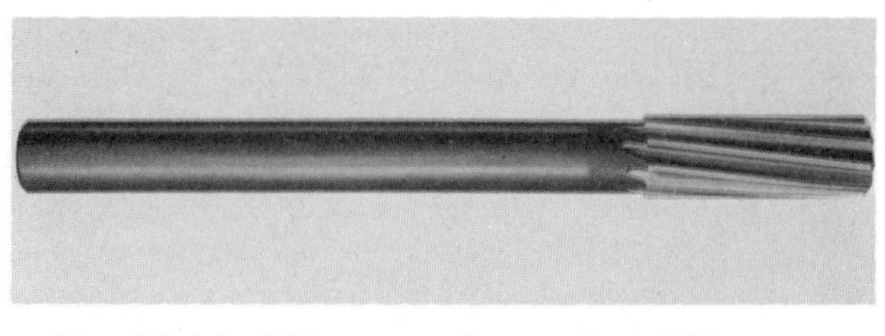

Fig. 15-11. This reamer is a typical HSS cutter.

Another important high-speed steel is molybdenum or "moly" high-speed steel. Moly steels replace two parts tungsten with one part molybdenum. A popular alloy is a 6-6-4-2 which contains 6 percent tungsten, 6 percent molybdenum, 4 percent chromium, and 2 percent vanadium. Another popular alloy contains 1.5 percent tungsten, 8 percent molybdenum, 4 percent chromium, and 1 percent vanadium. Moly tool steels have excellent cutting ability and toughness.

A third important high-speed steel is called cobalt high-speed steels. These alloys contain 2 to 15 percent cobalt along with tungsten, chromium, and vanadium. Cobalt high-speed steels have higher heat and abrasion resistance. They cost more and are more difficult to grind and heat treat, however. Cobalt high-speed steels are therefore used primarily for operations which produce high pressures and temperatures for the cutting tool.

Among the most popular high-speed steels today are several developed during the 1930s by leading steel companies and toolmakers.

1. M1 and M7 consisting of 9 percent molybdenum and 1.5 percent tungsten grades.
2. M2 made up of 6 percent tungsten, 5 percent molybdenum, and 2 percent vanadium.
3. M10, an 8 percent molybdenum grade with no tungsten.
4. Two carbon-high vanadium grades (M4 and T15).

Grades M1, M2, M7, and M10 are the most popular today. They exceed tungsten based HSSs in popularity by 20 times.

Cast Nonferrous Alloys. Cast alloys mainly contain cobalt, chromium, tungsten, and carbon with small amounts of carbide-producing elements—tantalum, molybdenum, or boron. In general, cast alloys contain 15-25 percent tungsten, 25-35 percent chromium, 40-50 percent cobalt, and 1-4 percent carbon. The material has a very high percentage of carbide and no iron. The material cannot be softened by heat treating and, therefore, cannot be machined to shape, but must be cast to its basic shape and then ground.

Cast alloys (cast carbides) can hold a good cutting edge to temperatures up to 1700°F (927°C). The tools operate best at high temperatures, about 1500°F (816°C). They lose efficiency at room (cold) temperatures. Cast nonferrous alloys are used at high cutting speeds with deep cuts at low feed rates. Cast alloys were replaced in popularity by the introduction of carbides.

Cemented Carbides. Carbides are made using the sintering process described in Chapter 14. They are materials with a very high carbide to matrix mix. Nearly 80 percent of the cutting tool is composed of carbides. The carbides are made from cobalt powders mixed with tungsten, tungsten-titanium, and tungsten-tantalum carbides. The materials are compacted and sintered at about 2500°F (1370°C).

The resulting sintered carbide material is ground to shape and attached to a tool blank. Formerly, this attachment was made by brazing. However, brazed tools have largely been replaced with mechanically clamped inserts. Fig. 15-12 shows a common carbide cutter with insert tools. These tools have superior hardness and will retain their cutting edge at temperatures over 2200°F (about 1200°C). Carbide tools may also be coated with a thin layer of ceramic materials. These materials may be titanium carbide, aluminum oxide, or titanium nitride. The resulting coated carbides are used for machining very hard materials.

Carbides have a high initial cost but can be operated at fast speeds. This allows carbides to be generally more cost efficient than high-speed steel tools for long production runs.

Sintered Oxides. Sintered oxides are usually compacted, sintered aluminum oxide powder with small amounts of other additives. The material, often called ceramic tooling, has excellent resistance to temperatures up to 2000°F (1090°C). Extremely brittle, its hardness is surpassed only by tungsten carbides and diamonds. Ceramic tools, because of their hardness, ability to resist high temperatures, and low resistance to the cutting action can be operated at two to three times the speed of carbides.

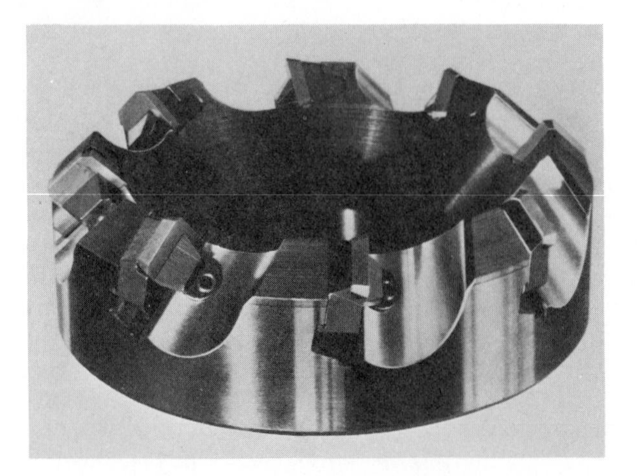

Fig. 15-12. This milling cutter is outfitted with carbide cutter inserts. (Greenleaf Corp.)

Fig. 15-13. A squaring shear uses a straight-stationary and a straight-moving blade to cut sheet material. (Cincinnati Inc.)

Due to their brittleness, sintered oxide tools must be securely held and used only on very rigid machines. This requirement greatly limits their use. Most modern machines cannot operate at the high speed required for cutting efficiency. Further, they are not rigid enough to keep the tool from being damaged. Whenever machine speeds and rigidity requirements are met, ceramic tooling's long life and high cutting speed makes it an excellent tool material.

Other Tool Materials. In addition to the five materials discussed, other materials are used for tools. Diamonds, the hardest known material, are used to cut very hard materials. Their brittleness and poor heat conductance limit them to applications requiring light cuts such as precision boring of holes.

Also, new tooling materials are constantly being developed. One result of this continuing search for better cutting tools is UCON, a non-carbide material developed by Union Carbide Corporation. It contains 50 percent columbium, 30 percent titanium, and 20 percent tungsten. UCON is specially manufactured to produce a hard, tough tool capable of withstanding cutting speeds two to three times that of carbides.

SHEARING TOOLS

Shearing, which will be more fully described in Chapter 23, separates material by using offset opposed edges. Three major types of opposed edges are used: The first, Fig. 15-13, has two straight blades. Whenever straight blades are used the process is called shearing.

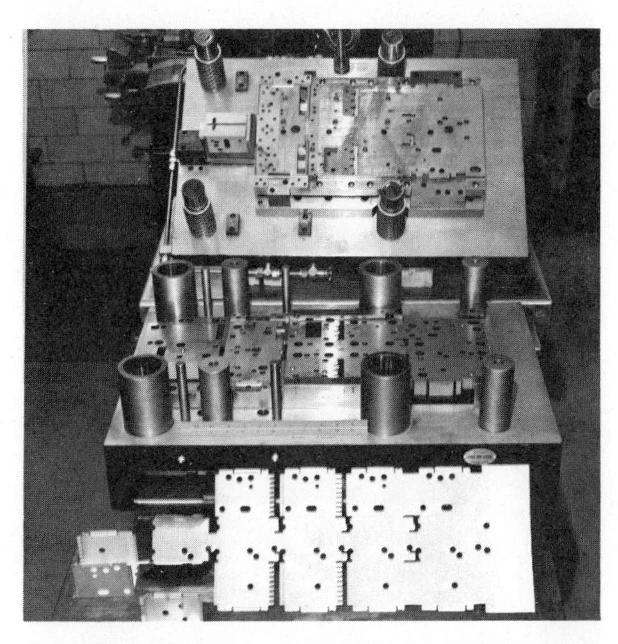

Fig. 15-14. A die set which shears the part then forms it. The material is fed from right to left through five stations to complete the process. (American Metal Stamping Assoc.)

A second type of opposed edges are the punches and dies used in a press. The shearing of material using the curved edges of punches and dies, as shown in Fig. 15-14, is called by more specific names: *blanking, piercing, notching, die cutting,* and *trimming.*

The third type is created by mated rolls or rotating blades. Fig. 15-15 shows sheet metal being slit or sheared using a set of slitting rolls.

A fourth method of shearing uses a single sharp edge. It depends upon movement between the cutting edge and the workpiece to separate the material. The process, as shown in Fig. 15-16, is called *slitting.* The material cut is usually thin stock. The process is often used to cut paper, plastic films, metal foils, and veneer.

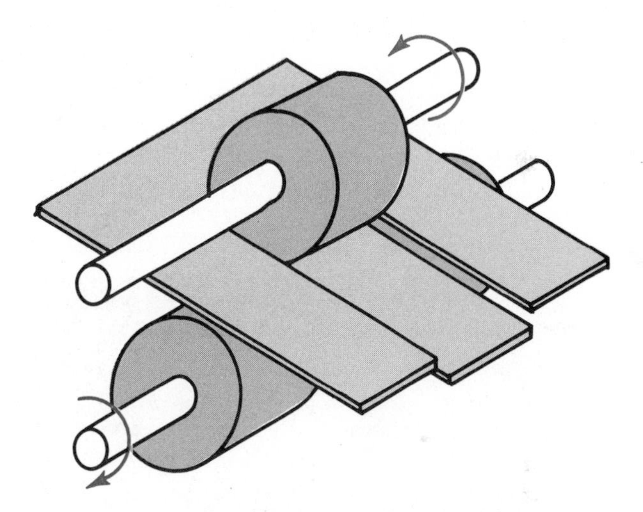

Fig. 15-15. A schematic of roll shearing. Opposed edges on rollers cut material.

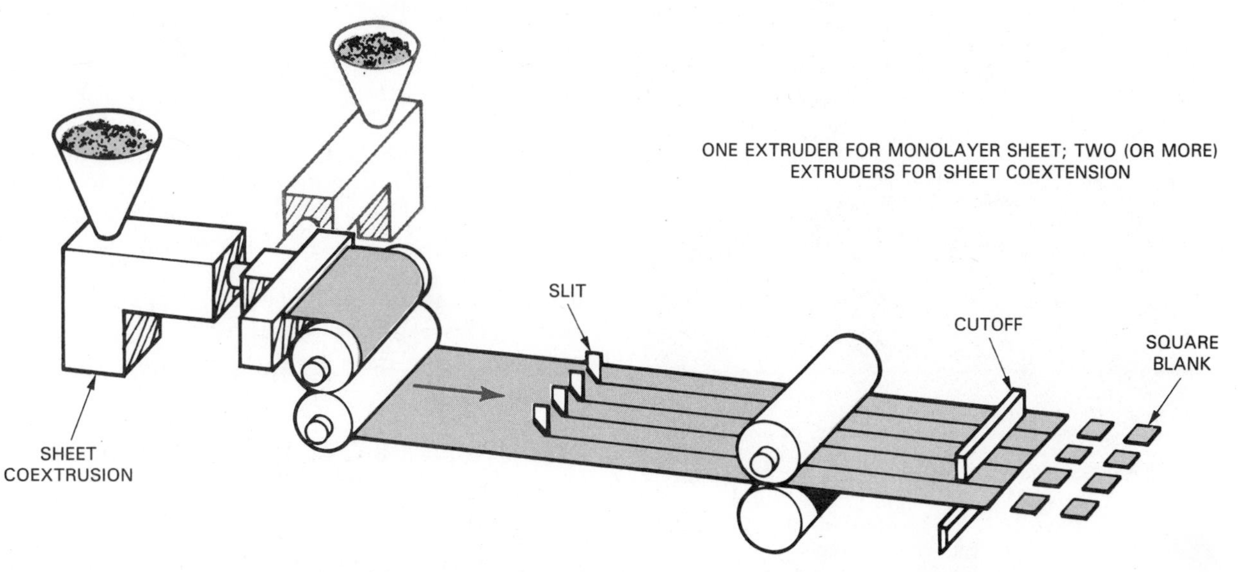

Fig. 15-16. The knife slitting process. (Dow Chemical Co.)

FLAME CUTTING ELEMENT

Flame cutting does not use a "tool" as many people define the term. The process, however, does have a cutting element in the burning gases. Most flame cutting torch tips, like the one pictured in Fig. 15-17, have a series of pre-heating openings through which oxygen and acetylene are fed. This mixture burns at about 1700°F (927°C) and heats the metal to be cut. In the center of the tip is a larger hole through which oxygen is delivered at high pressure (30-60 psi). When the oxygen is fed through the center hole onto the hot metal, rapid oxidation (burning) of the metal occurs. Thus the metal is separated by burning away material on a line between the part and the scrap.

OTHER CUTTING ELEMENTS

Other sources of cutting are used for processes usually called nontraditional machining. These processes, described in Chapter 22, use streams of electrons (electrical sparks), chemical actions, electrochemical action, high intensity beam of light, and sound waves. All these sources of cutting action are of recent origin. Each has its own special, and often limited, use.

Fig. 15-17. Flamecutting is a separating process that uses a flame as the cutting tool. (AIRCO Welding Products)

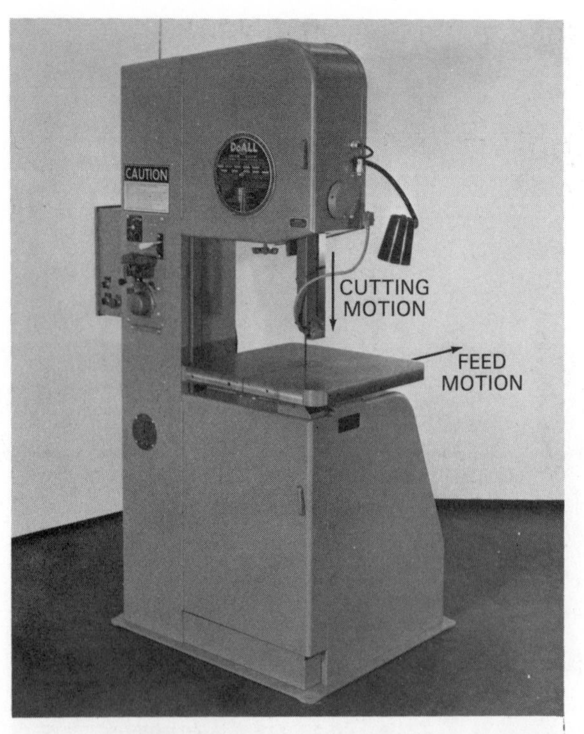

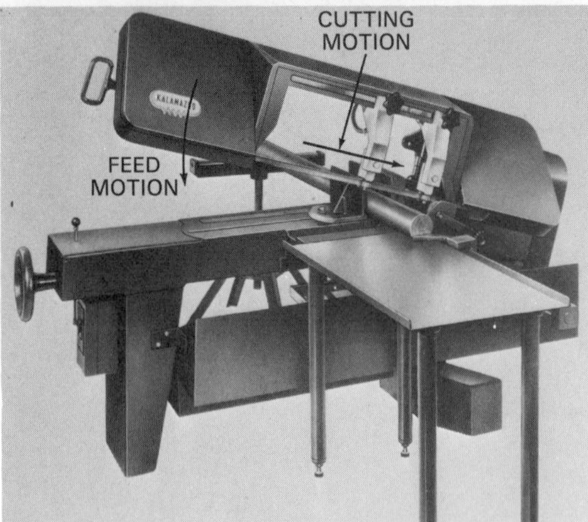

Fig. 15-18. A band saw has a linear cutting motion. Feed motion can be in the workpiece or in the tool. (DoAll Co., and KTS Industries)

MOVEMENT BETWEEN WORK AND THE CUTTING ELEMENT

A separating operation requires a cutting element to penetrate a workpiece. To accomplish this task

Fig. 15-19. A metal shaper has a reciprocating cutting motion. (Rockford Machine Tool Co.)

there must be motion involved. The cutting element, the work, or both must move to create the force required to complete the separating action. The motion generates force that causes the tool to penetrate the work and to overcome the forces holding the material together.

CUTTING MOTION

Movement which causes the cutting action is called *cutting motion*. It separates the unwanted material from the workpiece. Specifically, cutting motion produces the separating action which provides the desired size, shape, and surface finish.

Cutting motions can be one of three types: *linear, reciprocating* (back and forth), or *rotary*.

1. Linear cutting motion creates a chip or some form of waste material by a continuous, straight movement of the cutting element. Typical machines using a linear cutting motion are band saws, Fig. 15-18, broaches, and electrical discharge machines.
2. Reciprocating cutting motion creates a chip by a back and forth motion of the cutting element. Reciprocating motion is actually a series of linear motions with a return stroke between each cut. Typical machines using reciprocating cutting motion are power-driven hack saws, metal cutting shapers, Fig. 15-19, and metal-cutting planers.
3. Rotary or circular cutting motion creates a chip by using a rotating cutting element. Typical machines using rotary cutting motion are milling machines, wood-cutting table (variety) saws, boring machines, drill presses, and woodworking shapers, Fig. 15-20.

Measuring cutting motion

The rate at which the various cutting motions remove material is called *cutting speed*. Cutting speed is the rate at which chips move past the cutting tool. It is the distance the tool moves through the work or the work moves past the tool in a given period of time. Cutting speed is measured in either feet per minute (fpm), surface feet per minute (sfpm), or meters per minute (m/min).

It is quite easy to calculate the cutting speed of an operation. For rotating cutters or work, or for linear

Fig. 15-20. A milling machine table saw. Both milling machine and table saw have rotary or circular cutting motions. (Cincinnati Milacron and Delta International Machinery Co.)

cuts driven by a rotating wheel (a band saw), the formula is:

CS = Circumference × Revolutions per minute (in feet or meters)

$$\text{Circumference} = \frac{\pi D}{12} \text{ when D is given in inches}$$

$$= \frac{\pi Dm}{1000} \text{ Dm is always given in millimeters}$$

(Dm = Diameter – metric)

$$\pi = 3.14$$

Example. The cutting speed of a 14 in. wood cutting band saw driven by a 1725 rpm direct-drive motor is calculated as follows:

$$CS = \frac{\pi D}{12} \times \text{rpm}$$
$$= \frac{3.14 \times 14}{12} \times 1725$$
$$= 3.66 \times 1725$$
$$CS = 6313.50 \text{ fpm}$$

The formula may also be used to determine the correct machine speed (rpm) when the cutting speed is known. (Charts are available which give cutting speeds of common materials for the various tool materials – high-speed steel, cast carbides, carbides, etc.) For this application the formula is:

$$RPM = \text{Cutting Speed} \times \frac{12}{\pi D} \text{ (when D is given in inches)}$$

$$RPM = \text{Cutting Speed} \times \frac{1000}{\pi Dm} \text{ (when Dm is given in millimeters)}$$

Example: The cutting speed of mild steel using high-speed steel cutters is approximately 70. The correct rpm for drilling a 1/2 in. hole would be determined as follows:

$$RPM = CS \times \frac{12}{\pi D}$$
$$= 70 \times \frac{12}{3.14 \times .5}$$
$$= 70 \times 7.65$$

RPM = 535 (You would set the drill press to operate at about 535 rpm.)

Cutting speeds for reciprocating machines are calculated using a different formula. For hack saws, metalcutting shapers and planers and other reciprocating machines the following formula is used:

$$CS = \frac{\text{Length of the stroke}}{12} \times \text{strokes per minute}$$

(When length of stroke is given in inches)

To calculate the strokes per minute needed to generate a certain cutting speed is somewhat more difficult. Reciprocating machines cut only on the forward stroke. Also, the forward cutting stroke is slower than the return stroke. This ratio of cutting time (forward stroke) to noncutting (return stroke) time must be factored into the formula.

For mechanically driven shapers, the ratio is about 1.5 to 1. Hydraulically driven shapers have about a 2 to 1 ratio. The formula for strokes per minute needed to generate a certain cutting speed then is:

$$N = CS \times \frac{12}{L} \times \text{Cutting Ratio}$$ when Cutting Ratio is percent of the stroke time used for cutting

Example. For shaping mild steel using a hydraulic drive shaper with a 6 in. stroke and a HSS cutter (cutting speed = 70) the following procedure is used to determine the proper number of strokes per minute. (On hydraulic shapers, speed may be expressed in fpm.)

$$N = 70 \times \frac{12}{6} \times \frac{2^*}{3} = 70 \times 2 \times .667 = 93.3$$ strokes per minute

*Ratio 2 to 1. Cutting stroke is two units and return stroke is one unit. Total units is three.

The proper setting of the machine speed is very important. *Cutting speed* is the most important factor in tool life. If cutting speed is too fast, the tool will build up excessive temperatures and "burn." A too-slow cutting speed is inefficient. Time is wasted by not cutting away material as fast as possible.

FEED MOTION

Feed motion is the second motion between the cutter and the work. It brings new material into contact with the cutting element or tool. Cutting motion removes the chip. Feed motion brings new material into contact with the tool so that cutting can continue.

For example, if a piece of stock is placed against an idle band saw blade no cutting will take place. When the machine is turned on, the first tooth moving against the work will cut. It will produce a chip. This first tooth has cutting motion. The next tooth will not cut because it will pass through the area cut by the first tooth. It will only cut when the work is moved forward to force the second tooth to penetrate new material. The material must be fed into the blade if continuous cutting is to happen. There must be *feed motion* present if the cutting action is to continue. The drawing in Fig. 15-21 shows this combination of feed and cutting motions.

Measuring feed motion

Feed motion is a measurement of linear (straight-line) movement. It is the movement of the cutting element along or into the work in one cycle of the operation. For machines on which the work turns, or the cutter rotates, feed is often given as inches or millimeters per revolution. If multitooth cutters are used, feed is sometimes given as inches or millimeters or movement per tooth. When the cutting action is reciprocating, the feed is expressed as inches or millimeters of movement per stroke.

Feed rates are determined using simple formulas. For drilling and turning operations the feed rate is the linear distance the cutter or work advances with each revolution of the work or cutter. The formula to determine this rate is:

$$\text{Inches (millimeters) per revolution} = \frac{\text{In. (mm) per minute}}{\text{Revolutions per minute}}$$

$$\text{ipr} = \frac{\text{ipm}}{\text{rpm}}$$

For multitooth cutters such as milling cutters and circular saw blades the feed rate is the distance the work moves between the cutting action of each successive tooth. The feed rate is given in inches (millimeters) per tooth. The formula to determine this feed rate is:

$$\text{inches (mm) per tooth} = \frac{\text{inches per revolution}}{\text{number of teeth}}$$

For reciprocating machines, the feed rate is expressed in inches per stroke. This rate is generally directly set on the machine.

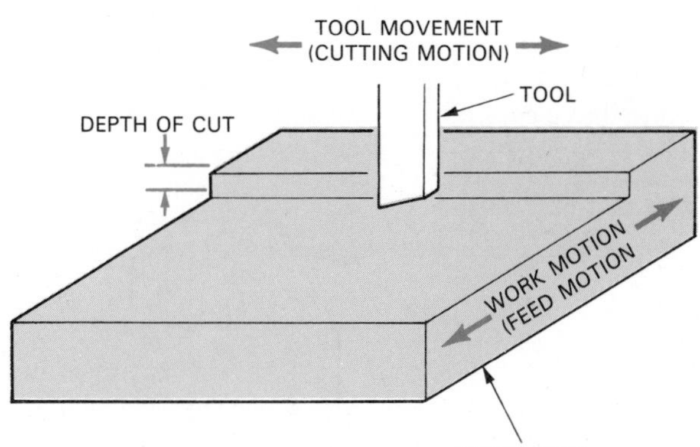

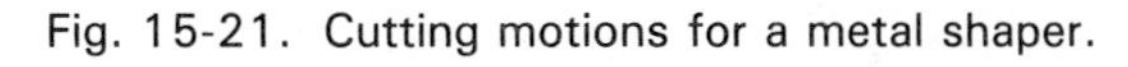
Fig. 15-21. Cutting motions for a metal shaper.

Feed is directly related to a number of factors. In general, it is reduced as:

1. Material hardness increases.
2. Speed is increased.
3. Tools become dull.
4. Tool temperature increases.
5. Less coolant is used.

Feed and surface finish

A general rule suggests that as feed rates are increased the quality of the surface will decrease. A single-point tool, as shown in Fig. 15-22, will leave wider and deeper grooves as feed is increased.

Rotating cutters cut in a series of arcs. As the feed is increased, the width and depth of the arcs or ripples increases. Fig. 15-23 shows this action which is called millmarks in woodworking.

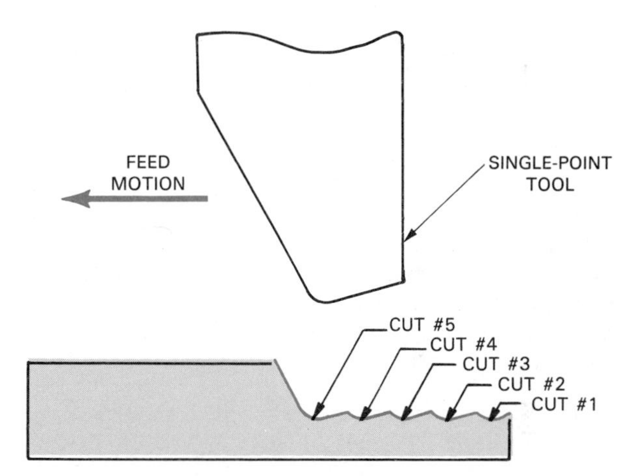

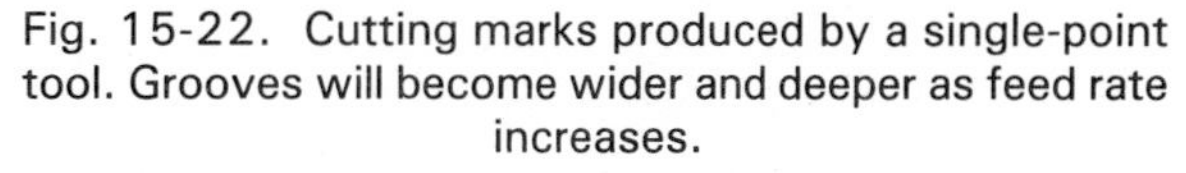
Fig. 15-22. Cutting marks produced by a single-point tool. Grooves will become wider and deeper as feed rate increases.

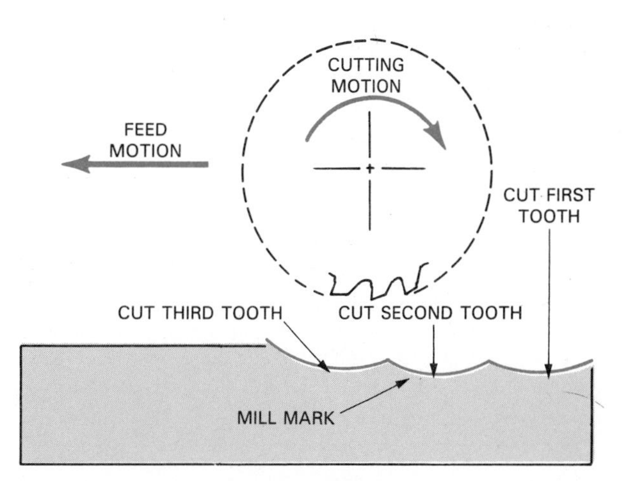

Fig. 15-23. Mill marks are produced in wood by rotary cutters.

Increased feed rate causes more material to be removed with each tooth or pass of the cutter. The increased amount of material being removed increases the force against the cutting element. The increased force may cause a less rigid tool or machine to produce chatter. Chatter is an uneven cut caused by a vibrating or bouncing tool.

The proper balance between cutting speed and feed rate, plus a third factor, depth of cut, is needed if smooth efficient separating is to take place.

DEPTH OF CUT

Depth of cut is a measurement of the distance a cutting element penetrates the work. The depth of cut for revolving or reciprocating tools used on flat work is the difference between the original thickness and the new thickness. This measurement is shown in Fig. 15-24.

With operations in which the tool is fed into revolving work, the depth of cut still is the distance the tool penetrates the work. However, the size reduction will be twice the depth of cut.

Depth of cut is usually given in thousandths of an inch or millimeters for metals and most synthetic materials. Wood, being much softer, can withstand much greater depth of cuts. Many joinery operations have depth of cuts as small as a fraction of an inch or several centimeters.

CUTTING AND FEED MOTIONS FOR COMMON MACHINES

All machines which use a cutting tool have a cutting motion and a feed motion. These motions may be linear, reciprocating (back and forth), or rotating (revolving). Cutting motions may be developed by the motion of either the work or the tool. Feed motion, likewise, may involve either work or tool movement.

This relationship of three types of motion applied to either the work or the tool to generate cutting and feed motions produces eight basic machine groups. In a general way, these groups are described in Fig. 15-25. Chapters 16-24 contain complete descriptions.

SUPPORTING WORK AND TOOLS

The third essential element in separating processes involves holding or supporting the work and the tool. Each of the three types of movement—rotating, linear, and reciprocating—involves special holding devices. Furthermore, holding of either the work or the tool presents its own holding or support needs. The supporting devices may be grouped into categories.

1. Rotating tools.
2. Reciprocating tools.
3. Linear-moving tools.
4. Rotating work.
5. Reciprocating or linear-moving work.

HOLDING ROTATING TOOLS

A rotating tool must be held so that it can revolve around an axis. This need may be accomplished by three major devices—an arbor, a chuck, or a cutter head.

An *arbor,* Fig. 15-26, requires a cutting tool with a hole in its center. The tool, which may be a cutter, a saw blade, or a grinding wheel, is placed on the arbor. The cutter is then secured by tightening an arbor nut. The nut will force collars or bushings tightly against the cutter thereby holding it in place.

A *chuck* is used to hold drills, taps, reamers, and router cutters. The chuck has two or three segments which grip the cutting device when the chuck is

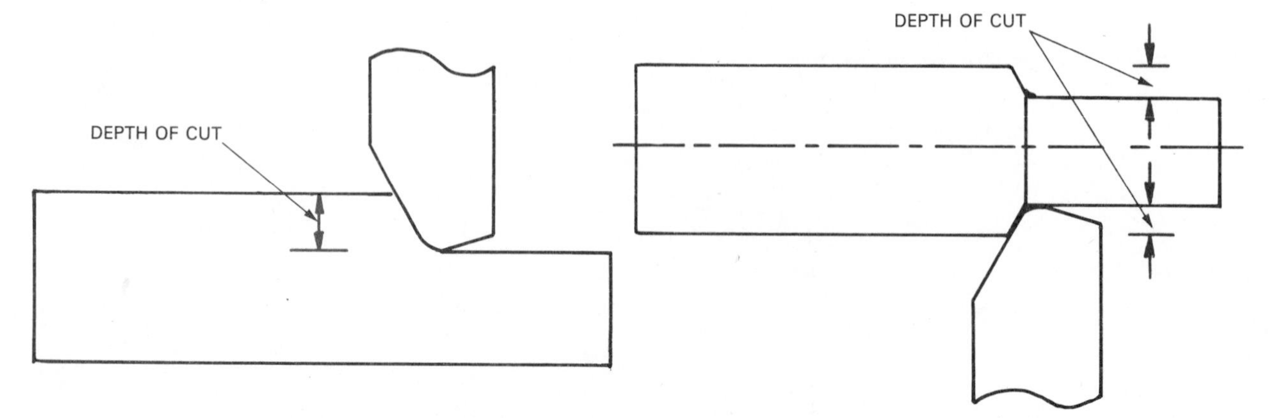

Fig. 15-24. Depth of cut (colored arrows) are shown on flat cuts, left, and cylindrical parts, right.

CUTTING ACTION	FEED MOTION	DIAGRAMS	TYPICAL OPERATION	TYPICAL MACHINES
Rotating work	Linear movement of the tool	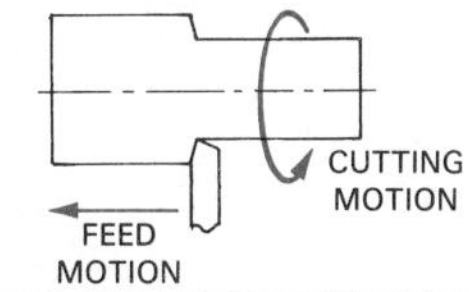 CUTTING MOTION FEED MOTION	Turning Boring Reaming	Lathe
Rotating tool	Linear movement of the tool	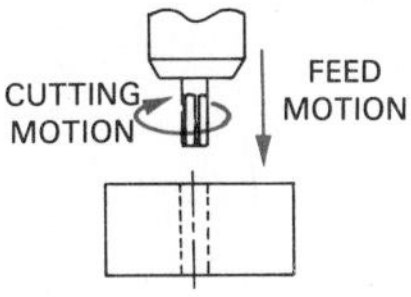 CUTTING MOTION FEED MOTION	Drilling Grinding Sawing	Drill press Cylindrical grinder Radial saw Cutoff saw
Rotating tool	Linear movement of the work	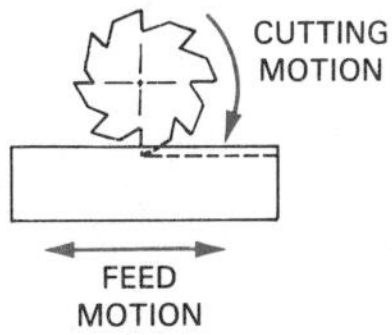 CUTTING MOTION FEED MOTION	Milling Shaping (wood) Routing Planing (wood) Jointing Sawing Sanding	Milling machine Wood shaper Router Surfacer Jointer Circular saw Drum sander Disc sander Spindle sander
Rotating tool	Reciprocating work	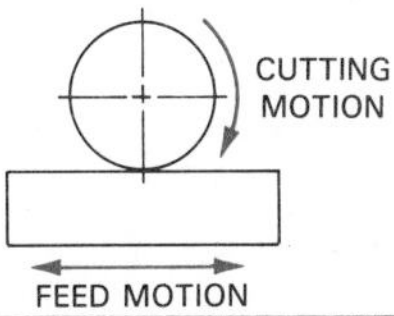 CUTTING MOTION FEED MOTION	Grinding	Surface grinder
Linear moving tool	Linear movement of the tool	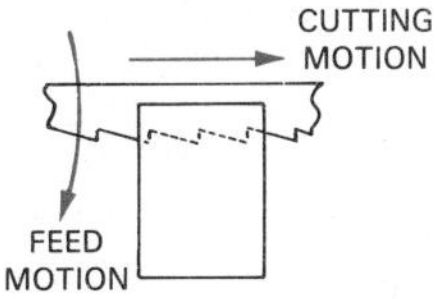 CUTTING MOTION FEED MOTION	Broaching Sawing Sanding	Broach Horizontal band saw Belt sander (portable)
Linear movement of the tool	Linear movement of the work	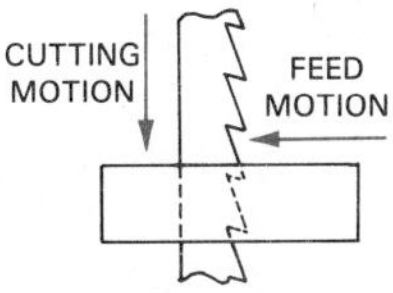 CUTTING MOTION FEED MOTION	Sawing Sanding	Vertical band saw Belt sander (stationary)
Reciprocating tool	Linear movement of the tool	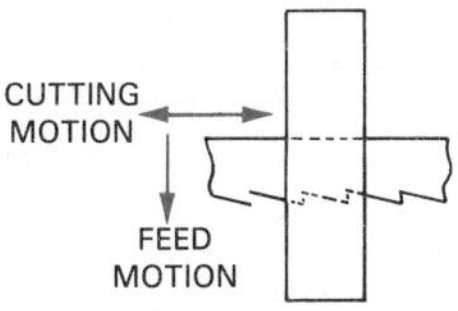 CUTTING MOTION FEED MOTION	Sawing	Hacksaw
Reciprocating tool	Linear movement of the work	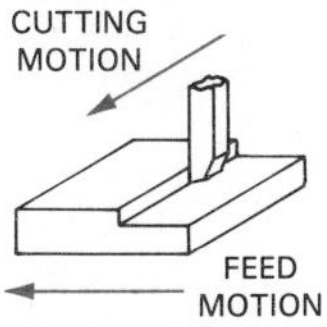 CUTTING MOTION FEED MOTION	Sawing Shaping (metal)	Jig saw Metal shaper
Reciprocating work	Linear movement of the tool	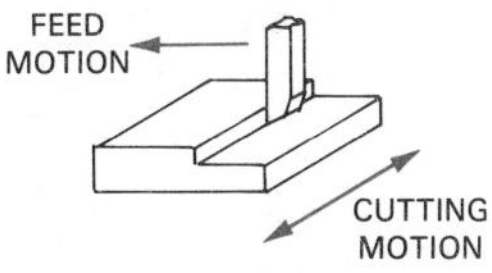 FEED MOTION CUTTING MOTION	Planing (metal)	Metal planer

Fig. 15-25. Machines can be grouped by cutting and feed motions.

Fig. 15-26. An arbor is used to hold a rotating multiple-point tool such as milling cutters. (POLAMCO Machine Tools)

Fig. 15-27. A chuck is used to hold drills, end-cutting milling cutters, router bits, and reamers.

tightened. Fig. 15-27 shows a typical chuck.

The third method of holding rotating cutting tools is a *cutter head.* This device, shown in Fig. 15-28, is widely used in woodworking machines. The head is a basic part of the machine. Removable knives or cutters are clamped into the head.

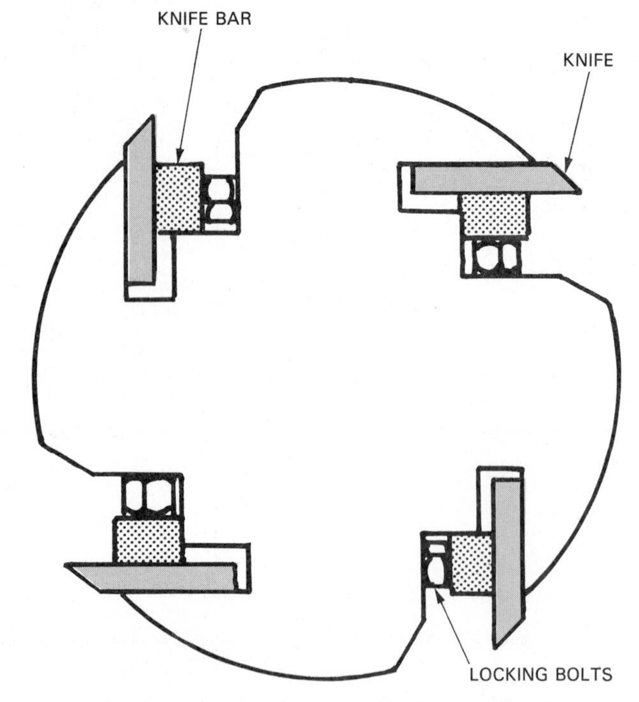

Fig. 15-28. A cutter head holds knives in jointers, planers, and molders.

Other devices used to hold rotating cutters are *expandable drums* which hold abrasive sleeves and *rotating discs* to which abrasives may be attached.

HOLDING RECIPROCATING TOOLS

Reciprocating tools may be either single-point or multipoint. Single-point reciprocating tools are used mainly on the metal shaper. They are secured by a *tool holder* as shown in Fig. 15-29. The hacksaw and sheet abrasives are the principle reciprocating multi-point tools.

HOLDING LINEAR MOVING TOOLS

A linear-moving tool is typically a continuous band. The band may be a *saw blade* or an *abrasive belt.* This type of tool, as can be seen in Fig. 15-30, is supported by two or more wheels. One wheel is power driven and causes the band to move.

A *broach* is another type of linear moving tool. It is held in a special machine and pushed or pulled through a hold to create its cutting action. (Sometimes the broach is stationary while the work is forced by the tool).

Fig. 15-29. Tool holders are used for gripping cutting tools on metal cutting shapers, planers, and lathes. (Rockford Machine Tool Co.)

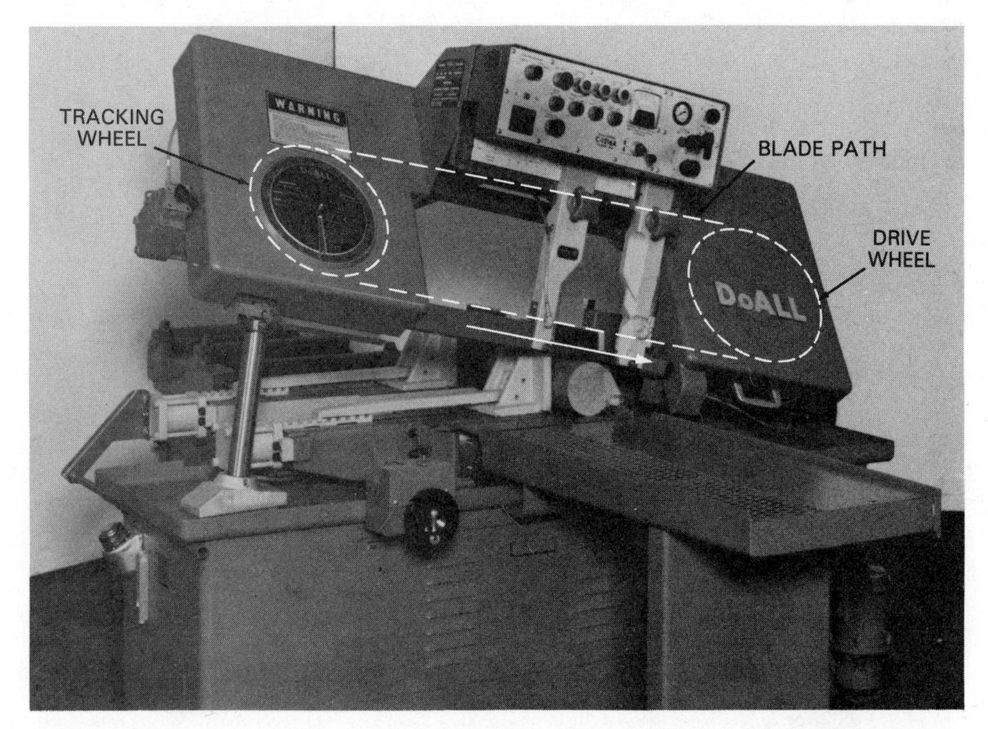

Fig. 15-30. Two wheels are used to hold and drive a band saw blade. (DoAll Co.)

HOLDING ROTATING WORK

Like a rotating cutter, rotating work must spin around an axis. The principal methods of rotating work are either to hold it between centers, or attach it to a chuck or *faceplate,* Fig. 15-31.

Material is held between centers when long cylindrical parts are being produced. A chuck or faceplate is used to hold discs, short, round parts, and rectangular or square material.

HOLDING RECIPROCATING OR LINEAR MOVING WORK

Work which reciprocates or moves in a linear mode must be supported. If the material is metal it is

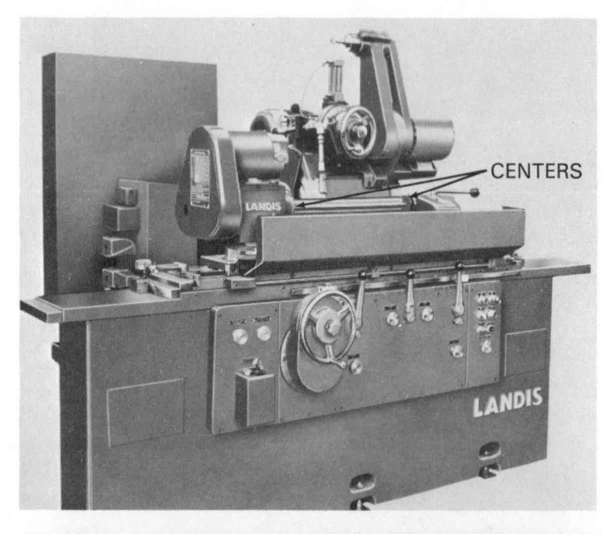

Fig. 15-31. Rotating work may be held by either of two ways. Top. Between centers. Bottom. By a chuck. (Landis Tool Co. and LeBlond Makino Machine Tool Co.)

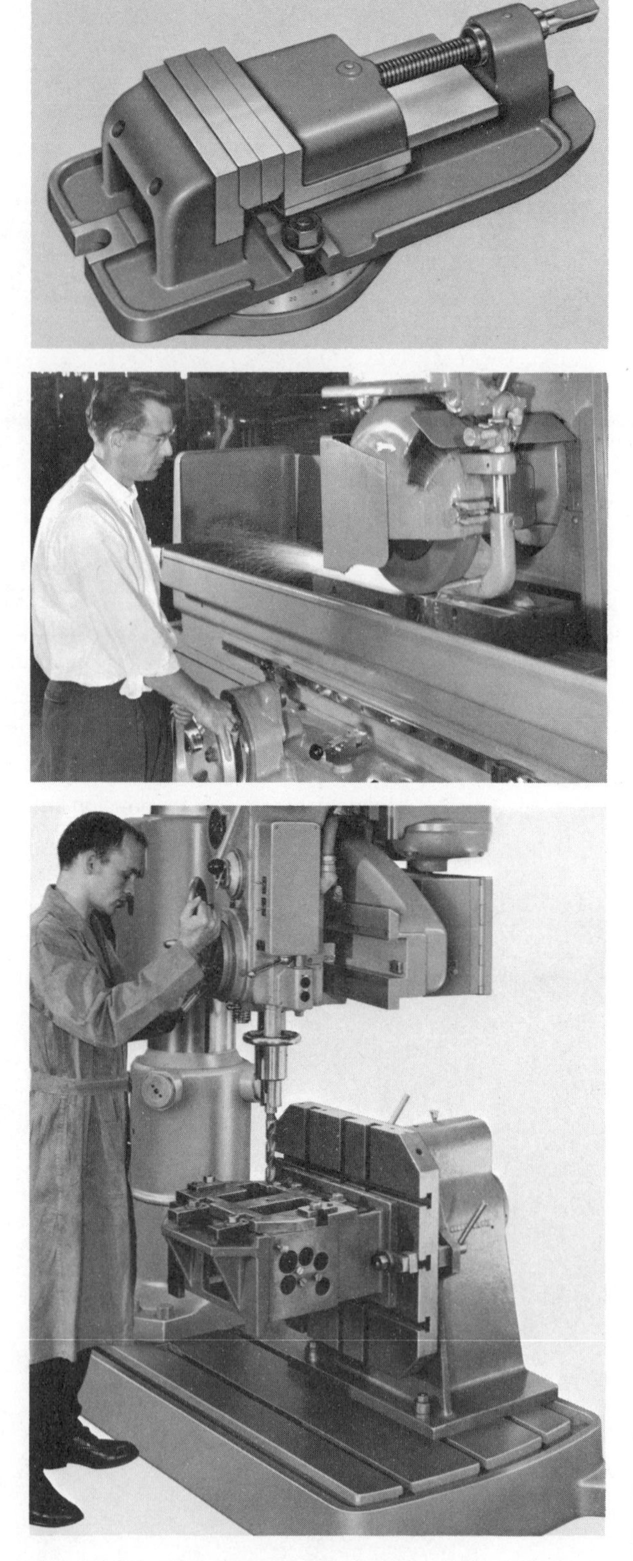

Fig. 15-32. Flat work can be held by several devices. Top. In vices. Middle. Magnetic tables. Bottom. By t-slot bolts. (The Norton Co., and LeBlond Makino Machine Tool Co.)

Fig. 15-33. Hand feeding material is a common practice in woodworking industries. (Delta International Machinery Corp.)

usually clamped into place. Typical clamping methods include vices, magnetic tables, or T-slot bolts, Fig. 15-32.

Material may simply be supported by a table and hand fed into the cutter. This is typical of many working machines. Fig. 15-33 shows an operation in which wood is being hand fed over a table into a rotating cutter.

SUMMARY

Separating processes must have a cutting element, motion between the cutting element and the work, and a means to hold or support the work and the cutting element. Typical cutting elements used in separating are tools, flames, and nontraditional means (sparks, chemicals, light, sound, etc.).

Motion between the cutting element and the work generates the force to produce a chip or fracture the material. This activity is called cutting motion. Another motion, called feed motion, brings new material into contact with the cutting element.

In separating, three types of motion are possible: rotating, reciprocating, and linear. The tool and the work may have any of these three motions during a specific separating operation. In fact, machines and separating acts may be grouped by the combination of tool and work motions present.

The various tool and work motions require special methods of support. Tools must be held so that they can rotate, reciprocate or move in a linear path. Likewise, the work must also be supported for one of the three modes of motion.

Separating is truly a group of very important processing techniques. Our manufacturing ability is based on our ability to:

1. Separate materials to produce precision parts.
2. Prepare molds, pattern, and dies for other processes.
3. Build machines for all types of work.

STUDY QUESTIONS – Chapter 15

1. Define separating.
2. List and define the three major groups of separating processes.
3. Name the three essential elements of all separating acts.
4. List and briefly describe the three major types of cutting elements.
5. Give examples of single-point tools, multiple-point tools, and random-point tools.
6. Briefly describe the cutting action of a tool.
7. What are Type 1, Type 2, and Type 3 chips?
8. Why is the shape of a cutting tool important?
9. What is the purpose of relief angles, cutting edge angles, and rake angles?
10. List and briefly describe five major tool materials.
11. List and describe the four major shearing tools.
12. Describe the basic types of cutting motions and give two examples of common machines that use each type.
13. Describe the process of determining cutting speed for rotating cutters, linear moving cutters, and reciprocating cutters.
14. What is meant by cutting motion and feed motion?
15. Define depth of cut.
16. List the combinations of tool and work movements used to produce the cutting and feed motions and give an example of a machine which uses each combination.
17. List the major tool-holding devices and give an example of a tool which is held by each type.

Cylindrical grinding is shown using a chuck to hold rotating workpiece. Tool is an abrasive wheel attached to an arbor. (Clausing Corp.)

Chapter 16

TURNING AND RELATED OPERATIONS

Turning includes those operations that produce a cylindrical or conical (cone-shaped) part. These are among the simplest of all machining operations. They involve the use of a rotating workpiece which creates the cutting motion. A single-point tool is generally used. Its motion provides the feed motion. The tool is usually fed at either right angles (90°) to or parallel with the work surface. Fig. 16-1 shows two typical turning operations: straight turning and facing. Note the direction of the cutting motion and feed motion. Also observe the depth of cut for both operations.

TYPES OF TURNING OPERATIONS

There are a number of turning operations which are common to the machining of metals, woods, and plastics. Among the most important are:

1. Straight turning.
2. Taper turning.
3. Contour turning.
4. Facing.
5. Forming.
6. Necking.
7. Parting.
8. Boring.
9. Threading.
10. Knurling.

On machines designed for turning some nonturning techniques may be performed. Included in this list are drilling, reaming, and grinding.

STRAIGHT, TAPER, AND CONTOUR TURNING

The basic turning operations are straight, taper, and contour turning. These operations are performed on the external (outside) surface of the workpiece. The work, as shown in Fig. 16-2, is rotated on an axis (centerline). The single-point tool is fed along the surface.

If the tool is fed parallel to the work axis, a uniform diameter workpiece will be produced. This type of turning is called *straight turning*. When the tool is fed in a straight line which is not parallel to the work axis, a taper is produced. The part will have a uniformly decreasing diameter along its length. This turning operation is called *taper turning*.

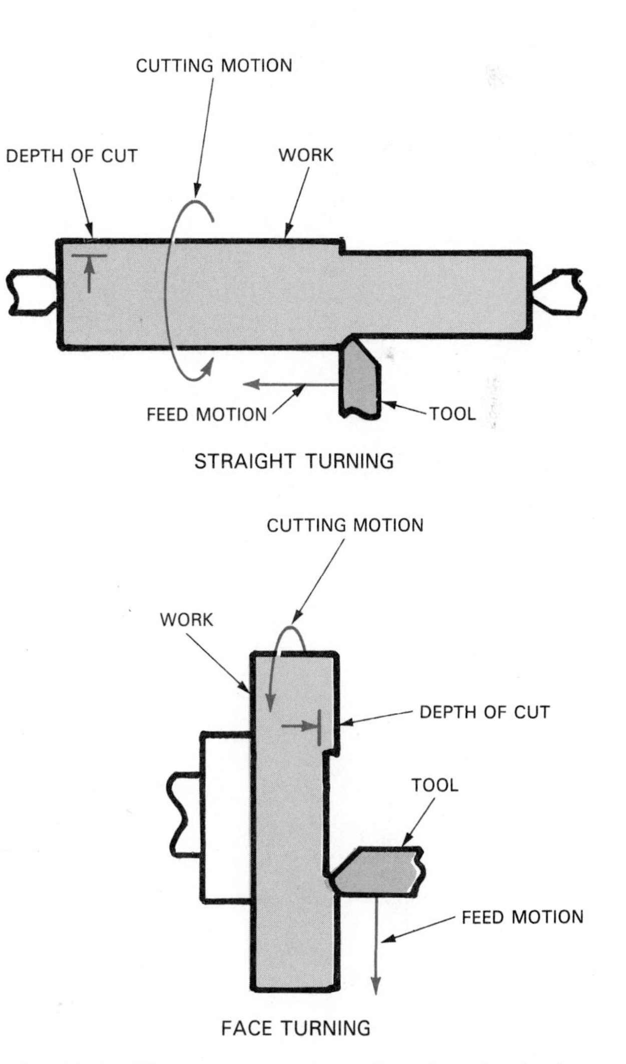

Fig. 16-1. There are two types of turning. Study the lathe terms; they will be used throughout this chapter.

Fig. 16-2. A straight turning operation. Tool motion is parallel to the axis (centerline) of the workpiece. (Clausing Corp.)

Contour turning is done by causing the tool to follow an irregular (curved) path. The part will have varying diameters along its length. Fig. 16-3 shows diagrams of straight, taper, and contour turning.

Most basic turning operations involve one or more roughing cuts which produce the correct size and shape. The roughing cuts use fairly heavy cut depths and fast feed rates. They are followed with lighter finishing cuts at slower feeds. These finishing cuts produce the specified surface finish on the part.

FACING

Some turning operations cut with the tool moving across the end of the rotating workpiece, Fig. 16-4. These operations, called facing operations, may be either end facing or shoulder facing.

End facing trues and smooths the end of the workpiece. The tool often engages the work at its center (axis) and is fed outward. The depth of cut is established when the tool penetrates the work. This penetration is caused by feeding the tool along the axis

Fig. 16-4. A typical facing operation is performed on a metal-cutting lathe. (LeBlond Makino Machine Tool Co.)

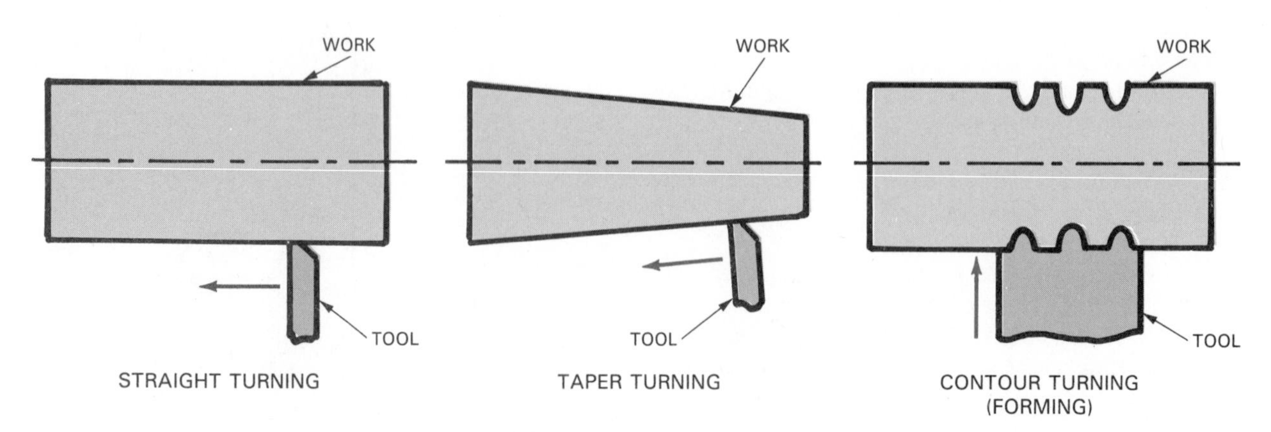

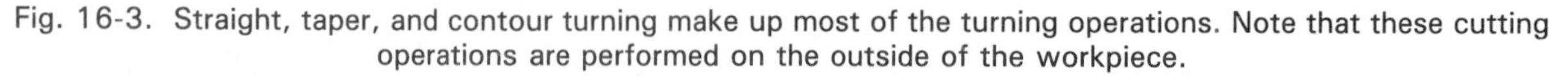

Fig. 16-3. Straight, taper, and contour turning make up most of the turning operations. Note that these cutting operations are performed on the outside of the workpiece.

(into the work). The depth of cut is equal to the amount the length of the part is reduced.

Shoulder facing trues and smooths a shoulder previously cut, cast, or formed on the part. The cutting takes place as the tool is fed outward from the bottom of the shoulder. End facing and shoulder facing are shown in Fig. 16-5.

FORMING

Forming involves feeding a special cutter into the revolving work. The shape of the single-point tool determines the form of the cut. Form turning may be either internal or external. External forming involves feeding the tool into the cylindrical surface of the workpiece. Internal forming requires the forming tool to be fed into the end of the work.

Wood is often formed and contour turned by using standard tools rather than specially shaped tools. These tools are usually hand fed into the work to shape the outside and produce internal cavities. Fig. 16-6 shows typical internal and external wood turning operations. High-production turnings may be done by automatic wood lathes and certain special forming tools.

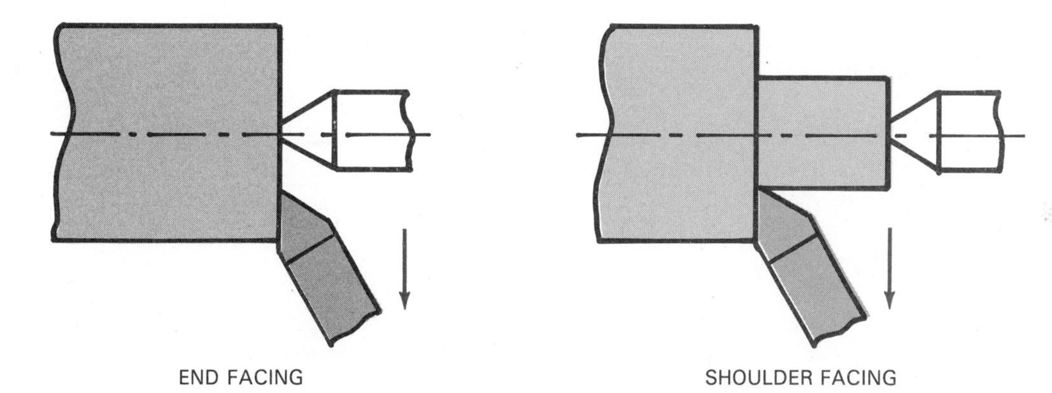

Fig. 16-5. Types of facing operations. Left. In end facing, the tool often starts the cut at the center and works to the outside. Right. The tool engages the work at the bottom of the shoulder.

Fig. 16-6. Left. Faceplate turning. Right. External contour turning. (Delta International Machinery Corp.)

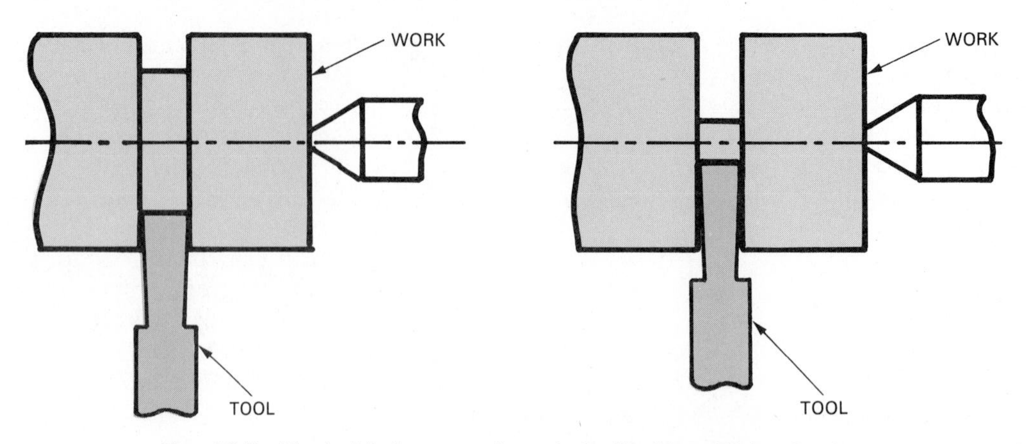

Fig. 16-7. Typical lathe operations. Left. Necking. Right. Parting.

NECKING AND PARTING

Necking involves the feeding of a square-pointed tool directly into the work. This operation cuts a shoulder or establishes a diameter.

If this operation is continued, the part will be cut off from the workpiece. When this happens the operation is called *parting*. Diagrams in Fig. 16-7 show typical necking and parting operations as they might be performed on wood, metal, or plastic.

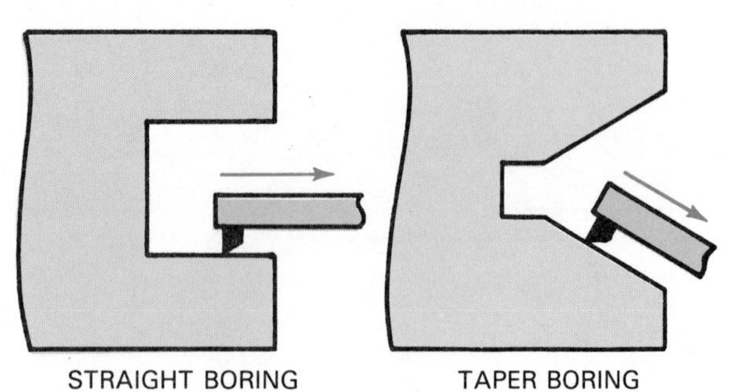

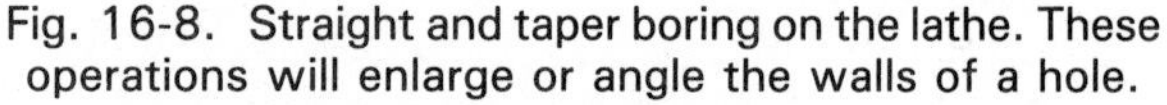

Fig. 16-8. Straight and taper boring on the lathe. These operations will enlarge or angle the walls of a hole.

BORING AND INTERNAL FORMING

Boring involves enlarging a hole in a part which is held in a chuck or on a faceplate. The goal of the operation is to insure that the hole is concentric (round) and is in the proper location. The boring operation corrects any errors caused by a drill which drifted off the centerline or produced an eccentric (out of round) hole.

Boring may also change a straight hole into a tapered hole. This operation requires the tool to be fed outward at an angle. Straight and taper boring operations are shown in Fig. 16-8.

Reaming, which is a nonturning operation, may also be used to accurately size a hole. This operation, when done on a lathe, involves feeding a stationary reamer into the revolving workpiece. Fig. 16-9 shows a typical reaming operation. Holes may be drilled the same way by replacing the reamer with a twist drill.

Fig. 16-9. Reaming on a lathe. Tool maintains straightline motion while workpiece rotates. (LeBlond Makino Machine Tool Co.)

THREADING

Most threads are not produced on a lathe. Threads on common bolts, threaded rods, and nuts are produced by roll forming (which was described in Chapter 11) or by *thread chasing*. However, when very accurate threads are needed, or when a special threaded part is required, lathe threading is often used. The cutting of threads requires a specially shaped tool which is fed along the work at a uniform (even) speed.

When cutting threads, the movement of the tool is controlled by the machine. A half-nut grips the lead screw of the lathe. As the lead screw turns, the carriage with the tool is uniformly moved along the work. Fig. 16-10 shows a typical threading operation.

KNURLING

Knurling is a cold-forming process often done on a lathe. The operation, Fig. 16-11, produces a roughened surface on the workpiece. A special tool is fed into and along the work. The knurling tool has two toothed rolls. When the rolls contact the work they revolve with it and cause the workpiece to be deformed. The resulting diamond- shaped pattern is seen on many adjusting knobs and locking devices.

THE BASIC TURNING MACHINE

The basic turning machine is called a lathe. These machines have some common elements. All lathes:
1. Have a basic structure.
2. Use a cutting tool.
3. Provide a way to support and move the tool.
4. Hold and rotate the work.

Fig. 16-10. Thread cutting on a lathe. Tool is attached to carriage which is moved along by lead screw. (LeBlond Makino Machine Tool Co.)

BASIC STRUCTURE OF LATHES

The basic structure of lathes involve four main parts. These parts, as seen in Fig. 16-12, are:
1. Bed.
2. Headstock.
3. Tailstock.
4. Tool holding device.

Fig. 16-11. A knurling operation. Tool has toothed rolls which produce a diamond pattern on the workpiece. (LeBlond Makino Machine Tool Co.)

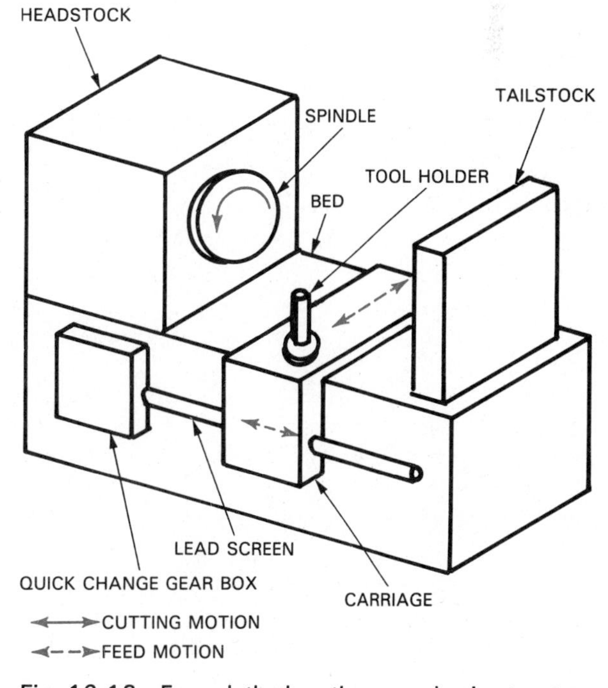

Fig. 16-12. Every lathe has the same basic structure shown here.

The *bed* of the lathe, Fig. 16-13, is the foundation of the machine. It provides a rigid base for the other main parts. The bed must resist vibration and deflection (twisting).

The bed of most lathes is made of cast iron and contains two parallel, machined top surfaces called *ways.* The ways allow the tailstock and the tool-holding device to be moved. The ways also insure that the headstock and the tailstock remain in alignment. The bed is generally attached to two legs or end housing units.

The *headstock* contains the drive or power element. It remains fixed and is almost always on the left end (facing the lathe) of the bed. The headstock generally contains a hollow spindle and a speed changing mechanism.

The hollow spindle is designed to hold devices that support and rotate the work. It is accurately positioned in the headstock and runs in heavy bearings. The diameter of the hollow spindle limits the size of stock which may be fed into a chuck from the headstock end of the lathe.

Spindle speed may be controlled by a gear box or pulleys. Most metal cutting lathes use a gear box. Wood (or speed) lathes usually use step pulleys or variable diameter (variable speed) pulleys to control spindle speed.

The *tailstock* is on the opposite end of the bed from the headstock. It is usually adjustable along the length of the bed. The spindle of the tailstock does not rotate. It can, however, be fed in and out of the tailstock. This allows clamping of stock between centers, or for drills and reamers to be fed into the work.

The *tool-holding element* varies greatly between types of lathes. Metal cutting lathes, wood lathes, and spinning lathes each have different tool-holding devices. These will be discussed later.

The *lathe size* is directly related to its basic structure. Size is expressed as *swing,* the diameter of work which can be accommodated above the ways. Thus, lathes are called 13 in. (33 cm), 16 in. (40.6 cm) etc. A second size measurement is often included to further specify lathe capacity. This dimension is the maximum length of stock which will fit between centers. (Some manufactures give the bed length instead.) Using this measurement, we might have 36 in. (1 m) lathes or 60 in. (1.5 m) lathes.

Fig. 16-13. A large lathe. Compare its main components with the drawing in Fig. 16-12. (LeBlond Makino Machine Tool Co.)

LATHE TOOLS

Lathe tools are the single-point devices which produce the chip. There are two major categories:
1. Metal-cutting lathe tools.
2. Wood-cutting lathe tools.

Metal-cutting lathe tools

Most metal-cutting lathe tools are either square pieces of high-speed steel or carbide-insert tools. These two generally have a side-cutting edge and an end-cutting edge. The edges are produced by grinding the proper relief, rake, and cutting-edge angles. These angles, as shown in Fig. 16-14, vary with the tool material and the material being cut.

Also, the shape of the tool will change for each of the various turning operations as described earlier. Typical turning tools are pictured in Fig. 16-15 and they include:
1. Right hand turning tools.
2. Left hand turning tools.
3. Round nose turning tool.
4. Left hand facing tool.
5. Right hand facing tool.
6. Threading tool.
7. Cutoff (parting) tool.

The uses of these basic tools are shown in the sketches contained in Fig. 16-16. Note that they may be used for both external and internal turning operations. These tools are placed about 5 degrees above center for straight turning. For taper turning, boring, and thread cutting, the tool is placed exactly on center.

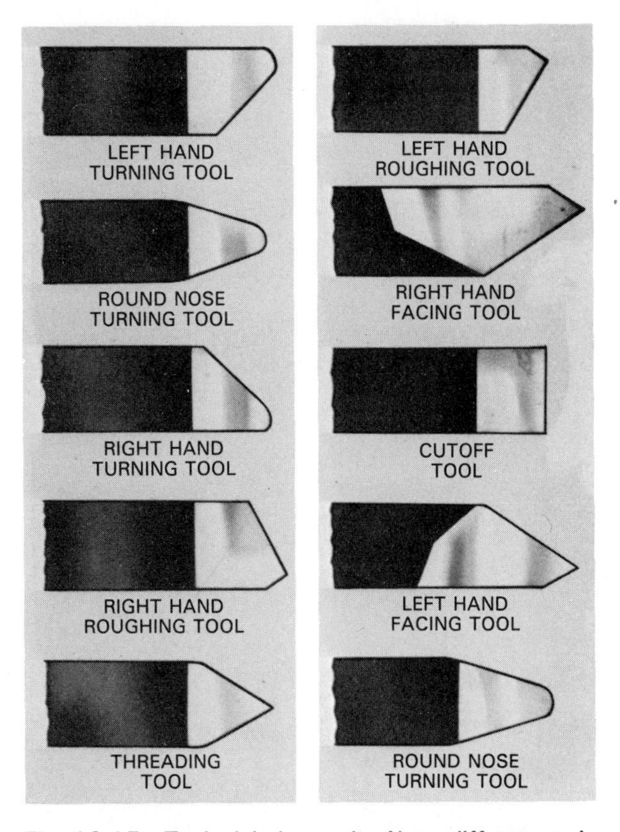

Fig. 16-15. Typical lathe tools. Note differences in shapes. (Armstrong Bros. Tool Co.)

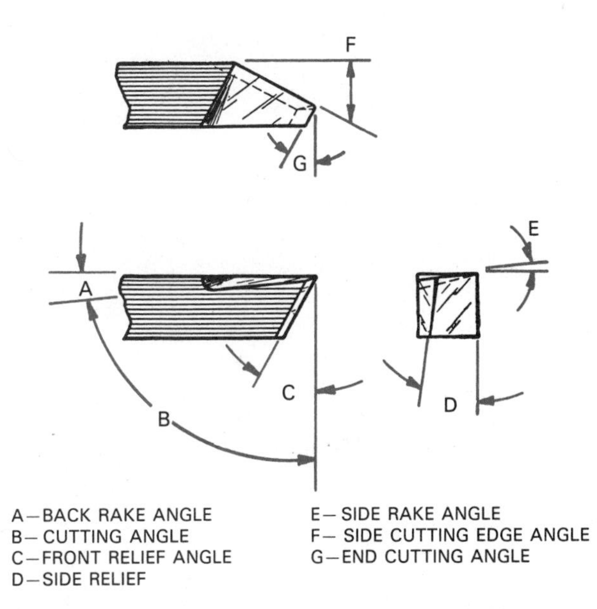

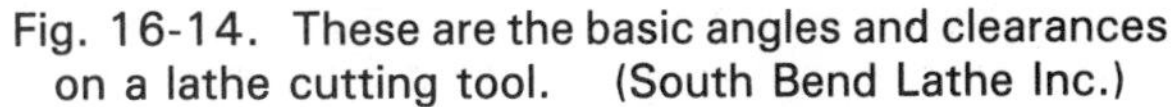

Fig. 16-14. These are the basic angles and clearances on a lathe cutting tool. (South Bend Lathe Inc.)

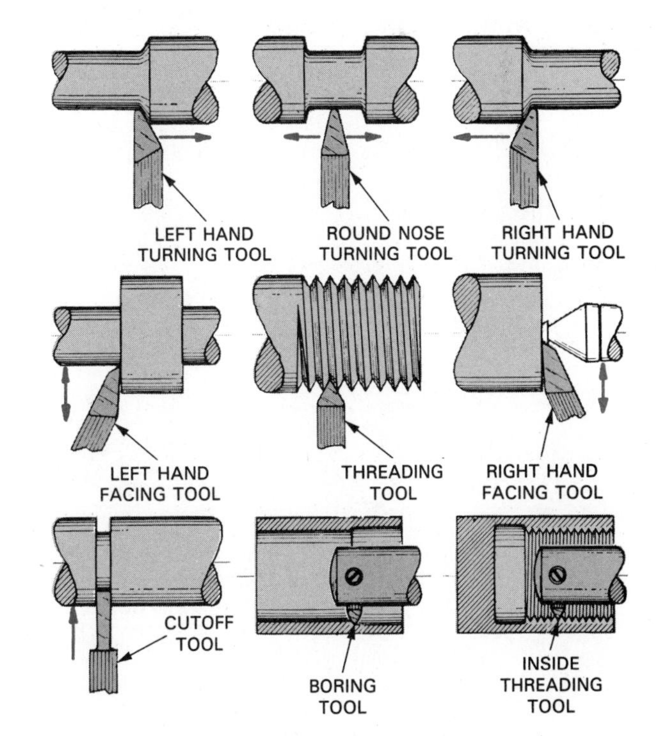

Fig. 16-16. Selected lathe tools and how they are used. (South Bend Lathe Inc.)

Wood cutting lathe tools

A completely different set of single-point tools are used for common wood turning. A set of wood turning tools would contain *gouges* for rough cutting stock; *skews* for smooth cutting; *parting tools* to establish diameters, cut recesses, and face ends; *spear point tools* to finish grooves, flutes and other shapes; and *round-nose tools* for scraping concave recesses and cylindrical surfaces. Fig. 16-17 shows the shape of these typical wood cutting lathe tools.

These tools are used to produce four types of cutting action, Fig. 16-18. These actions are:

1. Scraping (zero rake).
2. Cutting (30°rake).
3. Cutting (80° rake).
4. Shearing.

Hand-fed machines use the 30° rake cutting action while some automatic feed wood turning lathes use the 80° rake cutting action. A bail wood lathe (lathe for turning handles for buckets and for other small turnings) almost always uses an 80° rake action with the work turning up to 6000 rpm.

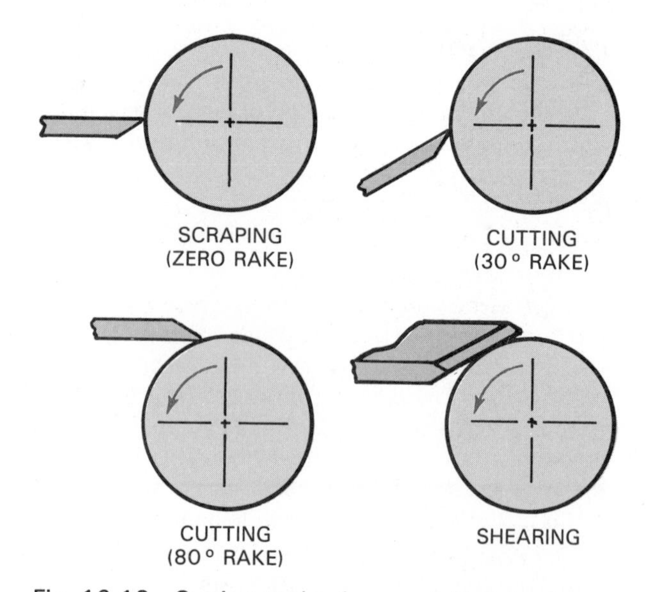

Fig. 16-18. Cutting angles for wood. Note tool position for various lathe operations.

HOLDING AND MOVING THE LATHE TOOL

There are a number of techniques for holding or supporting the lathe tool and for allowing it to move during cutting. First, the holding and supporting systems will be discussed.

Lathe tool holding devices

Metal cutting lathe tools must be securely held. Holding devices must reduce chatter (vibration) and allow the tool to be moved mechanically by the machine. Typically, the lathe tool is held in one of four common devices:

1. Lathe tool holder.
2. Boring bar.
3. Carbide tool holder.
4. High-production tool holders.

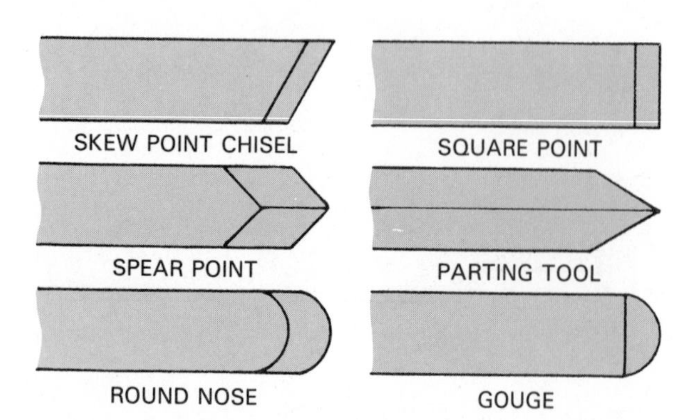

Fig. 16-17. Shapes of six common wood turning tools.

The *standard tool holder* is designed to hold a high-speed tool bit. This device also holds the tool at the correct cutting angle. Standard tool holders are available in straight, left-hand, and right-hand models. Modified tool holders hold cutoff, threading, and knurling tools. This method of holding tools is rapidly being replaced by quick-change holders. Fig. 16-19 shows some typical tool holders.

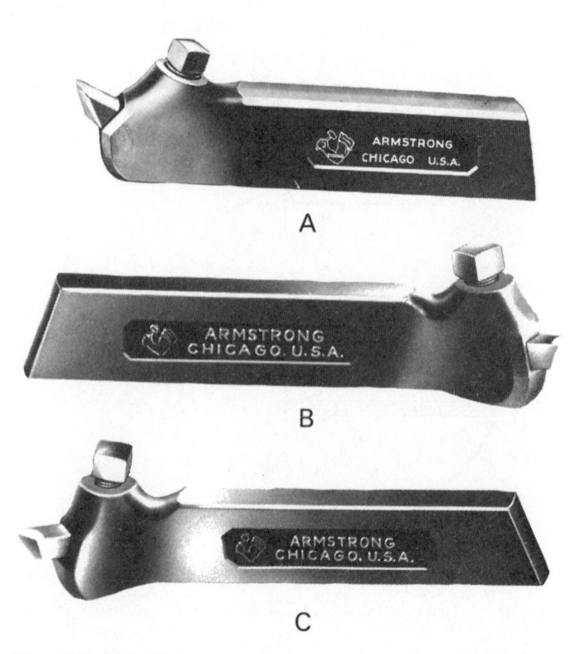

Fig. 16-19. Common metal lathe tool holders are shown with their cutter bits. A—Straight tool holder. B—Left-hand tool holder. C—Right-hand tool holder.

Fig. 16-20. Boring bars without cutter bits. Note set screws (at left) for holding cutters. (Armstrong Bros. Tool Co.)

A *boring bar,* as shown in Fig. 16-20, is a special tool holder for various internal turning operations. Boring, internal threading, and other similar operations can be performed by adjusting the angle of the bar and the tool which is placed in it.

The *carbide tool holder* is generally a steel part which will hold a carbide cutting tool insert. Fig. 16-21 shows several types of carbide tool holders. Carbide inserts may also be brazed on the cutting tool and then held by a standard tool holder.

High production machines use several types of tool holders to hold, index, and position the tool for cutting. These will be discussed later in the chapter under the production turning machine section.

Tool moving methods

The clamped or supported tool must be moved to generate the feed motion for the lathe cutting action. The tool must be capable of moving in both the Z axis (parallel with the work axis) and the X axis (90°

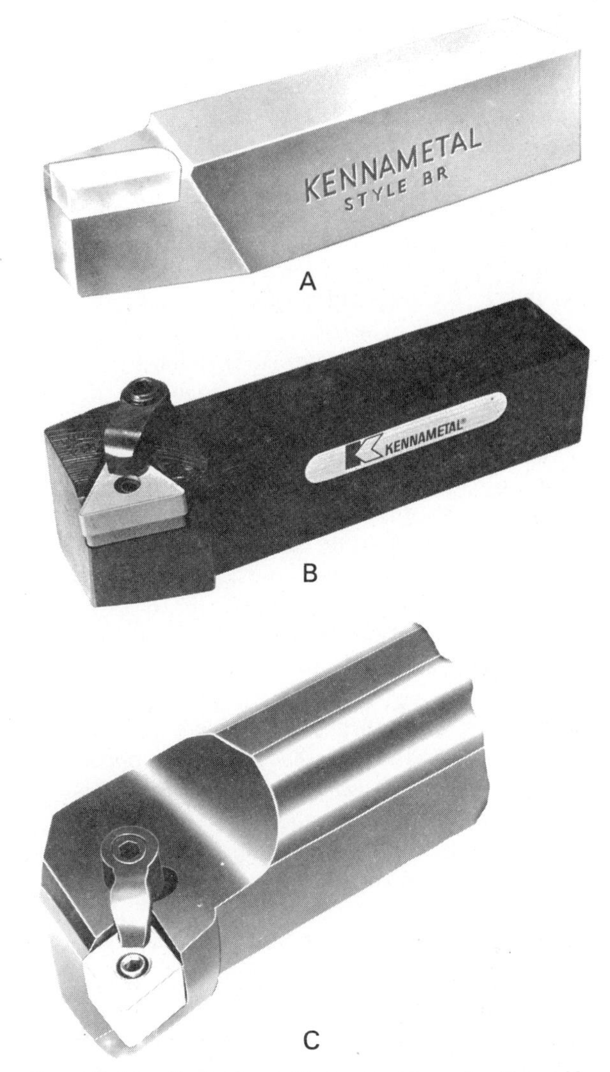

Fig. 16-21. Typical carbide tool holders. A—Brazed insert. B—Locked insert turning tool. C—Locked insert boring bar.

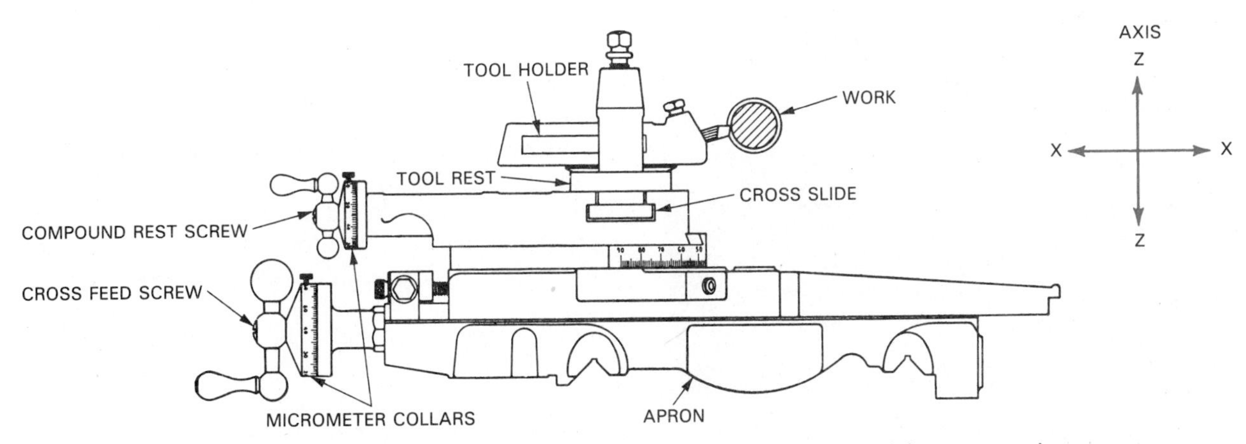

Fig. 16-22. The upper portion of the carriage shows the cross slide and the compound rest. (South Bend Lathe Co.)

or perpendicular to the work axis; movement of the cross-slide). To accomplish this, two separate tool-moving mechanisms are included on the metal lathe. These mechanisms are part of the carriage assembly.

The carriage, Fig. 16-22, contains a saddle, an apron, a compound rest, and a cross slide. The *saddle* supports the other parts and provides the Z axis movement. It moves along the ways of the lathe. The *apron* supports the controls which engage the automatic feed mechanisms for both the cross slide and the carriage. The *cross slide* travels along ways which are part of the saddle. This produces the X axis movement. The *compound rest* is attached to the cross slide and supports the tool holders. The compound rest pivots to provide movement for angled cuts.

The adjustment and feed of the cross slide, saddle, and the compound rest establish the depth of cut and provide the feed motion.

Woodworking lathes often use hand-held and hand-fed tools. A tool rest, as shown in Fig. 16-23, is used to support the tool and allows it to move, generating the feed motion.

An automatic wood turning technique uses a shaped knife which turns the entire length of the part in one operation. The lathe is called a *back-knife lathe.* The part turns away from the operator producing a downward action at the knife, Fig. 16-24. Production wood lathes are often automatically loaded from a hopper as shown in Fig. 16-25.

Fig. 16-23. Wood products are formed using a hand-held tool steadied on a tool rest. (Delta International Machinery Corp.)

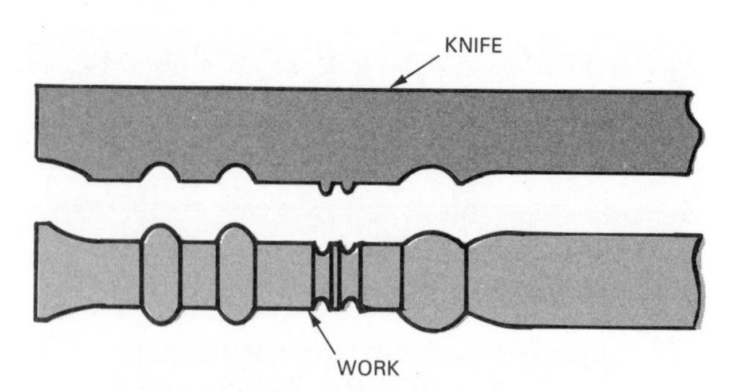

Fig. 16-24. Diagram of a back-knife lathe. Knife is shaped to contour of workpiece along its entire length.

Highly automatic wood lathes may use a multi-point cutter. Both the cutter and the work rotate. The rotating cutter is fed into the work by a cross slide. The part is shaped by a single movement of the cross slide. Fig. 16-26 shows the basic operation of an automatic shaping wood lathe.

Holding and rotating work

The last major element of a lathe is a mechanism to hold and rotate the workpiece. Most often, devices are attached to or inserted into the spindle to perform this task. The basic devices used are:

1. Centers.
2. Mandrels.
3. Chucks.
4. Collets.
5. Faceplates.

Centers

Workpieces which are relatively long compared to their diameters are turned between centers. One center is in the headstock spindle hole and may provide the drive to rotate the stock. The other center is in the tailstock hole and is dead (has no power). The stock to be turned must have extra length or otherwise account for holes needed for the lathe centers.

When metal is being turned, the headstock center does nothing more than support the stock. The work is rotated by means of a faceplate and lathe dog as seen in Fig. 16-27. The lathe dog is clamped to the work and inserted into a notch in the faceplate. The faceplate, being attached to the spindle, rotates when the spindle rotates and, in turn, rotates the work.

Wood lathes, however, use drive centers, Fig. 16-28. These centers not only support the work but, with the use of the four teeth, rotate it.

If a very long metal workpiece is being turned between centers, a *follow rest,* Fig. 16-29, is used. The rest is attached to the carriage or tool rest and supports the work near the cutting area. The follow rest

Fig. 16-25. Hopper-fed, back-knife wood turning lathe loads and unloads workpieces automatically. (Goodspeed Machine Co.)

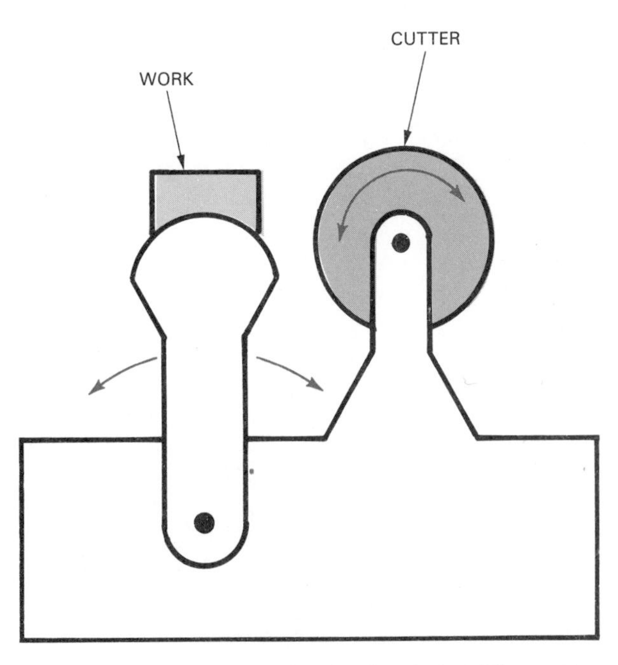

Fig. 16-26. An automatic shaping lathe. The cutter heads turn at about 3600 rpm and shape the full length of the work which turns slowly either clockwise or counterclockwise.

Fig. 16-27. Work mounted between centers with a faceplate and dog. A follow rest is also being used. (LeBlond Makino Machine Tool Co.)

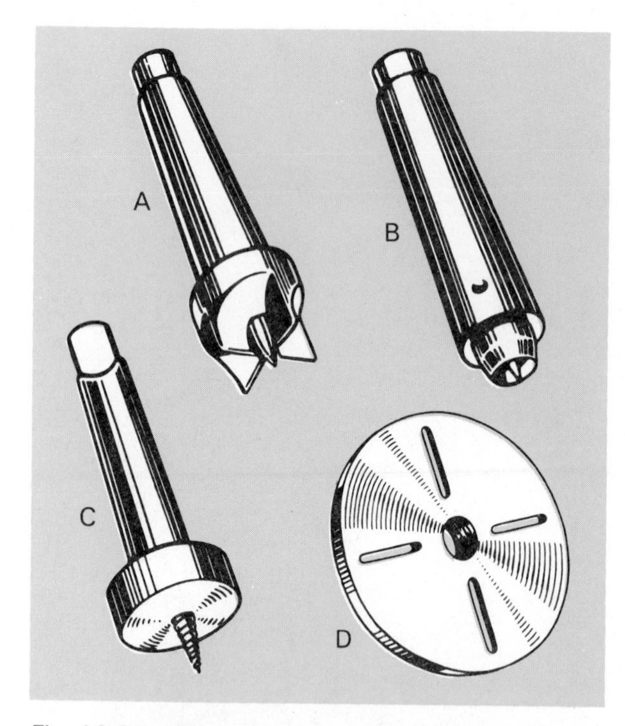

Fig. 16-28. Common woodworking lathe work holding devices. A—Drive center. B—Tailstock dead center. C-Screw center. D—Faceplate. (Delta International Machinery Corp.)

Fig. 16-29. Turning with a follow rest prevents long workpiece from flexing away from cutting tool. (LeBlond Makino Machine Tool Co.)

resists the cutting forces which tend to flex the part away from the tool.

Mandrels

Mandrels are used for between-center turning of hollow parts. There are several types. *Solid mandrels* are accurately ground, tapered (about .006 in. per foot) bars which can be placed between centers. A hollow disc-like part can be pressed on the mandrel and turned between centers like any other part.

Gang mandrels clamp the part between discs for turning. This increases the speed of the operation because each part does not have to be pressed on and off the mandrel. However, the diameter of the mandrel and the hole in the part must be about the same. the same.

Cone mandrels are similar to gang mandrels except they can clamp parts with a range of hole sizes. Fig. 16-30 shows the three types of mandrels.

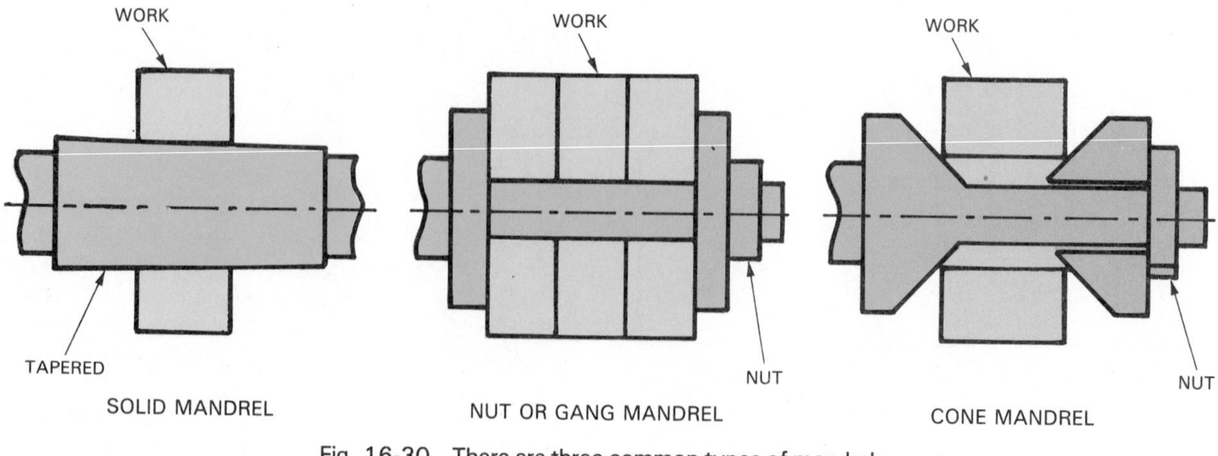

Fig. 16-30. There are three common types of mandrels.

Chucks

Lathe chucks are designed to hold a wide variety of sizes and shapes of work. Chucks grip one end of the part and allow machining of the extended portion of the work, Fig. 16-31.

There are two basic types of chucks: three-jaw self-centering chucks, and four-jaw independent chucks. The clamping jaws of a *three-jaw chuck* move inward or outward together. A wrench turns a bevel gear which operates a spiral cam. This cam moves all three jaws at once. The three-jaw chuck is used to hold round stock and center it at the axis of the lathe. A properly maintained three-jaw chuck will hold accuracy within 0.001 in.

The *four-jaw independent chuck* is designed to chuck irregularly shaped pieces and round stock off-center from the lathe axis. Each jaw is independently adjusted to properly position the part, Fig. 16-32.

Special chucks are also available. Jaws of a four-jaw combination chuck can be adjusted simultaneously or independently. Also, air or hydraulically operated chucks are used in high-production applications.

Collets

Collets are a holding device for stock which must be held very accurately, Fig. 16-33. They are thin steel tubular bushings which have been split into three equal segments along two-thirds of their length.

The internal shape of the collet may be designed to grip round, hexagonal, square, or other common shapes. The collet is placed in a sleeve in the lathe spindle. It is drawn into the sleeve with a draw bar. As it is drawn inward, the outside shape of the collet causes the sleeve to compress the collet segments. This creates a gripping action around the work.

Fig. 16-31. A part being held in a chuck. Jaws are adjustable. (Cushman Industries)

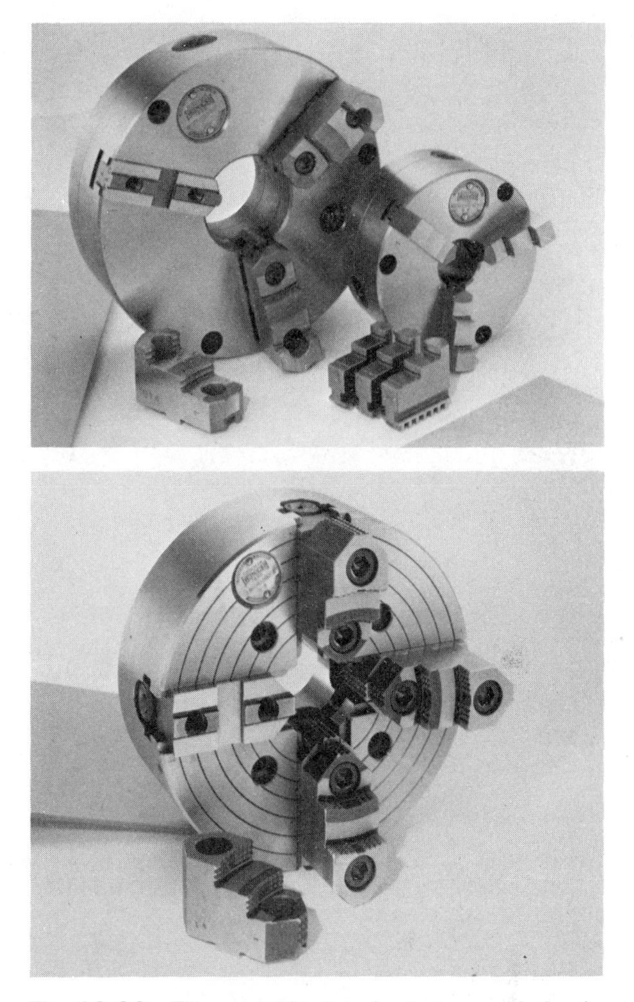

Fig. 16-32. These are the two basic types of chuck. Top. A three-jaw check. Bottom. A four-jaw independent chuck. (Cushman Industries)

Collets provide very accurate centering of work. However, the size of the stock must be within +0.002 in. to −0.005 in. of the collet size. This requirement limits collet usage to clamping drill rod, previously machined parts, extruded shapes, and cold collet stock. See Fig. 16-33.

Faceplates

Faceplates are used in both wood and metal turning operations. Metal may be clamped onto the face plate using a special fixture. This procedure lends itself to long-run operations on a single shaped part.

Many woodworking lathe operations use a faceplate. The work to be turned is attached to the faceplate (generally by first glueing a scrap piece to the back of the work). The faceplate is attached to the scrap piece with screws. The work is turned, then split off the scrap. This practice eliminates screw holes in

Fig. 16-33. A part is being removed from a lever-operated collet chuck. (Clausing Corp.)

work and allows for thin-bottomed products to be manufactured. Fig. 16-23 shows a typical faceplate turning operation.

TYPICAL TURNING MACHINES

Though there are a number of different turning machines, they may be grouped under three basic classes:

1. General-purpose lathes.
   a. Speed lathes.
   b. Engine lathes.
   c. Toolroom lathes.
2. Automatic turning machines.
   a. Automatic lathes.
   b. Turret lathes.
   c. Automatic bar machines.
   d. Turning centers.
3. Vertical boring and turning machines.

GENERAL-PURPOSE LATHES

Several turning machines are designed to be used in general or multipurpose situations. These lathes are quickly and easily set up for production. However, they usually require manual loading of stock, changing of spindle feeds, feeding of tools, and unloading of stock. General-purpose lathes also require skilled operators. Bench lathes and tracer lathes are also used in general-purpose situations.

Speed lathe

Speed lathes are the simplest turning machines. They generally have only a headstock, a tailstock and a simple tool rest. These are mounted on a lightweight bed. The speed lathe may be mounted on two legs or directly to a bench top.

Most speed lathes are driven by cone pulleys, variable-diameter pulleys, or variable-speed motors. Their speeds (revolutions per minute) are generally higher than other lathes. This is why they are called speed lathes.

The tools used on a speed lathe are held and fed by hand. The tool rest supports the tool and aids in the even movement of the tool by the operator. The most common speed lathe is the woodworking lathe, Fig. 16-34. Speed lathes are also used for metal spinning and polishing.

Engine lathe

The engine lathe is the common general-purpose metal cutting lathe. It contains all the basic features of a lathe: a headstock, a tailstock, a bed, and a tool holding device. The engine lathe is much heavier and more rigid that a speed lathe. Refer, again, to the turning operation in Fig. 16-2.

The engine lathe generally has a greater number of spindle speeds. Often the speeds are set by a gear box (gear-headed lathe). The tool may be hand or power fed on both the X axis and the Z axis.

Engine lathes, like speed lathes, can be used for both between-center turning and chuck (or faceplate) turning. Engine lathes are manufactured in a wide range of sizes from 9 to 50 in. (229 to 1270 mm) swing and bed lengths from 3 ft. to over 15 ft. (1 to 5 m). Some special lathes may have bed lengths up to 50 ft. (over 15 m).

Engine lathes may be modified by adding a tracer attachment. This allows the tool to follow a path set by a template. The tool movement, as shown in Fig. 16-35, is caused by hydraulic, mechanical, air, or electrical devices controlled by the stylus movement. These lathes are called *tracer lathes*. They can produce a specifically shaped part without complex hand adjustment or the cross slide and the carriage. The machine operator loads and unloads the work and starts the machine cycle. The trace attachment does the rest.

Fig. 16-34. A common wood lathe is typical of the type called speed lathes. (Delta International Machinery Corp.)

Another special engine lathe is the *bench lathe.* This is simply a small, lightweight engine lathe designed to be attached to a bench. It can perform all the operations of the engine lathe. It is, however, limited to about a 10 in. (254 mm) swing and a 3 ft. (about 1 m) bed.

Toolroom lathes

Toolroom lathes are very similar to engine lathes. They differ in that they are designed to produce much more accurate cuts. They, also, have a wider range of speeds and feeds available. Toolroom (or toolmaker's) lathes are primarily used to produce the accurate cuts required by tool and die work.

PRODUCTION TURNING MACHINES

As machining needs increased, more complex turning machines were developed. Most of them evolved from the basic elements of the engine lathe. The newer machines are designed to produce parts more efficiently. They are easier to load, machine parts more rapidly, and can often use a series of tools in sequence (order).

Most of these machines automatically feed tools into the work and then withdraw them after the desired cut is completed. Features of each machine will be discussed in the following sections.

Automatic lathes

Automatic lathes have the same basic features as general-purpose lathes. These include a bed, headstock, tool support and moving systems, and, in some cases, a tailstock. The main additional features of an automatic lathe is a method or mechanism for feeding several tools into the work, one at a time.

Look at the simplified drawing in Fig. 16-36. As it shows, tools are mounted on blocks in front and in back of the work. The tools in the front are fed along the work to turn tapers, curves, and straight surfaces. The tools in the back are fed into the work for facing, necking, and form turning.

Automatic lathes may have automatic part loading and unloading. Machines with this feature are said to be *hopper fed.* (Fig. 16-25 showed a hopper-fed back-knife wood lathe which is used to produce small diameter turnings.)

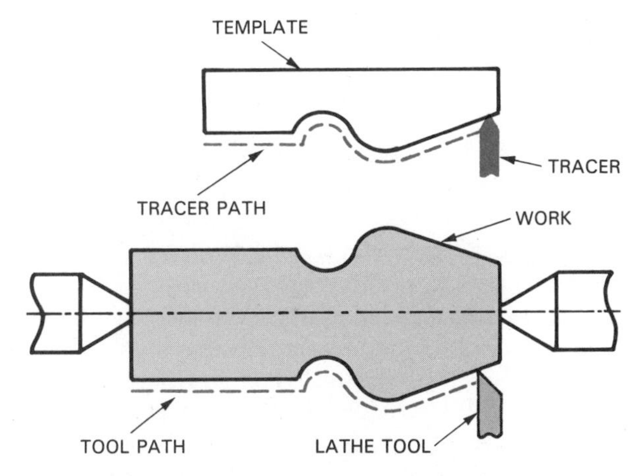

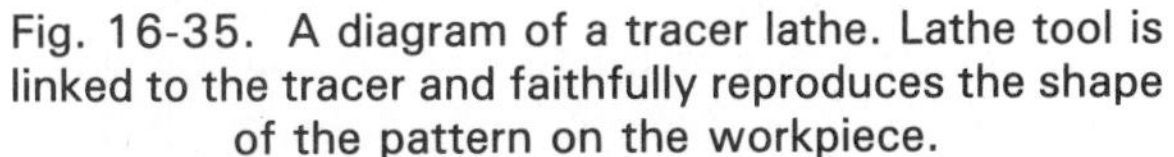

Fig. 16-35. A diagram of a tracer lathe. Lathe tool is linked to the tracer and faithfully reproduces the shape of the pattern on the workpiece.

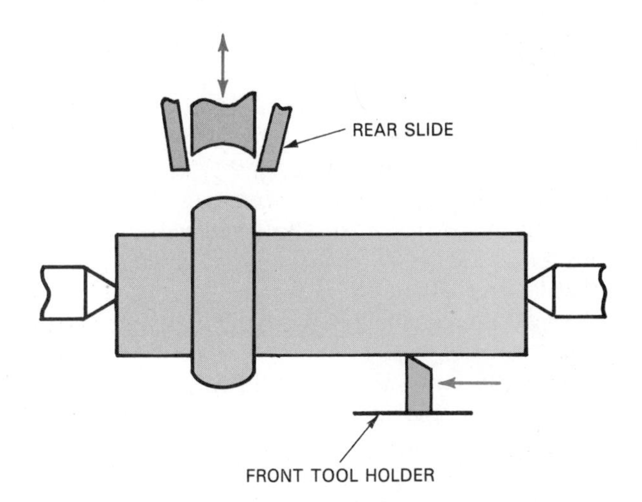

Fig. 16-36. A diagram of an automatic lathe. Tools are mounted front and back.

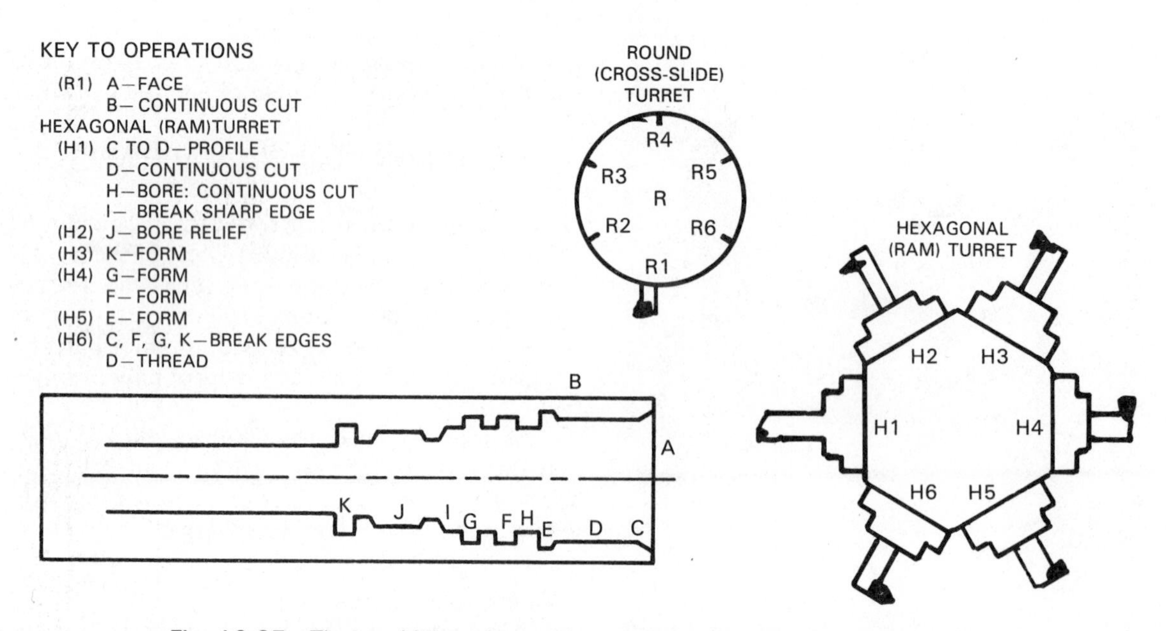

Fig. 16-37. The machining of a part by a turret lathe. Note key to operations performed on part. (Jones & Lamson)

Special automatic lathes are also built. These machines are often called *chucking machines.* Also, vertical automatic lathes are built. They reach maximum speed and feed rates rapidly. They can make several different cuts at once which increases their productivity. The operator of the vertical automatic lathe loads and unloads parts. The automatic features control the remaining turning tasks. One type has no tailstock and is used for turning operations requiring a three or four-jaw chuck.

Turret lathe

A turret lathe is semiautomatic. It is capable of producing accurate parts faster and in larger quantities than an engine lathe. The turret lathe can also perform several operations on a part because it is not limited to a single tool, Fig. 16-37. The tools for several operations are mounted in a special toolholding device called a turret. The turret is a six or eight-sided device which replaces the tailstock. Each side may hold a cutting tool. A four-sided turret may also be placed on the cross slide to add additional cutting tools. Fig. 16-38 is a simplified drawing of a typical turret lathe.

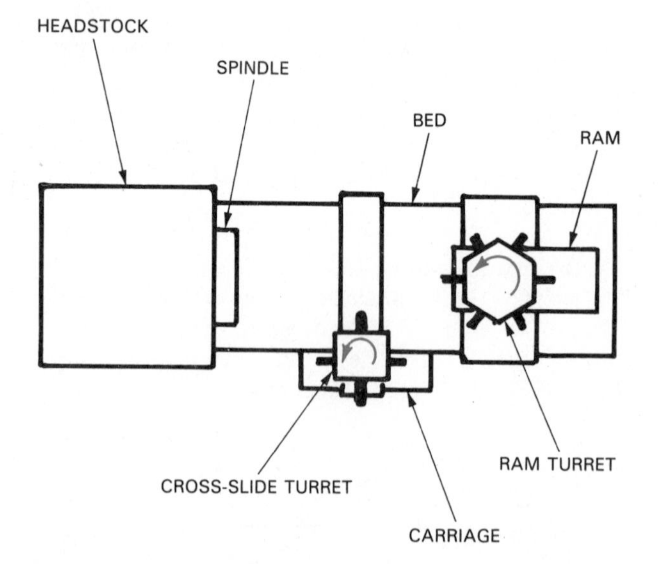

Fig. 16-38. A diagram of a turret lathe. One turret replaces tailstock.

Turret lathes are built in two basic forms: *horizontal* and *vertical.*

Horizontal turret lathes use either of two basic turrets: *ram* type and *saddle* type. These names come from the system used to move the turret.

Ram type turret lathes have a saddle which is held in a fixed position. The turret is mounted on a sliding member (ram) which moves forward to feed the tool into the work. It then moves back where the turret can be rotated to position the next cutting tool.

Saddle-type turret lathes move the entire saddle to bring the turret forward for cutting operations. Ram-type machines are used for lighter work because the tools are held less rigidly. However, ram machines operate more rapidly. Ram and saddle turrets are shown in Fig. 16-39.

Turret lathes may be totally operated by a worker or may be highly automatic. For shorter production

A

B

Fig. 16-39. Types of horizontal turret lathes. A—Ram type. B—Saddle type. (Jones & Lamson)

runs the operator may chuck the work, index and feed the turrets, and remove the completed part. For longer runs, automatic turret lathes which chuck the parts, rotate the turrets, feed the cutting tools, and change spindle speed are often used. The greater expense for the machine and its setup is overcome by higher production rates. In many cases, a worker can operate two automatic machines at once.

Vertical turret lathes are primarily for large-diameter and heavy parts. The weight is more evenly distributed on the table than could be accomplished on the spindle of a horizontal lathe.

It has a rotating table or chuck and a turret above the table. The turret can be moved up and down as well as sideways. Fig. 16-40 shows a typical machine. Vertical turret lathes are used primarily for work

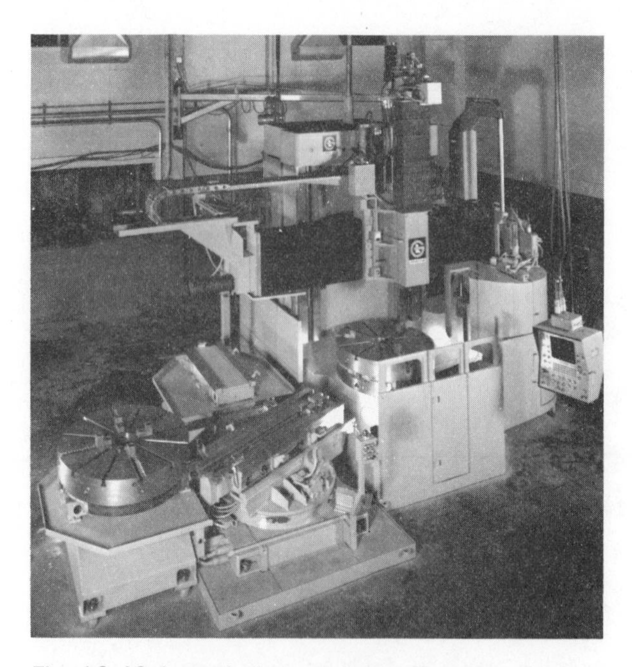

Fig. 16-40.A vertical turret lathe. Rotating chuck or table can support heavier weights than a horizontal spindle. (Giddings & Lewis)

which can be held in a chuck. The machine is usually used for boring operations. Bar stock cannot be turned on a vertical turret lathe.

Automatic bar machines

Automatic bar machine is a new name for the automatic screw machine. It was first designed for high-volume production of screws and other threaded fasteners. Now most screws and bolts are produced by *thread rolling machines.*

Automatic bar machines combine the best features of the automatic lathe and the turret lathe. They turn bar stock by:

1. Feeding the stock into position.
2. Automatically indexing tools into the machining position.
3. Adjusting speeds and feeds for each operation.
4. Feeding and withdrawing the tools.
5. Cutting the part off the bar.
6. Feeding the stock into position for the next cycle.

Automatic bar machines are of three major types: single-spindle, Swiss-type, and multiple-spindle.

Single-spindle machines

Single-spindle machines, Fig. 16-41, are used to machine single round, square, hexagonal, or other shapes of bars. The bars are fed through the hollow spindle into a collet. When they are in proper machining position, they are automatically clamped by the collet.

Tools in a six-station turret in the tailstock position are used in the machining act. The tools in the turret will machine (drill, bore, tap, ream, and shoulder) details into the end of the bar.

The first tool is fed into the rotating stock. When its machining operation is completed the tool is withdrawn. The vertical (upright) turret rotates and brings the second tool into machining position. The second tool progresses through the machining cycle. The other four tools, likewise, are brought into position and used. The last machining operation parts or cuts off the machined portion of the bar.

If additional cuts are needed, turrets may be mounted in front and behind the workpiece on the cross slide. These turrets hold tools which will machine detail along the length of the bar.

A series of trip "dogs" determine which turret and tool will be used in any given step. Cams are used to set the distance the tool will be fed into the work. These dogs and cams insure that the tools are used in proper order and that they produce the desired cut.

When all the machining acts are completed the finished portion of the bar is cut off. The stock is fed through the collet into position to produce another part. The total cycle is repeated and a second piece is cut off the bar. This feeding of the bar, machining, cutting off cycle is repeated until the entire bar is changed into a number of machined parts. A new bar is then inserted into the machine.

Swiss-type

The *Swiss-type automatic bar machine* is also a single-spindle machine. It was designed for more ac-

Fig. 16-41. A single-spindle automatic screw machine. A six-station turret is located in the tailstock position. Note second turret behind chuck. It performs operations along the length of the workpiece. (POLAMCO Machinery Co.)

curate machining. The Swiss-type machine is different from the other automatic bar machines in that it does not use a turret to hold the tool. Instead, five tool slides are mounted around the bar. The two horizontal slides machine the diameter of the stock. The other three are used for special operations such as shouldering, knurling, parting, and chamfering.

The Swiss-type bar machine is the only turning machine in which the headstock both rotates the stock and produces longitudinal (along the axis of rotation) movement. During the forward movement the workpiece is supported by a guide bushing. This bushing is positioned close to the tool and keeps the work from deflecting. The stationary tools and guide bushings enable the Swiss-type machine to produce very accurate parts.

Multiple-spindle

Multiple-spindle automatic bar machines, seen in Fig. 16-42 are used when greater output is needed.

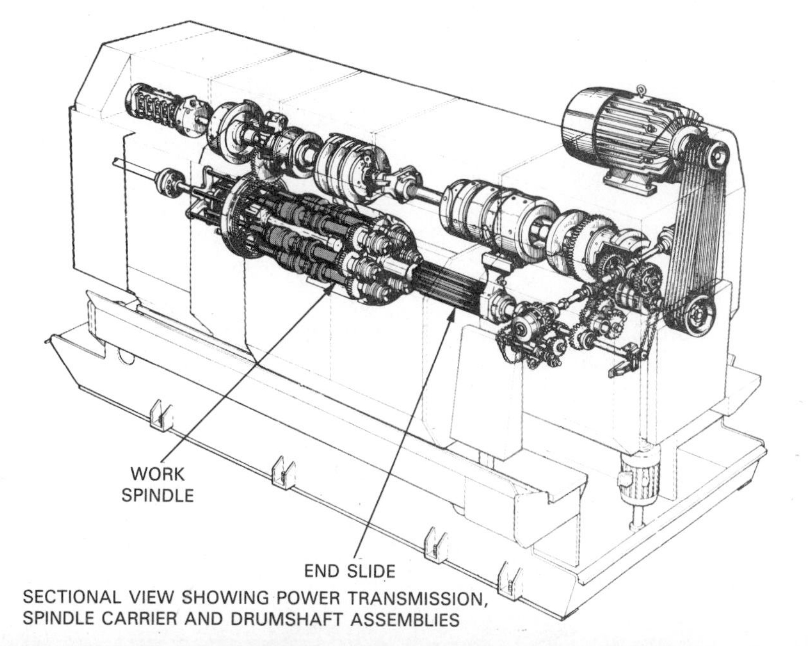

Fig. 16-42. Multiple-spindle automatic bar machine. Top. Schematic. Bottom. Close-up of operations. (Cone Blanchard Machine Co.)

Single-spindle bar machines make a series of cuts on a single bar. Multiple-bar machines can machine four to eight bars at one time. Typically, machines are built with four, five, six, or eight spindles.

Multiple-spindle machines are able to clamp and rotate bar stock in each of their spindles. At each spindle location there may be a tool on an end tool slide and one on a cross slide. The end slide will move forward to cause the tools to machine the end of the bars. The cross slides hold tools that machine details along the length of the bars.

How it works

When a multiple bar machine is operating:

1. The end and cross slides move forward machining the ends and sides of the bars being held and rotated in the spindles.
2. The end and cross slides move backward away from the bars.
3. The spindles are rotated clockwise.
4. Steps one through three are repeated.
5. At the last step the part is cut off.

Fig. 16-43 shows the operations performed at each of the eight stations of an eight spindle automatic bar machine. Note that stations 2, 3, 4, 6, 7, and 8 use end and cross slide/spindle operations.

NC/CNC LATHES

Programmable lathes are taking on more of the work of the turning machine. These lathes involve two groups:

1. Numerically controlled (NC) lathes.
2. Computer numerically controlled (CNC) lathes.

The numerically controlled lathe is the older of the two types. The action of the lathe is controlled by a unit which reads a punched tape. The actual control unit and tape are described in detail in Chapter 35. The holes in the tape indicate speed, feeds, and depth of cuts. The control unit "reads" the tape and transmits directions by way of electrical impulses to devices which adjust the machine's speed and location and movement (feed) of the tool.

Numerically controlled lathes are available in three basic types.

1. *Center-type machine.* This machine is designed to turn stock between centers and is used primarily

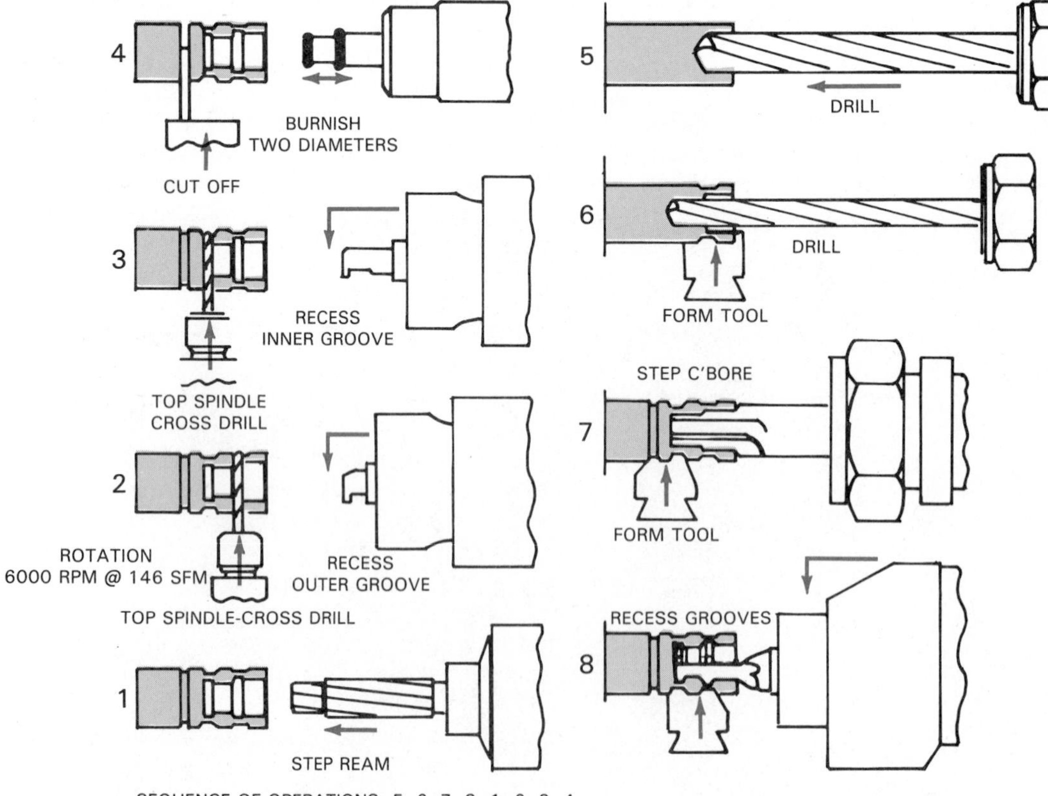

Fig. 16-43. These drawings represent the operation of an eight spindle automatic bar machine. (Cone Blanchard Machine Co.)

for outside-diameter operations. It has high spindle speeds and is available in long bed lengths. The cutting tools are held in a single cross slide or in a turret.
2. *Chucking-type machine.* This machine is designed to hold the workpiece in a chuck and perform both outside and inside-diameter operations. The spindle speed is slower than center-type machines and the bed is shorter. Tools are held in one or more heavy-duty slides or turrets.
3. *Universal machine.* This machine is a modification of either a chucking or a center-type machine. It may be a center-type machine with a chuck or a chucking-type machine with a tailstock. The tools are often held in a cross-slide turret for outside diameter operations and end turrets for inside diameter tasks.

NC lathes can increase productivity (parts produced for each unit of labor worked), reduce operator skill requirements, and minimize inspections, special tooling, and material handling. However, NC lathes are more expensive. Also, additional skills are needed to program and maintain the more complex machine.

A more recent addition to the list of turning machines is the CNC (computer numerically controlled) lathe. This machine works on the same principle as the NC lathe. A control unit reads data and adjusts the operation of the lathe. The difference lies in the source of the data. While the NC lathe used a punched paper tape, the CNC lathe uses a computer program for its data source. The computer program is easier to correct and modify. It also allows the machine direct use of data from computer-aided design computers.

VERTICAL BORING AND TURNING MACHINE

The vertical boring machine is constructed somewhat like the vertical turret lathe. The machine, as diagramed in Fig. 16-44, has a rotating table which holds the work. The cutting tools are held stationary in heads which move on cross rails. The boring machine is used when:

1. Large holes (over 12 in. in diameter) must be bored.
2. The material to be turned is too large for horizontal turning.

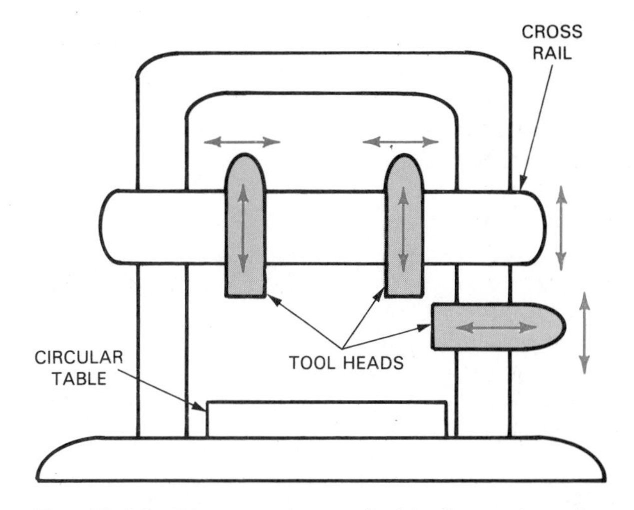

Fig. 16-44. Diagram of a vertical boring and turning machine. Table may be up to 40 ft. (12 m) in diameter.

SUMMARY

Turning machines are important machine tools. They may be used to turn, face, form, neck, part, bore, thread, and knurl various materials. Most woods and metals and many plastics can be turned.

Typical turning machines have a basic structure which includes a bed, a headstock, a tailstock, and a tool support or holding system. They must also hold and rotate the work.

Many different tools may be used to produce the desired results. Most often, specially shaped single-point tools are used in turning. However, drills, reamers, and taps may be used on turning machines.

Programmable lathes are taking on more of the work of turning machines. Older numerically controlled (NC) machines used a punched tape read by a control unit. A new generation of machines (CNC) is controlled by a computer program.

The most common turning machines are general-purpose lathes and automatic turning machines. The general-purpose machines include speed, engine, and toolroom lathes. Common automatic machines are the automatic lathes, turret lathes, and automatic bar and chucking machines. Also, special turning operations are performed on vertical boring and turning machines.

STUDY QUESTIONS — Chapter 16

1. Describe straight, taper, and contour turning.
2. Describe end and shoulder facing.
3. Describe necking and parting.
4. Describe boring and internal forming.
5. Describe threading and knurling.
6. List and describe the four parts which make up the basic structure of a lathe.
7. List the common metal turning tools.
8. List the common woodworking lathe tools.
9. Describe the operation of back-knife and shaping wood lathes.
10. List and describe the major work-holding devices.
11. List and describe the common general-purpose lathes.
12. List and describe the common automatic lathes.
13. Describe a vertical boring and turning machine.

Chapter 17

MILLING AND RELATED OPERATIONS

Milling includes all the operations which produce a flat or curved surface by progressively forming and removing chips. Milling generally uses a multipoint tool which is rotated to create the cutting motion. The feed motion is produced by feeding the workpiece into the cutter in a straight line. Fig. 17-1 shows the typical components of a milling operation.

Milling techniques may be used for shaping, sizing, and producing surface finishes on wood, metal, or plastic materials. When performed on metal the process is usually called milling. Similar processes used to machine wood are called *planing, jointing, molding, shaping, routing,* and *carving*. Sheet plastic material may be machined by all of these processes.

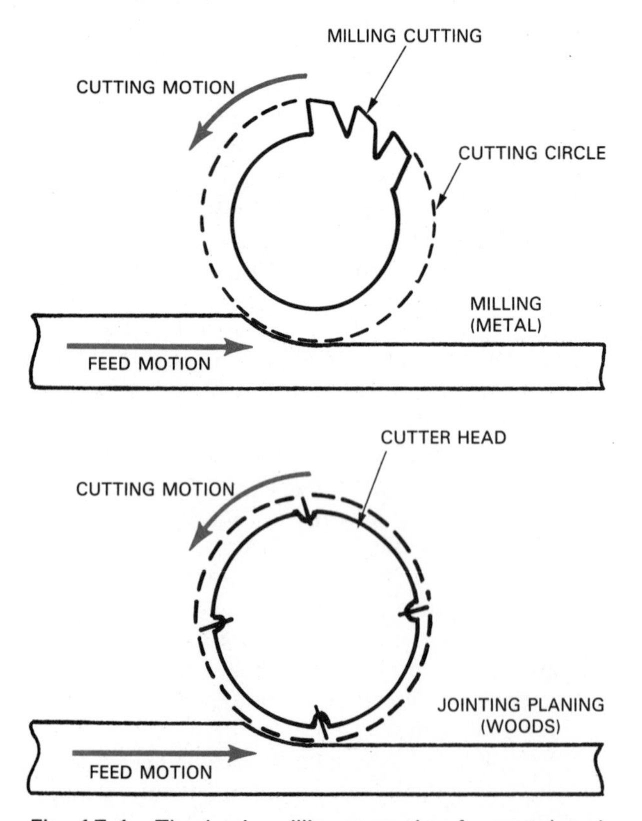

Fig. 17-1. The basic milling operation for metal and wood. Flat or curved surfaces are formed by progressively removing chips.

MILLING METALS

Most milling of metal is done on machine tools called milling machines. To understand this basic machining practice better, the following topics will be discussed:

1. The basic milling operation.
2. Types of milling operations.
3. Surface generating methods.
4. The milling machine.

THE BASIC MILLING OPERATION

The basic milling operation must have two major elements if the process is to be effective. These are:

1. A rotating cutter.
2. A workpiece moving in a straight line.

The cutter

The rotating milling cutter may be held in either a horizontal or a vertical position as shown in Fig. 17-2. Horizontal cutters are usually placed on an arbor. The arbor extends through the center hole of the cutter and clamps the cutter in position.

Vertical cutters are usually secured in chucks or collets. These devices are attached to the end of a spindle.

A number of different cutters are made for both horizontal and vertical milling operations. Each cutter is designed to produce a particular type of cut. Fig. 17-3 shows some typical cuts.

Milling cutters are grouped into three major types. These, shown in Fig. 17-4, are:

1. *Arbor cutters.* Cutters that have a hole in their center for mounting on an arbor.
2. *Shank cutters:* Cutters with a straight or tapered shank which may be placed in a spindle, chuck, or collet.
3. *Face cutters:* Cutters which can be bolted or otherwise held on the end of a spindle.

Fig. 17-2. A—Horizontal milling. Cutter is held by an arbor. B—Vertical milling. Cutter is held by a chuck or collet. (Cincinnati Milacron)

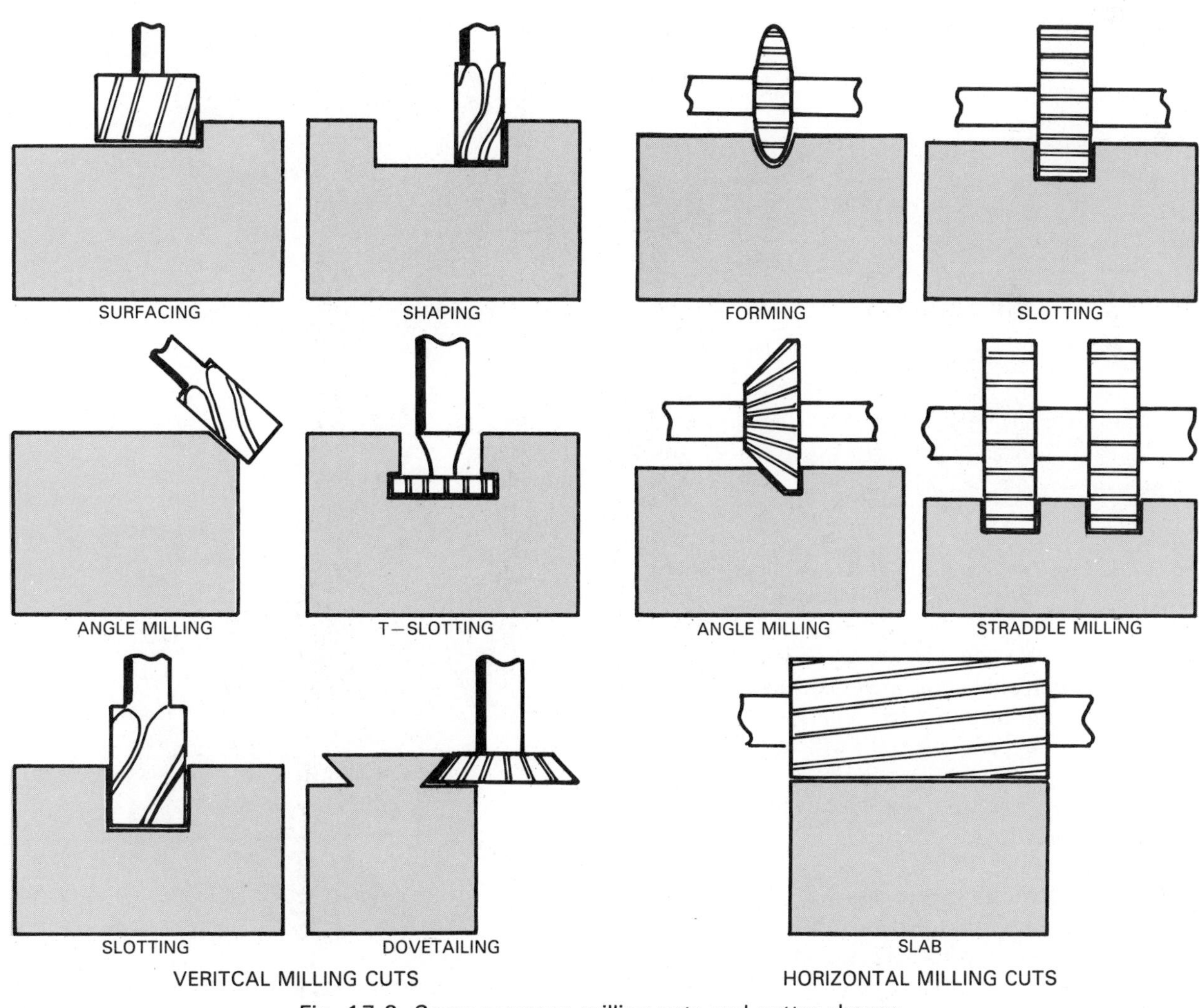

Fig. 17-3. Some common milling cuts and cutter shapes.

A

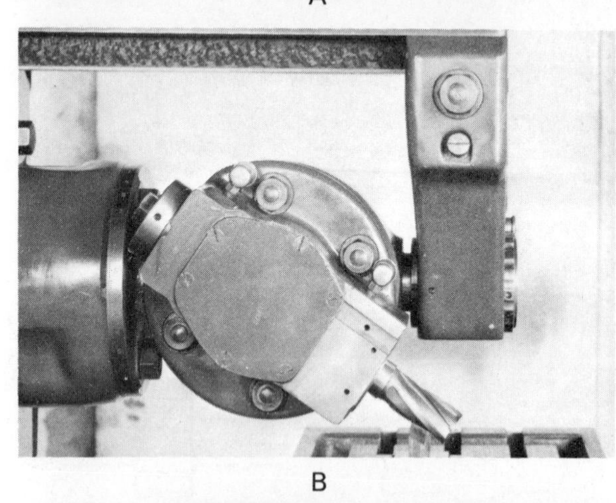

B

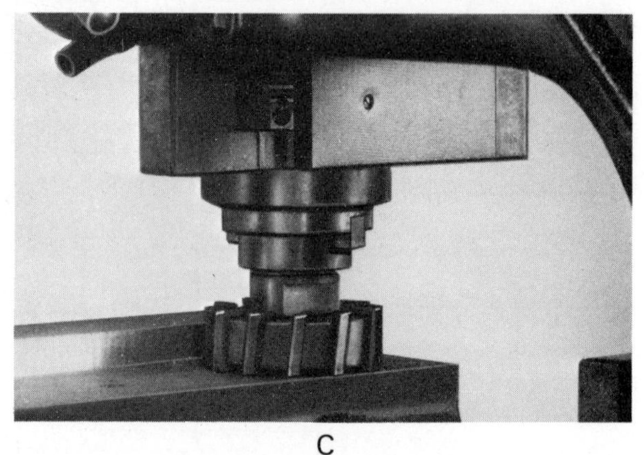

C

Fig. 17-4. Types of milling cutters. A—Arbor. B—Shank. C—Face. (Cincinnati Milacron)

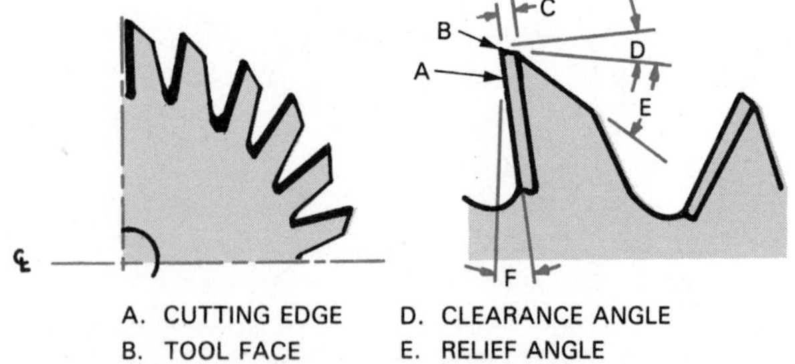

Fig. 17-5. Clearances, relief angles, and other features on a milling cutter are important to their operation.

Milling cutters, like most metal-cutting tools, may be made of high-carbon steel, high-speed steels, or carbides. The shape of the teeth will vary according to the type of cutter and the material to be machined. Fig. 17-5 shows a section of a typical cutter with important features indicated.

Holding and moving the work

The second important feature of the basic milling operation is a method to hold and move the work. The milling machine has a table which will move on three axes or directions. It will move left and right

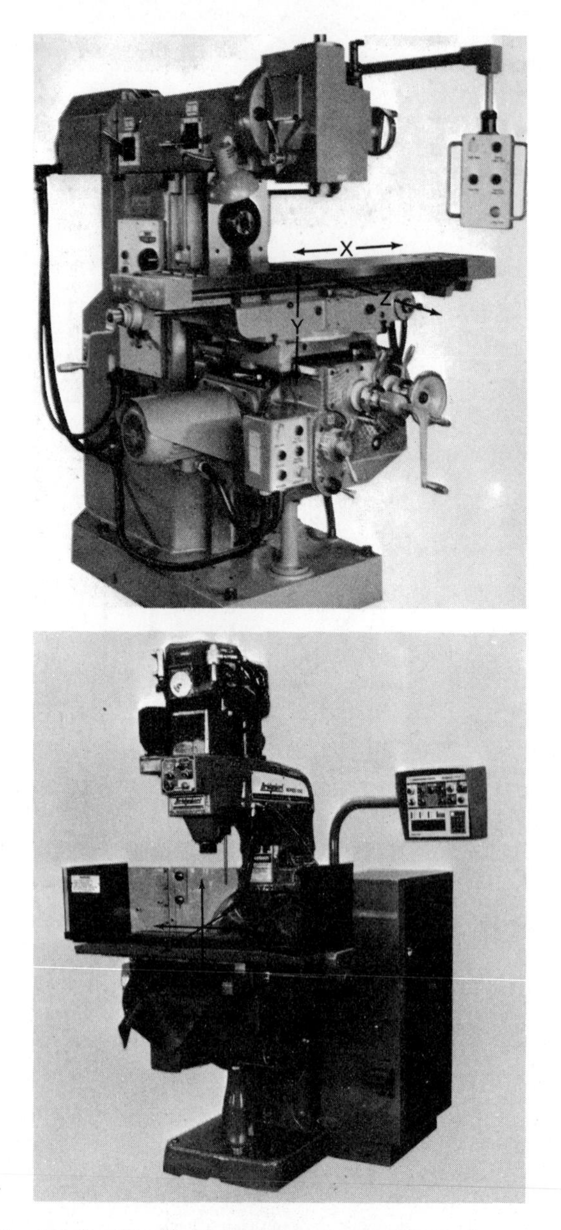

Fig. 17-6. The X, Y, and Z axes. The X axis is always the left to right horizontal motion of the table.

and in-out under the cutter to generate the feed motion. The table will also move up and down to set the dept of cut. On some milling machine spindles will move up and down.

These axes are designated by X, Y, and Z, depending on the orientation of spindle. Milling machines have either horizontal or vertical spindles, Fig. 17-6. The table motion parallel to the spindle is on the Z axis. Horizontal motion parallel with the work holder, and therefore the table, is designated the X axis. Table or spindle movement perpendicular to both the X and Z axes is the Y axis. These movements are summarized in Fig. 17-7.

The work must be held to the table during the cutting action. Typically, a vice is used. See Fig. 17-8. Special holding devices (jigs, index heads, etc.) and bolts in the table's T-slots can also be used.

	HORIZONTAL SPINDLE MACHINES	VERTICAL SPINDLE MACHINES
X Axis	Horizontal movement of the table	Horizontal movement of the table
Z Axis	In-out movement of the table	Up-down movement of the spindle (up-down movement of the table is on the W axis)
Y Axis	Up-down movement of the table	In-out movement of of the table

Fig. 17-7. Y and Z axes designations depend on the orientation of the spindle. Chart summarizes designations.

TYPES OF MILLING OPERATIONS

All basic milling operations are included under two major headings: *peripheral milling* and *face milling.*

Peripheral milling

In peripheral milling, cutting teeth are on the outer edge (periphery) of the milling cutter. The cutter rotates on an axis parallel to the surface being machined. The movement of the table under the cutter can produce both flat and curved surfaces. Also, form cutters may be used to produce a desired cross section. Peripheral milling is sometimes called *slab milling.* This type of milling, shown in Fig. 17-9, is usually done on horizontal (spindle) machines.

Fig. 17-9. Peripheral milling is being done on a horizontal milling machine. (Cincinnati Milacron)

A

B

Fig. 17-8. Parts may be held for milling by several devices. A—Vices. B—Fixtures. (Cincinnati Milacron)

Face milling

Face milling, as shown in Fig. 17-10, is used to produce a flat surface at right angles to the cutter axis. In face milling, teeth on both the side and the periphery of the cutter are used. The periphery teeth remove the majority of the material. The side teeth smooth the machined surface. Both vertical and horizontal milling machines can be used in face milling.

METHODS OF MILLING SURFACES

The final surface of the part may be produced or generated by two basic milling methods. These methods are *conventional* or *up milling* and *climb* or *down milling.*

Up milling

In up milling, the cutter rotates against the direction the work is being fed. This action is shown in Fig. 17-2A. It produces a chip that is thin where the cutter tooth first engages the work but much thicker as the teeth completes its cut.

The action of up milling causes the work to be lifted as the cut is being made. This lifting action removes all looseness from the table ways and feed system. The result is a very smooth cut. However, the part must be securely clamped or the lift will pull the work from its holding device.

Fig. 17-10. Face milling on a vertical milling machine. Cutters have teeth on two different planes at right angles to each other. (Cincinnati Milacron)

Up milling is also used to machine castings, forgings, and other materials which have an abrasive outer layer. Since the chip is started at the bottom of the cut and extends upward, the cutter is not seriously dulled by the abrasive skin of the work.

Down milling

In down milling, the work is fed in the direction of the cutter rotation. This procedure causes the cutter tooth to take its maximum bite at the start of the cut. The chip becomes thinner as the cut progresses, Fig. 17-11.

The cutting action tends to pull the work into the cutter. If there is excessive play in the table and feed mechanism, the cut will be less than satisfactory. Vibration will be present and chatter (rippled cuts) will occur. However, the cutting action forces the work down onto the table or into the vice. Therefore, less clamping pressure is needed. A well adjusted milling machine with little wear will produce very good cuts using the down milling technique.

Down milling will produce a better finish on hard steel. Teeth marks are less likely to appear. Also, chips are more readily rejected in down milling and are less likely to be carried along by the cutter, Fig. 17-12.

THE MILLING MACHINE

The basic milling machine has six major parts as shown in Fig. 17-13. The components include:

1. Frame.
   a. Base.
   b. Column.
   c. Overarm.
2. Spindle.
3. Knee.
4. Saddle.
5. Table.
6. Feed mechanisms.

Frame

The frame produces the basic structure for the milling machine. It includes a base to provide stability for the machine, a column to provide the height for the machine, and an overarm to support the machine's spindle or arbor.

Spindle

The spindle is the device which transmits power from the machine motor to the cutter and provides rotary motion for the cutter. The spindle may be placed horizontally or vertically. Horizontal spindles drive arbors and hold face cutters. Vertical spindles may hold the cutters themselves. They may also accept chucks and collets which, in turn, hold the cut-

Fig. 17-11. Climb or down milling refers to milling done with cutting action in the same direction as the feed. (Cincinnati Milacron)

ters. The position of the spindle often gives the milling machine its name; that is, a horizontal milling machine or a vertical milling machine.

Knee

The knee of the milling machine supports the saddle and the table. The knee is connected to the column and can move up and down on a set of ways. This movement raises and lowers the table and often sets the depth of cut.

Saddle

The saddle is the machine element between the knee and the table. The saddle moves in and out along ways on the knee. This movement provides for positioning the work under the cutter.

Table

The table is attached to the saddle. It moves right and left along a set of ways on the saddle. This movement provides the feed motion. The table contains T-slots which enable vices, index head, fixtures, and, even the part itself, to be bolted to it.

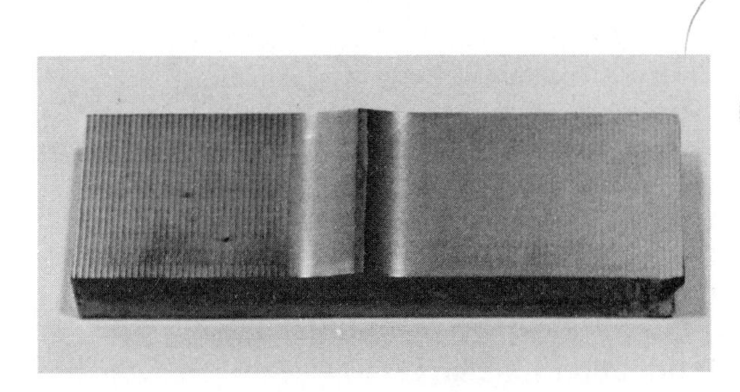

Fig. 17-12. Comparison of milled surfaces. Left. Surface produced by conventional milling. Right. Surface produced by climb milling. (Cincinnati Milacron)

Feed mechanism

All milling machines contain systems for moving the knee, saddle, and the table. This movement may be manual or automatic. One, two, or all three elements may be moved at once enabling the milling machine to produce complex cuts.

In the case of the vertical mill, the spindle may also be moved up and down giving a second movement along the Y axis.

Milling machine sizes

Many but not all milling machines are sized by a set standard. The sizes for general purpose machines are indicated by numbers from 1 to 6. The numbers are used to indicate the approximate longitudinal (lengthwise) table travel. The following chart indicates this relationship:

SIZE (Number designation)	1	2	3	4	5	6
TABLE TRAVEL (inches)	22	28	34	42	50	60

These size numbers do not relate to the power or strength (rigidity) of the machines. Often machines

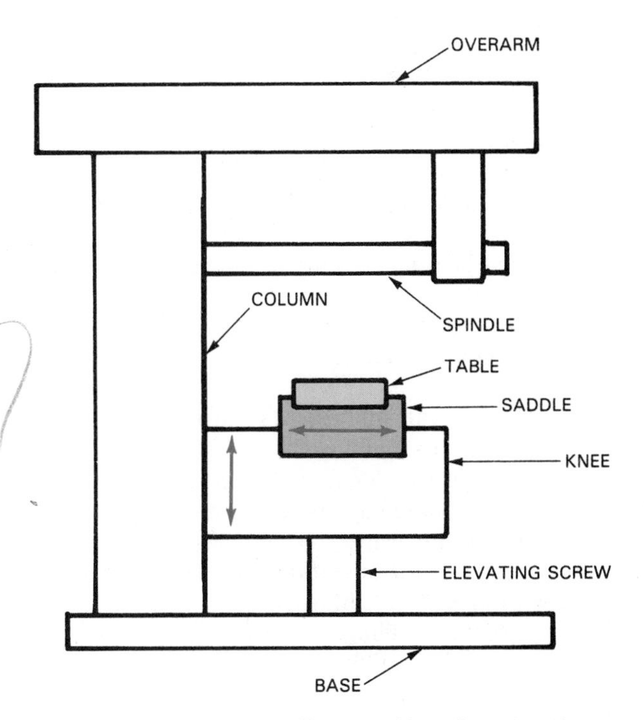

Fig. 17-13. Parts of a milling machine. Compare these with the photographs in Fig. 17-14.

Fig. 17-14. Left. A standard milling machine. Right. A numerically controlled horizontal milling machine. These are typical of horizontal milling machines. (Cincinnati Milacron and Giddings & Lewis)

of a certain size will be made in light, medium and heavy-duty models. The heavier duty a machine is the more it will weigh, the more power it will require, and the more it will cost.

TYPES OF MILLING MACHINES

Milling machines may be separated into two basic categories: *general purpose* machines and *production* machines. Included are these machines:

1. General purpose machines.
   a. Horizontal knee-and-column milling machines.
   b. Universal knee-and-column milling machines.
   c. Vertical knee-and-column milling machines.
2. Production machines.
   a. Fixed bed-type milling machines.
   b. Planer-type milling machines.
   c. Tracer milling machines.
   d. Machining centers.

General purpose milling machines

General purpose milling machines are knee-and-column machines as described earlier in this chapter. These machines are capable of doing a number of varied operations, hence the name general purpose. Common among the operations that can be done on them with proper attachments are:

1. Plane surface milling.
2. Curved surface milling.
3. Drilling and boring.
4. Slotting.
5. Gear cutting.

Horizontal or *plain* knee-and-column machines have a horizontal drive spindle. Arbors and adapter are mounted in the spindle. Arbors and adapter are mounted in the spindle. These cutter-holding devices are often rigidly supported by the overarm. Fig. 17-14 shows typical horizontal milling machines.

Universal knee-and-column milling machines are very much like plain milling machines. The main difference is that the table can be swiveled on the horizontal plane. This allows cutting a helix similar to those found on drills, cams, special gears, and milling cutters. The universal milling machine, like the one shown in Fig. 17-15, is almost always used in tool rooms. They produce accurate, complex cuts on short runs or single parts, such as special tools, dies, and fixtures.

Vertical knee-and-column milling machines have a head that is at right angles to the table surface, Fig. 17-16. This head contains both the power unit and the machine spindle. This arrangement permits more efficient face and end milling and also produces more accurate drilling and boring. On some machines the vertical spindle can be fed up and down and the entire head swiveled around its horizontal axis. These additional motions add greater variety to the possible machining operations.

Production machines

Production type milling machines are designed to machine material more rapidly with less operator attention than general purpose machines. Production

Fig. 17-15. A universal milling machine. Its table can be switched to different horizontal angles. (Cincinnati Milacron)

machines, however, are less flexible and require more effort to change setups from one job to another. This feature is not particularly a disadvantage since production or manufacturing milling machines are designed for long runs with infrequent changes. Moreover, special cutters and fixtures for locating and holding parts are used to increase the efficiency of the machines.

Bed-type machines are a type of production mill. A solid bed supports the table and provides a base for the power and spindle units. Bed-type machines, as shown in Fig. 17-17, can only move the table back and forth (lengthwise) along the bed. The movement of the cutter across the work after each cut is accomplished by moving the spindle in and out of the machine spindle carrier. The carrier is raised and lowered to set the depth of cut.

Often the actions of the machines are automatically controlled. The table will move rapidly forward until the work contacts the cutter. The table will then

Fig. 17-17. Bed-type milling machines support the table with a solid bed (base). (Cincinnati Milacron)

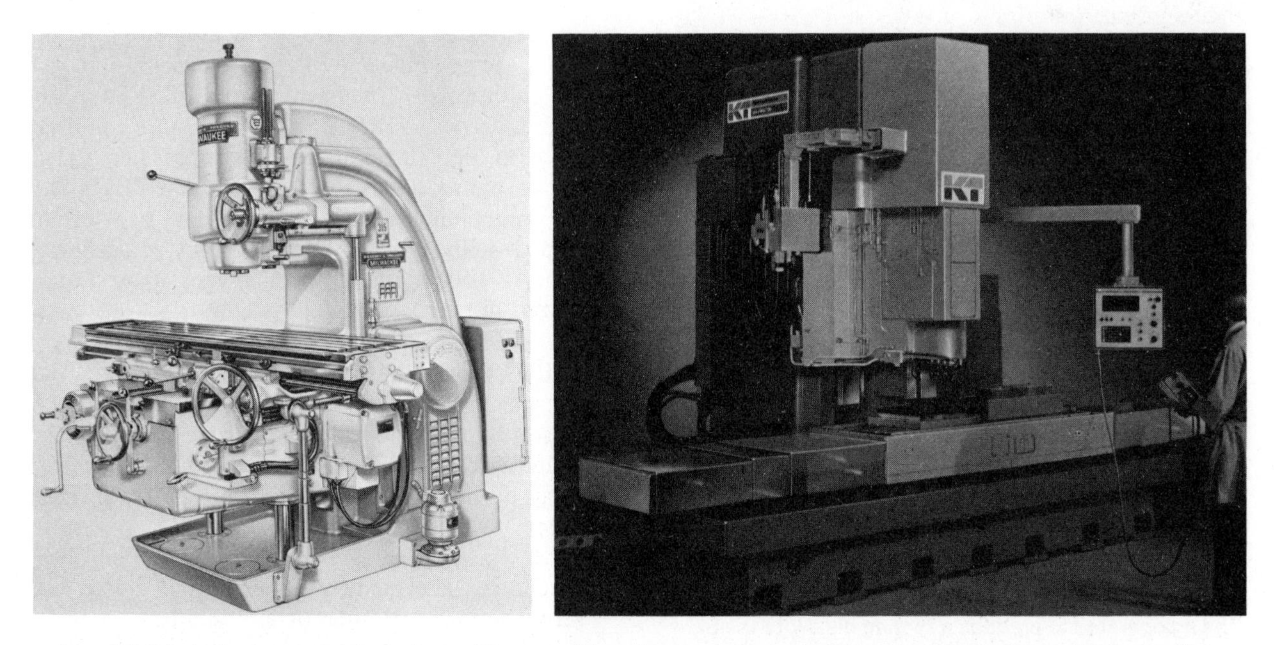

Fig. 17-16. Left. A standard vertical milling machine. Right. A numerically controlled vertical milling machine. Head is at right angles to the table surface. (Cincinnati Milacron and Kearney & Trecker)

slowly feed the work under the cutter. The table will rapidly move back after the cut. The spindle will only rotate during the actual cut and stop so that the work can be moved backward under it after each cut. The spindle is also fed a set distance across the work before each successive cut.

The *planer-type* milling machine is very similar to the bed-type machine. The main difference lies in the way the spindles are arranged. The bed-type machines have spindles which are fed out of the spindle carriers to move the cutter across the work. The planer-type machine moves the spindles like a metal planer (see Chapter 19). The spindles move across a cross rail to position themselves for each cut. The table moves lengthwise under the cutter to produce the feed motion for the cut.

A *tracer* milling machine may be a standard machine with a trace attachment. Like the trace lathe described in Chapter 16, it uses a stylus to determine the path of the cutter. The stylus follows a pattern which dictates the movement of the cutter on the X, Y, and Z axes. Tracer mills can produce curved and irregular shapes with relative ease. Numerically controlled machines can be programmed to accomplish similar tasks. A tracer mill may sometimes be called a *die-sinking* mill. It is used to reproduce cavities in various dies as shown in Fig. 17-18.

Fig. 17-18. A vertical mill with a tracing attachment is being operated to produce a forging die. Note the pattern to the right of the die. (Forging Industry Assoc.)

Machining centers

Machining centers are a new class of machine tools that perform a wide variety of milling and other operations. A machining center, Fig. 17-19, is:

1. Multifunctional (does many functions or operations).
2. Numerically or computer numerically controlled (NC or CNC).
3. Has automatic tool changers.
4. Uses rotating tools (cutters, drills, reamers, taps, etc.).

Machining centers were introduced in the late 1950s as a new, versatile machine tool. Today they are one of the most common machines found in modern manufacturing plants.

Machining centers were first used for producing small lots of parts requiring several different operations. The ability to program the machine through NC or CNC makes the task of completing multiple operations relatively easy and highly efficient. The machines are now being used for longer production runs.

Machining centers may use either vertical or horizontal spindle machines. The type will vary with:

1. Type of workpiece, (size, shape, and variety).
2. Number of parts to be produced.
3. Number of different operations to be used.

Vertical-spindle machines are used on flat parts when most of the operations are to be completed on the face of the part. With special part holders, small multisided parts can be machined. The machine is easy to load and has a wide open area for setup, Fig. 17-19A.

Horizontal-spindle machining centers, Fig. 17-19B, are more flexible. They also are available in a wider range of sizes and can, therefore, machine larger parts. Additionally, they can be controlled on more axes and, because of this, can produce more complex parts. Also, they can easily machine multisided parts.

The machines have heavy-duty spindle drives for heavy machining applications. They are often equipped with random-selection tool magazines, automatic tool-changing mechanisms, and computerized numerical control systems.

MILLING MACHINE ACCESSORIES

Most milling machine accessories are called work-holding devices. They are designed to hold the work securely during the milling operations. The devices include:

1. Vices which are most common on general purpose machines. They may be opened and closed by screw action, hydraulic or air cylinders, or with quick-acting cams or toggle mechanisms.
2. Chucks similar to those used on lathes.

A

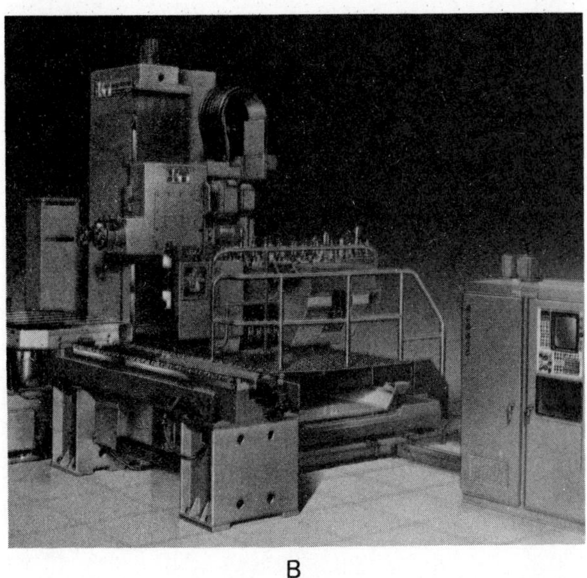

B

Fig. 17-19. A—A vertical-spindle machining center. B—A horizontal-spindle machining center. Note the numerical control units and the tool cassette holding several rotary tools.
(Giddings & Lewis and Kearney & Trecker Corp.)

3. Fixtures that are specifically designed to hold a particular part. The fixtures are bolted to the table.
4. Rotary table (circular milling attachment) which is a round table that can be pivoted on its axis. The rotary table is attached to the standard milling machine table. This attachment allows indexing a part so that cuts may be made at specific locations around the workpiece.

Fig. 17-20. An index head is being used to mill a helical cut. (Cincinnati Milacron)

5. Dividing or index head which is used to divide a circle into equal parts, Fig. 17-20. The device allows indexing a round part so teeth or other features may be uniformly cut on the circumference.

MILLING-TYPE WOODWORKING OPERATIONS

There are a number of woodworking operations which fit the operating definition of milling. They:
1. Use multitooth cutters.
2. Support the work on a table.
3. Rotate the cutter to generate the cutting motion.
4. Move the work in a straight line to generate feed motions.

The most common woodworking operations (which may also be used on selected plastics, particularly acrylics) are jointing, planing, molding, shaping, and routing.

All the milling-type machining operations involve removing excess wood (or plastic) in the form of single chips. These chips are formed by individual knives or cutting edges in a head or cutter. The typical arrangement of these knives or cutting edges is shown in the planer head for Fig. 17-1. As the stock is fed past this revolving head, a series of arc-shaped cuts are made. These cuts leave high spots much like waves in the ocean. The ripples resulting from cutting a flat surface with a rotating cutter are called mill marks and are similar to the machining marks shown in Fig. 17-12. The greater the feed rate the greater will be the width and depth of the mill marks. Therefore, high cutter head speeds and slow feeds will produce the least noticeable mill mark. The material will require only a little sanding to remove these marks. Similar mill marks are produced on metals and plastics machined on milling machines.

Jointing

Jointing is the woodworking process which trues or smooths a surface of a board. It uses a single cutter head located between an infeed and an outfeed table. As shown in Fig. 17-21, the cutter head, with usually three or four knives, is located so that it is always the same height as the outfeed table. The infeed table is adjusted to set the depth of cut. As the board is hand or mechanically fed over the cutter head, it is turned (made flat).

Boards may also be fed over a standard jointer with their edges against the table to produce a straight, true edge. Also, special jointers are made to produce straight edges for gluing (glue jointers) and on veneers (traveling-head veneer jointers).

PLANING

Planing is the process which produces one or more smooth sides on lumber. It is almost always done with the grain of the wood. The machine used is called a *surfacer*. The three major types of surfacers are:

1. Single surfacer.
2. Double surfacer.
3. Planer-matcher.

Single surfacer

The single surfacer, Fig. 17-22, is the simplest of the planing machines. The machine contains six major parts, Fig. 17-23:

1. A *table* or lower platen over which the lumber passes.
2. *Infeed rolls* which grip and feed the lumber into the cutter head.
3. A *chipbreaker* ahead of the cutter head which holds the workpiece down, directs the flow of the chips, and prevents advancing chipping (chipping far ahead of the cutter).
4. The *cutter head* which contains a number of knives (from three to twenty). Newer models have short, separated knives or spiral knives to reduce noise and increase cutting efficiency. The rotating cutter head produces the chip.
5. A *pressure bar* behind the cutter head to hold down the work as it passes from the cutting zone.
6. *Outfeed rolls* which grip and pull the workpiece from the machine.

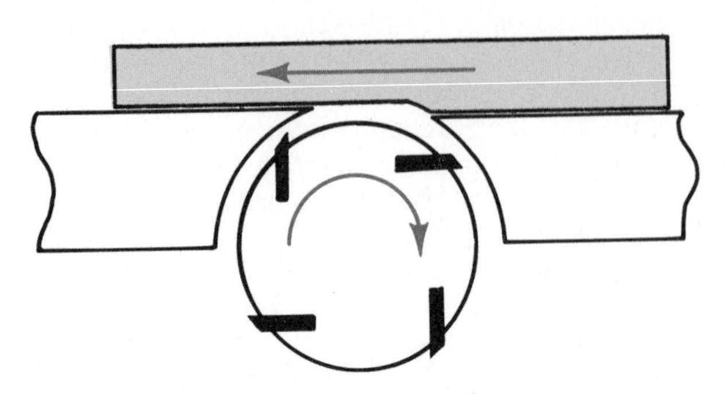

Fig. 17-21. A simple sketch of a jointer. Cutter rotates against feed direction.

Double surfacers

Double surfacers are designed to mill both sides of a board in a single pass. The machine has a cutter head located above the work as in a single surfacer. A second cutter head is located below the work. Fig. 17-24 shows a conventional double surfacer. Other models are made. One, the double-thickness planer, has chip breakers and pressure bars with each cutter head.

Planer-matcher

A planer-matcher has two heads to surface the two faces of the board. In addition, two more cutter heads, located after the thickness heads, surface the edges of the board. The planer-matcher can true all four major surfaces of a board in one pass.

Some machines add two additional horizontal cutter heads. Called *profile heads,* they produce a pat-

Fig. 17-22. A single-head planer has one cutter head. (Delta International Machinery Corp.)

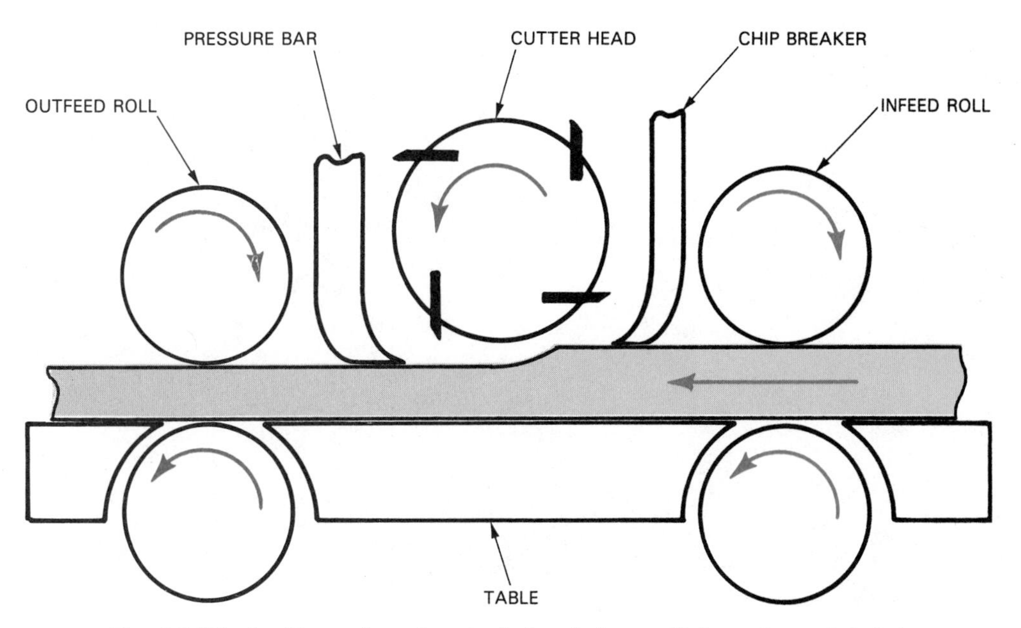

Fig. 17-23. An illustration of a single-head planer. Main parts are labeled.

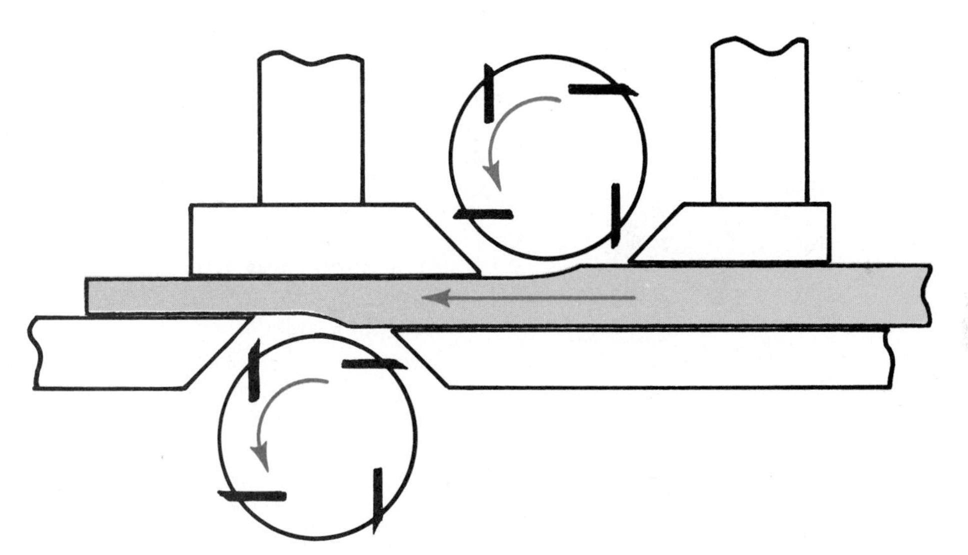

Fig. 17-24. A schematic of a double-head planer. Both sides of workpiece are planed in one pass.

tern on the faces of the lumber as it leaves the machine. Fig. 17-25 shows a planer-matcher with profile heads.

MOLDING

Molding is very much like the planer-matcher operation. The simplest molding machine has a top cutter head, followed by two side cutter heads which are staggered, followed by a bottom cutter head. The side heads can be tilted to produce bevels and other angled cuts.

The molding machine produces decorative shapes on material used to finish off cabinets, door and window frames, and furniture. These shaped pieces, called molding or trim, can be produced in an endless variety of cross sections.

SHAPING

Shaping is a process which produces a profile (shape) on the edge or end of a rectangular part or on the edge of a round flat member. The simplest shaper, as shown in Fig. 17-26, is a single-spindle

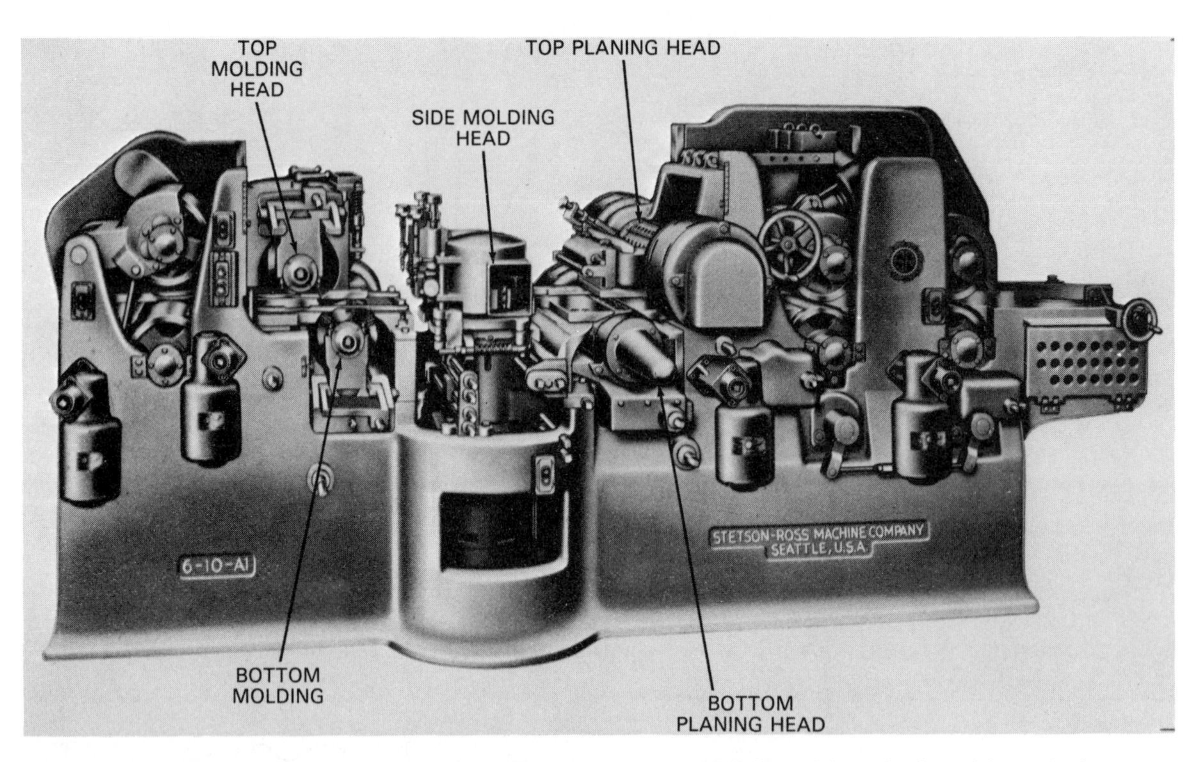

Fig. 17-25. Cutaway view of a planer-matcher which can true all four sides of a board in a single pass. (Stetson-Ross Machinery Co.)

Fig. 17-26. A single-spindle shaper is being used to machine a pattern on a cabinet door frame. (Delta International Machinery Corp.)

machine. A cutter of the desired shape is placed on the spindle. The work is then passed against the cutter to produce the required contour. Fig. 17-27 shows typical shaper cutters.

Double-spindle shapers are used to overcome chipping caused by single-spindle shapers. The best cutting action is produced when the cutter is rotating so its cut is made up the curve of the part. With a single-spindle shaper, the shaping action around a circle is with the downward slope of a curve during one half of the cut. This causes excessive chipping. The double-spindle shaper has cutters rotating in opposite directions. One-half of the curved cut is made on the first spindle while the other half uses the second spindle. Look carefully at Fig. 17-28 to see how this is done.

ROUTING

Routing is very similar to vertical milling. A small-diameter cutter is placed in the vertical spindle. This spindle is held by an overarm, hence the name *overarm router*. The cutter is lowered into the work. The work is then moved on the table to produce the desired cavity or cut.

The work is guided by one of several ways. These, as shown in Fig. 17-29, are:

1. A fence along which the work travels.

Fig. 17-27. Some typical shaper cutters and the shape they produce. (Delta International Machinery Co.)

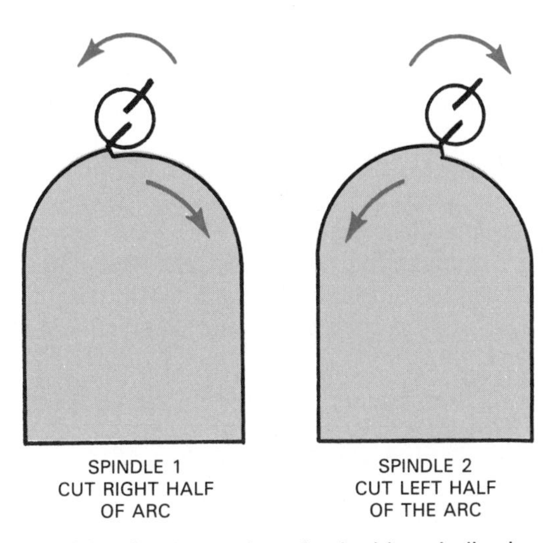

Fig. 17-28. Cutting action of a double-spindle shaper.

2. An external guide pin which positions the work.
3. A guide pin following a template attached to the underside of the work. Simple cuts may be made with a hand router but this machine has only limited industrial applications.

CARVING

A machine which also uses small cutters in vertical spindles is a *carver.* The carver is used to produce up to 30 duplicate shapes at one time. The machine works on the principle of the pantograph. A stylus is used to follow the contour and shape of the master template. As the stylus moves on the three axes (up-down, forward-back, right-left), the carver heads move in unison. The result is a number of parts which duplicate the master template. The master may be flat and clamped in a set position or round and rotated on an axis. The carver may be used to produce round parts like table legs or flat carvings like drawer fronts.

A

B

Fig. 17-29. Cutting with an overarm router requires using various accessories. A—A fence. B—Guide pin and template. (Delta International Machinery Corp.)

Steel turbine diffusers are being precision milled on a rotating table. (Cincinnati Milacron)

SUMMARY

Milling, broadly defined, is machining material using a stationary rotating cutter under which the material is fed. The rotating cutter produces the cutting motion. The movement of the workpiece on a table produces the feed motion.

Milling may be done on metals, woods, and selected sheet plastics. Typically, metals are milled using general and production milling machines. Milling operations on wood uses jointers, surfacers, molders, shapers, routers, and carvers.

All of these machines must provide a means of holding and rotating the cutter and a way to hold and move the work. The difference among the various machines basically lies in the way they accomplish these tasks and in the material they will machine.

STUDY QUESTIONS – Chapter 17

1. Describe the basic milling operation.
2. What are the two common milling cutter positions?
3. List and describe the three common types of milling cutters.
4. What are the common devices used to hold workpieces for milling?
5. What are the three axes in the movement of the workpiece/cutting tool and on what does their designation depend?
6. Describe peripheral milling and face milling.
7. Describe conventional and climb milling.
8. List and describe the major parts of a milling machine.
9. List and describe the major types of general purpose milling machines.
10. List and describe the major types of production milling machines.
11. Describe a machining center.
12. List and describe the major milling type woodworking machines.

Chapter 18

SAWING, BROACHING, AND FILING

A number of important manufacturing processes involves passing a continuous series of fine, evenly spaced teeth over a workpiece. The teeth generally are arranged in a narrow line. These processes may be grouped under three basic types:

1. Sawing—separating or cutting the workpiece into other parts using a narrow cutting device.
2. Broaching—shaping internal or external surfaces or round, flat, or contoured surfaces using a straight-line cutting device.
3. Filing—shaping and smoothing a surface using a wider surface-cutting device.

These definitions are vague. A closer look at each will add clarity and understanding.

SAWING

Sawing involves cutting off portions of a material. The cutting element for most sawing operations is a thin blade with evenly spaced teeth. The teeth separate the material by making a series of narrow cuts. Each tooth removes a very small chip, Fig. 18-1. The action of a large number of teeth passing a point in a given length of time makes sawing a relatively fast operation.

Since there are other ways to cut off material (parting on the lathe, flame cutting, shearing, etc.), it is the use of a thin blade and multiple, evenly-spaced teeth which makes sawing differnt. It is the narrow passage of the blade through the material that is unique. This path is called the *kerf.*

Understanding sawing begins with the study of saw blades and sawing machines. These two elements must work together.

TYPES OF SAW BLADES

Three major types of blades are widely used in industry. They are:

1. Straight blades.
2. Continuous band blades.
3. Circular or disc blades.

Straight blades are thin strips of steel with teeth along one edge. Common examples of straight blades are hacksaw, jig saw, and saber saw blades. Straight blades are used in machines which have a reciprocating (back and forth or up and down) cutting motion.

Continuous band blades are longer straight blades which have been end-welded to form a loop. Teeth are usually on only one edge. However, some applications, such as lumber mill head rigs (band saws), use band saw blades with teeth on both edges. The cutting action produced is a continuous straight-line cut.

Circular saw blades are thin discs of steel with teeth on their outer edges. The cutting action is a continuous arc or curved cut.

Special purpose blades include abrasive cutoff wheels, friction discs, and hole saws. Abrasive cutoff wheels are discs of abrasive material which will cut the material using the random cutting edges of the abrasive. Abrasive wheels are widely used to cut

Fig. 18-1. A close-up view of the cutting action of saw blade teeth. (DoAll Co.)

ceramic materials (brick and stone).

Friction discs are steel discs which cut material using only frictional heat. A disc turning at high speeds will build up enough heat at its point of contact to melt its way through a material. Friction discs are used to cut large I-beams and other structural members. Similar results can be achieved using band friction blades on high-speed band saws.

Hole saws are special saws designed to cut circular holes in wood, metal, or plastic. They are cup shaped with teeth on the rim. The saw cuts with a continuous revolving action.

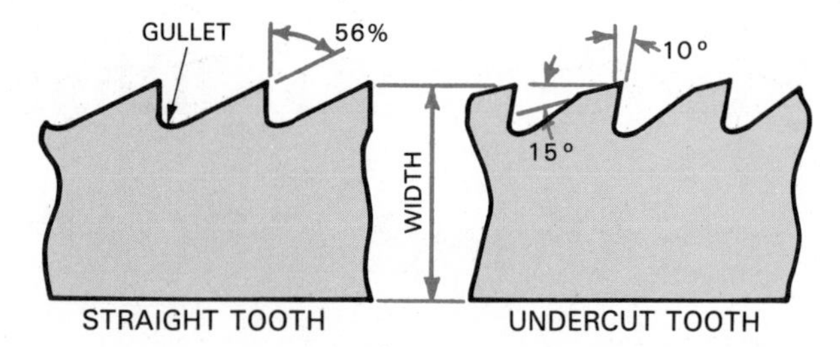

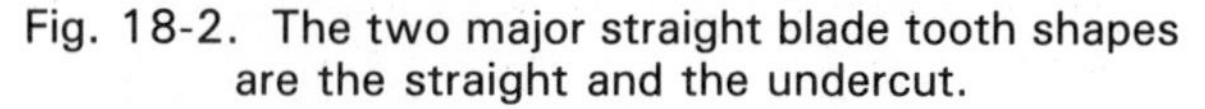
Fig. 18-2. The two major straight blade tooth shapes are the straight and the undercut.

SAW BLADE FEATURES

All saw blades have certain features which must be considered in their selection. These basic features include:

1. Material.
2. Tooth form.
3. Tooth set.
4. Tooth spacing.
5. Blade size.

Materials

Saws are made of a number of different materials. Typically, high carbon steel, high speed steel,and cemented carbides are used. Some blades are uniform throughout. They are made of the same material. Most cheaper wood cutting blades are of this type. Other blades have a body of a tough, flexible alloy. The tooth area is a hard but more brittle material welded to the body. Many metal-cutting hacksaw and band saw blades use this construction. Other blades have carbide teeth brazed onto a steel blade body. High quality wood cutting circular saw blades are of this type.

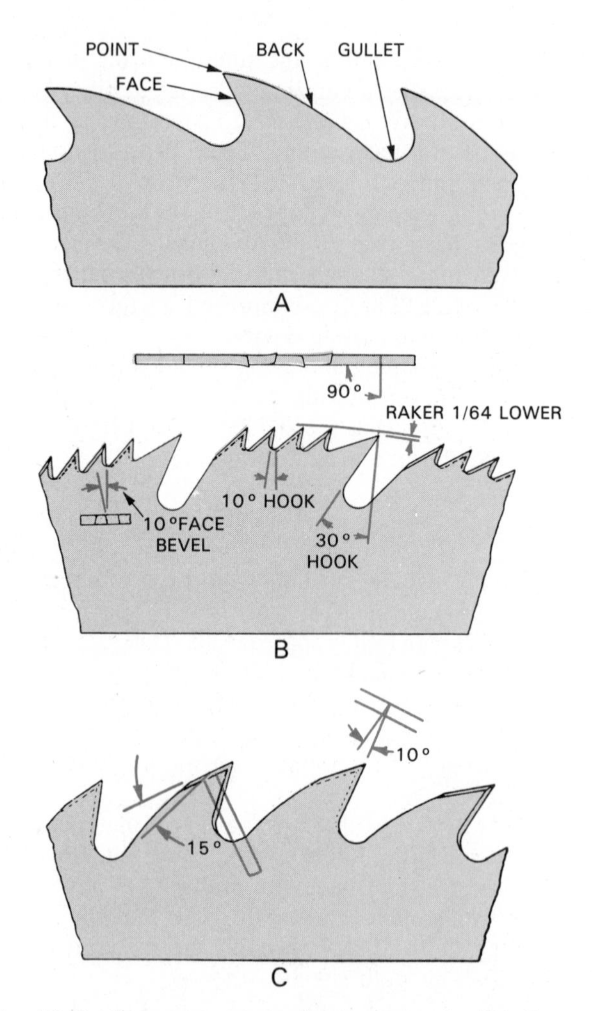

Fig. 18-3. Common circular blade types. A—Rip. B—Planer combination. C—Flat-ground combination. Crosscut blades are the fourth type and have teeth like the four uniform teeth in each set shown in view B. (Simonds Cutting Tools)

Tooth form

Many different tooth forms are used in saw blades. Straight and band blades use two main forms: straight tooth and undercut face tooth, Fig. 18-2. The straight tooth form is most common. The undercut face tooth is limited to some very coarse (teeth widely spaced) blades.

Circular saw blades have a variety of tooth forms. The typical woodworking tooth forms, shown in Fig. 18-3 are the crosscut, rip, combination, and planer types. Note that all the tooth forms have rake and clearance angles similar to single-point tools.

Tooth set

As a saw blade passes through a material its teeth produce a kerf. This kerf must be wider than the saw body. If it is not, the blade will bind and generate frictional heat from rubbing on the kerf walls. The technique used to deflect the teeth to cut a kerf wider than the blade body is called *setting*.

There are three major ways to set straight and band

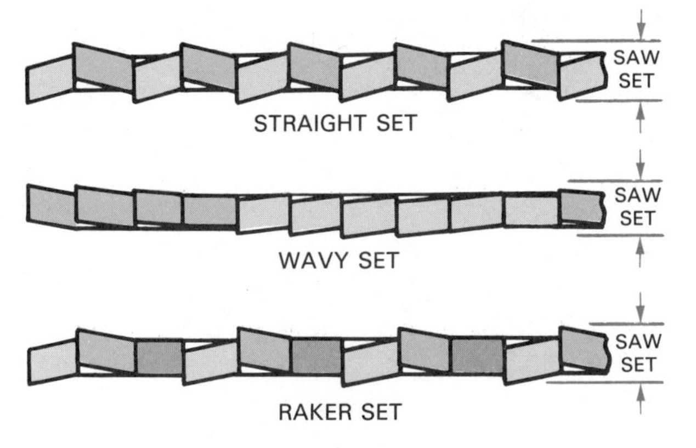

Fig. 18-4. These types of saw sets are common on straight and band saw blades.

blade teeth. As seen in Fig. 18-4, these are:

1. *Straight* set. The teeth are set (bent) evenly in alternating directions.
2. *Wave* set. The teeth are bent varying amounts to produce a "wave" of teeth from one side of the blade to the other.
3. *Raker* set. In each set of three teeth, one is set to the right, one to the left, and one has no set. This set is repeated in the same order for the length of the blade.

The straight and raker techniques are also used on circular blades. In addition, a method called *hollow grinding* is used. This process grinds a taper on the blade from the teeth to a point near the arbor hole. This action reduces the blade thickness between the teeth and the blade body. Fig. 18-5 shows hollow grinding.

Tooth spacing

Tooth spacing is a very important factor in saw efficiency. Teeth must be close enough to insure that two or three teeth are in contact with the work, as shown in Fig. 18-6. If only one tooth is in contact at a time the work can be caught in the tooth gullet (space between teeth). The result can be snagging of the tooth. This could cause damage to both the work and the blade. However, there must be enough space between the teeth to carry off the chips.

The fineness or coarseness of straight and band blades is given in pitch or number of teeth per inch

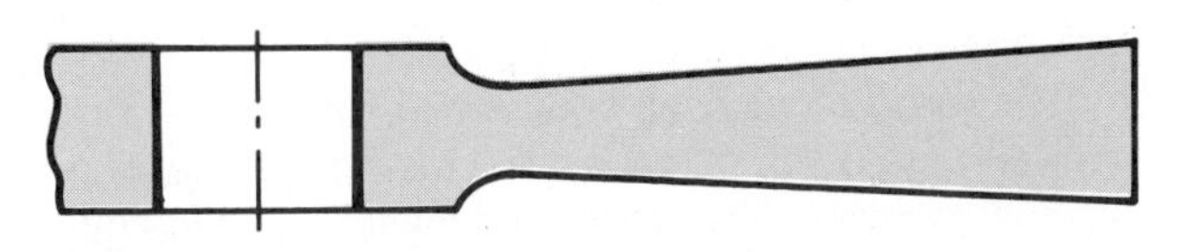

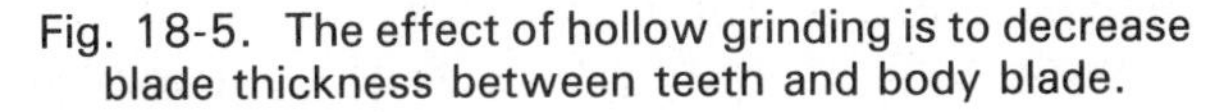

Fig. 18-5. The effect of hollow grinding is to decrease blade thickness between teeth and body blade.

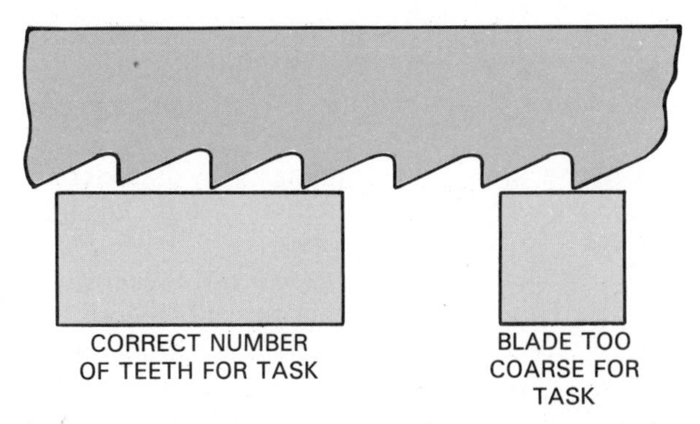

Fig. 18-6. Three saw teeth must be in contact with the work. Otherwise, teeth can be stripped from blade.

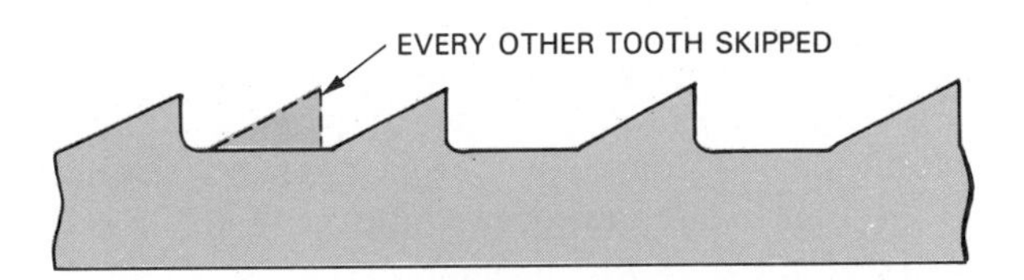

Fig. 18-7. Skip or buttress blade showing the "skipped" tooth.

of blade. There is generally a direct relationship between number and size of teeth. A blade will have a few large teeth per inch or more small teeth per inch.

The exception to this rule is the *buttress* or *skip tooth* as seen in Fig. 18-7. Actually, the teeth are smaller than would be expected from the pitch. This is accomplished by leaving out (skipping) every other tooth. The wider gullet allows for greater chip carrying capacity. Skip tooth blades are widely used in cutting wood, plastics, and free machining of metals.

Circular saw blade teeth are carefully spaced for fine cutoff; smaller closely spaced teeth are used. Coarser cuts, such as those for ripping wood, call for larger, more widely spaced teeth. Grinding wheels which also can be used as circular blades, have unevenly (randomly) spaced pieces of grit (abrasive) instead of teeth.

Blade size

The last factor considered in selecting blades is the blade size. This must be matched to the machine and the work being done.

Hacksaw blades vary from 0.025 in. thick, 1/2 in. wide, and 10 in. long to 1/8 in. thick 4 1/2 in. wide and 36 in. long. The smaller blades are for hand hacksaws while the larger blades are used on sawing machines.

Band saw blades can range in width of 1/16 in. to several inches. Narrower blades often come in long rolls. They are cut to the proper length and butt welded to form the band. Very wide blades (6–18 in.), like those used in sawmills, are manufactured from a thick blade stock. Fig. 18-8 shows a large band saw blade.

Circular saws are sized primarily by diameter and then by gauge (thickness). Usually, as the diameter increases so does the thickness.

The wider and thicker a blade the straighter will be the cut. This rule is important for cutoff work. The reverse must be considered as contour cutting is done. A wide, thick blade will not cut a very sharp curve. In fact, the sharpness of the radius that can be cut is a direct function of the blade width. Fig. 18-9 is a chart which shows the minimum radius standard width blades will cut.

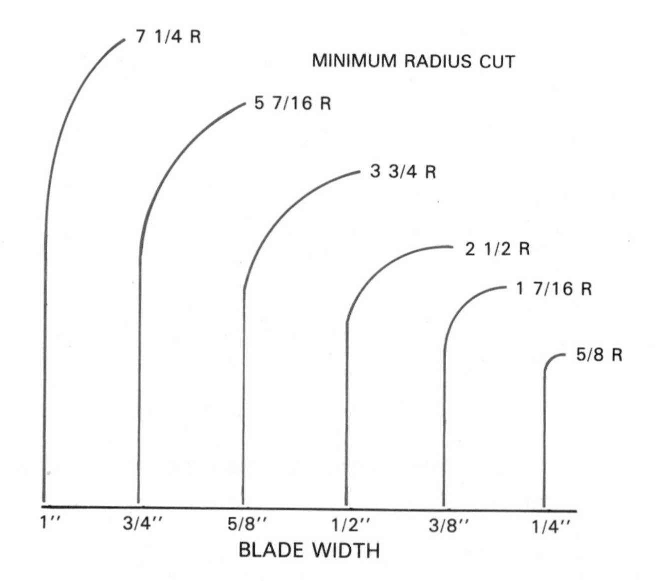

Fig. 18-9. Chart shows minimum radii which can be cut with standard width band saw blades.

Sawing machines

The various sawing machines for wood, metal, and plastic may be grouped by the type of blades they use:

1. Straight blade sawing machines.
2. Band blade sawing machines.
3. Circular blade sawing machines.

Straight blade sawing machines

Straight blade sawing machines use a reciprocating (back and forth or up and down) motion. The machine generally applies pressure on the forward or downward stroke. Pressure is often released on the return stroke to reduce wear on the teeth.

Fig. 18-8. This large continuous (band) blade is used in saw mills. (Weyerhaeuser)

The two major machines using straight blades are the metal cutting hack and the jig or scroll saw which will cut wood, plastic, and thin metals. Both of the machines have a C-frame. The blade is held across the open end of the "C."

Hacksaws, as shown in Fig. 18-10, are composed of a bed, a vise to support and clamp the work, a frame and blade assembly, a power mechanism, and often a feed mechanism. The blade is tightly held in a frame which is moved back and forth across the clamped work. Some machines use the weight of the frame-blade assembly to feed the blade into the work. A cam action will lift the blade up for the return stroke. More expensive, heavy duty machines use a feed mechanism. This system, usually hydraulic, applies force on the cutting stroke and lifts the blade for its return. The cutting and feed motions are both produced by the blade. The reciprocating action of the blade produces the cutting motion. The downward movement of the frame/blade mechanism generates the feed motion.

The hacksaw is somewhat inefficient. Cutting takes place in only one direction. For this reason other types of saws, to be discussed later, have largely replaced power hacksaws for cutting metal.

The *scroll or jig saw,* is designed for making fine, curved cuts on thin material. This machine uses an upright reciprocating blade. The blade is held in upper and lower chucks. The lower chuck is moved up and down by the power mechanism of the machine. The

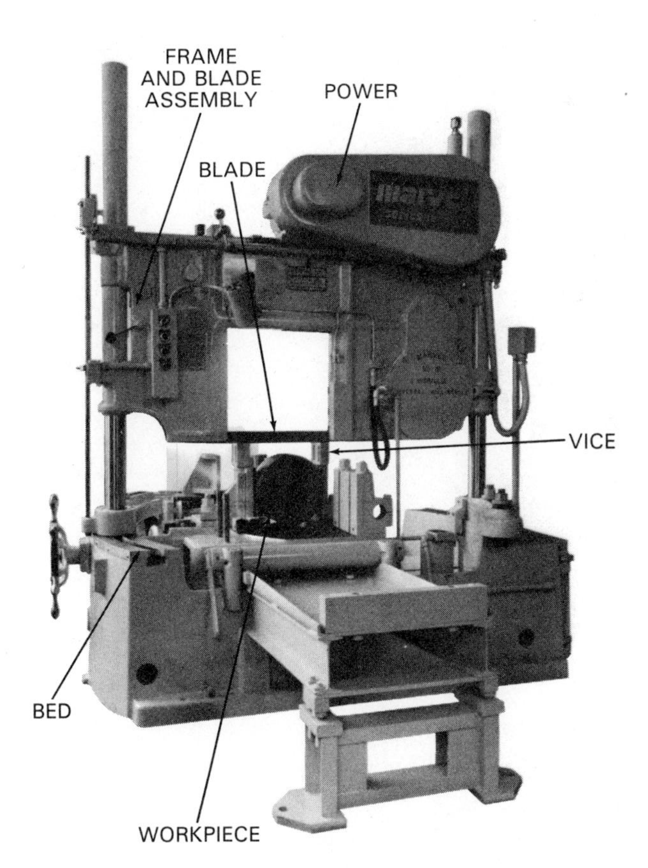

Fig. 18-10. A power hacksaw. Hydraulic system lifts blade. (Armstrong-Blum Mfg. Co.)

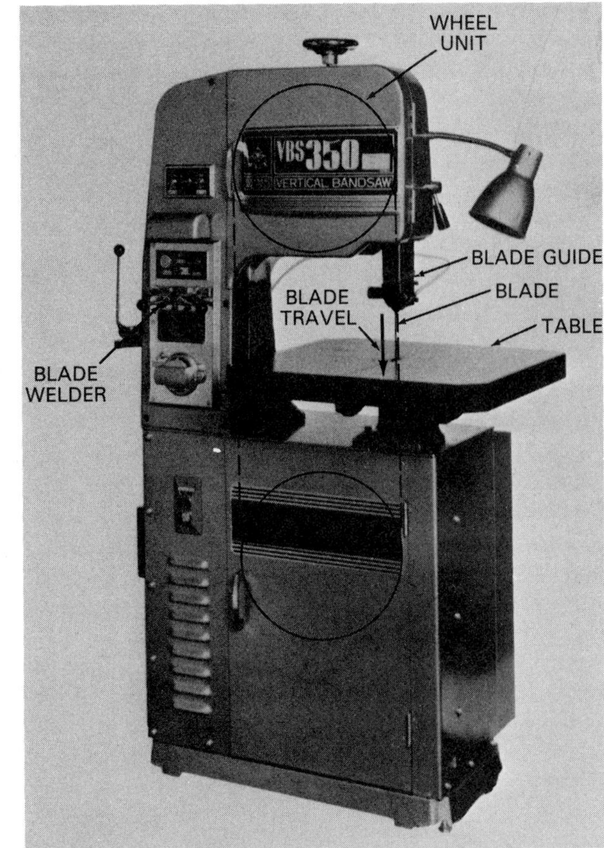

Fig. 18-11. Vertical band saw. Superimposed drawing shows location of wheel units which support and drive the blade. (Jet Equipment)

upper chuck is spring-loaded and maintains tension on the blade. The material is cut by hand feeding the stock as the reciprocating blade produces the cutting action. The linear movement of the stock into the blade produces the feed motion. Scroll saws are used in many hand-crafting operations including jewelry making, custom wood working, and some pattern-making.

Band blade sawing machines

The principal band blade machine is called a band saw. Band saws were first developed as a woodworking machine. Later the metal cutting band saw was developed. Two principal types of band saws are in use today:

1. Vertical band saw.
2. Horizontal band saw.

All band saws use a continuous saw blade which tracks over a pair of wheels. One of these wheels is powered while the other is used to adjust the blade track. The typical band saw has a frame, wheel units, blade guides, and a table or bed.

Vertical band saws, Fig. 18-11, were the first to be developed. They are driven so that the blade passes downward through the work and the table. The cutting action is generated by the linear (straight line) blade motion. Feed motion is produced by moving the work on the table into the blade.

In many applications the downward force of the blade motion is enough to hold the work on the table. No additional clamping is needed. However, fences and miter gages may be used to guide the work for straight cuts. Work may also be hand fed for straight cuts. Hard materials, like steel, may be power fed as shown in Fig. 18-12.

Special fixtures may be used to cut tapers, circles, and arcs. These devices add to the flexibility of the machine and increase part accuracy.

Vertical band saws are built for either wood or metal cutting applications. Wood cutting machines are lighter weight with higher cutting speeds. They also use a different, lighter blade guide system.

Special operations may be performed on the vertical band saw. A band made up of short pieces (segments) of files may be used for continuous filing.

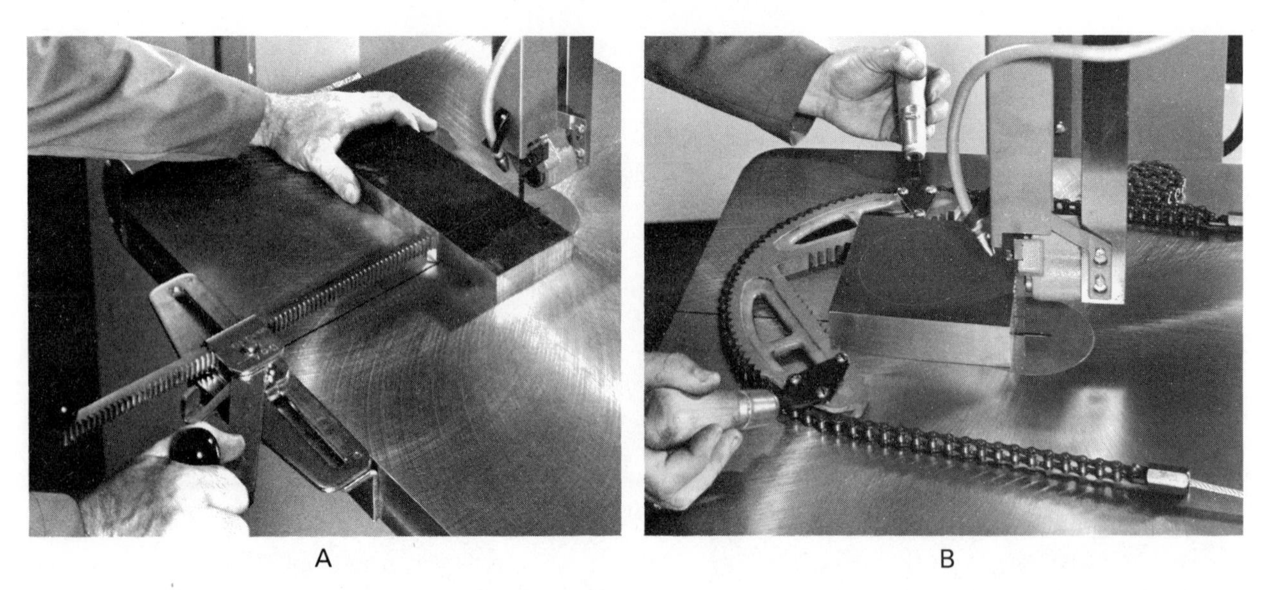

Fig. 18-12. Two methods to power feed using mechanical forces. A—Ratchet feed. B—Weight-type feed and contouring attachment. (DoAll Co.)

Also, endless emery cloth belts, backed with a rigid plate, may be used for polishing. Finally, high speed band friction cutting may be used on thin nonferrous metals and some plastic materials.

Horizontal band saws combine the holding and clamping features of the power hacksaw with the continuous cutting action of the band saw. Again, the cutting action is produced by the linear movement of the blade across the stock. The feed motion, however, is also produced by the blade/frame assembly. As this unit moves downward, new material is brought into contact with the blade. Fig. 18-13 shows a typical horizontal metal cutting band saw.

Circular blade sawing machines

Circular blade sawing machines are of two basic types. The first type operates on the principal of the milling machine. The blade is mounted on a stationary

Fig. 18-13. A horizontal band saw is used primarily for stock cutoff.

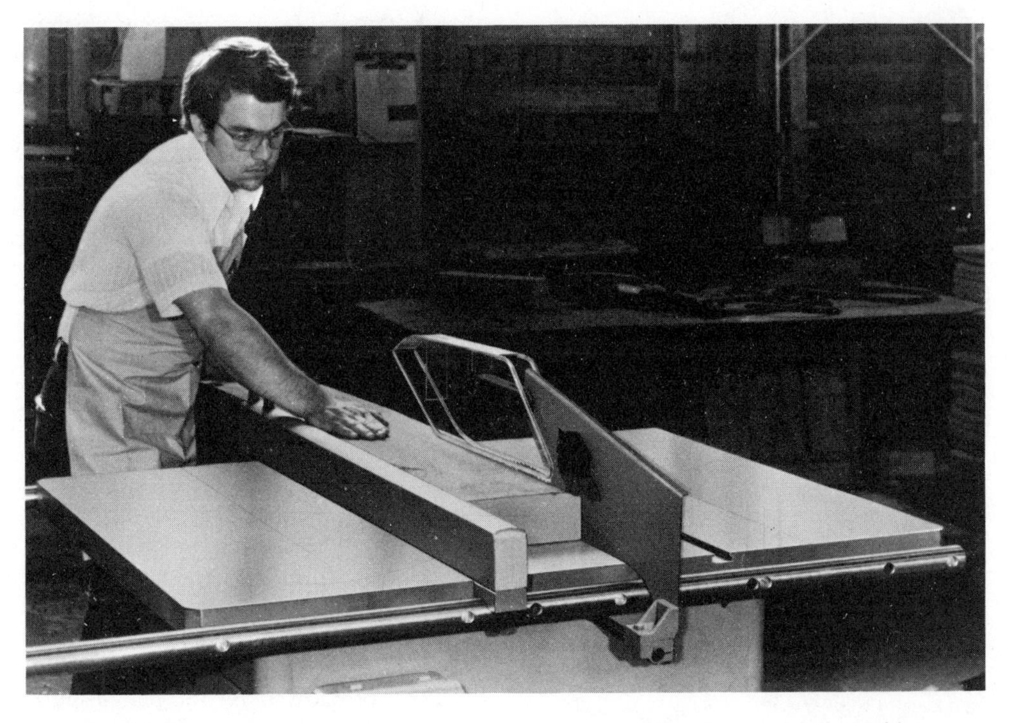

Fig. 18-14. A variety saw is the most common circular blade sawing machine.

rotating arbor. The rotation of the circular saw blade generates the cutting motion. The work is fed into the rotating cutter to produce the feed motion.

The most common machine of this type is the woodworking table saw or variety saw, Fig. 18-14. The blade is below the table. The arbor holding the blade can be raised and lowered to set the depth of cut. Also, the arbor assembly can be tilted to produce angled cuts.

The work rests on a table and is fed into the work with the aid of a rip fence or miter gage, Fig. 18-15. Power feed units may also be used to move the stock

Fig. 18-15. Rip fences are used to guide stock being cut along its length. Miter gages are used to cut across the width.

at a more uniform rate.

The radial saw also has a rotating blade to produce the cutting motion. However, the feed motion is produced by moving the blade through the work. This movement may be accomplished by either moving the blade-arbor assembly horizontally along a track or by pivoting the assembly from a fixed point.

Radial arm saws, Fig. 18-16, use a horizontal track to move the rotating blade through the stock. The overarm can be pivoted to produce angled cuts across the face of the work. Also, the arbor can be pivoted to produce angled cuts on the end of the stock.

Radial arm saws are used mostly for cutting wood. They may also be used on construction sites to cut nonferrous metals such as aluminum siding.

Fig. 18-16. A radial arm saw. Workpiece remains stationary and blade provides both feed motion and cutting motion.

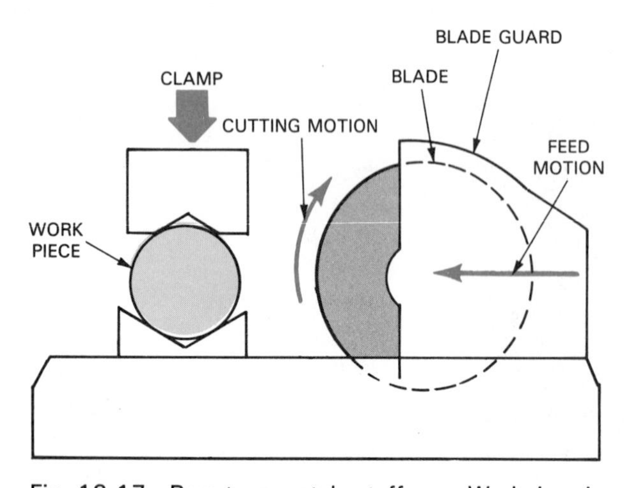

Fig. 18-17. Ram-type metal cutoff saw. Workpiece is stationary; ram provides feed motion at the arbor.

A horizontal-feed circular saw which uses the same sawing principle as the radial arm saw is shown in Fig. 18-17. This metal cutting saw uses a ram-type feed to move the blade-arbor assembly.

Pivot-type machines are generally cutoff machines. They are designed to only cut stock to length. They allow the blade to be fed downward into the work as the rotating blade cuts the materal. A typical cutoff machine is shown in Fig. 18-18.

Abrasive cutoff machines

Abrasive cutoff machines are a type of circular blade sawing machine. An abrasive wheel is used to replace the multitooth saw for cutoff operations, Fig. 18-19. The abrasive wheel, however, is a random-tooth or point cutter.

Abrasive sawing uses speeds up to 18,000 sfpm. It is able to cut harder materials than other blades and is one of the cheapest ways to cut metal. The finished cut is generally very straight and smooth.

Four basic techniques, as shown in Fig. 18-20 are used to perform abrasive sawing. These are:

1. Pivot or chop stroke for general-purpose cutting.
2. Rotating work for cutting thicker material.
3. Oscillating wheel for deep, fast cuts.
4. Horizontal feed for long cuts in flat stock.

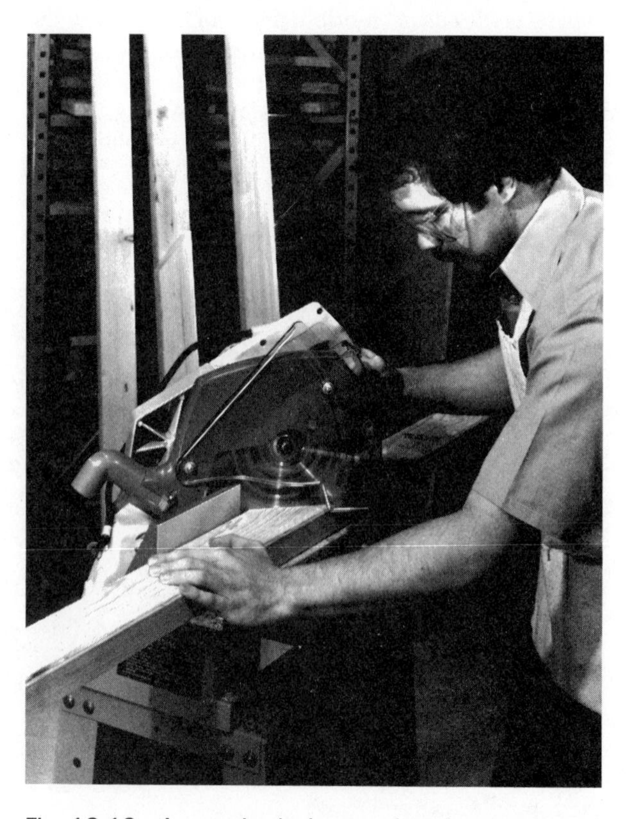

Fig. 18-18. A motorized miter saw is a pivot type cutoff machine. (Delta International Machinery Corp.)

Fig. 18-19. An abrasive cutoff operation. Cutoff blade is barely visible above the workpiece.

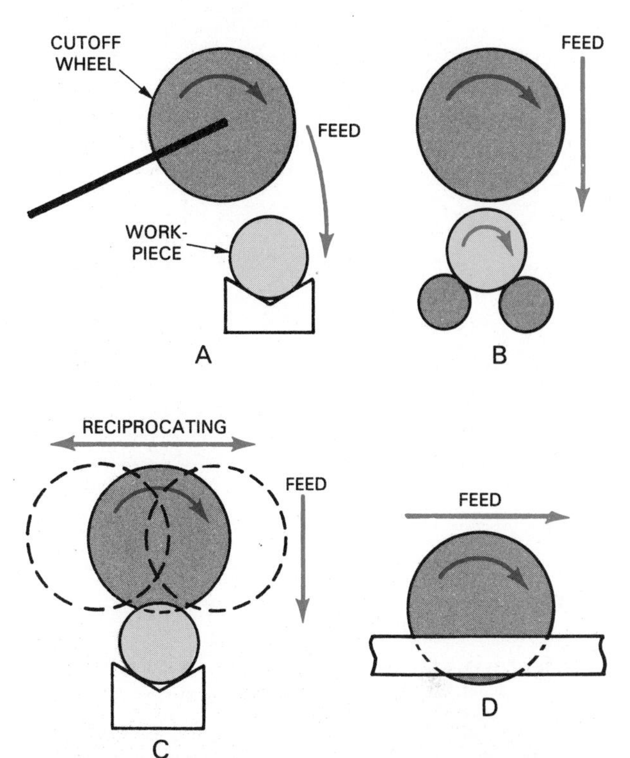

Fig. 18-20. Types of abrasive cutoff machines. A—Pivot or chop stroke cutoff machine. B—Rotary cutoff machine. Workpieces also rotate during the cut. C—Oscillating wheel cutoff machine. D—Horizontal cutoff machine.

Chop stroke machines are the simplest type and are used for cutting bars or tubes from 2 to 4 in. in diameter. The workpiece is held in a fixed position while a pivoted wheel up to 26 in. in diameter travels downward to make the cut.

Rotary machines rotate both the wheel and the workpiece. Thus, it is possible to cut solid, round stock twice the diameter that could be cut by a stationary workpiece. This method is preferred for cutting bars over 8 in. in diameter. Some machines use oscillation and controlled vibration of the wheel along with the cutting action to provide chip clearance and help cool the cutting edges.

Oscillating machines have the sawing action of a straight blade. They work well on larger workpieces—up to 12 in.—either round or square. Amplitude (distance traveled) of the oscillation and frequency (length of time to complete one oscillation cycle) can be varied over a wide range.

Horizontal-traverse machines may have either a wheel that moves across the work or a fixed wheel against which the work is moved. The type with a traversing (moving) wheel is used for cutting medium or large plates and slabs up to about 3 in. thick. The fixed spindle machine (wheel does not travel) is commonly used to cut glass and nonmetallic materials including brick, tile, and refractories.

Hot-sawing or *friction-saving machines* are also adaptations of circular sawing machines. Frictional heat generated by a rotating disc cuts off the material. The heat melts the material in contact with the disc. The disc does not melt because only a very small part of it is in contact with the work at any one time. The rest of the disc is being cooled as it rotates free of the work. Hot sawing is generally used to cut thin nonferrous metals.

The action leaves heavy burrs caused by the melting action of the "cut." Also, the cut is less accurate than other methods of sawing materials.

BROACHING

Broaching, an internal or external machining process for flat, round, or contoured shapes, uses an action similar to straight-blade sawing. The tool, called a broach, has a series of teeth of gradually increasing height spaced along its length. The broach is pushed or pulled along the exterior surface or through an internal cavity. External broaching is called *surface broaching*. Using a broach on an inside surface is called *internal* or *hole broaching*.

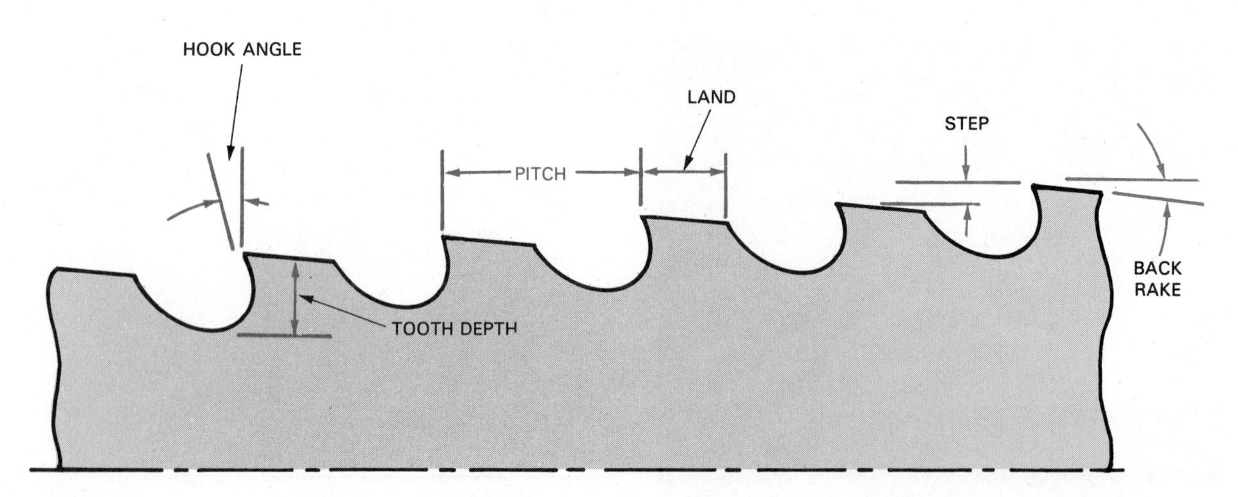

Fig. 18-21. Cross section of a broach. Each successive tooth is generally higher than the one preceding it.

BROACHES

There is wide variety in the design of broaches. However, there are basic commonalities too. All broaches have a series of single teeth spaced along a rigid bar or plate. The cutting edge of each succeeding tooth is slightly higher than the one preceding. This increase in the tooth height, called its *step,* sets the depth of cut for each tooth. The difference in height between the first tooth and the last tooth determines the total depth of cut for the broach. A typical cross section of a broach is shown in Fig. 18-21.

A broach usually has several cutting zones along its length. The first set of teeth is generally *roughing teeth.* These teeth have a depth of cut up to .006 in. Following the roughing teeth are semi-finishing teeth, then finishing teeth. The depth of cut for finishing teeth will be at least .001 in. It is necessary to match the depth of cut with the material being machined. Cuts which are too large generate stress on the teeth and workpiece. Very small cuts cause a rubbing action rather than good cutting action.

BROACHING MACHINES

A number of machines use the many broaches available. These machines have a device to hold the work, a broaching tool, a mechanism to drive the broach or the work, and a supporting frame. These machines are grouped into four basic types according to their operating principles. Included are:

1. *Pull broaching machines.* The broach is pulled across or through the work. The work is stationary. The cutting motion is created by the movement of the broach. Feed is produced by the increasing height of the teeth on the broaching tool.
2. *Push broaching machines.* The broach is pushed across or through the stationary work. The cutting and feed motion is produced in the same way as in pull broaching.
3. *Surface broaching machines.* A surface broach is moved across a flat workpiece or the work is moved across the broach. The movement of either the work or the broach generates the cutting motion.
4. *Continuous broaching machines.* The work is moved in a straight line or rotated against a stationary broach. The moving workpiece generates the cutting motion.

Fig. 18-22 shows simple sketches of the typical broaching actions. These actions may be on either horizontal or vertical machines which usually have hydraulic drives.

Broaching is used to produce a number of specific shapes. Included are:

1. Shaped holes from round holes.
2. Keyways in holes.
3. Finished holes to accurate limits.
4. Finished flat surfaces.
5. Straight or helical splines.

Broaching produces a roughing and finishing cut in a single pass. The process produces parts at high production rates on either internal or external surfaces. Also, any shape a broach can be made into may be produced on the part. However, broaches are expensive and require rigid setup. The operation can only be used where the broach can pass completely through an opening or over a surface.

FILING

Filing is often a hand operation which is not within the scope of this book. However, some machine filing is done.

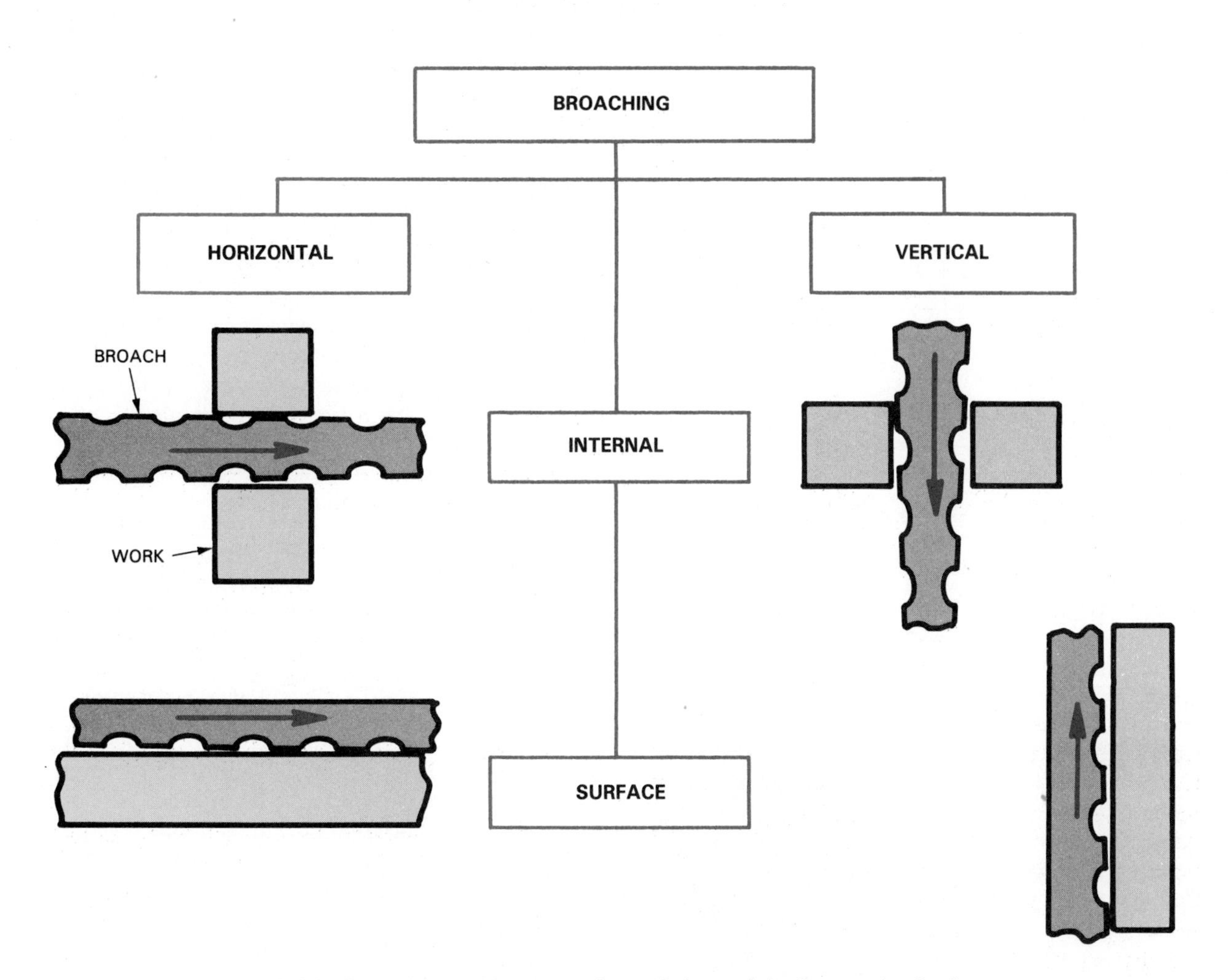

Fig. 18-22. Typical broaching operations. Only straight-line motion is shown.

Filing removes material in the same basic way a saw does. A series of teeth remove very small chips. The primary difference lies in the width of the cutting tool. Saws and saw teeth are very narrow and are intended to cut deeply into the material. By contrast, files are generally wide. They are designed to remove material from the surface of a part.

TYPES OF FILES

Files are organized into types according to:

1. Type or cut of the teeth. The four major cuts of files are single-cut, double-cut, rasp, and vixen-cut.
2. The coarseness of the teeth.
3. Shape of the file cross section.
4. Construction of the file.

Single-cut files have evenly spaced parallel teeth across the width of the file. These teeth are placed at a 65 to 80 degree angle.

Double-cut files have two sets of teeth across the file. The first series is placed at a 40-45 degree angle while a second, coarser set is placed at an opposite angle. The angle for the second set of teeth may be anywhere from 10 to 80 degrees.

Rasp-cut teeth are short, individual teeth raised out of the file body. These teeth are arranged in rows across the file. *Vixen-cut* files have a series of curved parallel teeth.

The coarseness of the file refers to the size of the teeth. They range from very small to large. From finest to coarsest they are called: dead smooth, smooth, second cut, bastard, coarse, and rough.

Files are also available in several shapes, Fig. 18-23. The most common are flat, square, 3-square, half-round, and round. Flat files may have teeth on all four surfaces or only on the faces. Flat files with no teeth on the edges are called safe edge files.

File construction may be of three major types:

1. Single units for hand filing or die-filing machines.
2. Band segments for band-filing (saws) machines.
3. Discs for disc-filing machines.

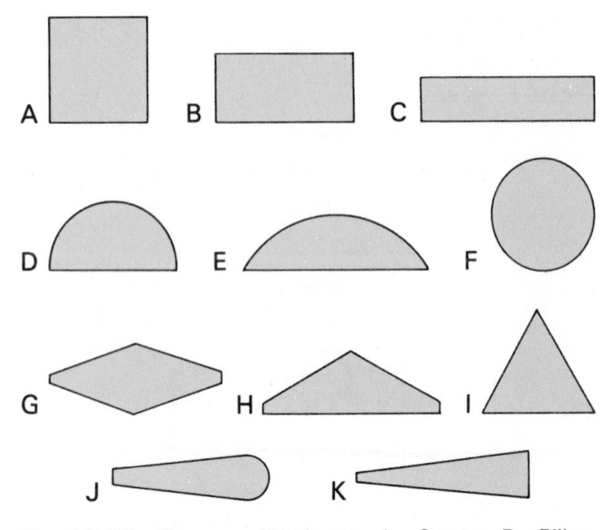

Fig. 18-23. Common file shapes. A—Square. B—Pillar. C—Mill. D—Half-round. E—Marking. F—Round. G—Slitting. H—Cant saw. I—3-square. J—Crosscut. K—Knife. (Nicholson File Co.)

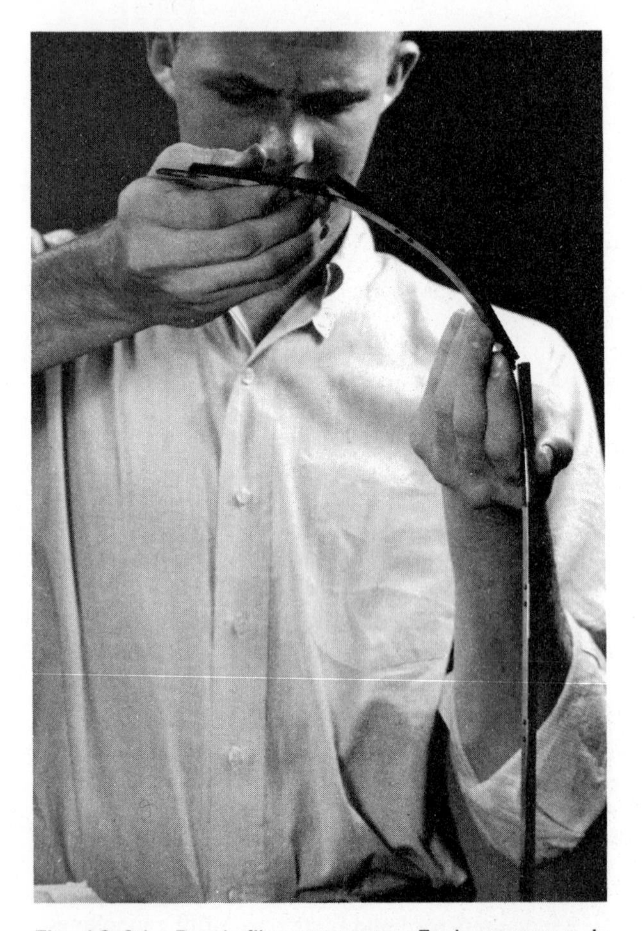

Fig. 18-24. Band file segments. Each segment is riveted to a flexible steel band so band can flex and curve independently of other segments. (DoAll Co.)

FILING MACHINES

To speed up the hand filing operations, three types of filing machines have been developed. The *die filing machine* operates much like a jig saw. The file extends up through a table. The file reciprocates (moves up and down). As it moves it is held rigid by a roller guide on an overarm.

The band-filing machine is an adapted band saw. As described earlier, file segments are connected to form a band. The band then travels around two wheels to create a continuous downward filing action. Fig. 18-24 shows band-file segments while Fig. 18-25 shows a band-filing machine in operation.

A disc file operates like a disc sander. A rotating file disc provides the filing action.

All three filing machines require a skilled operator to produce accurate work. Therefore, most filing machines are used in tool and die work and in maintenance shops.

SUMMARY

Sawing, broaching, and filing remove stock using a series of teeth on a cutter. Each tooth removes a small chip. The action of the many teeth passing a point produces the total cutting action.

Sawing separates the workpiece into other parts, using a narrow cutting device. Other devices are also used to cut off material but the saw uses a thin blade with multiple, evenly-spaced teeth. This makes it different from other types of cutoff machines.

Broaching shapes internal or external surfaces using a straight-line cutting device. In filing operations, a surface is smoothed or shaped with a wider surface-cutting device.

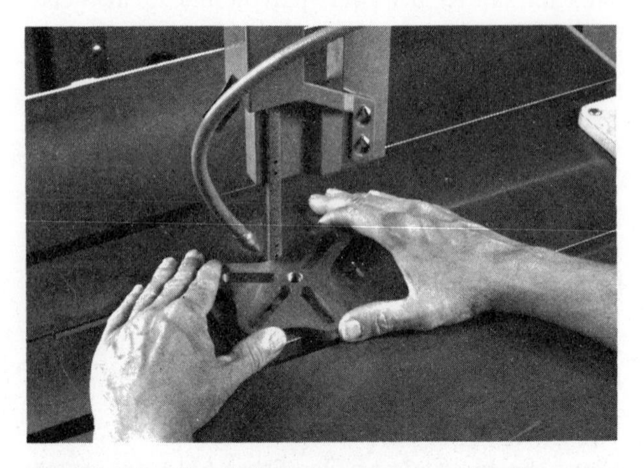

Fig. 18-25. Band filing operation. Band moves downward, traveling around a pair of wheels, one of which drive the file band. (DoAll Co.)

Flat cutters are used in sawing, broaching, and filing. Band and circular cutters are common in sawing and filing operations.

STUDY QUESTIONS – Chapter 18

1. Define sawing, broaching, and filing.
2. What is a saw kerf?
3. List and describe the three major types of saw blades.
4. What is meant by tooth or saw set?
5. List and describe the three major ways to accomplish saw set.
6. What happens if too coarse a blade is used in cutting hard material?
7. Describe the power hacksaw and scroll saw.
8. List and describe the types of band saws.
9. List and describe the types of circular sawing machines.
10. Describe the basic abrasive sawing techniques.
11. List and describe the four basic types of broaching machines.
12. List the four ways in which files are classified.
13. Name 11 common file shapes.
14. List and describe the three common filing machines.

Typical metal sawing operation. This power hacksaw cuts with a reciprocating motion. Machine can be set for automatic or manual operation.

Chapter 19

SHAPING AND PLANING METALS

Shaping and planing metals are entirely different processes than shaping and planing woods. The woodworking processes, as described in Chapter 17, were closely related to milling metals. They used a rotating multiple-point cutter and linear feed of the work. Metal shaping and planing have several features which make them unique processes. These features include:

1. Reciprocating cutting motion.
2. Linear feed motion.
3. Single-point tool.

The difference between metal shaping and metal planing lies in the element which reciprocates to cause the cutting motion and the element that moves linearly to produce the feed motion. The shaper tool reciprocates and the work feeds in a linear direction. The planer generally reciprocates the work on a moving table to cause the cutting motion. The tool feeds across the work to produce feed motion.

Both machines have common elements which, as seen in Fig. 19-1, include:

1. A table.
2. A method of securing the work to the table.
3. A single-point tool and holder.
4. An adjustable tool slide.

However, the machines differ in other ways which make it necessary to discuss the two machines separately.

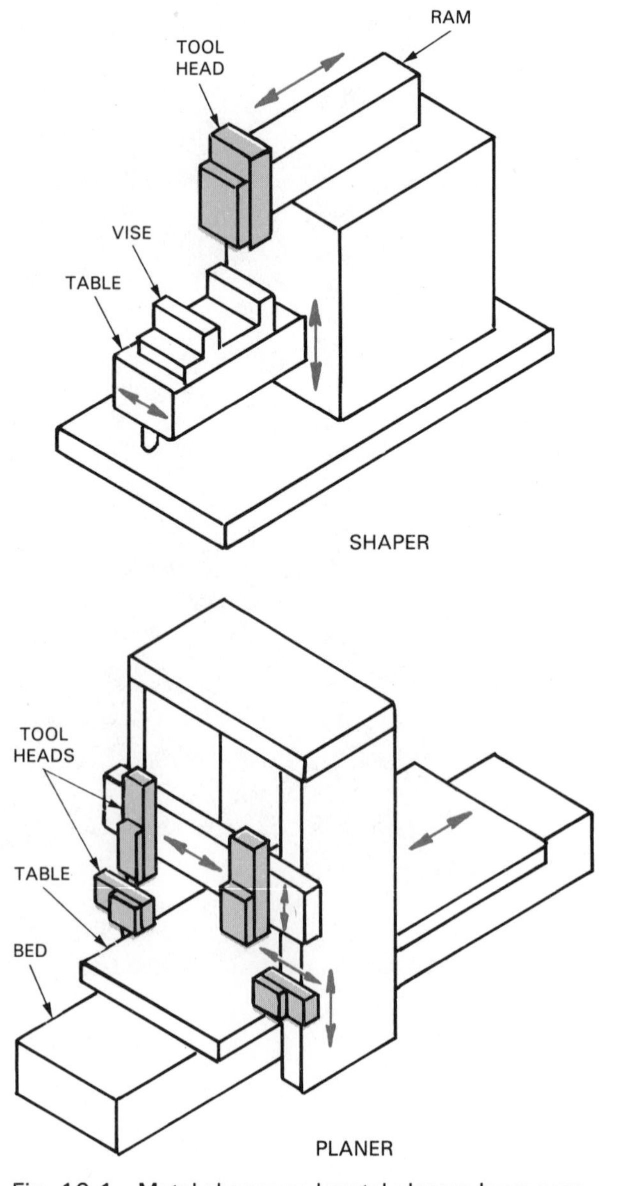

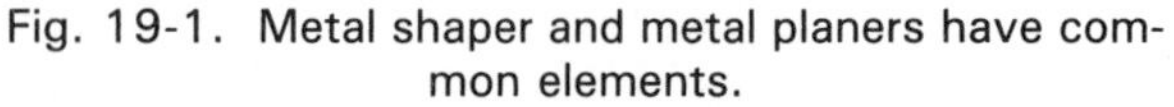
Fig. 19-1. Metal shaper and metal planers have common elements.

SHAPERS

Metal shapers are designed to produce flat surfaces on the workpiece. These surfaces may be horizontal, vertical, or angled. With a special tracer attachment, shapers may be modified to produce a curved surface.

The shaper cuts by pushing or pulling a single-point tool across a workpiece. The tool movement is straight-line. At the end of the stroke, the machine cycles the tool back to its original position. No cutting is done during the return stroke. The work is fed at right angles to the cutting line during the back

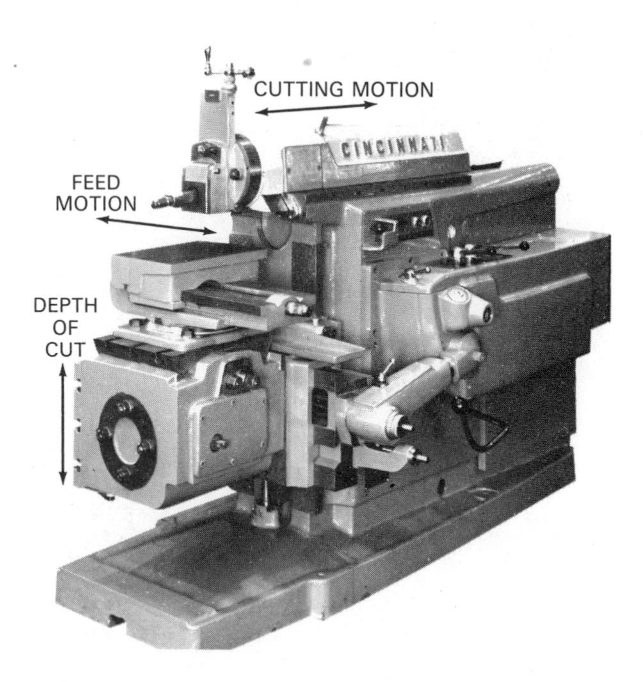

Fig. 19-2. Cutting and feed motions of a metal shaper. Cutting action produces flat surfaces. (Cincinnati, Inc.)

stroke, however. This action brings new material into the tool path. Fig. 19-2 shows the cutting and feed actions of a shaper.

Types of shapers

Shapers are made in several general designs. These include:

1. Horizontal shapers.
   a. Push cut.
   b. Pull cut.
2. Vertical shaper.
   a. Keyseaters.
   b. Slotters.
3. Special purpose (gear cutters) shapers.

All these machines consist of several basic parts. Among these parts is the *ram.* The ram moves the tool during its cutting and return strokes. Generally the return stroke is more rapid than the cutting stroke. The rapid return increases the speed of the operation.

Shapers also have a tool-securing mechanism. As shown in Fig. 19-3, the tool mechanism generally has a *tool slide* at the end of the ram, a *clapper box,* and a *toolholder.* The tool slide allows the tool to be moved to set the depth of cut. The clapper box permits the tool to pivot upward on the return stroke. The latter action eliminates unnecessary rubbing of the tool against the work on the return stroke.

All of these devices are secured to a *tool head* which can pivot. This allows for positioning the tool for flat

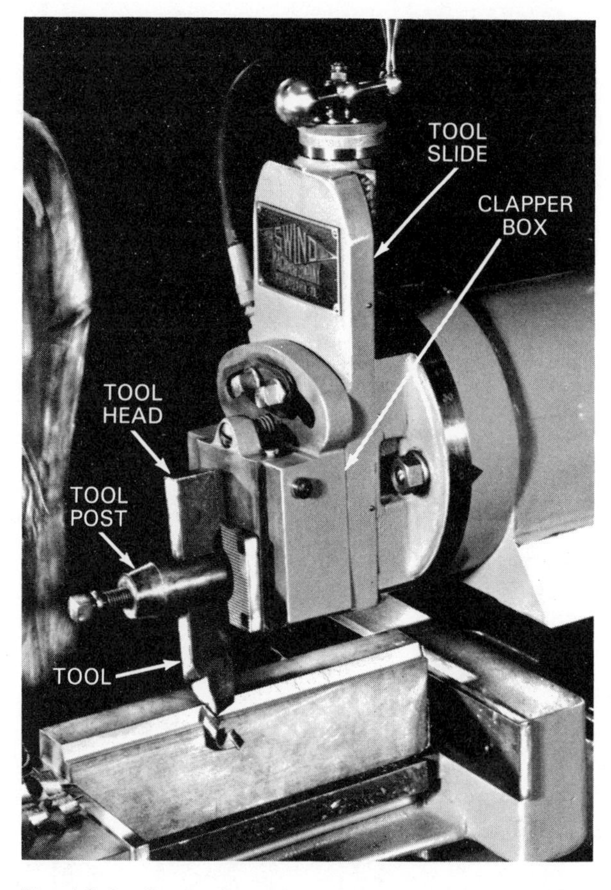

Fig. 19-3. Parts of a tool mechanism. Note the adjustment points. (Rockford Machine Tool Co.)

cuts on the face or the side of the work and for various angled cuts. Fig. 19-4 diagrams some typical shaper cuts.

Another important part of a shaper is the table. The table serves two main purposes. It secures the workholding device. The device may be a vise, a fixture, or bolts to name a few possibilities. Also, the table moves along crossrails to feed the work.

Shapers have a heavy cast frame and base to hold the operating systems. Within the frame are various drive mechanisms. There are systems to cause the ram to reciprocate and the table to feed the work.

Horizontal shapers

Horizontal shapers have a ram which is parallel with the base of the machine. The most common horizontal shaper is the *push type.* It cuts on the push or forward stroke and feeds the work on the back stroke. Fig. 19-2 showed a push type horizontal shaper in use.

Horizontal shapers can have a tracer attachment which automatically changes the height of the tool. This allows for machining a contoured shape. Such

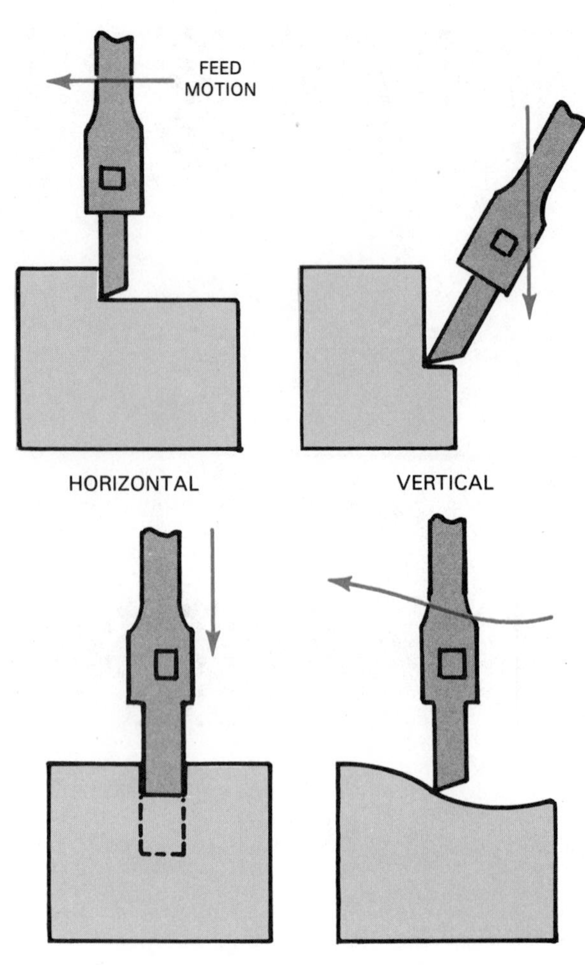

Fig. 19-4. Typical shaper cuts. Contour cuts require special attachment.

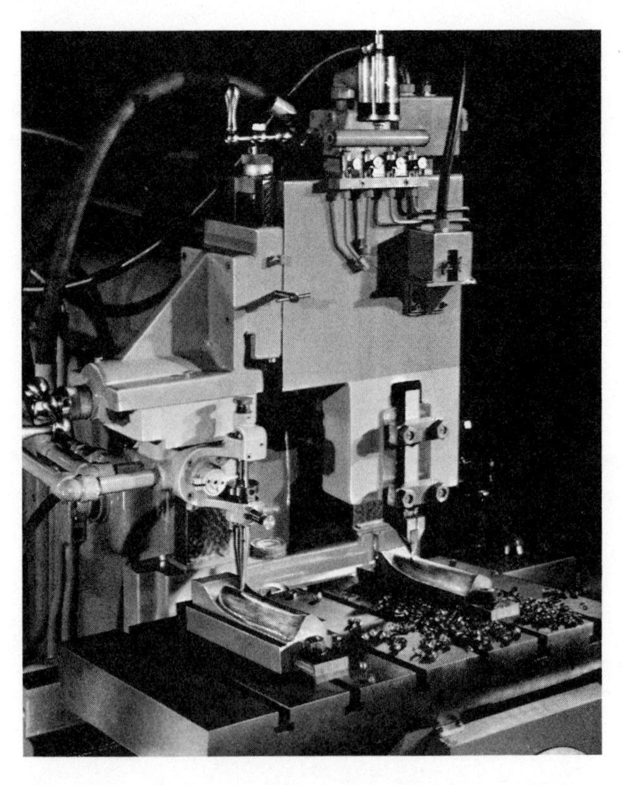

Fig. 19-5. A shaper with a tracer attachment. Tracer attachment is at left. (Rockford Machine Tool Co.)

a machine is shown in Fig. 19-5.

Some horizontal shapers cut on the back movement of the ram. These machines are called *pull* or *draw-cut shapers*. They are designed for heavier cuts with longer ram movements. The inward movement makes it easier to hold the work and reduce tool vibration and chatter.

Vertical shapers

Vertical shapers have a ram at right angles with the base of the machine. They often have round tables which rotate and move horizontally along the crossrails. The most common vertical shaper is also called a *slotter*. It is used to machine curved surfaces, arcs, and internal surfaces.

A special vertical shaper is called a *keyseater*. This machine is designed for machining keyways in gear, wheel, and pulley arbor holes. The machine uses a vertical reciprocating cutting bar to cut the keyway. The table moves inward to produce the feed motion as shown in Fig. 19-6.

Special purpose shapers

Special purpose shapers are designed primarily for gear cutting operations, Fig. 19-7. These machines use a tool the shape of the gear tooth. The gear blank is clamped to the rotary table. The vertical ram causes the tool to cut the gear teeth. After each cut the table is rotated or indexed so that the next tool shape may be cut on the gear. Fig. 19-8 shows a gear cutting shaper.

Gears may also be produced by casting, molding, extrusion, powder metallurgy, roll forming, and machining (milling machines, gear hobs).

Shaper sizes

The sizes of shapers are determined by the maximum length of their strokes. Horizontal push-cut shapers can have up to 36 in. (914 mm) strokes while pull-type machines can have strokes as high as 72 in. (1829 mm).

Most shapers have a table cross-feed equal to the maximum stroke. A shaper with a 24 in. stroke can usually machine a flat surface which is 24 in. by 24 in.

Vertical shapers are also sized by the length of the

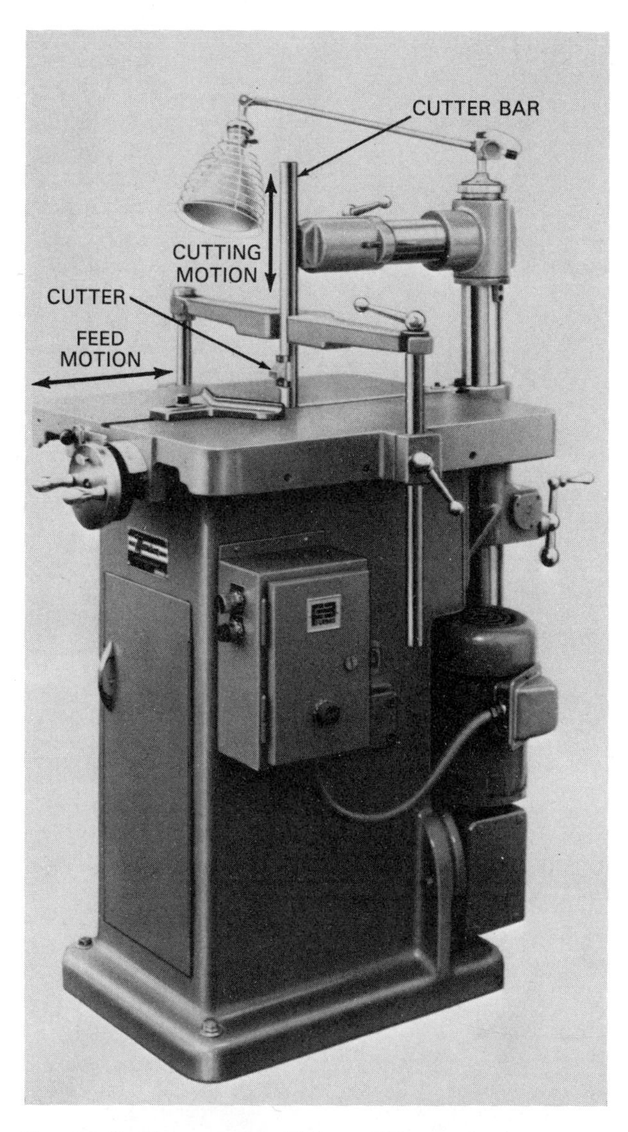

Fig. 19-6. A keyseater. This machine tool cuts on the downstroke. (D.C. Morrison Co.)

Fig. 19-8. This close-up shows a gear shaper setup and a shaped gear. (Fellows Corp.)

ram. The diameter of the table is usually given as a second size factor.

PLANERS

The planer is a machine which also uses a reciprocating cutting motion. The machine generally moves the work against a stationary single-point tool. The movement of the work is linear. Like the shaper, the feed motion occurs on the back stroke. During

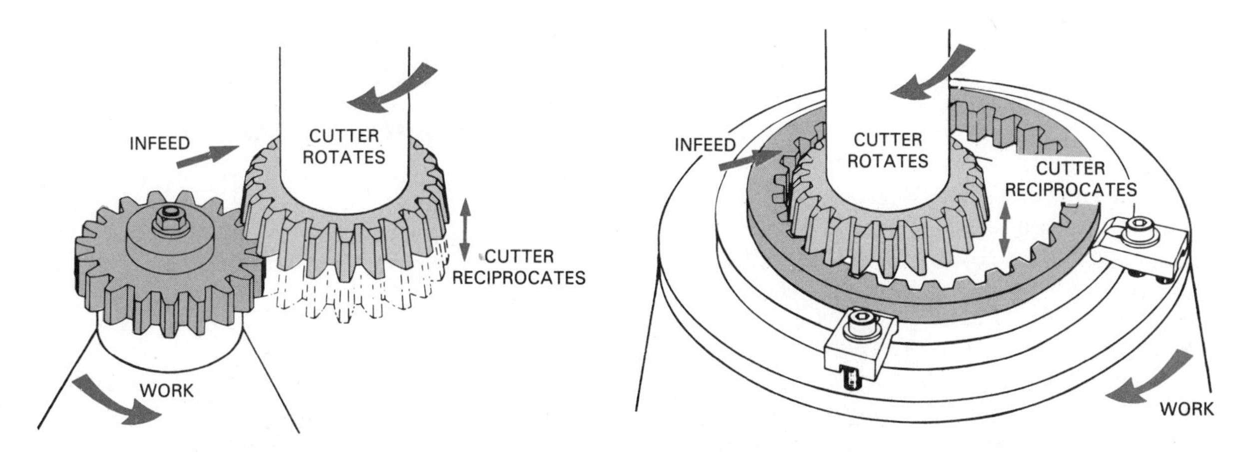

Fig. 19-7. Diagrams for shaping gears. Left. External gear. Right. Internal gear. (Fellows Corp.)

Fig. 19-9. Close-up of planing action. Cutting motion is in the table which is holding the workpiece. Feed motion is in the tool (top). (Cincinnati, Inc.)

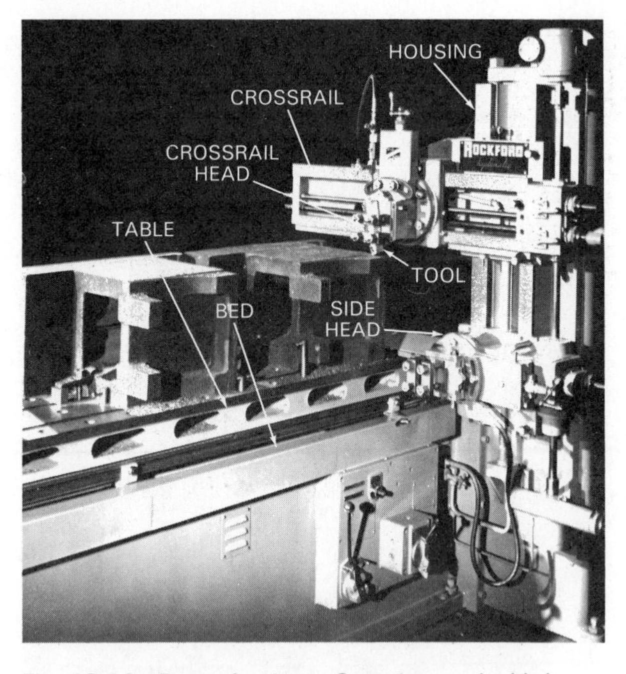

Fig. 19-10. Parts of a planer. Some have a double housing supporting both ends of the crossrail and three or four heads. (Rockford Machine Tool Co.)

this time the tool is indexed over for the next cut. Fig. 19-9 shows a close-up of the planer cutting action.

Planers, as seen in Fig. 19-10, have several basic parts. These include:

1. A table which holds the work and usually provides the reciprocating motion.
2. One or more tool heads which are located above or at the side of the work.
3. A bed on which the table may move.
4. A *crossrail* to support the tool heads.
5. A column or housing which supports the crossrail. Sometimes two are used.
6. A heavy-duty tool that will withstand the forces generated by deep, long cuts.

Planers are designed to do horizontal, vertical, and angled cuts on work too large for a shaper. Typically these machines are of four basic types:

1. Open-side planers.
2. Double-housing planers.
3. Ease or plate planers.
4. Pit-type planers.

Open-side and *double-housing* planers are the basic and most common types. They operate on the reciprocating work-stationary tool principle. The main difference between the two is that the open-side planer has only one column (or housing). This allows greater access to the tools and also allows wider work to be clamped to the table. Fig. 19-11 shows an open-side planer with three tool heads. Each of these heads will move during the back stroke to provide feed motion. The tool head also has depth of cut adjustments.

Fig. 19-11. An open-side planer. The crossrail is supported on only one side. Tools are thus more accessible and larger workpieces can be machined. (Rockford Machine Tool Co.)

The double-housing planer has a column on each side to provide added rigidity. It generally can handle heavier cuts. A typical double-housing planer is shown in Fig. 19-12.

Fig. 19-12. A huge double-housing planer. Two columns support crossrail. (Mesto Machine Co.)

Plate or *edge planers* are specially designed machines. They are used for planing edges of metal plates. With this planer the work is held stationary on the table. The tool reciprocates on a side-mounted tool carriage. The tool is reversed at the end of each stroke so that it cuts in both directions.

Pit planers are used for very large work. The large, heavy piece is clamped to a stationary table. The column-housing assembly holding the tool heads reciprocates on tracks or ways beside the table. The tool heads reverse at the end of each unit. This allows the tools to cut in both directions.

Other types of planers include:

1. The *convertible open-side* planer. This is an adaptation of the open-side planer with a removable housing. The housing is attached to the left side of the bed. It supports the outer end of the crossrail. Another head may be mounted to it.
2. *Adjustable convertible open-side* planer. Also a variation of the open-side planer, it has a removable housing on the left-hand side. The housing is mounted on a runway at right angles to the table travel. Its position can be adjusted to suit the size of the workpiece.
3. *Milling* planers. These are more versatile than the standard planer, being made with various combinations or planing, milling, boring, and drilling heads. They are made in double-housing, open-side, and convertible types.
4. *Double-cut* planers. This type incorporates the ability to cut on both the forward and reverse strokes. The head is designed to rotate over a limited arc to switch the tool bit whenever direction of cut changes. Edge and pit planers are of this type.

Planer size

Sizes of planers are designated in different ways depending on the type. The open-side planer uses planing width (maximum horizontal movement of the tool head), maximum distance from the table to the rail, and the maximum table travel, for example: 24" x 36" x 10'.

Double-housing and pit planers substitute the distance between the vertical columns for the planing width. Size designation might be 42" x 42" x 10'.

Plate planers are designated by the maximum size of plate that can be machined.

SUMMARY

Shapers and planers operate with a reciprocating cutting motion. They basically produce straight cuts along or across the workpiece. Shapers are used for smaller parts while planers can handle very large, heavy workpieces. Both shapers and planers use single-point cutting tools which remove the excess material in a series of linear cuts.

STUDY QUESTIONS – Chapter 19

1. What is the difference between metal shaping and metal planing?
2. List and describe the major types of metal shapers.
3. What are the basic parts of a metal shaper?
4. List and describe the four basic types of metal planers.
5. List the major parts of a metal planer.

Chapter 20

DRILLING, BORING, REAMING, AND TAPPING

Many manufactured parts and products require machine operations that put holes in them. These holes may be part of the functional design of the product or are for product assembly with fasteners such as screws, bolts, rivets, shafts, tabs, and tapered pins.

Holes may be produced using a number of manufacturing processes. They may be made by punching, making cores in casting operations, flame cutting, ultrasonic machining, drilling, fly cutting, hole sawing, boring, and electrical discharge machining (EDM). This chapter will discuss those hole-producing operations which are performed on drilling and boring machines.

HOLE MACHINING OPERATIONS

Operations which may be completed on drilling and boring machines, Fig. 20-1, are:

1. Drilling.
2. Counter drilling.
3. Step drilling.
4. Boring.
5. Counterboring.
6. Countersinking.
7. Reaming.
8. Spotfacing.
9. Tapping.

Drilling is the easiest way to produce a hole in solid material. It uses a rotary end-cutting tool with one or more cutting lips. This tool is called a drill or a bit. When drilling is done to enlarge a hole the process is called *counter drilling* or *core drilling.*

Drilling may also be done with special drills having two or more diameters. This operation, called *step drilling,* will produce a hole with the same diameters as the drill.

Boring is the machining technique which accurately enlarges the diameter of a hole. Boring trues a drilled hole and removes errors caused by the drilling operations. Boring insures the holes are round and straight.

If a hole is enlarged only partway along its length, the operation is called *counterboring.* A counterbore is often used to allow bolt heads, nuts, and round head screws to be set below the surface of the material. A counterbore may also be used as a seat for a rod or shaft.

Countersinking involves machining a beveled surface into the end of a hole. Most often, a countersink will produce an angled opening to accept flat head wood and machine screws. Most common countersink angles are 60, 72, 82, and 90 degrees.

Reaming, like boring, enlarges a hole. It is used primarily to produce accurate hole sizes with good surface finish. The reaming process removes little material. Often, a hole is drilled and bored to produce the approximate size, then reamed for final size and finish.

Spotfacing is a special operation designed to smooth the surface around a hole. Spotfacing produces a smooth, flat seat for bolt heads, nuts, and shaft shoulders.

Tapping is another special operation which produces threads in a hole. The operation is difficult because of several factors. Included in these are the amount of material to be removed, the torque required to produce the chip, and the problem of chip removal.

CUTTING AND FEED MOTIONS

All of these processes may be performed by feeding a rotating multipoint tool or lip cutter into the work. The turning cutter produces the cutting motion while the linear movement of the cutter generates the feed motion. This relationship is shown in Fig. 20-2.

In some cases other combinations of feed and cutting motions are used. As described in Chapter 16, drilling and reaming on a lathe involve feeding a stationary tool (drill or reamer) into the revolving work. In other cases, work is fed into a revolving tool.

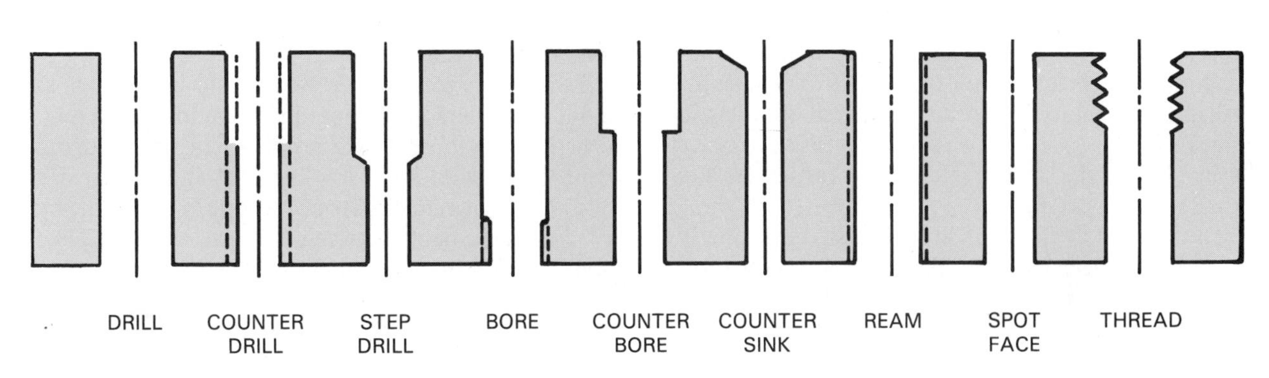

Fig. 20-1. This is how working drawings represent types of drilling operations.

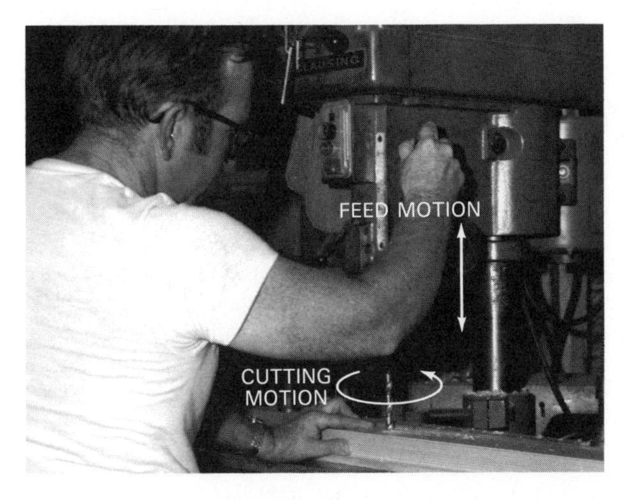

Fig. 20-2. Note the cutting and feed motions for drilling and boring operations. (Andersen Corp.)

HOLE-MACHINING TOOLS

There are a number of common hole-machining tools. These include:
1. Drills and bits.
2. Special hole cutters.
3. Boring tools.
4. Countersinks.
5. Reamers.
6. Taps.

DRILLS AND BITS

A number of drills and bits can be used to produce holes in metal, plastic, wood, and ceramic materials. The most commonly used are:
1. Twist drills.
2. Gun drills.
3. Spade drills and bits.
4. Spur bits.
5. Forstner bits.

Twist drills

Twist drills are the most commonly used hole-producing tool. They contain five basic features: a body, a point, cutting lips or edges, flutes, and a shank. See Fig. 20-3.

The body is the main portion of the drill. It contains the flutes which carry the chips up and away from the cutting lips. The point, created by grinding the cutting edges, does the cutting. The shank is the part of the drill which is held by the drilling machine. As seen in Fig. 20-4, the shank may be a straight,

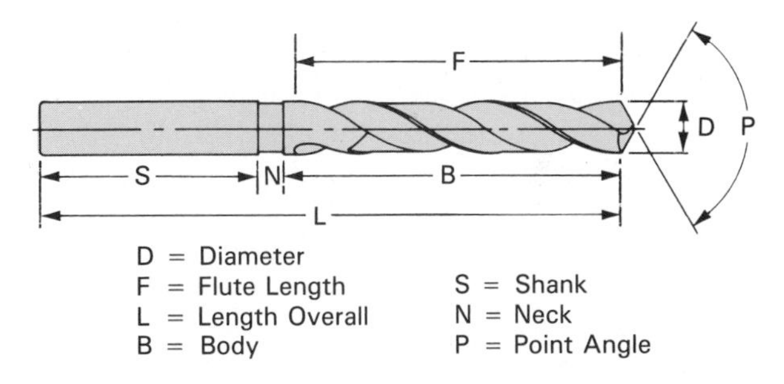

Fig. 20-3. Parts of a drill. The body contains flutes (spiraled grooves). (National Twist Drill)

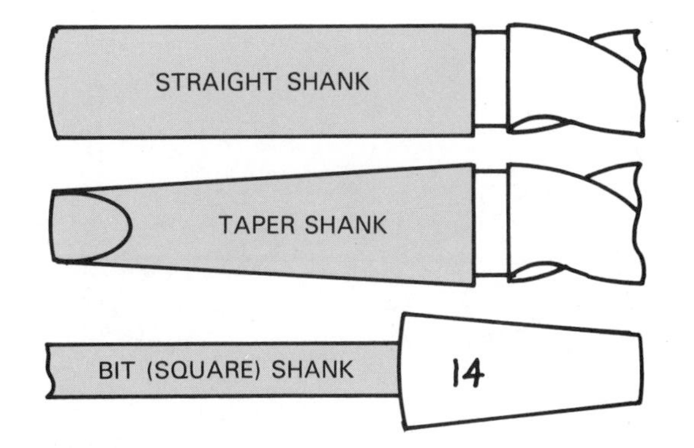

Fig. 20-4. The drill shank is the part held by the drilling machine or tool. There are three types of shanks.

taper, or bit type. The straight shank fits into a standard drill chuck found on drill presses and portable drills. Taper shanks fit into the tapered spindles in large drill presses and lathe tailstocks. Bit shanks are designed to fit the hand braces used by carpenters and other woodworkers.

The body and point of a drill must be properly shaped for efficient cutting. Fig. 20-5 shows the standard twist drill elements. These include:

1. *Cutting lip* or *edge.* These are like single-point cutting tools. The rack angles behind. The cutting edge is formed by the flute helix (spiral) angle built into the drill. The rake angle cannot be varied by normal grinding. The normal 25 deg. helix angle of most drills sets the rake angle.
2. *Chisel edge.* The coming together of the cutting edges forms a chisel edge between them. The edge is at a 118 deg. angle for normal drills. The chisel edge is inefficient and often causes the drill to start off-center. Some drills are ground to reduce the chisel edge. This action, however, often weakens the point.
3. *Margins.* The margin is the cylindrical part of each land which is not cut away to provide clearance. The margins provide the full diameter of the drill. The rest of the body of the drill has a smaller diameter. This difference in diameter is called *body clearance.*

   The margins reduce the friction between the drill and the walls of the hole. They also allow cutting fluid to flow to the point of the drill.
4. *Flutes.* The flutes provide a path for chips to move away from the cutting lip. The number of flutes also determines the number of cutting lips a twist drill will have. Most common twist drills have two flutes but three and four flute drills are available. These drills are generally used to enlarge cored, punched, or drilled holes. They cannot drill holes in solid material because their cutting edges do not extend to the center of the drill.

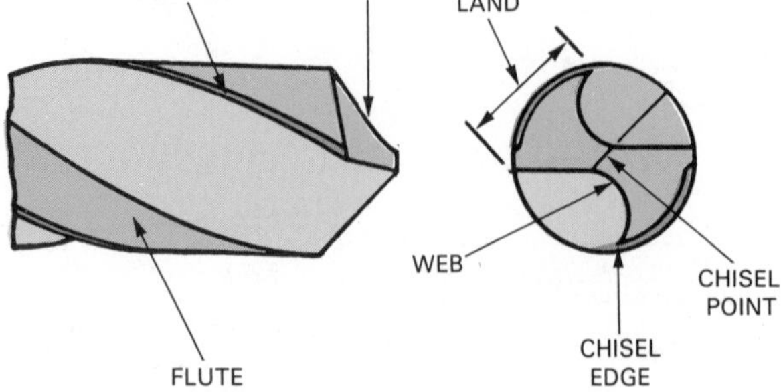

Fig. 20-5. Twist drill elements. (National Twist Drill)

Twist drill sizes

Twist drill sizes are expressed in four basic systems: numbers, letters, fractional inches, and millimeters. *Fractional inch drills* are produced in sizes ranging from 1/64 in. to 3 1/3 in. The drills increase in size by 1/64 in. increments from 1/64 in. to 1 1/4 in.; by 1/32 in. increments between 1 1/4 in. and 1 1/2 in.; and by 1/16 in. increments between 1 1/2 in. and 3 1/2 in.

Letter sized drills range from A to Z. An "A" drill has a 0.234 in. diameter while a "Z" has a 0.413 in. diameter. *Number sized drills* range from 1 (the largest) to 80. A No. 1 drill is 0.228 in. in diameter. The diameter of a No. 80 drill is 0.0135 in.

The number and letter size drills fill gaps in the fractional-inch drill set. This allows a wide variety of holes to be drilled with each being only a few thousandths of an inch smaller than the next drill.

Straight shank metric drills start at 0.20 mm and increase to 100 mm in diameter. Taper shank metric drills range from 3.00 mm to 100 mm in diameter.

The length of drills varies with the diameter. The larger the diameter the longer the drill. Also, most sizes of straight shank drills are available in both short and long lengths. Special twist drills will drill two or more diameters at a time. Fig. 20-6 shows two special-purpose multicut drills.

GUN DRILLS

Gun drills are single lip drills built for deep hole drilling. They usually have a single flute parallel with the axis of the drill, Fig. 20-7. The single cutting lip is usually a carbide insert at the end of this flute. A hole also runs the length of the drill through which coolant is forced to increase the efficiency of the drilling operation.

Gun drills are used to produce holes that are not possible with standard drills. Holes from 0.075 in. to 2.000 in. in diameter and up to 150 in. deep are possible. The holes produced are true and tolerances in the tenths are possible. Also good surface finish is obtained.

Fig. 20-6. Two common multidiameter drills. Top. This type will drill and countersink a hole in one operation. Bottom. A subland drill produces a two-diameter hole. (National Twist Drill)

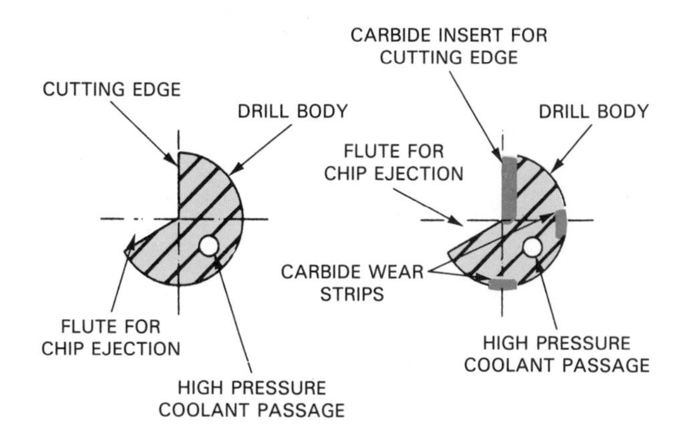

Fig. 20-7. Cross sections of gun drills. Left. High speed. Right. Carbide tipped.

SPADE BITS AND DRILLS

Spade bits are often used to produce larger-diameter holes. They have been used for drilling holes in wood for a number of years. The bits themselves are a single piece of metal with a shank and flat spade on the end. A pointed extension of the wood cutting spade insures that the bit enters the material properly. The spade portion is a two-lip cutter much like a twist drill, Fig. 20-8.

Metal cutting spade drills consist of two main parts: a toolholder and an interchangeable blade. The tool holder holds the blade during drilling. It may also have a hole through which fluid can be pumped to aid cooling and chip removal.

The blades do the actual cutting. They are available in a number of shapes and sizes. The width of blades generally start at 1 in. and can range up to 15 in.

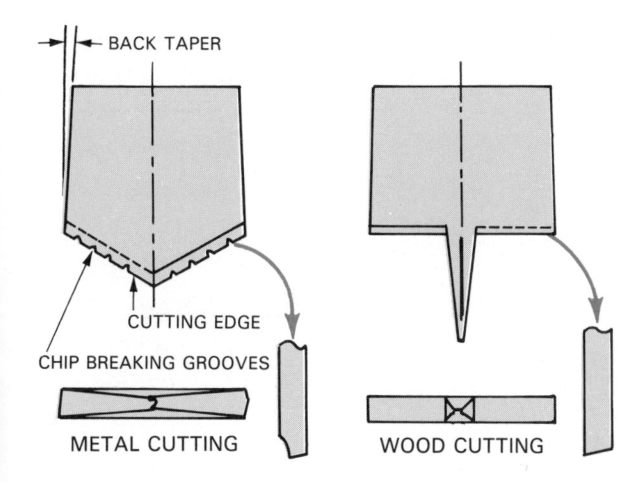

Fig. 20-8. Metal and wood spade drills have flat bodies and two-lipped cutters.

SPUR BITS

Spur bits are designed for cutting clean holes in wood. The bit, Fig. 20-9, has three distinct features:

1. A point which guides the bit into the wood.
2. A spur along the circumference of the drill to cut the fibers around the edge of the hole.
3. Cutting lips to cut the material between point and spur cut.

Spur bits produce a high quality hole. First the point directs the bit. Then the spur scores and cuts the fiber around the edge of the hole. Finally the mass of the wood is removed between the spur cut and the center of the hole.

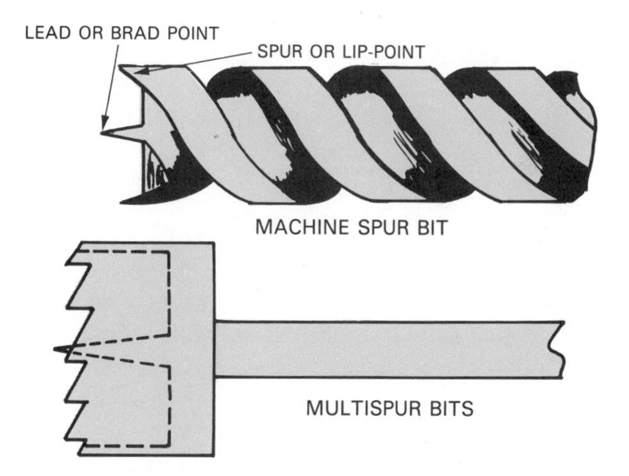

Fig. 20-9. Machine and multispur bits are used for drilling large holes in wood.

FORSTNER BITS

Forstner bits, as seen in Fig. 20-10, are another wood-cutting tool. They are designed to produce an almost flat-bottomed hole.

The forstner bit has a sharp knife edge around its circumference. This edge scores and cuts the wood

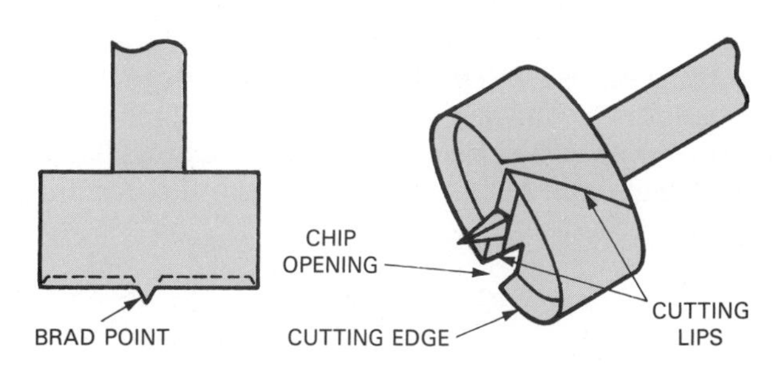

Fig. 20-10. A forstner bit is used to drill flat-bottomed holes in wood.

much like the spur on the spur bit. The bit then removes the material from within the scored circle with two cutting lips.

SPECIAL HOLE CUTTERS

There are two special hole cutters which deserve attention, Fig. 20-11. The first is the *hole saw* or *hole cutter*. This tool, introduced in Chapter 18, is a combination drill and saw.

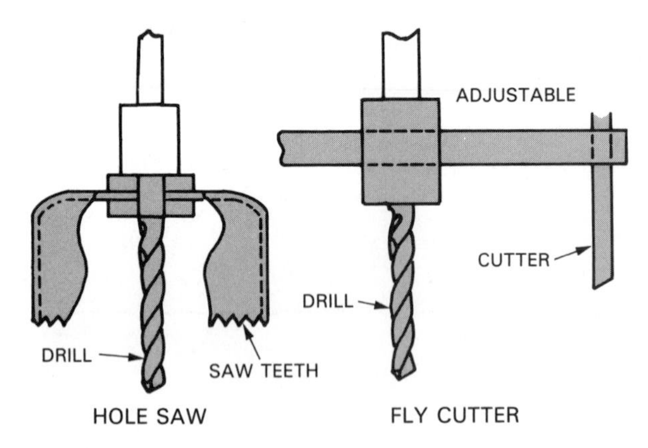

Fig. 20-11. Left. Hole saw. Right. Fly cutters. They are designed to cut large holes in wood, metal, and plastic materials.

The hole cutter has a drill in its center. This drill enters the work and provides an axis around which the saw rotates. The saw is a cup-shaped steel part with teeth on its lip (rim). The saw cuts a round opening as it enters the work. When the hole is cut the cutter is raised and the scrap is removed. The scrap will be a disc with a hole in its center.

Hole cutters are available in a wide variety of sizes. The various sizes may be placed on the drill-arbor assembly to cut the size hole wanted.

The other special hole cutter is called a *fly cutter*. Like the hole saw, it has a drill in the center. The drill provides a pivot point for an arm. On the end of the arm is a single-point cutting tool much like a lathe tool. When rotated, this tool will cut a circular groove through the work. The diameter of the hole to be cut is adjusted by moving the toolholder along the radial arm. Any size hole between the cutter minimum and maximum can be cut by carefully positioning the toolholder.

BORING TOOLS

Boring tools are generally single-point. They may be made from high-speed steel, carbides, or ceramic materials. Typical boring tools, as seen in Fig. 20-12, are boring bars, single-tool boring heads, multiple-tool boring heads, and counterboring tools.

The normal operations performed with boring tools are shown in Fig. 20-13. These include:

1. *Boring*. Correcting the location, size, and roundness of holes.
2. *Facing and spotfacing*. Producing a flat surface at the end of a hole.
3. *Turning*. Correcting the roundness or size of an external round surface.
4. *Grooving*. Producing a slot on the wall of a hole or on an external round surface.
5. *Undercutting*. Enlarging the diameter of a hole below its top end.
6. *Counterboring*. Enlarging the diameter of a portion of a hole starting at its end.

REAMERS

Reamers are multipoint cutting tools designed to finish a hole already drilled or bored. Reamed holes maintain very close tolerances and have good surface finish.

Common reamers may be grouped into three major types, Fig. 20-14. These are:

1. Solid.
2. Shell.
3. Adjustable.

SOLID REAMERS

Solid reamers are made from solid bar stock. They may be wholly high-speed steel or high-carbon steel

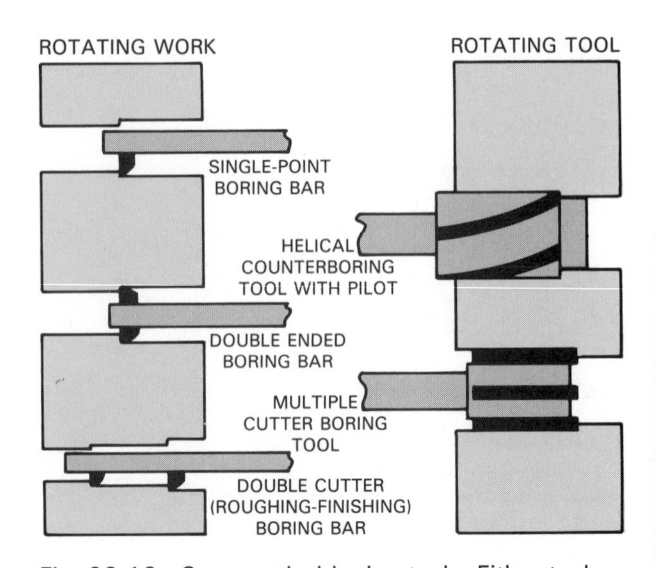

Fig. 20-12. Some typical boring tools. Either tool or work may rotate.

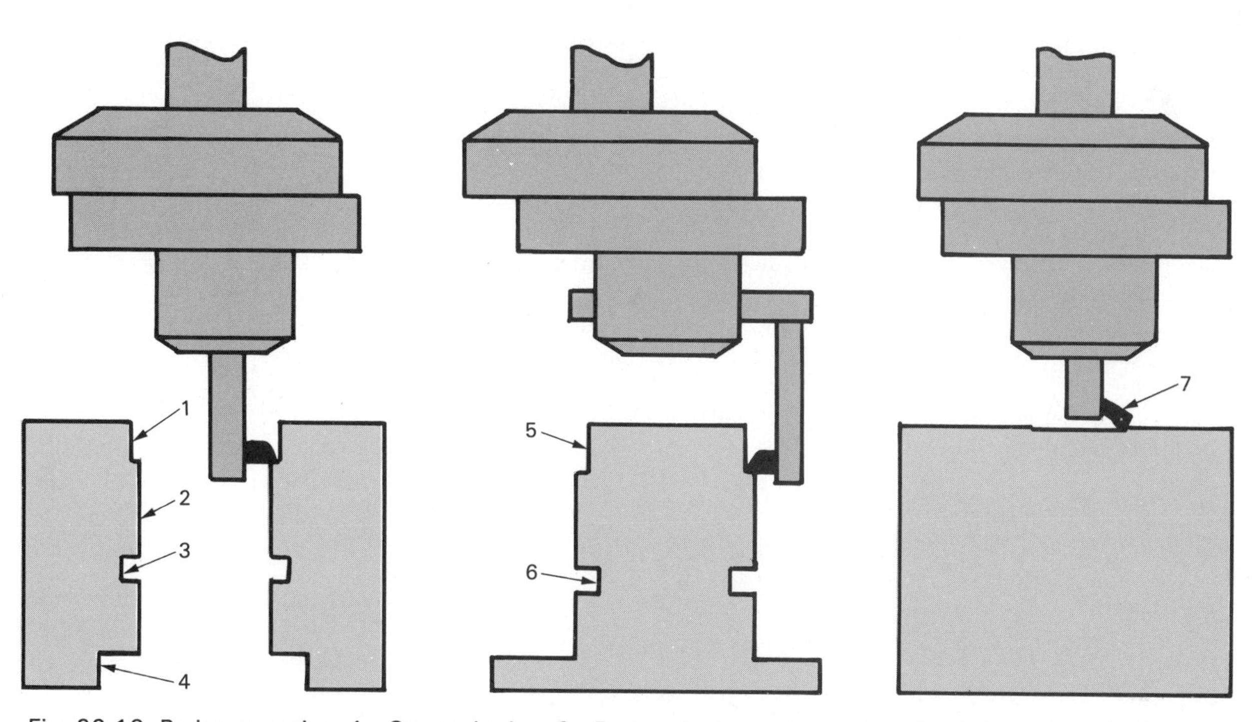

Fig. 20-13. Boring operation. 1—Counterboring. 2—Boring. 3—Internal grooving. 4—Undercutting. 5—Turning. 6—External grooving. 7—Facing.

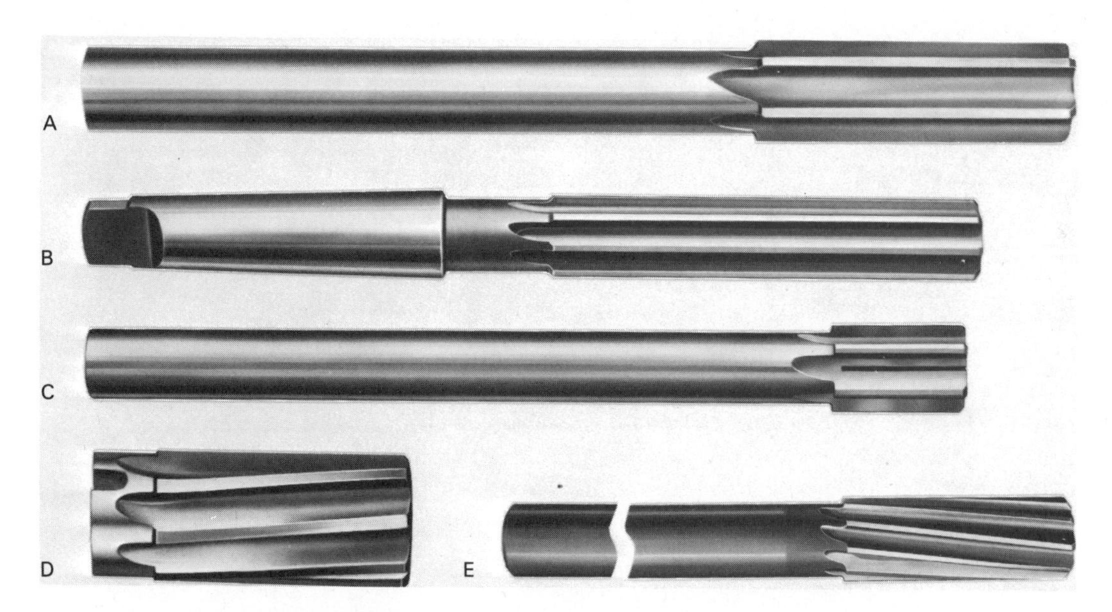

Fig. 20-14. Types of reamers. A—Straight flute chucking. B—Taper shank straight flute. C—Expansion chucking. D—Shell. E—Helical flute chucking. (Morse Cutting Tools)

with brazed-carbide cutting edges.

There are several types of solid reamers. The *straight reamer* may be a hand reamer which is turned with a wrench and removes small amounts of material. Or it may be a machine or chucking reamer designed to be chucked in a drill press or lathe.

Straight reamer cutting edges may run straight down the length of the reamer (straight flute) or may wrap around the body of the reamer (helical flute). The flutes may be right hand or left hand. Many straight reamers taper slightly along the first part of their length. This allows for easy starting of the reamer in the hole.

The *rose-chucking* reamer looks much like a straight reamer. However, it has no relief angle back of the cutting edges. It also has bevel cutting edges on the end of the reamer. These edges actually do the cutting. Rose-chucking reamers are designed to take heavy roughing cuts. Their main use is in enlarging cored holes in castings. The resulting reamed hole is not particularly smooth.

Tapered reamers are designed to produce a standard tapered hole. These commonly are either Morse or Brown and Sharp tapers. Tapered reamers are widely used to produce tapered holes for assembly operations using tapered pins.

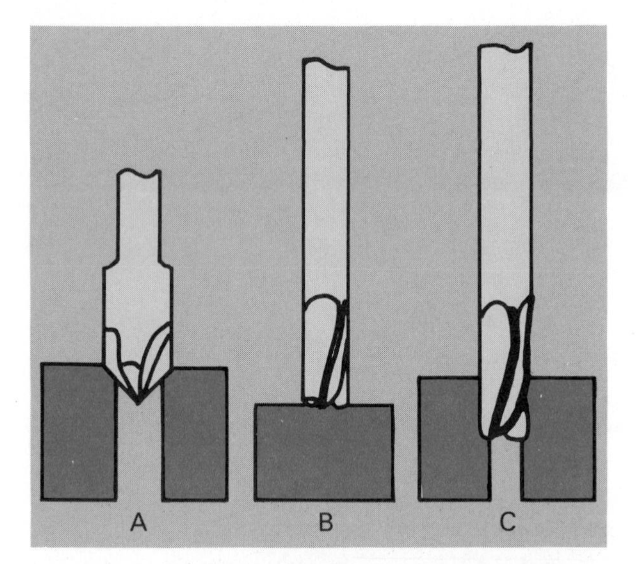

Fig. 20-15. Three types of boring operations for modifying holes. A—Countersinking. B—Spotfacing. C—Counterboring.

SHELL REAMERS

A shell reamer is used on larger holes, usually over 3/4 in. diameter. It has two main parts: an arbor and a shell. The shell section, round and hollow, is attached to the arbor.

Whenever the shell is dull it may be replaced on the same arbor. Shell reamers are also available in adjustable models. These have a series of adjustable cutting blades.

ADJUSTABLE REAMERS

Adjustable or expansion reamers allow for reaming almost any size hole from 3/8 to 2 1/2 in. Each reamer will adjust over a limited range of sizes. For example, a 1/2 in. adjustable reamer will ream holes ranging from 15/32 to 17/32 in.

An adjustable reamer has separate cutting blades set in tapered slots. The blades are moved along these slots and locked in place by nuts at each end. The farther the blades are forced toward the tip, the smaller the reamer diameter becomes.

COUNTERSINKS

Closely related to reamers and boring tools in cutting action are countersinks, counterboring, and combination countersink-counterboring tools, Fig. 20-15. These tools may produce one of three basic cuts in wood, metal, plastic, and some ceramic materials. The cuts are:

1. Countersinking for accepting flat head screws.
2. Counterboring to recess screw and bolt heads and provide seats for rods, shafts, and dowels.
3. Spotfacing to provide a flat surface for bolt heads, screw heads, or nuts.

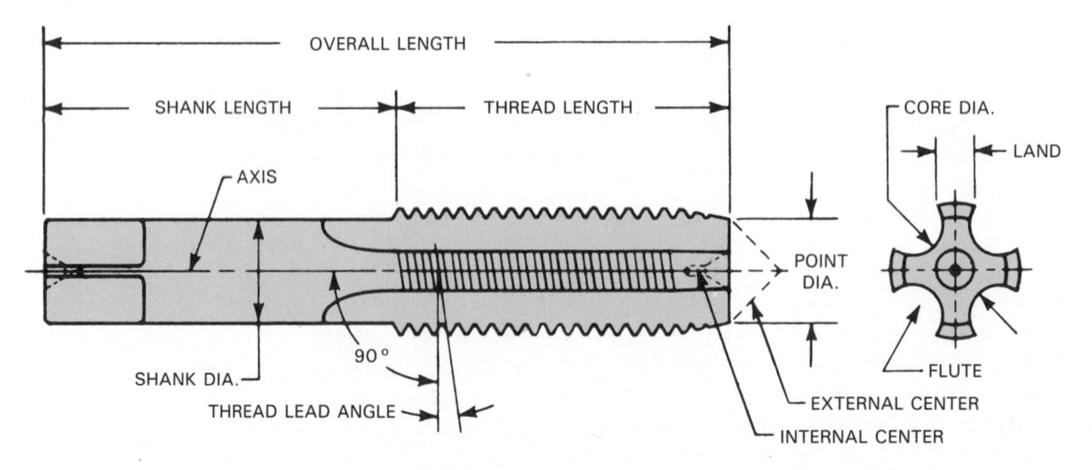

Fig. 20-16. A schematic drawing of a hand tap. (National Twist Drill)

TAPS

Taps, used to cut threads in a hole, have shanks and rows of cutting teeth separated by flutes. As the tap is fed into the hole, the angle of the teeth cause it to cut spiral threads.

Tapping is often a process with several steps. First a hole of proper size must be punched, drilled, or otherwise formed. Drilled holes require a twist drill size that will give proper thread height in the hole. For example, to obtain about 75 percent of full threads in a 1/4-20 threaded hole (accepts a 1/4 diameter x 20 threads per inch bolt), a 13/64 in. hole is needed. This drill size is called the *tap drill size* and is found on a tap drill chart.

After the tap drill hole is produced, it is often countersunk to make tap entry easier. Finally, the hole is threaded using one of several types of taps.

There are a number of types of taps. The major ones are *hand* taps, *serial* taps, and *tapper* taps.

Hand taps

Hand taps, shown in Fig. 20-16, have short shanks with squared ends. They are made in sets of three for each standard size: a *taper,* a *plug,* and a *bottoming* tap.

Each of these taps is identical except for the length of the chamfer at the end. On taper taps, the chamfer extends for the first eight to ten threads to make them easier to start. However, taper taps cannot cut threads to the bottom of a blind hole. The first three to five threads on plug taps are tapered while on bottoming taps only the first thread is tapered.

Hand taps may be used for both hand and machine applications. Machines generally use only the plug-type tap.

Serial taps

Serial taps are also made in sets of three, all designed to be used in threading a hole. The taps are numbered 1, 2, and 3 and have one two, and three rings. The No. 1 tap is used first since it has the smallest major and pitch diameter. This tap makes the first roughing cut. Tap No. 2 cuts a fuller thread while No. 3 finishes the job.

Serial taps are used to thread tough materials or to cut threads which require excessive material removal (such as Acme Threads).

Helical-fluted taps

Helical or spiral-fluted taps have flutes that wrap around the tap in an upward spiral. This helps draw the chips out of the hole. They are especially useful when tapping a blind hole in brass, aluminum, copper, and other nonferrous metals. The tap's cutting angle causes the chips to feed naturally up the spiraled flutes thus preventing clogging of the tap.

Tapper taps

Tapper taps are designed for production tapping of nuts. The first 10 to 15 threads are chamfered for ease of starting. Tapper taps have long shanks which allow the finished nut to be forced off the tap and onto the shank. Bent shank tapper taps are used on automatic tapping machines intended for high-volume production of nuts.

THE BASIC DRILLING MACHINE

The basic drilling machine is designed to provide two types of motion to the cutting tool. It can rotate the tool to create the cutting motion. The tool can also be fed linearly (up-down, forward-back) to generate the feed motion.

The basic drilling machine has four parts. These, as seen in Fig. 20-17, are:

1. A base to support the machine.
2. A support column to hold the power head and allow the head and table to be adjusted along it.
3. A power head to rotate and feed the tool.
4. Table to support the work.

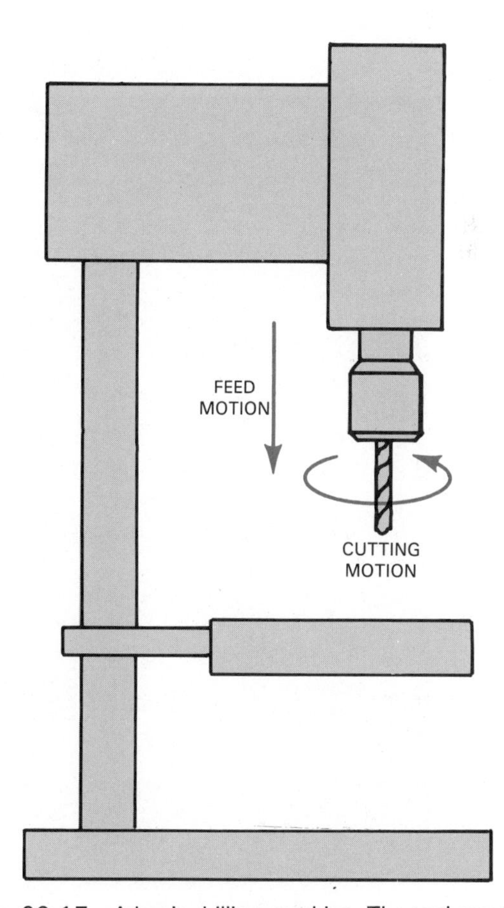

Fig. 20-17. A basic drilling machine. The tool provides rotary and linear motion.

HOLE-PRODUCING MACHINES

A number of hole-producing machines contain the elements of the basic drilling machine. These are:
1. Drill press.
2. Radial drilling machine.
3. Multispindle drilling machines.
4. Boring machines.
5. Tapping machines.

Drill press

The common *vertical drill press* is very much like the basic drilling machine described above. It contains a column that supports the table and the spindle head.

The spindle head has two main parts: a spindle which rotates the tool and a nonrotating quill which moves up and down providing feed motion. The spindle is inside the quill. As the quill moves up and down for the feed motion, the spindle continues to rotate.

Drill presses may hold the drill, reamer, or other tool in a chuck attached to the spindle. Larger tools may be inserted into the tapered hole in the spindle of larger drill presses. Not all drill presses will accept taper shank tools. Some will only hold straight shank tools in a chuck.

Drill presses are classified as *sensitive* and *power fed.* Sensitive drill presses are hand fed. The hand, then, is sensitive to the drilling action. Power fed drill presses feed at a set rate and are not sensitive to the drilling action. Sensitive drill presses are usually used for light drilling operations. Power fed units are used for heavier work and in high-production applications. Fig. 20-18 shows a typical vertical drill press in use. These machines are made in bench and floor models. Their size is equal to twice the drill-to-column distance.

Radial drilling machine

The head of the radial drilling machine can be moved along an overarm. The overarm also swings on the column. These movements allow the cutting tool to be positioned over the workpiece. This is different from the vertical drill press which requires aligning the work under the tool.

Radial drilling machines, like the one shown in Fig. 20-19, are used on heavy work which cannot be easily moved. These may be *plain* machines which drill only in the vertical position or they may be *semi-universal* allowing the head to pivot to drill angled holes in the vertical plane. Fully *universal* models have

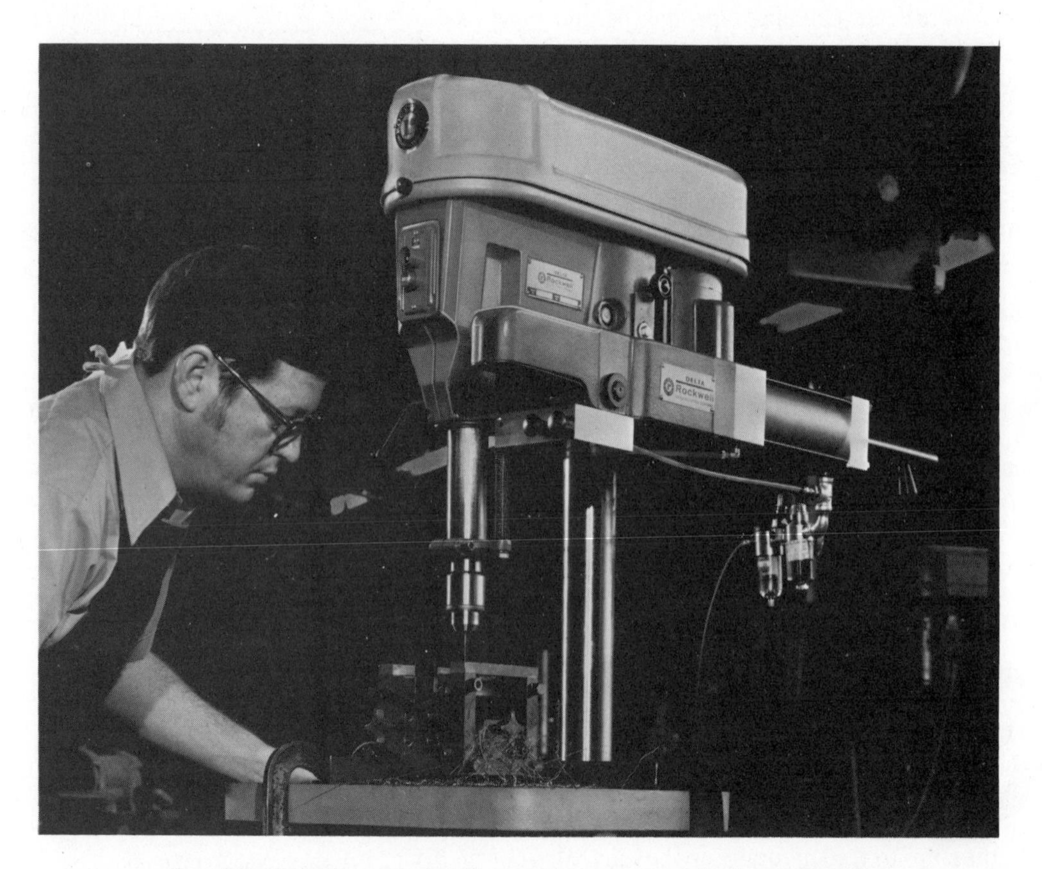

Fig. 20-18. This vertical drill press has a power feed attachment.

Fig. 20-19. Radial drill press. Rotational and linear motion allows drill head to be moved over the workpiece. (Leblond Makino Machine Tool Co.)

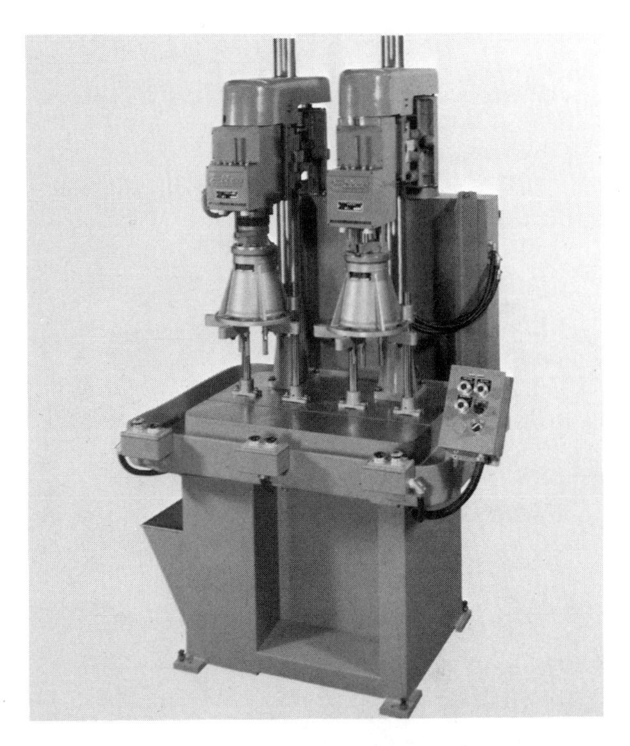

Fig. 20-20. This machine is typical of the two-head gang drill press with multiple spindle heads. (Ettco Tool Co.)

radial arms which can rotate around the horizontal axis. A universal radial drilling machine can drill a hole at any angle.

Multispindle drilling machine

Some drilling applications require the drilling of several holes of the same or different sizes. To increase the productivity of such operations, several multispindle drilling machines have been developed. These include:

1. Gang drilling machines.
2. Turret machine.
3. Multiple-spindle drilling machine.

Gang drilling machines, Fig. 20-20, are basically two or more upright drill presses in a row. They usually have a single base and table. The work is passed from one drilling station to another to complete several tasks.

Turret drilling machines use a single table, column, and power head. A turret, Fig. 20-21, is attached to the power head. The turret is indexed to its several positions to complete the work on a particular part. Turret machines may be numerically controlled (NC) or hand operated.

Multiple-spindle drilling machines have a group of spindles. These spindles may be driven by one or more

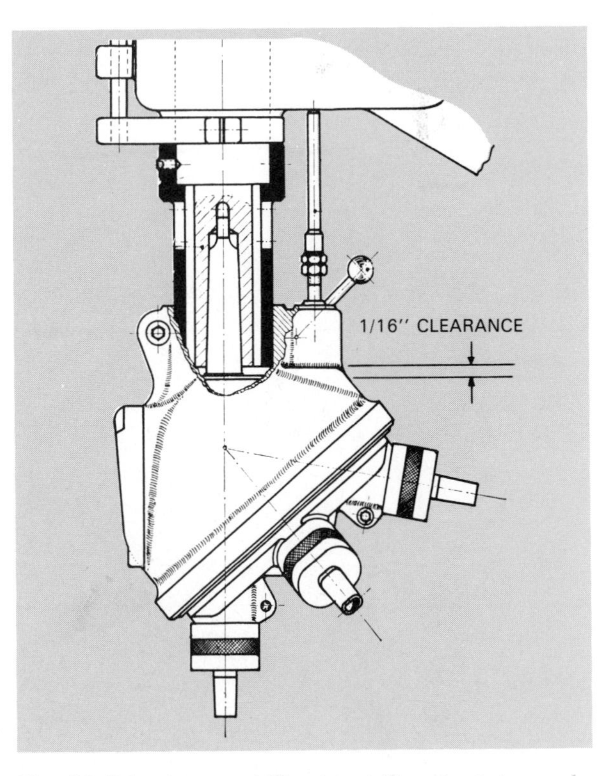

Fig. 20-21. A turret drilling head. Turret rotates on an indexing head. (Jersey Mfg. Co.)

power heads. The machine, Fig. 20-22, will drill all the holes it is set up for in one cycle. Often the machine uses a table feed to eliminate the problems associated in feeding all the spindles into the work. Fig. 20-23 shows both horizontal and vertical table-fed multiple-spindle woodworking drilling machines.

Fig. 20-22. A closeup of a multiple-spindle drilling head. Holes are drilled on the left and tapped on the right. Note the rotating table which moves the part from the drilling station to the tapping station. (Zagar Inc.)

Boring machines

Several machines have been developed to do boring operations. Important among these are the jig borer, the vertical boring machine, and the horizontal boring machine.

The *jig borer* is much like the vertical milling machine described in Chapter 17. It is used to accurately locate and bore holes in gages, dies, jigs, and fixtures.

The jig borer, Fig. 20-24, is a precision machine with very accurate controls. The machine is capable of positioning parts within tolerances of ±0.0001 (±0.003 mm). The jig borer has two sets of direct-reading dials: one moves the table in or away from the column, and the other moves the table left or right. The operator sets these dials to match the drawing.

The *vertical boring machine* was basically described in Chapter 16. The machine as diagramed in Fig. 20-25, is built on the same structural concept as the double-housing planer. It has two side columns, a crossrail, toolheads on the crossrail, two on the columns, and a circular table.

The part to be machined is placed on the table. The toolheads are positioned for the desired operation. Vertical boring machines are used to bore large holes in parts which, because of their weight or shape, cannot be turned on horizontal lathes. These heavy-duty machines are like vertical lathes and are called *vertical boring and turning machines.*

Small precision production vertical boring machines are designed to do boring quickly and accurately. The

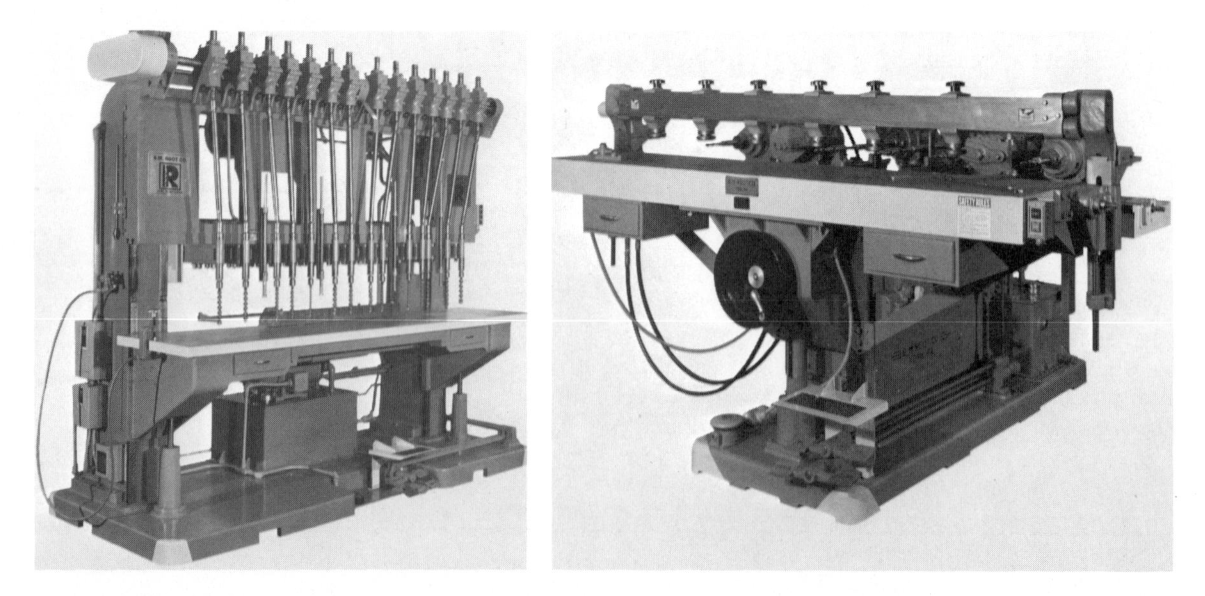

Fig. 20-23. The furniture industry uses both vertical and horizontal table-fed multiple-spindle drilling machines. Left. Horizontal spindle machine. Right. Table-fed horizontal spindle machine. (B.M. Root Co.)

Fig. 20-24. A jig borer in use. It positions work automatically. (LeBlond Makino Machine Tool Co.)

machines may have several spindles allowing the machining of several parts at a time. Or they can perform a different boring operation on a part at each station.

Horizontal boring machines, like the vertical machines, are of two types. The heavy-duty machines are like a horizontal milling machine. They are designed to bore holes in the horizontal surface of a part. It consists of a table which can move in two directions in the horizontal plane, a toolhead which can move vertically, and a rotating spindle which can be fed horizontally. These features may be seen on the machine shown in Fig. 20-26.

The precision horizontal boring machine also bores holes in a horizontal surface. It is used for lighter work, however, and produces very accurate work. In some cases, the work is rotated by a chuck against a stationary boring head.

TAPPING MACHINES

Tapping may be done on engine, turret, and automatic lathes. Also, drill presses can be used—in fact, some tapping machines are modified drill presses. Tap holders and reversing mechanisms have been added to increase their efficiency. Tapping accessories may be attached to standard drilling machines. These attachments turn the tap one way as

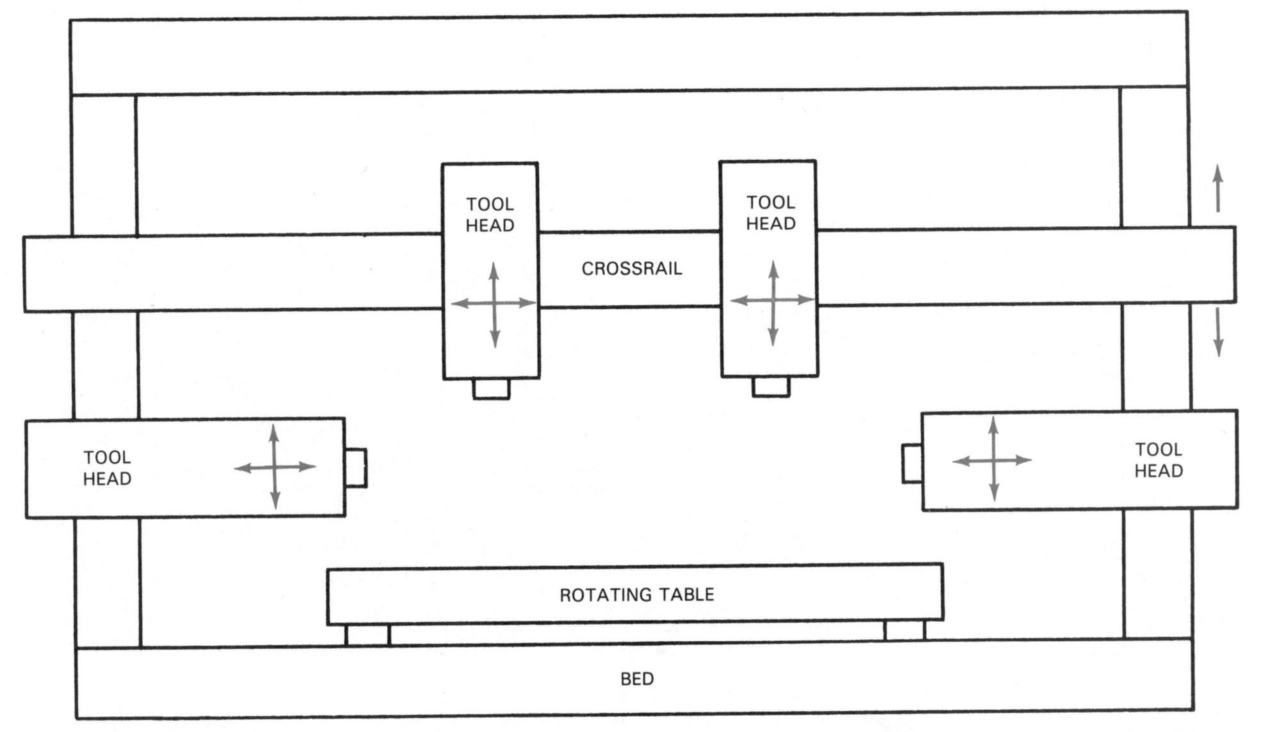

Fig. 20-25. Vertical boring and burning machines have two toolheads on the crossrail and one on each column.

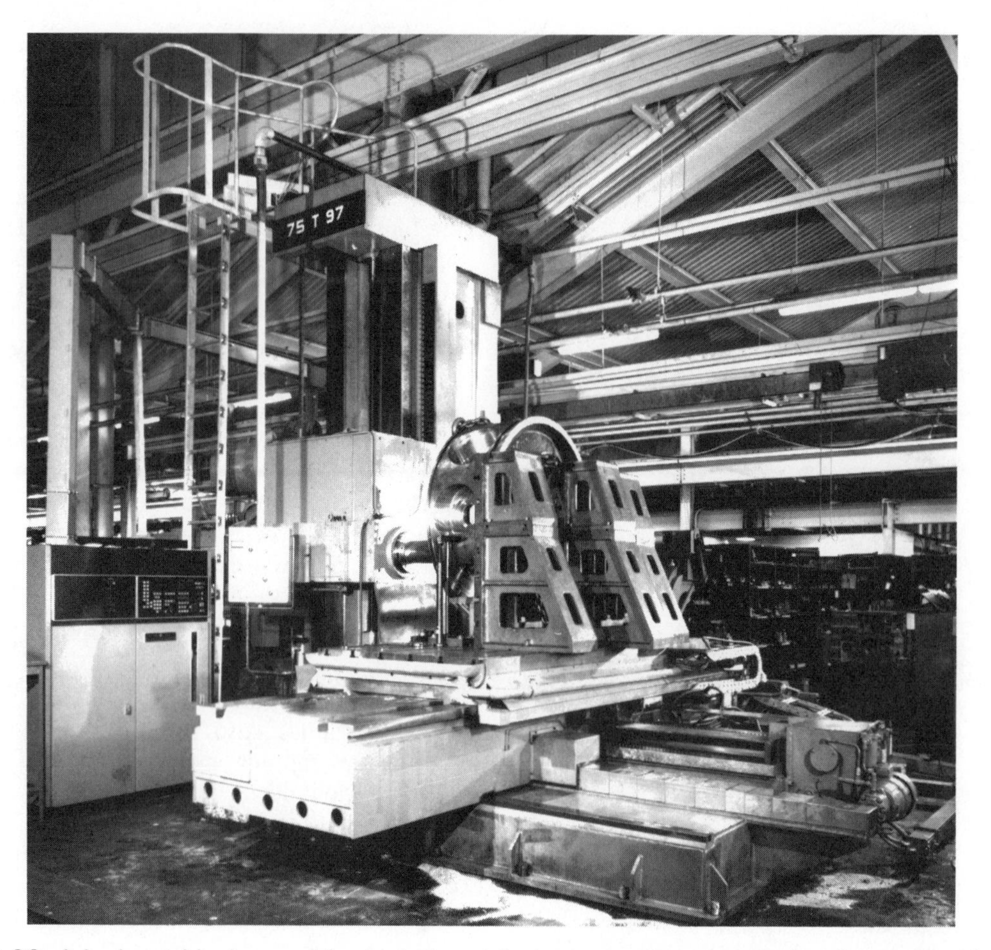

Fig. 20-26. A horizontal boring machine is designed for heavy cutting operations. (Lucas Machine Co.)

it is lowered and reverse the rotation as it is withdrawn.

However, special machines are generally used for tapping large quantities of nuts. Gang and multiple-spindle machines use straight-shank tapper taps. The threaded portion is forced through nut blanks, one after another. Each finished nut slides up the long shank of the tapper taps. When the shank is full, the tap is removed from the machine. Completed nuts slide off the shank.

Another multiple-spindle machine also cuts threads on a nut. It then reverses and backs out of the nut. The nut blanks are automatically loaded into the machine while finished nuts are ejected.

The other important tapping machine uses a bent tap. Blanks are hopper fed into position. The tapper tap is turned and forced through the nut. The finished nut slides onto the shank. The process is repeated. Each new nut that slides onto the tapper tap shank forces another nut around the tap bend and off the shank. The ejected nut falls into a bin. Fig. 20-27 shows an automatic tapper.

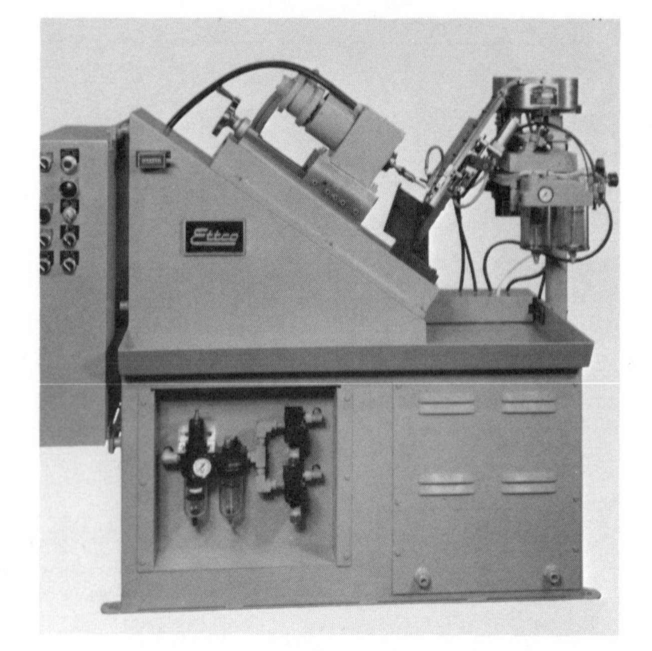

Fig. 20-27. A hopper-fed automatic tapping machine. (Ettco Tool Co.)

SUMMARY

The use of a rotating cutter fed into the work is a common manufacturing practice. Typically these operations are called drilling, boring, reaming, and taping. Drilling produces a hole in solid material. Boring trues the hole by insuring that it is round and in the correct location. Reaming accurately sizes and provides a high quality finish on the hole. Tapping produces threads in a hole.

Each of these operations has its own cutting tools and offers its own special equipment. The proper selection of the tools and equipment insures efficient production.

STUDY QUESTIONS – Chapter 20

1. List and describe nine major hole-machining operations.
2. How are cutting and feed motions produced for the common hole-machining operations?
3. List and describe the major hole-machining tools.
4. List and describe the major types of reamers.
5. What are the four basic parts of a drilling machine?
6. List and describe the five major types of drilling machines.
7. Tapping can be accomplished on a number of different rotating machines. List them.

Production drilling operation. Five holes are being drilled simultaneously. (Delta Machine Tool Co.)

Chapter 21

ABRASIVE MACHINING OPERATIONS

Using abrasives to remove material is a common manufacturing process. Small particles of natural or synthetic material are used to produce a cutting action. These small particles form a random multipoint cutting tool. Each point produces a very small chip during the cutting process.

PROCESSES

The four most common abrasive machining processes are:

1. Grinding processes.
2. Sanding processes.
3. Loose-media processes.
4. High-finish processes.

GRINDING

Grinding is the use of abrasive particles bonded to form a wheel. The wheel is a multipoint cutting tool which provides a constant supply of new, sharp, individual cutting surfaces. The three basic types are: *rough grinding, precision grinding,* and *abrasive machining.*

Rough grinding

Rough grinding involves the use of grinding wheels to remove excess materials from castings, forgings, and weldments. Flash, the excess materials from casting gating systems, and extra weld material, are typical of the types of material removed. Rough grinding also is used to remove sharp edges and corners, burrs, and other unwanted projections. Rough grinding, as its name implies, does not produce an accurately sized part but simply removes unwanted material. Most rough grinding, as shown in Fig. 21-1, is a hand operation using simple grinding machines.

Precision grinding

Precision grinding is a group of processes which produce parts to the correct tolerance and surface finish. It is also used to machine materials that are too hard to cut using other machining techniques. Many types of grinding machines can do precision grinding. Fig. 21-2 shows a typical operation.

Abrasive machining

Abrasive machining rapidly removes material to produce a part of proper size and shape. Abrasive machining of metals is most often a grinding technique. However, coated abrasives (abrasive papers and cloths) are used in abrasive machining of woods. See Fig. 21-3.

SANDING

Sanding processes use abrasive paper and abrasive cloth to remove material. These materials, called *coating abrasives,* are ceramic grains bonded onto a backing of paper or cloth. They are then fabricated into sheets, sleeves or drums, belts, or discs.

Sanding processes include rough sanding, finish sanding, and abrasive machining operations. Rough sanding serves the same function as rough grinding. It quickly removes unwanted projections and sharp

Fig. 21-1. Rough grinding a ceramic pump plunger is done with a diamond wheel. (Adolph Coors Co.)

Fig. 21-2. This multiwheel precision grinder is being used for grinding engine crankshafts. (Landis Tool Co.)

edges. The emphasis is on speed rather than accuracy.

Finish sanding produces a smooth surface. Most finish sanding is associated with woodworking. The finish sanding activities remove machining marks and small surface defects in preparation for application of finishes.

Fig. 21-3. Particleboard panels are being machined by this wide-belt sander. (American Forest Products Industries)

Abrasive machining using coated abrasives rapidly removes material (usually wood) to produce a desired size or shape.

LOOSE MEDIA

Loose-media abrasive processes use individual abrasive materials or shaped media to smooth surfaces or deburr a part. Typically, the abrasive material and the parts are placed in a container. The container is rotated (tumbled) or vibrated. The motion causes the abrasive to rub the surface of the parts, producing a fine cutting and smoothing action.

OTHER ABRASIVE MACHINING

Several specialized abrasive processes are designed to produce a very accurate part or a very smooth surface. These, to be discussed later, are *boring, lapping,* and *super finishing.*

ABRASIVES

Abrasives are the basic material for all grinding, sanding, loose media abrading, and high finish operations. Most abrasives are crushed ceramic particles, Fig. 21-4. The particles have been graded to size so that uniform cutting occurs.

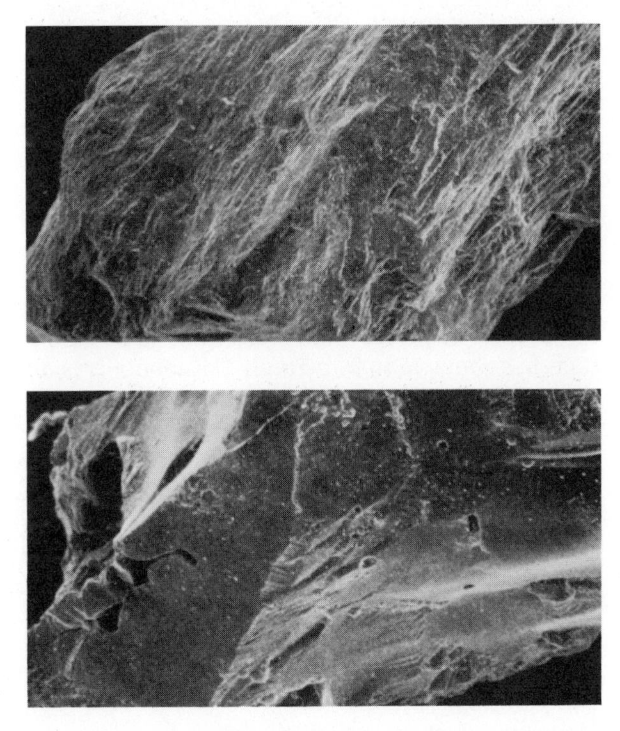

Fig. 21-4. Top. Alumina-zirconia. Bottom. Aluminum oxide. These abrasive grains have been magnified 93 times. (The 3M Co.)

Abrasives are always harder than the material they are to cut. All abrasives can be grouped under two major classes, *natural* and *synthetic* (manufactured).

NATURAL ABRASIVES

Until almost 1900, natural abrasives only were available to manufacturers. These abrasives are minerals extracted from the earth. The most important natural abrasives are:

1. Sandstone or solid quartz.
2. Garnet.
3. Emery.
4. Corundum.
5. Diamonds.

Sandstone and quartz

Sandstone and solid quartz are little used in modern industrial application. Some hand operated grinders have sandstone wheels. Also, flint (a quartz compound), a grayish white particle, is used for inexpensive abrasive papers. Flint abrasive paper is the common household sandpaper. Flint is also used in air (sand) blasting operations designed to clean a wide variety of materials and objects.

Garnet

Garnet is any of several silicate minerals which may be used as gemstones and abrasives. The main garnet mineral used for abrasives contains aluminum, iron, and silicon oxides. This reddish colored abrasive becomes wedge-shaped grains when crushed. Garnet is harder than glass and is one of the most widely used natural abrasives. It is used principally in coated abrasives.

Emery and corundum

Emery and corundum are both made up of aluminum oxide and iron oxide. Emery contains about 50 to 60 percent aluminum oxide while corundum has about 75 to 90 percent. Corundum crystals are much larger than emery crystals. Therefore corundum is used for rough, fast cutting. Emery produces a finer finish. Neither material is widely used in commercial applications because of the variation in their aluminum oxide content.

Emery and corundum are used for sharpening stones. They often take the name of the location where they are found. "Arkansas stone" is made of emery while "India stone" is made of corundum.

Diamonds

Diamonds, either natural or manufactured, are the hardest of all materials. Natural stones not suitable for gems are crushed for abrasives. Synthetic diamonds are manufactured specifically as abrasives. They are widely used for sharpening carbide and cermamic tools. They also are used for dressing, shaping, and trueing grinding wheels. Diamonds are the most expensive abrasive. They are, therefore, used only when another abrasive cannot do the desired job.

MANUFACTURED ABRASIVES

The development of the electric furnace in the late 1800s opened the way for manufactured abrasives. Today there are six principal synthetic abrasives. These are:

1. Silicon carbide.
2. Aluminum oxide.
3. Aluminum oxide/zirconium oxide cofusion.
4. Sintered bauxite.
5. Cubic boron nitride (CNB).
6. Diamonds.

Silicon carbide

Silicon carbide was first discovered during an attempt to produce precious gems using an electric furnace. In 1891, the first silicon carbide was produced in a crude carbon-arc furnace. The resulting cyrstals approached the hardness of diamonds.

Silicon carbide is produced by subjecting silica sand, petroleum coke, sawdust, and salt to high temperatures (4200°F; 2315°C). After cooling, the resulting material is broken up, graded, and crushed. The very sharp grains are then sized before being converted into abrasive products.

Silicon carbide, being extremely hard, is also very brittle. This characteristic somewhat limits its use.

Aluminum oxide

Aluminum oxide was first manufactured commercially a few years after silicon carbide. The abrasive is made by heating bauxite ore with small amounts of coke and iron filings. (Bauxite is the same ore which produces pure aluminum.) The bauxite, coke, and iron mixture melts and fuses at about 3510°F (1930°C). The refined aluminum oxide forms a glassy mass called a pig. The pig is crushed and the particles are graded to size.

Aluminum oxide, slightly softer but tougher than silicon carbide, is the most widely used abrasive in industry.

Cofusions

Cofusions of *aluminum oxide* and *zirconium oxide* improve the grinding rates and the durability of the abrasive. *Sintered bauxite* is usually produced in extruded shapes. It is effective in the grinding of stainless steels.

CNB-cubic boron nitride

Cubic boron nitride (CNB) is produced under extremely high pressures and temperatures. The diamond-like material forms cubic crystals. CNB is used to machine hard-to-grind steels including hardened tool steel and die steel. Cubic boron nitride works better with hard materials than with softer steel.

Manufactured diamonds

Manufactured diamonds were discussed briefly under the natural abrasive section. Manufactured diamonds are carbon structures produced under very high temperatures and pressures. They were used as abrasives to grind ceramics, stone, glass, and cemented carbides. Also, hard-to-grind materials are machined with diamonds whenever another abrasive cannot do the job economically.

ABRASIVE TOOLS

Abrasive grains are generally bonded to form an abrasive cutting tool. An exception is loose-media abrasive machining. Abrasive tools include two major types:

1. Bonded abrasives.
2. Coated abrasives.

BONDED ABRASIVES

Abrasive wheels are the most common of the bonded abrasives. Abrasive stones are the other major type.

Abrasive wheels are a major type of cutting tool. Their cutting action depends upon the abrasive used, abrasive grain size, bonding material, wheel grade, and wheel structure.

Abrasive used

The types of abrasives have already been discussed. The most common abrasives used in grinding wheels for industrial applications are aluminum oxide and silicon carbide. Emery and diamonds are used to a limited degree.

Grain size

After the abrasive is crushed and impurities removed, it is graded by size. Coarser grains are screened and specified according to their screen size. (An 80-grit abrasive has a grain size just able to pass through a screen which has 80 openings per linear inch. That is, the grain is able to pass through an opening 1/80 in. square.)

Standard screened grain sizes for aluminum oxide and silicon carbide are 4 to 220. Finer grains are separated by floating them on air or floating them through a liquid sedimentation process. These very fine abrasives are called *flours.* Flours can range as fine as 600 grit – 1/600 inch square. Three grain sizes are shown in Fig. 21-5.

Bonding material

Grinding wheels and some sharpening stones are manufactured by bonding abrasive grains together in

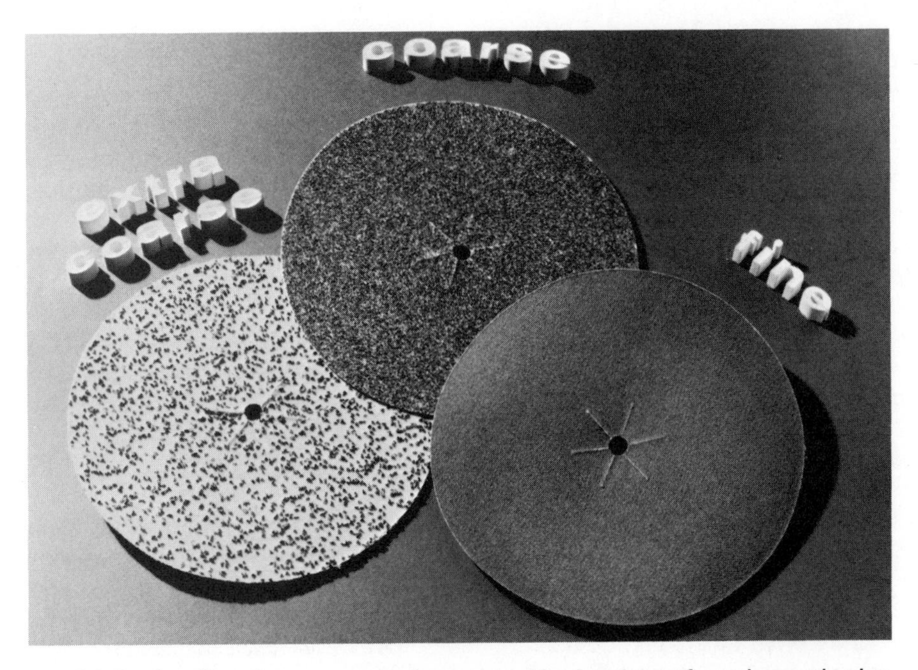

Fig. 21-5. Sanding discs are manufactured with abrasives of varying grain size.

a desired shape. A major factor in determining the cutting action of the grinding wheel is the bonding material. The four major bonding materials are:

1. **Vitrified bond** – A clay bonding material heated to form a glass-like material. The resulting wheels have high strength and porosity. They are not affected by oils, water, or acids. Vitrified bonded wheels are the most common type in use, accounting for nearly 75 percent of all grinding wheels.
2. **Resinoid bond** – A synthetic plastic adhesive compound. The abrasive grains are mixed with a thermosetting synthetic-resin and a liquid solvent. The mixture is molded to shape, then baked. The resinoid bond is very hard, strong, and fairly flexible. Resinoid bonded wheels can be turned at high speeds and still maintain cool cutting temperatures.
3. **Rubber bonds** – A vulcanized (heat treated) rubber material. The rubber-abrasive mixture is heated and rolled to thickness. The wheels are then cut to size and vulcanized (heat cured) under pressure. Rubber bonded wheels are strong and flexible. They can be produced in very thin cross sections. Rubber bonded wheels are used in high-speed grinding and cutoff operations.
4. **Shellac bonds** – A natural shellac material. The shellac and abrasive grains are mixed, rolled or pressed into shape, and then baked. The resulting wheel is very strong. It is used for cool cutting with high finishes. Shellac bonded wheels are used to grind hardened steel such as camshafts.
5. **Metal bonds** – Metal bonds are used in two basic applications – to produce diamond wheels for grinding ceramic materials and also wheels for the electrostatic grinding process. These are aluminum oxide or diamond wheels that must conduct electricity. The metal bond allows for this action.

There are also several other bonding materials which are less used. the *silicate bond* grinding wheel may be used where the heat of grinding is at a minimum. An example would be the grinding of edged tools. *Oxychloride bond* has limited use in certain wheels and segments, particularly on disc grinders.

Wheel grade

The grade of the wheel is a measure of how strongly the grains are held by the bonding agent. The bonds are formed by the bonding agent surrounding the grains. Links (called post) connect the bonded grains. The size and strength of these posts can be increased by using more bonding agent.

A wheel's overall strength or ability to hold the grains is called its hardness. Hard wheels hold the abrasive grains in place strongly. Soft wheels allow the grains to be easily pried away, Fig. 21-6.

Wheels which are too hard will hold onto abrasive

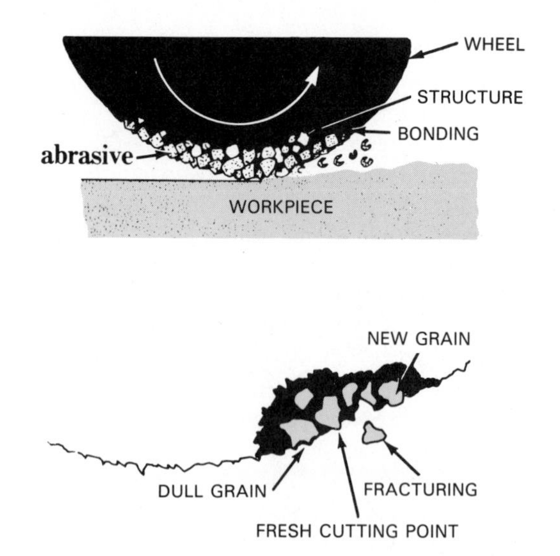

Fig. 21-6. Top. Grinding wheels are bonded abrasives. Bottom. Bonding allows dull grains to be fractured from the wheel. (The Norton Co.)

grains even after they are dull. Too soft a wheel will wear away too quickly.

In general, hard grades of wheels are designed to remove large amounts of stock rapidly, to grind small areas of contact, to remove stock from soft materials, and to hold the correct shape on precise operations.

Soft grades are meant for light stock removal, for grinding where there are large areas of contact, for removing stock from hard materials and expensive parts, where grinding conditions are light, or where continuous wheel wear is desirable to maintain a wheel surface with sharp abrasive grits.

Wheel structure

A grinding wheel cuts as a series of exposed small cutting edges – the abrasive grains – contact the workpiece. The number of cutting edges in contact with the work at any one time is related to the spacing of the grains. This spacing, plus the size of the voids (open space) between grains, is called the wheel structure.

A grinding wheel must have voids to allow for the grains to cut and chips to be carried away. If the voids are too small, the chip will stick in the space causing "loading up" of the wheel. A loaded wheel will generate excess frictional heat. Also, its cutting action is reduced. The wheel structure, therefore, must be matched to the material and desired cutting action. Fig. 21-7 shows the voids in a grinding wheel.

Grinding wheel shapes

Grinding wheels take a number of standard shapes and face contours. Fig. 21-8 shows the common grind-

Structure number

Dense	*to*	*Open*
0, 1, 2, 3, 4, 5, 6, 7, 8, 9, 10, 11, 12		

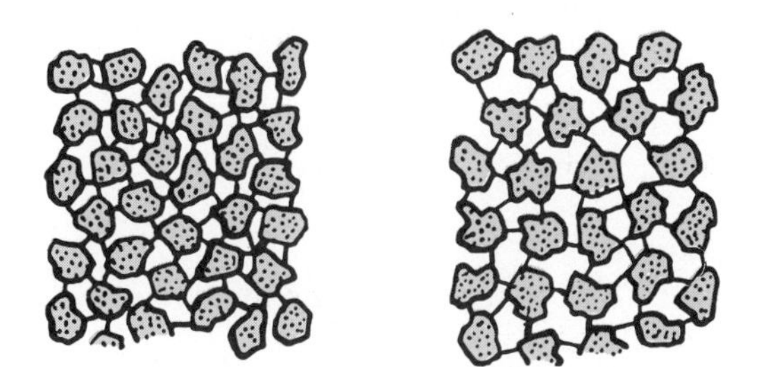

Fig. 21-7. Diagram shows the spacing of identical sized grains in two grinding wheels. (The Norton Co.)

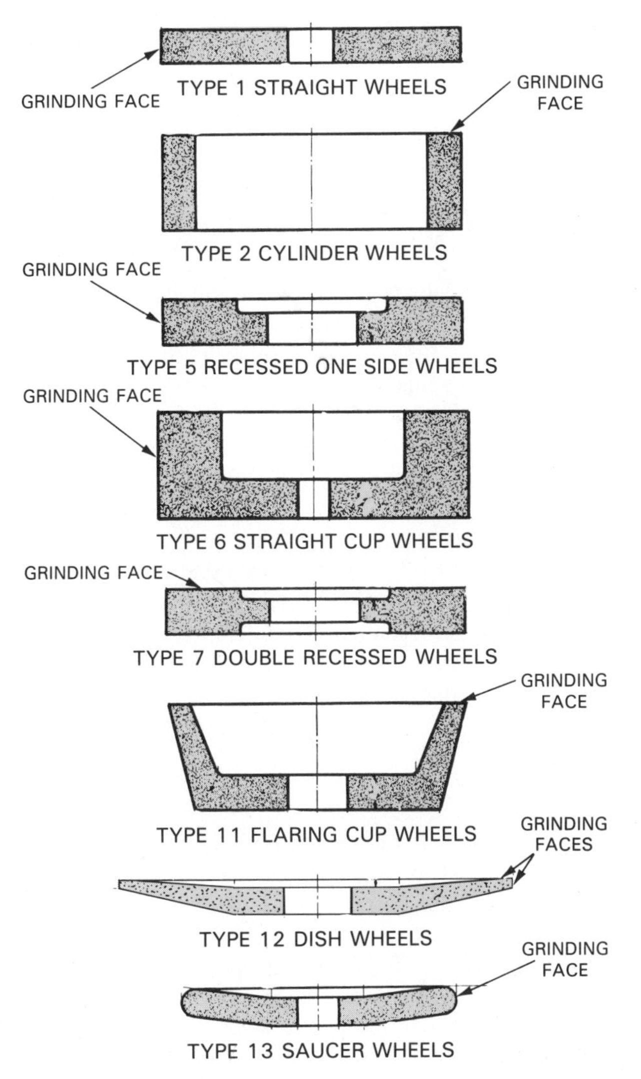

Fig. 21-8. Standard grinding wheel shapes. Note also the differences in face contours. (The Norton Co.)

ing wheel shapes. In addition, straight wheels can be purchased with a variety of face contours. Fig. 21-9 shows 12 common contours.

Grinding wheel selection

To match the grinding wheel to the job, abrasive type and size, bonding agent, and wheel grade and structure must be considered. In making this selection the following criteria apply:

1. Material to be ground. Hard materials require a soft bond (grade) so dull grains will break away. Hard grade wheels may be used on soft materials because the abrasive grains will stay sharper longer.
2. Type of operation. Different grinding operations exert varying pressures on wheels. Flat grinding and grinding of internal surface both allow a sizable part of the wheel to contact the work. Therefore, unit pressures are low. Grinding outside diameters of round stock requires a very small part of the wheel to be in contact with the part.

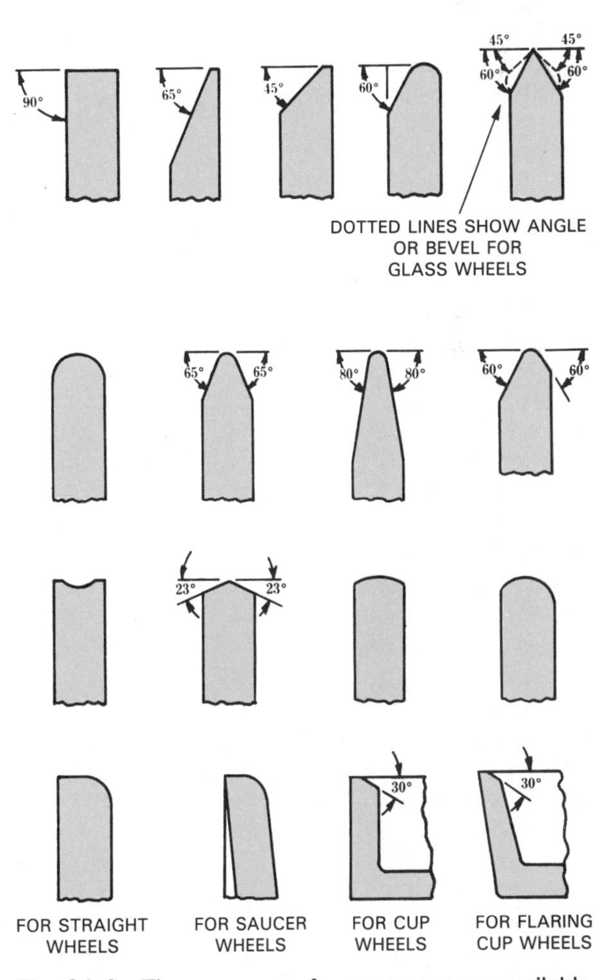

Fig. 21-9. These common face contours are available on straight grinding wheels. (The Norton Co.)

This produces high pressures on the wheel. A harder grade wheel is needed for external cylindrical grinding than for surface or internal cylindrical grinding.

3. Machine. Each machine has its own speed, feed, and other characteristics. These factors will play a roll in wheel selection.
4. Task to be performed. The type of grind required and the finish specified will help indicate the type of wheel required.

To help in wheel selection, a standard code has been developed. This code, illustrated in Fig. 21-10, specifies the manufacturer, abrasive type and grain size, wheel grade, and structure and type of bond.

COATED ABRASIVES

Coated abrasives are used for abrasive machining or sanding. Coated abrasives are generally used for two major tasks:

1. To reduce a rough, machined surface to a relatively smooth, flat surface.
2. To prepare material for application of finishing materials.

The selection of coated abrasives requires consideration of five basic factors. These factors are:

1. Abrasive.
2. Backing material.
3. Bond.
4. Distribution and orientation of abrasive grains.
5. Product form.

Abrasives

There are six commercially important minerals used for coated abrasives. These are:

1. Natural minerals.
   a. Flint (solid quartz).
   b. Garnet.
   c. Emery.
2. Manufactured abrasives.
   a. Silicon carbide.
   b. Aluminum oxide.
3. Crocus, a natural mineral which can also be manufactured.

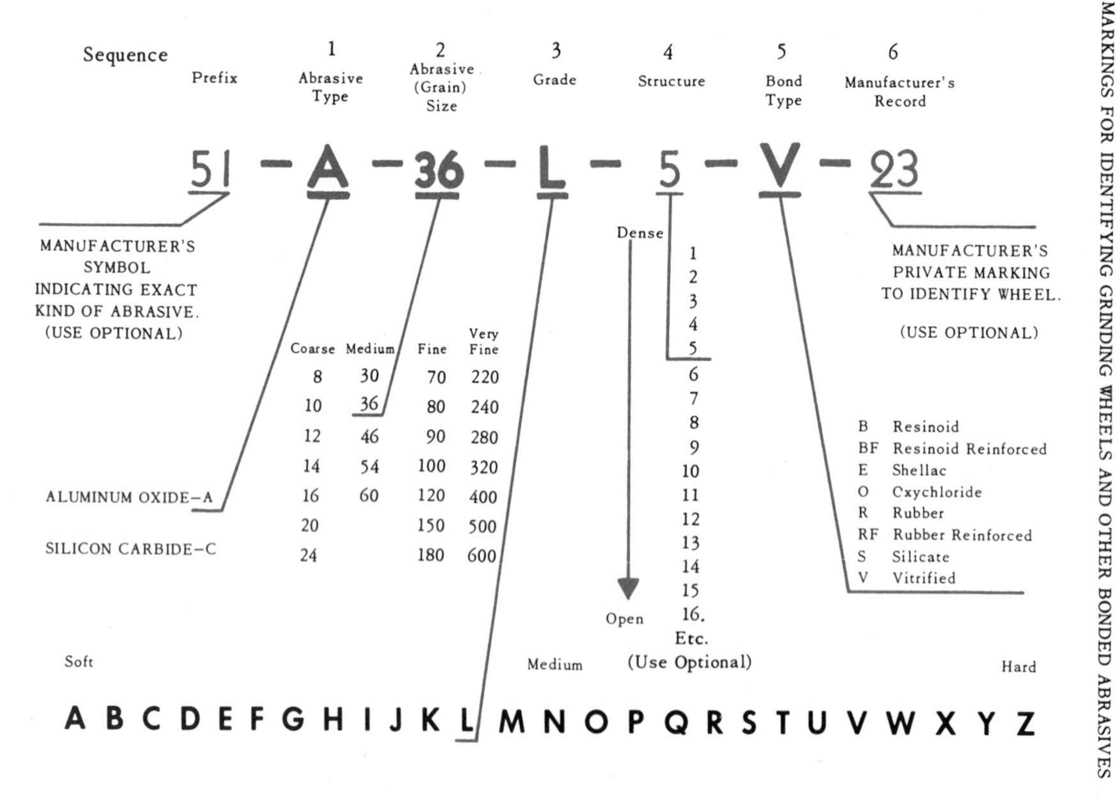

Fig. 21-10. Marking system for standard grinding wheels. Diamond and cubic boron nitride wheels are coded using another system. (Grinding Wheel Institute).

Emery and crocus (purple iron oxide) are used primarily for polishing and buffing metals. Flint, as discussed earlier, is used in common sandpaper. It has little industrial use.

Garnet and aluminum oxide are the principal minerals used for coated abrasives for sanding operations. Silicon carbide is used for light sanding.

The abrasives are sized using the same methods described under grinding. Their size is specified by the mesh (screen) size.

Backing materials

Fig. 21-11 shows the use of backing material in the basic construction of coated abrasives. The major types of backing are paper, cloth, and a combination of paper and cloth. *Paper backings* are available in four weights, each designated by a letter: A, C, D, and E. "A" is the lightest while "E" is the heaviest. A, C, and D papers are generally sheet stock abrasives for hand sanding. The papers are also used on reciprocating and orbital sanding machines. An E weight paper is used for sanding belts, drum sander wraps, and sanding discs.

Cloth backings are used on belts and disc for heavy-duty sanding applications. The belts are stronger and less likely to tear.

Combination paper and cloth backs are widely used for drum sander wraps. They have much higher resistance to breakage than E weight paper. Combination backs are being replaced by a vulcanized fiber back. This material is made of cotton cellulose treated with zinc chloride. The cellulose gelatinizes the cotton fibers. They are then vulcanized (treated with heat and sulphur) and converted into sheets.

Bonding material

Coated abrasives use a two-layer bond. The first layer is called the "maker coat." The maker coat is used for holding the abrasive particles as they are applied to the backing. The second coat, the "sizer" coat, holds and bonds the abrasive in its original position.

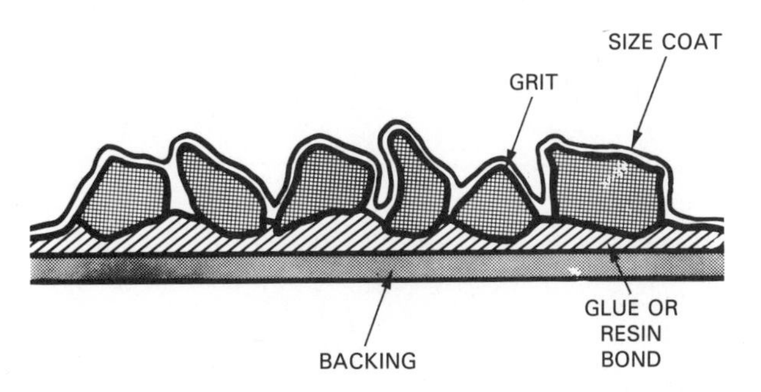

Fig. 21-11. This enlargement shows the composition of a typical piece of coated abrasive. (The 3M Co.)

The bonding materials may be hide glues, urea resins or phenolic resins. All-glue bonding is cheaper but has lower abrasive holding strength. Resin bonds do not retain a tackiness. They, therefore, have less tendency to fill and load up.

A resin sizer is sometimes placed over a glue maker coat. The advantage of this is that the abrasive has a better cutting action than when glue bonded but produces a better final finish than resin bond. Other bonding materials are varnish and vinyl acetate.

Generally, sheet goods are manufactured with hide glues. Resin bonding agents are used for wet sanding and machine sanding where heat buildup could be a problem.

Distribution and orientation of abrasive grains

The control of the distribution and the orientation of abrasive grains is critical for high quality coated abrasives. To accomplish this, one of two basic manufacturing processes may be used, Fig. 21-12.

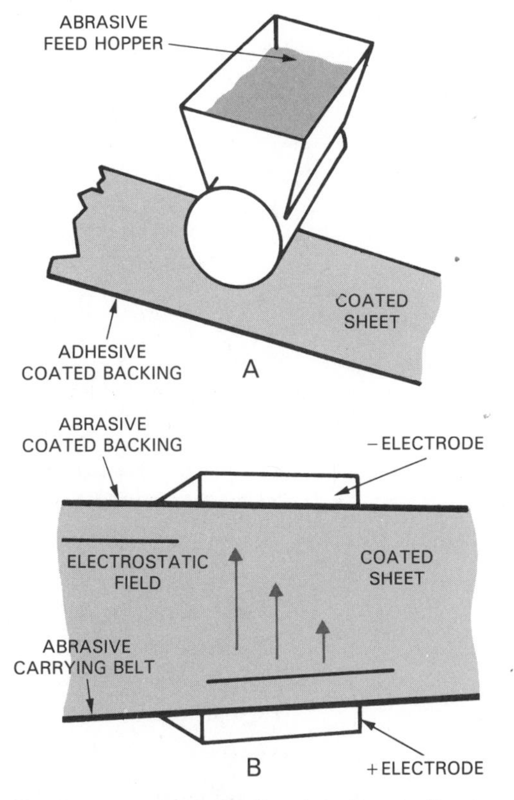

Fig. 21-12. The methods for abrasive application for coated abrasives. A—Gravity coating. Abrasive grains are dropped from hopper onto the coated backing. B—Electrostatic coating. Adhesive-coated backing and abrasive grains pass through an electrically charged field together. As grains are propelled they become embedded. (The 3M Co.)

These are the gravity method and the electrostatic method.

The *gravity* method distributes the abrasive on an adhesive-coated strip of backing material. The strip is vibrated so particles stand on their long axis.

The more common is the *electrostatic method* which involves passing the coated backing strip under an electrically charged plate. The abrasive particles are given the opposite electrical charge. Since unlike electrical charges attract, the particles are drawn upward and become embedded in the maker coat adhesive. Both methods apply a sizer coat of bonding material. The material is then heat cured.

The density (closeness) of the abrasive particles is also important. If the abrasive completely covers the adhesive side of the paper, the material is a *closed-coat abrasive.* When open spaces between the abrasives exist, it is called an *open-coat abrasive.* Open-coat abrasives have about 60 percent as many abrasive grains as closed-coat products. Open-coat abrasives are more flexible and less likely to clog or load up.

Product form

The typical product forms for abrasives are sheets, belts, rolls, and discs. Special forms such as cones, coils, and contour head wheel strips are also produced.

ABRASIVE MACHINING MACHINES

There are a number of machine tools that do abrasive machining. In order to describe them we will divide them into four major types:

1. Grinding machines.
2. Sanding and polishing machines.
3. Lapping, honing, and superfinish machines.
4. Loose-media machines.

GRINDING MACHINES

Grinding machines, like grinding operations, may be separated into precision grinding machines and rough grinding machines.

There are three basic precision grinding methods, as shown in Fig. 21-13. These are:

1. Cylindrical grinding.
2. Surface grinding.
3. Tool grinding.

Cylindrical grinding

Cylindrical grinding involves grinding a round surface or workpiece. These grinding operations, as illustrated in Fig. 21-14, may be on either internal or external surfaces. External surfaces may be ground straight, tapered, or formed. Internal surfaces may

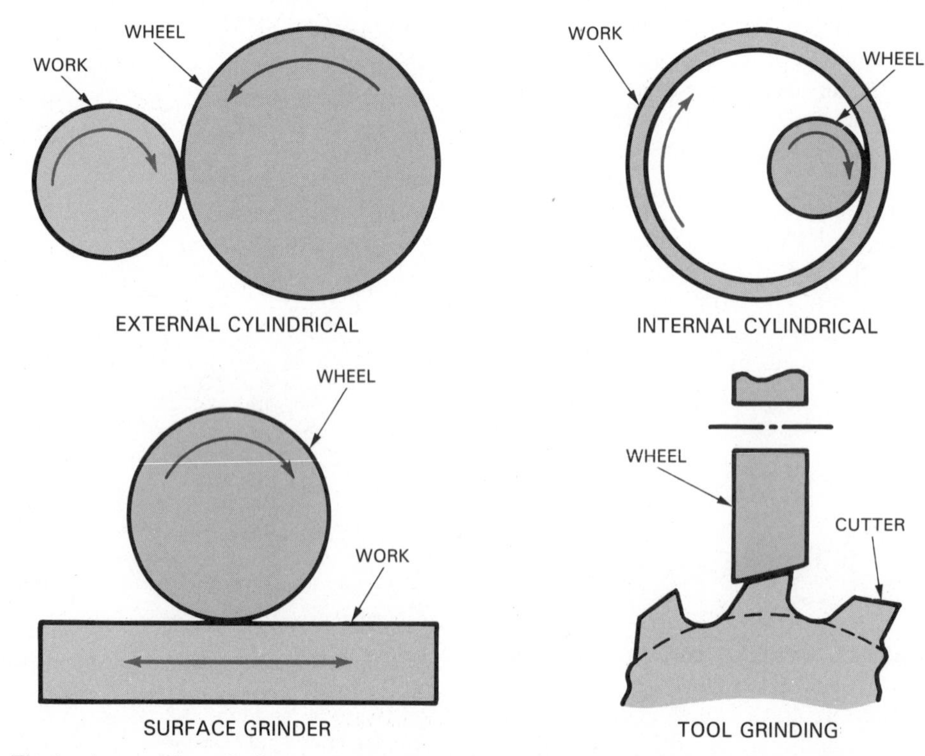

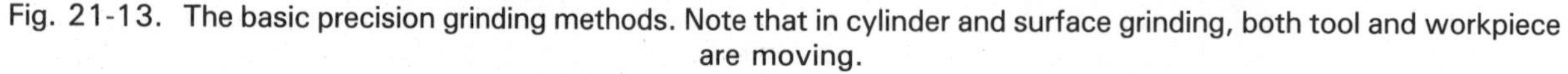
Fig. 21-13. The basic precision grinding methods. Note that in cylinder and surface grinding, both tool and workpiece are moving.

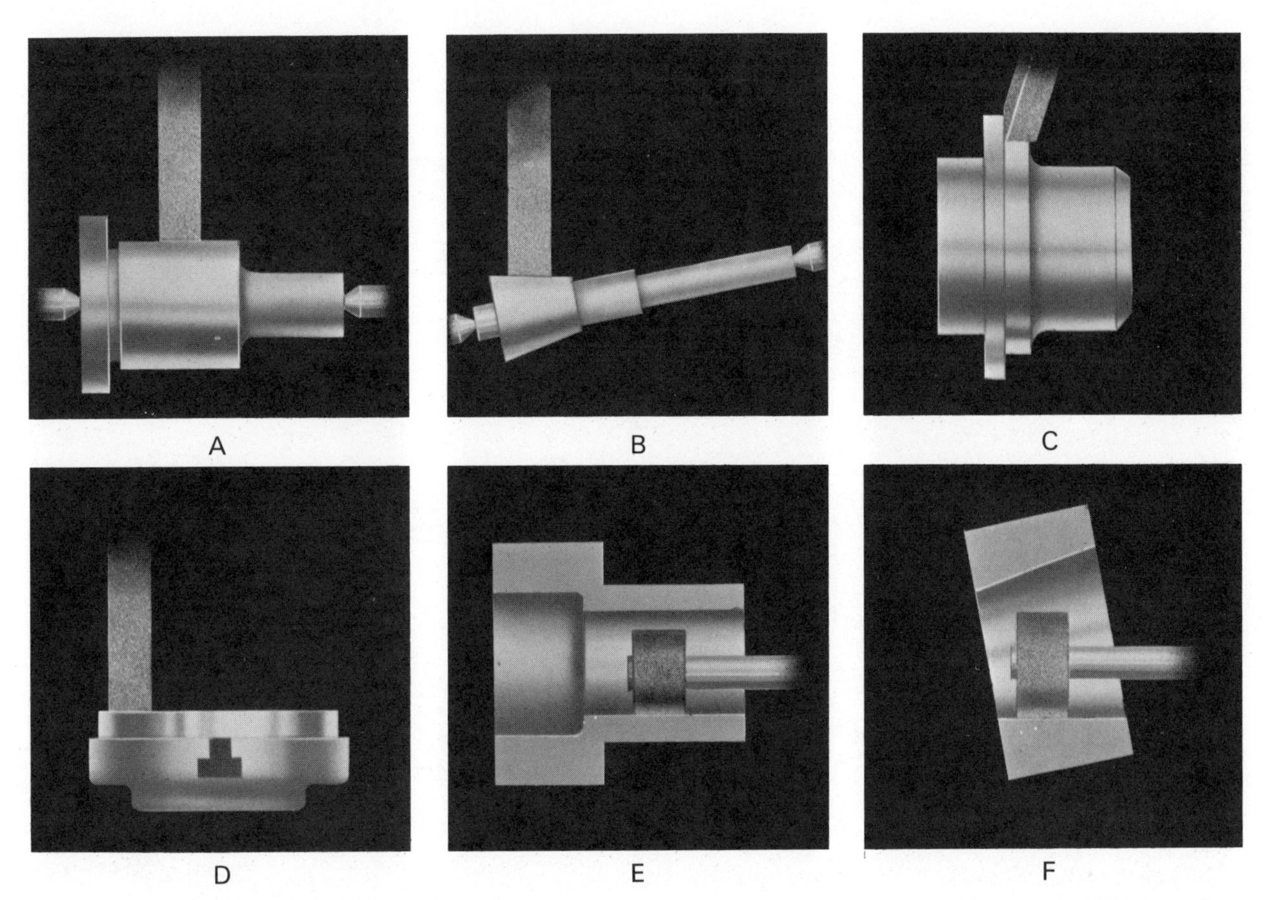

Fig. 21-14. Types of cylindrical grinding. A—External straight. B—External tapered. C—Shoulder. D—Face. E—Internal straight. F—Internal tapered. (Landis Tool Co.)

also be straight, taper, or form ground. In addition, *blind hole,* which is a type of straight grinding, may be cylindrically ground.

Cylindrical grinding generally has two actions. First, the rotating grinding wheel is fed into the revolving work or the work is fed into the wheel to obtain the depth of cut. Then the wheel is moved along the work or the work across the wheel. This action produces both the feed and cutting motions. The rotating wheel produces the cutting action and the linear movement of the work or the grinding wheel generates the feed motion.

Cylindrical grinding, as just described, may be done on one of three basic machines:

1. Plain cylindrical grinder.
2. Universal cylindrical grinder.
3. Centerless cylindrical grinder.

Plain cylindrical grinder

The *plain cylindrical grinder* holds and rotates the stock between centers much like an engine lathe. The tailstock adjusts to accommodate various lengths of work. Generally the wheel is adjustable only inward to set the depth of cut. The table traverses (moves at right angle to the wheel) for feed motion. The center-type cylindrical grinder, Fig. 21-15, will grind straight or taper forms. Examples of typical plain grinding applications are shown in Fig. 21-16.

When special shapes are to be ground in large quantities, *form* grinding may be used. First the grinding wheel is shaped, possibly by *crushing* during which a carbide roll is forced against the slowly revolving wheel. This action crushes the wheel to shape.

Another shaping method is by diamond trueing. This process uses a shaped roll with embedded diamonds. The roll is rotated and fed into the revolving grinding wheel, thus shaping the wheel.

Finally the wheel may be template dressed. A single diamond dressing tool may follow a template to form the grinding wheel. The newly formed wheel can then be used to grind the desired form.

Universal cylindrical grinder

The *universal cylindrical grinder* will hold a workpiece between centers like a plain grinder. In addition, material may be held in a chuck. Added flexibility is

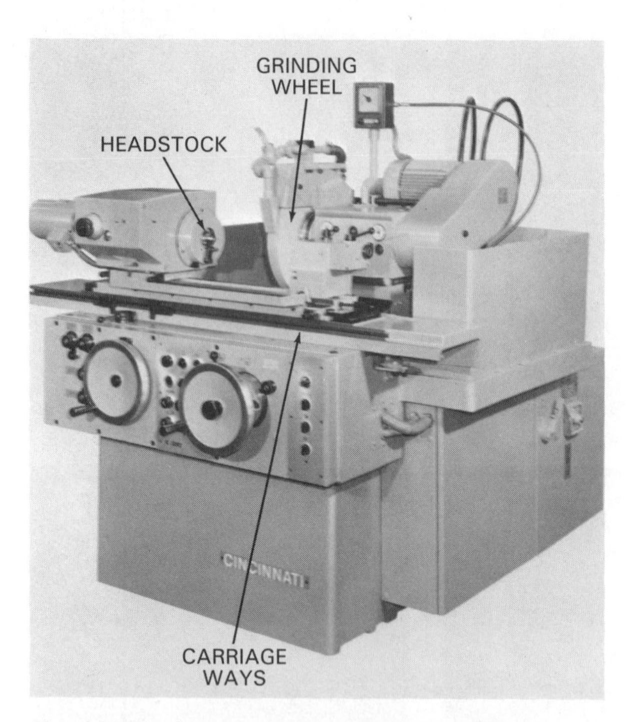

Fig. 21-15. A center type grinding machine grinds straight or taper forms. (Cincinnati Milacron)

obtained from a headstock that can swivel on a vertical axis. The wheelhead can also be swiveled and traversed on any angle.

Universal grinders like the one in Fig. 21-17 can do both internal and external grinding. Internal grinding requires an internal grinding unit shown in the upper right of the photo. The universal grinder can perform a wide variety of grinding operations. A sample of these was shown in Fig. 21-13.

Centerless grinder

The *centerless grinder* can grind both internal and external surfaces without chucking or holding the work between centers. The work lies on a work rest. Two revolving grinding wheels machine the work.

The principle of centerless grinding is shown in Fig. 21-18. The grinding wheel turns at regular grinding speed. It does the actual grinding. The regulating wheel turns much slower. It is set at an angle to the grinding wheel. The regulating wheel turns and feeds the workpiece. Fig. 21-19 shows two views of a centerless grinder.

Internal centerless grinding, Fig. 21-20, requires three external rolls and an internal grinding wheel. The grinding wheel grinds the internal surface as it is moved inward (traverses).

Special cylindrical grinders have been developed to grind threads and, through use of multiple wheels,

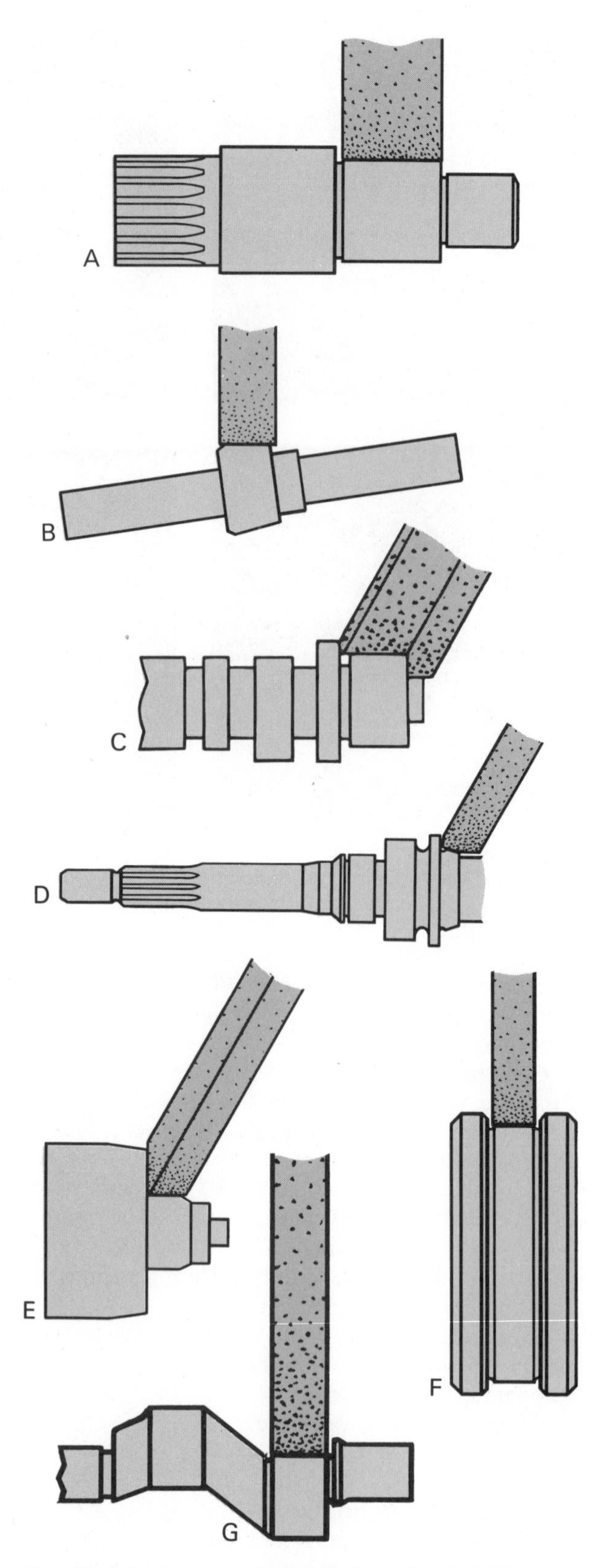

Fig. 21-16. Typical plane grinding. A—Straight. B—Tapered. C—Formed. D—Bevel. E—Shoulder. F—Straight recess. G—Cam. (Landis Tool Co.)

Fig. 21-17. A universal grinder can do both internal and external grinding. Internal grinding unit (arrow) swings out of the way when not in use.

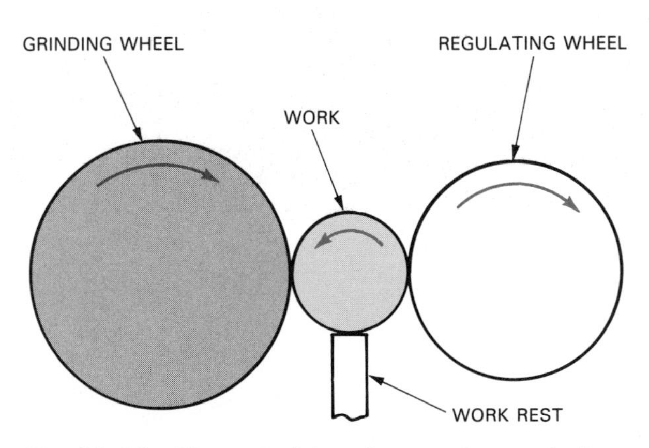

Fig. 21-18. The principle of centerless grinding. Workpiece rotates between two grinding wheels.

A

B

Fig. 21-19. A close-up view of the centerless grinding operation. A—Stopped. B—In action. (Landis Tool Co. and The Norton Co.)

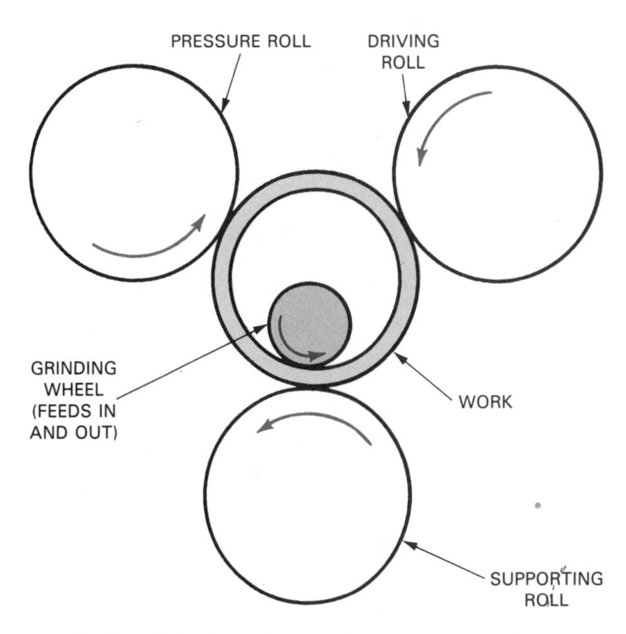

Fig.21-20. Principle of internal centerless grinding. One roller rotates the workpiece, a second roller supports the workpiece, and a third one applies pressure.

Fig. 21-21, to grind camshafts, crankshafts, and similar products.

Surface grinding

Surface grinding involves the grinding of a flat surface. Typically, surface grinding machines a planed or formed surface, Fig. 21-22.

Surface grinders and their operation can be divided into three major categories (Fig. 21-23):

1. Horizontal-spindle grinders.
2. Vertical-spindle grinders.
3. Disc grinders.

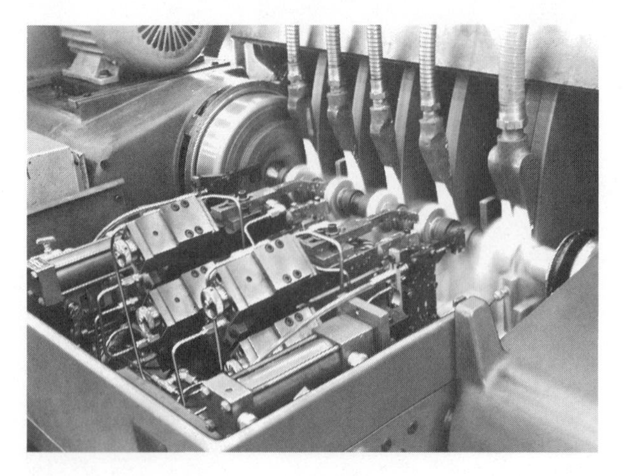

Fig. 21-21. Camshaft is being ground on a multi-wheel grinder. (The Landis Tool Co.)

Fig. 21-22. Surface grinding. Such machines produce flat surfaces. (The Norton Co.)

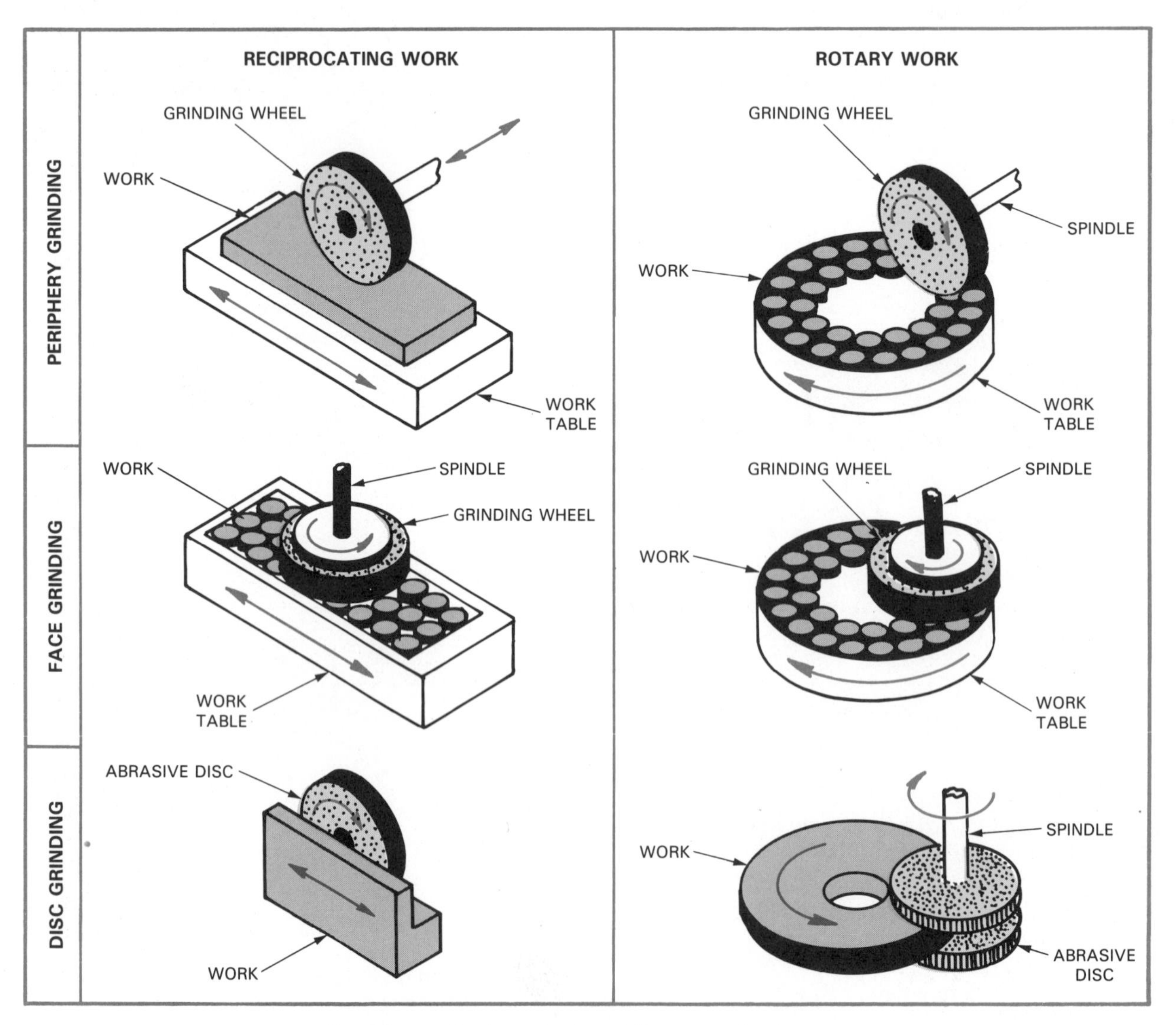

Fig. 21-23. Surface grinders fall into these basic types.

Horizontal-spindle machines

A horizontal-spindle machine has one or more grinding wheel mounted on its shaft. The grinding takes place as the peripheral (outer circumference) surface of the wheel contacts the work. Typical applications for this type of machine include grinding flat surfaces, tapers, slots, surfaces next to shoulders, and angles.

Horizontal-spindle machines are available with two types of tables: horizontal and rotary, Fig. 21-24. The horizontal table has a reciprocating (moves back and forth) motion under the rotating grinding wheel. The part is held by a magnetic chuck (table), a vacuum chuck, or in a special fixture. The rotating table turns under the rotating grinding wheel. Parts held on the table will contact the wheel as the table rotates. Light, medium, and heavy-duty horizontal table models are produced. They vary in size and power and will accommodate increasingly larger parts and cuts.

Vertical-spindle machines

Vertical-spindle machines use the face of a cup or cylinder grinding wheel in their operation. They may use a reciprocating or rotating table or a feed-through conveyor to hold and support the workpieces. The reciprocating and rotary tables work in the same manner as they do with horizontal-spindle machines. The feed-through conveyor replaces the table and allows parts to be continuously fed under the grinding wheel.

Vertical-spindle surface grinding machines are generally used to machine a flat surface in small to moderately large parts. Most vertical-spindle grinding operations are not designed to give highly accurate dimensional control; nor is a high-quality finish always desired. Often, high-volume production is the most important goal. However, both accuracy and fine surface finishes are possible on certain vertical-shaft surface grinders.

Disc grinders

Disc grinders are either vertical or horizontal shaft machines capable of high volume production. They can maintain close tolerances and produce good surface finishes at these high rates. These machines are the most efficient grinders available to machine large

Fig. 21-24. There are two basic methods for holding parts for surface grinding. A—Magnetic chuck on horizontal table. B—A rotary table. (The Norton Co., Cincinnati Milacron)

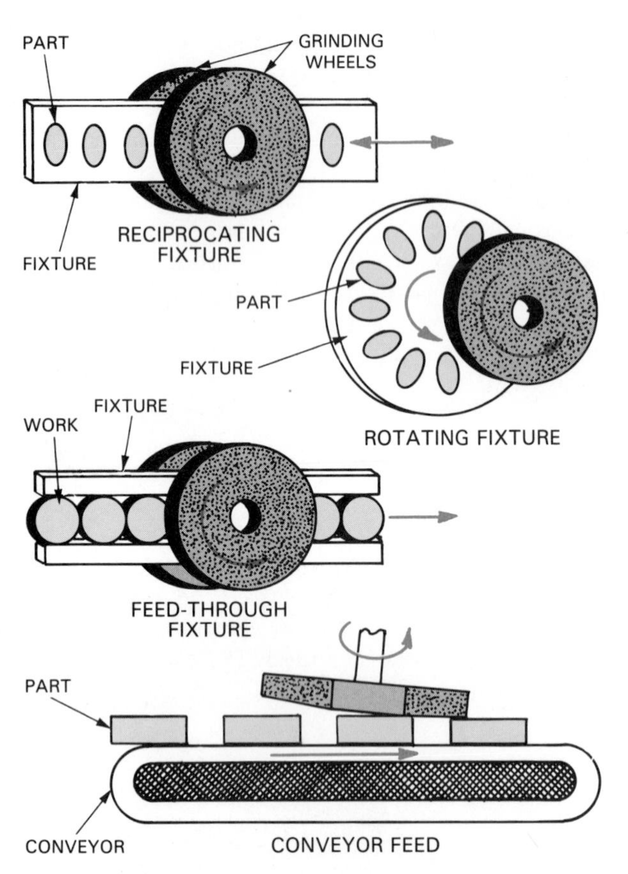

Fig. 21-25. Parts feeding systems for disc and vertical shaft surface grinding.

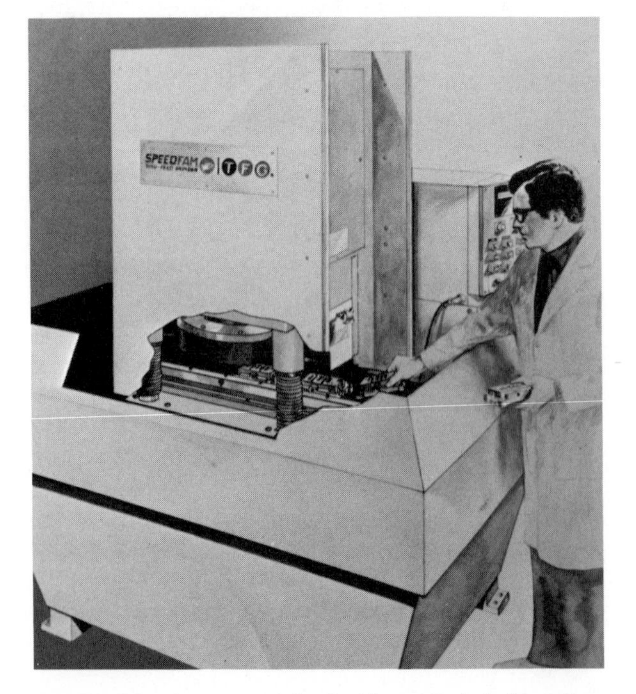

Fig. 21-26. A conveyor feeds this vertical shaft surface grinder. (Speedfam Corp.)

numbers of parts with flat surfaces.

Parts may be machined on one or both opposing surfaces at once. Special feeding mechanisms enable parts to be machined continuously, Fig. 21-25.

Vertical-shaft models use the side of the wheel for grinding. The grinding wheel may be solid or segmented. The vertical shaft surface grinding wheel contacts a large area of the work, Fig. 21-26. Grinding is rapid but leaves circular marks. Fig. 21-27 is a diagram of the grinding actions of reciprocating and rotary table surface grinders both vertical and horizontal spindles.

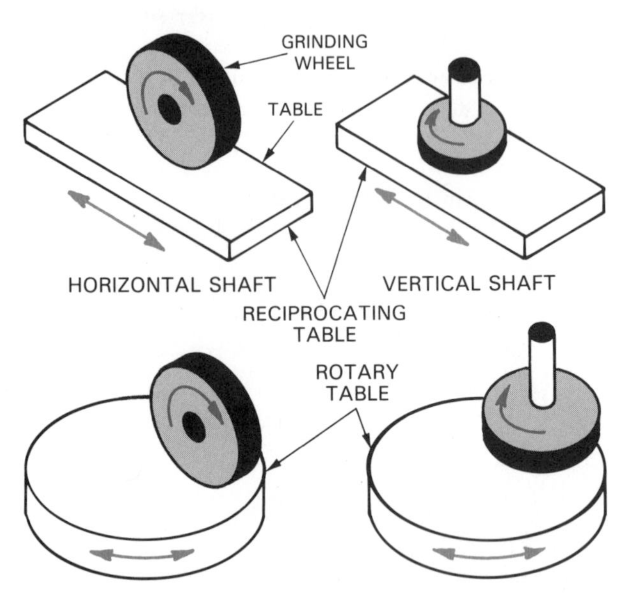

Fig. 21-27. There are four basic cutting actions for vertical and horizontal reciprocating and rotary table surface grinders.

Tool grinding

Tool and cutter grinding is the process of finishing and sharpening various tools and cutters. Typically, the grinder is used to sharpen milling cutters, drills, reamers, taps, hobbing cutters, etc.

Universal grinding machines may be used for cutter sharpening. Also, special cutter grinders are made. The three in Fig. 21-28 shows typical setups.

Rough-grinding machines

Rough grinding or nonprecision grinding quickly removes unwanted material. This process is sometimes called *snag grinding*.

Major machines used in rough grinding are pedestal or bench grinders, swing-frame grinders, and portable grinders, Fig. 21-29. Small parts which can be easily held are snag ground on pedestal (or stand) grinders.

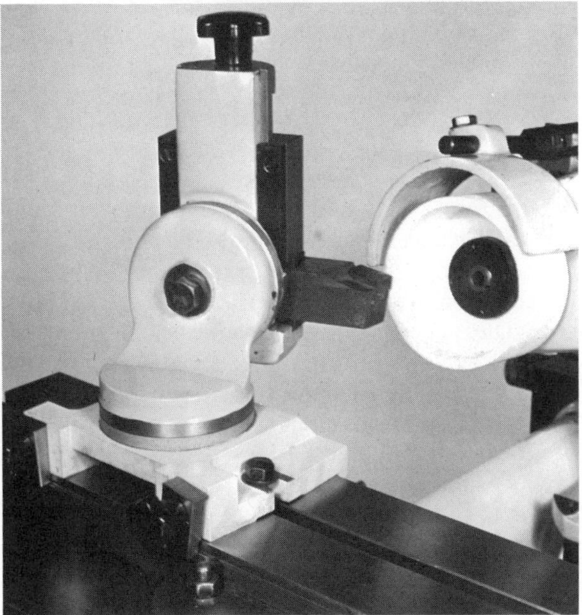

Fig. 21-28. Typical universal grinding machine. These are common cutter grinding setups. (POLAMCO Machine Co.)

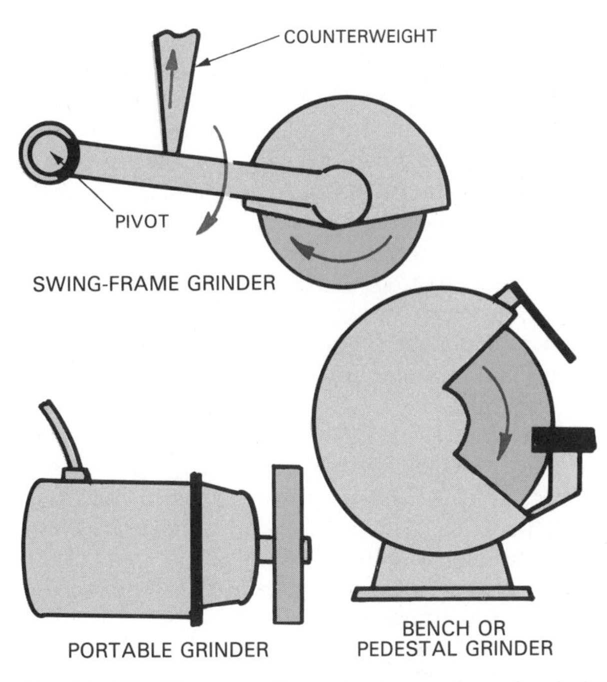

Fig. 21-29. These are the major types of rough grinding machines.

These machines may have wheels up to 36 in. in diameter.

A swing-frame grinder may be used on objects too heavy to easily move. The frame is supported from a column which allows for easy movement. The wheel is positioned above the excess material and then lowered. The depth and area of the grind is controlled by the operator.

Portable grinders are used for more varying grinding operations. They may be used for grinding defective spots, casting parting lines, and weld areas.

SANDING AND POLISHING MACHINES

Coated abrasives are widely used in the processing of wood, metals, and plastics. These machines are used to sand or polish:

1. Flat surfaces.
2. Contoured surfaces.

Flat surface sanding

The sanding of a flat surface requires a moving coated abrasive to generate the cutting action. The part is moved to produce feed motion.

The three major machines that produce these motions are the disc sander, the drum sander, and the belt sander.

The *disc sander* uses an abrasive disc adhered (glued) to a metal disc. This disc is rotated by a power-driven shaft. The disc sander has some limited industrial use in sanding end grain of wood. Since the direction of rotation is across the faces or edges of wood, disc sanders will not produce adequate surface smoothness. Fig. 21-30 shows a typical disc sander.

The *drum sander* has one or more carefully balanced cylinders with an arbor through its center. Coated abrasives are spiral-wound around the drum. The material to be sanded is passed on a moving belt or on rollers under the revolving drums. As the drums revolve they move slightly back and forth on their axes. This keeps the abrasive from loading and wearing unevenly.

Many drum sanders have several drums. Each drum will have a slightly finer abrasive on it. Therefore, the first drum rough sands the work while the last one does the finish sanding. A typical commercial drum sander is shown in Fig. 21-31.

Belt sanders for machining flat surfaces are of three major types. These are:

1. Edge sanders.
2. Wide-belt sanders.
3. Stroke sanders.

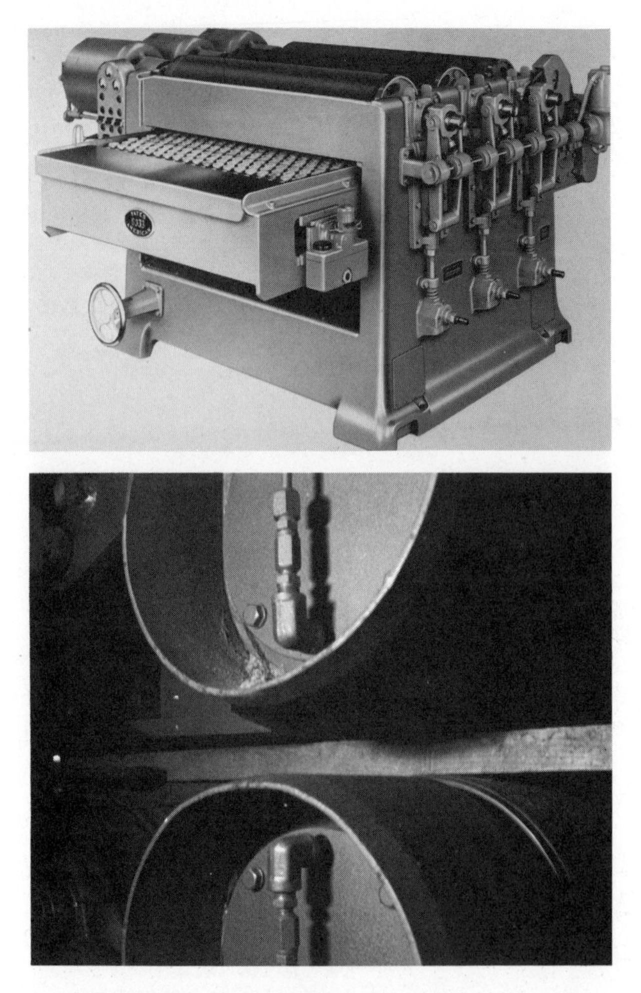

Fig. 21-31. Top. Drum sander with its covers removed. Bottom. A board passing through on a horizontal drum sander. (Yates-American Machine Co.)

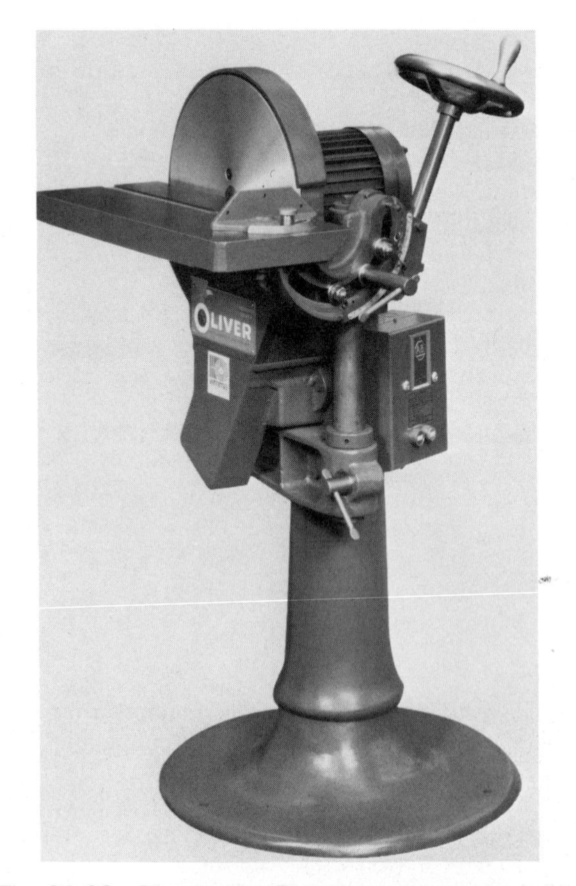

Fig. 21-30. Disc sander. Disc size can range from 30 to 36 in. in diameter. (Oliver Machinery Co.)

The *edge-belt sander,* as its name suggests, is designed to sand edges. The belt moves around two vertical drive and tracking wheels. The belt is backed at the sanding area by a platen (flat piece). The edge of the material is forced against the belt at the platen area. The movement of the abrasive belt and the force against the platen cause the work to be sanded.

The belt often oscillates (moves up and down) to distribute wear and reduce loading. Also the table may be tilted to sand angled edges. A typical edge belt sander is diagramed in Fig. 21-32.

The *wide-belt sander* operates much like the drum sander. The drum has been replaced by belts. The belts pass around and over either a rubber roll or a platen (flat piece). The work is fed under the moving abrasive belts on a continuous feed belt. The sanding occurs as the rubber roll or the platen forces the belt against the work. As with the drum sander, the belt

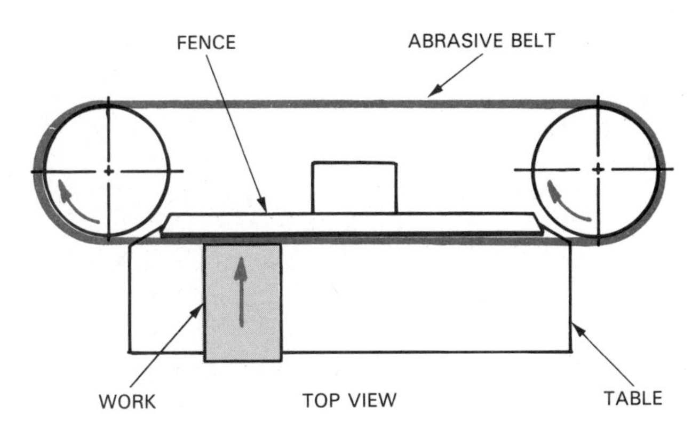

Fig. 21-32. A simplified diagram of top of an edge belt sander. Workpiece is pushed against fence as abrasive belt moves past.

also moves left and right to reduce wear and loading.

Wide-belt sanders, Fig. 21-33, may have one or more belts. These belts can be as wide as 62 in. and as long as 20 ft.

Stroke sanders, Fig. 21-34, are used primarily for final sanding before applying finish. They use a belt moving around two horizontal drums. The work is placed on a table below the moving abrasive belt. A pressure pad is brought down against the back of the abrasive belt. This forces the belt down to the work where sanding occurs.

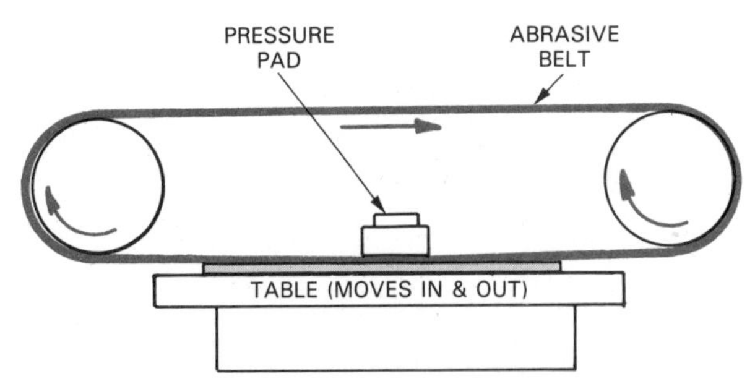

Fig. 21-34. Simplified diagram of a stroke sander. Operator controls pressure on workpiece.

Other belt sanders are similar to grinding machines discussed earlier. Common examples of these machines include centerless, surface, swing, and stationary platen sanders, Fig. 21-35.

Contour sanding machines

A number of techniques are used to sand and polish a contoured surface. The most common tools are abrasive wheels, abrasive belts, and abrasive sleeves or drums.

Abrasive belt contour sanding and polishing may use a contact wheel, a flexible belt, or a formed wheel. These techniques are shown in Fig. 21-36. Also the edge-belt sander may use a formed wheel or a formed

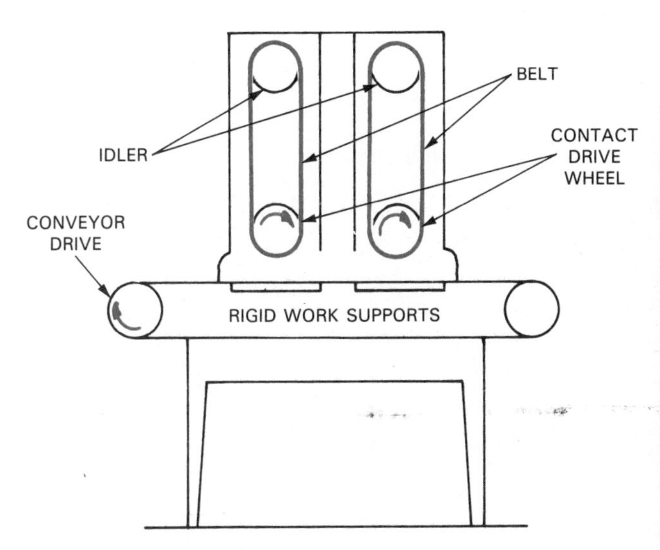

For use on flat surfaces, including larger sheets and panels, where a succession of abrasive grades is used. Uses: simultaneous roughing, blending, finishing, polishing and precision sizing.

Fig. 21-33. Left. Diagram of a wide-belt sander. Right. A wide-belt sander with its cover and belts removed. (The 3M Co. and Yates-American Machine Co.)

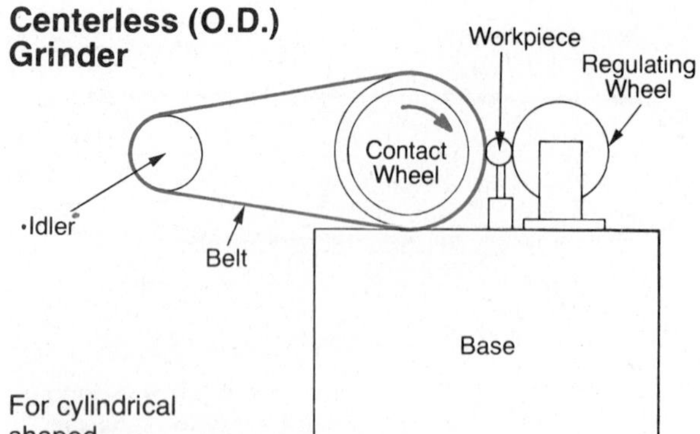

For cylindrical shaped workpieces, such as rod or tubing of varying diameters and lengths, often requiring precision finishes. Uses: roughing, blending, finishing, polishing and precision sizing.

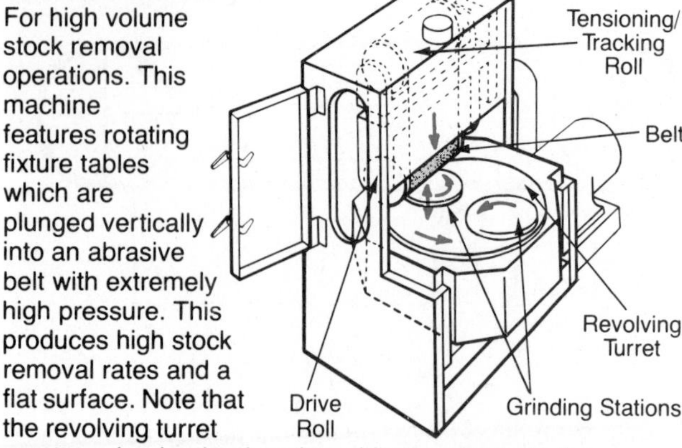

For high volume stock removal operations. This machine features rotating fixture tables which are plunged vertically into an abrasive belt with extremely high pressure. This produces high stock removal rates and a flat surface. Note that the revolving turret creates a load-unload station while the alternate fixture table is in a grind cycle. Uses: heavy stock removal.

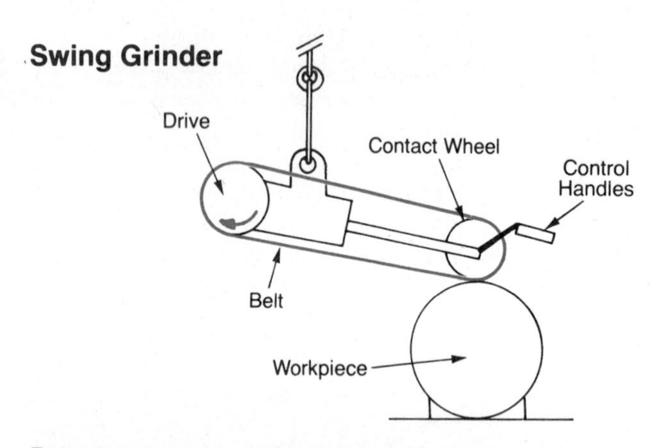

For use on larger workpieces that cannot be brought to a floor grinder. Available in 1-15 H.P. ratings. Uses: roughing, blending, finishing.

Stationary Platen Grinder

Belt
Idler
Platen
Work Table
Drive

For use where a true, flat surface must be developed. Uses: roughing, blending, finishing, polishing, and precision sizing.

Fig. 21-35. These belt sanding and grinding machines are commonly used. (The 3M Co.)

platen to produce the desired contour.

Abrasive sleeves on a spindle sander, Fig. 21-37, may be used to sand along a curved edge or end. An abrasive drum on a drill press performs the same task.

Abrasive wheels may also sand contoured surfaces. They are usually used to sand a contour on an edge or end of a part. They may be abrasive impregnated formed wheels, brush-backed sanding wheels, or wheels made of many abrasive strips, Fig. 21-38.

LOOSE-MEDIA ABRASIVE MACHINES

Loose-media abrasive processes include abrasive machining and abrasive finishing. Abrasive machining is most often done using the abrasive jet process. The major abrasive finishing processes using loose media are barrel finishing and vibration finishing.

Abrasive jet machining

Abrasive jet machining is used on hard, brittle, and/or fragile materials. It is similar to sand blasting except the machining process is closely controlled. Finer abrasive grains are used and their velocities are carefully regulated.

The cutting action is produced by propelling the grains at very high speeds (500-1000 fps) against the work. Air or carbon dioxide (CO_2) is generally used to carry the abrasive grains.

For most machining operations aluminum oxide or silicon carbide powders are used. Sodium bicarbonate

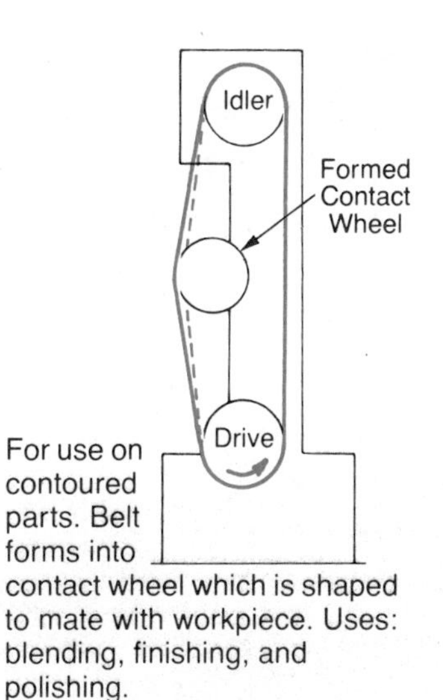

Fig. 21-36. A formed wheel sanding and grinding machine may be either horizontal or vertical. (The 3M Co.)

Fig. 21-37. An oscillating single spindle sander and metal grinding machine. As spindle rotates it moves up and down. (Oliver Machinery Co.)

Fig. 21-38. A brush back contour sanding wheel sands along curved edges or ends. (The 3M Co.)

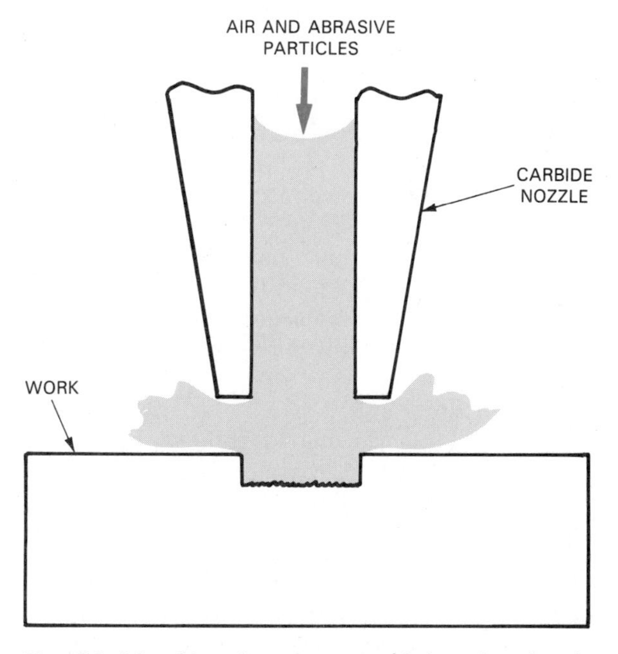

Fig. 21-39. Abrasive jet machining bombards workpiece with abrasive particles carried by compressed air.

or dolomite are often used for cleaning, polishing, or etching.

Fig. 21-39 shows a typical abrasive jet cutting operation. The process is used for machining materials, frosting glass, etching decorative patterns, cutting holes and shapes in thin materials, shaping crystalline material, and deburring parts.

The process is limited in its applications by its slow cutting rate. Abrasive jet machining is used when conventional processes will not work.

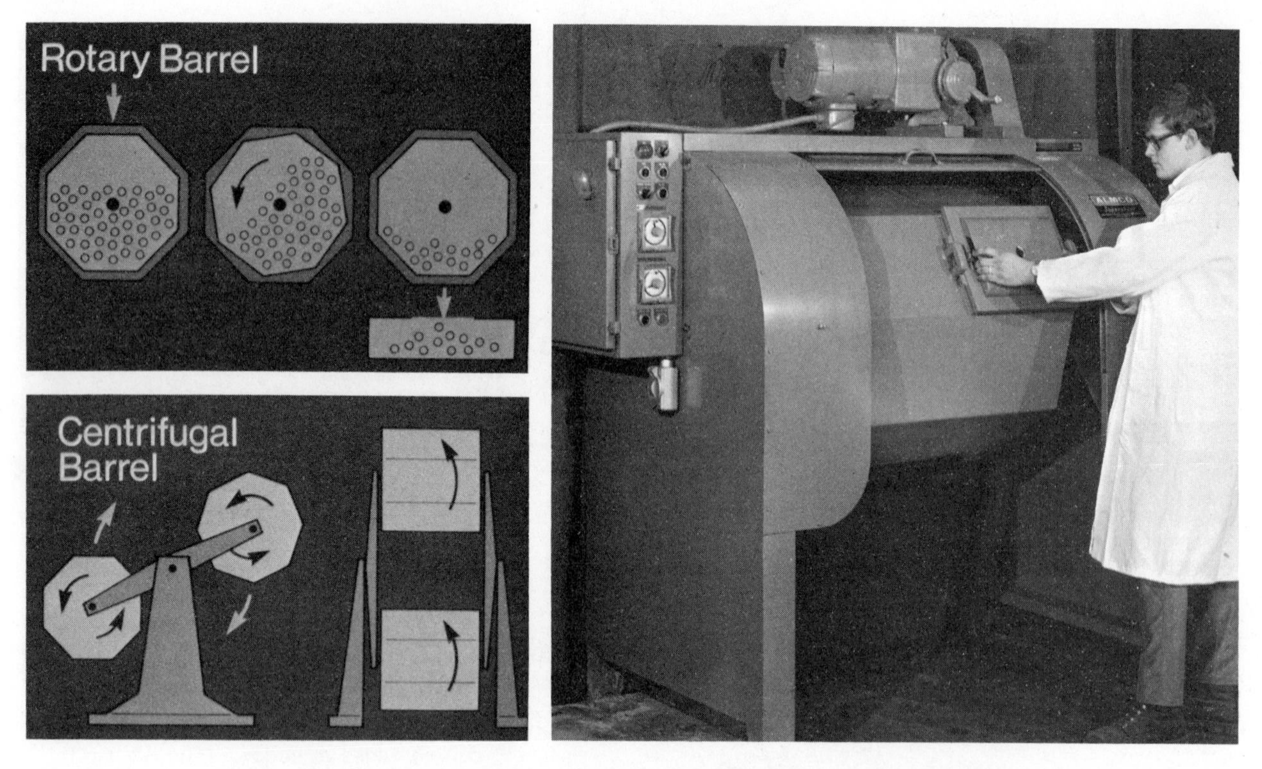

Fig. 21-40. Rotary and centrifugal barrels are used for barrel finishing. (Ransburg Corp. and ALMCO Div., King-Seeley Thermos Co.)

Barrel finishing

Barrel finishing involves placing the workpieces and an appropriate abrasive media into a six or eight-sided barrel. When the barrel is rotated, the abrading action between the abrasive media and the work produces the finish, Fig. 21-40.

The media may be stone, bonded abrasive shapes, abrasive grains, or hardened steel shapes. Barrel finishing is used to smooth fired ceramic materials, metal parts, and small wood shapes, Fig. 21-41.

Fig. 21-41. Bisque fired china cups are tumbled with round stones to smooth them before final decoration and glaze is applied. (Syracuse China Co.)

Vibration finishing

Vibration finishing requires that the workpieces and an abrasive be placed in a vibratory finishing machine, Fig. 21-42. The materials are vibrated causing an abrading action. Vibration finishing can finish holes and slots not reached by barrel finishing.

HIGH FINISH PROCESSES

Often very accurate smooth finishes are desired. They can be obtained using one of three techniques: honing, lapping, or superfinishing.

Honing

Honing is an abrasive machining technique chiefly used on internal cylindrical surfaces. Typical of these

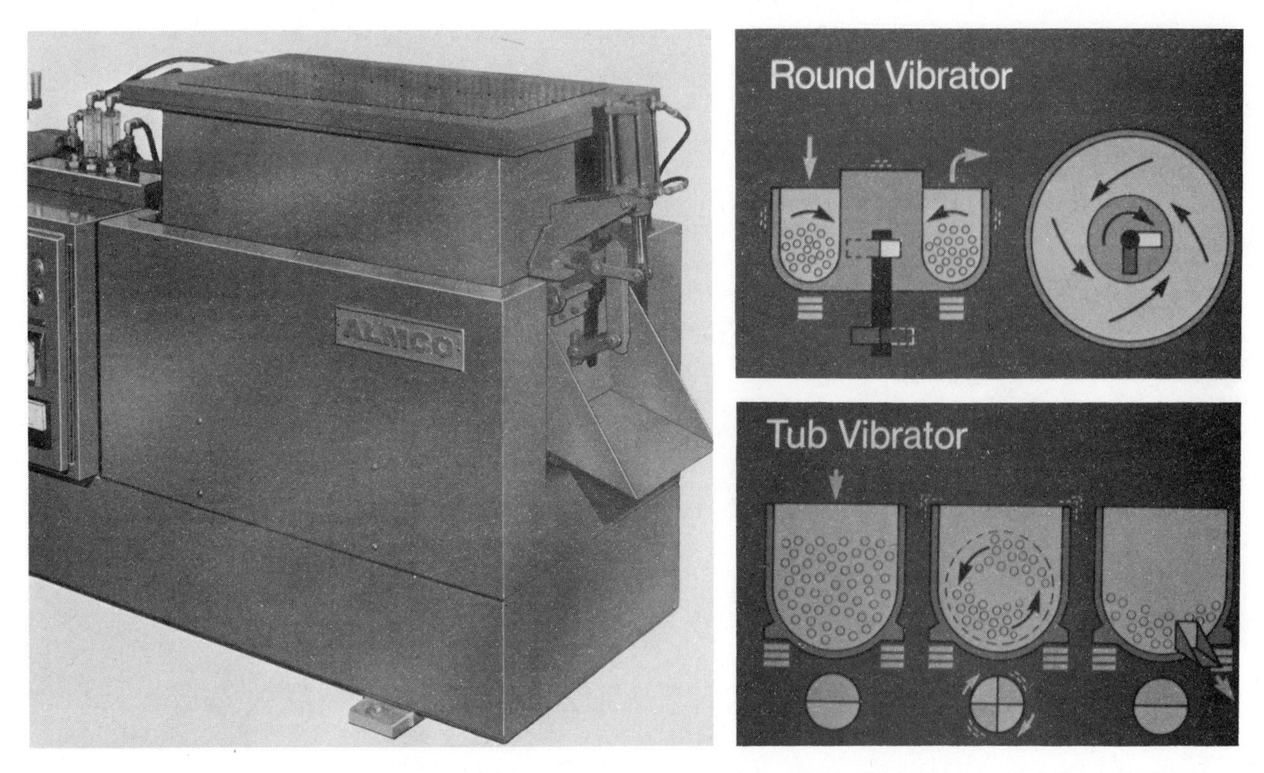

Fig. 21-42. Left. Commercial vibratory tub-type finishing machine. Right. Round and tub (vibrator) finishing machines. (ALMCO Division-King-Seeley Thermos Co. and Ransburg Corp.)

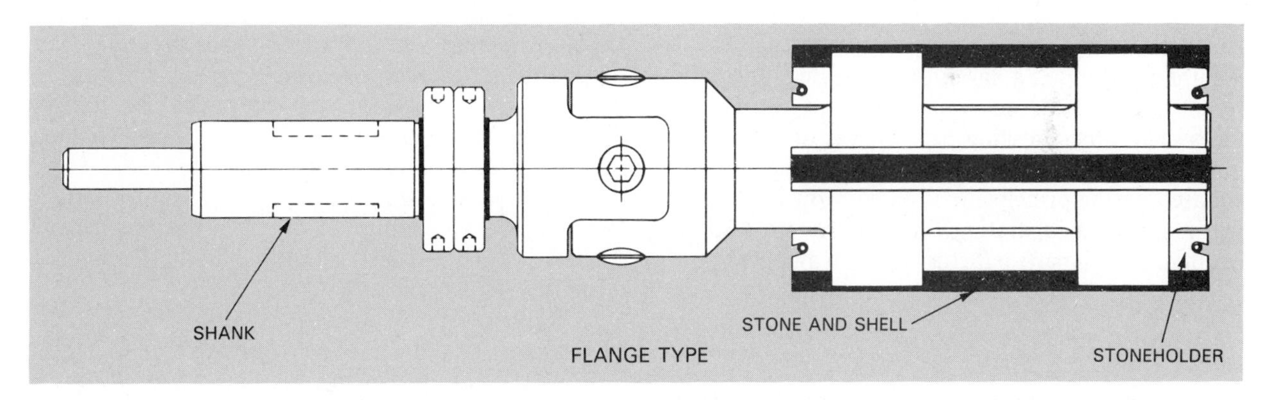

Fig. 21-43. A standard hone like this is used for abrasive machining of internal cylindrical surfaces. (General Hone Corp.)

surfaces are hydraulic cylinders, gun barrels, and engine cylinders.

Honing uses a fine abrasive stone, Fig. 21-43. The rotating stone is placed in the hole then caused to rotate and reciprocate, Fig. 21-44. An exact amount of pressure is applied to the stone as it rotates in the hole. When the hole is honed to the proper diameter the cutting action stops.

Honing is done at slow speeds and produces accurate results. Producing parts within tolerances of 0.0001 in. to 0.0003 in. is within the range of a honing operation.

Lapping

Lapping is a process designed to remove scratches. It will not remove material rapidly. Maximum material removal seldom reaches 0.001 in.

Lapping uses abrasive materials embedded in a soft material that ranges from cloth to copper. The part to be polished is placed on the lap along with a

Fig. 21-44. A—Horizontal honing machine. B—Schematic of the honing action. (General Hone Corp.)

liquid—usually oil. The lap is rotated or vibrated causing the abrasives to polish the part.

Superfinishing

Superfinishing is a variation of honing. It is different in that it is applied primarily to external surfaces. The superfinishing process was developed to improve the wear characteristics of mating parts. It is based on the principle that oil of proper viscosity will maintain a separating film if parts are smooth enough. Rough parts puncture the film and break the lubricating barrier. The honing stick is designed to abrade away the points, Fig. 21-45.

Superfinishing uses a honing stick which is forced against the part with light, controlled pressure. The pressure used ranges between 10 to 40 psi. The part is then slowly rotated (50-60 rpm) while the hone is oscillated. During the superfinishing process coolant floods the area.

The basic superfinishing equipment for cylindrical parts is shown in Fig. 21-46. This equipment will remove very little material. Typically the diameter is

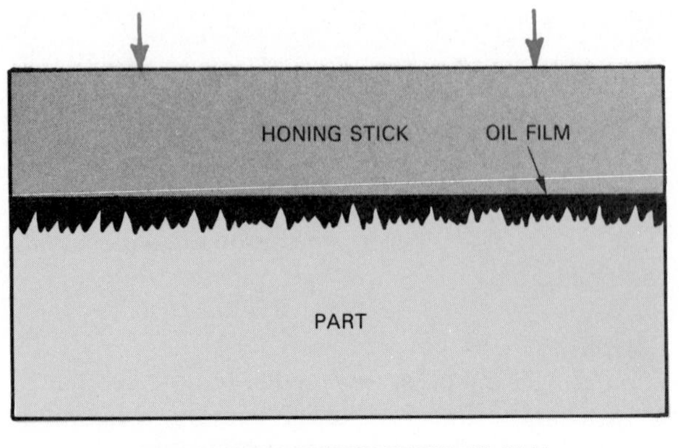

SHARP POINTS PENETRATE THE OIL FILM AND ARE REMOVED BY THE HONING STICK

Fig. 21-45. Principle of superfinishing. Honing over oil film abrades high points only.

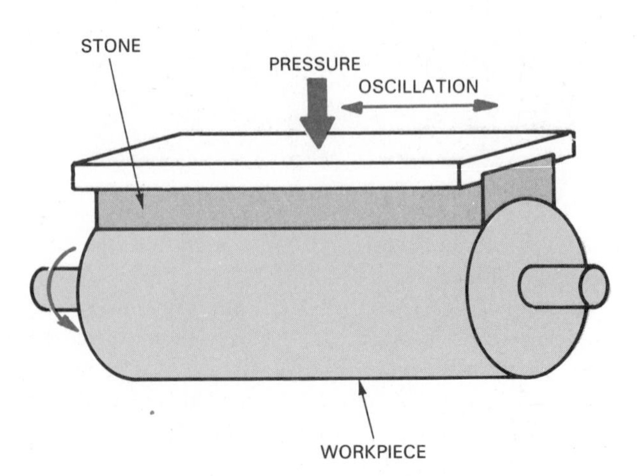

Fig. 21-46. Schematic of a superfinishing or microstoning operation for machining cylindrical parts.

reduced 0.0001 in. to 0.0004 in. The resulting finish, which is extremely smooth, can range from 1 to 80μ in. (micro inch, 10^{-6} inches).

Superfinishing may also be used on flat surfaces. A flat-surface superfinishing machine has two spindles. The upper spindle holds the stone while the lower spindle carries the work. The spindles, as can be seen in Fig. 21-45, are parallel but offset. As they rotate, a very flat surface is produced.

SUMMARY

Abrasive machining uses abrasive grains as loose media, bonded abrasives, or coated abrasives. In most cases the grains form a random multipoint cutting tool. These tools are formed into wheels for grinding; stones for honing and superfinishing; and sheets, belts; and sleeves for sanding.

Abrasive machining may be used to quickly remove stock. These actions are called rough grinding or sanding. Abrasive machining may also be used to shape parts quickly and accurately. Finally, abrasive machining through precision grinding and finish sanding operations may produce a smooth, accurate part.

Abrasive machining may be used on flat surfaces, external cylindrical shapes, and internal cavities. These processes may shape and smooth flat, tapered, and formed surfaces. Abrasive machining practices are essential in producing accurate parts with proper surface finishes.

For smooth and accurate finishes, three processes are used. They are: honing, lapping, and superfinishing.

STUDY QUESTIONS — Chapter 21

1. List the four common types of abrasive machining processes.
2. List and describe the three basic types of grinding.
3. Give three examples each of natural abrasives and synthetic abrasives.
4. What are bonded abrasives and coated abrasives?
5. Describe bonded abrasives in terms of:
   a. The abrasive used.
   b. Grain size.
   c. Bonding material.
   d. Wheel grade.
   e. Wheel structure.
   f. Wheel shape.
6. What information is given in a grinding wheel marking code?
7. What are the two major tasks for which coated abrasives are used?
8. List the six important coated abrasive minerals and their typical uses.
9. List and describe the backing materials commonly used for coated abrasives.
10. Describe the two methods by which coated abrasives are manufactured.
11. Describe the difference between an open-coat and a closed-coat abrasive.
12. List and describe the major grinding machines.
13. List and describe the major sanding and polishing machines.
14. List and describe the major lapping, honing, and superfinishing machines.
15. List and describe the major loose-media machines.

Chapter 22

THERMAL AND CHEMICAL MACHINING

A number of material removal processes using heat (thermal) or chemicals have been developed. These processes are usually called *nontraditional machining* because they use techniques which do not fit into the common machining groups. They are not the typical turning, drilling, milling, shaping, and planing operations. The most common of these operations which cut and shape materials using thermal or chemical actions are:

1. Electrical discharge machining (EDM).
2. Electrochemical machining. (ECM).
3. Chemical machining.
4. Laser machining.
5. Electron beam machining.
6. Flame cutting.
7. Plasma-arc cutting.

Each of these operations uses electrical, chemical, electrochemical, and/or thermal actions to do the cutting or machining. No mechanical force is used.

ELECTRICAL DISCHARGE MACHINING

Electrical discharge machining (EDM) is used to accurately remove material from a metal workpiece. This process is the oldest of the nontraditional machining systems. It was developed during World War II to remove broken taps from holes. Today it has wider application. The EDM process, however, will only work on conductors of electricity (metals). EDM cannot be used to machine polymers or ceramics, Fig. 22-1.

Electrical discharge machining is sometimes called *spark machining*. This name comes from the fact that the metal removal is actually done by an electrical spark, Fig. 22-2.

The process involves a rapidly repeating electrical discharge (spark) between the electrode (tool) and the work. The spark melts and vaporizes the metal workpiece. The resulting metal particle or "chip" is flushed away from the gap between the work and the electrode by a nonconducting liquid, Fig. 22-3. This liquid is usually a light oil and is called a *dielectric*. The entire cutting action takes place in a container of this dielectric.

As can be seen in Fig. 22-4, an EDM machine has several major parts. These include:

1. An electrode in the shape of the cavity which will be produced in the workpiece.
2. A device to hold the work under the electrode.
3. A dielectric container and system to circulate (pump) the dielectric medium onto the cutting surface.

Fig. 22-1. An EDM (electrical discharge machine) is useful in machining materials which conduct electricity. (Elox Div., Colt Industries)

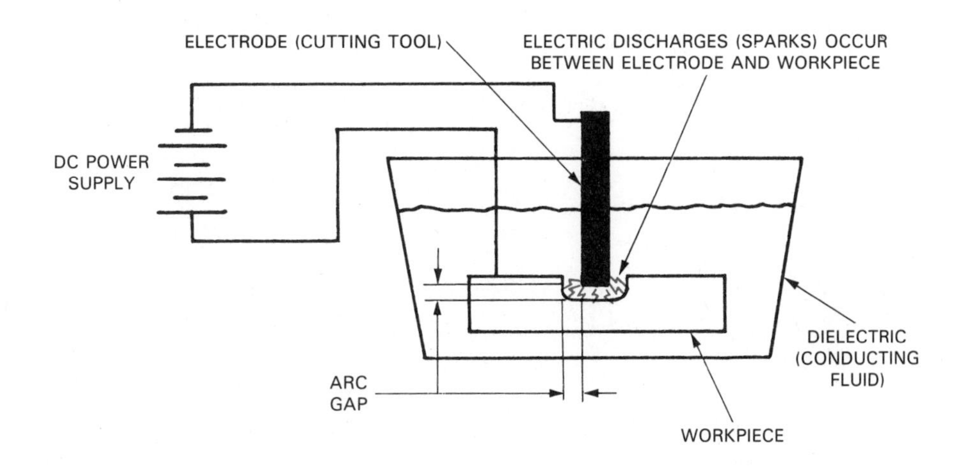

Fig. 22-2. Simple diagram represents the entire EDM process. (Caterpillar Tractor Co.)

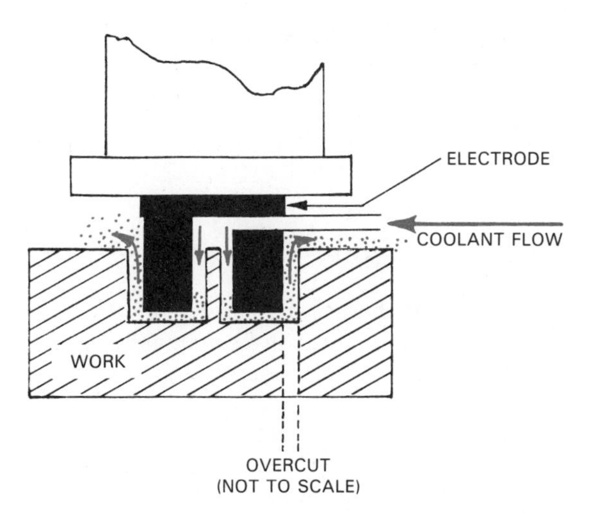

Fig. 22-3. Chips or particles are flushed away from the EDM cut by a coolant. (Elox Div., Colt Industries)

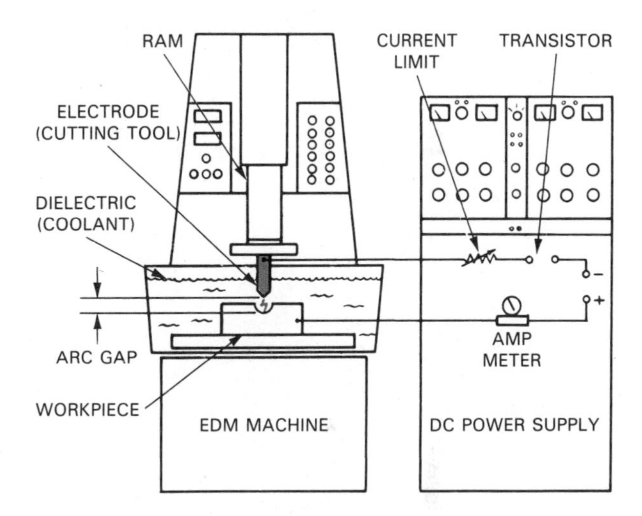

Fig. 22-4. Diagram of an electrical discharge machine. Voltage is controlled by a direct current power supply. Electrode, which may be male or female, is always shaped like the part it is to produce. (Elox Div., Colt Industries)

4. A system to move the electrode toward the work and maintain the proper arc (spark) gap.
5. A system to generate pulsating direct electrical current. This current jumps between the negatively charged tool to the positively charged work.

The spark produced is contained in a very small area. It produces very high temperatures (about 10,000°C) and high pressures. As the sparks naturally move from different areas of the electrode to the work, very even, smooth cuts are possible. Any shape that can be made into an electrode can be reproduced in the work.

ELECTRICAL DISCHARGE GRINDING

Special EDMs have been built for special machining activities. Electrical discharge grinding (EDG), Fig. 22-5, removes materials by using a rotating tool. The rapid, repeating spark discharges between the tool and the workpiece causing the "cut." The dielectric covers the work and flows between the tool and work. This flow removes the chip.

EDG is used to accurately machine hard materials. However, it is generally used to shape an outside surface rather than a cavity as in EDM. Typical uses are machining forming tools and carbide tool bits.

ELECTRICAL DISCHARGE SAWING

Electrical discharge sawing (EDS) combines the motion of a band saw with the EDM principle. A high speed (5000+ fpm) knife-edge band "blade" is fed into the work. A continuous spark discharges between

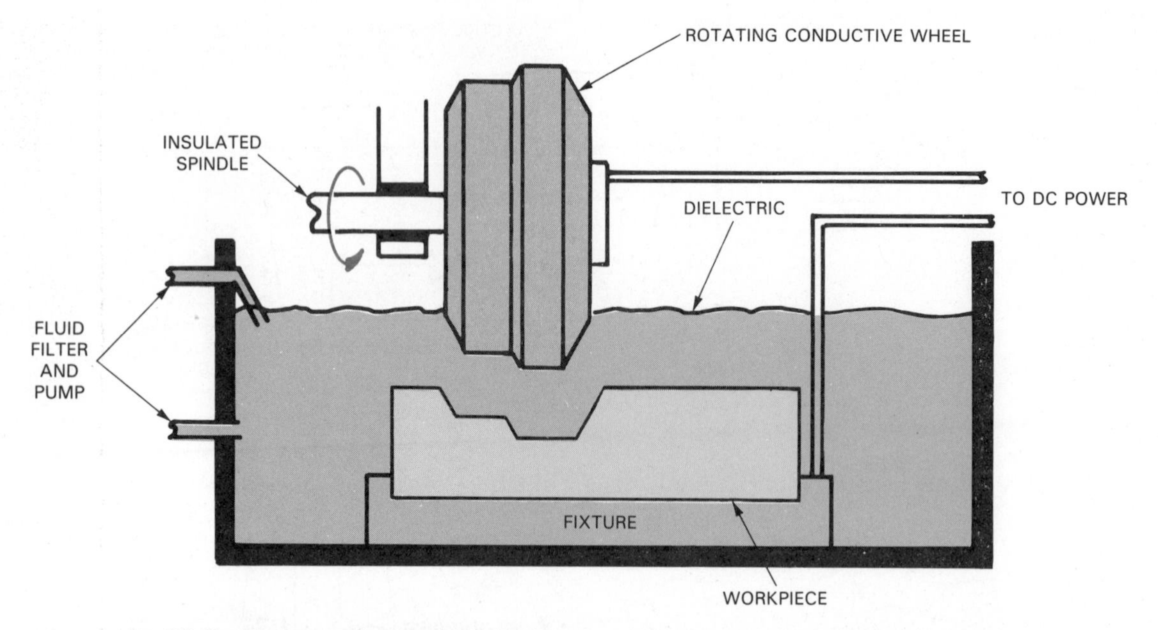

Fig. 22-5. Diagram of electrical discharge grinding operation. Dielectric between workpiece and conductive wheel floats chip away from the machining area.

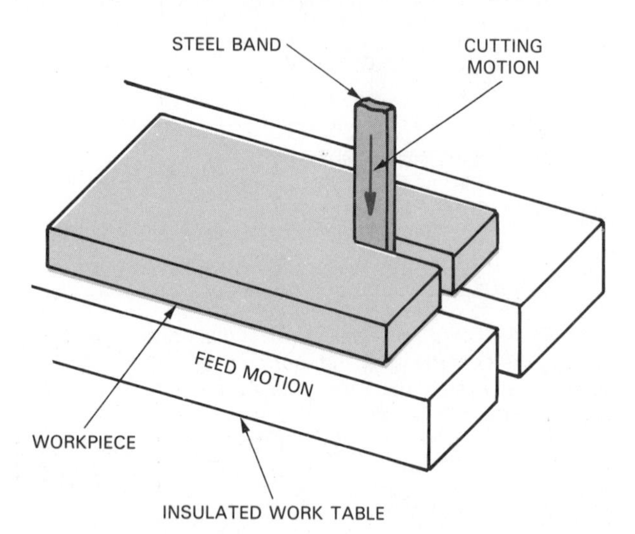

Fig. 22-6. Diagram of electrical discharge sawing (EDS). This process is useful in cutting fragile material which would be damaged by traditional machining.

the blade and the work vaporizing the material at the machining line. No dielectric is used but water is used to cool the band. Fig. 22-6 is a diagram of a typical EDS operation.

Electrical discharge sawing has great advantages over other processes in cutting fragile materials. Typical applications are cutting aluminum and titanium honeycomb panels and thinwall tubing. The cutting action neither bends nor leaves a burr, nor crushes the material, Fig. 22-7.

ELECTRICAL DISCHARGE WIRE CUTTING

Electrical discharge wire cutting (EDWC) uses a taut, traveling, conductive wire—usually copper or brass—as the "saw." The process is also called traveling wire EDM.

The small-diameter (0.002 to 0.010 in.) wire moves into the work. The spark between it and the work produces the EDM action. A dielectric produces the same results as the standard EDM process.

Often the path of the wire is numerically controlled. A wire cutting EDM is shown in Fig. 22-8. It produces a wide variety of parts, Fig. 22-9. These include punches, dies, and stripper plates for forming and shearing operations.

ELECTROCHEMICAL MACHINING

Electrochemical machining (ECM), Fig. 22-10 is, in many ways, like electroplating. The process involves bringing a hollow electrode (tool) very near (0.001 to 0.010 in.) the work. Low-voltage direct current, is applied between the electrode and work. The electrode is given a negative charge while the work is positively charged.

The entire work area is covered with a rapidly flowing electrolyte (liquid which will conduct electricity). As the electrical current flows from the tool to the work, electrons are removed from the surface of the workpiece. The resulting positively charged atom (atom with a shortage of electrons) is drawn off the

A B C

Fig. 22-7. A—An EDS operation cuts a thin "slice" from honeycombed metal block. B—A part cut by the EDS process. C—The EDS blade. (DoAll Co.)

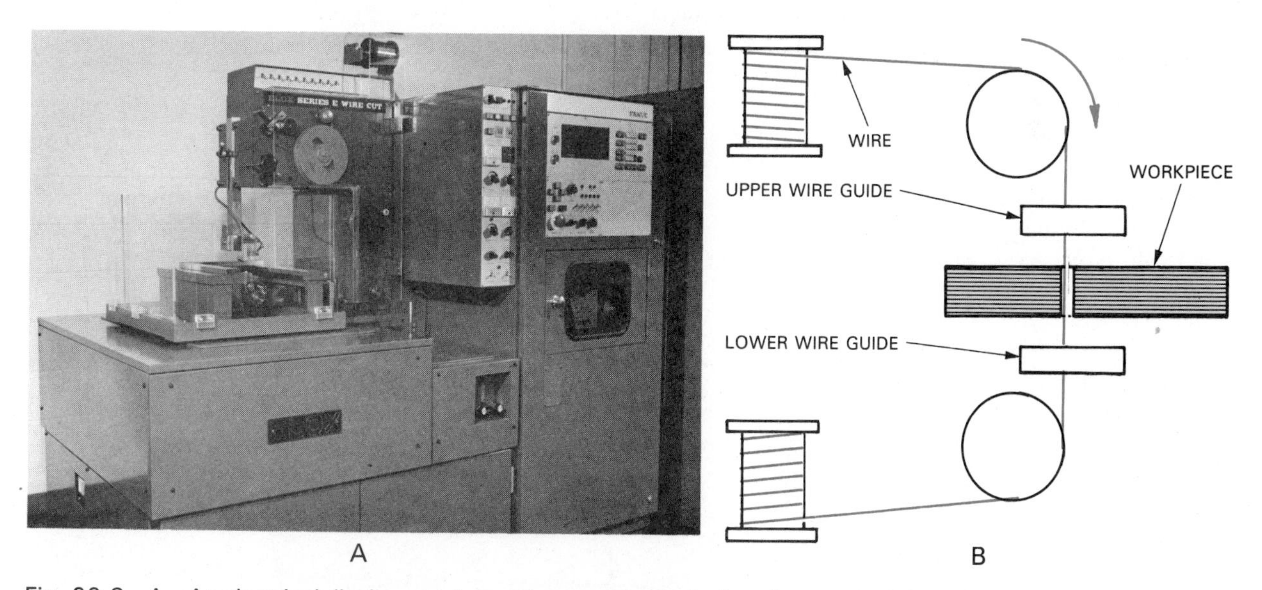

A B

Fig. 22-8. A—An electrical discharge wire cutting machine is designed to cut straight-sided parts made from hard materials. B—Diagram of its operation. (Elox Div., Colt Industries)

work toward the electrode (tool). The ion would be deposited on the electrode as in electroplating. Instead, the fast-flowing electrolyte washes it away.

ECM removes metal one atomic layer at a time. The electrons are removed by electrical current. Then the positively charged ions are drawn away from the work

Fig. 22-9. These are some typical parts produced by a numerically controlled EDWC machine. (Elox Div., Colt Industries)

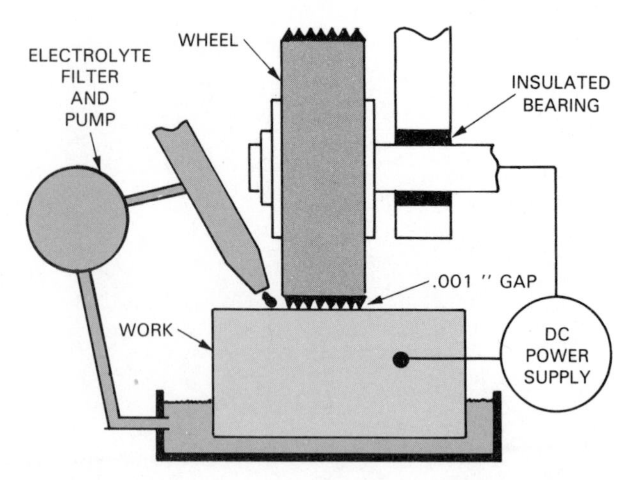

Fig. 22-11. Diagram of electrochemical grinding. Wheel is metal bonded and studded with diamond grit.

by the negatively charged tool. The ion is swept away before it can reach the electrode. The electrode is then fed farther into the work and the action is repeated.

ECM is an alternative to EDM for machining very hard materials or for producing complex cavities. EDM will produce more accurate cuts but ECM is much faster.

ELECTROCHEMICAL GRINDING

A modification of electrochemical machining is electrochemical grinding (ECG) or *electrolytic grinding*. This process, as shown in Fig. 22-11, uses a metal-bonded, diamond-grit grinding wheel as an electrode.

The metal bond of the wheel acts as the cathode (negatively charged element). The diamonds are insulators and brush away the ions formed in the process. The diamonds also will cut if the wheel contacts the work. However, in normal operations the grip of the wheel produces very few chips. Over 90 percent of the material is removed by the electrolytic action.

Electrochemical grinding is not as economical as regular grinding. It is, therefore, used for shaping and/or sharpening very hard materials. Electrochemical grinding is generally reserved for grinding carbide cutting tools. The electrochemical action can shape carbides quickly and produces a smooth surface. Also, the diamond cutting wheels are subjected to very little wear in electrochemical grinding.

CHEMICAL MACHINING

Chemical machining involves removing material with a chemical reagent. This technque is the simplest of the nontraditional machining operations. It uses the same process which causes metal to corrode. There are two basic types of chemical machining: *chemical blanking* and *chemical milling* or *contour forming*.

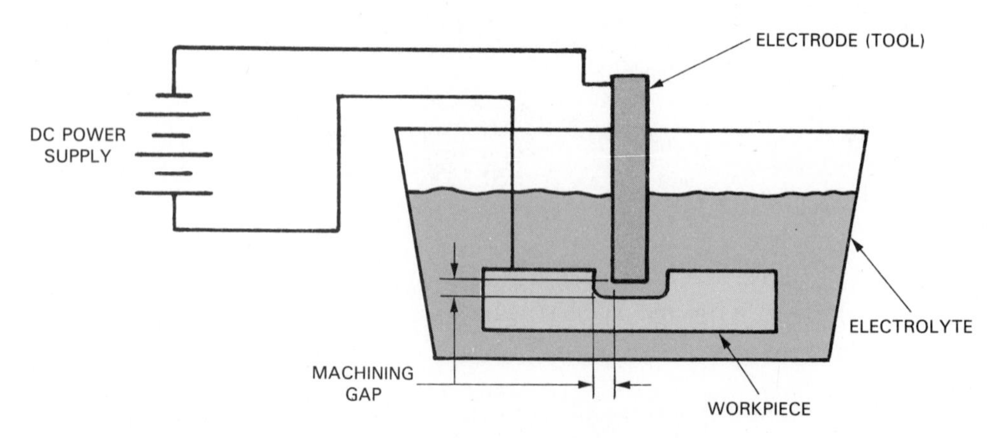

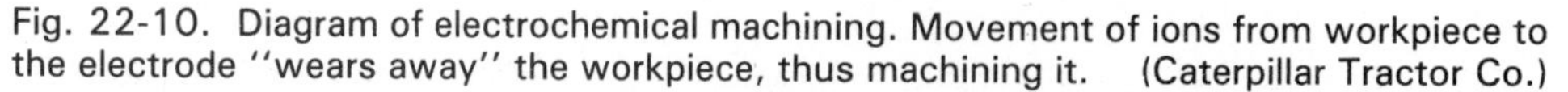

Fig. 22-10. Diagram of electrochemical machining. Movement of ions from workpiece to the electrode "wears away" the workpiece, thus machining it. (Caterpillar Tractor Co.)

Both processes may follow some basic steps.

1. An accurate drawing of the workpiece is made. This "artwork" is draw up to 100 times larger than the final size.
2. The artwork is then photographically reduced. The resulting negative will be the exact size of the part. Reducing the artwork sharpens the lines and reduces any errors.
3. The metal is completely cleaned. It is then coated with a photoresist.
4. The resist is allowed to dry and is baked.
5. The negative from Step 2 is placed over the coated workpiece.
6. A powerful light exposes the resist through the negative.
7. The resist is developed. This leaves a chemically resistant image on the metal. The image is in the shape of the part. Any area to be machined will not have a resist coating.
8. The metal is coated with or placed in an etching solution.
9. The solution is allowed to etch away the unwanted material.

CHEMICAL BLANKING

Chemical blanking uses the same procedure to cut metal parts from thin sheets. The metal is left in contact with the etching solution until it "eats" through the sheet. The part is then removed from the scrap.

Chemical blanking is used for short-run operations. It is more economical for these applications because it avoids the high cost of blanking and shearing dies. Also, chemical blanking produces highly accurate parts not possible with hand-cutting operations.

CHEMICAL MILLING

Chemical milling involves removing large amounts of material from part or all of a workpiece. Generally it is used on complex shapes. The process includes cleaning the part, masking off the area not to be machined, etching the unmasked areas, and removing the mask.

The mask may be a photosensitive material like that used in chemical blanking. This type of mask is widely used. However, masks of vinyl or rubber are also used. Using these masks involves coating the part with the masking material and then cutting openings to expose the areas to be machined.

The depth of the chemical milling "cut" is regulated by the length of time the reagent is left on the part. When proper milling action is completed, the reagent is washed off and the mask removed.

Chemical milling is widely used in the aircraft industry where cavities are produced in parts to reduce weight. Also, the parts have very complex shapes making traditional milling difficult.

LASER BEAM MACHINING

Laser beam machining (LBM) removes material through intense, concentrated heat. The process uses a highly focused beam of monochromatic (one color) light to produce the heat. In fact, laser is an acronym for light amplification by stimulated emission of radiation.

Laser machining uses this very strong light to melt and vaporize the material being cut away. Lasers can machine all known materials and can work through transparent materials. Fig. 22-12 shows the operation

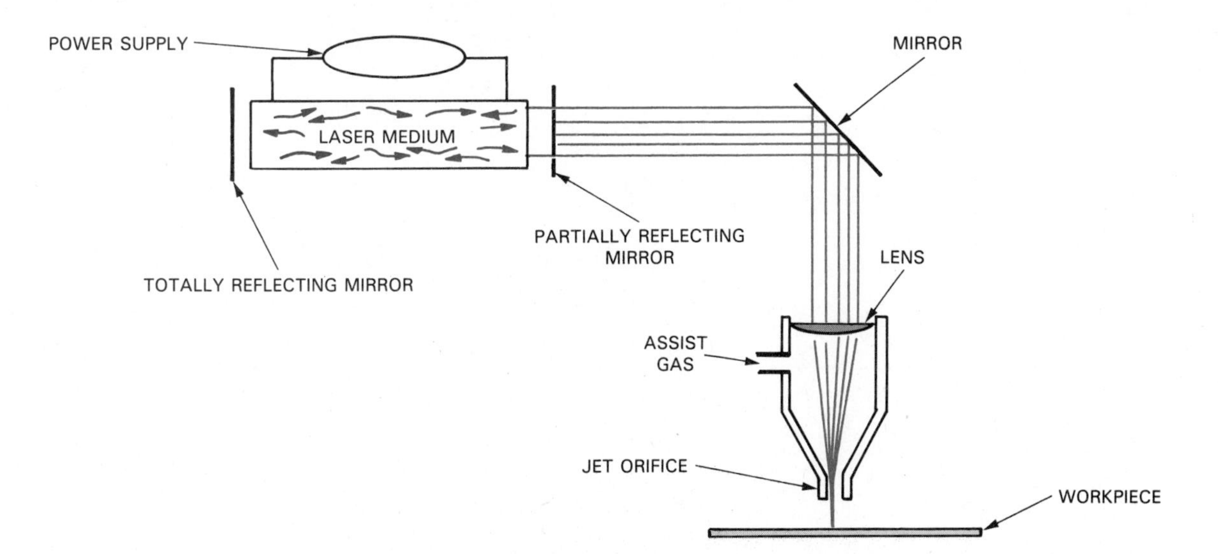

Fig. 22-12. Diagram of a laser machining operation. Concentrated beam of light vaporizes metal on workpiece.

of a typical laser.

The effectiveness of laser machining is determined by three material properties:

1. Thermal conductivity—the properties which relate to the flow of heat in a material.
2. Optical reflectivity—the properties which relate to the absorption or reflection of light.
3. Specific heat—the properties which relate to the amount of heat required to change a material from one state to another (i.e. solid to liquid).

Laser machining works best on materials which do not conduct heat away too rapidly, which absorb light energy, and which do not have high specific heat levels, Fig. 22-13. Some of these features are improved by coating the material with graphite to increase light absorption. The laser may be gas assisted. Oxygen may be sprayed on the cutting area to increase the burning action.

Use of laser machining is limited because of high equipment costs and low operating efficiency. Its use for cutting, drilling, slotting, welding, scribing, and heat treating has grown recently. The process is well suited to microdrilling—drilling very small diameter holes. Also, lasers can be used to mark or scribe ceramic materials without causing heat shock or induced fracture.

Lasers are finding increasing use in welding. These uses will be discussed in Chapter 27.

ELECTRON BEAM MACHINING

Electron beam machining (EBM) is another thermal machining process. The process has grown out of electron beam welding. However, the applications are much wider. Electron beams can make holes in metals and plastics, harden metals, cure paints, and a variety of other processes.

Electron beam machining uses high-speed beams of electrons to perform the cutting. When the electrons strike the workpiece, the energy is changed into heat (thermal energy). The electron beam, like the light beam of the laser, melts and vaporizes the material. Lower-power beams are used to produce heat for hardening and curing operations.

Electron beam machining, Fig. 22-14, is usually carried out in a vacuum. The electron beam is generated by a triode (electronic component). The beam is electronically changed to a pulsating (on/off) beam. This pulsating beam of electrons is necessary for drilling. Computer controls control workpiece movement, the focusing of the electron beam, and beam direction.

Electron beam machining is very accurate. It can produce very small holes and shapes (micromachining). The process does require expensive machines and skilled operators which presents economic disadvantages. Also the need for working in a vacuum restricts the size of the workpiece.

Fig. 22-13. General purpose laser cutting system in operation. Note the cut out blank at left. (Coherent)

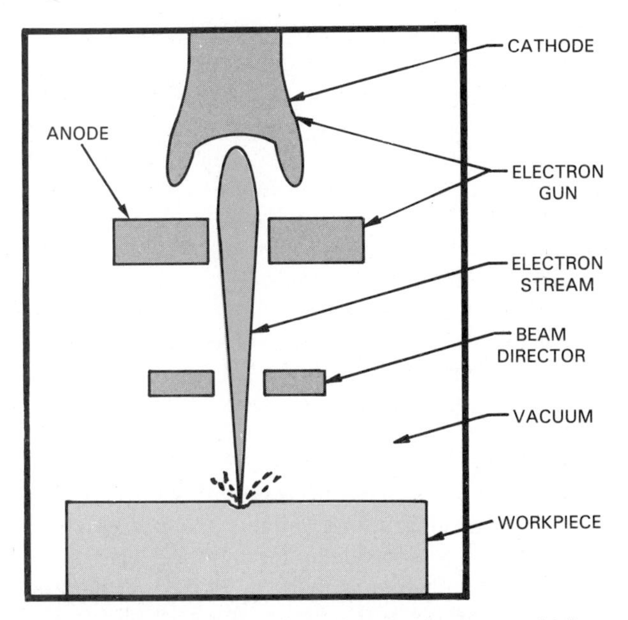

Fig. 22-14. Diagram of an electron beam machining operation. This process is similar to laser machining.

FLAME CUTTING

Flame cutting is not usually called a nontraditional machining technique. It does, however, shape parts without mechanical force. Burning gases cut metals and, thus, shape them.

Flame cutting uses one or more cutting torches to generate the heat needed. Most torches use a mixture of acetylene and oxygen. A preheating flame first heats the steel. Then a blast of oxygen is added. The flame cutting torch is shown in Fig. 22-15.

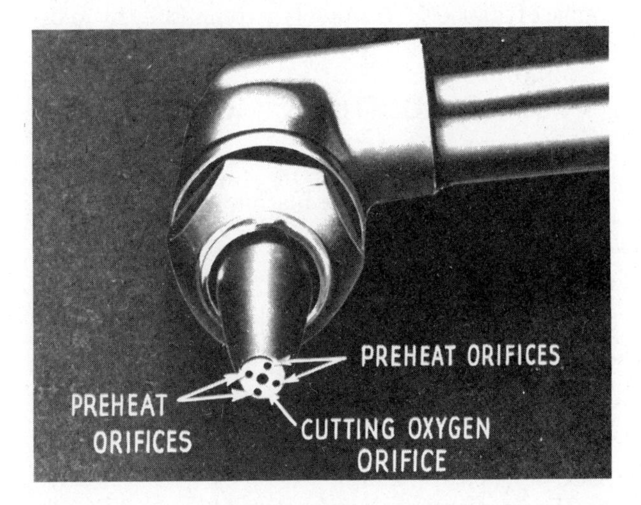

Fig. 22-15. Flame cutting torch tip. Note center orifice. It is for the oxygen which is introduced to the superheated workpiece.

Iron and steel have an attraction to oxygen. The metal naturally forms iron oxide (rust). At normal temperatures this action is relatively slow. At high temperatures the oxidizing action is rapid. When steel is red hot it burns to iron oxide almost immediately.

Flame cutting is a very common way to cut steel. Rough cuts may be done with a hand-held torch, Fig. 22-16. More accurate cuts are done with a torch cutting machine. These machines may have several torches so that several parts may be cut at one operation.

The movement of the cutting torches may be controlled in several ways:

1. Tracer units which follow a template.
2. Electric eyes may follow lines on a drawing.
3. Numerical (tape) control units.

Each of these units activates electronic or electrical controls which move the cutting heads over the work. The result is quick, accurate cutting of steel sheet, plate, and structural shapes.

PLASMA ARC CUTTING

Plasma arc cutting uses matter in its "plasma" state. Plasma is a gas which has been broken down into free electrons, neutrons, and ions. In physics, plasma means "a stream of ionized particles."

The plasma torch uses a tungsten electrode and a water-cooled copper nozzle. A gas, as shown in Fig. 22-17, is forced to flow around the electrode and through the nozzle. While the gas is flowing, electric

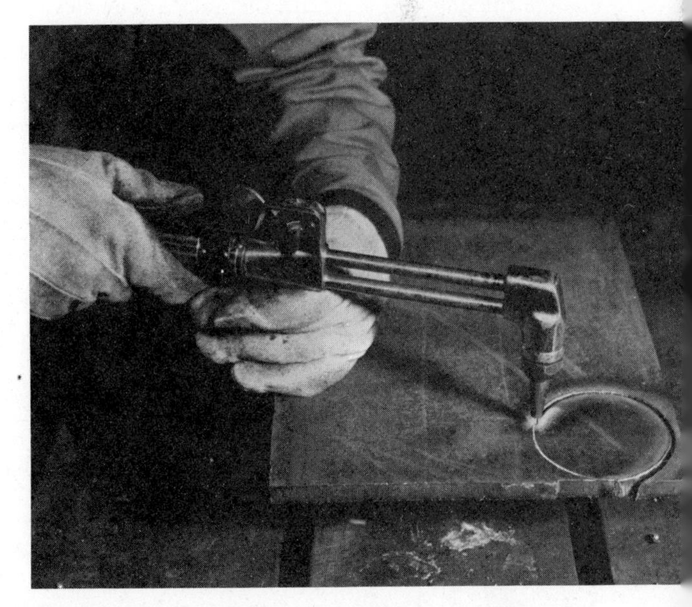

Fig. 22-16. A hand-operated flame cutting operation. This method is satisfactory where close tolerances need not be maintained.

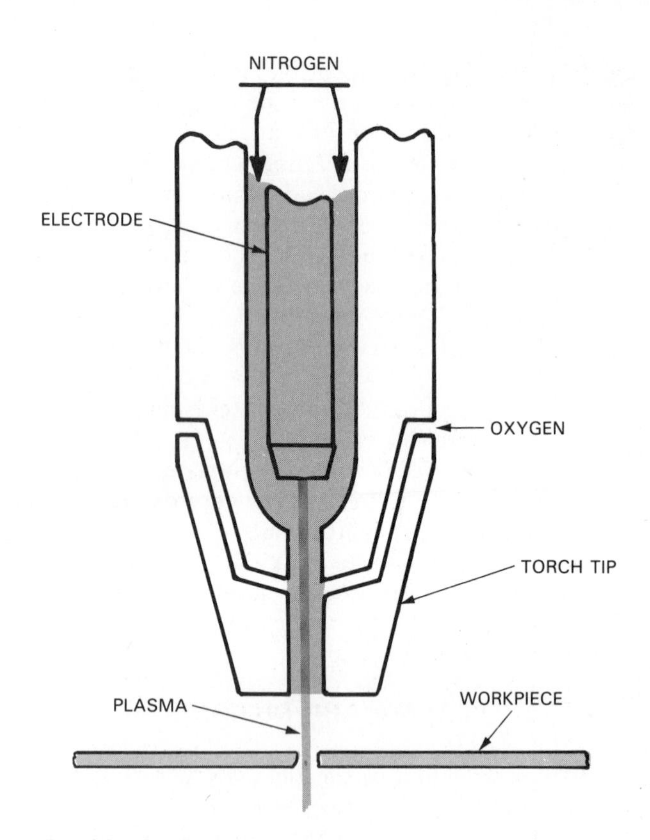

Fig. 22-17. Diagram of a plasma arc cutting torch. Note method for combining nitrogen and oxygen at torch tip.

sparks are caused to jump from the electrode in a continuous stream to the copper nozzle. These sparks change the gas into a stream of ionized particles.

The production of the plasma takes high levels of electrical energy. This energy is released as the atoms recombine while the plasma strikes the workpiece. The released energy can produce temperatures in the 50,000°F (27,500°C) range.

The gas used will vary with the material being cut. Typically, nitrogen, hydrogen, argon, or mixtures of nitrogen and hydrogen and of argon and hydrogen are used. Oxygen can also be used but it causes electrode corrosion.

Plasma arc cutting cuts any metallic material rapidly, leaving a narrow kerf. However, plasma arc cutting equipment is expensive. Also, since the cut is produced by melting the material, there is always the danger of damaging the surface. This damage is principally caused by either oxidation or overheating.

SUMMARY

A number of machining processes do not use mechanical force applied to a cutting tool. These processes are often called nontraditional machining. Material is removed by chemical action or by heat.

There are several common machining processes in this grouping. Electrical discharge machining uses a rapidly repeating spark which leaps the gap between a shaped electrode and the workpiece to shape the part. Electrical discharge grinding is a variation of EDM which uses a rotating abrasive wheel and a dielectric to float away the chips. Electrical discharge sawing combines the motion of a band saw with the electric spark of EDM. In electrical discharge wire cutting, a taut, traveling, conductive wire gives off sparks to cut material.

Basically the process of electrochemical machining (ECM) involves bringing a hollow electrode (tool) close to the work while the latter is immersed in a rapidly flowing electrolyte. A modification of ECM, electrochemical grinding, adapts the process to a rotating diamond studded grinding wheel (electrode).

Chemical machining involves removing material with a chemical. The two processes are called chemical blanking and chemical milling (often called contour forming).

Laser beam machining (LBM) through concentration of a beam of light creates intense heat. The heat is so strong that it vaporizes the metal being cut in a very narrow kerf.

Electron beam machining (EBM) uses high-speed beams of electrons to do cutting. As the electrons strike the workpiece, their energy is turned to heat energy which is so intense that the metal is vaporized.

Flame cutting is an old process borrowed from oxyacetylene welding. Heat and oxygen perform the cutting action.

Plasma arc cutting uses high levels of electrical energy to cut metal. Streams of ionized particles bombard the workpiece and create temperatures in the range of 50,000°F. Cutting is accomplished by ionizing the metal in the kerf.

STUDY QUESTIONS – Chapter 22

1. Describe:
   a. Electrical discharge machining.
   b. Electrical discharge grinding.
   c. Electrical discharge sawing.
   d. Electrical discharge wire cutting.
2. Describe electrochemical machining.
3. Describe electrochemical grinding.
4. Describe:
   a. Chemical machining.
   b. Chemical blanking.
   c. Chemical milling.
5. Diagram and explain laser machining.
6. Describe the electron beam machining process.
7. Describe flame cutting and plasma arc cutting.

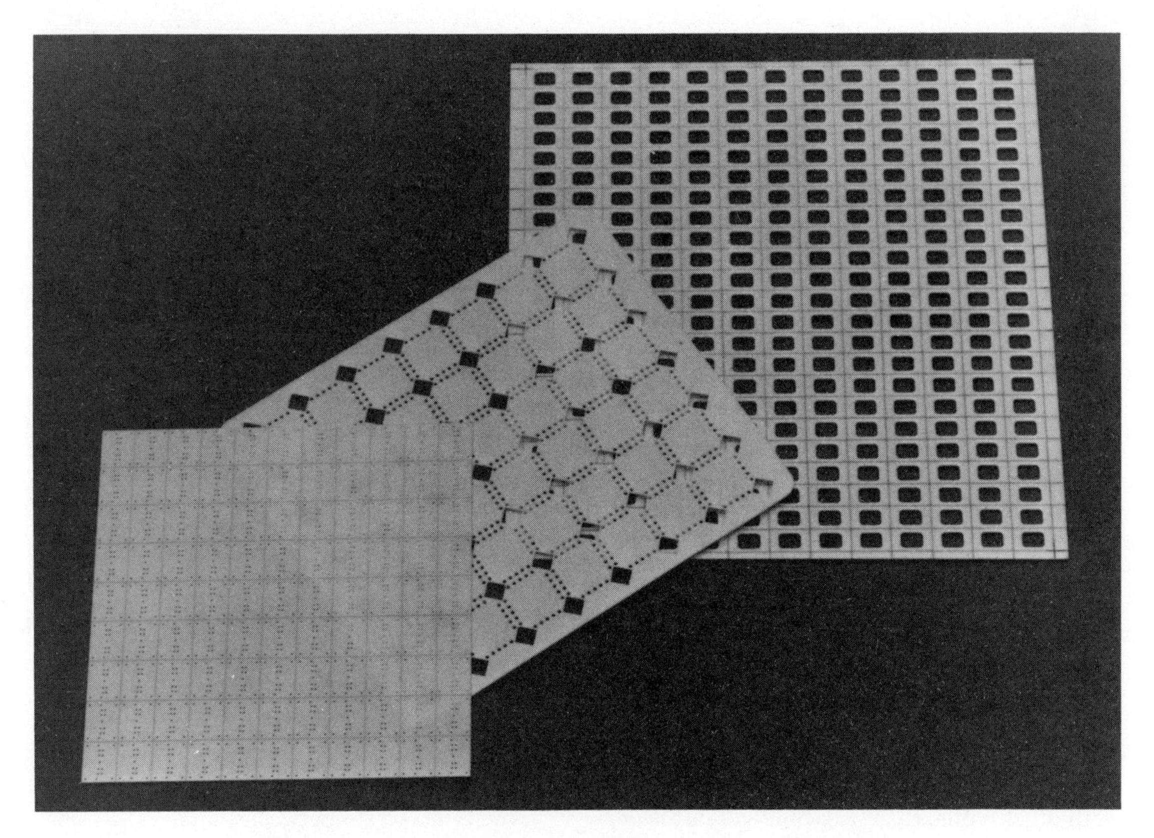

These ceramic parts were shaped by lasers. (Coherent)

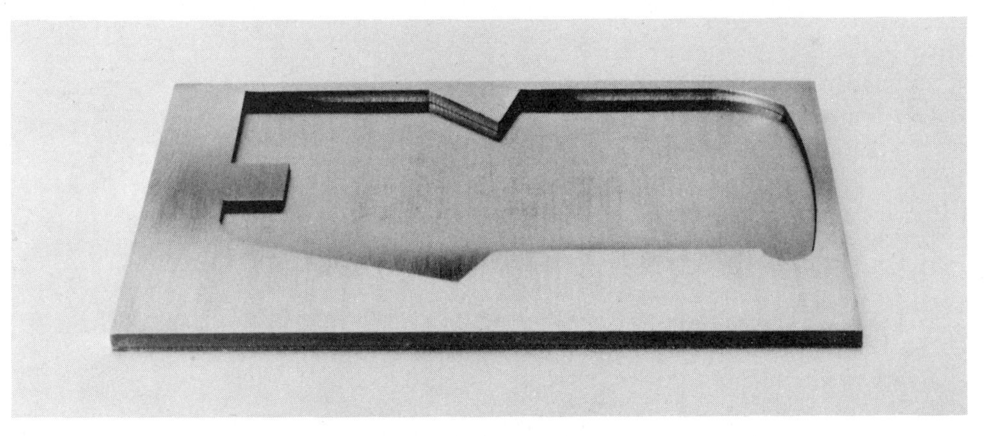

A laser beam shaped this metal part.

Chapter 23

SHEARING

Most sheet material is cut to shape by breaking the unwanted material (scrap) away from the desired shape (part). The processes used in this controlled breaking can be grouped under the general term, shearing.

Shearing can be used on a variety of material. Most commonly it is used to cut sheet metal, wood veneer, paper, cloth, leather, and plastic sheet stock.

SHEARING OPERATIONS

Shearing may also be used in a more specific sense, along with other terms, to describe individual shearing operations. These operations include:

1. **Shearing and slitting.** These involve cutting across or along sheet or coil material to produce smaller standard shaped sheets, Fig. 23-1. Shearing is used to cut across the sheet. The product of a shearing operation is square, rectangular, or angular stock which is ready for use or further processing.

   Slitting cuts along the length of the material. It is commonly used to cut several narrower strips from a wide coil of material. Slitting may produce strips narrower than an inch wide.
2. **Blanking, cutoff, and parting.** Shaped parts are cut in a single stroke with a punch and die, Fig. 23-2. Blanking cuts all sides of the part with one cut. A blanking operation removes the part, leaving scrap completely around the resulting hole in the sheet material.

   Cutoff produces the part by cutting completely across a sheet. Cutoff uses a single cut and produces no waste between parts.

   Parting uses two cuts across the sheet to produce the part. There is also waste as each part is produced. Often blanking, cutoff, and parting are

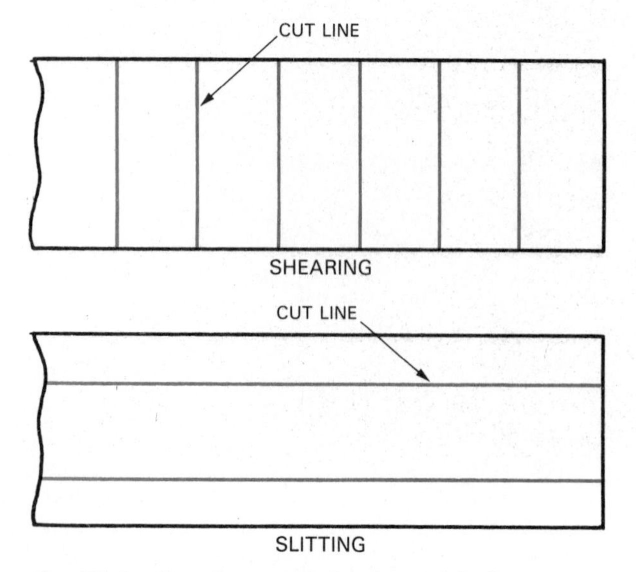

Fig. 23-1. Shearing and slitting are straight-line cuts on sheet material.

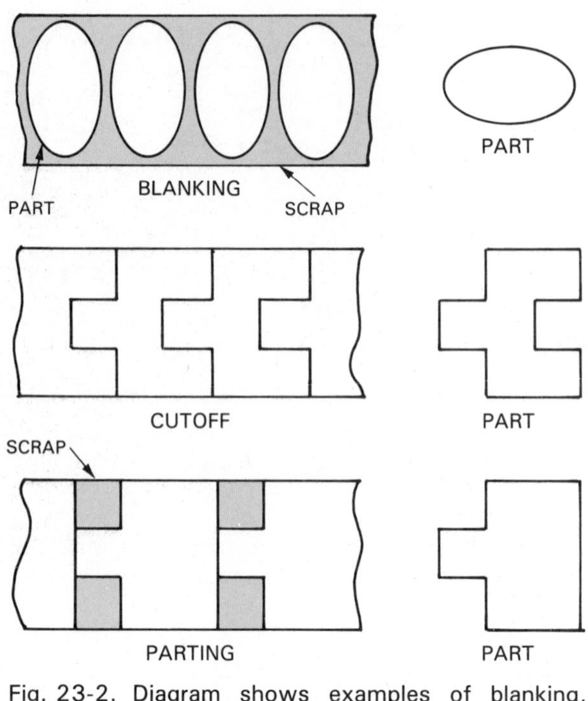

Fig. 23-2. Diagram shows examples of blanking, cutoff, and parting operations.

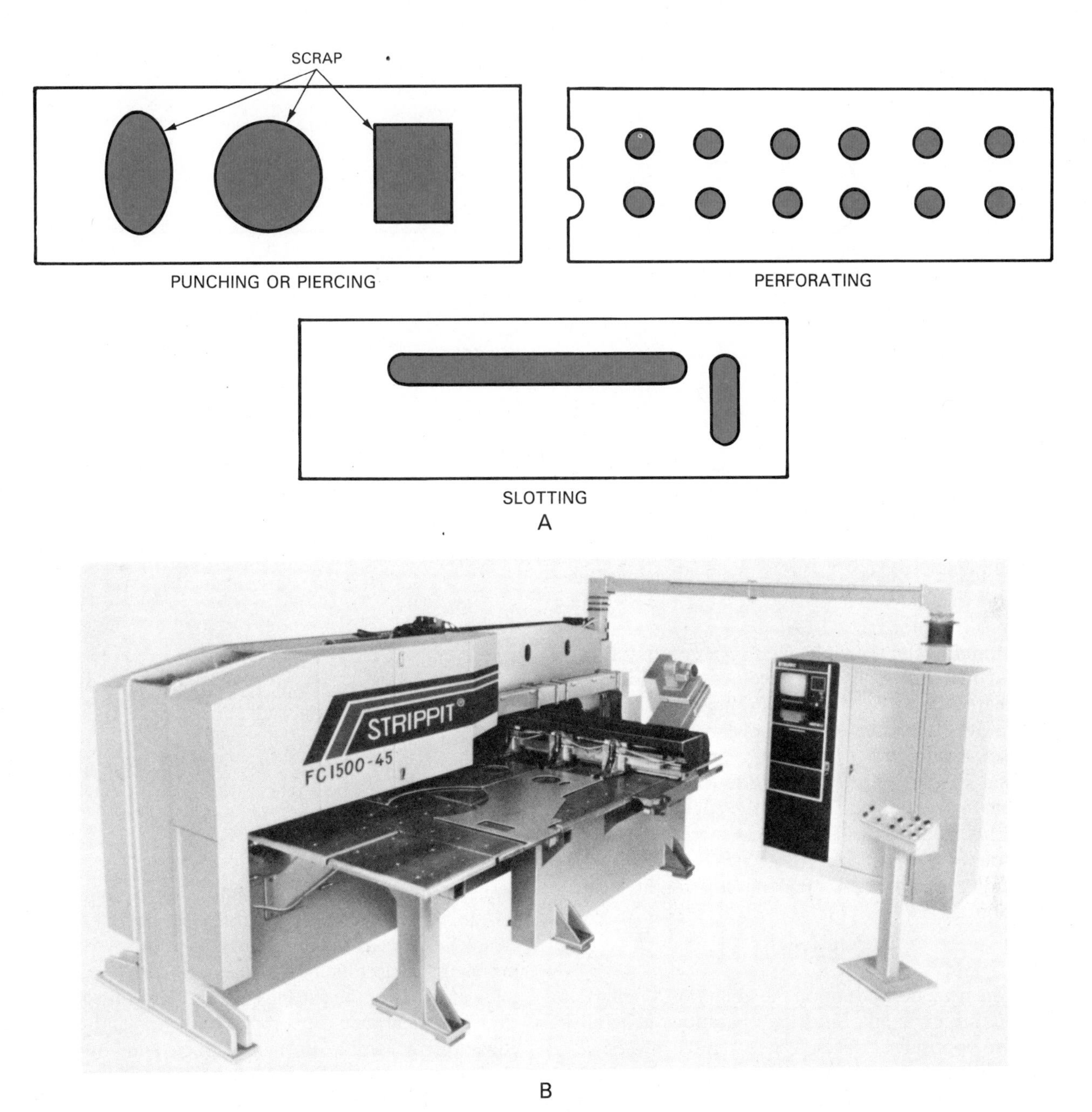

Fig. 23-3. A—Punching and piercing, perforating, and slotting produce holes in sheet stock. B—Automated holemaking machine is capable of making a series of shearing operations without an operator. (Strippit, Inc.)

but the first step in producing a part. Several other shearing and forming operations may follow to create the final shape.

3. **Punching or piercing, perforating, and slotting.** These three operations produce holes in sheet rock. While similar in that they produce holes, they are each unique, Fig. 23-3. Punching or piercing produces a round or shaped hole in sheet rock.

   Perforating also produces round or shaped holes. This process, however, produces many identical holes in a pattern. The holes are generally evenly spaced along and across the sheet.

   Slotting produces an elongated hole. The slot is almost always the same width along its length.
4. **Notching.** This is an operation that removes material along the edge of a sheet, Fig. 23-4. Notches may be any shape or size.

   Lancing. A cut part way across a sheet or a

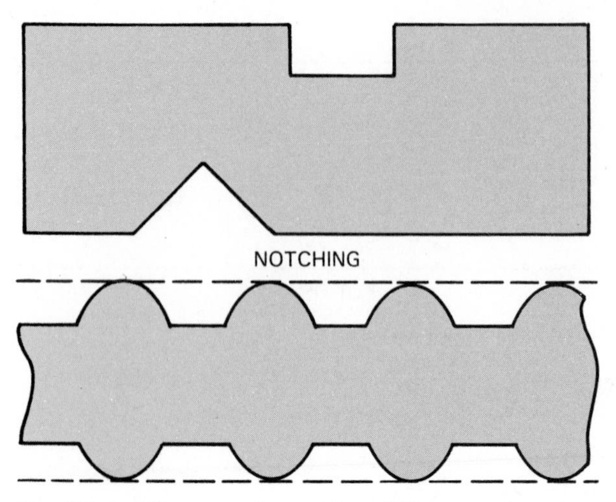

Fig. 23-4. Diagram of notching. This operation cuts away parts of the edges of material.

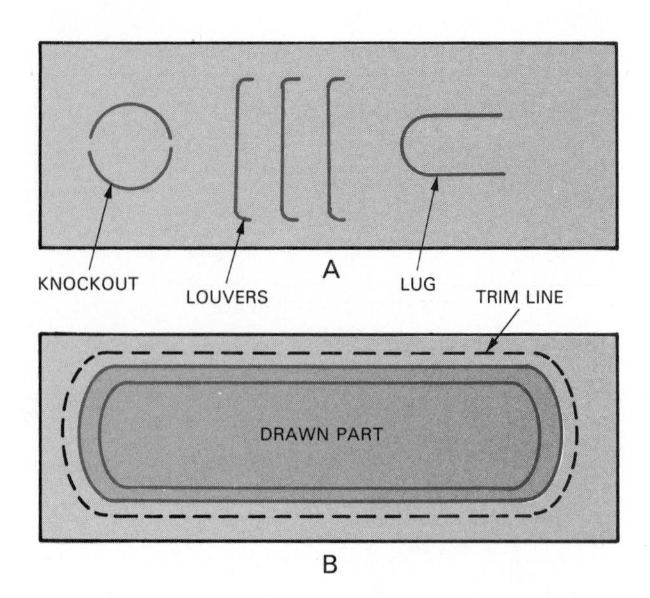

Fig. 23-5. Lancing and trimming produce cuts partially across or holes partially through the workpiece. A—Notching. B—Trimming.

partially cut hole in the sheet is called lancing. It is often the first step in producing tabs, louvers, and air deflectors, Fig. 23-5A.

5. **Trimming and Shaving.** These operations cut excess material from a formed, cast, molded, or sheared part, Fig. 23-5B. Trimming is used to remove flash from a casting or molded part. It is also used to trim a flange on a drawn part to its final size or a drawn cup to its finished height. Shaving cuts a thin strip off the edges of sheared or blanked parts. This process increases the accuracy of the part size. Also, shaving improves the quality of the edge, making it straighter and smoother.

SHEARING

Shearing depends upon the ability of a material to fracture or break. Shearing operations generally use two opposing surfaces. The material is placed between these surfaces. One of the surfaces is movable and applies the force necessary to shear the material.

The three common shearing devices are opposed blades, rotary cutters, and punches with mated dies. Punch and dies and blades shear the material in four basic steps, as shown in Fig. 23-6. These steps are:

1. The movable blade or die is brought into contact with the workpiece.
2. Pressure is applied. The blade or die penetrates the workpiece.
3. As penetration continues, fractures appear in two areas:
   a. On top of the workpiece around the punch or along the blade.
   b. On the bottom around the die opening or along the lower blade.
4. As additional pressure is exerted, the two fracture areas connect and the part is severed.

Rotary cutters produce all four steps as the stock is pulled under the two slitting rolls.

Shearing may be improved by shaping the punch or upper blade. The shape is designed to cause the die or blade to progressively engage the part. These shapes, as seen in Fig. 23-7, reduce the maximum force needed to shear the stock. However, this technique, called *angular shear,* causes distortion.

Another important factor in efficient shearing is the clearance between the punch and die or the two shearing blades. Too much clearance will cause the stock to be drawn over the cutting edge and bent (distorted). Too little clearance causes high tool wear and rough edges on the work. Both too much and too little clearance increase the load required to shear the material. Fig. 23-8 shows the effects of clearance on shearing.

SHEARING MACHINES AND TOOLING

As just described, shearing is primarily done using one of three major cutting tools.

1. Blades.
2. Punches and dies.
3. Rotary cutters.

BLADES

The most common shearing machine using blades is the *squaring shear.* This machine, as shown in Fig.

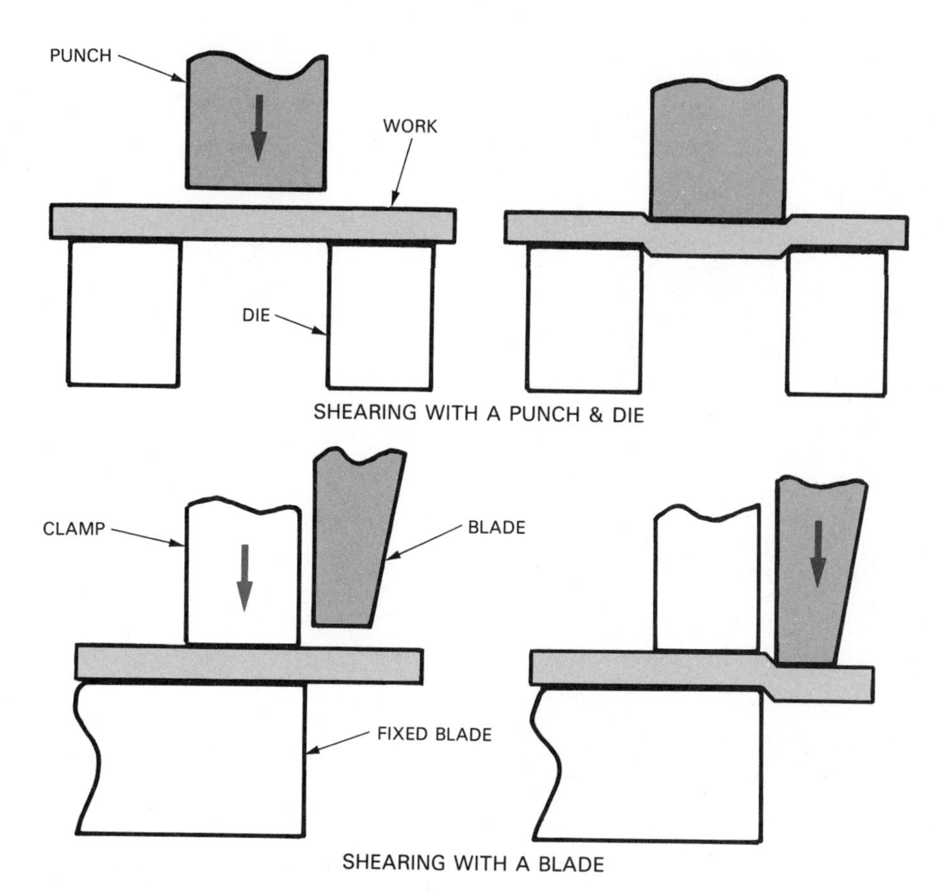

Fig. 23-6. Shearing is done with a punch and die or a blade.

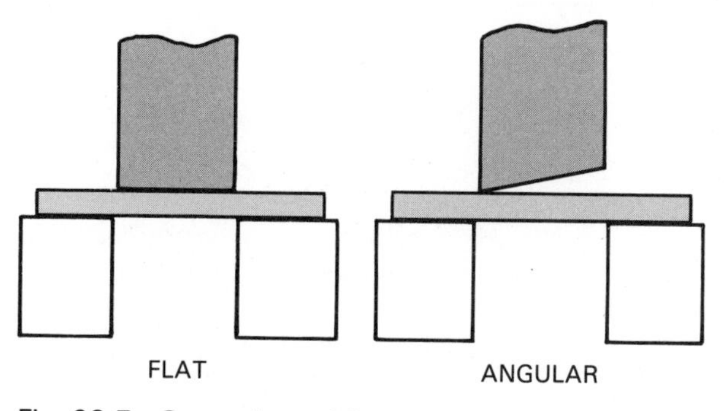

Fig. 23-7. Comparison of flat and angular shearing. Flat shearing has shear engaging all of the part at once. In angular shearing, contact of the shear with the part is progressive.

23-9, is used to cut straight-sided shapes from sheet or plate metal. Similar machines are used to cut veneers and paper.

Squaring shears have several basic parts: a shear table, upper and lower knives, a stock hold-down, a back gage, and an upper blade power unit.

The typical cycle for a cut includes:

1. The back gage is adjusted to set the size of the

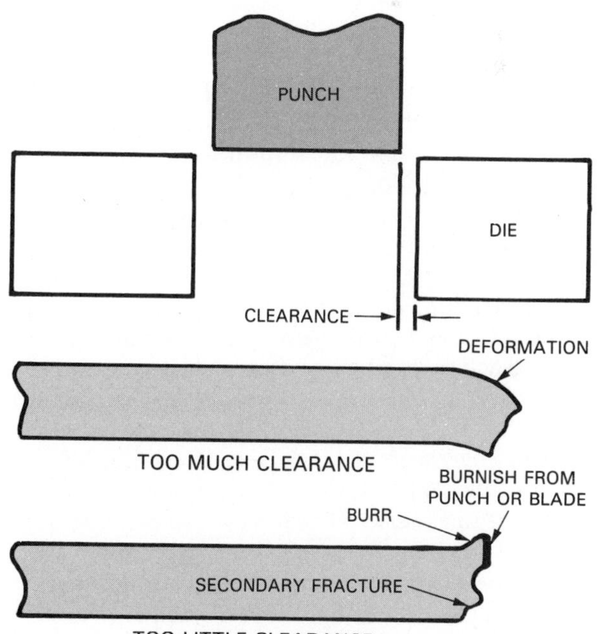

Fig. 23-8. Shear clearance is important. Too little causes burring and secondary fractures. Too much deforms edge of the workpiece.

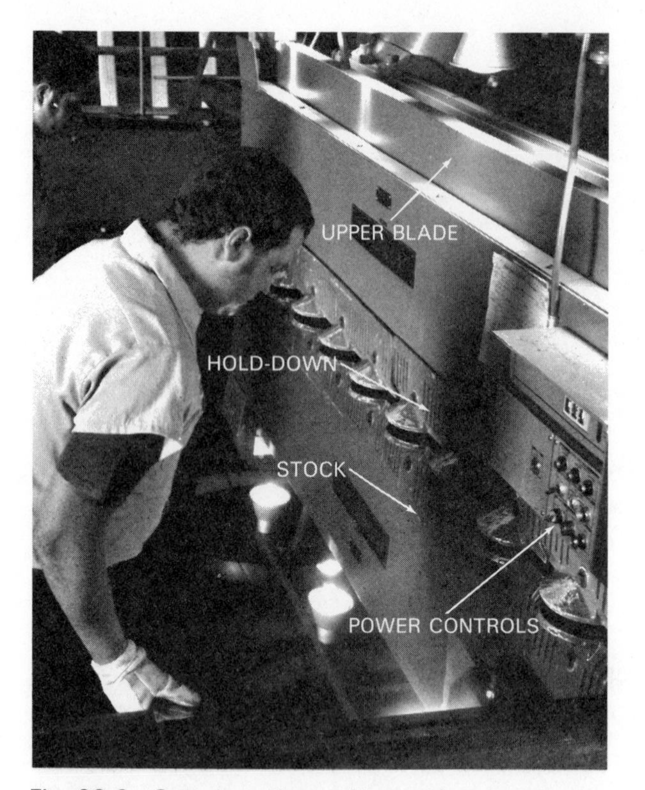

Fig. 23-9. Squaring shear. Note that machine is reflected in polished surface of workpiece. (Cincinnati Inc.)

finished workpiece. Many modern shears allow the operator to set the back gage from the front of the machine.

2. The stock, positioned on the shear table, moves into the machine until it is against the back gage.
3. The hold-down moves down to firmly hold the stock in position.
4. The upper blade moves down to shear the material. The blade is positioned in the machine so that one end penetrates the stock first. It then continues its penetration, shearing across the sheet as it moves further downward. This action is much like the cutting action of a pair of scissors. The angle of the blade is called its rake angle. This angle is increased as the thickness capacity of the machine increases.

Squaring shears are available in cutting lengths from 2 ft. to 20 ft. or more. They are built to cut many different thicknesses of material. The heaviest duty machines can cut up to 1 1/2 in. mild steel plate.

Paper and veneer cutters are similar to a squaring shear. Their main differences are in their lack of a lower blade and in the blade motion. The blade moves downward and across the sheets finishing its cut against a nylon or wood insert in the shear table. The cutting action of these shears is much like cutting bread or meat on a cutting block. The knife is forced down and slides across to cut the fibers of the material.

Rotary cutters

Rotary cutters are of two major types: rotary shears and slitting rolls. *Rotary shears,* Fig. 23-10, are used primarily to produce round and irregular shapes. Two revolving circular cutters are used.

The circular shape of the cutters allows the stock to be easily moved right or left. This freedom of movement permits cutting irregular shapes with small radii. However, the need to hand guide the material limits the use of rotary shears to low-volume production applications.

A special rotary shear, Fig. 23-11, called a *ring and circle shear,* has a pivot pad which holds the stock. The rotary cutters can be adjusted to any distance from the pad. As the rotary cutters engage the stock, the stock is rotated around the pad pivot point and cut. The output of this machine is a round disc.

Slitting rolls cut with the same action as rotary shears. However, slitting rolls are designed to make one or more straight cuts on a sheet of coiled material.

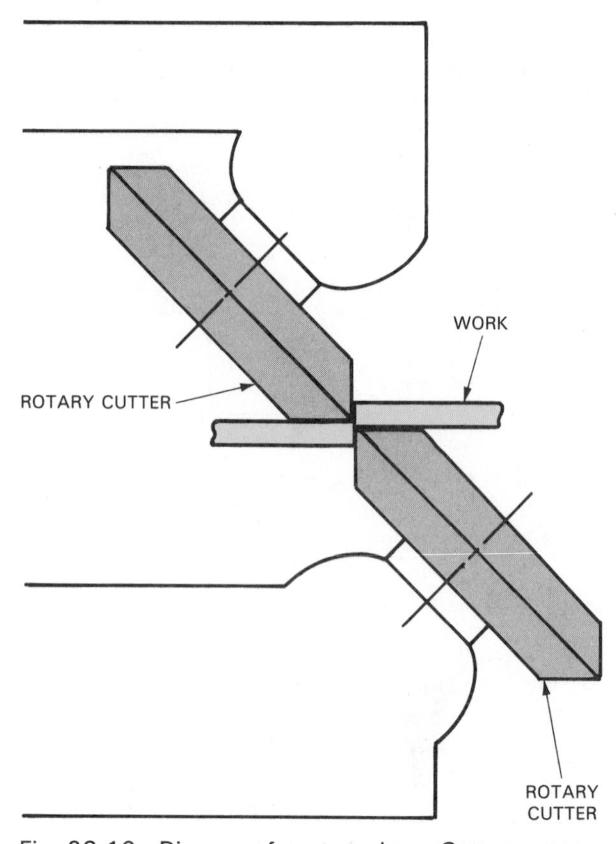

Fig. 23-10. Diagram of a rotary shear. Cutters are opposed to each other.

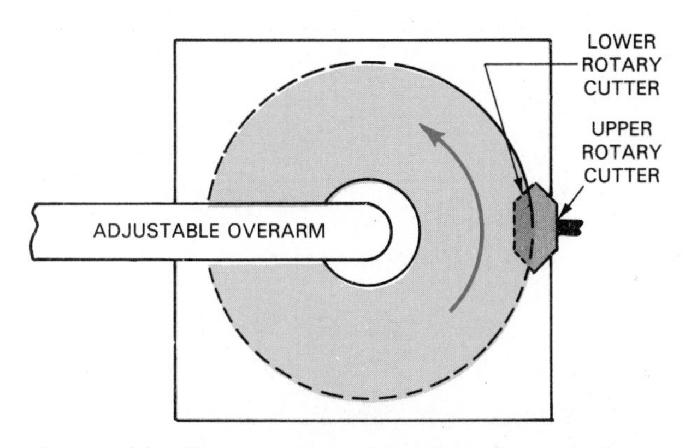

Fig. 23-11. Ring and circle shear. Pair of cutter wheels do cutting as workpiece is rotated.

A typical slitting roll machine is called a *shear slitting machine,* Fig. 23-12. The mated rolls can be used to cut metals, plastic materials, paper, fabric, and other thin sheet materials. The cutting action is like a scissors.

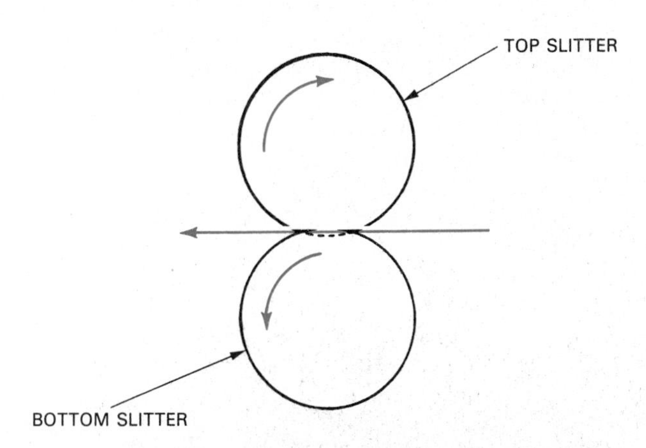

Fig. 23-12. Diagram of shear slitting. Rotating slitters work like the blades of a scissors. (Dow Chemical Co.)

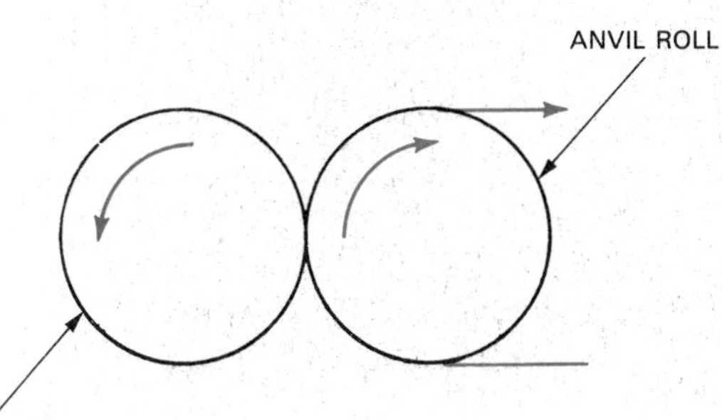

Fig. 23-13. Diagram of score slitting. Slitter wheel has a bead which scores and slits as it rotates and as material moves past it. Anvil wheel is smooth. (Dow Chemical Co.)

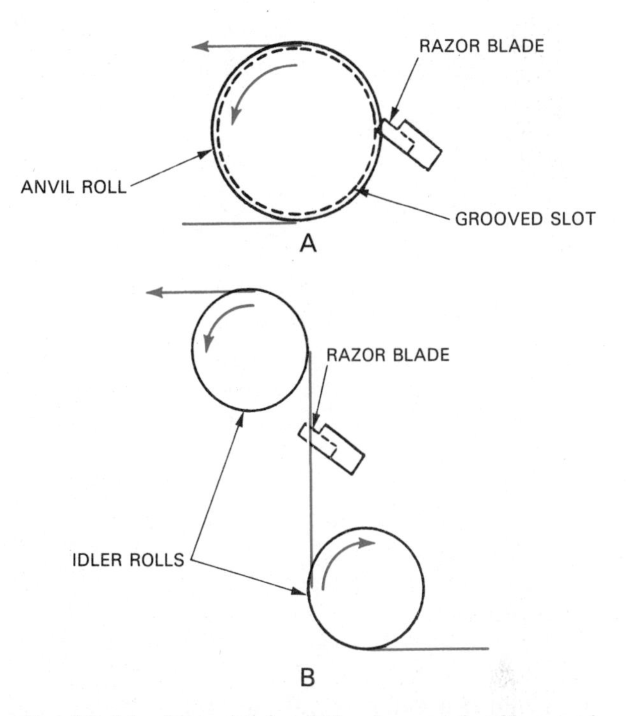

Fig. 23-14. Razor blade slitting is used for films and plastics because it produces clean, cut edge. Two systems are shown. A—One-roller system. Roller is anvil for slitting operation. B—Sheet material is carried over two rollers and slitting takes place somewhere in between.

Score slitting is used to cut pressure-sensitive tapes, waxed paper, and gummed tape. The rolls are smooth with a bead on one of them, Fig. 23-13. The bead scores and slits the material. Score slitting rolls are inexpensive and simple. This allows for rapid changing of strip widths. They require considerable maintenance to maintain a quality cut, however. For this reason, score cutting is not widely used.

A third roll slitting technique is *razor blade slitting.* The material passes over a roll while a blade which fits into a groove in the roll slits the material. A similar process uses a blade which is held between two rolls. As the material passes over the rolls, it is cut. Fig. 23-14 shows both types of razor blade slitting. These processes are widely used to cut plastic films such as vinyl, saran, polyester, and thin acetate.

Another roll cutting technique is called *sheeting.* It is used to cut thin plastic films and metal foils. The process, as shown in Fig. 23-15, uses a knife mounted across a roll. Each revolution of the rolls causes a standard length of the material to be cut off. The sheets are then moved away from the machine by conveyors.

Punches and dies

Punches and dies are used for a number of shearing operations, Fig. 23-16A. A typical punch and die

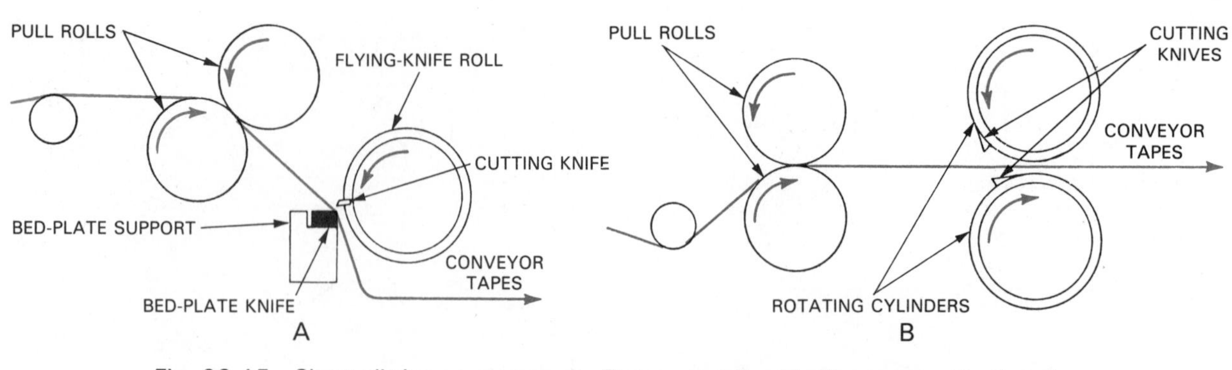

Fig. 23-15. Sheet slitting processes. A—Rotary and fixed-knife process. B—Rotating synchronized-knife process. (Dow Chemical Co.)

set, Fig. 23-16B, includes:

1. A punch holder which attaches to the press ram.
2. A die shoe which attaches to the bolster plate of the press.
3. Guide pins which keep the punch holder and die shoe in alignment.
4. A punch in the shape of the part or hole.
5. A die which mates with the punch.

Included in many punch and die sets is a stock hold-down. Also, a set of ejector pins are commonly used to push the part or scrap out of the die.

Punches and dies may be used for conventional cutting and for shaving. As shown in Fig. 23-17, conventional cutting shears material away from the sheet to produce a hole, slit, or other cut. Shaving trues up a previously produced cut. Typically, shaving removes material equal to no more than 10 percent of the stock thickness.

Sharp-edged punches may also be used against a solid block of wood or other soft material. The operation, called *dinking* or *die cutting,* is used to cut paper, leather, rubber, plastics, and other similar materials. Fig. 23-18 shows a typical dinking operation.

To increase production, several die sets may be mounted together in a single press. The combination is called a *progressive die set.* It is designed to complete one operation at each die station as the workpiece progresses through the dies.

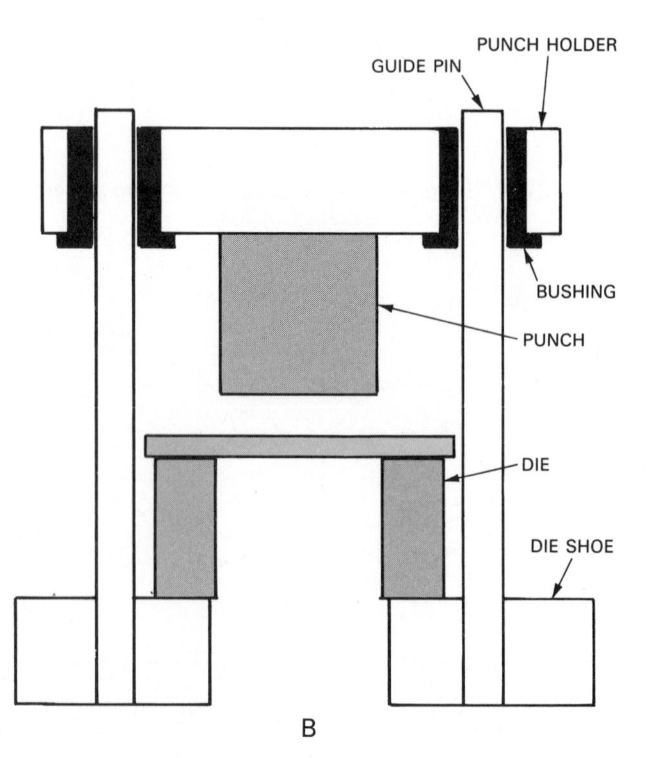

Fig. 23-16. A—Modern hole punching machine can handle thicknesses up to 20 ga. Turret handles up to 20 stations. B—Diagram of a single station of a punch and die set. Stock is inserted over die. (Strippit, Inc.)

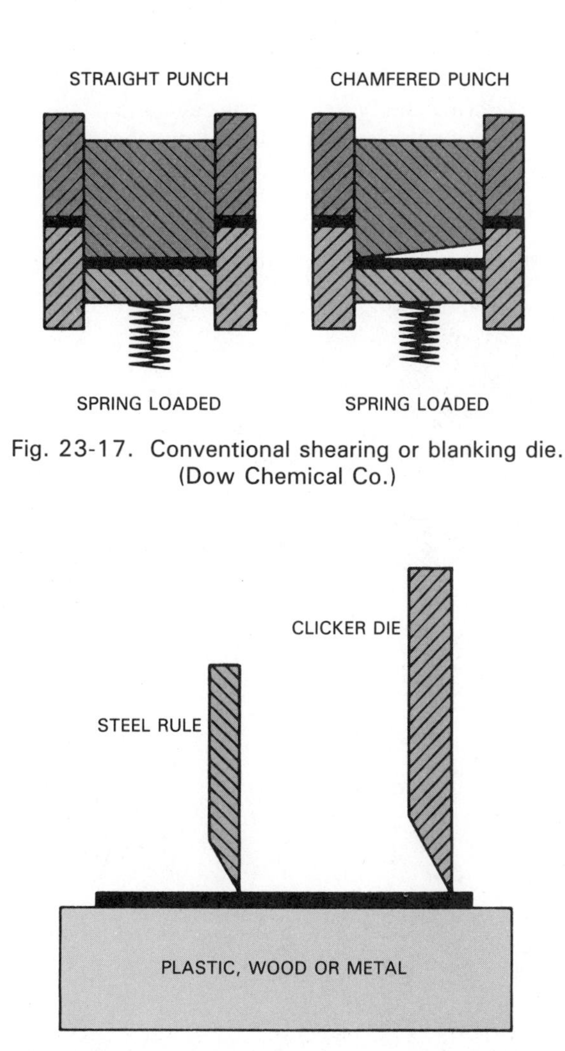

Fig. 23-17. Conventional shearing or blanking die. (Dow Chemical Co.)

Fig. 23-18. Dinking die cuts material against backing of plastic, wood, or metal. (Dow Chemical Co.)

The work is fed and located so that, as material moves through each station of the progressive die set, the part takes shape. A typical progressive die is shown in Fig. 23-19. Notice how the part is produced by a series of independent operations.

Progressive dies can include individual dies which pierce, notch, lance, blank, bend, draw or otherwise shear and form the part. Progressive dies are used to produce a wide variety of parts out of sheet stock.

SUMMARY

Shearing is a series of operations involving the controlled fracture of stock. It involves producing external shapes to internal openings in sheet stock.

Shearing uses blades, rotary cutters, and punch and die sets. The device selected depends upon the operation to be completed. Blades are used primarily for shearing stock along straight lines. Rotary cutters are

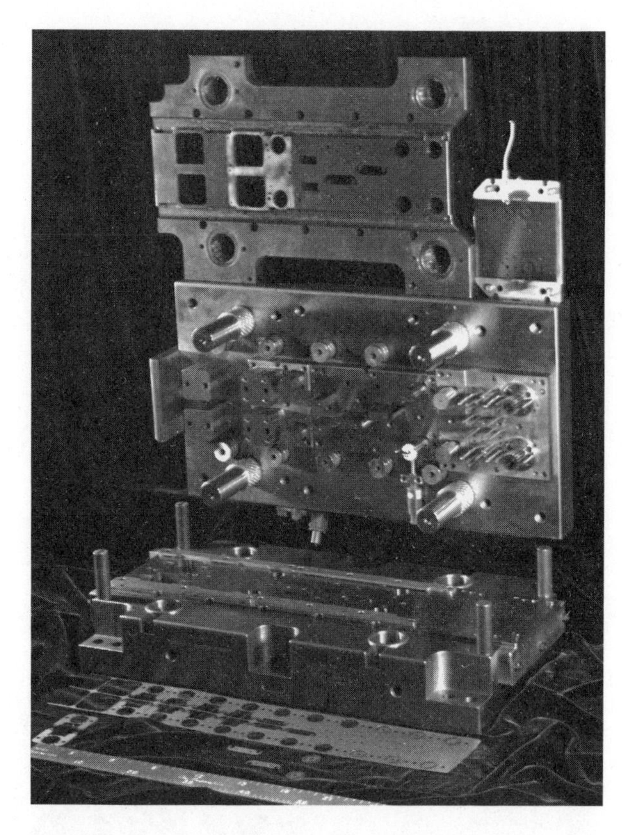

Fig. 23-19. A progressive die. Note the part-in-work shown below. (Oberg Mfg. Co.)

used for limited production of irregular shapes and for producing round discs. They also are widely used in slitting operations. Punches and dies are used for blanking, cutoff, parting, piercing, perforating, slotting, notching, lancing, and shaving.

Combined with other forming dies, punches and dies are used in progressive die sets. These sets produce complex parts through a series of operations.

Shearing is a most important manufacturing process. Large quantities of sheet material are sheared during the manufacture into usable products.

STUDY QUESTIONS – Chapter 23

1. Define shearing.
2. What is the difference between shearing and slitting?
3. Define blanking, cutoff, and parting.
4. Describe punching or piercing, perforating, and slotting.
5. What are notching and lancing?
6. What is the difference between trimming and shaving?
7. Describe the shearing act.
8. What are the three common shearing devices?

9. Describe the four basic steps in blade shearing.
10. What happens in shearing if:
    a. There is too much clearance between the opposing edges?
    b. There is too little clearance?
11. Describe the basic shearing machines.
12. Describe the cutting action of punches and dies.
13. What is dinking or die cutting?

Worker adjusts a shearing and forming die set. (American Metal Stamping Assn.)

View of factory floor. Unit in foreground is a squaring shears which is designed to cut sheet metal. Table supports workpiece while a knife in the heavy frame above the table shears away excess material.

SECTION 5
CONDITIONING

Conditioning processes are those operations which alter the internal structure of a material to improve mechanical and physical properties. These processes fall into three categories: thermal, mechanical, and chemical. They include the following operations:

- Annealing—for softening materials and relieving stress.
- Normalizing—which causes steel to return to its "normal" internal structure.
- Hardening—used to increase material's hardness, strength, and wear resistance.
- Tempering—for relieving internal stress caused by hardening.
- Drying—to remove excess moisture.
- Firing—to change the internal structure of ceramics.
- Work hardening or shot peening—to improve internal structure or relieve stress.
- Catalytic action or polymerization—chemical reactions that change internal structure of plastic materials.

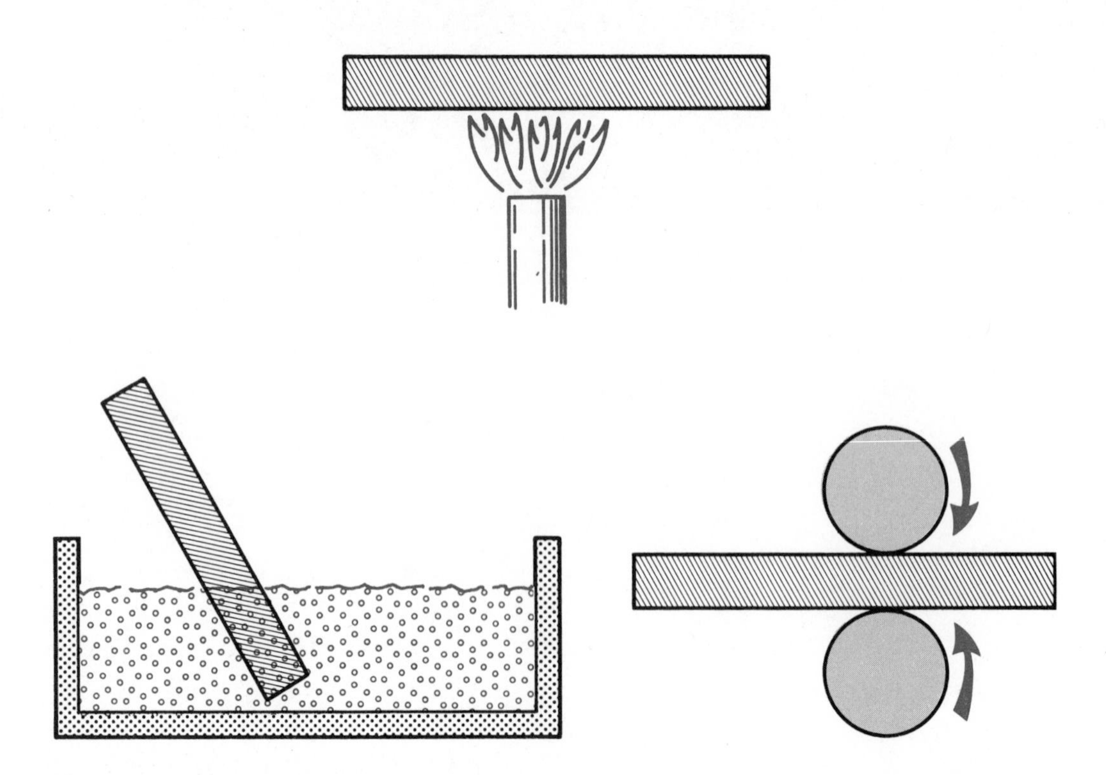

Chapter 24

INTRODUCTION TO CONDITIONING

Materials do not always have the strength, toughness, hardness, or other mechanical properties needed to serve their intended purposes. These properties can often be changed and improved. Operations which create these changes are called *conditioning processes.*

Specifically, conditioning processes are those processes which alter the internal structure of a material. They change the material's mechanical and physical properties. Typical properties changed in conditioning, Fig. 24-1, include:

1. **Strength** – the ability to resist breaking or rupturing. This resistance is often measured by the force needed to break the material. A material has several strengths, Fig. 24-2:
   a. **Tensile strength** – resistance to fracturing, (breaking) by forces pulling on the material.
   b. **Compressive strength** – resistance to fracturing by forces pressing down on the material.
   c. **Shear strength** – resistance to fracturing by offset forces applying loads in opposite directions.
   d. **Torsion strength** – resistance to fracturing from forces twisting the material.
2. **Elasticity** – the ability of a material to be stretched or bent without undergoing permanent change. An elastic material will return to its original size and shape after the load is removed. Springs are highly elastic. Stiffness is the opposite of elasticity.
3. **Ductility** – the ability of a material to flow (plastic flow) under load without fracturing. The opposite of ductility is brittleness.
4. **Toughness** – the ability of a material to absorb energy (blows, etc.) without rupturing. Toughness is related to strength and ductility.

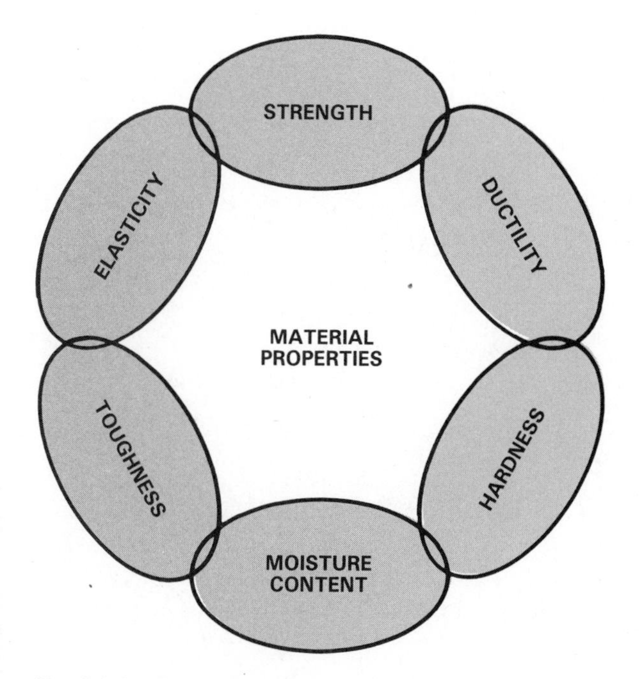

Fig. 24-1. Properties of materials. Any of these can be changed through conditionings.

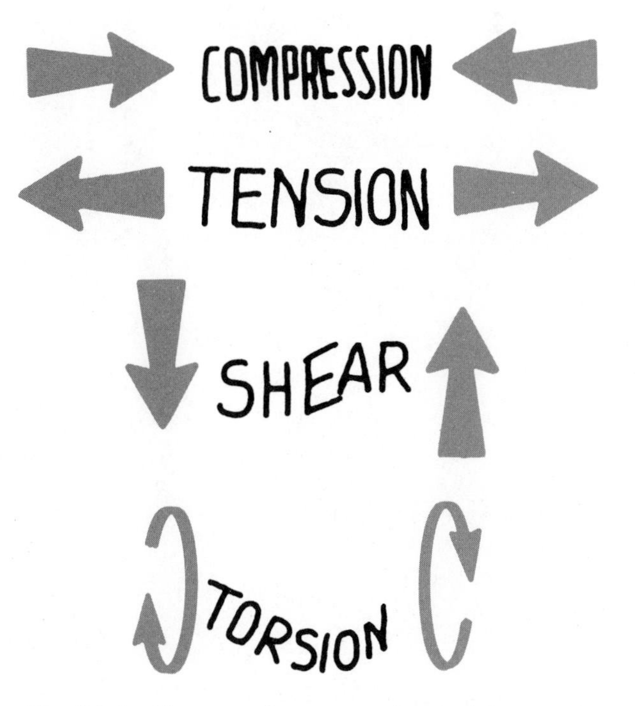

Fig. 24-2. There are four types of material strength.

5. **Hardness**—the ability to resist scratching and denting. Hardness is a measure of the wear resistance of a material. Generally the harder a material the more brittle or less elastic it is.
6. **Moisture content**—The amount of water or other fluids contained in the material. Conditioning processes greatly reduce the moisture content of most ceramics and woods.

Conditioning processes are performed for three basic reasons: to improve the forming and machining properties of the material, to remove internal stress and strain which has built up during other processing acts, and to add desired mechanical and physical properties to the material.

To accomplish these tasks it is essential to:

1. Establish the material properties that are needed.
2. Determine the changes of internal structure that are needed to develop the necessary properties.
3. Select a conditioning process that can properly change the internal structure.

TYPES OF CONDITIONING PROCESSES

An important factor in conditioning materials is the process used. The needed property is developed in the material by using the correct conditioning process. All conditioning processes are grouped into three categories: thermal (Using heat), mechanical (using physical forces), and chemical (using chemical reactions).

THERMAL PROCESSES

Heat is a common conditioner of materials. It may be used to change the internal structure of the material, remove built-in stress, or remove excess moisture. Typical thermal conditioning processes include heat treating metals and ceramics, firing ceramic products, and drying wood.

The most common thermal conditioning operations, as seen in Fig. 24-3, are:

1. **Annealing**—a process to soften materials and/or relieve stress.
2. **Normalizing**—a process used to cause steel to return to its "normal" internal (pearlitic) structure.
3. **Hardening**—a process used to increase a material's hardness, strength, and wear resistance.
4. **Tempering**—a process to relieve internal stress, usually caused during hardening.
5. **Drying**—a process to cause an even removal of excess moisture in a material.
6. **Firing**—a process to change the internal structure of ceramic materials to increase their hardness.

Fig. 24-3. Heat is used to anneal, normalize, harden, temper, dry, and fire. This view is of a continuous heat-treating furnace. (Drever Co.)

MECHANICAL PROCESSES

Another technique to condition material uses physical force. The material is flexed, compressed, pulled, or struck to cause internal structural changes. Materials as they are processed will be changed.

Forming (forging, drawing, bending, etc.) and machining operations apply mechanical forces to a material. These forces cause the internal structure of the material to change. It usually becomes harder and more brittle. This action is called *work hardening.* Often work hardening is not desired. The material will need to be thermal conditioned to remove the hardness. However, in some cases, the hardness developed while working the material is wanted.

Materials, especially metals, are sometimes mechanically struck to improve their properties. Shot may be thrown, as shown in Fig. 24-4, to improve material properties. The process, called *shot peening,* will relieve stress and cause work hardening. The process, however, is widely used to improve the surface finish. The shot removes very small surface defects by evenly compressing the outside layers of the material.

Fig. 24-4. Shot "thown" against a workpiece will mechanically condition it. (Wheelabrator-Frye Inc.)

CHEMICAL PROCESSES

The use of techniques to cause chemical reactions is the third way to condition materials. This type is commonly associated with plastics. Catalysts (compounds which cause chemical actions to take place) are added to thermosetting liquids. The catalysts cause the material to polymerize (form long organic chemical chains).

Also, radiation from a gamma source (gamma rays) can cause a plastic material to polymerize. In both cases, the material changes. Short organic chains are combined into longer more stable chains. The material takes on a new "set" or structure.

SUMMARY

Conditioning processes are used to change the internal properties of a material. Through conditioning, materials develop a different set of properties. The strength, elasticity, ductility, toughness, hardness, and moisture content may be changed.

The changes are caused by applying heat, using mechanical forces, or causing chemical reaction to take place. Through these thermal, mechanical, and chemical conditioning processes, a material may be changed to suit its use better.

STUDY QUESTIONS – Chapter 24

1. What are the typical properties changed by conditioning processes?
2. Define these properties.
3. List and define the four main types of material strength.
4. Why are conditioning processes used?
5. List and describe the three major types of conditioning processes.

Chapter 25

CONDITIONING MATERIALS

Chapter 24 introduced conditioning in broad terms. The properties affected were presented. This chapter will take a closer look at thermal, mechanical, and chemical conditioning.

THERMAL CONDITIONING

Thermal conditioning processes use heat to change the internal condition of a material. Properties are altered to "cause a better fit between a material and its use." To be called thermal conditioning, the process must use controlled heating and cooling to cause the internal changes.

To understand thermal conditioning, answers to three basic questions must be understood.

1. What happens inside the material?
2. How is the necessary heat generated?
3. What conditioning processes use heat?

WHAT HAPPENS INSIDE THE MATERIAL?

Chapter 3 presented the concept that each material has a basic structure. What makes materials different can largely be explained by their structures. Each structure provides a material with certain characteristics. One structure produces a harder material while another provides elasticity. Still another provides ductility.

The whole reason for conditioning processes is to create a material structure which has the best possible properties. These properties will change as the use for the materials changes. The conditioning processes to be used are selected because of their ability to produce the desired property change.

Thermal conditioning of metals

Most metals can be thermally conditioned. The common metal conditioning processes are called *heat-treating*. All heat-treating practices are carefully controlled activities. Time, temperature, and rate of temperature change are controlled. Also, the atmosphere in the heat-treating furnace is often controlled. It is through this careful control that the properties of the metal are changed to meet product needs. Proper heat-treating is a powerful process because it develops material properties that match the intended material use.

Theory of heat-treating. Metals are made up of an orderly three-dimensional arrangement of molecules called crystals. These crystals may be bonded together in a number of ways. The bonded crystals make up the microstructure or grain structure of the metal.

Heat-treating is used to modify the grain structure for one or more reasons. These reasons include:

1. Relieving internal stress.
2. Producing a refined or uniform grain size.
3. Altering (changing) surface chemistry.
4. Strengthing the material.

The basis of heat-treating lies in the ability of a metal to have more than one crystal structure. Materials which can exist in two or more crystal structures are called *allotropes* or *allotropic materials*. Most iron alloys are allotropic. Fig. 25-1 shows the various stages iron passes through as it cools and changes from a liquid to a room-temperature solid.

Nonallotropic materials have only one possible crystal structure. All nonferrous metals and alloys and a very few ferrous alloys are nonallotropic.

Nonallotropic materials are not widely heat-treated. The heat-treating that is done primarily relieves stress and softens work-hardened pieces.

Allotropic materials are widely heat-treated. In fact, over 90 percent of all heat-treating involves iron alloys. Iron has several phases (grain structures) which it passes through when it is heated and cooled.

The chart in Fig. 25-2 shows the extra heat absorbed as steel passes through its phases. These points on the chart are called *critical points*. Allotropic (grain structure) changes take place at each of the critical points. The properties of the metal also change at each critical point.

Steel must be heated above the lower critical point if hardening is to take place. In some cases, it must be heated above the upper point.

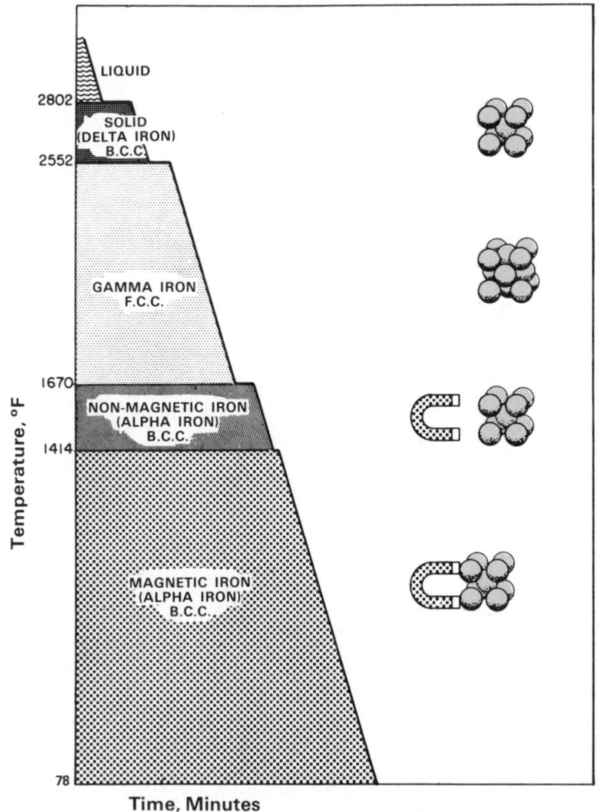

Fig. 25-1. Changes take place in pure iron as it cools to room temperature. (Caterpillar Inc.)

The actual changes that take place are very complex and beyond the scope of this discussion. In summary, steel heated above the Ar_3 point is called *austenite.* It is a solid solution of carbon in gamma iron. The iron is nonmagnetic and has a face-centered cubic lattice, Fig. 25-3A.

As the steel cools below Ar_3, a new grain structure starts to form. A body-centered cubic lattice, as shown in Fig. 25-3B, appears. This new form is called alpha iron or *ferrite.* The ferrite iron cannot hold as much carbon as the austenite (gamma) iron. At this point the iron is a mixture of austenite and ferrite iron.

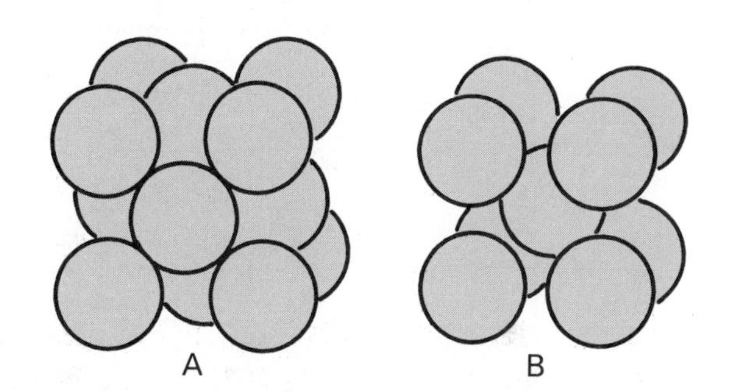

Fig. 25-3. Models of atomic structure of steel. A—Face-centered cubic lattice (FCC). B—Body-centered cubic lattice (BBC) structures. (Caterpillar Inc.)

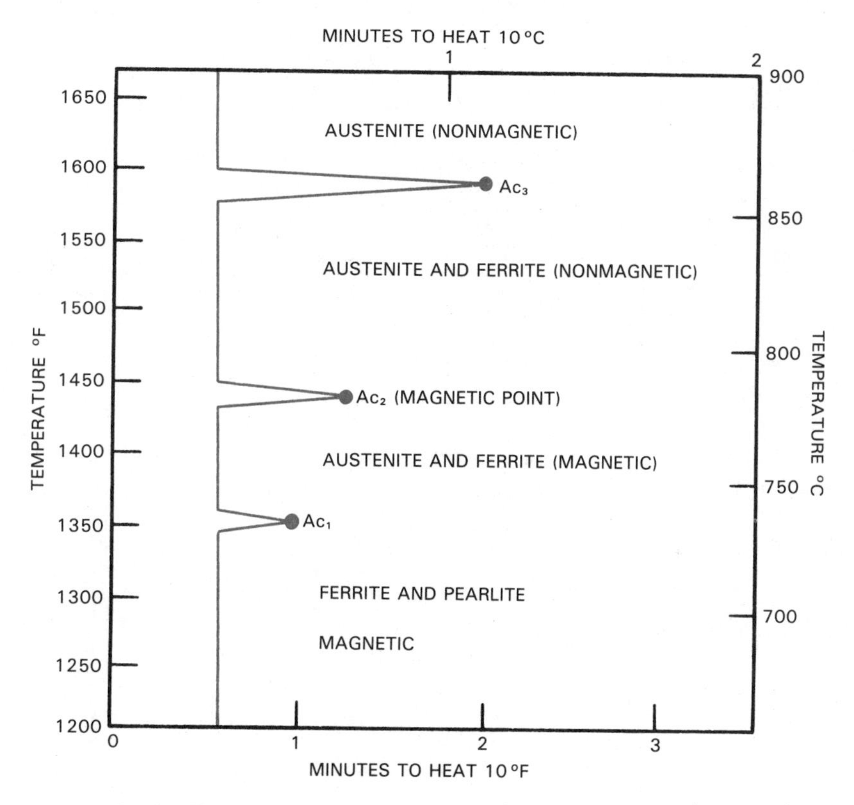

Fig. 25-2. Time and temperature chart for a common carbon steel.

Below the Ar_1 line the remaining austenite changes into pearlite. The pearlite is a combination of alternating layers of ferrite (alpha iron) and iron carbide (iron-carbon compound). Fig. 25-4 shows the relationship of the allotropic stages to carbon content and temperature.

The two-step hardening process. Steels are often heat-treated to increase their hardness and wear resistance. The typical process involves two steps:

1. The steel is heated above its critical temperature and rapidly cooled (quenched).

 This step produces martensite if the steel has a

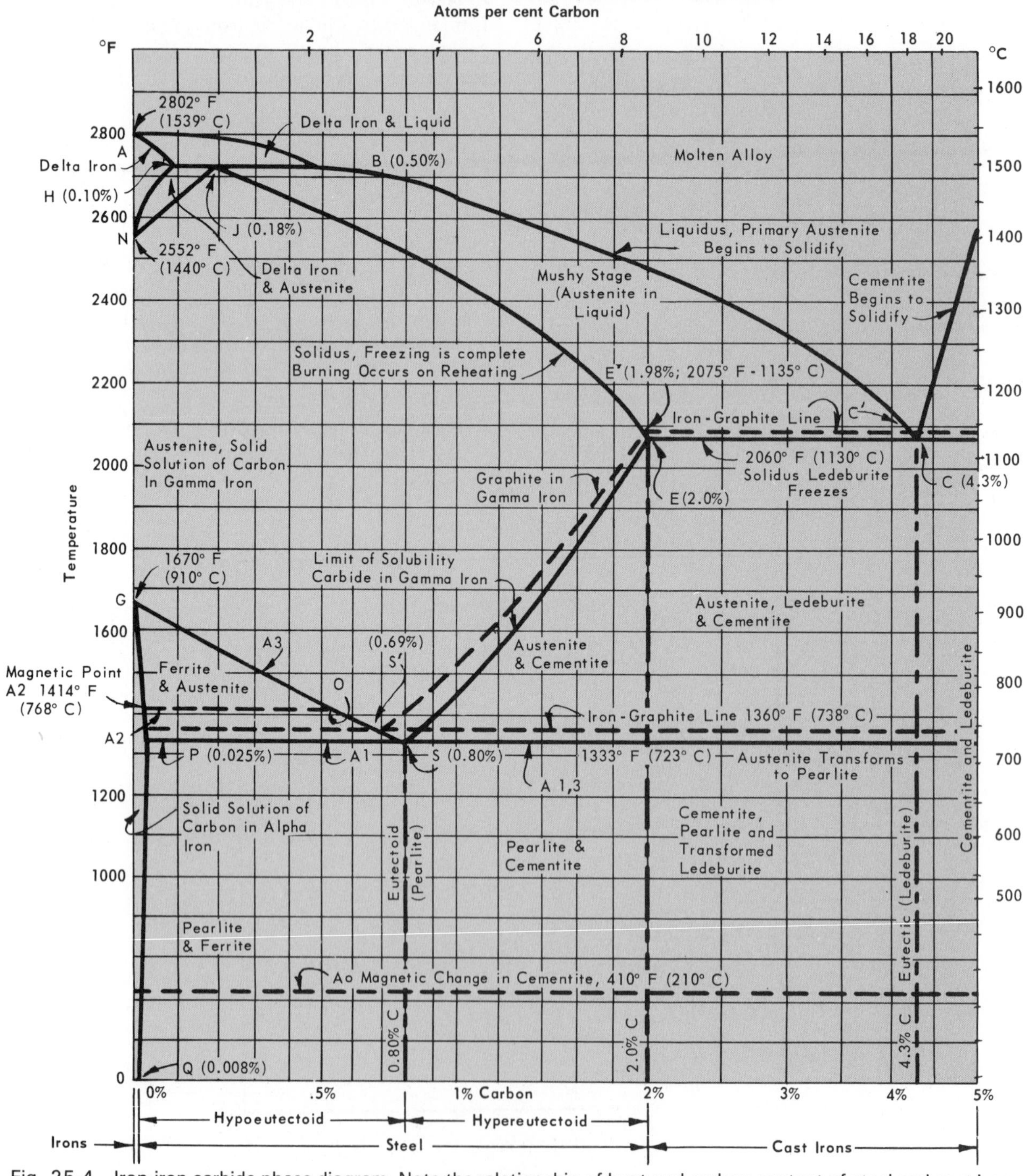

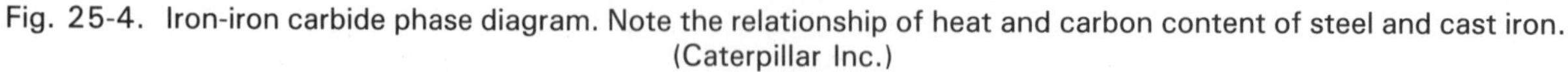

Fig. 25-4. Iron-iron carbide phase diagram. Note the relationship of heat and carbon content of steel and cast iron. (Caterpillar Inc.)

carbon content between 0.35 to 2.0 percent. (Below 0.35 the steel cannot be hardened). Martensite is a solid solution of iron and carbon. The solution is unstable because it has more carbon than it can easily hold. It is supersaturated with carbon.

Also, martensite has an unstable (abnormal) grain structure. The material wants to go to its stable body-centered cubic structure.

The result of the hardening step is a material which is hard, brittle, and unstable. The material will distort and is likely to fail under use.

2. Step two involves "tempering" the material. The steel is heated gradually to a predetermined temperature. It is then quenched. Tempering allows the material to take on its normal lattice structure. Stress is removed. The material becomes less brittle and somewhat softer.

Thermal conditioning other materials

Almost all metals and ceramic may be thermally conditioned to remove stress. This process, called "annealing," involves heating the material to near its lower critical temperature. The material is then allowed to cool very slowly. The result is a stress-free structure.

Ceramic materials are "fired" to change the internal structure. Ceramic is made of clay and fluxes. At temperatures above red hot, the fluxes turn to glass. This glass forms a cement which bonds the clay into a strong body. The glass also fills the pores in the ware giving it a very solid structure. After firing, the ceramic material has a new internal structure.

Wood materials may be dried to change their properties. Properly controlled kiln drying removes excess moisture and relieves internal stress in the wood. The material is less subject to checking and warp. Kiln drying also "sets" resins in the wood. It kills fungi and inserts which can stain or destroy the material. Dried wood has increased strength and nail-holding power.

In general, all thermal conditioning processes produce a more suitable material. The material should serve its purpose better after thermal treatment.

GENERATING HEAT FOR THERMAL CONDITIONING

Heat for thermal conditioning operation can be generated in many ways. The most common are:

1. Furnaces.
2. Induction coils.
3. Flames.
4. Kilns and ovens.

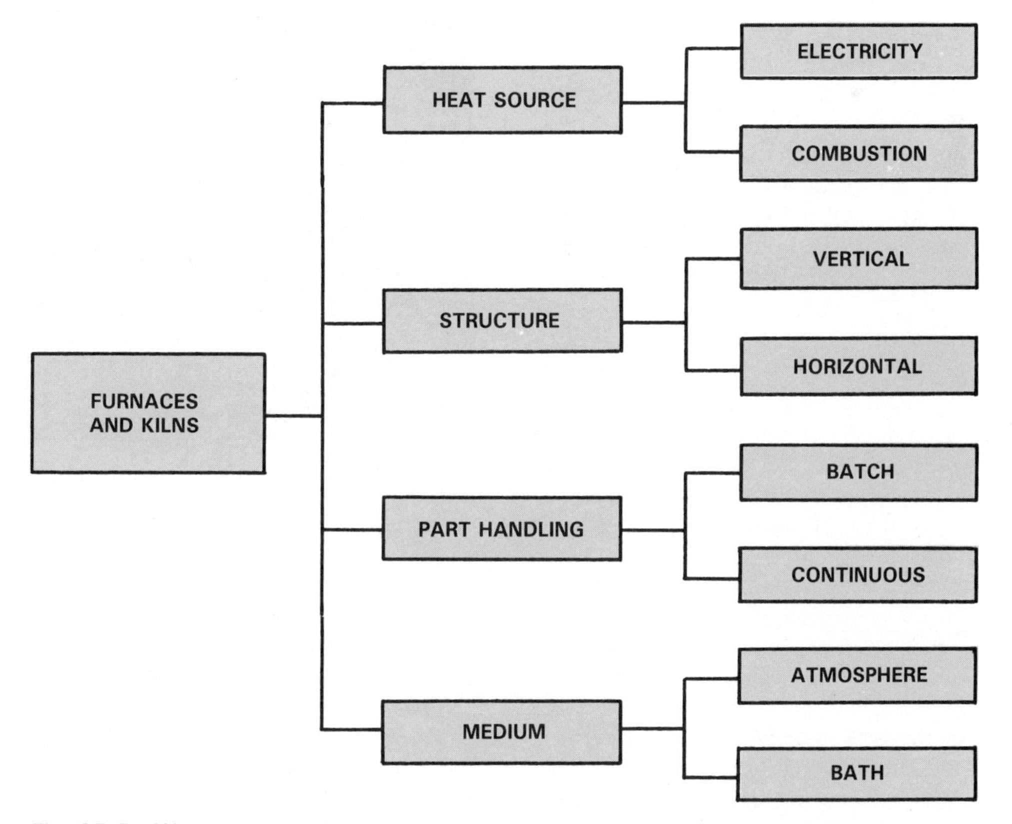

Fig. 25-5. Ways to group furnaces and kilns. There are four types and eight variations.

Furnaces

Carefully controlled thermal conditioning requires a reliable heat source. The most common source is a furnace. Furnaces are generally closed containers or boxes. They have a refractory (high-temperature resistant) lining and a method to control the temperature. Many furnaces also have a system to control the atmosphere within the furnace.

Furnaces may be grouped by four methods. See Fig. 25-5.

1. The means of heating.
2. The structure of the furnace.
3. The way the parts are handled.
4. The medium in which the part is treated.

Heating method

There are two principal means of heating furnaces, one is by burning of fuels (or *combustion heating)* and the other is through *electrical resistance.*

Combustion heating depends upon burning oil or gas to generate heat. Burning, of course, requires the presence of oxygen.

If combustion takes place in the same chamber with the parts, problems can develop. Too much oxygen in the hot atmosphere will cause the carbon in the outer layer of the steel part to burn away. This leaves a low carbon, softer, outer layer on the part. Too little oxygen leaves excess fuel in the furnace atmosphere. Elements in the fuel will combine with the parts to form iron carbide.

These problems are avoided when the flame is kept outside the part chamber. The internal atmosphere can then be an inert gas such as nitrogen or argon.

Electric furnaces generally use heating coils (resistance heating elements) to generate the heat. These coils may be set in the sides of the treating chamber or may be located away from the chamber.

Furnace structure

Basically, heat-treating furnaces are either vertical or horizontal in structure, Fig. 25-6. *Horizontal* furnaces generally are square containers often called box furnaces. A door on one end allows loading and unloading of parts. Typical horizontal furnace types include the standard box furnace and the car-bottom furnace, Fig. 25-7.

Standard box furnaces are loaded through a door in one end. They may be direct-fired hearth furnaces in which fuels burn over the parts or indirect-fired box furnaces which separate the heating and combustion chambers, Fig. 25-8.

Car bottom furnaces use flat cars on which the work is loaded. The cars roll into the furnace. This type is generally used for longer parts which standard box furnaces cannot handle.

Fig. 25-6. These furnaces represent two different types. Top. Horizontal, continuous annealing furnace. Bottom. Vertical continuous annealing furnace. (Drever Co. and U.S. Steel Corp.)

Electric furnaces may also be box type. These are either directly heated or have separate chambers for the heating coils and the parts.

Vertical furnaces are designed to heat treat long parts and are often shaped like a long, round cylinder. In some cases, they may be set so their tops are at floor level. This type is called a *pit furnace.*

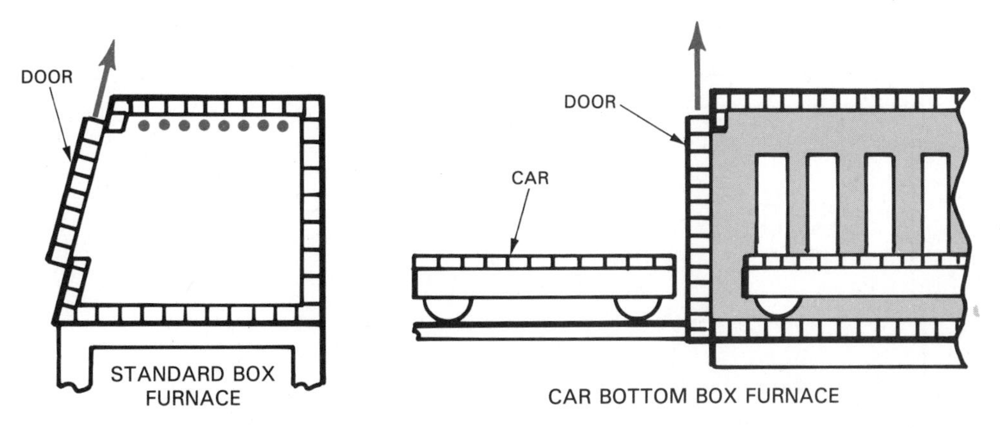

Fig. 25-7. Box furnaces may be direct or indirect fired. Car bottom types are used for large parts.

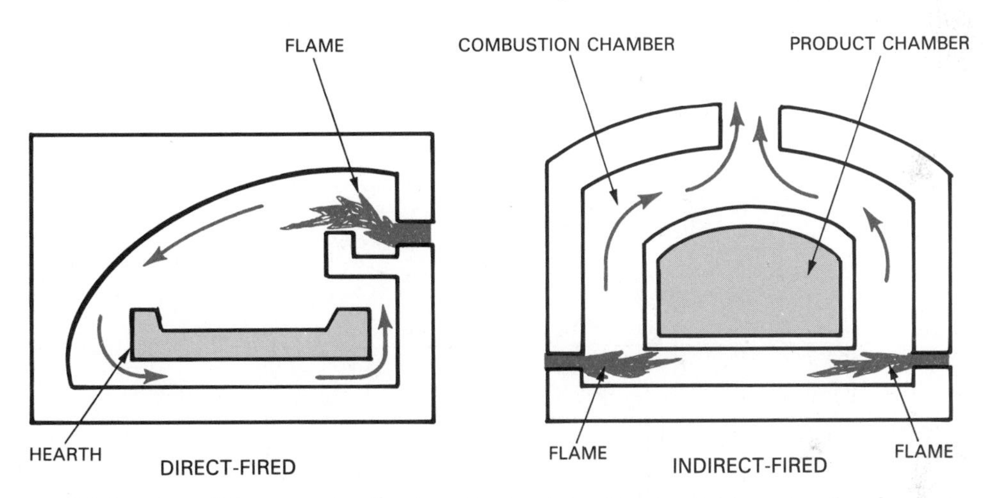

Fig. 25-8. Note differences in direct and indirect-fired horizontal heat-treating furnaces.

Another type of vertical furnace is shaped like an upside-down, open can or box. Called a *bell furnace,* it is lowered over the parts and heat is applied. Bell furnaces, Fig. 25-9, are often used for annealing processes.

Part handling

The third classification of furnaces involves the way they handle the work. Batch furnaces heat-treat a group of parts at one time. All the furnaces shown so far are this type.

However, parts can often be continuously heat-treated. Parts can be fed in one end of the furnace and come out heat-treated at the other. Belt or chain conveyors may move the parts through the furnace. Other furnaces have a rotary table or hearth. The parts are loaded in one door and removed after one rotation from an adjacaent door. One type of furnace uses a mechanical or hydraulic ram to move the parts through the furnace. The parts are placed on a tray. The tray is pushed into the furnace by the ram. Conditioned parts are forced out the far end of the furnace by the incoming tray. Called *continuous furnaces,* they are shown in Fig. 25-10.

Medium within the furnace

Parts may be heated in different gaseous atmosphere or mediums. These furnaces make up the fourth group. The atmosphere may be air or a controlled atmosphere of inert (nonreacting) gas.

Heat-treating may also take place in a molten bath. Often, liquid salts or lead are used. A liquid bath allows for:

1. Rapid and uniform heating.
2. Close temperature control.
3. Exclusion (absence) of air.

Bath furnaces, seen in Fig. 25-11, may be either combustion fired or electrically heated.

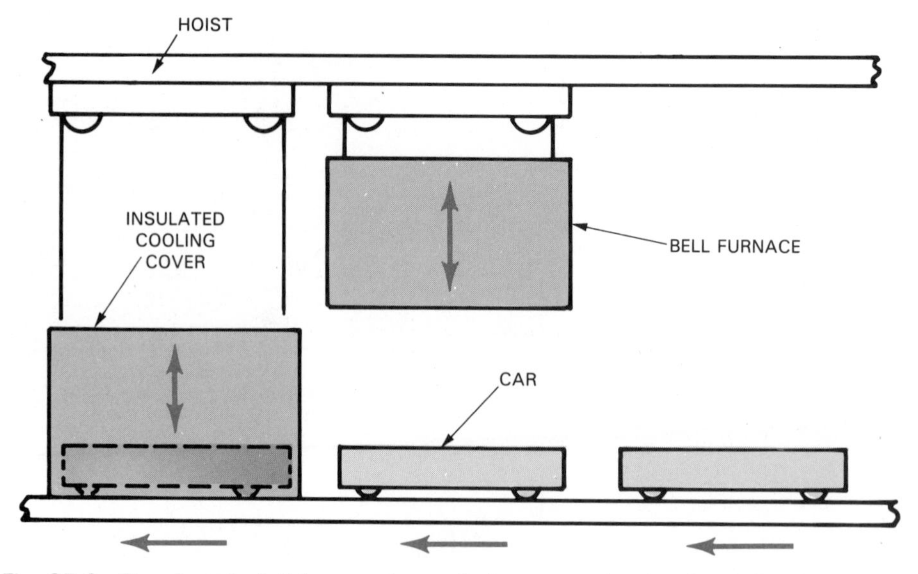

Fig. 25-9. Drawing of a bell furnace shows the furnace and an insulated chamber in which parts slowly cool after heat treating.

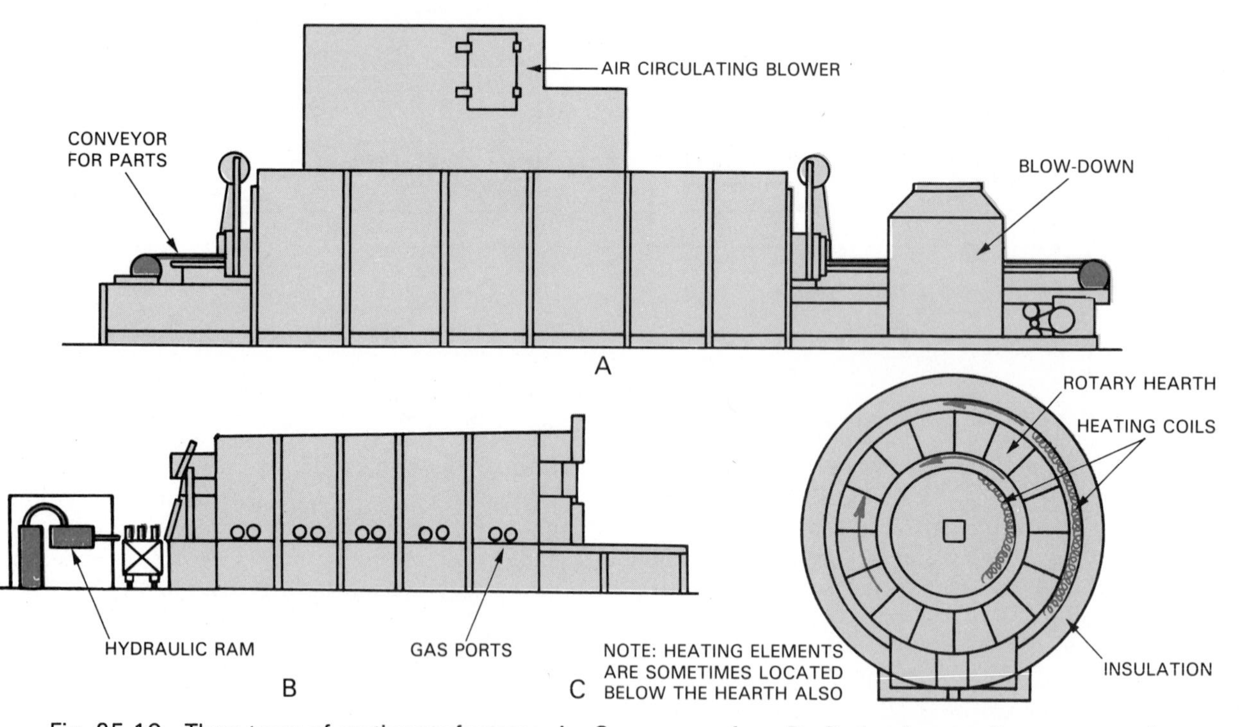

Fig. 25-10. Three types of continuous furnaces. A—Conveyor surface. B—Pusher furnace. Parts are moved by a ram. C—Rotary furnace. Note heating elements. (Caterpillar Inc.)

Induction coils

A second way to generate heat for thermal conditioning is by induction coils. The workpiece is placed inside a coiled electrical conductor or the coil is placed inside or on the part. Then high frequency alternating current is applied to the coil. The expanding and collapsing magnetic field around the coil induces eddy currents in the work. These eddy currents generate heat and so the part is heated to the proper conditioning temperature.

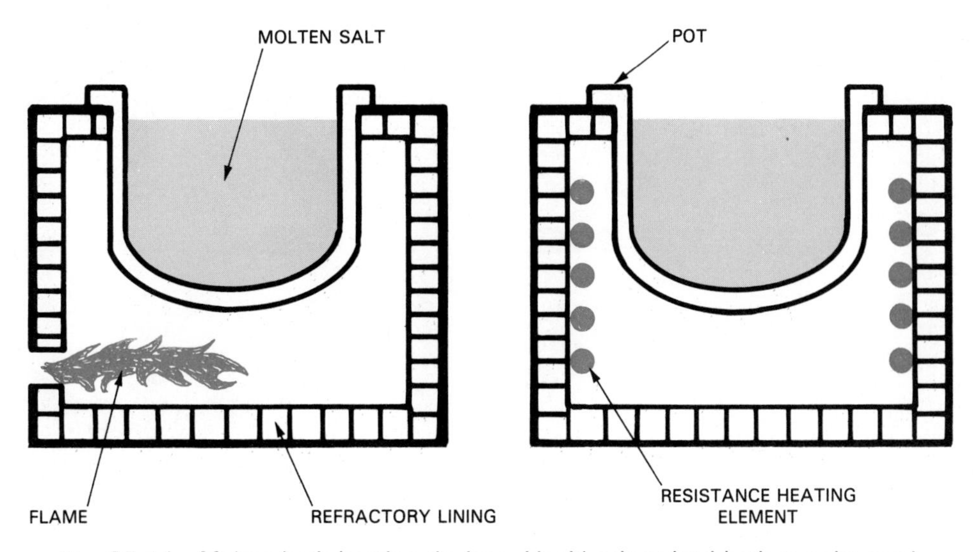

Fig. 25-11. Molten bath heating devices. Liquid salt or lead bath may be used.

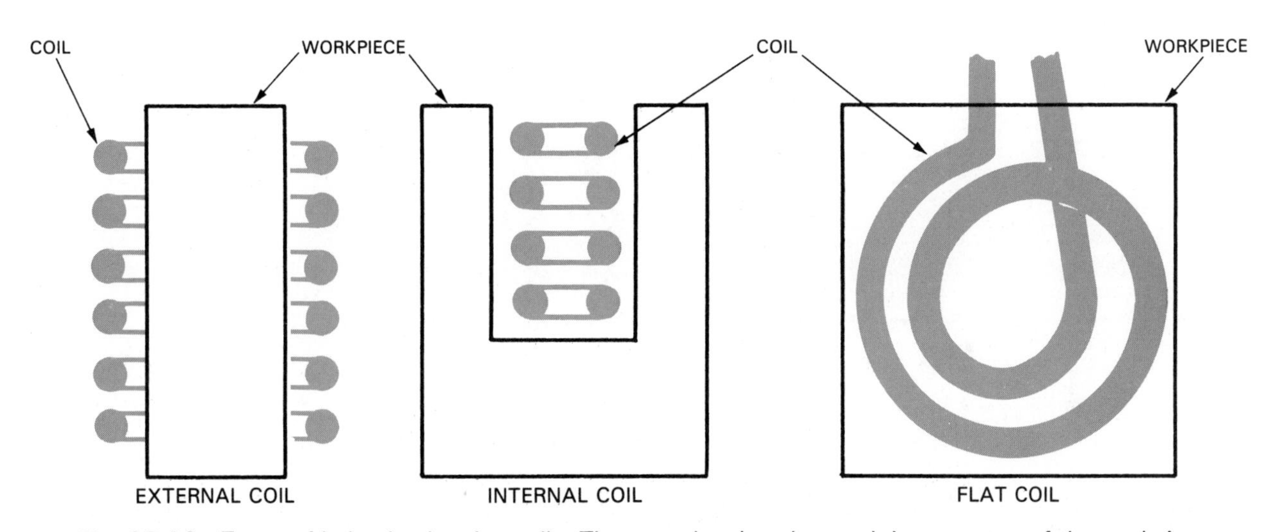

Fig. 25-12. Types of induction heating coils. They may be placed around, in, or on top of the workpiece.

Induction heating, as shown in Fig. 25-12, is used for surface (case) hardening of parts such as camshafts and crankshafts. The system is quick and effective.

Outside combustion

Combustion outside a furnace may also be used for thermal conditioning. The most common process of this type is flame hardening. The technique, shown in Fig. 25-13, quickly heats the surface of the part. It is then rapidly cooled (quenched) causing surface hardening with a ductile core or back.

Kilns and ovens

Kilns and ovens are special-purpose furnaces. They may be directly or indirectly heated with electricity or gas. They are also available as batch or continuous models.

The typical uses of these devices are for firing ceramic products, annealing glass, drying wood, and other similar operations. Fig. 25-14 shows a continuous ceramic kiln with its drying, firing, aging, and cooling stages. The ware move through the kiln on special cars, Fig. 25-15.

THERMAL CONDITIONING PROCESSES

The thermal conditioning processes include four main groups. They are arranged by what they do:

1. Harden material.
2. Soften materials.
3. Relieve stress in materials.
4. Dry materials.

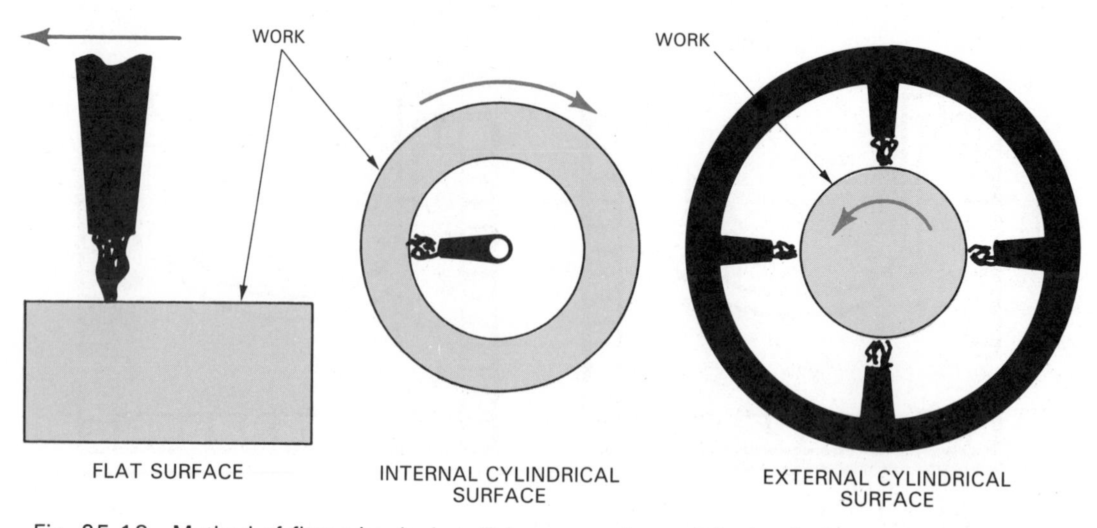

Fig. 25-13. Method of flame hardening. This process is good for hardening workpiece surface.

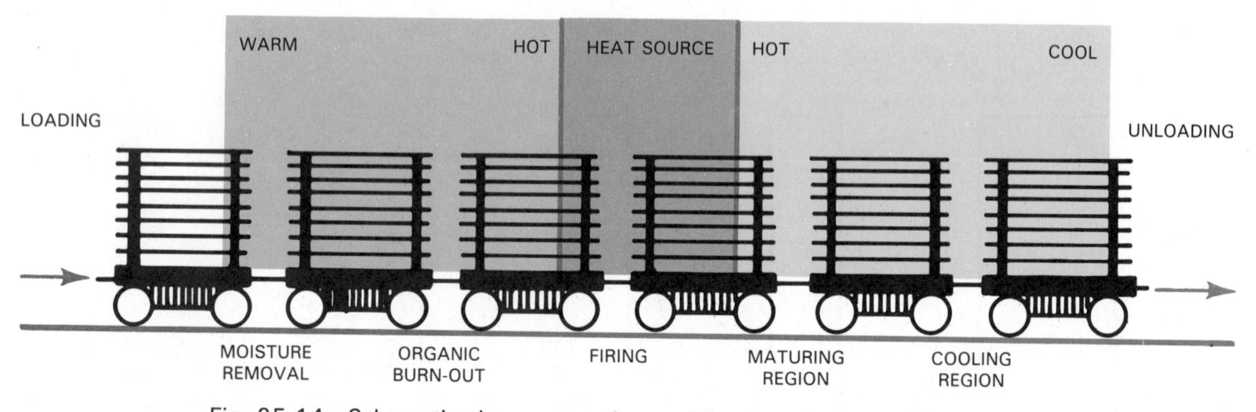

Fig. 25-14. Schematic shows a continuous kiln. Note its several heat zones.

Fig. 25-15. Kiln cars are carefully loaded with ceramic ware prior to firing. (Syracuse China Co.)

Hardening processes

Materials are frequently hardened to increase ability to resist scratching and denting. There are a great many ways to do this. The most common processes are shown in Fig. 25-16.

Full hardening

Full hardening, or more commonly called just hardening, involves treating the entire material. The goal is to increase the hardness, tensile and fatigue strength, toughness, and wear resistance. This type of hardening includes quench hardening of metals and firing of clays.

Quench hardening, as described earlier, is at least a two-step process. There is a hardening step followed by a stress-relieving step. The hardening step involves heating a piece of steel to a temperature within or above its critical range. The material is then rapidly cooled in a liquid (quench). The quenching medium may be oil, water, or brine (salt water).

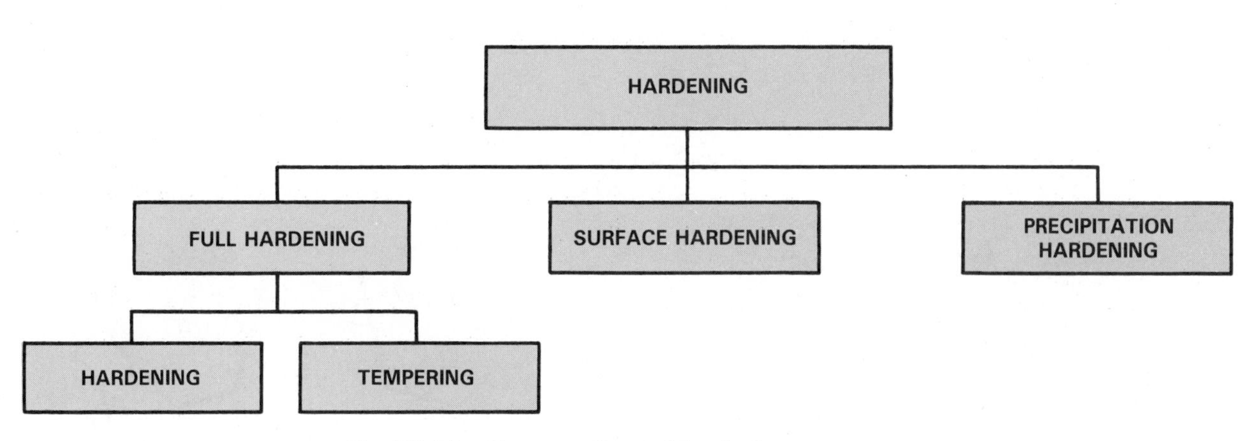

Fig. 25-16. Common thermal hardening processes.

There are several important conditions for proper hardening.

1. The material must be heat-treatable. Steel is a combination of iron, carbon, and other elements. Steels below 25 points (1 point = 0.01 percent) carbon is called low-carbon steel and will not harden. Medium carbon steels are between 30-60 points and can be toughened by heat-treating. They will not, however, become very much harder. High carbon steels are above 80 points carbon and will react well to hardening processes.
2. The material must be heated correctly. Heating will take place from the outside inward. The heating rate must be slow enough to allow the center of the part to obtain the correct temperature without overheating the outside.
3. The material must be quenched properly. The quenching medium is important. Very rapid cooling is necessary to produce the best hardening results. Medium carbon steels can be quenched in water. However, oil, brine, and brine or water sprays are used when more rapid cooling is needed. Most tool steels use an oil quench.

Steels which have been rapidly quenched are generally brittle. The material must be further treated to bring it into a service condition. This further treatment is called *tempering* or *drawing*.

Tempering reduces the material's hardness and brittleness while increasing its ductility and toughness. The general practice of tempering requires the heating of the material to its tempering temperature and holding it there until the part is uniformly heated. The tempering temperature varies with each alloy.

The heated part is then quenched. Generally, it is rapidly cooled to room temperature. However, in some cases interrupted quenching is used. The heated part is first cooled to a set temperature in a salt bath and allowed to soak (remain at this set temperature). Later it is quenched to room temperature. The most common interrupted quench processes are austempering and martempering. These processes are used to produce special physical properties.

Ceramic firing is another hardening process. As described earlier the material is heated to change internal structure of the materials. The clay ware is heated in several stages. At temperatures between 212°F (100°C) and 1022°F (550°C) all remaining water is driven off. The second step, betwen 752°F (400°C) and 1032°F (556°C) organic impurities are burned off. Finally between 1652°F (900°C) and 2552°F (1400°C) the hardening process takes place. The final temperature is maintained over time. This allows for maturing. The fluxes turn to glass and flow around the heat-hardened clay particles. After maturing, the clay ware is allowed to slowly cool. Rapid cooling will cause the product to crack.

The sintering process described in the chapter on Powdered Metals is another ceramic firing operations. The metal powders are bonded by a thermal conditioning process.

Glass, another ceramic, is often tempered. The material is heated to near its melting point and quickly cooled. The result is increased product strength. For sheet glass the tempering lehr (oven) is often a part of the overall forming machine as seen in Fig. 25-17.

Surface hardening

In many cases, a fully hardened part is not desirable. What is needed is a ductile part with a hard, wear resistant surface. This need calls for a surface hardening technique. These techniques, sometimes called case hardening, include:

1. Material-adding treatments.
   a. Carburizing.
   b. Cyaniding.
   c. Carbonitriding.
   d. Nitriding.
   e. Chromizing.

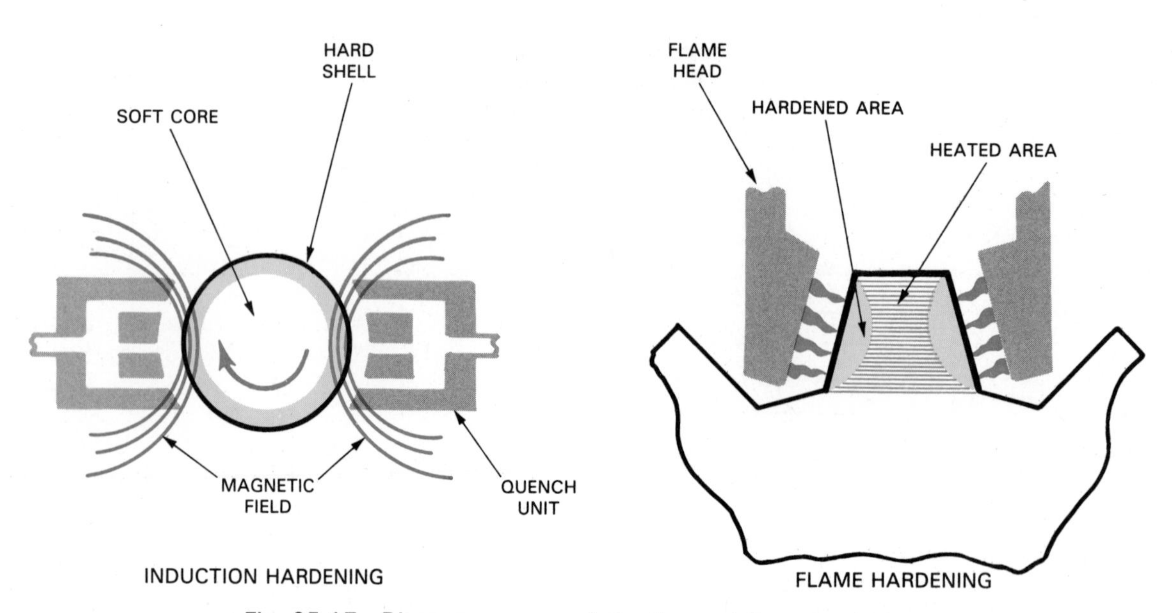

Fig. 25-17. Diagrams compare induction and flame hardening.

2. Surface treatments.
   a. Induction hardening.
   b. Flame hardening.

Some case hardening techniques involve the addition of carbon or nitrogen or both to the surface layers of the steel part. This modified case or shell can, then, be hardened by heating and quenching.

Carburizing is the oldest method. It involves placing low carbon steel in contact with a high carbon medium. The steel is then heated to or above its critical temperature. At this temperature the iron attracts carbon. The carbon is absorbed into the outer shell of the steel. If the temperature is maintained, it will diffuse (spread) inward. The result is a high carbon steel shell or case around a low carbon steel core. The thickness of the case will vary with the time and temperature of the process.

Carburizing can be done by several methods including:

1. Pack carburizing—heating the steel in a container of charcoal or coke.
2. Liquid carburizing—heating the steel in a cyanide salt bath.
3. Gas carburizing—heating the steel in an atmosphere of carbon monoxide, hydrogen, and nitrogen or in natural gas or propane.

The carburizing process produces steel with varying carbon content and case thicknesses. Carbon content of the case can range up to 1.2 percent and case thicknesses of 0.160 in. are possible. This type of material may require several heat treatments to refine the grain structure of the core and to produce an acceptable hardened case.

Cyaniding is similar to the liquid carburizing process. The cyaniding technique uses a molten salt bath containing sodium cyanide, sodium carbonate and sodium chloride. When heated between 1450°F and 1650°F (788°C and 899°C) the bath breaks up into free nitrogen and carbon. Steel in the bath will absorb these elements, forming a hard carbide-nitride case. The heated part is quenched after a proper soak time to form the hardened case. The quenched parts are then thoroughly cleaned.

The cases formed by cyaniding will vary up to 0.015 in. thick. The process is, therefore, well suited for case hardening small parts.

Carbonitriding involves holding a heated part in a gaseous atmosphere of carbon and nitrogen. The steel absorbs the carbon and nitrogen which, when cooled, produces a hard shell. The addition of nitrogen increases the ability to harden the case.

Nitriding is used on a steel that has been previously hardened and tempered. The steel is then heated to about 1000°F (538°C) in an atmosphere of ammonia gas. The nitrogen from the ammonia gas is absorbed by the steel. This forms a very hard case of nitrides.

The nitriding process works best with steels containing alloys of chromium, molybdenum, aluminum, and vanadium. These elements have a high tendency to form the necessary nitrides.

The treated part is allowed to air cool since it was already heat-treated. The nitriding process causes little change in the core material and the case is highly resistant to corrosion.

Chromizing is a process which produces a

chromium-enriched shell on the part. This shell is highly resistant to heat, wear, or corrosion. There are a number of techniques to introduce the chromium to the steel. All of them, however, use high temperatures which can cause distortion, internal stress, and dimensional changes.

The process involves heating the steel in a pack of powdered chromium halide salts, or in a salt bath containing chromium and chromium halides, or in a gas produced from a reaction between acids and chromium. The steel absorbs the chromium, forming the desired case which is hardened by rapid cooling.

Hardened cases or shells may be produced by rapidly heating and cooling high carbon steel. The rapid heating allows only the outside layers of the material to reach the proper hardening temperature before quick cooling takes place.

The most common hardening methods, which have been described previously, are *induction hardening* and *flame hardening*. These processes, as shown in Fig. 25-18, produce hard cases on ductile cores. The case depth can be easily controlled and the resulting surface is scale free.

Precipitation hardening

Precipitation hardening or age hardening is used to harden nonferrous (not iron base) metals. Most nonferrous metals are alloys of two or more base metals. They may be seen as a solid solution. One metal is "dissolved" in another.

To harden these materials, the alloy is heated until both (in the case of two metals) elements are melted. The temperature is held above its solubility line until a uniform solid solution is produced. The metal (solid solution) is rapidly cooled to room temperature. This action leaves the material in an unstable state. The alloy is soft but has more particles in it than it can naturally hold. Over time, these particles precipitate out from the solid solution. The precipitated particles form connectors between the crystals of the alloy. This action reduces slippage between the crystals and, therefore, increases the hardness of the material.

If the precipitation occurs at room temperature, the process is called *natural aging*. Artificial aging can take place if heat is applied to speed up the process.

Many alloys of aluminum, copper, magnesium, and nickel respond to precipitation hardening. Also, some stainless steels will age harden.

Softening processes

Often, metals become hard as they are worked. To allow for continued working (machining or forming) the material requires softening. This proces is called *annealing*.

Annealing requires the metal to be heated above the Ac_3 critical temperature. The material is held at this temperature until it is uniformly heated. The metal is allowed to slowly cool to room temperature.

Maximum softness and ductility is achieved when the metal is cooled very slowly. Again, control is the key—controlled cooling.

Annealing may be used to produce desired physical and mechanical properties. Specific grain structures may be developed using this process.

Normalizing is a process similar to annealing. It

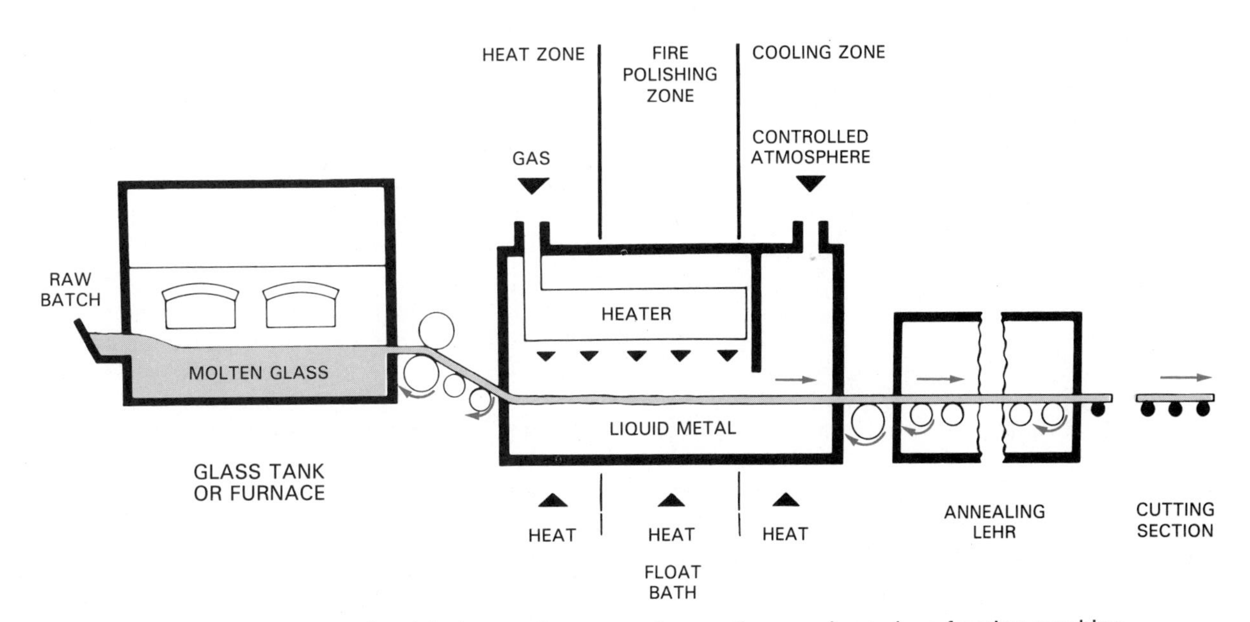

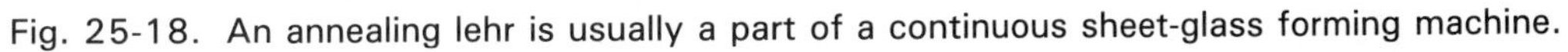
Fig. 25-18. An annealing lehr is usually a part of a continuous sheet-glass forming machine.

involves heating the metal 100-200°F (38-93°C) above the hardening and annealing temperatures. The metal is then slowly cooled in still air. The result is a soft material with a refined grain structure. Normalizing is used instead of annealing when a finer-grained, more uniform structure is wanted.

Normalized metals are harder and generally have higher strength properties than annealed metals.

Stress-relieving processes

Stress relieving is used to remove internal stress caused by machining, forming, and welding operations. This stress can cause a part to distort or fail.

Stress relieving does not change the grain structure of the material. Annealing and normalizing does.

Stress relieving involves heating the part to a specific temperature. The actual temperature varies with the metal. Common stress-relieving temperatures range from 600-1200°F (316-649°C). After it has been heated, the part is allowed to cool slowly.

Glass is almost always stress relieved immediately after it is formed. Uneven cooling that usually is part of the forming process produces internal strain. If the strain is allowed to remain, the glass will fracture. An annealing process of even heating and cooling is used to remove the internal stress.

DRYING PROCESSES

The fourth type of thermal conditioning process is drying. As described earlier it is usually associated with the wood products industry. Kiln drying is a controlled removal of excess moisture from the wood. This process improves the properties of the material for construction, furniture making, and other related applications.

CHEMICAL CONDITIONING

Other conditioning processes use chemical actions rather than heat. These are primarily polymerization actions with natural or synthetic plastic materials. A chemical action takes place which produces a new internal structure. This structure is produced as smaller molecules unite (polymerize) to create more complex molecules.

The drying of enamels and polyurethane finishes are examples of chemical conditioning. The curing (hardening) of liquid casting plastics is another example.

Often the chemical conditioning process is started or speeded up by a catalyst. This material is a chemical which causes a chemical reaction but does not enter into the reaction itself.

The preparation of casting acrylic or waterproof glue involves adding a catalyst. This starts the polymerization process. A liquid is changed into a solid. Molecules unite in a nonreversible action.

Glass is also chemically strengthened or conditioned. Chemicals are used to treat a series of special glass compositions. The result is an ultra-high-strength product which can withstand flexure forces up to 100,000 psi.

These glass materials find wide use in lenses for safety glasses, industrial tubing, and tableware. The chemical treatment permits the manufacture of lightweight strong glass products.

MECHANICAL CONDITIONING

Many materials change their internal structure as they are worked. Rolling, forging, drawing, machining, and many other processes cause material to work harden. Usually this action is undesirable.

One process uses mechanical forces to change the nature of the material. This process is called *shot peeening*. It is a cold working act which uses a stream of metal particles called "shot." The shot is thrown at the part at very high speed.

Shot peening is used to increase the fatigue resistance of a part. The shot sets up compressive forces on the outside layer of the material. This layer resists fractures caused by tensile forces inside the parts. Shot peening is used widely to condition springs.

Other processes can be used to condition materials. Some metal alloys can only be conditioned by mechanical working. Others, such as piano wire and wire for bridge cable, are work hardened during their drawing process.

SUMMARY

Conditioning processes are used to change the mechanical and physical properties of material. These processes may be primarily thermal, mechanical, or chemical in nature.

The process may condition the entire part or just its surface. Also, the material may be hardened, softened, stress relieved, or dried. The intended end result is a material which is well suited to its use.

STUDY QUESTIONS — Chapter 25

1. How are most metals conditioned?
2. What factors are controlled during thermal conditioning?
3. What is an allotropic material?
4. What is a "critical point"?

5. Describe the two-step hardening process.
6. What is firing?
7. Why is wood thermal conditioned?
8. List and describe the four ways of generating heat for thermal conditioning.
9. What are the two main sources of heat in furnaces?
10. List and describe the thermal conditioning processes.
11. List and describe the major surface hardening processes.
12. List and describe the softening processes.
13. Describe chemical conditioning.
14. Describe mechanical conditioning.

Arc welding process is used for assembly of huge portable crane. Note huge jig used to hold heavy parts during assembly.

SECTION 6
ASSEMBLING

Assembling processes temporarily or permanently join two or more parts into assemblies or finished products.

Assembly techniques explained in this section include:

- Mechanical—using screws, bolts, nails, rivets, staples, or clips to grip or squeeze the parts together.
- Bonding—use of adhesives (adhesion) or heat and pressure (cohesion) to join parts.

Chapter 26

INTRODUCTION TO ASSEMBLING

It is difficult to name more than a handful of products having only one part. Most products are assembled, (put together) from several mating parts or components. Even the common lead pencil, Fig. 26-1, has five parts. The techniques used to join parts into products are called *assembling processes*. By definition, these are processes which temporarily or permanently join two or more parts into assemblies or products.

The assembly of parts into products is as old as civilization. Cave dwellers sewed animal skins with animal tendon "thread." By the high point of the ancient Egyptian civilization, glue from animal parts and milk was invented. The early Greeks had developed metal clamps by 600 BC. Complex wood cabinet joints were widely used during the dark ages. Machines to make nails were introduced in the late 1700s and welding was attempted by 1890.

Today, many complex assembly techniques are available to the manufacturer. These methods of temporarily or permanently combining parts may be grouped under two major headings. These groups, as shown in Fig. 26-2, are:

1. Mechanical assembly processes.
2. Bonding processes.

MECHANICAL ASSEMBLY

Mechanical fastening uses physical force to hold two or more parts together. These parts may be held in place by either a fastener or by the parts themselves. Typical mechanical fasteners are screws, bolts, nails, staples, cotter pins, rivets, and retaining clips. The fasteners grip or squeeze the parts and hold them in position.

Mechanical fasteners are designed and selected for one of three basic uses. They may be used for assemblies which are:

1. Temporary—can be easily assembled, dissassembled, or adjusted.
2. Semipermanent—can be disassembled though this is not normally done.
3. Permanent—disassembly destroys the fastener and, sometimes, the part.

Parts may also be designed to position and hold one another. The components may be bent or machined to grip their mating part to form an assembly. Typically, this technique uses a joint or seam. Sheet metal parts, as shown in Fig. 26-3, can be seamed together. Pipe and tubing can be compressed around a shaft.

Parts may also be designed to be pressed together. The friction between surfaces will hold them in place. Press fits, as this joint is called, are widely used. Bearings are generally pressed into their opening or onto a shaft. Pins are often driven into holes. These are but two examples of press fits.

The parts may be designed to hook together. Velcro, the self-closing fabric fastening material, is a good example. Also, T-slots, tabs, and other such shapes locate and hold mating components.

BONDING

Parts may also be assembled by the use of bonding techniques. There are two major bonding methods: cohesion and adhesion.

Cohesive forces hold the millions of molecules in a part together. They are the forces that create the internal structure of the material. These same forces

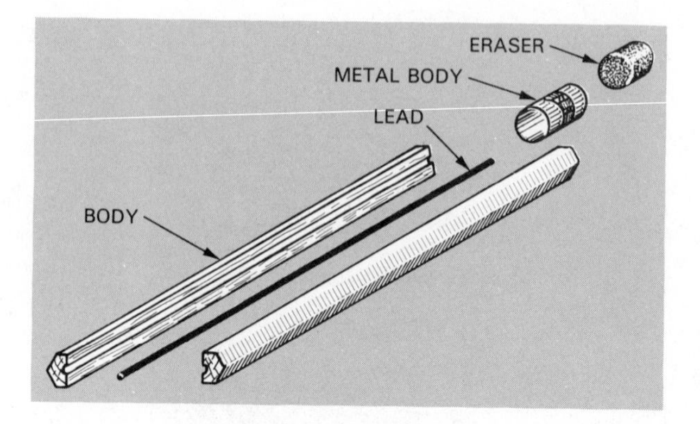

Fig. 26-1. A common wood pencil is an assembly of at least five parts.

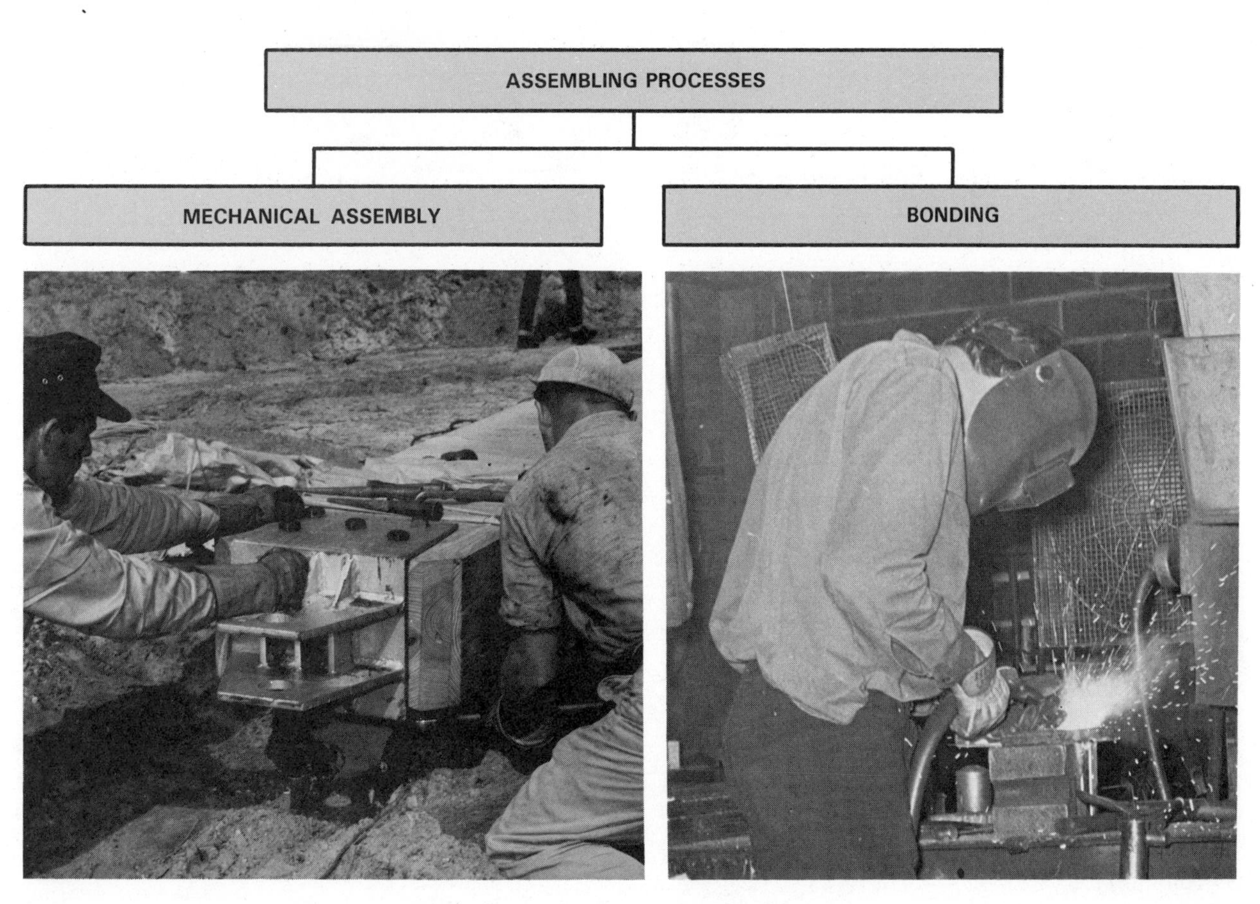

Fig. 26-2. Types of assembling processes.

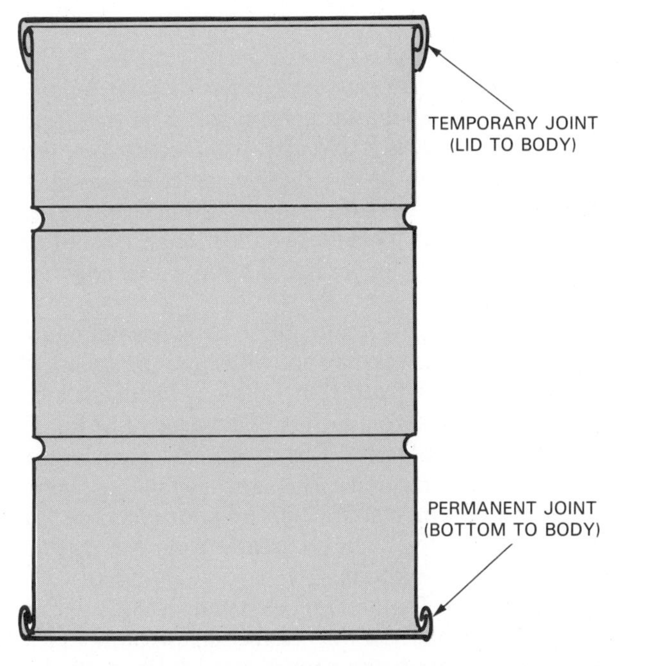

Fig. 26-3. Metal can have permanent mechanical seams at the bottom and temporary assemblies to hold the plastic snap lid.

are used to cause the surface molecules of two or more parts to "grip" one another. Heat and/or pressure is used to bring the parts together. The result is a single piece made from the several parts. The area of joining becomes a uniform or monolithic layer. The parts become one. Cohesive bonding is developed through a group of processes generally called welding.

Adhesive forces are used to hold two different materials together. Adhesive forces cause one material to bond or grip another material. An adhesive material can be applied between two parts. It grips the parts and holds them in place. Adhesives are often called glues or cements. They are widely used to assemble a full range of metallic, polymeric, and ceramic materials. Many parts that were once bonded with cohesive forces (welded) or by mechanical means are now glued.

JOINTS

When two or more parts are assembled they form a structure. The area where the parts meet is usually called a joint. While there are a great number of different joints, they may be grouped into five major

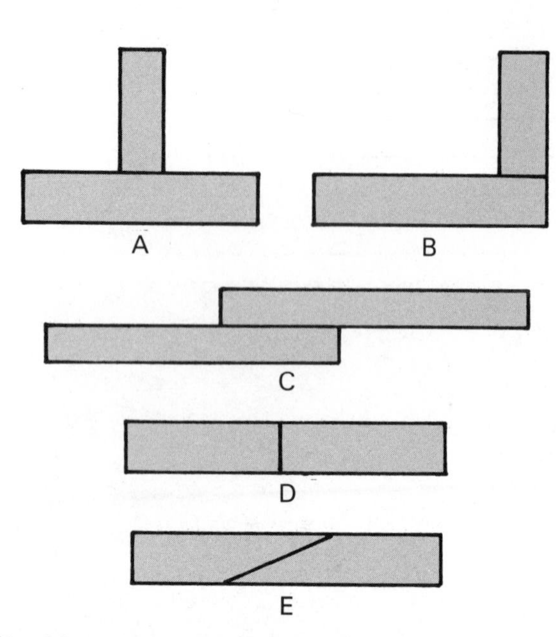

Fig. 26-4. Common joints used to join parts. A—T-joint. B—Corner joint. C—Lap joint. D—Butt joint. E—Skarf joint.

categories, Fig. 26-4.
1. T-joint.
2. Corner joint.
3. Butt joint.
4. Lap joint.
5. Skarf joint.

Each joint has its own use. These uses are easy to see. The T-joint and the corner joint join parts which are at an angle to one another. Often this angle is 90 deg. but any angle may be used.

Butt, lap, and skarf joints serve similar purposes. They join parallel parts. The butt joint provides a smooth joint but is the weakest of the three. The lap joint is strongest but is not a smooth transition. The skarf joint is a compromise between the two. It is harder to make but provides a smooth, reasonably strong joint.

Joints are chosen to suit their use. Appearance and strength are important factors. The joint must be attractive and able to withstand the stress applied to it. Typical stresses on joints, shown in Fig. 26-5, are:
1. *Tensile stress*—forces at right angles to the joint.
2. *Shear stress*—forces parallel to the joint.
3. *Cleavage stress*—forces at an angle greater than zero and less than 90 deg. to the joint.
4. *Peel stress*—forces trying to roll one part away from another.
5. *Torsion stress*—forces which rotate or twist the joined parts in opposite directions.
6. *Compression stress*—forces which squeeze the bonding materials.

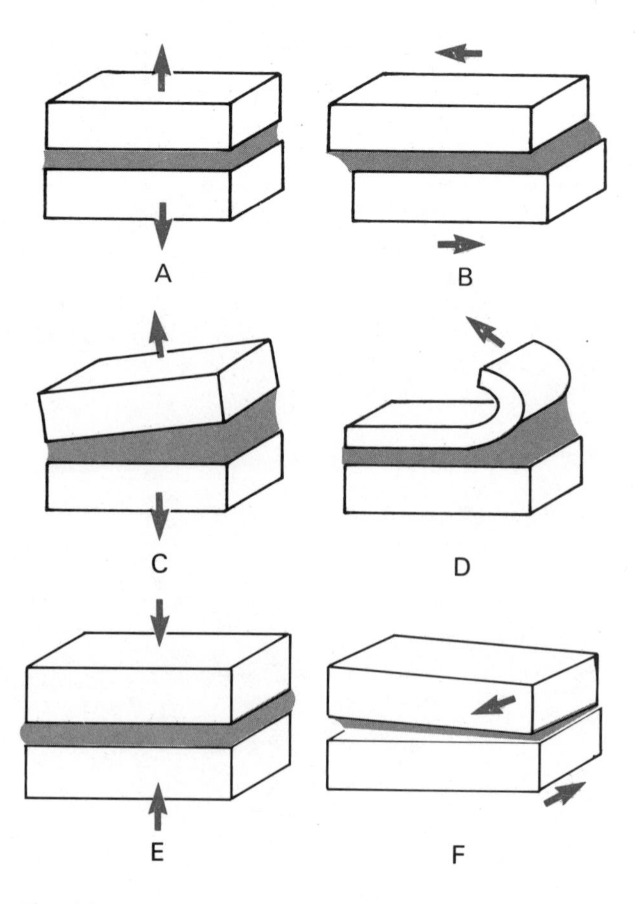

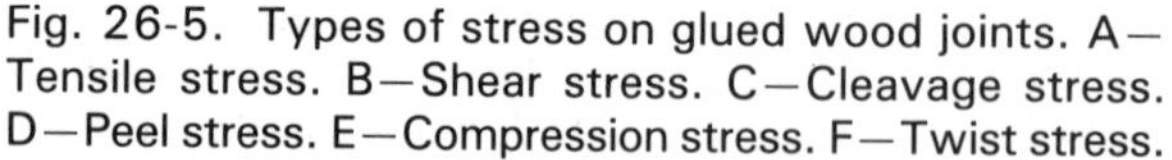

Fig. 26-5. Types of stress on glued wood joints. A—Tensile stress. B—Shear stress. C—Cleavage stress. D—Peel stress. E—Compression stress. F—Twist stress.

These forces can cause two types of joint failure, Fig. 26-6. Either the fastener or adhesive may fail.

When fastener is too weak it will fracture (break). Bonds which hold the adhesive material may also break. Each part in the assembly will have a coating of adhesive. But the adhesive itself has ruptured or broken. This type of failure is often called *glue line failure.*

The other type of failure involves material fatigue. The parts themselves may not withstand the stress applied. The parts fracture or allow a fastener to be pulled from or through them. The heat of welding, brazing and soldering processes can change the strength of the materials. The parts may become more brittle. Also, the material can be poorly chosen for its task. The stress can be greater than the natural strength of the material.

The proper selection of the fastening technique, the joint, and the material will produce an assembly which is able to withstand the applied stress or load. To increase the strength of an assembly, joints are often modified. Surface areas for adhesion are increased.

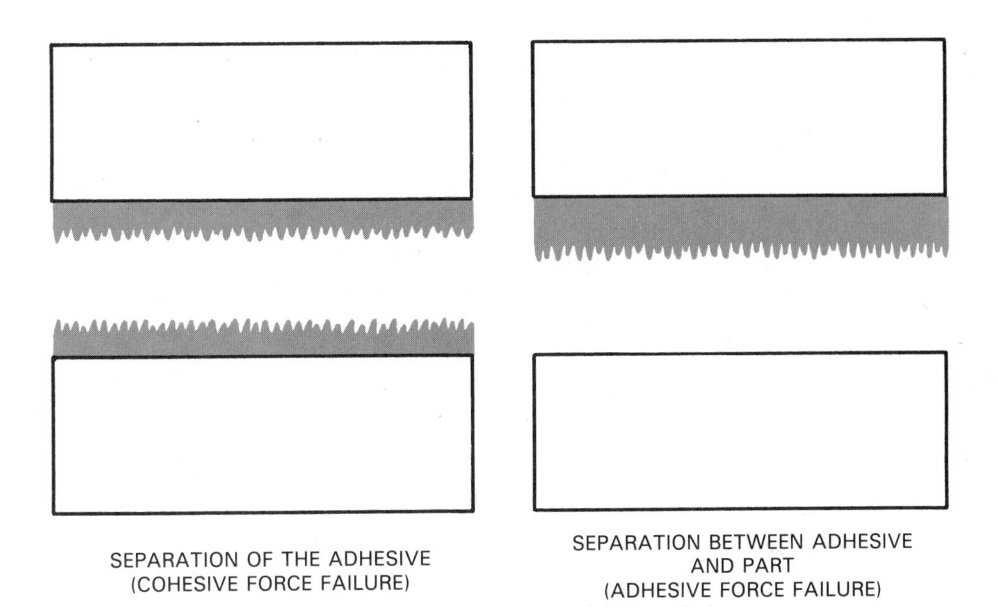

Fig. 26-6. There are two types of adhesive joint failure.

The joint is designed to transfer the load from the adhesive or fastener to the material itself.

Fig. 26-7 shows several modified woodworking corner joints. In several cases, the parts are made to interlock and increase the bonding surfaces. Consider how each of these are an improvement of the common butt corner joint.

JOINT PREPARATION

Materials must be prepared for assembly. The parts must fit tightly. The contact surfaces, in most cases, must be clean. Chemicals (usually acids) or abrasive materials may be used to remove grease, oil, and oxides. Dust and dirt can be wiped off.

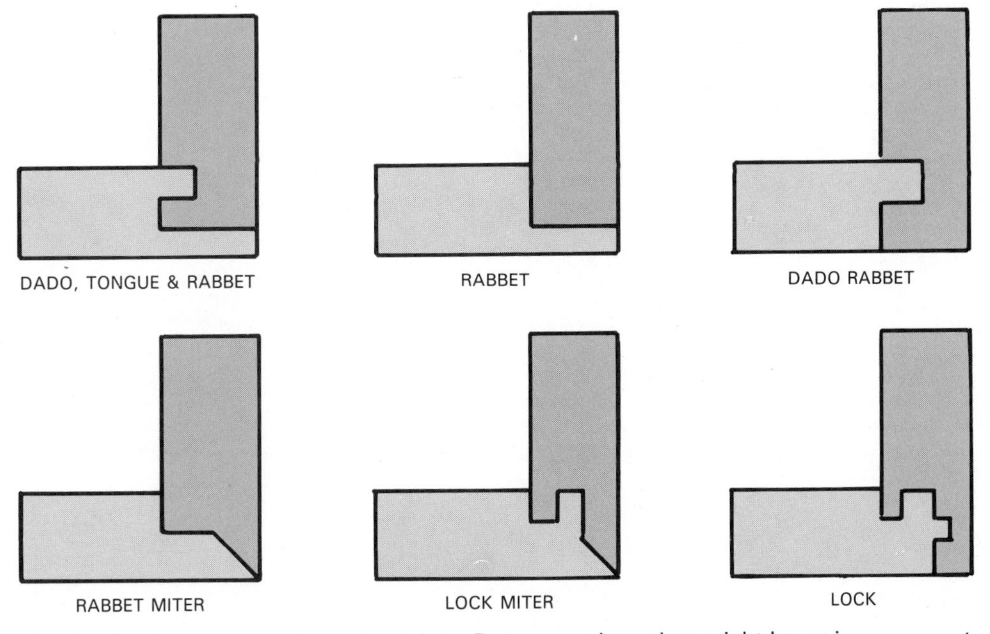

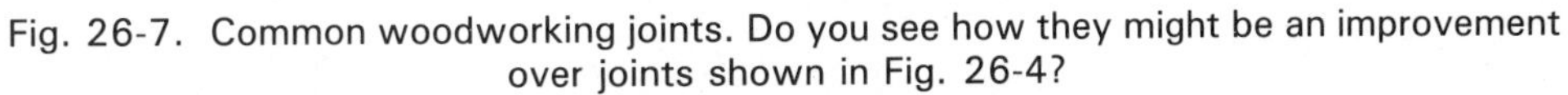

Fig. 26-7. Common woodworking joints. Do you see how they might be an improvement over joints shown in Fig. 26-4?

Special allowances for the joining materials are also considered. Holes for fasteners may need countersinking to accept the fastener head. Weld joints often are V-grooved. The groove allows for weld penetration and weld metal buildup.

Careful preparation of joining surfaces will increase the strength of the assembly. As with all processes, selection and preparation will insure desired results.

SUMMARY

Parts may be assembled using mechanical assembly processes and bonding processes. The mechanical processes include the use of fasteners or mechanical fits or forces. Bonding practices include use of cohesive and adhesives forces. They bond material using heat and/or pressure or by using an adhesive material.

The assembly must resist the forces placed upon it. Proper selection of the joint design, the fastening technique, the base materials, and the fastening material (adhesive, fastener) will insure a good assembly.

STUDY QUESTIONS – Chapter 26

1. Define assembling.
2. Define mechanical fastening.
3. What are the three types of assemblies produced by mechanical fastening?
4. What is cohesive force? Adhesive force?
5. List and describe the major types of joints.
6. List and describe the typical types of stress which can be exerted on a joint.
7. What are the two types of failure that can cause a joint to separate?

Chapter 27

WELDING PROCESSES

Welding is an important group of assembly processes. These processes use heat, pressure, or both, to permanently join two or more workpieces. Most welding processes assemble metal parts but plastics can also be welded.

The result of the welding process is a cohesive bond between the parts. The parts become fused into a single, uniform material. The molecules of the parts are brought so closely together by the heat and/or pressure that they form cohesive bonds. These bonds between the parts are exactly like the bonds that hold in place the molecules within the part.

WELDING CATEGORIES

The various welding processes differ primarily in the way they use heat and pressure. There are a number of ways to group these processes. The American Welding Society uses a six-category system: brazing, gas welding, resistance welding, arc welding, solid-state welding, and other processes. This system deals primarily with the method of generating heat and is based on metal welding processes. A more useful grouping accounts for all materials:

1. Heat welding processes.
2. Pressure welding processes.
3. Heat and pressure welding processes.

HEAT WELDING PROCESSES

The two major types of welding processes which use heat and no additional pressure are *gas welding* and *arc welding*. Gas welding, as the name implies, uses burning gases to generate heat. Arc welding uses an electrical spark (arc) as a heat source. A third process, *thermit welding,* is used to repair broken casting and fasten very large assemblies.

Gas welding

The most widely used gas welding technique is *oxyacetylene welding*. The heat is developed by combustion (burning) of acetylene with oxygen. These two gases are stored in high-pressure tanks. Each tank is equipped with regulators which control the pressure of the gas leaving the tanks. The gas flows through hoses from the tanks to the welding torch. The torch, Fig. 27-1, controls the flow of each gas and mixes them. When lit, a flame is maintained by the torch tip.

The flame can be adjusted to produce three types of flames. These, as shown in Fig. 27-2, are:

1. Neutral flame.
2. Oxidizing flame.
3. Carburizing or reducing flame.

A neutral flame uses about equal amounts of oxygen and acetylene. It is used for most welding applications. The neutral flame has the least chemical effect on the heated metal parts.

Oxidizing flames are produced by a higher ratio of oxygen to acetylene. The principal use for an oxidizing flame is for welding copper and certain alloys of copper. The most important of these alloys are brass (copper-zinc alloys) and bronze (copper-tin alloys).

A reducing flame has a smaller percentage of acetylene. This flame is used to weld low-carbon steels, nickel, monel, and for some flame surface-hardening processes.

Gas welding produces a bond by melting the two parts together. The metal must flow across a gap between the parts. This causes the joint area to be thinner. To overcome this action, a filler rod is usually used. The rod is also melted into the joint area. Refer to Fig. 27-3.

Gas welding rods

The most commonly used gas welding rods are:

1. Steel rods for welding iron and steel.
2. Cast iron rods for welding cast iron.
3. Brass (or bronze) rods for welding brass or for brazing (see Chapter 28).
4. Aluminum rods for welding aluminum and aluminum alloys.
5. Hard surface rods for producing hard surface coatings.

The melting of the rod and the base metal create

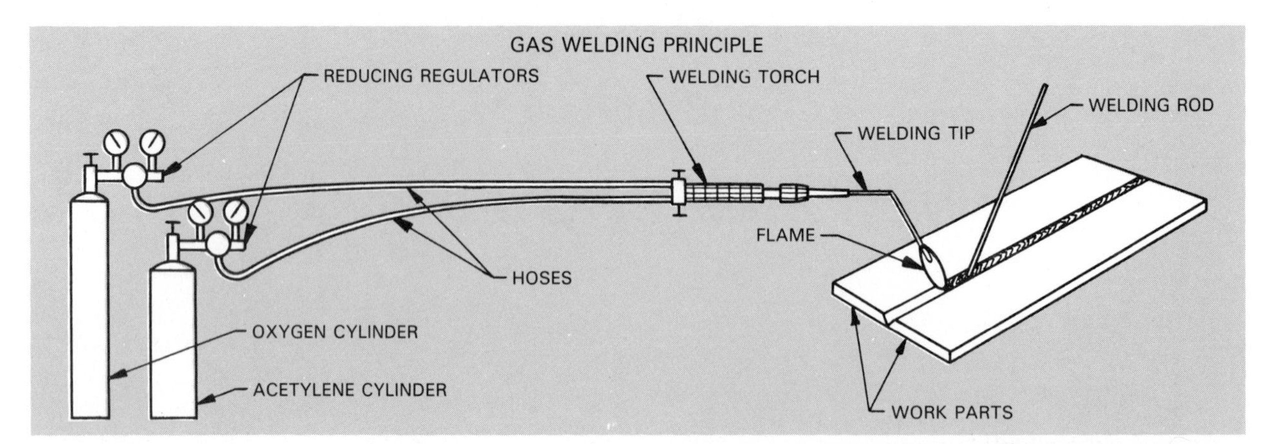

Fig. 27-1. A typical gas welding system. Torch flame will produce a temperature of about 6300 °F (about 3500 °C) at the inner cone. (General Motors Corp.)

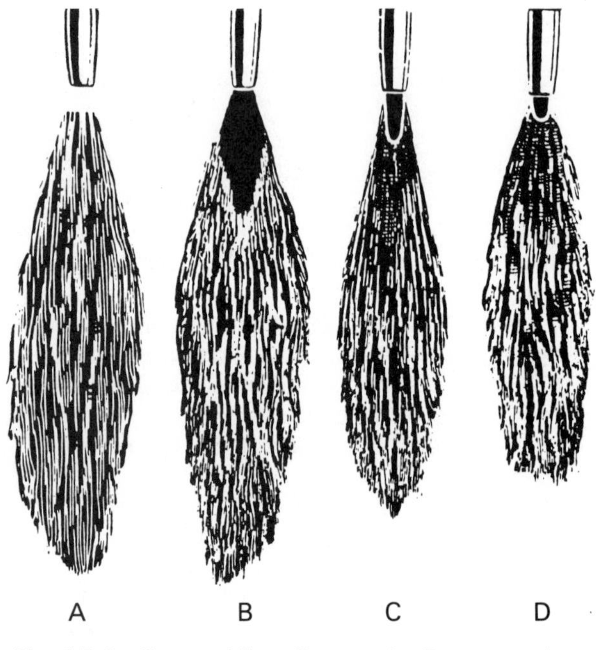

Fig. 27-2. Gas welding flames. A—Pure acetylene flame. B—Reducing flame. C—Neutral flame. D—Oxidizing flame. (Modern Engineering Co.)

Fig. 27-3. Gas welding aluminum using a filler rod. Rod provides extra metal for added strength. It also improves weld appearance.

a puddle of molten metal. The flame is moved slowly as the desired puddle size is developed. If the flame is directed toward unmelted base metal and moved away from the puddle, the act is called *forehand welding,* Fig. 27-4. *Backhand welding* also moves the tip away from the puddle. However the flame is directed toward the puddle.

Forehand welding preheats the metal before it is melted. A wider joint area is produced.

Backhand welding keeps the puddle hot longer. Also, the weld area cools more slowly. The slow cooling tends to anneal the weld area and removes internal stress. Backhand welding is often used in welding cast iron and thick steel members.

Multiple pass welds, Fig. 27-5, may also be used for thick weld areas. The first pass will insure good weld penetration. Later passes build up and complete the weld.

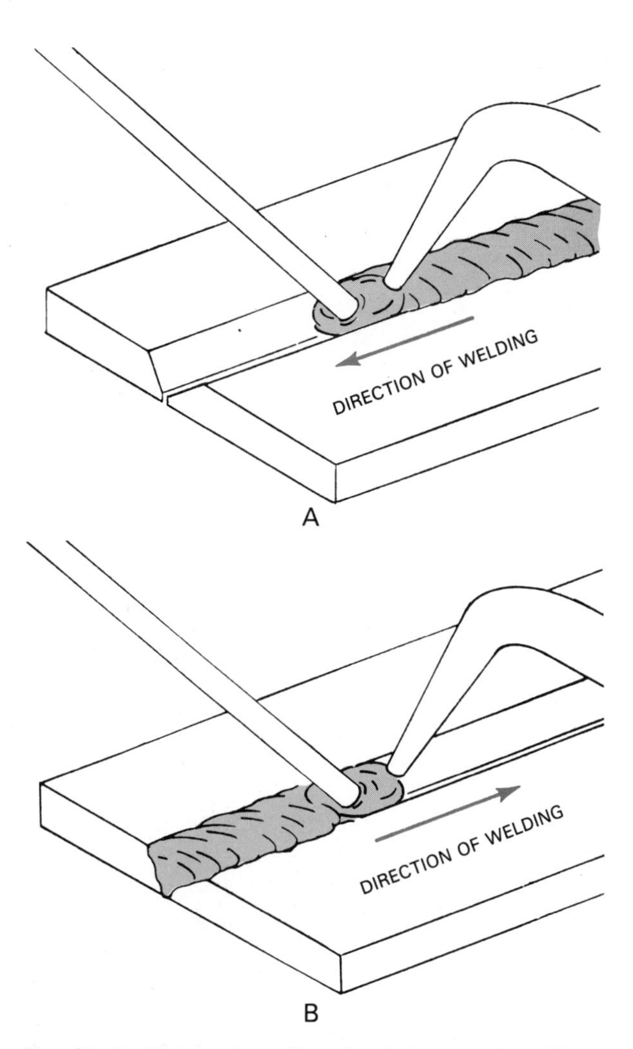

Fig. 27-4. Types of welding. A—In forehand welding, also called puddle or ripple welding, rod is moved ahead of the torch tip in the direction of the weld being made. B—In backhand welding, the torch tip moves ahead of the welding rod in the direction of the weld. (AIRCO Welding Products)

Acetylene is a very dangerous gas. It is explosive over a wide range of mixtures with oxygen or air. To overcome this hazard, other gases are sometimes used. An important substitute for acetylene is MAPP (methylacetylene propadiene). This gas is less explosive than acetylene and can be stored in ordinary pressure tanks.

Oxygen and hydrogen may also be used for welding. Oxyhydrogen gas burns at about 3600°F (1980°C) which is much lower than oxyacetylene. This cooler flame is used to weld thin sheet metal and low-melting alloys. Oxyhydrogen welding uses the same equipment as oxyacetylene welding.

Gas welding is not widely used in volume production application. It is relatively slow. Also the metal

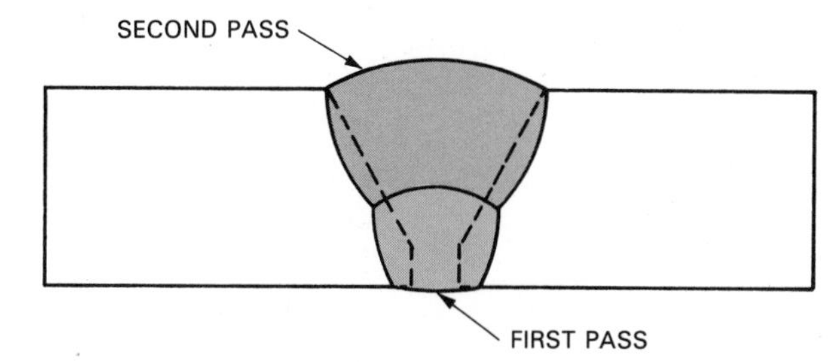

Fig. 27-5. Cross section of multipass welding. A number of passes may be needed.

is often effected by the long heating period required to produce the weld. The material can be warped or distorted by this prolonged heat.

Thermoplastic materials may also be welded using burning gases. The material is prepared like metals and a plastic filler rod is melted and bonds the base materials together. Like gas welding of metals, the quality of the weld is largely determined by the operator's skill.

Arc welding

Arc welding overcomes some of the disadvantages of gas welding. Heat, in the range of 10,000°F (5538°C), is rapidly produced by an electric arc. The arc is developed between a rod or electrode and the work. The spark carries high current at a relatively low voltage. The basic electric arc system is shown in Fig. 27-6. As can be seen, the basic equipment includes:

1. An alternating current (ac) or direct current (dc) power supply.
2. An electrode holder.
3. A lead from the power supply to the electrode holder.
4. A lead from the power supply to the work.

The circuit is completed when a spark (arc) is produced between the electrode and the work. Current flows around the circuit to maintain the arc. When the electrode is scratched on the work, an arc starts. The arc continues as long as the proper gap is maintained between the work and the electrode. If the electrode is moved away from the work, the circuit will be broken and the arc stops.

Arc welding involves two major processes: consumable electrode and nonconsumable electrode. Their subdivisions are:

1. Consumable electrode arc welding.
   a. Shielded metal arc welding.
   b. Flux cored arc welding.
   c. Gas metal arc welding.
   d. Submerged arc welding.
   e. Electroslag welding.

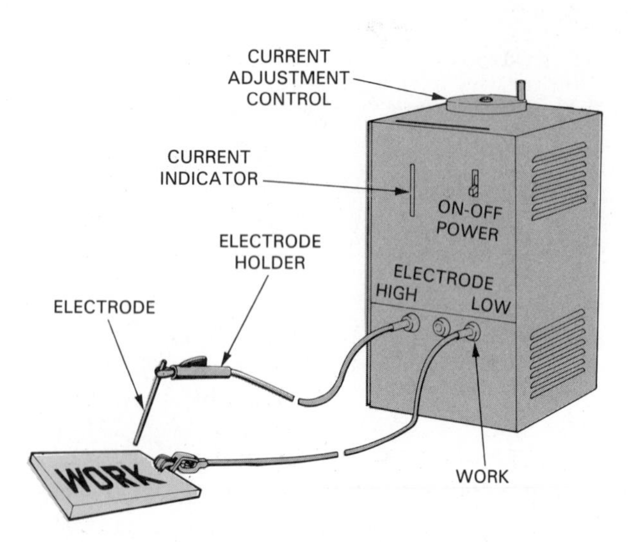

Fig. 27-6. A basic arc welding setup. Electrode acts like a switch. (Miller Electric Co.)

2. Nonconsumable electrode arc welding.
   a. Gas tungsten arc welding.
   b. Plasma arc welding.
   c. Stud welding (stud is both electrode and part).

All these techniques have one feature in common. The welding heat is produced by an electric arc. Each technique has one or more unique features which make it a separate process.

Shielded Metal Arc Welding

All molten metals are affected by the atmosphere. Oxygen and hydrogen are absorbed from a normal atmosphere. Upon cooling, the molten metal has less strength and may be brittle. Since welding involves molten metal, the atmosphere around the weld area is important. Often, the weld needs to be shielded from the normal atmosphere.

Shielded metal arc welding, Fig. 27-7, uses an electrode which is a filler rod with a chemical coating. Some of the coating breaks into a cloud of protective gas when the arc is struck. This cloud provides a shield around the weld area.

This type of arc welding is used in a number of applications. Most important of these are general repair work, on-site building construction, and low-volume production.

Flux cored arc welding

Flux cored arc welding is very similar to the gas shielded process. The main difference is that gas shielded arc welding uses a flux coated stick while flux cored arc welding uses a hollow core tubular wire electrode. The core is filled with a chemical compound. During welding the electrode is automatically fed into the weld area.

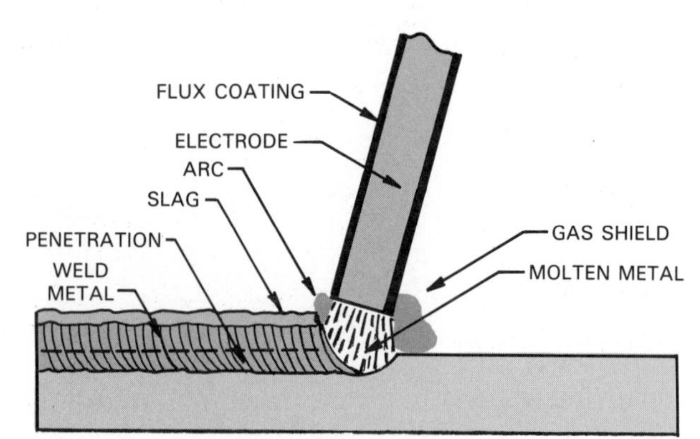

Fig. 27-7. Diagram of the shielded metal arc welding process. Shielding is formed as the chemical coating of the filler rod turns to a cloud of gas.

The process just described is called self shielding. A second variation is called external shielding. In this process, additional shielding is developed by an external gas—usually CO_2 (carbon dioxide). In both cases, the gas shields the weld area against possible oxidation from normal atmosphere. Fig. 27-8 shows flux core arc welding.

This process was developed to allow faster welding than the gas-shielded technique. The coil electrode does not have to be constantly inserted in the rod holder. Rather it is continuously fed from a moisture-protected container to the weld area.

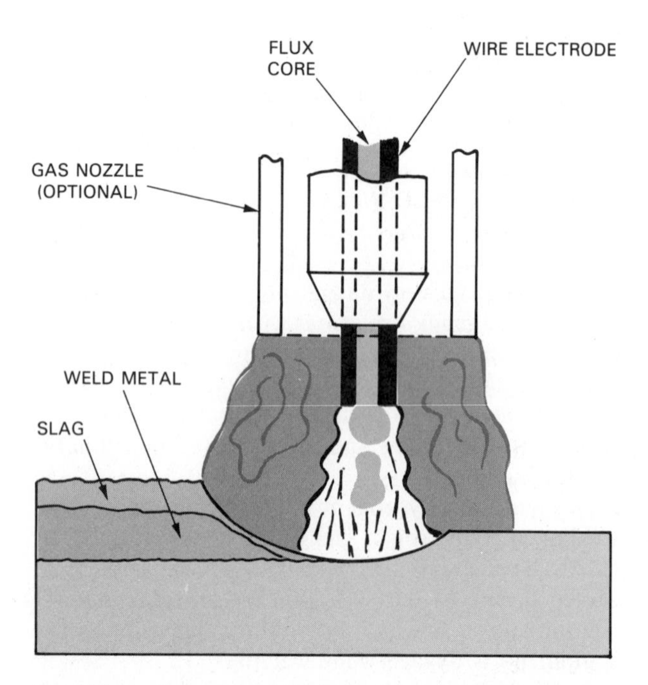

Fig. 27-8. Diagram of flux cored arc welding. As the electrode melts, the internal flux changes to a gas shield.

Gas metal arc welding

Often called MIG (metal inert gas) welding, gas metal arc welding has two main features:

1. A continuously fed electrode and an external inert gas shield.
2. An arc produced between the electrode and the work. The arc melts the electrode which is automatically fed through the holder, Fig. 27-9.

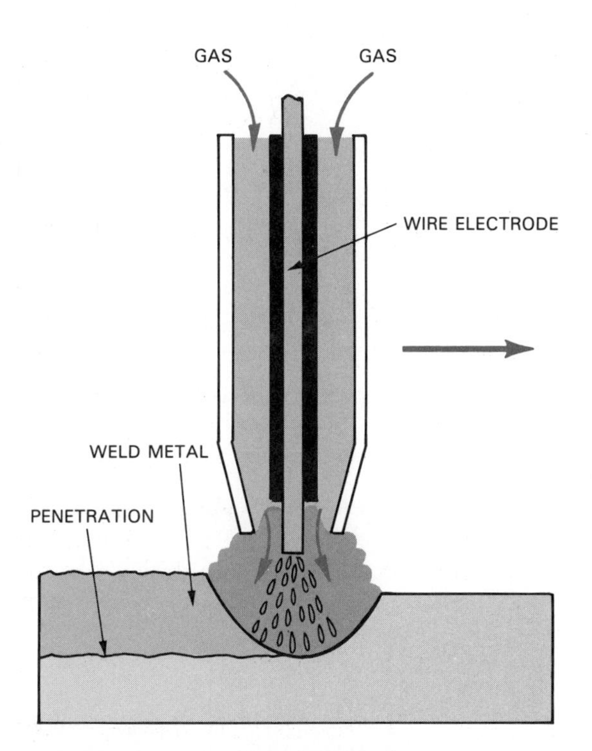

Fig. 27-9. Gas metal arc welding. Argon, helium, or a mixture of the two gases covers the weld area.

MIG welding does not produce slag (impurities) on top of the weld bead. Thus, the operator can make multiple-pass welds without removing slag between passes. MIG welding may use either manual (operator-controlled electrode movement) or automatic welding equipment. MIG welding works well on aluminum magnesium, copper, and steel.

Submerged arc welding

Submerged arc welding shields the weld area under a layer of fusible granular material. The filler metal is obtained from the consumable bare wire electrode. The shield material is applied directly ahead of the weld, Fig. 27-10.

Submerged arc welding is fast, has high penetration and deposition rates, and produces smooth welds. The process works on flat and horizontal surfaces where the flux will not fall off.

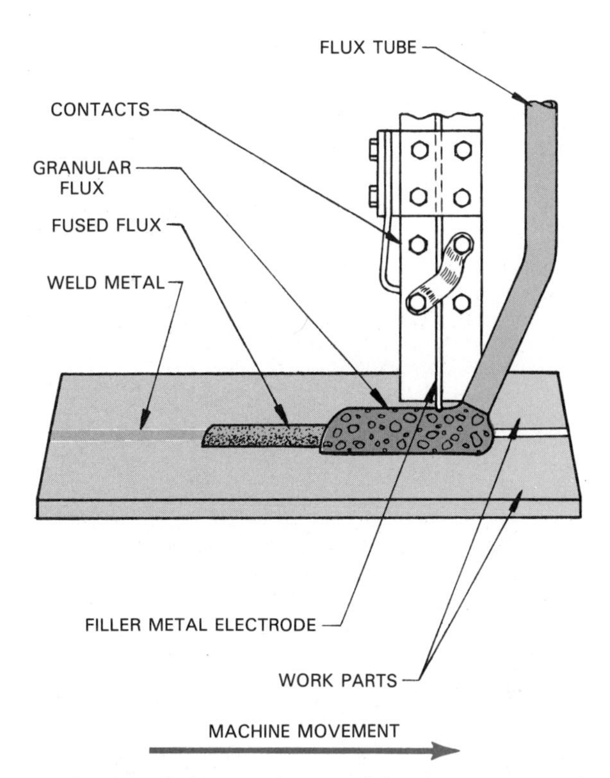

Fig. 27-10. Submerged arc welding is an automatic welding process. (General Motors Corp.)

Electroslag welding

Electroslag welding is not actually an arc welding process. It uses the same equipment, however. The heat for welding is produced by the resistance of the slag to electric current.

Electroslag welding is started with an arc between the electrode and the work. The arc is quickly put out by the slag as it melts. Additional heat is produced by current flowing through the molten slag. This heat melts the electrode as it is automatically fed into the weld area. The weld is shielded from oxidation by the molten slag.

The slag and molten weld material are contained in the weld area by water-cooled shoes or dams. As the electrode moves, the shoes also move. This action contains the molten weld material until it cools.

Electroslag welding, as shown in Fig. 27-11, is used to fabricate thick workpieces (weldments) and for butt welding plate metal. Material thicknesses from 1/2 to 12 in. can be successfully welded.

Gas tungsten arc welding

Gas tungsten arc welding is one of the nonconsumable electrode processes. This process produces an arc between the work and one or more tungsten electrodes. The weld area is protected by an inert gas which is usually argon, helium, or a mixture of both.

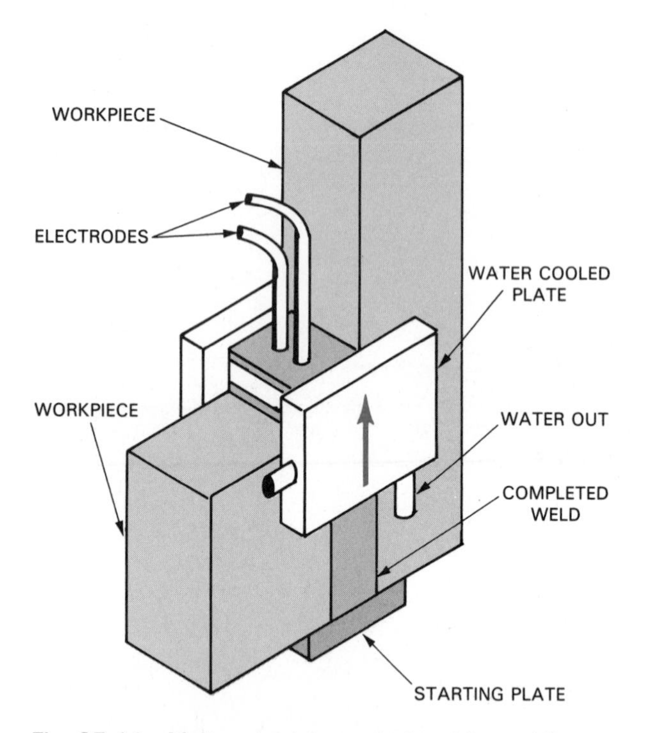

Fig. 27-11. Molten slag (electroslag) welding, while not an arc welding process, uses the same equipment.

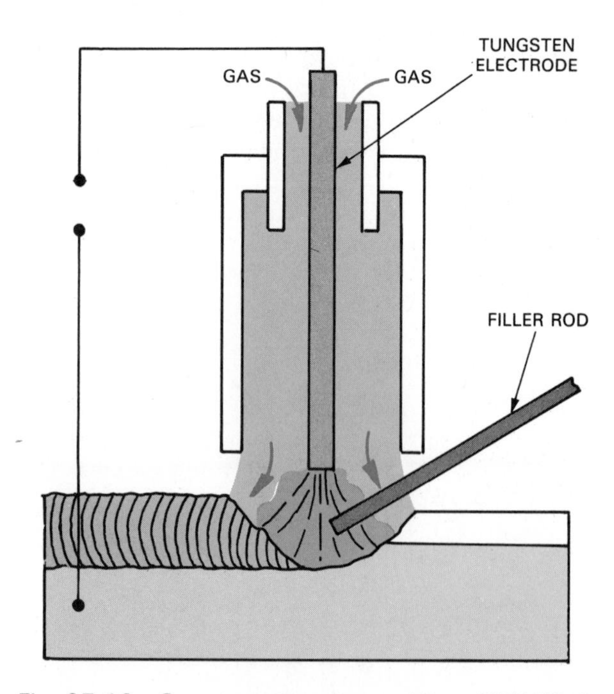

Fig. 27-12. Gas tungsten arc welding (GTAW) is popularly known as "TIG" (for tungsten inert gas).

The heat generated by the arc will melt the base metal but not the electrode. Filler metal is added from outside the arc. This technique reduces the splatter since the filler rod does not cross the arc.

Gas tungsten arc welding (GTAW) is employed for flat or vertical welding. Its widest use is in welding nonferrous metals. Aluminum, magnesium, nickel alloys, and copper alloys are commonly welded using the GTAW process, Fig. 27-12.

Plasma arc welding

Plasma arc welding, Fig. 27-13, also uses a tungsten electrode and an inert gas. However, the arc is used to heat and ionize a gas. The ionized gas will easily conduct current. This gas is then forced through a nozzle. The result is a narrow column of very hot ionized gas which conducts the arc. Temperatures can reach beyond 24,000°F (13 316°C). Plasma arcs can also be used for cutting. The high temperatures and concentrated arc will quickly burn through the metal.

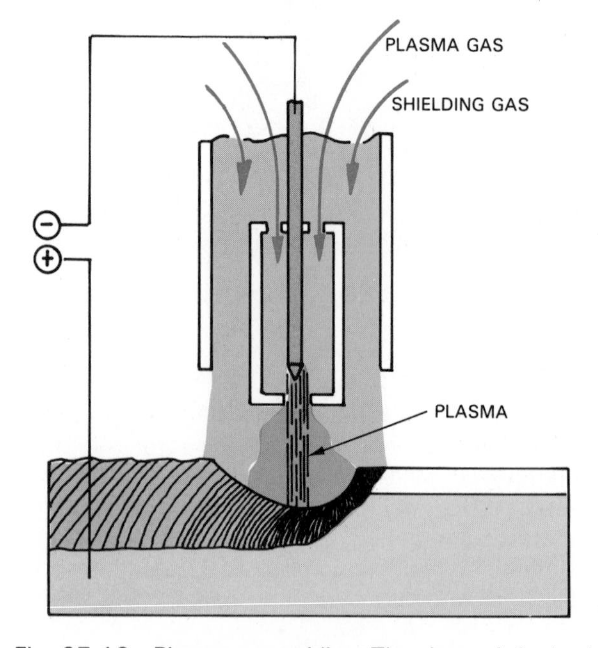

Fig. 27-13. Plasma arc welding. The plasma is ionized (atoms have lost or gained one or more electrons) as a result of super heating.

Stud welding

Stud welding is used to attach studs, bolts, pins, or hangers to metal plate. The process uses a special stud gun shown in Fig. 27-14.

The stud is made to function as an electrode. It is put in the gun and brought into contact with the work. Current is applied as the stud is moved away from the work. An arc develops between the stud and the work which melts the end of the stud and the work below it. The current is automatically shut off. The molten end of the stud is pressed into the molten section of the work by a spring in the gun. The molten areas

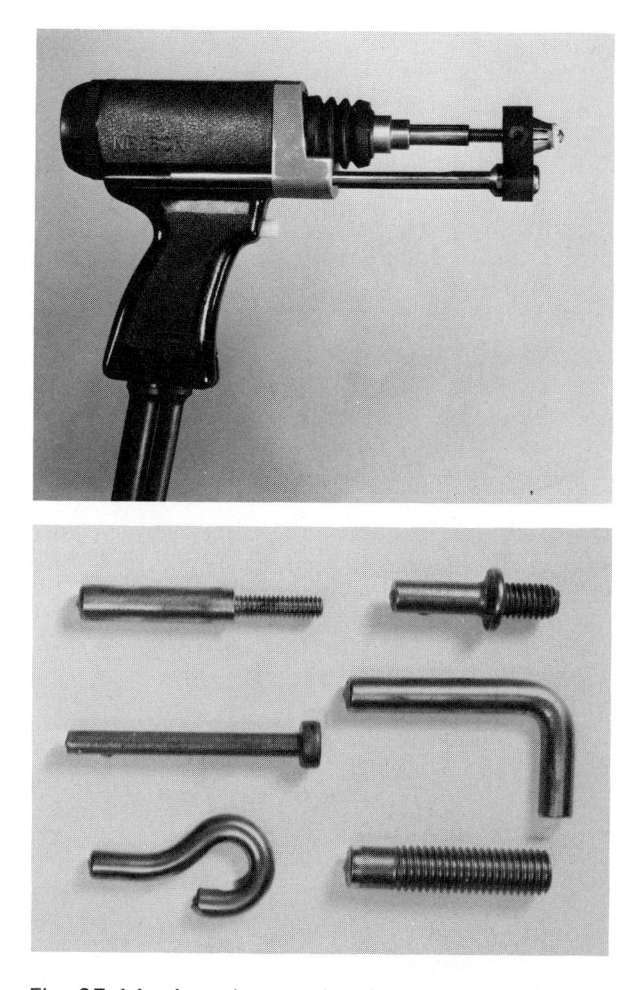

Fig. 27-14. A stud gun and typical devices which can be welded with it. The devices are used as electrodes. (Nelson Div., TRW)

solidify, welding the stud to the work, Fig. 27-15.

Stud welding can be semiautomatic or automatic and can usually be switched from one mode to the other. The process can weld studs vertically, horizontally, or overhead.

Thermit welding

Thermit welding finds its primary uses in repair work and for assembling parts too large to weld by other processes. Actually, thermit welding is a combination of casting and welding.

First a mold is made. Wax is used to form the shape of the weld bead. The wax is placed over the crack or area to be joined. Then a wax riser and sprue are attached. Foundry sand is packed around the wax patterns to form a mold. Heat is applied to melt the wax and preheat the parts to be bonded.

The metal which will become the weld material is melted. The melting process involves a chemical reac-

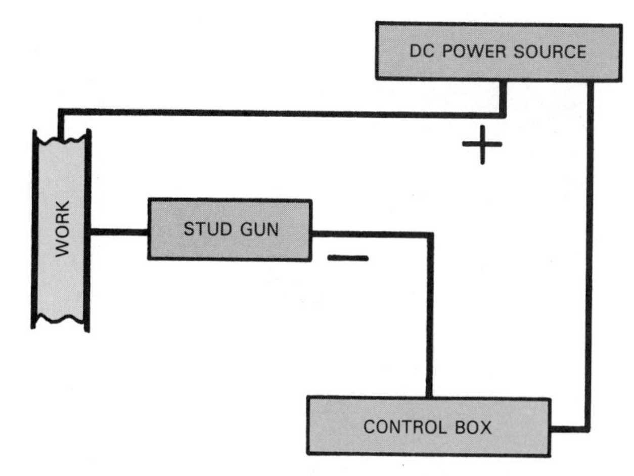

Fig. 27-15. The stud welding circuit is like any arc welding circuit. (Nelson Div., TRW)

tion between pure aluminum and the oxide of the weld material—iron, copper, nickle, or manganese. The reaction produces pure metal from the oxide and aluminum oxide (slag). The thermite mixture (aluminum and metal oxide) may be placed in the mold then ignited by a chemical. Or it may be placed in a crucible and lit. The resulting chemical action takes about 30 seconds and reaches about 4500°F (2482°C) for iron to over 9000°F (4982°C) for manganese.

If the mixture was originally in the weld zone, it is allowed to cool. Otherwise, it is poured from the crucible into the mold. The hot iron from the thermite mixture melts the base metal. Upon cooling, a fusion bond is formed.

There is no limit to the size of the weld that can be made. The thermite reaction can melt well over a ton of material. The process does, however, require building a mold for each weld. This requirement restricts the use of the process to applications where other welding processes are not economical or practical.

PRESSURE WELDING PROCESSES

Parts may be forced to fuse to one another by applying large amounts of pressure. The pressure breaks down any oxide layers on the parts. Also, the molecules on the surfaces are forced closely together. The result is a cohesive bond without changing the surfaces into a liquid, Fig. 27-16.

This major pressure welding process is called *cold welding*. Mechanical force creates the necessary bonds. Heat is not generated from an outside force nor required by the welding process, itself.

Cold welding will only work on material which is

Fig. 27-16. A cross section of a cold welded joint. Note "flash" extending on each side of the weld. (Kelsey-Hayes Co.)

highly ductile and does not work harden easily. Therefore, nonferrous metals are most successfully bonded by this method. Pressure may be supplied by rolls, punches and dies, or explosives, Fig. 27-17.

Roll bonding uses two rolls to force the layers to the base metal. Simple cold welding presses, Fig. 27-18, squeeze the parts together to form the bond.

Explosive welding or cladding uses an explosion as the bonding force. The metal is placed between an explosive mat and a support. A charge is exploded above the mat. The shock wave is transmitted through the mat, forcing the metal layers together.

Cold welding is also used to bond materials which cannot be welded easily using heat. Low melting point metals and metals with different melting points are often cold welded.

Copper can be bonded to other metals. Also aluminum wire and rod can be butt welded. Thin aluminum or copper sheets which are hard to heat-weld can be cold welded.

HEAT AND PRESSURE WELDING

Many welding processes use a combination of heat and pressure. Heat is used to change the surfaces to

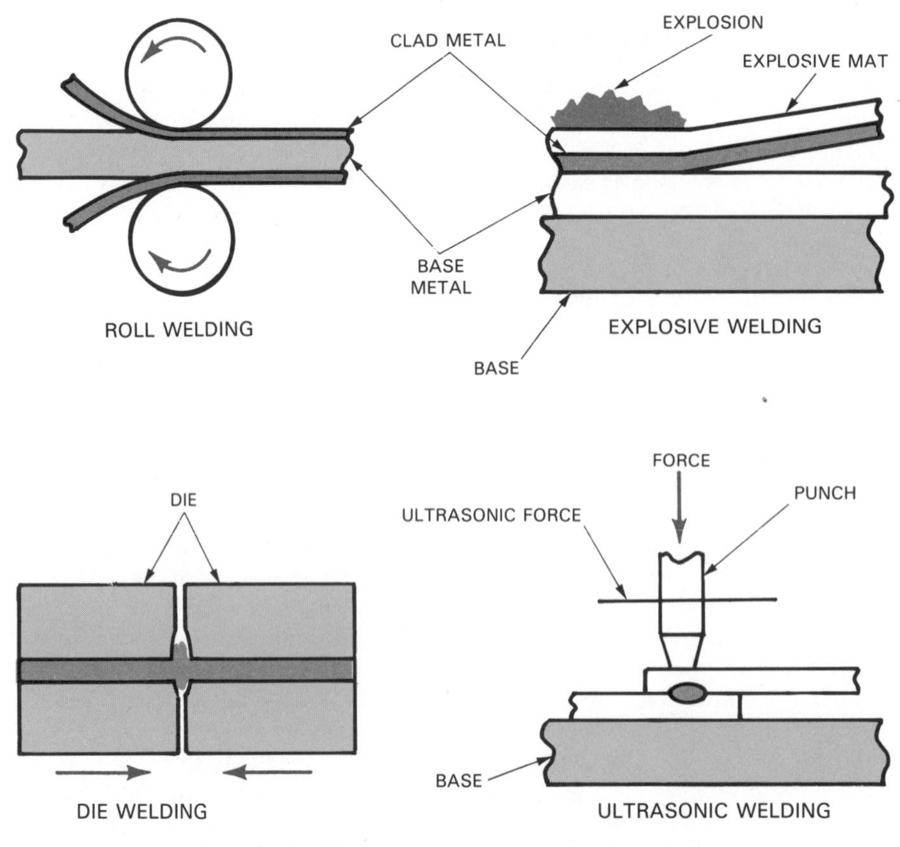

Fig. 27-17. There are four types of cold welding.

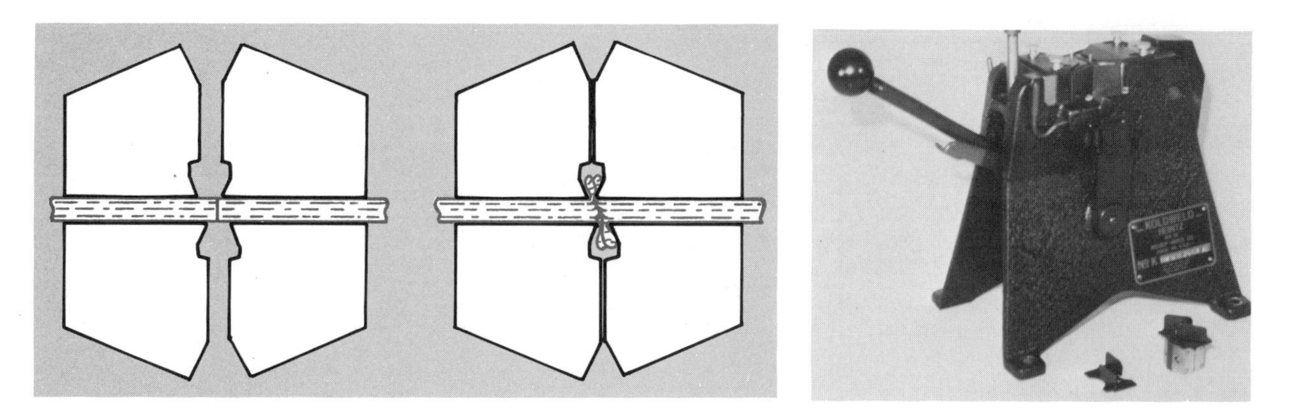

Fig. 27-18. Dies and a cold weld machine are used to produce cold welds. (Kelsey-Hayes Co.)

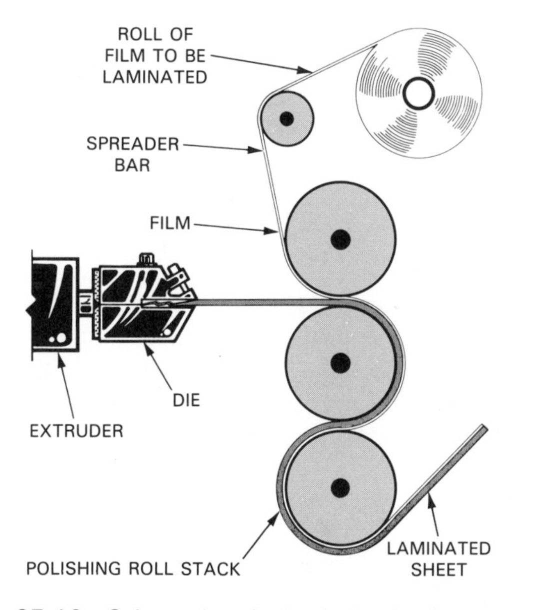

Fig. 27-19. Schematic of plastic lamination setup. Bonding of film to a thicker base is done wihtout using adhesives. (Dow Chemical Co.)

their molten state. Pressure is then used to force them together. The most common of these welding processes are:

1. Forge welding.
2. Roll laminating.
3. Resistance welding.
4. Induction welding.
5. Energy beam welding.
6. Friction welding.
7. Diffusion bonding.

Forge welding

Forge welding is much like cold welding. Both are solid state welding processes; neither requires the material to be melted. The main difference is that, in forge welding, the parts are heated first. Then they are forced together.

Forge welding was one of the earliest types of welding to be developed. It was practiced by the blacksmith of earlier times. Forge welding has only limited use today and has been largely replaced by more modern processes.

Roll laminating

Roll laminating is a plastic bonding process which is very much like forge welding. The plastic film is preheated and then brought together on a combining roll. The hot film fuses together producing a two or three-ply lamination.

The lamination, as shown in Fig. 27-19, is often done with similar films. It is possible to laminate a thin, expensive film on a heavier gage base. This allows for wider use of expensive films. The final laminate is much cheaper than an exotic film of the same thickness. Laminates with high-temperature and high-strength outer layers are used in aerospace and biomedical applications.

Resistance welding

Resistance welding encompasses a group of processes which:

1. Generate heat through the resistance of the weld material to electric current.
2. Use pressure to secure the weld.

Resistance welding techniques include a number of variations. The major ones are:

1. Spot welding.
2. Projection welding.
3. Seam welding.
4. Flash welding.
5. Upset welding.
6. Thermal sealing.

All resistance welding requires a basic electrical circuit, Fig. 27-20. It includes a power source, a

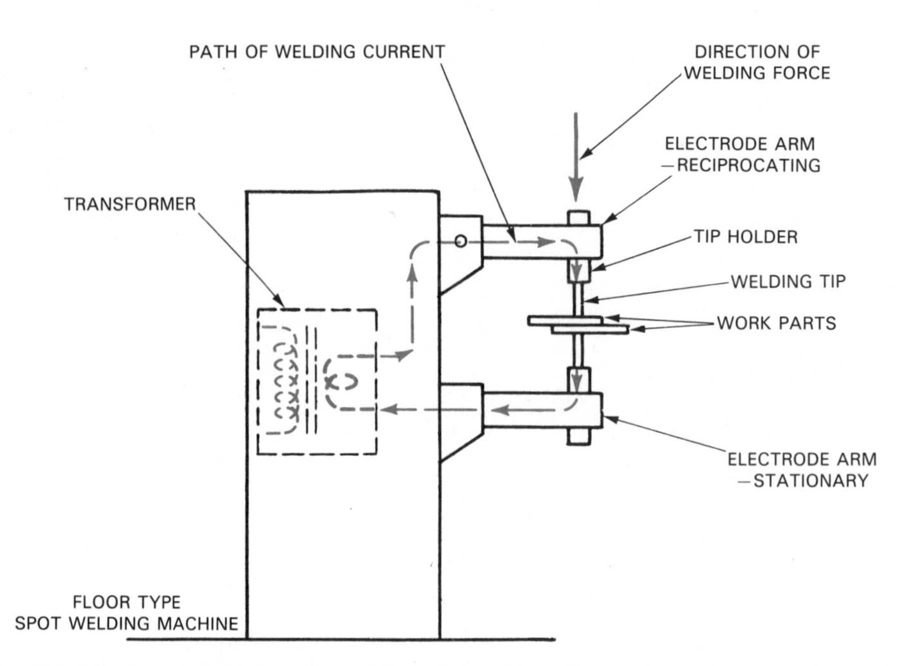

Fig. 27-20. A typical resistance welding circuit. Transformer is used to step down the voltage. (General Motors Corp.)

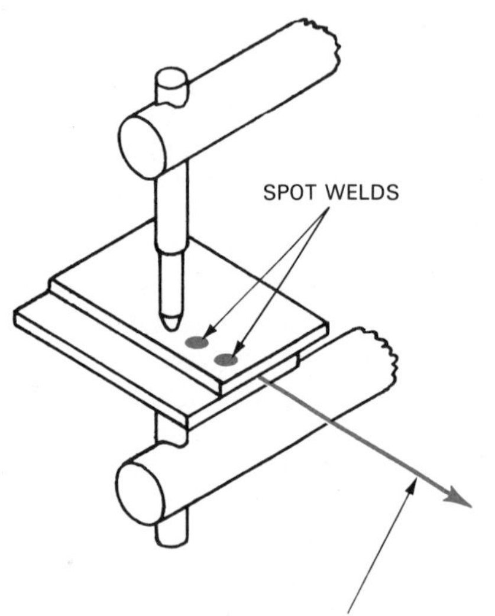

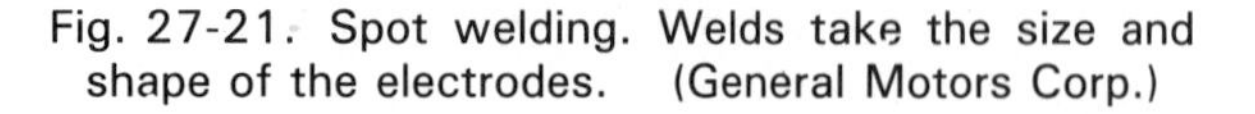

Fig. 27-21. Spot welding. Welds take the size and shape of the electrodes. (General Motors Corp.)

transformer, a timer, a switch, and a set of electrodes or dies. The transformer produces low-voltage, high-amperage electrical current from the power source. The switch starts the weld cycle and the timer stops it. The electrodes or dies conduct the current and the pressure to the work. The current heats the work to its molten state while the pressure squeezes the work together.

Resistance welding processes are generally used for high-production applications. Band and rod steel products are assembled using resistance welding. So are automobiles, appliances, tubing, and other similar products.

Spot welding

One of the most common resistance welding processes is spot welding, Fig. 27-21. Probably the simplest of this family of processes, it is used to assemble sheet metal parts such as car body parts, refrigerator cases, and metal cabinets. Wrought iron (band and rod steel) products are also often spot welded.

The basic process involves four major steps:

1. Parts are placed between electrodes and pressure is applied (squeeze time).
2. The electrical circuit is closed and current flows between the electrodes (weld time). Heat melts the metal directly between the electrodes.
3. Current is stopped and pressure is maintained until the weld cools (hold time).
4. The electrodes are released and the work is removed (off time).

Spot welding produces a nugget of fused metal which bonds the parts together, Fig. 27-22. A good weld will not pull apart. If the parts are separated the sheet will tear around the weld nugget. To increase the holding power and reduce the change for sheet tearing, a series of spot welds are usually used.

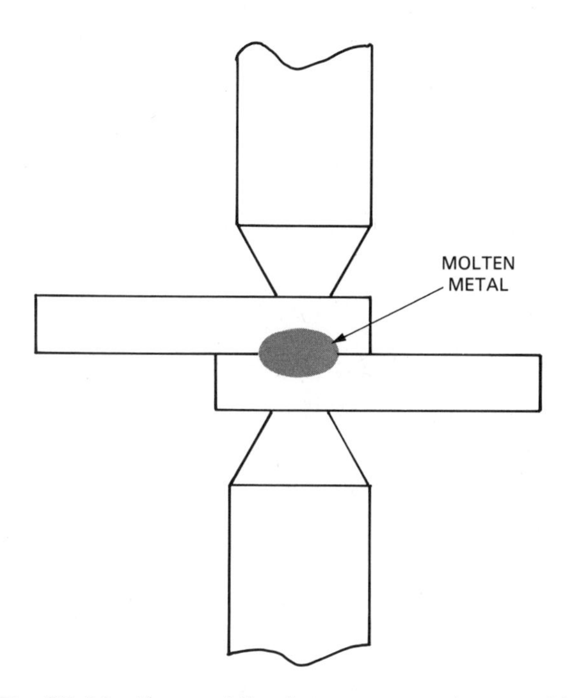

Fig. 27-22. Spot welding fuses two metal parts with a nugget of molten metal.

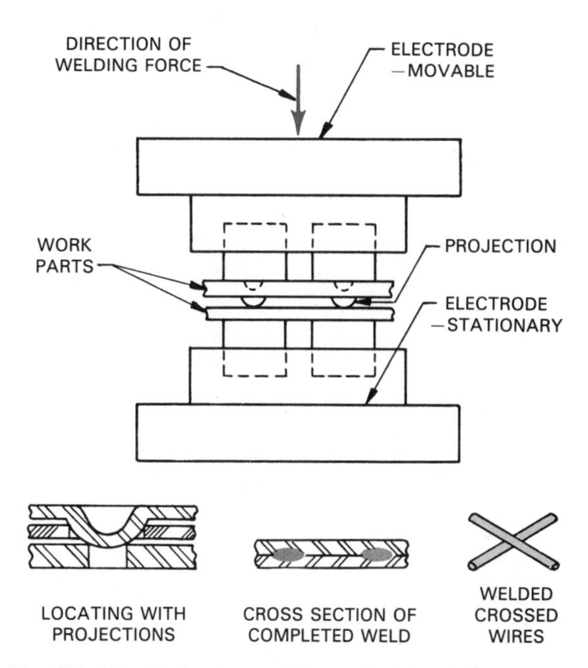

Fig. 27-23. Projection welding principle. This process, a variation of spot welding, can produce more welds per unit than a spot welder. (General Motors Corp.)

Spot welding is a rapid fastening process. It can be either fully automated or manually operated. Today, robots on continuous assembly lines complete many spot welds.

The major disadvantage of spot welding is its appearance. The material is slightly forged during the hold time (period after the molten nugget is formed and solidified). The clamp pressure produces a slight indentation in the material.

Projection welding

Projection welding, Fig. 27-23, is a variation of spot welding. The difference lies in the design of the parts to be welded. They are shaped or formed to have dimples or projections. These projections concentrate the electrical current between the electrodes and therefore the heat. The projection ares melt and fuse rapidly.

Projection welding can make several welds at one time. The electrode can cover an area with several projections. When the electrical current flows, it will be concentrated at each dimple. This activity melts the nuggets of metal where each projection contacts the base metal producing multiple welds.

Projection welding finds wide application in welding wire into screens, attaching wire grills to frames, and attaching threaded studs or nuts to sheet metal base material. In the last application, the stud or nut becomes a projection. The current flowing through it melts its end and the base material at the point of contact. Special threaded fasteners are available for projection welding applications.

Seam welding

Seam welding involves a series of spot welds which either overlap or are very close together. Seam welding uses basic resistance welding equipment. Its major difference from other resistance welding is in the rotating disc or wheel electrodes.

The material is passed between the rotating electrodes. The discs grip and force the work together. Current passes between the discs creating a series of spot welds. The area of the weld is covered with a coolant. The spacing of the welds is determined by the length of the on and off cycles of the welder. Short off cycles will cause overlapped welds. Longer off cycles will space the welds apart. When cycles are widely spaced, the process is often called *roll spot* or *stitch welding.*

Seam welding, as pictured in Fig. 27-24, may be used to produce lap, flanged, or butt seams. When the welds are overlapped the product can be used for liquid or pressure resistant applications. Typical of these uses are gasoline cans and tanks, engine mufflers, and air and water tanks. Seam welding is also used to butt join two sheets or plates. The most widely used application is probably in producing continuous welded pipe.

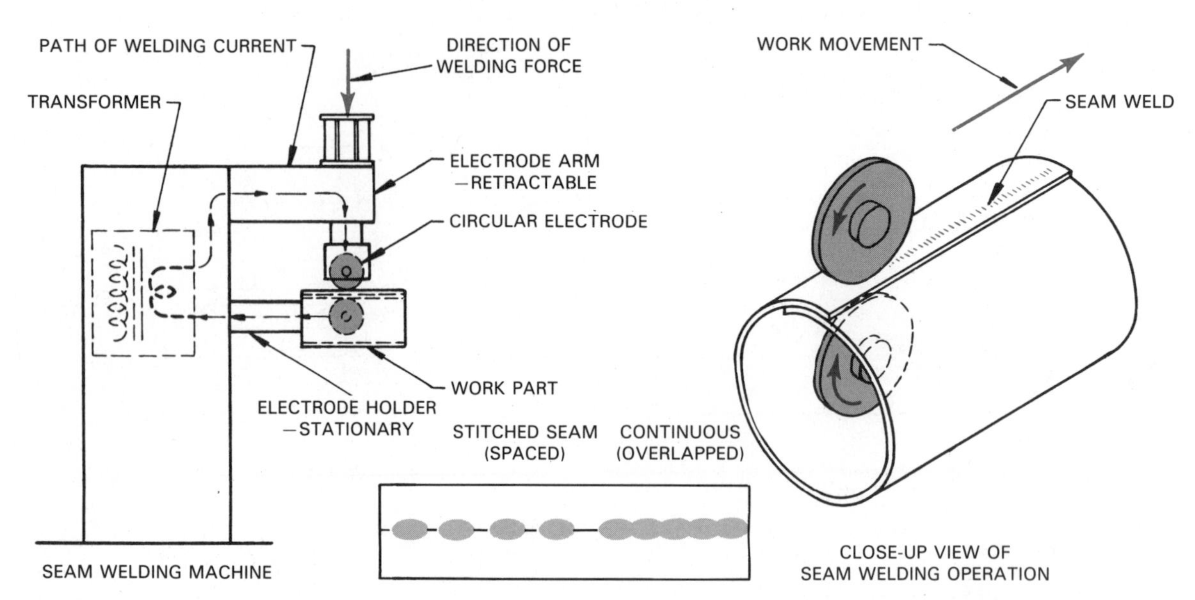

Fig. 27-24. Seam welding is a type of resistance welding which uses discs or wheels as the electrodes.

Flash and upset welding

Flash and upset welding are different processes but similar in result. They both use similar equipment and are used to weld bars, sheets, and strips.

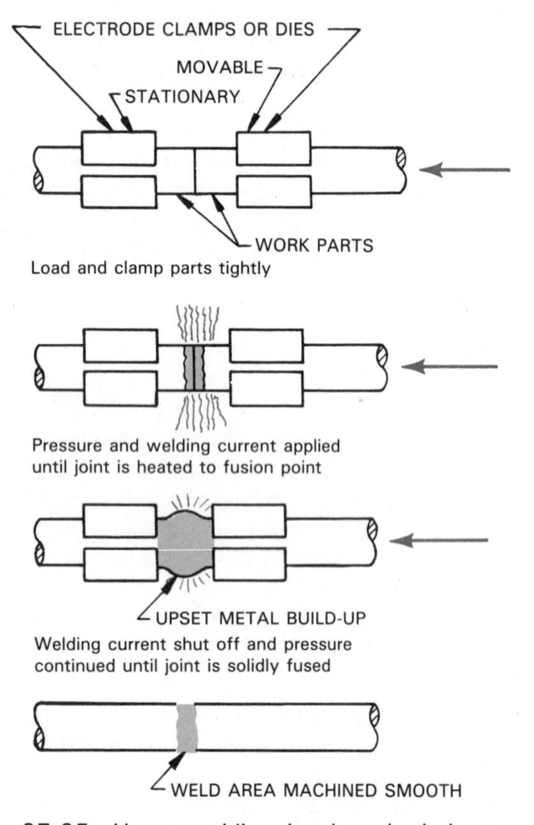

Fig. 27-25. Upset welding involves both heat and pressure. (General Motors Corp.)

In upset welding, Fig. 27-25, the workpieces are tightly butted. Electrical current is then passed through the work. Resistance to the current causes heating of the work. The greatest resistance is at the point where the two parts meet. This high resistance zone melts first. The parts are, then, forced more tightly together causing both bonding and an upset forging action. The result is a welded seam which has greater thickness or diameter than the base materials.

A process similar to upset welding is used for thermoplastic material, Fig. 27-26. The plastic parts are placed against a heated tool. When they are soft, the tool is removed. The parts are forced together creating the bond. Parts, from small cigarette lighter bodies to large diameter pipe, may be assembled using this process.

Flash or percussion welding holds the parts slightly apart, Fig. 27-27. The electrical current jumps the gap, creating high temperatures. The heat melts the ends of the parts. They are then forced together creating the weld.

Flash welding is often chosen over upset welding. The flash process produces a stronger weld with less electrical power. Also, upsetting at the joint area is less. A further advantage is that less heat is generated in the parts themselves. This factor reduces chances of distortion of the welded parts.

Thermal sealing

Thermal sealing is another resistance welding process used for plastic materials. The process bonds thermoplastic materials to form a liquid and air-tight seal.

Thermal sealing requires forcing the material together, then heating and cooling it. The hot plastic

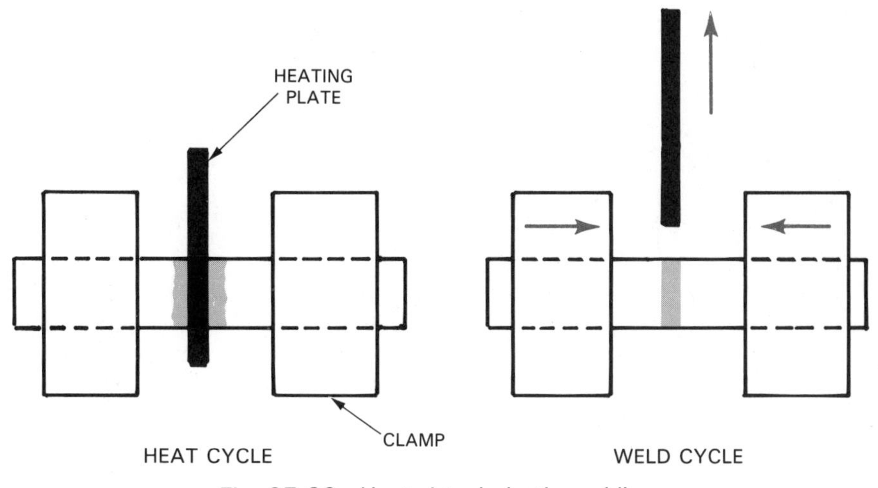

Fig. 27-26. Heated-tool plastic welding.

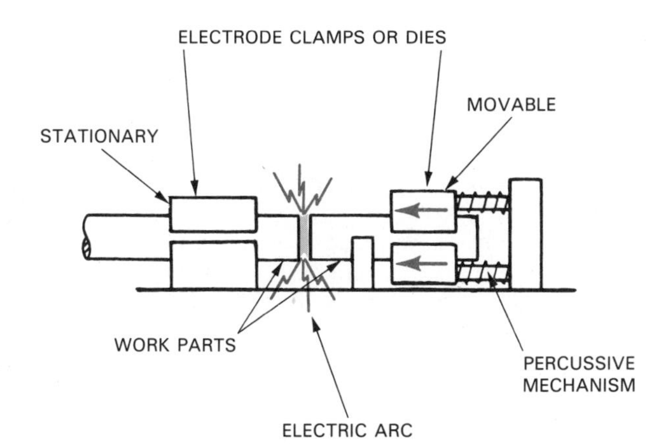

Fig. 27-27. Flash or percussion welding is similar to stud welding. (General Motors Corp.)

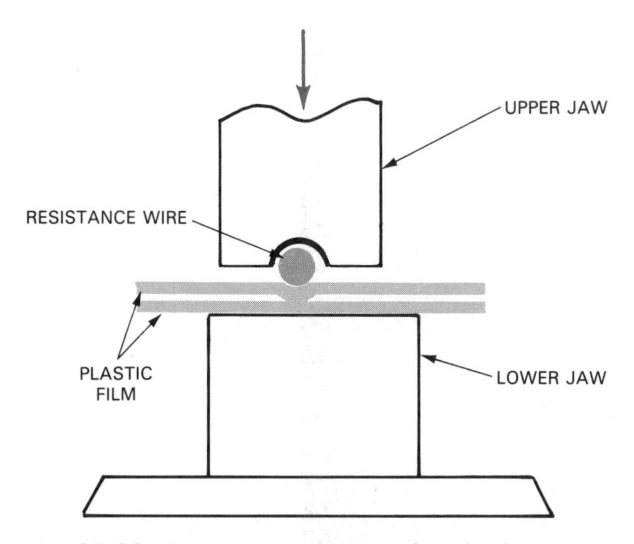

Fig. 27-28. Heat sealing process for plastic uses a resistance wire which carries a pulse current.

will fuse and cohere if under pressure as it cools.

Another common thermal sealing process is *impulse* sealing of plastic films and tubes. The material, as shown in Fig. 27-28, is placed between two gripping bars. The upper bar has an electrical resistance wire in it. A pulse of current is caused to flow through the wire. The wire becomes hot and melts the plastic along the length of the wire. The current is stopped. The plastic is held under pressure as the weld line cools.

Other thermal sealing practices use hot jaws or dies and rotary bands or wheels. The jaws or rotary devices apply heat and pressure. Rotary devices are used for high-volume sealing of films. Often, knives are located between two mated rolls which slit the products or packages apart as they are sealed.

Thermal sealing is used to produce bags and seal-shrink plastic packages.

Induction welding

Induction welding is completed by placing the parts inside a coil of wire called an *induction coil.* Alternating electrical current is caused to flow in the coil. The magnetic field around the coil rises and falls as the alternating current changes direction. This changing magnetic field induces an electrical current in the work. The induced current experiences resistance which heats up the workpieces. Pressure is applied by rolls or contacts. Induction welding is used to weld pipe, expanded metal, and structural shapes.

A similar process, called *electromagnetic bonding,* can be used to weld thermoplastic materials, Fig. 27-29. the process uses a plastic with very small magnetic particles in its structure. The parts to be bonded are placed inside the coil while uniform pressure is applied. Electrical current is then caused to flow in the coil. The magnetic particles are heated

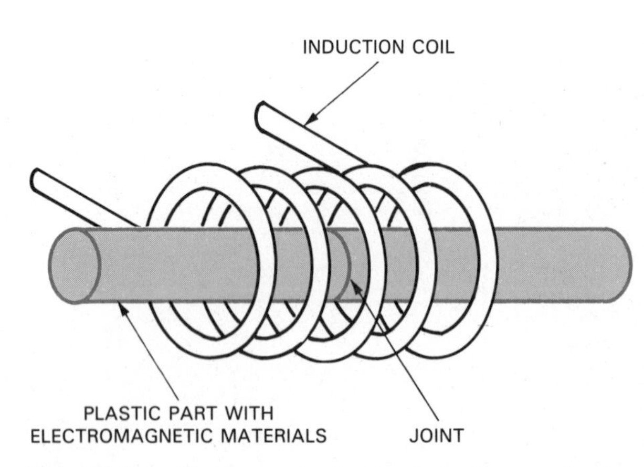

Fig. 27-29. Electromagnetic bonding. Current in the coil induces magnetic field and heats up the workpieces.

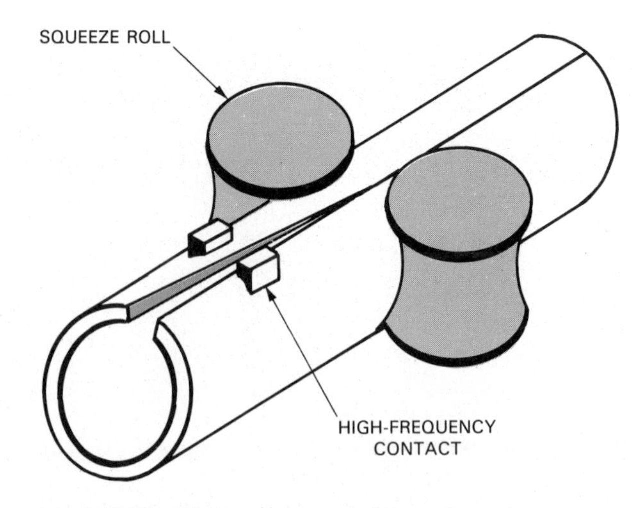

Fig. 27-30. High frequency welding of metal tubing applies current directly to the workpieces. Edges are heated and then forced together.

by induced currents. The heated particles, in turn, melt their host plastic material. The result is a polymer-to-polymer (cohesive) bond.

Electromagnetic bonding is used for mass production of parts. The process is particularly good for irregular-shaped parts. The low contact pressure reduces distortion.

Another similar process uses high frequency alternating current to bond metal workpieces. However, the current is directly applied to the work. Frequencies up to 500,000 cycles per second (Hz) are used, as opposed to 400 Hz for induction welding.

High frequency current flows near the surface and along the path of least resistance. This path, Fig. 27-30, is along the edges of the weldments.

Plastic material can also be sealed using high frequency currents. The process causes radio waves to pass through the material which normally would not conduct electricity (a dielectric). The resistance to conduction causes heating.

This process is often called dielectric heating or sealing. It involves a radio wave generator and dies or strips to direct the waves to the weld area. Similar equipment is used to emboss texture plastics and to heat cure wood adhesives. (See Chapter 28.)

Energy beam welding

Beams of energy may be used to weld materials together. The major sources of the energy are electron beams and light (LASER) beams. The processes are often called high energy beam welding.

Electron beam welding

Electron beam welding works on the same principle as the vacuum tube. A vacuum tube has a cathode which emits electrons and an anode which absorbs the electrons.

An electron beam welding system has four major parts, Fig. 27-31.

1. An electron beam gun with a power supply.
2. Vacuum pumping system.
3. Work holding and moving system.
4. Beam alignment system.

The welding takes place in a vacuum chamber as the cathode emits a stream of electrons. They are accelerated and focused by a beam alignment system. Metal, melted by the intense beam of electrons striking the work, flows into the joint area. Either the gun or the work is moved along the weld area. The movement allows the molten weld material to cool and produces a continuous weld.

Special electron beam welding systems have been developed to work outside a vacuum. These systems overcome the size limitations of a vacuum chamber. However more X-rays are given off and the gun must be within 1/2 in. of the surface. Also wider weld joints are produced.

Electron beam welding gives deep penetraion and a narrow weld joint. No filler rod is used. This process is used for welding thin materials, dissimilar materials, and high temperature or hardened materials.

Laser beam welding

Laser welding uses focused coherent (held together) monochromatic (one color) light on the material to be welded, Fig. 27-32. The light heats the surface of the material and moves down through the weld area by thermal conduction.

The heating power of the light beam must be higher than the melting point of the material. The extra heat is used to raise the temperature of the base material. However, the heat must not be so high that it

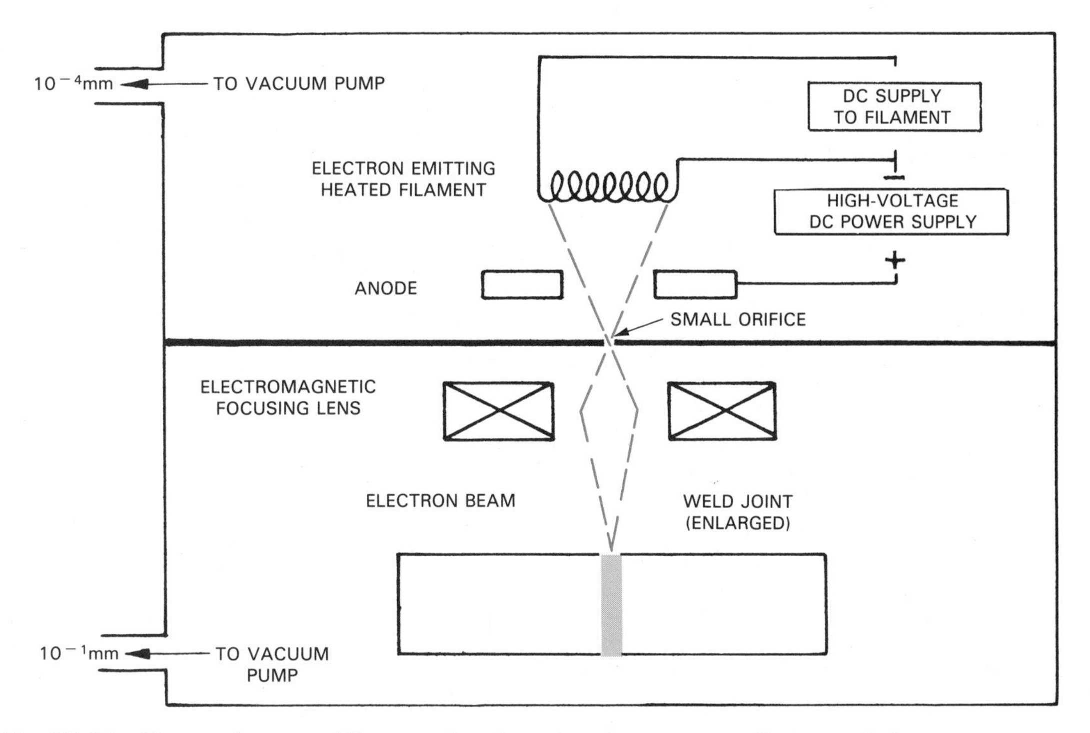

Fig. 27-31. Electron beam welding usually takes place in a vacuum. Streams of electrons are concentrated on the weld area. (General Motors Corp.)

Fig. 27-32. A CO_2 laser is welding nickel plated cold-rolled battery cans. (Coherent)

vaporizes the surface metal.

Laser welding uses intense energy focused on a tiny spot. This action allows for the welding of high-melting-point materials and refractories. Most of these materials cannot be welded by any other process.

Laser welding is slow because of the pulse rate of the light. Also, only thin materials can be welded. Material penetration is limited to about 1/8 in.

Friction welding

Friction or inertia welding, Fig. 27-33, is done with the heat produced by one workpiece surface rubbing against another. The typical friction welding process holds one part stationary. The second part is held against the first part and rotated. Frictional heat melts the weld surfaces. The pressure insures that a bond is obtained, Fig. 27-34.

The parts to be welded must be clean. Also, at least one of them must be cylindrical so it can be held in a chuck and rotated.

Frictional welding can be used to bond steels, nonferrous metals, and dissimilar metals. An important application is in welding superalloy turbine wheels to steel shafts and engine valve heads to their shafts. Frictional welding can also be used to bond thermoplastic materials.

Frictional or inertia welding is fast. The weld zone is rapidly heated by the rubbing of the material. Likewise, it is rapidly cooled by the cold base material behind the weld area. Only a small portion of each part is heated, therefore, distortion and stress are minimized.

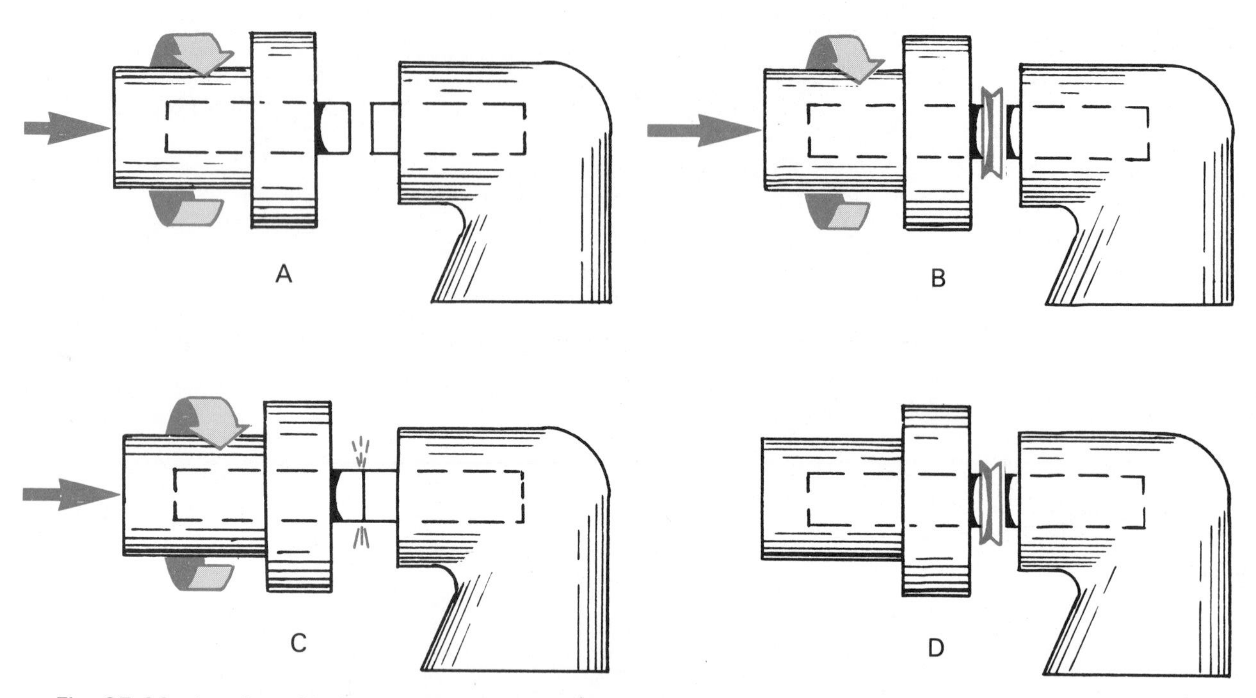

Fig. 27-33. Inertia or friction welding. A—One workpiece is stationary while the other spins. B—Workpiece ends are brought together with light pressure. C—As heat of friction brings the metal surfaces to a plastic state, greater pressure is applied. D—Completed weld. (Caterpillar Inc.)

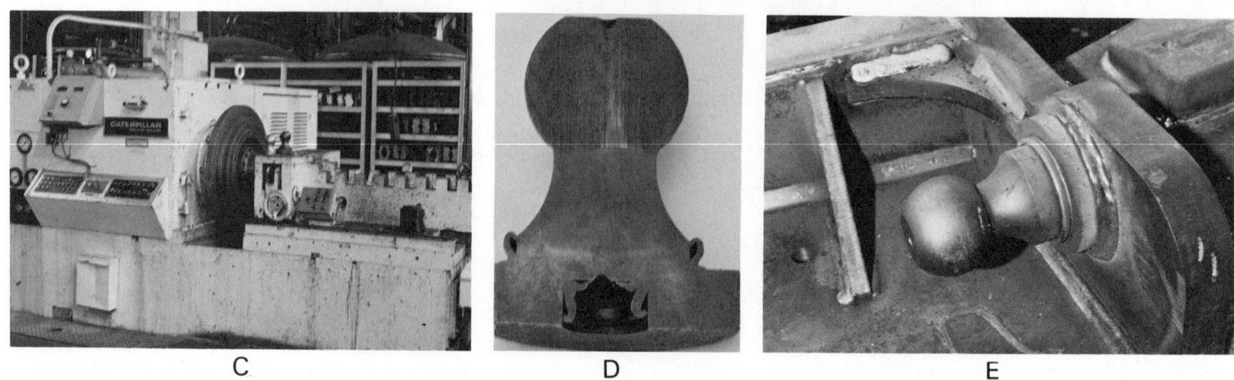

Fig. 27-34. The inertia welding process. A—Parts to be welded. B—Parts are loaded in the machine. C—The machine rotates one part against the other. D—The part is welded. E—The finished part is later welded to a larger assembly. (Caterpillar Inc.)

Ultrasonic welding

A process similar in theory to frictional welding is *ultrasonic welding*. The difference lies in the method of generating the frictional movement. Ultrasonic welding uses sound waves to vibrate two pieces which are closely held together. The vibration causes the parts to rub each other creating the welding heat.

Both metals and plastics can be welded using ultrasonic techniques. The resulting weld is as strong, or stronger, than the base material.

Diffusion bonding

Diffusion bonding uses heat and pressure to bond metal. The metal parts must be clean and flat. The workpieces are heated and pressure is applied. As the metals are squeezed together the atoms of the two pieces of material diffuse (cross the joint lines). No filler rod is applied.

Diffusion bonding will work on almost all metals. Even two different metals can be bonded together. However extremely smooth and flat surfaces are required. Also the cycle time is quite slow. The bond must "grow" between the materials. This slow action makes diffusion bonding appropriate only for materials which are difficult to bond by other means. It is not a high production process.

SUMMARY

Welding is a widely used process to permanently bond metals and plastics. It can be done using heat, pressure, or a combination of both. The result of the welding process is cohesive bond between the parts.

Welding processes use flames, electric arcs, mechanical force, friction, light, sound and radio waves, chemical reactions, electromagnetic forces, heated bars, and electron beams to create the needed heat and/or pressure. Each process is selected for use by considering its effect on the parts, its speed, and its accuracy.

STUDY QUESTIONS — Chapter 27

1. Define welding.
2. List and briefly describe the three major groups of welding processes.
3. List and describe the two major heat welding processes.
4. What are the major components of a gas welding system?
5. List and describe the major types of oxyacetylene flames.
6. What is the major disadvantage to using acetylene as a welding gas?
7. What are the major components in an arc welding system?
8. Describe the major consumable electrode arc welding processes.
9. Name and describe the major nonconsumable electrode arc welding processes.
10. List and describe the major pressure welding processes.
11. What types of material can be pressure welded?
12. List and describe the heat and pressure welding processes.
13. What are the major types of resistance welding techniques?
14. Describe induction welding.
15. Describe energy beam welding.
16. Describe inertia welding.

Chapter 28

ADHESIVE BONDING

Manufacturers are always looking for new technology for quicker and better processes. This is particularly true for assembling processes. Adhesives or "glues" have long been used for wood products. However, a wide range of adhesives and adhesive bonding processes have been recently developed to bond metals, ceramics, and polymeric materials. But what is an adhesive?

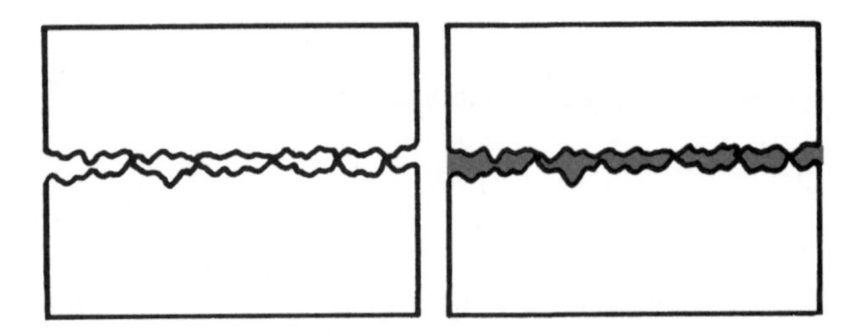

Fig. 28-1. Adhesives fill the spaces between parts, left, to create solid surface contact, right.

ADHESIVES

Adhesives are a group of materials which have some tackiness or sticking power. They are used to bond two or more base materials together. When two parts are placed together, they actually touch each other in a very few places as shown in Fig. 28-1. Over 95 percent of the surface area is separated by voids. When these areas are filled with a liquid material, it is much harder to separate the parts. That is why wet materials stick together better than dry materials. The materials are even harder to separate if the voids are filled with a material which adheres (clings, sticks) to the base materials. A material which adheres to another material is called an adhesive. These adhesives may be grouped in several ways, Fig. 28-2.

First, they may be arranged by their use. There are two major divisions:

1. Nonstructural.
2. Structural.

Nonstructural adhesives are those which do not have to support loads. Most household adhesives (glues and cements) are this type.

Structural adhesives can withstand loads over a long period of time. These have significant industrial applications. There are five main groups:

1. Thermoplastic adhesives.
2. Thermosetting adhesives.
3. Adhesive alloys.

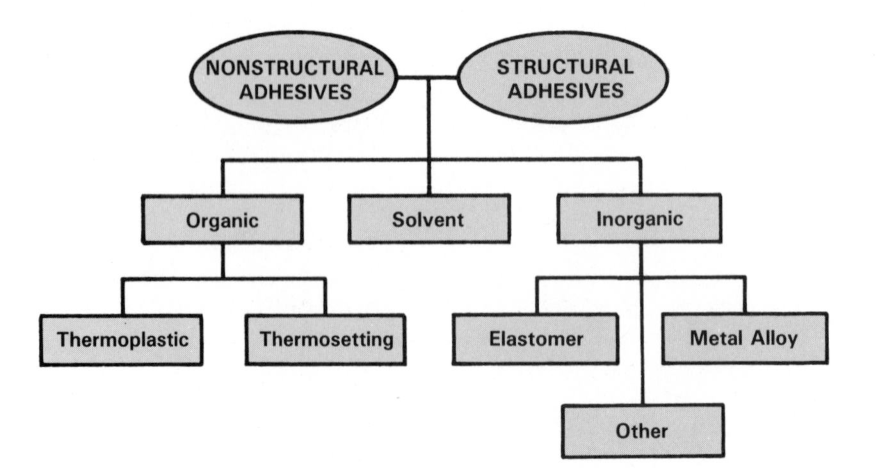

Fig. 28-2. Types of adhesives. This is common method of grouping them.

ADHESIVE	TYPE	CURING METHOD	DESCRIPTION	COMMON MATERIALS BONDED
Urea Formaldehyde	Two-part: Resin and hardener	Pressure, heat	Low cost, cures at room or elevated temperatures, water resistant	Wood Plywood
Resorcinol Phenol-resorcinols	Two-part: Liquid, resin with powdered hardener	Heat, radio frequencies	Expensive, waterproof, strong	Wood Plastic-to-wood core Metal-to-wood core
Polyester	Two-part: liquid or hot melts	Room temperature or elevated temperatures	Flexible, expensive	Foils Plastics Plastic Laminates
Epoxy	Two-part: liquid or pastes	Room or elevated temperatures	Very versatile, strong except to peel. Moisture resistant	Metals Ceramics Wood Glass Rubber

Fig. 28-3. Types of thermoplastic adhesives.

4. Elastomers.
5. Metal alloys.

The first four can be called *organic adhesives.* All are made from once-living matter. Metal alloys are *inorganic adhesives.*

In addition to structural and nonstructural adhesives there is a third group. This type uses a *solvent* for adhesion. A liquid, the solvent can dissolve the base material. Once dissolved (soft), the materials are pressed together. As the solvent evaporates or is absorbed the parts bond. Solvent adhesion is actually like a welding process. The base material is liquefied and forced together. Upon drying, the materials are actually bonded by cohesive bonds.

ORGANIC ADHESIVES

Organic adhesives can be traced back to living things. They are compounds in which carbon and hydrogen form their base. Some organic adhesives are *natural* products. The oldest type of adhesives, they are made from starch, dextrin, animal products, casein, and soya flour. Early craftspersons used animal hide, fish, and casein (milk) glues to assemble wood parts. Few modern applications remain for these products.

Synthetic (fabricated by humans) organic adhesives have replaced natural glues for most industrial applications. These products have two major features:

1. They will adhere to base materials.
2. They change from liquids to solids to form the bond (except pressure-sensitive adhesives).

Synthetic adhesives are complex combinations of several different materials. All have an *adhesive material* or base that will form the bond to hold the workpieces together. Many have a *hardener* which will cause the curing action to take place. Often a *catalyst* is included to speed up the curing action. Almost all adhesives have a *solvent* which suspends the adhesive materials and other components. It allows the adhesive to be spread.

Other materials are added to improve the adhesive. *Fillers* add desired properties such as strength and workability. *Diluents* like solvents, are used to make the glue easier to spread.

Synthetic adhesive materials are made up of polymers. (A discussion of monomers and polymers is included in Chapter 7.) These adhesive polymers may be thermosetting, thermoplastic, or elastomers.

Thermoplastic adhesives

Thermoplastic adhesives are polymers which have a basic tackiness. They will soften with heat and, often, when exposed to water. The "glue" joint produced by a thermoplastic is somewhat flexible or soft. The adhesive will allow the base materials to creep (slide across one another) if subjected to continued strain.

Thermoplastic adhesives are available in several forms. The most common are solvent solutions, water based emulsions, and hot melts. Solvent solutions and water-based products have wide application for bonding porous materials. They are particularly used for adhering wood, paper, and plastic foam material. Each of these materials allows the solvent or base material to be absorbed. This leaves the adhesive material in the joint area, forming the necessary bond. Descriptions are contained in Fig. 28-3.

Both the solvent and water-based material will be softened by their base liquid. Therefore, water-based adhesives have poor water resistance. Also, any solvent which will carry an adhesive will also weaken adhesive bonds formed between parts.

Hot-melt thermoplastics are materials which will soften under heat and harden when cooled. They are available in thin strips, films, sticks, pellets, and powders. The bonding action of solvent-based, emulsion, and hot-melt glues is shown in Fig. 28-4. Hot melts are used to bond paper, metal foil, and plastic film. Wide application for hot-melt adhesives is found in the packaging industry.

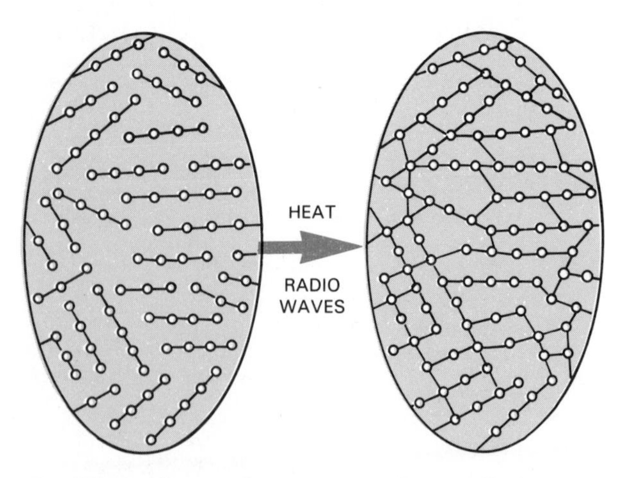

Fig. 28-5. Heat, radio waves, or other catalysts cause thermosetting organic chains, left, to unite, right, to form permanent bonds.

Thermosetting plastics

Thermosetting adhesives develop their adhesive bonding characteristics during a curing process. The material polymerizes into a long-chain organic molecule. Adhesive bonds are developed, Fig. 28-5, as the chains are formed.

Thermoplastic adhesives will become soft when heated or subjected to their solvents; thermosetting plastics will not. Once a thermosetting adhesive is cured a strong somewhat brittle joint is produced. The joint will withstand long continuous loads. However sharp blows may break the bonds.

The curing of thermoset adhesives is more complex than that of thermoplastics. It is not enough to allow the solvent to be absorbed. Polymerization must take place. The polymerization can be caused by several actions. These include:

1. *Plural components:* The chemical action is started by mixing the adhesive material with a curing agent. A catalyst and/or heat is often used to speed the action. Ureas and some phenolics and epoxies fall into this category.
2. *Heat:* The chemical action is started by an outside heat source. The material is premixed. Certain phenolics and epoxies are heat-activated adhesives.
3. *Moisture:* The chemical action is started by moisture in the air or base material. These adhesives are produced in liquid and paste form. They are quick-set materials which can be hand clamped. Silicones and urethanes are examples of moisture cure adhesives.

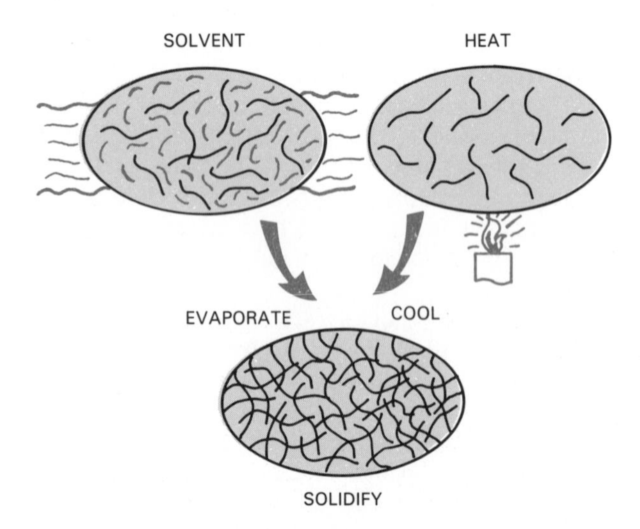

Fig. 28-4. Thermoplastic materials may be broken down by solvents or heat but form new bonds as they cool or the solvent evaporates.

Thermosetting adhesives may be used on a wide variety of materials. Fig. 28-6 lists some common thermosetting adhesives and their applications.

Adhesive alloys

Adhesive alloys are a mixture of adhesive materials. These are almost always thermosetting materials. They are compounded to provide greater strength, creep resistance, and resistance to the environment. Many are epoxies with other materials added. Typical adhesive alloys are epoxy/nylon, epoxy/phenolic, and epoxy/polysulfide materials. Also phenolics with neoprene and vinyl are adhesive alloys.

Elastomers

Elastomers are nonstructural adhesives with wide industrial applications. They are natural or synthetic rubber material. Elastomers can be stretched over twice their normal length.

Elastomers adhesives are usually solvent based liquids. They have high tack but low creep resistance. Typical elastomers are the contact cements and adhesives for bonding paneling, drywall, flooring, and ceramic tile. Pressure-sensitive tapes and papers are coated with elastomers.

ADHESIVE	TYPE	CURING METHOD	DESCRIPTION	COMMON MATERIALS BONDED
Cellulosic Cellulose acetate Cellulose nitrates	Solvent solutions	Solvent evaporation	Water resistant Bonds to many surfaces Low cost	Leather, paper, wood, nonwoven fabrics, glass
Polyvinyls Polyvinyl chlorides Polyvinyl acetate Polyvinyl acetals	Solvent solutions Films	Solvent evaporation Heat and pressure	Low cost, room temperature curing Not water resistant, flexible bonds	Porous materials like wood and paper
Acrylic	Solvent solutions, emulsions or catalyzed mixtures	Solvent evaporation heat or	Resistant to ultraviolet light-poor heat resistance Colorless	Glass, plastic metal foils
Polyamide	Solids, films, solvent solutions	Heat and pressure	Flexible, resistant to water and oil, expensive	Metals, plastic films, papers (metal cans, shoe soles)

Fig. 28-6. Types of thermosetting adhesives.

Metal alloys

Metal alloys are the principal inorganic adhesive. The typical metal alloy adhesive processes are soldering and brazing. Many individuals consider these processes closely related with welding. However they do not melt the base metal as in welding. These materials primarily stick to the base metals creating a bond between them.

Soldering and brazing are the same basic process. They use melted metal alloys as a bonding agent. The difference between them lies in the temperature and the alloy used.

Soldering is done below 800°F (430°C) while brazing takes place above 800°F. Soldering uses a tin-lead alloy (solder) as a bonding agent in most cases. Cadmium, antimony, silver, zinc, or copper may also be added. These elements are added to improve physical properties or reduce the price.

Solder contains various combinations of tin and lead. Common combinations are 50-50, 60-40, and 40-60. The first number is always the percentage of tin in the alloy. Solder is an eutectic. In its lowest melting form (63-37), it melts and completely solidifies at a temperature below either base element in the alloy. The common solders just listed melt and flow at 370°F, 470°F, and 460°F respectively. This is well below lead's 620°F and near tin's 450°F melting point. The 50-50 solder is the eutectic form for solder.

Brazing also uses a metal alloy which melts below the melting point of the base metal. Also, like soldering, brazing uses capillary attraction to draw the braze metal into the joint.

Several metals and metal alloys are used as braze materials. The most common of these are:

1. Copper.
2. Brass and bronze (copper alloys) (1600°-2250°F).
3. Silver alloys.
4. Aluminum alloys.

These materials melt in a range from 1050°F (566°C) for aluminum alloys to 1981°F (1083°C) for pure copper. Solder is most often sold in wire coils. The solder usually is hollow and filled with flux. Braze material is manufactured as solid rods or preformed discs, rods, rings, or other special shapes. A description of soldering and brazing is included later in this chapter.

ORGANIC ADHESIVE BONDING PROCESSES

The bonding of parts using organic adhesive generally follows a common pattern, Fig. 28-7. This involves:

1. Selecting the joint and adhesive to be used.
2. Preparing the material to be joined.
3. Applying the adhesive.
4. Assembling and clamping parts.
5. Curing the adhesive.
6. Unclamping and inspecting the assembly.

SELECTING JOINTS AND ADHESIVE

Poor assembly results will surely occur if the joints and adhesives are not carefully matched with their application. Joints must be chosen to minimize the load

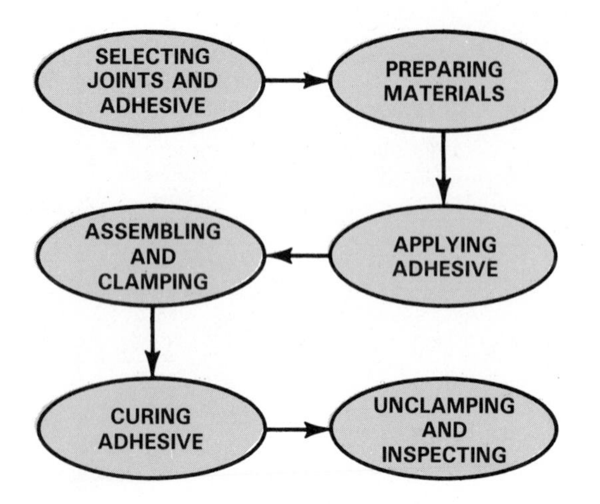

Fig. 28-7. Steps in adhesive bonding.

on the adhesive. A number of common joints are discussed in Chapter 26.

Also, the adhesive must be matched to the application. Conditions with high load on the adhesive require a thermosetting or alloy adhesive. If the parts must be able to adjust to expansion and contraction, a thermoplastic adhesive may be required. Applications with excess heat, moisture, or solvents suggest the use of a thermoset. Nonload applications where clamping is difficult may require an elastomer.

To make a proper selection, the technical data sheets of several adhesives must be studied. This information, coupled with an understanding of joint construction, provides the basis for adhesive bonding.

PREPARING THE MATERIAL TO BE BONDED

A properly prepared bonding surface is essential for a good adhesive bond. The preparation includes two main considerations: fit and cleanliness.

The first step is to produce a *good fit* between the surfaces to be mated. This may mean additional machining or forming to insure that the shape and smoothness of the parts are correct.

The second step is to clean the parts. Wood and plastic parts may need to be dusted. Metal and plastic parts often require chemical or mechanical cleaning. Solvent or acid baths are used to remove oil, grease, and other unwanted materials. Sandblasting, wire brushing, or sanding may be used to clean parts. Both chemical and mechanical means may be used to roughen the surfaces. Adhesives will bond better to a slightly rough surface. However, uneven, poorly mated surfaces produce a weak bond.

APPLYING THE ADHESIVE

There are a number of ways to apply adhesives. Sheet and film adhesives are placed between the joints. Hot melt adhesives may be extruded from a gun or spread along the joint area.

Liquid adhesives may be spread in a number of ways. Low-volume applications allow the spreading from a squeeze bottle or a brush. Larger areas may be spread with a nap roller or a trowel. In many industrial applications, either spray equipment or roll applicators, Fig. 28-8, are used.

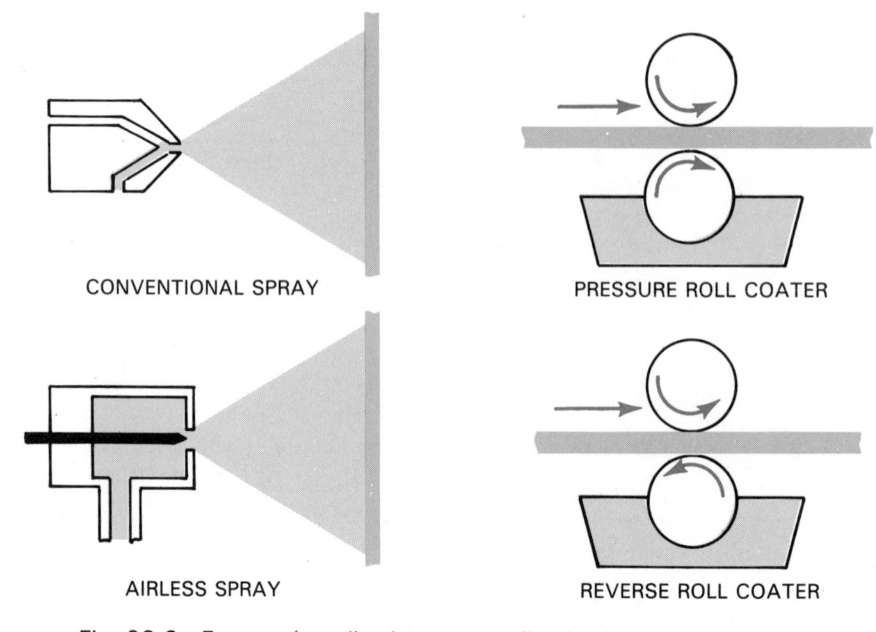

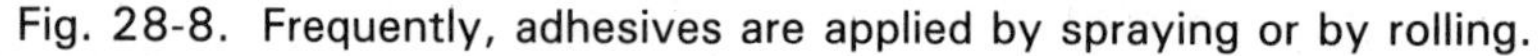
Fig. 28-8. Frequently, adhesives are applied by spraying or by rolling.

The purpose of all these techniques is to apply a uniform layer of adhesive to the workpieces. Glue line thickness is controlled by the amount of glue applied to the mating parts. As a general rule, about five times the desired glue line thickness must be laid down. Also, it is considered better to spread a thin coat on both parts rather than a thick coat on one part.

ASSEMBLING AND CLAMPING PARTS

Most adhesives allow for sliding the parts into position during the clamp phase. Pressure-sensitive adhesives and contact cement do not. These adhesives immediately establish the bonds and do not allow shifting of the parts once brought together. Also, they do not require a clamping period. Pressure need only be applied long enough to squeeze the surfaces together. Often a roll type clamp, Fig. 28-9, is used.

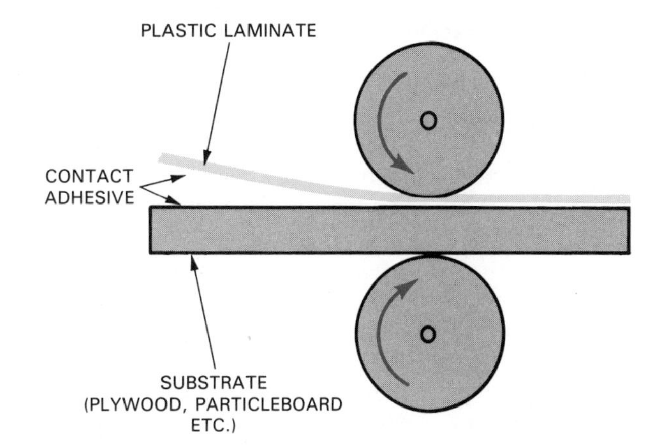

Fig. 28-9. Roll clamping of plastic laminate provides the temporary pressure needed for adhering pressure-sensitive adhesives and contact cement.

Most adhesives require clamping. This may be done in presses or by using gluing clamps. Bar clamps, C-clamps, and hand screw clamps are widely used for wood cabinet, furniture, and sash application. Hydraulic and pneumatic cylinders and toggle clamps are also used. Large, flat parts may be clamped in air bag presses. The bags are filled with compressed air which causes them to expand. The bag usually presses down on a platen between the parts and the bag. By eliminating the platen, the bag may be used to clamp curved pieces. The bag then takes on the contour of the part.

The clamping system provides the pressure required during the curing cycles for the adhesive. Without pressure, a weak joint will develop. If excess pressure is used, the adhesive will be forced from the joint. Wrong pressure or insufficient glue application will result in a starved joint. Obviously, a weak joint is the result. See Fig. 28-10.

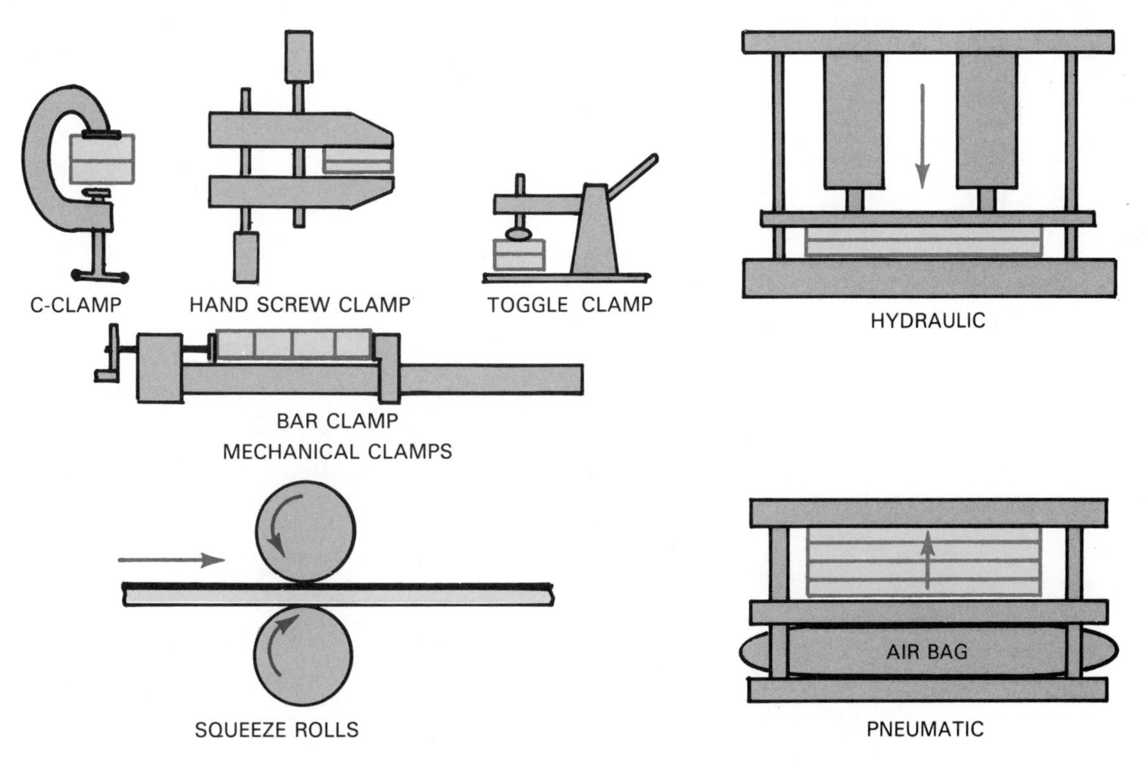

Fig. 28-10. Types of clamps used in adhesive fastening.

CURING THE ADHESIVE

Many adhesives cure naturally during the clamp cycle. Most thermoplastics and many elastomers cure by giving up their solvents. Many elastomers cure before the mating parts are squeezed together.

Thermosetting and adhesive alloys (which are also thermosets) cure by polymerization. This action can, in many cases, be speeded up by the application of heat. Plywood is cured in heated presses, Fig. 28-11. Electrostatic glue curers (wood welders) are used to cure wood joints using high frequency radio waves which pass through the wood. The wood, having high resistance to the waves, heats up. Curing is accelerated.

High frequency or RF flue curing can use a hand gun, Fig. 28-12, to cure standard joints. The generator can also be attached to metal strips on the faces of forms to cure bent wood laminations.

Ovens and induction heaters, Fig. 28-13, may be used to cure metal parts. Care must be given to insure even heating of the parts since uneven heating causes the parts to expand and contract at different rates. This weakens or breaks the adhesive bonds.

UNCLAMPING AND INSPECTING THE ASSEMBLY

After the correct clamp time has passed, the parts are removed from the clamps. Warm assemblies are allowed to cool. Joints are inspected to insure that:

1. A good adhesive-to-base material bond has developed.

Fig. 28-11. This giant press clamps and glue cures plywood. (Weyerhaeuser Co.)

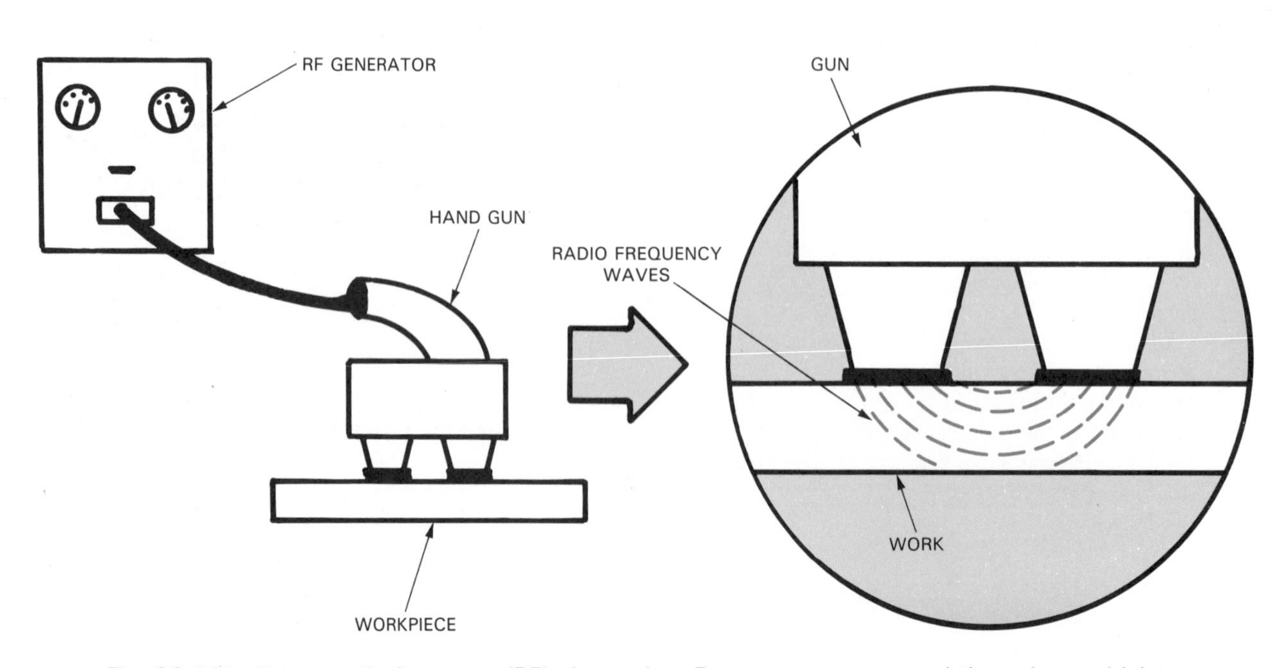

Fig. 28-12. High or radio frequency (RF) glue curing. Frequency waves travel through wood joints.

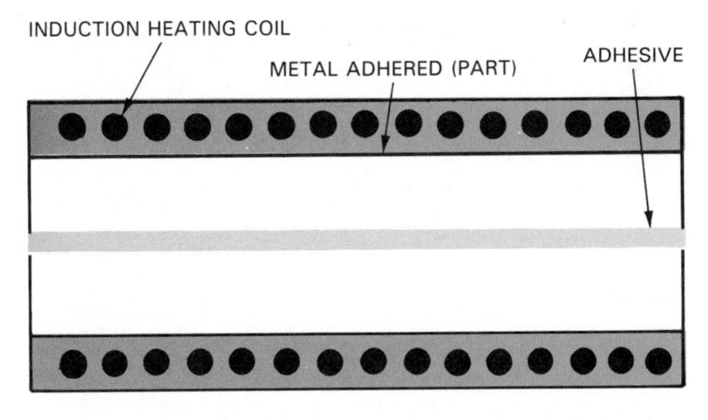

Fig. 28-13. Adhesives may be cured using induction heaters.

2. The adhesive is not itself cracked.

Adhesive bonding is becoming a major technique for assembling products. The aerospace industry has been in the forefront in developing the procedures. Many automotive and appliance components are now fabricated using adhesives. Adhesives are replacing welding and mechanical fasteners in many assembly applications.

BRAZING AND SOLDERING

Brazing and soldering are the principle adhesive techniques using a metal alloy as a bonding agent. The practices use a heat source to melt the alloy and cause it to flow in the joint area. There are three major types of brazing and soldering processes:

1. Soft soldering—the bonding of two metals using a tin-lead alloy which melts below 800°F (427°C).
2. Brazing or hard soldering—the bonding of two metals using a metal alloy which melts above 800°F.
3. Braze welding—the bonding of two metals with a bead of metal alloy usually bronze.

None of these processes melt the base metal. They form a bond through three basic actions.

1. The bonding alloy penetrates (flows into) the voids or open spaces in the base metal crystalline structure. When the solder or braze metal cools, it forms its own grain structure. This structure extends across the joint forming bonds or fingers.
2. A certain amount of diffusion occurs between the base metal and the bonding alloy. A new alloy is developed at the surface. A more complete discussion of diffusion is included in the section on *diffusion welding* in Chapter 27.
3. Atoms of one element are attracted to those of others. Solder and brazing alloys are chosen because their atoms are strongly attracted to many other metals.

Brazing and many soldering applications depend upon spreading the bonding alloy to all parts of the joint. The solder or braze material is melted and made to flow between the two tightly fitted parts by a natural physical force. This force, called *capillary action,* is the same force that causes water to rise in a thin fiber or tube.

Bronze welding and some soft soldering applications use a mass of material to create the bond. The metal alloy is melted and forms a bead or glob. The bonding material bridges the joint to form the bond.

Soldering

Soft soldering is usually called soldering. A tin-lead alloy called solder is the bonding agent. The solder is melted and flows over the joint area. It is chiefly the solder's natural attraction to the base metal that bonds the assembly together.

Soldering requires a basic six step process.

1. The parts to be soldered are chemically or mechanically cleaned. The surfaces are etched, sanded, or scraped.
2. A *flux* is applied. The flux is a chemical which removes oxides (tarnish) and helps the solder to flow. Fluxes are either acids or organic rosins. Rosin fluxes are used in electrical and electronic applications. Acid fluxes are appropriate for most other situations. Flux may be added first or may be contained in the hollow core of the solder. Often both flux application methods are used.
3. The parts are clamped or held in position. (Steps 2 and 3 can be reversed.)
4. The part to be soldered is preheated. A soldering gun, Fig. 28-14, or a torch may be used. Other heat

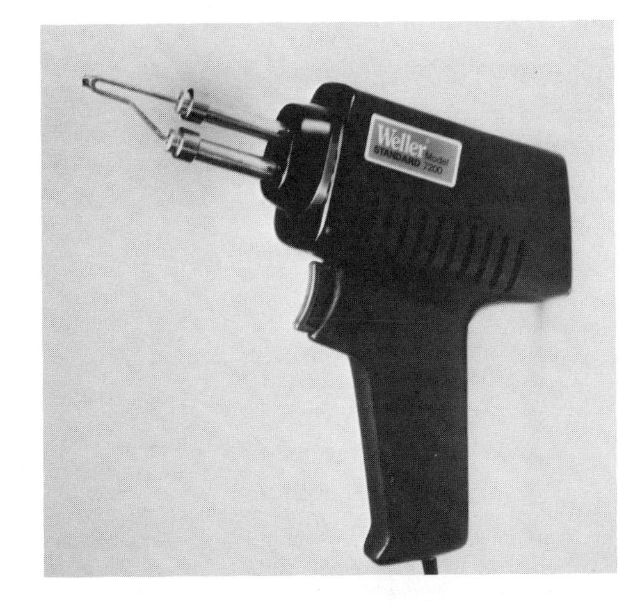

Fig. 28-14. A typical soldering gun. Tip temperatures vary between about 350° and 800°F. (Weller)

sources are furnaces, electromagnetic coils, dip baths, ultrasonic generators, and electrical current (resistance heating).
5. The solder is melted into the joint and allowed to cool.
6. Excess flux is cleaned from the assembly. This is often essential because the flux must be chemically active to remove oxide. It is, therefore, somewhat corrosive to the assembled parts.

BRAZING

Brazing is very similar to soldering. The parts are first chemically or mechanically cleaned. The surfaces to be bonded are flux coated. The base material is heated; then the brazing alloy is applied. The joint is cooled and cleaned, Fig. 28-15.

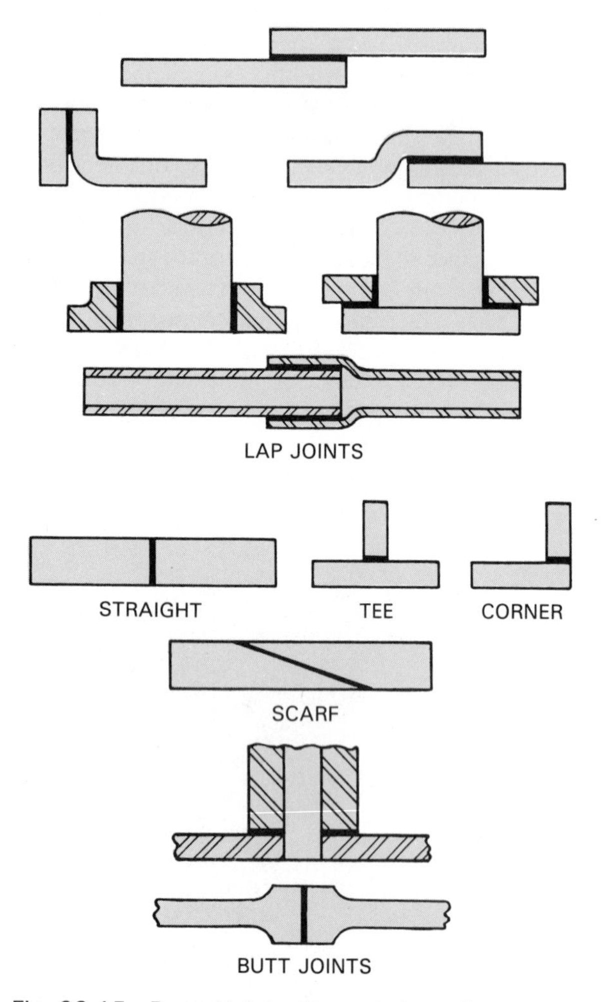

Fig. 28-15. Brazed joints. Strength depends upon surface area, clearance between parts being joined, type of joint, and material being brazed. Strongest joints are those in which the brazing filler is drawn in by capillary action.

There are a number of brazing techniques. The most important are:
1. Torch brazing.
2. Induction brazing.
3. Furnace brazing.
4. Dip brazing.
5. Resistance brazing.

Torch brazing

Torch brazing is the oldest and probably the most widely used brazing technique. It is adaptable to most jobs. Little special equipment is required.

Torch brazing uses the same basic equipment as gas welding: compressed gas, regulators, and a torch. Oxyacetylene, MAPP, or other gas systems can be used.

Two basic arrangements of torches can be used.
1. Hand held torch. This method, as shown in Fig. 28-16 is the most widely used torch brazing system. The operator holds the torch and directs the flame. The parts are heated to the proper temperature. Then a flux coated brazing rod is hand fed into the joint area. The heat first melts the flux. The liquid flux cleans the base metal and promotes the flow of the braze metal. The heat also melts the brazing rod. The braze metal flows between the

Fig. 28-16. Torch brazing uses standard oxyacetylene equipment. In this unit, the torch is hand-held. (AIRCO)

parts by capillary action, Fig. 28-17. As the torch is moved along or removed from the joint area, the braze metal cools.

2. Positioned or automatic torches. Two or more stationary torches are used in this system. Parts are moved from one torch station to another. One torch or set of torches will preheat the part. Another will provide the brazing heat. Still another could be used for post heat. This station would allow for slower cooling, Fig. 28-18. The assemblies could be manually moved from the various stations.

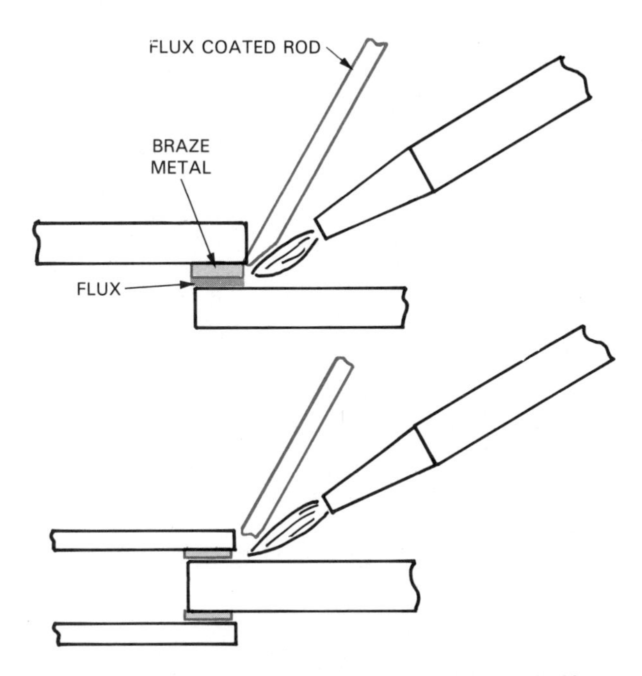

Fig. 28-17. Torch brazing with a hand-held torch. Heat melts the brazing rod and capillary action draws melted metal into the joint.

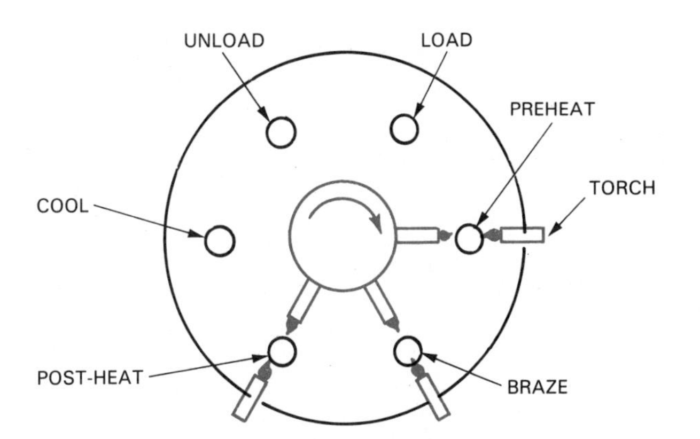

Fig. 28-18. Automatic torch brazing with three stations. Torches are stationary; parts move from one station to another.

Also, automatic brazing of small parts is possible. Programmed machines can load the parts, move them to the various stations and remove the completed assemblies. Automatic brazing usually uses a preformed disc or strip of flux coated braze material. The material is placed between the parts at the loading station. The material is automatically melted as the parts reach the proper braze temperature.

Induction brazing

Induction heating is used for many thermal conditioning processes such as hardening, tempering, and annealing. It can also be used to produce heat for brazing.

Induction brazing uses a coil of wire through which a high frequency current is passed. The magnetic field around the coil induces eddy currents in the workpiece which is placed near or inside the coil. The eddy currents generate the heat needed to melt the braze material.

Induction brazing, Fig. 28-19, requires very accurate part location to insure proper heating. The clamping mechanisms, which must be kept well away from the induction coil, are usually made of non-magnetic materials such as ceramics or aluminum. The braze material is often a preformed shape placed at the joint area during the original clamping stage.

With induction brazing, the area heated can be carefully controlled. This reduces unwanted material characteristics produced by heating large areas as in flame brazing. The induction process is also fast and produces very uniform bonds.

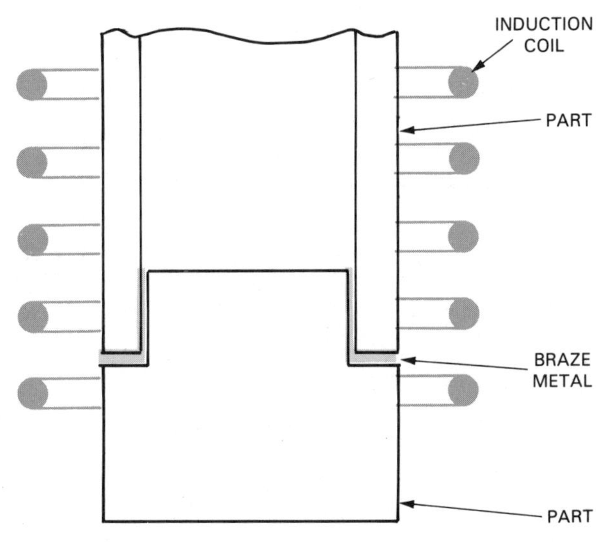

Fig. 28-19. Induction brazing is shown in cutaway. Eddy currents, induced by coil surrounding the workpiece, provide brazing heat.

Furnace brazing

Furnace brazing is a high production process well suited for mass production. The process produces the greatest quantity of brazed hardware today.

Furnace brazing, Fig. 28-20, is generally a continuous operation. The clean parts and preformed flux-coated braze materials are placed together. They are then fed into a furnace on a continuous belt. The atmosphere and temperature in the furnace are carefully controlled. Often, this atmosphere reduces oxidation and increases braze material flow.

Batch type furnaces may also be used. Parts are loaded on racks and placed in the furnace.

The use of fixtures is held to a minimum in furnace brazing. The fixtures absorb heat energy and must be cooled after each cycle through the furnace. Whenever possible, braze joints for furnace brazing are designed to be self-clamping. Grooves, shoulders, and notches, which locate and hold mating parts before brazing, are used.

Dip brazing

Dip brazing is another major brazing technique. This process uses a bath of molten salt to heat the parts. The cleaned parts and preformed braze material are clamped together and the clamped assembly is lowered into the salt bath. The hot salt heats the assembly until the braze material melts. The parts are then removed and washed thoroughly. The rinses remove excess flux and salt.

Dip brazing sometimes uses a bath of molten braze metal instead of salts. The bath furnishes both the necessary heat and the braze metal. This technique is used only for small parts. The entire part is coated and is, therefore, wasteful of the braze material.

Resistance brazing

Resistance brazing is much like spot welding. The parts, along with preformed braze material, are held between two electrodes. Electrical current is passed between the electrodes. The processes differ at this point. Spot welding uses highly conductive electrodes, usually copper. The current passes easily through the electrodes. The electrical resistance of the parts to be welded generate the welding temperature.

In resistance brazing, graphite or carbon electrodes are used. Their high resistance to electricity causes them to heat up. The hot electrodes bring the metal to the proper brazing temperature. The electrical current and cycle time are adjusted to control the temperature. Resistance brazing finds many applications in the electrical equipment industry. Conductors, terminals, and cable connections are often resistance brazed.

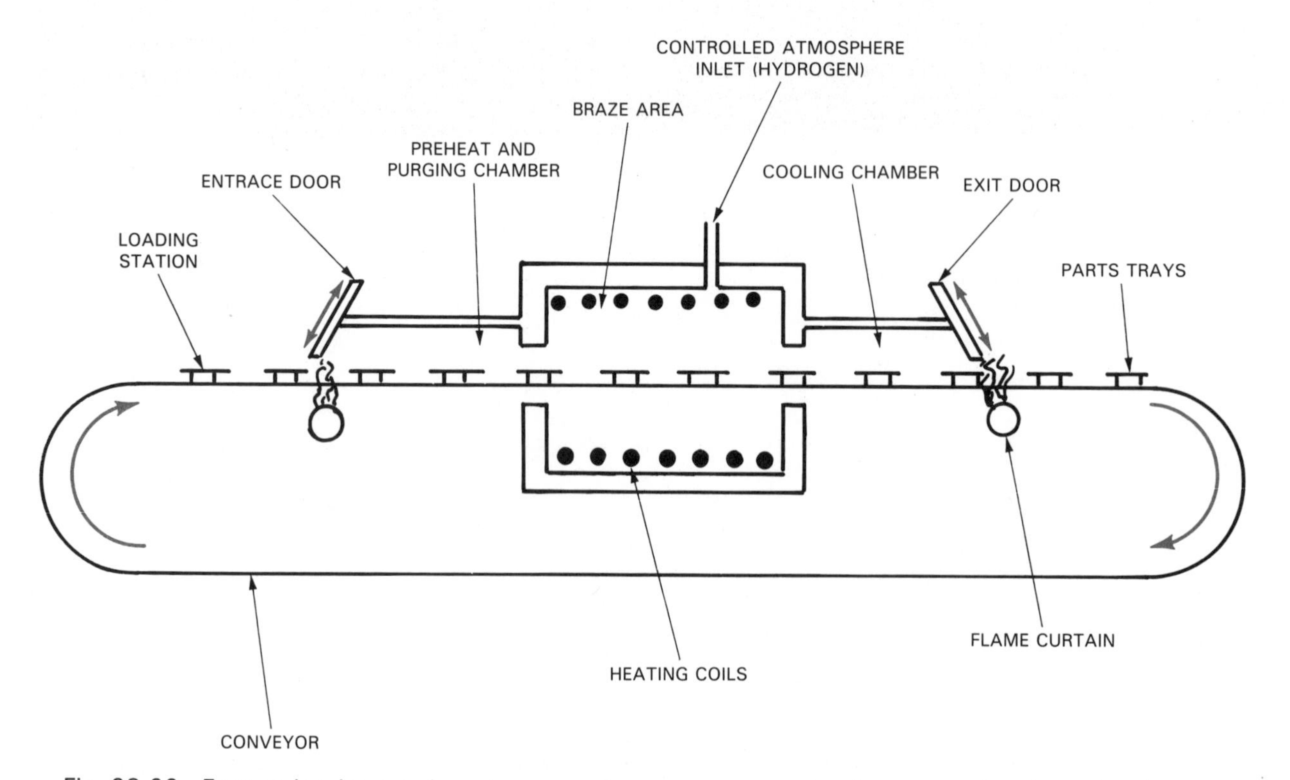

Fig. 28-20. Furnace brazing may be set up around a conveyor operation (as shown above) or a batch type furnace is used. Parts are loaded on racks and placed in the furnace.

BRAZE WELDING

Braze welding requires the addition of enough braze material to form a bead between the mating parts. The process which is also called bronze welding uses gas welding equipment and techniques. The joint areas are prepared. Then the parts are cleaned and positioned for welding. Flux coated brazing material is melted in the joint area. Often a first pass is used to coat the joint surfaces, Fig. 28-21. Later passes fill the area between the parts.

The brazed assembly is carefully cleaned after welding. Any remaining flux must be removed.

Braze welding is widely used for maintenance and repair work. Cast, malleable, and wrought iron members are easily assembled with braze welding techniques.

SOLVENT ASSEMBLY

Some parts may be assembled using a solvent. The base parts are softened with a material which will dissolve it. The soft parts are, then, pressed together. As the solvent evaporates, the parts become united.

All solvent bonding requires that:

1. The joints fit well.
2. The joints be free of dirt and other unwanted materials.
3. The solvent be evenly applied.
4. The parts be assembled before the joint areas harden.
5. The parts be held rigid while the joint cures (dries).

Solvents are used to assemble clay products. Slip (clay suspended in water) is used to coat and soften the parts. Refer again to Fig. 28-20. The parts are then assembled. Complex assemblies may be built using solvent assembly practices. Handles are attached to cups using solvent assembly. Some sanitary ware (bathroom fixtures) are also put together using this procedure. Likewise, acrylic sheet (Plexiglas) and plastic plumbing pipe can be assembled using solvents.

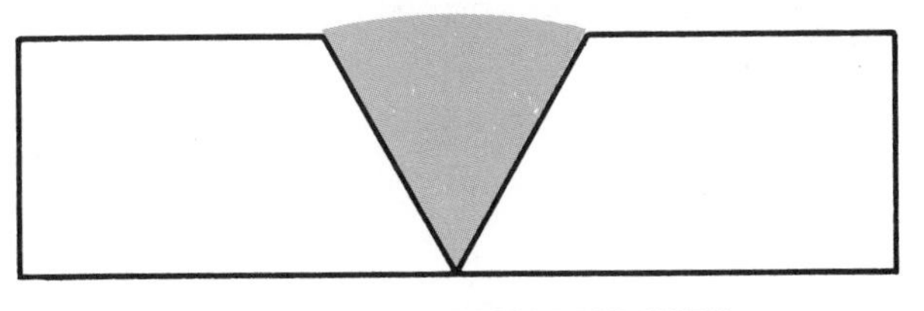

Fig. 28-21. Braze welding often uses two or more passes to complete the weld.

SUMMARY

Adhesives are commonly used to assemble parts. The materials which bond components using primarily adhesion are organic and inorganic adhesives. Solvents may also be used to soften surfaces which are then pressed together.

The common adhesive materials are thermoplastic adhesive, thermosetting adhesives, adhesive alloys, elastomers, and metallic alloys. These adhesives can also be grouped as structural or nonstructural adhesives. Structural adhesives are used where prolonged loads are expected.

Adhesive processes require that the material be prepared, the bonding material be applied, and the joint be clamped and cured. Typical adhesive bonding practices using these major stages are organic adhesive bonding, and solvent bonding.

STUDY QUESTIONS – Chapter 28

1. What is an adhesive?
2. What are the two classes of adhesives?
3. List the major types of organic adhesives.
4. How do adhesives set or solidify?
5. What are the major metal alloys which are used as adhesives?
6. Describe soldering and brazing.
7. List the steps followed by organic adhesive processes.
8. List and describe the common brazing processes.
9. Explain solvent assembly.

The beams in this church were produced by bending and gluing boards together.

Chapter 29

MECHANICAL FASTENING

Welding and adhesive bonding are commonly used methods for assembling parts. The third major assembling technique uses mechanical means. Mechanical fastening practices as shown in Fig. 29-1, include two major headings:

1. Mechanical forces.
2. Mechanical fasteners.

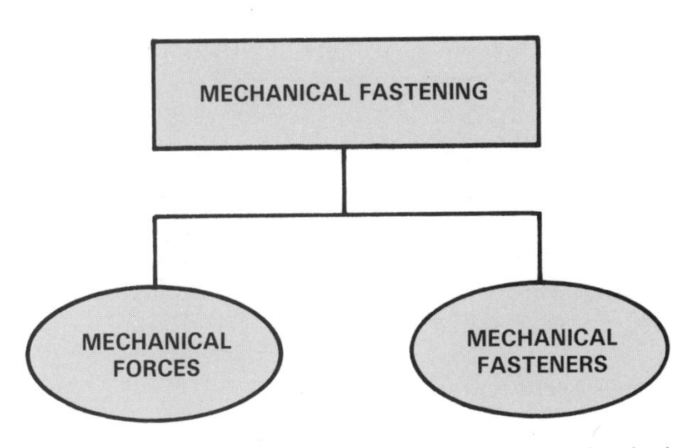

Fig. 29-1. There are two major types of mechanical fastening.

ASSEMBLING WITH MECHANICAL FORCES

Mechanical assembly forces hold the parts without the use of outside assembly devices. The internal mechanical forces of the material are used to position and hold the parts in place. Primarily this group of practices either uses a special *shape* built into the part or a *tight fit* between parts for assembly purposes. The parts could be considered self-assembling.

PHYSICAL STRUCTURES

Parts may be designed so that they interlock to hold and position themselves. The typical physical structure assembly practices are:

1. Seams.
2. Tabs and slots.
3. Beads, dimples, and flanges.

Seams

Seams are methods of assembly which interlock the edges of a material. Sheet metal is assembled with seams. Common seams, Fig. 29-2, are:

1. Single lock or grooved. This seam is a standard or locked pipe seam and provides a smooth surface on one side.
2. Double lock. This seam is a modified grooved seam which provides greater strength.
3. Pittsburg. This seam uses a connector strip to assemble two parts. It is widely used for on-site assembly of heating and air conditioning ducts.

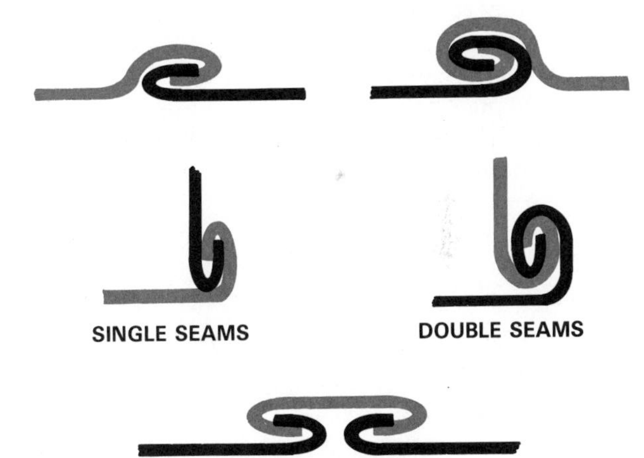

Fig. 29-2. Typical sheet metal seams.

Tabs and slots

Tabs and slots temporarily or permanently assemble sheet metal or plastic parts. The tabs protrude from the part, often fitting into a hole or a slot.

The common shelf bracket is a good example of a tab and slot assembly. The tabs on the shelf brackets lock into slots on the standard to produce a temporary assembly. In another technique, rectangular tabs are placed through slots and twisted. This action produces a permanent tab and slot assembly, Fig. 29-3.

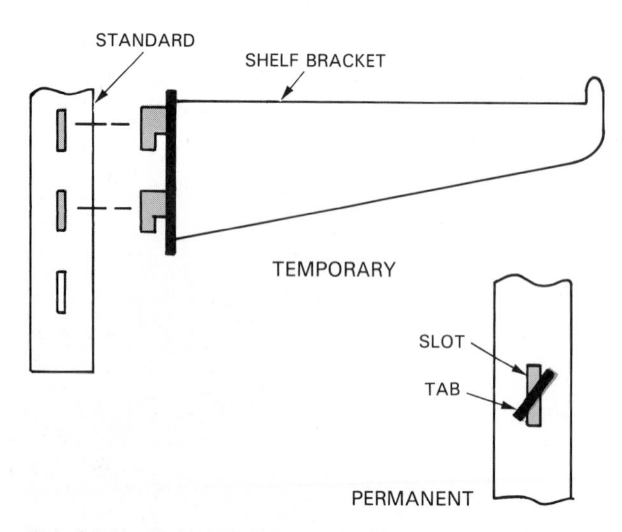

Fig. 29-3. Tab and slot assemblies can be either temporary or permanent.

Beads, dimples, and flanges

Sheet metal and plastic parts may be shaped so that they grip one another without producing a folded seam. Common designs include beads, dimples, and flanges.

A *bead* is a recessed groove around a cylinder, Fig. 29-4. The bead is usually produced after the two parts are assembled. Sheet metal parts can have roll formed beads. Tubing can be beaded to hold it onto a shaft. The bead may be roll formed or produced by high-energy-rate forming practices. (See Chapter 11.)

Dimples are recesses formed on a part. They hold parts together much like beads. Parts are often locked together by a punch or chisel mark, Fig. 29-5. The assembly is placed on a mandrel or other smooth surface. The surface is struck with a punch or die. The

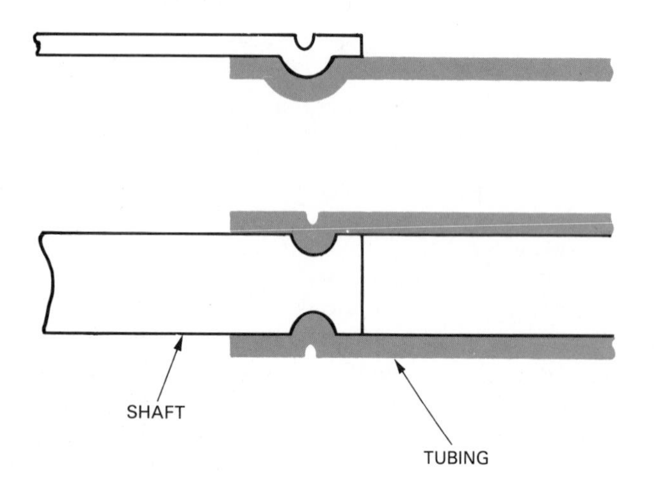

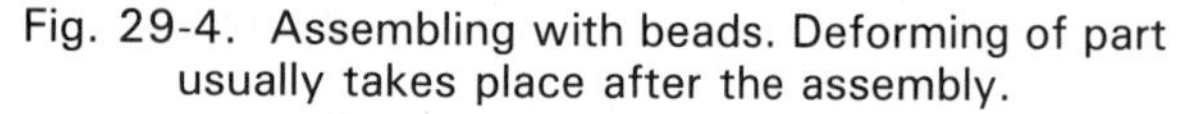

Fig. 29-4. Assembling with beads. Deforming of part usually takes place after the assembly.

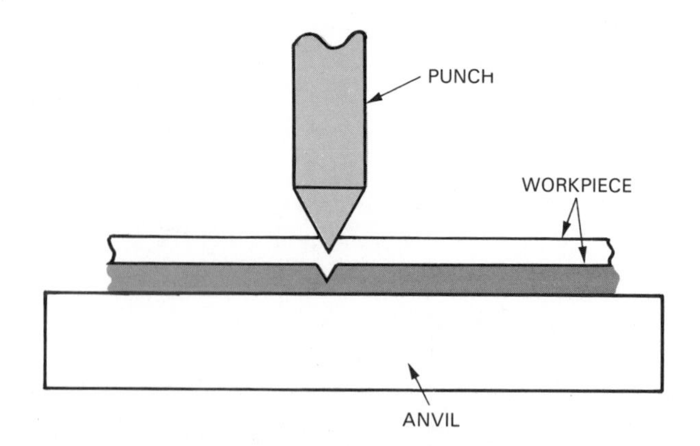

Fig. 29-5. A punch can be used to produce a dimple.

resulting indentation or mark locks the parts together. Like beads, dimples produce a permanent assembly.

Flanges are shaped edges often used for temporary assembly of plastic parts to mating pieces. Snap closures for containers such as margarine tubs and coffee cans use flanges. Zip-Loc™ bags use a double flange to grip the beaded strip, Fig. 29-6.

INTERFERENCE FITS

Two mating pieces do not always need special shapes to lock them together. Friction between the mating surfaces is used to hold the parts together. This

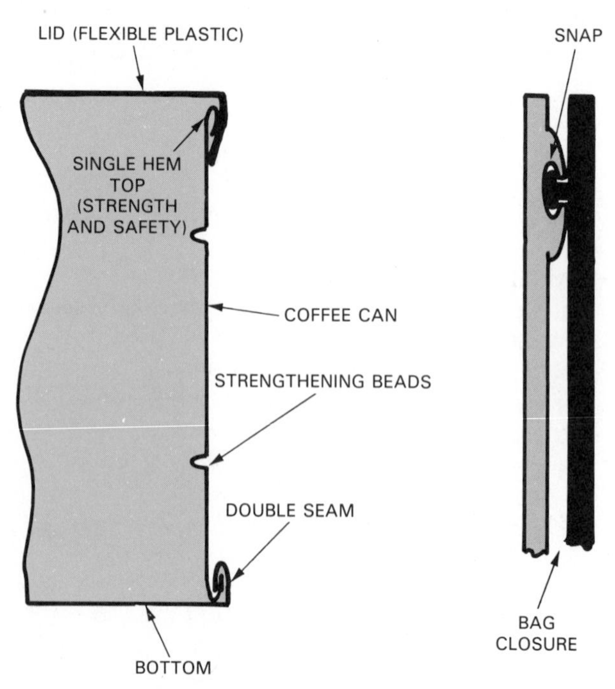

Fig. 29-6. Flange assemblies are commonly used on plastic lids and plastic bag closures.

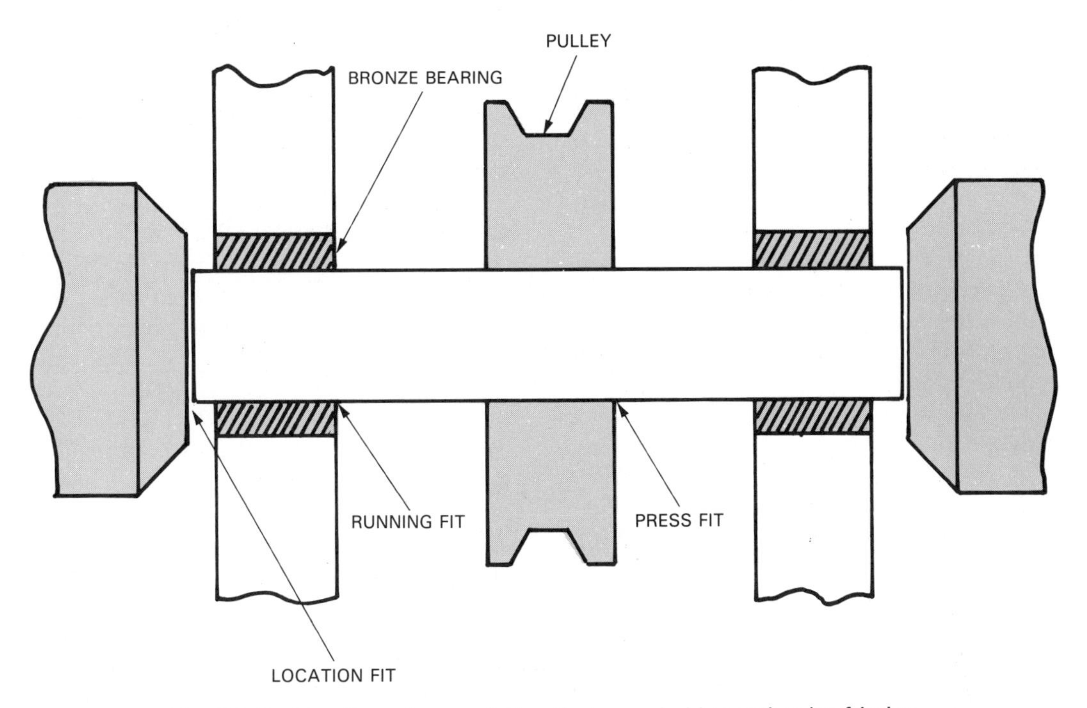

Fig. 29-7. The three types of fits. Parts are held together by friction.

type of assembly technique is called an *interference fit*.

There are three major types of fits used where two parts meet, Fig. 29-7:

1. Running or sliding (clearance) fit—parts fit together with enough allowance to permit free turning or sliding of one part in the other. Additional mechanical fastening, bonding, or location fits are required to keep the parts in place. For example, shafts must be welded, bonded, or bolted at some other location to keep it in a bronze bearing.
2. Location fit—positions one part in its proper relationship with another part. The fit may allow movement (clearance fit) or be very accurate (interference fit).
3. Interference fit—one part is tightly gripped by another. Interference fits may be either force fits or shrink fits.

   Force fits require a shaped part to be forced into a shaped hole. Often round shafts are forced into a round hole. The shaft is slightly larger than the hole. The difference in diameters vary by application. The greater the diameter difference the greater the stress placed on the mating parts. Also, greater pressures are required to force the shaft into the hole.

   To produce a *shrink fit,* the part with the hole must be heated. The heated material expands, as does the diameter of the hole. The parts are assembled before the hot part cools and shrinks around the shaft.

   Interference fits require close control of the size of the shaft and the hole. Excess variation in the sizes will greatly affect the quality of the fit.

ASSEMBLING WITH FASTENERS

Parts may be assembled with a device which is separate from the parts being united. These devices, called fasteners, grip and hold the parts in position. The most common fasteners are:

1. Threaded fasteners.
2. Rivets.
3. Pins and nails.
4. Retaining rings.
5. Stitches.
6. Quick-operating devices.

THREADED FASTENERS

Threaded fasteners include screws, bolts, and nuts. Often, lock washers and standard washers are used with them.

Selecting threaded fasteners

Threaded fasteners are made in a wide range of sizes, shapes, and materials. The common types are:

1. Wood screws.
2. Sheet metal screws.
3. Machine screws.

4. Bolts.
5. Nuts and internal threads.
6. Special fasteners.

Wood screws, as the name implies, are used to assemble wood parts. The common wood screws are identified by length, diameter, head shape, and material.

They vary in length from 1/4 in. to 4 in. Their diameter is specified by gauge size from 0, the smallest, to 24, the largest. Common gauge sizes for screws ranging from 3/4 in. to 2 in. long are No. 6 (.138 in.), No. 8 (.164 in.), No. 10 (.190 in.), No. 12 (.216 in.), and No. 14 (.242 in.). The wood screw has a smooth shank followed by a threaded portion, as given in Fig. 29-8.

Screws are available in a variety of head shapes as shown in Fig. 29-9. The three most often used for wood screws are the flat head, round or truss head, and oval head. The oval head screw is a decorative shape used with a finishing washer. Most wood screws are slotted to accept either a standard or a Phillips screwdriver.

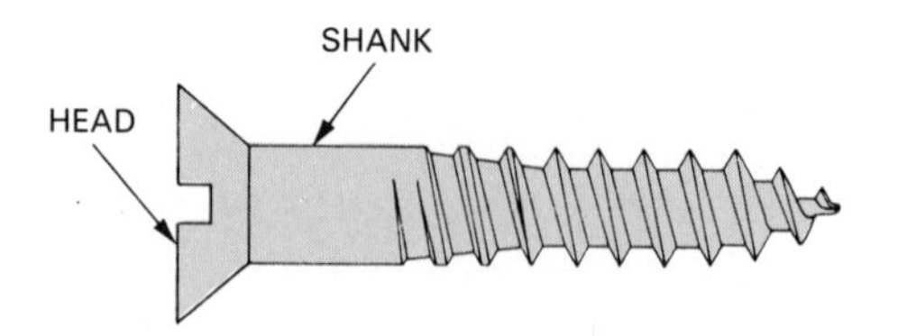

Fig. 29-8. A wood screw is identified by length, diameter, head shape, and the material. (General Motors Corp.)

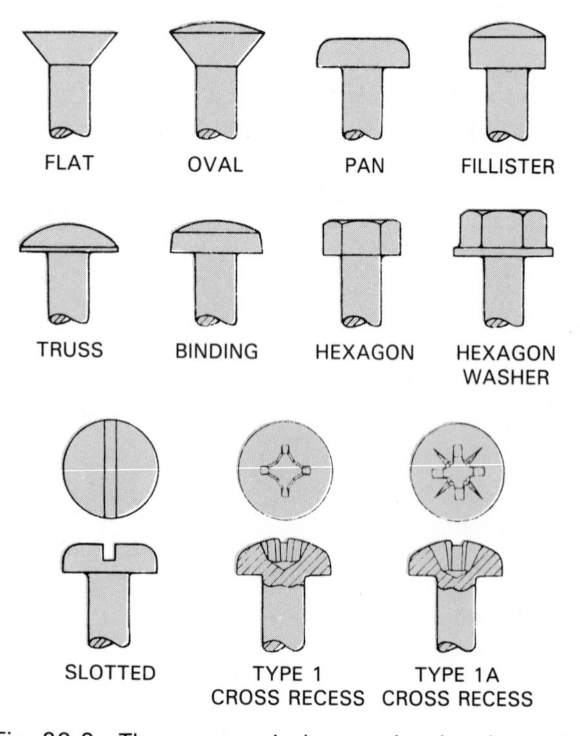

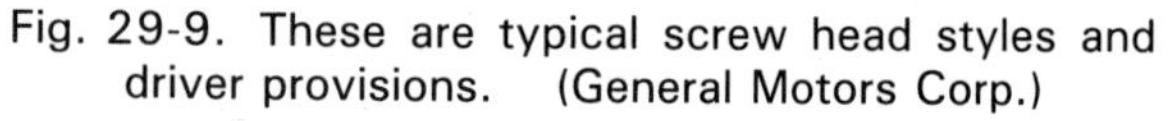

Fig. 29-9. These are typical screw head styles and driver provisions. (General Motors Corp.)

Wood screws are generally manufactured in standard steel, plated steel, or brass. Standard steel screws will rust in an open environment but plated ones will not. Brass screws are used where rust is a serious problem. Many marine applications require brass screws.

Consider typical wood screw specification: 1 1/2 x 12 F.H. cadmium plated steel wood screw. This specification gives the length (1 1/2), diameter (12), head shape (Flat head or F.H.), and material. If no special slot shape is specified, a standard slot is assumed.

Sheet metal screws are used for assembling sheet metal, plastic, and wood parts. Like wood screws, their size is given by length and gauge number. Sheet metal screws, however, have different threads. The threads extend the full length of the screw and may be standard or self-tapping. The self-tapping screws cut threads in the parts as they are set (inserted).

Most sheet metal screws have either a round or pan head. They generally have standard slots but are available with Phillips or other special drives.

Machine screws have a uniform diameter over their length. They are designed to be used with a nut or a threaded hole. The threads extend over the full length of the screw. Machine screw size is listed by the diameter, length, and number of threads per inch. Both flat head and round head shapes are available. Special shapes are manufactured for specific applications. Steel, aluminum, and brass machine screws are commonly available. Plastic machine screws can be purchased for special uses.

A typical machine screw size specification might be 10-24 x 1 1/4. This means that the screw is a No. 10 (diameter) screw. It is 1 1/4 in. long and has 24 threads per inch. The relationship between diameter and number of threads is standard. The common standards are National Coarse (NC) and National Fine (NF). A 1/4 in. screw will have 20 threads per inch if it has National Coarse threads and 24 threads per inch if National Fine.

Like machine screws, *bolts* generally have a uniform diameter. The exception is a lag bolt (or screw) which has a tapered, threaded diameter like a wood screw. Bolts come in a wide variety of lengths and diameters. Their threads are manufactured to a standard. National Coarse and National Fine are common bolt threads. Other thread standards are used for specialty bolts.

Bolts generally are designed to be gripped with a wrench. However, stove bolts have machine screw

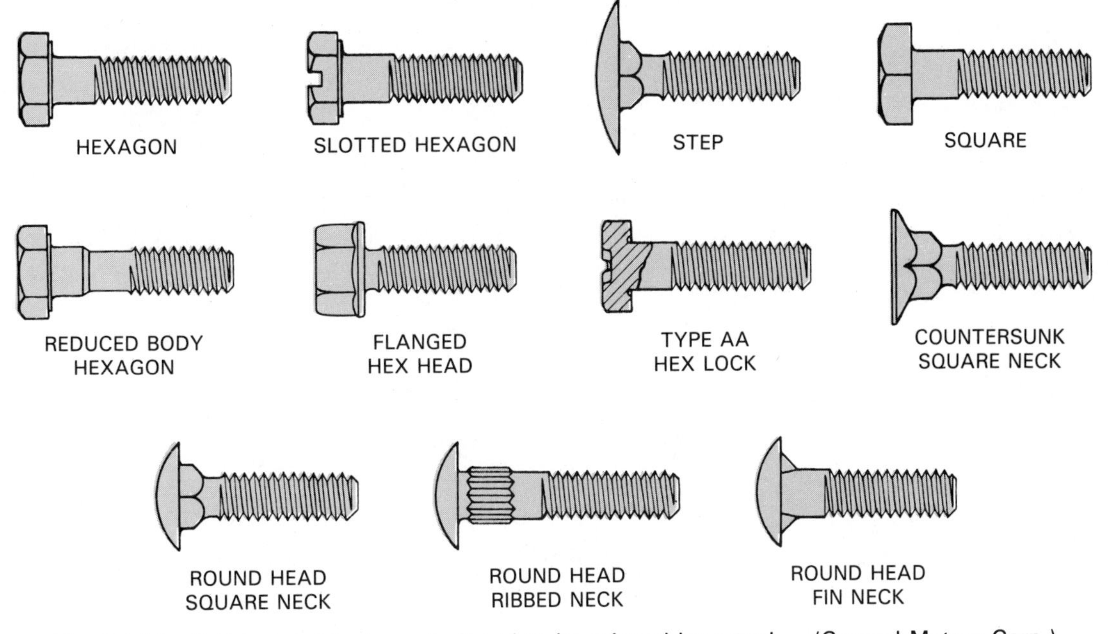

Fig. 29-10. Bolt types. Most are designed to be gripped by a tool. (General Motors Corp.)

type heads and carriage bolts have a smooth head, Fig. 29-10. With the exception of lag bolts, they are designed to be used with a nut or threaded part.

Bolts are sized by length, diameter, and number of threads. A 1/2-13 x 6 in. hex head machine bolt is 1/2 in. in diameter, 6 in. long, has 13 threads per in. and has a hex head which accepts a common box or open-end wrench.

Nuts and internal threads are used with machine screws and various types of bolts. Common nuts are square or hexagonal (six sides). Special types of nuts are also made. As shown in Fig. 29-11, they include:

1. Wing nuts—designed to be easily tightened by hand.

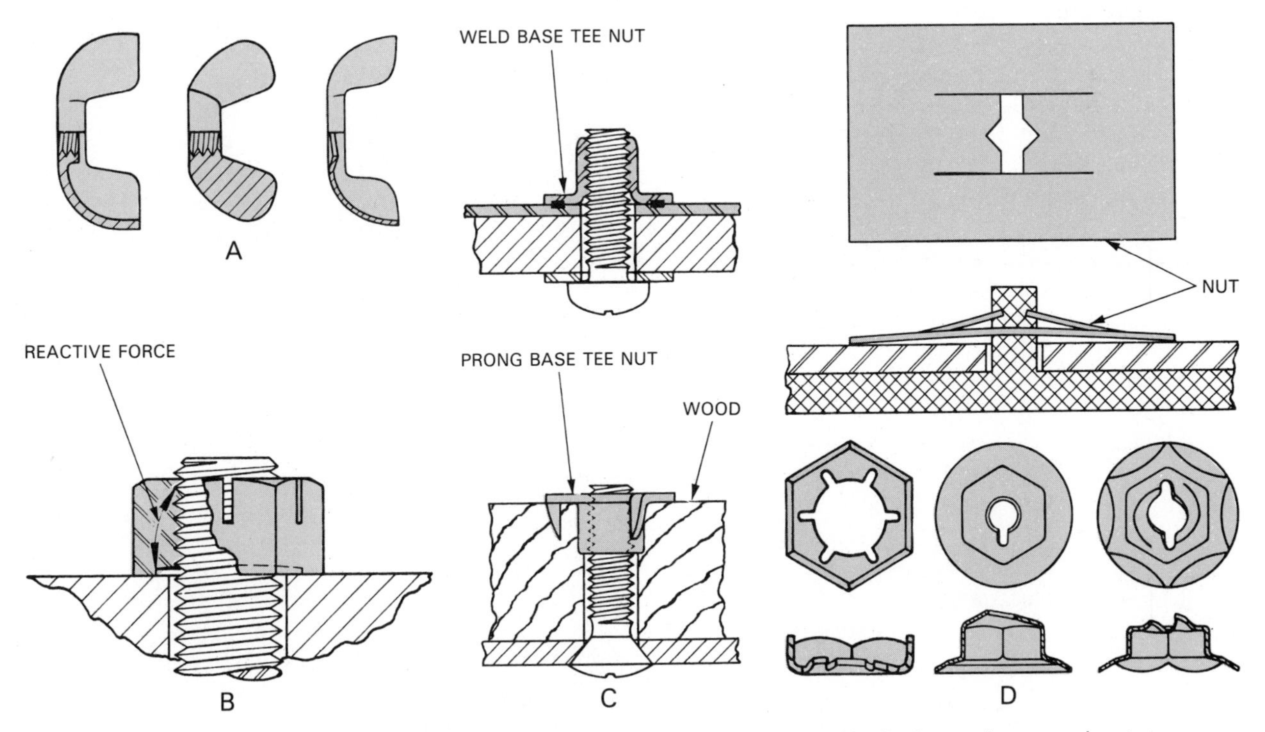

Fig. 29-11. Special nuts. A—Wing nut. B—Lock nuts. C—Inserts. D—Spring and stamped nuts. (General Motors Corp.)

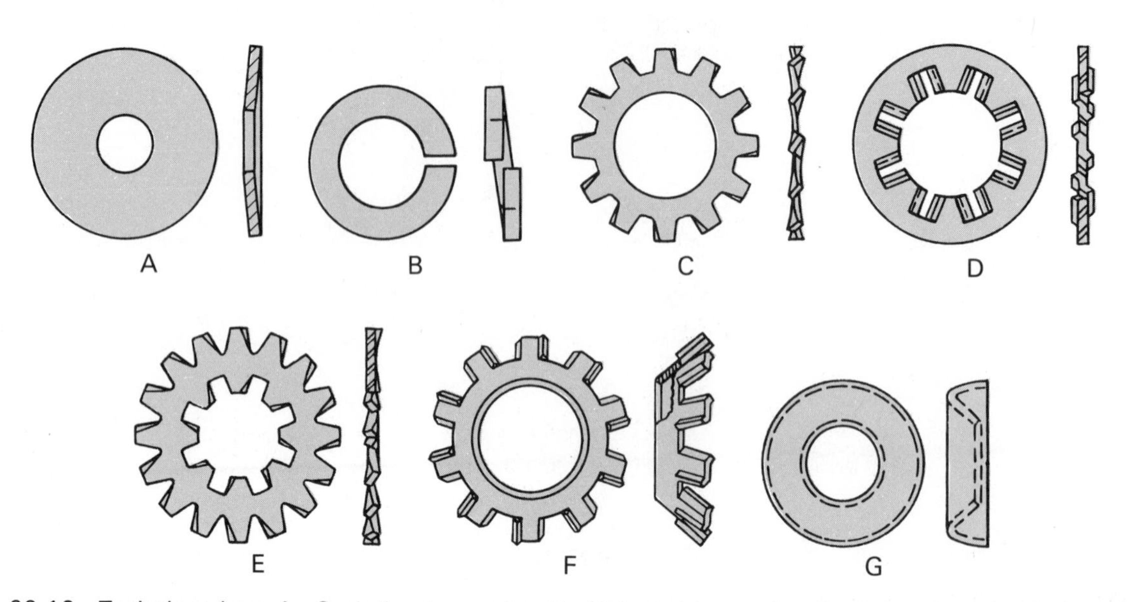

Fig. 29-12. Typical washers. A—Conical spring washer. B—Helical spring washer. C—External toothed lock washer. D—Internal toothed lock washer. E—Internal-external toothed lock washer. F—Countersunk external tooth lock washer. G—Finish washer.

2. Lock nuts—designed to resist vibration which could loosen the assembly.
3. Inserts—threaded metal parts which are placed in holes in wood, plastic, or soft metal.
4. Spring action—stamped sheet metal nuts which grip the bolt threads.

Bolts may also be used with a threaded part in place of a nut. The threaded hole in the part acts like a nut.

Washers are often used along with bolts and nuts, Fig. 29-12. Standard washers distribute the force applied to the surface. They also keep the turning bolt head and nut from marring the surface. Lock washers keep the nut or bolt from vibrating loose.

Special threaded fasteners

A common type of special fastener is the threaded stud. Studs have threads at both ends. One end is screwed permanently into a fixed part. The exposed end accepts a nut, Fig. 29-13. The stud may have machine threads on both ends or machine threads on one end and coarse or wood threads on the other.

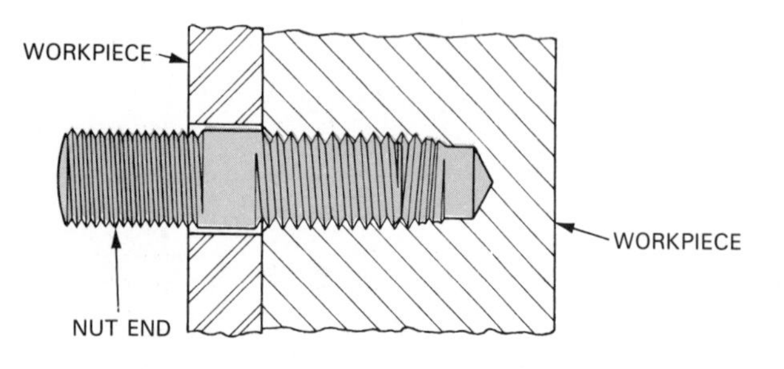

Fig. 29-13. Cutaway shows stud application. (General Motors Corp.)

Wooden legs are often attached to furniture using threaded studs. The wood screw end is threaded into the leg. The machine screw end is then fastened to a hanger bracket attached to the main frame of the furniture.

Installing threaded fasteners

Threaded fasteners require one or more loose pieces (inactive) and one solid piece (active). The loose piece(s) are gripped between the fastener head and the active piece. The active piece is drawn toward the head as the fastener is turned (set). This movement applies force between the head and the active part.

The active part can be either a nut or the second piece being assembled. In all cases the fastener must fit through a hole in the inactive (first) part. If it does not, threads will be cut in the inactive part. In this case, the parts will never be drawn together. This is like holding two nuts apart and threading a bolt into them. As long as the nuts are not allowed to turn independently the bolt will not pull them together. It will only lock against the first nut. Continued torque on the bolt may cause it to shear off.

This is the reason wood screws are twisted off when two parts are not properly drilled. The shank hole must extend completely through the inactive part. Otherwise, threads are cut in it and the two-nut principle takes over.

Fig. 29-14 shows the proper application of threaded fasteners. However, even proper preparation (hole

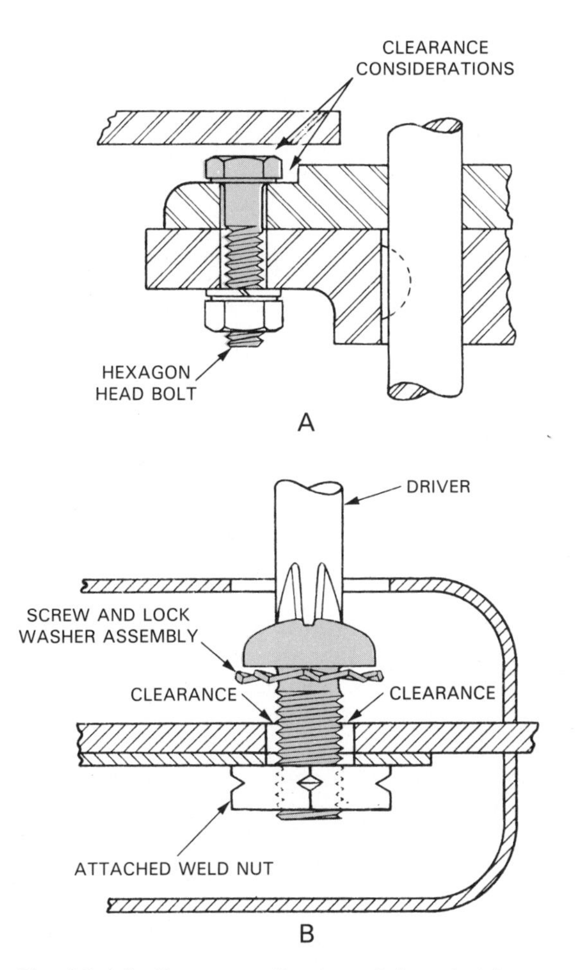

Fig. 29-14. Proper application of threaded fasteners. A—Hexagon bolts. B—Screw and washers. (General Motors Corp.)

sizes) is not enough. Installation forces cannot exceed the shear strength of the material. The resistance between the fastener and the parts may be so great that the parts will shear off. This is a particular problem in setting aluminum and brass wood screws. The thread (pilot) hole must be large enough to easily accept the screw. The screw may also need lubrication (wax or soap). Gentle continuous force (torque) must be applied. Sharp or rapidly applied forces can easily exceed the screw's tensile strength.

RIVETS

Rivets are metal pins which are used to permanently assemble parts. Rivets are generally one piece fasteners which have a head and a body.

The major types of rivets, shown in Fig. 29-15, are:

1. Solid—a rivet with a shaped head and a solid shank. It is usually used for heavy duty riveting in construction and on agricultural equipment.

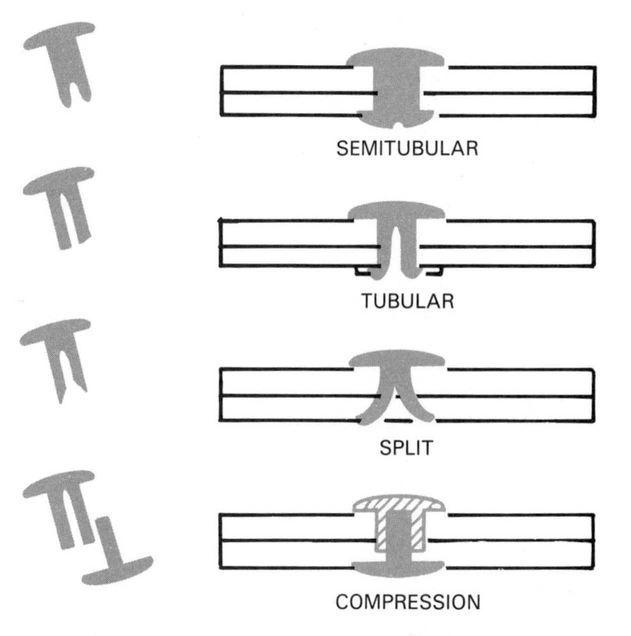

Fig. 29-15. There are four major types of rivets.

2. Semitubular—a rivet with a shaped head and a partially hollow shank. The hole in the shank is not deeper than the shank diameter. Clinching semitubular rivets take less force than solid rivets.
3. Tubular—a rivet that has a shaped head and a hollow shank. The shank can be used to punch its own hole in paper, leather, fabric, plastic, and other soft material.
4. Split shank—a rivet with a shaped head and a shank that is divided into two parts or legs. The pronged shank can produce its own hole in soft metals, plastics, leather, and fabrics.
5. Compression—a rivet made of two parts—a solid and a tubular rivet. The solid rivet has a slightly larger diameter than the hole in the tubular rivet. When pressed together a tight riveted joint is obtained. These rivets are used to assemble knife handles.
6. Blind—a rivet which can be clinched from the head end. Blind rivets are used where it is not possible or desirable to form the clinch end. Blind rivets are used in assembly and repair of semitrailers and aircraft. The familiar pop rivet is a blind rivet. As shown in Fig. 29-16, blind rivets may be clinched by explosive or mechanical means. Rivets are available in several head shapes, Fig. 29-17.

The riveting process may be manual or automatic. Automatic machines feed the rivets from a hopper to the rivet station. In either case, the steps involved include:

1. The rivet is placed in a hole or forced through the material.

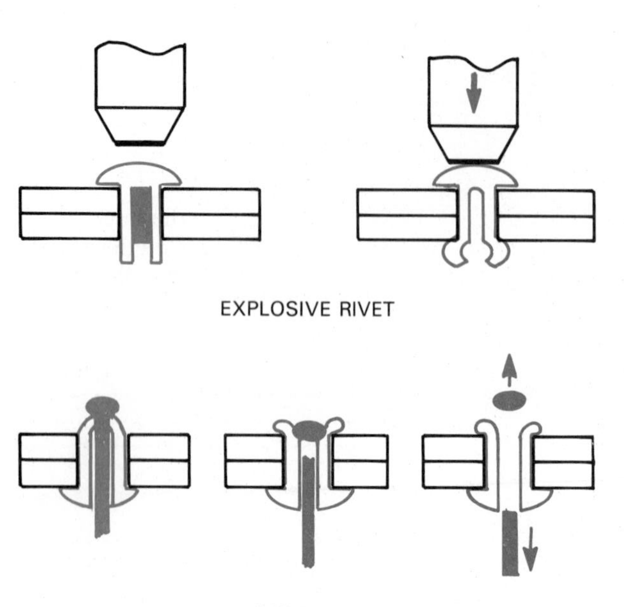

Fig. 29-16. Explosive and pop rivets are types of blind rivets.

2. The two layers of material are forced tightly together.
3. The rivet is upset thereby enlarging its diameter to fill the hole.
4. A head is formed (clinched) on the shank end of the rivet. The head should be one and one-half times the shank diameter.

The riveting may be done by a squeezing, hammering, or squeezing-twisting action. See Fig. 29-18.

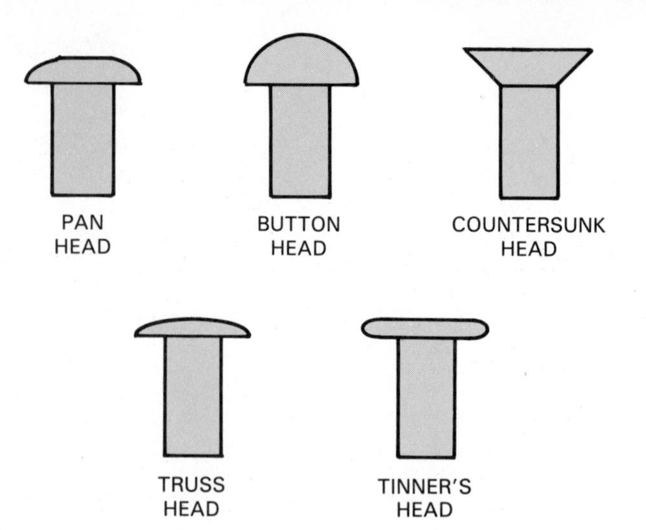

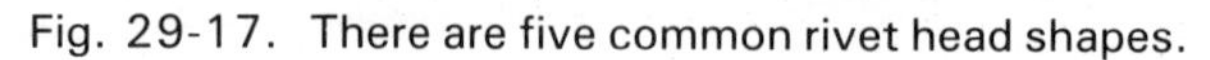

Fig. 29-17. There are five common rivet head shapes.

PINS AND NAILS

Pins and nails are another means to mechanically fasten parts together. Pins are used in predrilled holes while nails make their own holes.

Pins key two parts together. They are often used to hold a shaft in a hole or a pulley on a shaft. They are also used to stop nuts or washers from coming off a shaft. The principal types are taper pins, roll pins, and cotter pins.

Taper pins gradually change diameters along their length. When forced into a hole they create an interference fit.

Roll pins are a cylinder of high carbon. The cylinder is compressed as it is driven into a hole. Again, an interference fit is produced.

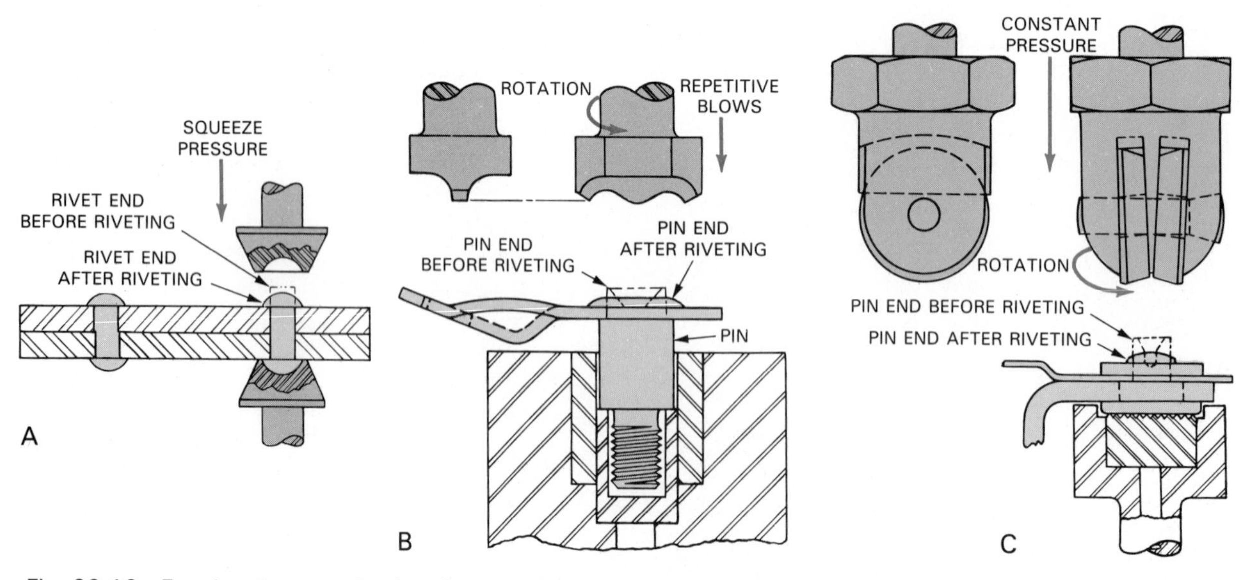

Fig. 29-18. Forming rivets can be done by several methods. A—Squeezing. B—Hammering. C—Squeezing-twisting. (General Motors Corp.)

Cotter pins are shaped and doubled wire strips. They are placed through a hole and the ends are bent in opposite directions to keep the pin from falling out. Cotter pins are used to keep nuts on a shaft or washers and spacers in place.

Nails hold two workpieces together. They are usually used to secure one part to a second part. The first part may be almost any material through which the nail can be driven or that has a hole in it. The second part usually is a fibrous material. Nails are generally used on wood or wood products. The primary types of nails are shown in Fig. 29-19. The common, box, and wire nails have similarly shaped heads. The relationship between the length and diameter varies. Common nails are thicker than box nails. Wire nails are usually shorter and are available in several diameters (wire gauge).

Finish nails and wire brads have the same head shape but finish nails are larger than brads. Wire brads, like wire nails, are available in several thicknesses.

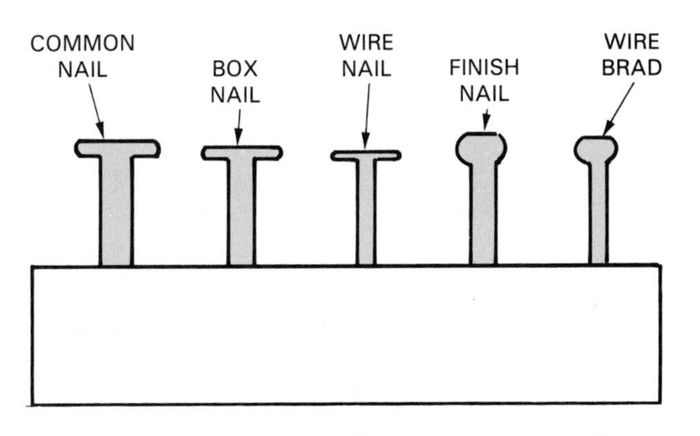

Fig. 29-19. Types of nails. Note that common nails are thicker than box nails.

Common nails, box nails, and finish nails have only one diameter for each length. The diameter is a standard for all nail manufacturers. The penny (d) is the unit of measure for length. The lengths vary from 2d or 1 in. to 60d or 6 in. The most common sizes are: 4d–1 1/2 in. long, 6d–2 in. long, 8d–2 1/2 in. long, 10d–3 in. long, and 11d–3 1/2 in. long. The shorter nails are used in cabinet and furniture manufacture. Construction workers always use the longer nails.

Many special purpose nails are also available. Nails with rings or barbs are used to nail floor underlayment and drywall. Large headed galvanized nails are used for roofing. Staples hold wire in place. Hardened steel nails can be driven into concrete, cement blocks, and brick.

Nails are selected for each specific use. At least three factors are considered:

1. Length.
2. Head shape.
3. Holding power.

The *length* of the nail is related to the thickness of the parts. A nail generally is driven through the thinner piece into the thicker member. The nail should be about three times as long as the thickness of the thinner member, Fig. 29-20.

Large diameter heads are used to hold soft materials like shingles and drywall. Smaller heads will hold solid wood. The ball-shaped heads of brads and finish nails can be driven below the surface.

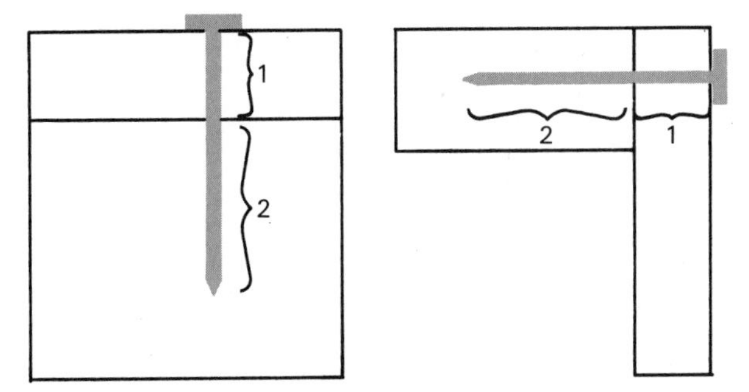

Fig. 29-20. Nail length should be three times the thickness of the thinner piece.

The *holding power* can be increased by the surface treatment of the nail. Coated and galvanized nails have a rougher surface. So do ring nails. Such nails resist being pulled from the stock.

Keys are special pins which hold wheels or pulleys or shafts. As shown in Fig. 29-21, they fit in a groove in the shaft and the wheel. They absorb the forces which try to make the wheel turn on the shaft. The most common are square and woodruff keys.

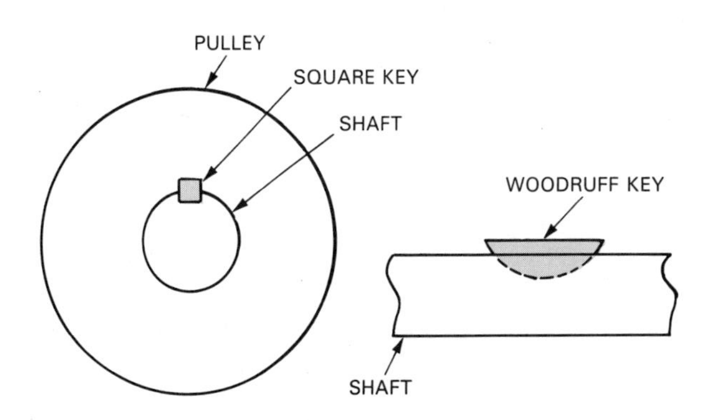

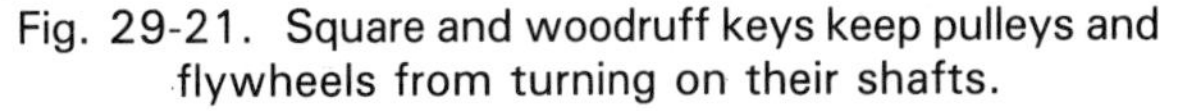
Fig. 29-21. Square and woodruff keys keep pulleys and flywheels from turning on their shafts.

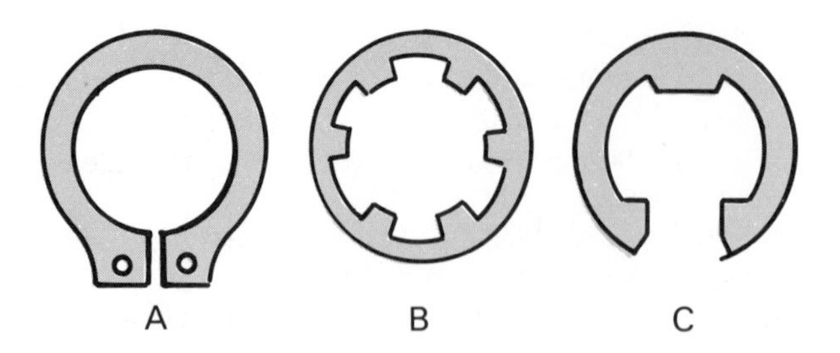

Fig. 29-22. Common retaining rings. A—External snap ring. B—External self-locking ring. C—External E type.

RETAINING RINGS AND CLIPS

Retaining rings are stamped steel parts. They either fit into a groove or are pressed on the shaft, Fig. 29-22. A retaining ring can serve several purposes including:

1. Providing a shoulder for a spring or other component.
2. Holding a shaft in a hole.
3. Holding components on a shaft.
4. Taking up end play in an assembly.

STITCHING

Two common types of stitching are used in industrial applications. Fabric parts are sewn using two interlocking threads. Paper and cloth bags are also sewn products.

Wire stitching is used to combine two or more pieces of metal, plastic, fabric, or paper, Fig. 29-23. Wire stitching forms a staple from a continuous wire. The wire "U" is driven through the parts and usually is clinched on the back side. Wire stitches, as shown in Fig. 29-24, are of five basic types:

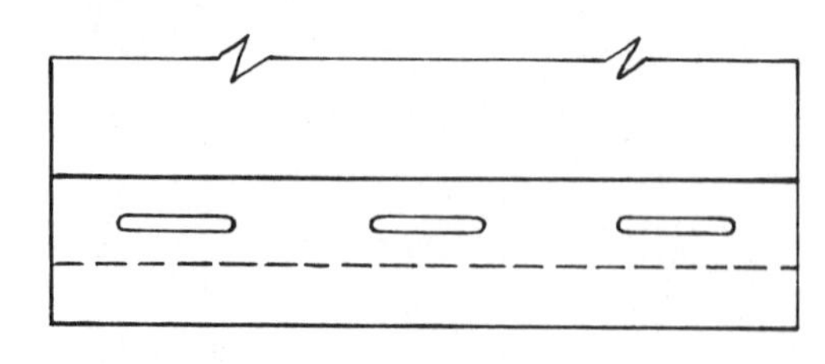

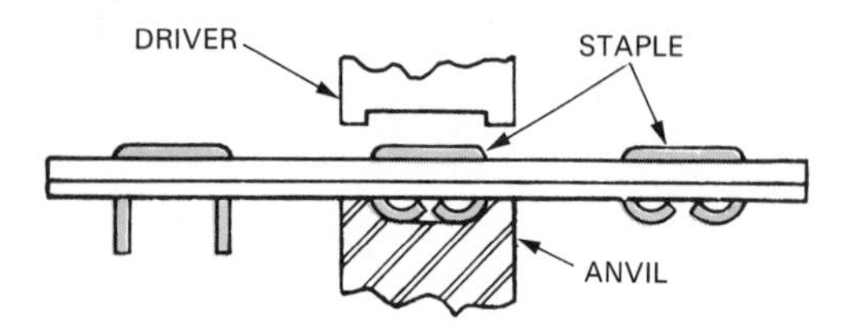

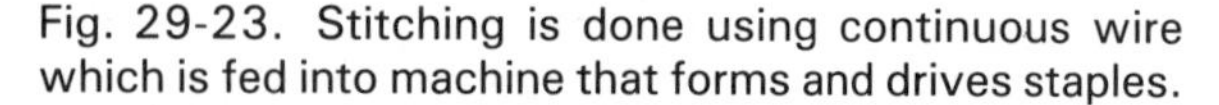
Fig. 29-23. Stitching is done using continuous wire which is fed into machine that forms and drives staples.

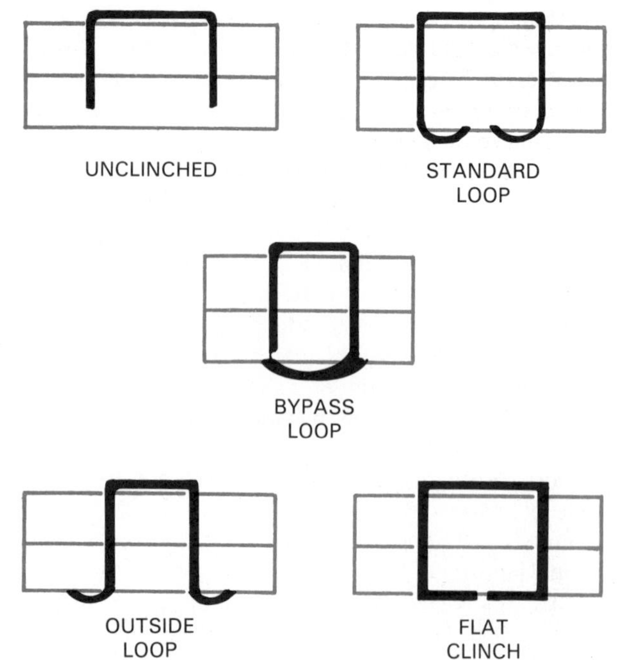

Fig. 29-24. Types of wire stitches.

1. Unclinched.
2. Standard loop.
3. By-pass loop.
4. Outside loop.
5. Flat clinch.

The unclinched stitch is used to attach thin materials to a thick base material. Polyethylene film is stitched (stapled) to studs. Wood trim is attached to homes using this process. The standard loop is used for normal work. Bypass and outside loop stitches provide a stronger assembly. The flat clinch is used when appearance is important.

A wide range of material can be stitched including aluminum, sheet copper, steel and brass, paper, cork, leather, fiber board products, and wood.

QUICK-ACTING DEVICES

Many quick-acting mechanical fasteners are available. These include magnetic catches, spring clips, lever controlled hooks, and other similar devices. They are designed for special purposes. They generally are used for temporary fastening like cabinet catches, luggage latches, and electronic case closures.

SUMMARY

Mechanical fastening involves the use of mechanical force or mechanical fasteners. Seams, tabs, slots, beads, dimples, and flanges are used as part of the assembly designs. Also, parts may be designed to be pressed together. Force fit and shrink fits are common pressed or interference fits.

Mechanical fasteners are also used for temporary and permanent assembly practices. The most common fasteners are threaded devices, rivets, pins and nails, retaining rings, stitches, and quick-acting devices.

STUDY QUESTIONS – Chapter 29

1. What are the two major groups of mechanical fastening processes?
2. List the ways that mechanical force is used in fastening.
3. What are the major types of fits?
4. List the six common types of fasteners.
5. List and describe the common types of threaded fasteners.
6. Describe the proper installation of threaded fasteners.
7. List and describe the types of rivets.
8. List the steps in riveting.
9. List and describe the major types of nails and pins.
10. What are retaining rings and clips?
11. What is stitching?

A threaded fastener is being installed on a aircraft wing. (Boeing Co.)

SECTION 7
FINISHING

Finishing processes are designed to alter or treat the surfaces of products to protect them from the environment or improve appearance. These processes include two basic categories treated in the following three chapters:

- Converted surfaces—in which a thin layer of the part's surface is changed chemically.
- Coatings—in which a layer of protective material (either organic or nonorganic) is applied to the product.

Chapter 30

INTRODUCTION TO FINISHING PROCESSES

Almost all products on the market have been surface treated during manufacture. Their exterior surfaces have been smoothed and finished by some process. The finish is applied for improved appearance and for protection against acids, chemicals, water and oil in the environment. The product is also protected from wear. These environmental elements can quickly discolor and/or destroy the product.

The name "finishing," is applied to the process which protects and/or beautifies a surface. This term includes all surface-decorating and surface-protecting treatments.

Finishing practices involve three basic steps, as shown in Fig. 30-1:

1. Selecting a finishing material.
2. Preparing the material to be finished.
3. Applying the finishing material or coating.

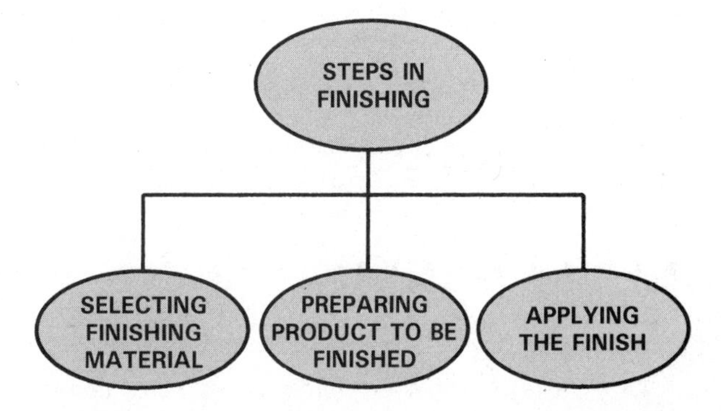

Fig. 30-1. Three steps are required to produce a finish on a product.

FINISHING MATERIALS

While finishes cover a wide variety of materials, they can be grouped into two large categories: converted surfaces and coatings. See Fig. 30-2.

Converted surface finishes are thin layers of material which are actually part of the metal product. The material is chemically treated to change the nature of the outside layer of the part. The result is a chemical compound which will resist reactive elements in the environment. The coating is provided by a reaction between a chemical and the metal atoms on the part's surface.

Converted coatings may be natural or artificial. Aluminum will develop an oxide when exposed to the natural environmment. This oxide coating resists further attack by environmental elements.

Coatings can also be developed through controlled processes. Phosphate coatings and black iron oxide layers are developed to protect steel products. Anodizing produces a thick, uniform oxide layer on aluminum.

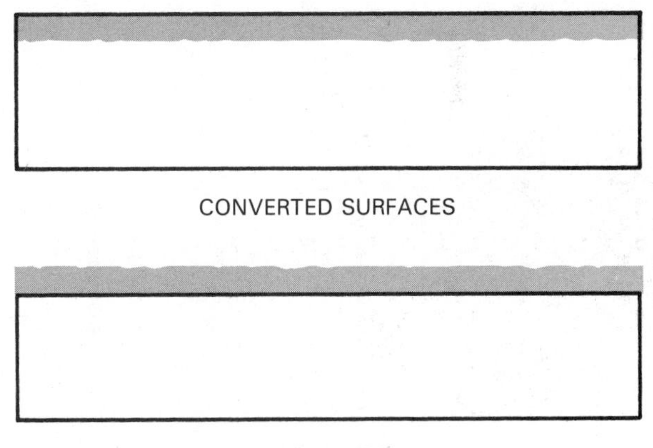

Fig. 30-2. Finishes may be either of these types. Top. Converted surface. Bottom. Coating.

Coatings involve the application of a layer of protective material on the base material. They are either inorganic or organic.

Inorganic coatings are metal or ceramic layers. They may be applied to metal, ceramic, or plastic products. Organic materials are natural or synthetic

Fig. 30-3. Final smoothing and cleaning may be done with coated abrasives. Here a belt sander is employed to smooth silverware finish.

hydrocarbons. The most common organic coatings are shellac, varnish, enamel, lacquer, vinyl, silicone, and epoxy. For additional information on finishing materials, see Chapters 31 and 32.

PREPARING MATERIAL FOR FINISHING

Many parts need to be smoothed and cleaned before finish materials can be applied. Scratches, burrs, and dents may need to be removed. Oil, grease, and other materials used during the manufacturing process need to be cleaned away. Surface oxides and other impurities may also need to be removed.

Three types of process are used in the material preparation stage of finishing:

1. Mechanical processes.
2. Chemical processes.
3. Electrochemical processes.

MECHANICAL PREPARATION PROCESSES

Mechanical processes use a rubbing action to remove unwanted materials from the surface of the product. Mechanical means also are used to smooth the material. The most common mechanical preparation processes are:

1. Abrasive cleaning.
2. Tumbling and vibrating.
3. Media blasting.
4. Burnishing.
5. Brushing.

Abrasive cleaning

Abrasive cleaning uses either a coated abrasive (abrasive paper or cloth) or an abrasive on a pad. The common abrasive processes are sanding, polishing, and buffing.

Sanding uses a coated abrasive to prepare the surfaces. The abrasives may be belts, discs, or sheets, Fig. 30-3. Sheet abrasives are often used on oscillating or pad sanders. A contour sander which uses a series of abrasive strips backed up with brushes is used to clean and smooth shaped edges. Many of the machines described in Chapter 21, Abrasive Machining Operations, can be used for surface preparation.

Polishing and buffing generally require cloth pads or wheels. Abrasive material (buffing compound) is applied to the moving pad or wheel against which the work is then pressed.

The term polishing is used when removal is a prime objective. Polishing removes surface defects.

Buffing removes only scratches left by sanding and polishing. Buffing is intended to produce a surface with no visible defects or scratches.

Abrasive processes are used to prepare metallic, ceramic, and polymeric materials for finishing. These materials can be sanded, buffed, and polished. Ceramics are often sanded before firing to remove defects and seam lines.

Tumbling and vibrating

Parts may be placed with an abrasive or nonabrasive media. Motion causes the parts to rub together. This removes roughness, burrs, sharp edges, and surface impurities such as scale.

Common processes using this type of rubbing action are tumbling and vibrating. *Vibratory finishing* uses a steel tub lined with rubber or polyurethane. The parts to be prepared are placed in the tub along with a medium. The medium may be abrasive particles or nonabrasive materials such as steel shot, smooth rock, round-end steel rods, or ceramic shapes. (Nonabrasive particles actually burnish the surface. This will be discussed later in this chapter.)

After the material is loaded, the tub is vibrated. This motion creates a rubbing action between the parts and the medium.

Vibratory finishing machines, Fig. 30-4, come in a number of sizes. They extend from about 1 cu. ft. to over 75 cu. ft. They also are built to vibrate in the range of 1200 to 2400 times a minute.

In *barrel finishing,* a revolving tub or barrel creates the rubbing action. The parts and the medium are tumbled together to smooth the parts.

Metal parts and fired ceramic ware can both be finished using vibratory and barrel finishing machines. Their surfaces are smoothed, burnished, and stress relieved by the processes.

Media blasting

Media may be impelled (thrown or shot) against a surface to clean and prepare it for finish. Media

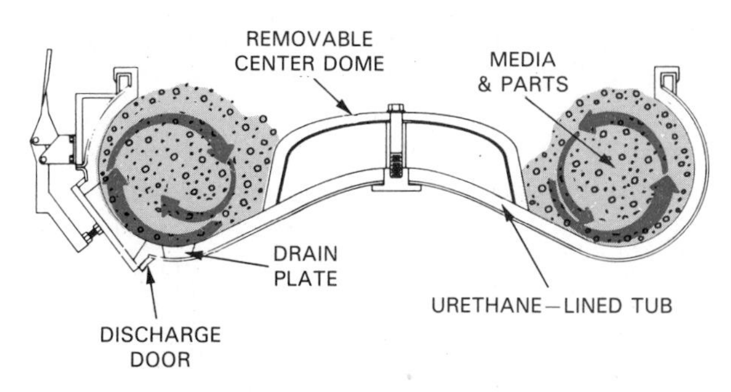

Fig. 30-4. Schematic shows simplified details of a round bowl vibratory finishing machine. (ALMCO Div., King-Seeley Thermos Co.)

blasting may be divided into three major categories: dry blasting, wet blasting, and shot peening.

Dry blasting, Fig. 30-5, uses high-velocity air to force abrasive particles against the surface. These particles will remove oxides and corrosion, old paint or other finishes, burrs, and sharp edges. It will also etch the surface for better paint adhesion.

Fig. 30-5. A structural beam enters a dry-media descaling machine. (Pangborn Corp.)

Wet blasting works on the same principle as dry blasting. However, the abrasive is mixed with a liquid, usually water. The abrasive-liquid slurry is then impelled against the part surface. Wet blasting controls dust and permits use of finer abrasives.

Shot peening throws fine steel shot against the part. The shot conditions the surface as was discussed in Chapter 25. It can also be used to produce a hammered finish on soft materials.

Burnishing

Burnishing uses a nonabrasive device to wipe a surface. The wiping action causes the surface of the material to be smoothed by creating a "plastic" flow of material.

Barrel burnishing is used for metal and ceramic parts. Smooth tools can also be used to burnish metals and plastics. Burnishing is often done in conjunction with metal spinning. It also is used to smooth ceramic ware as shown in Fig. 30-6.

Brushing

Wire and fiber brushes can be used to remove loose particles on the surface of parts. Usually, wire wheels are used. These brushes, as shown in Fig. 30-7, are available in a wide variety of shapes. They are produced in sizes from small brushes for hand-held tools to single brushes weighing over 1000 lb.

The rubbing action of the brush bristles scrape off loose surface materials. This leaves a clean surface ready to accept a finish.

Fig. 30-6. Burnishing has the effect of rearranging the direction of the grain on the surface of a product. Top. Grain orientation is "helter skelter" before burnishing. Bottom. After burnishing of the product, grain orientation is changed so grains are parallel to one another.

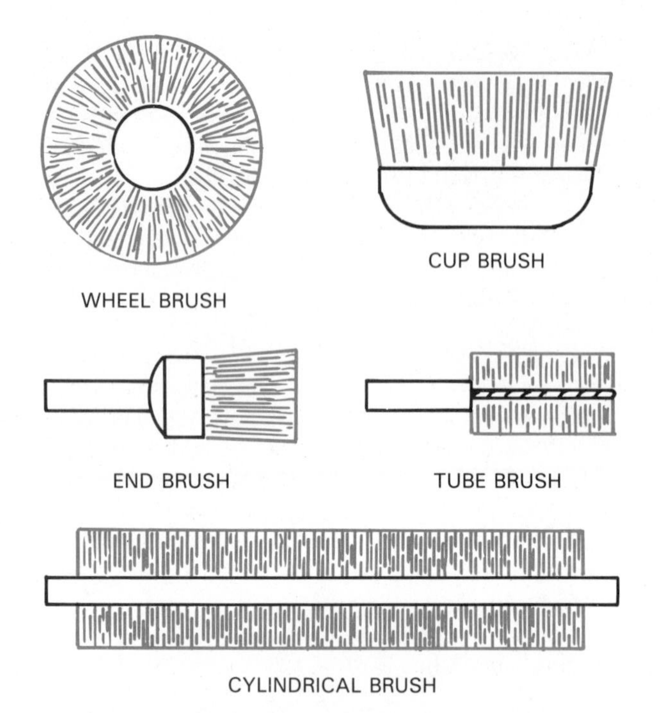

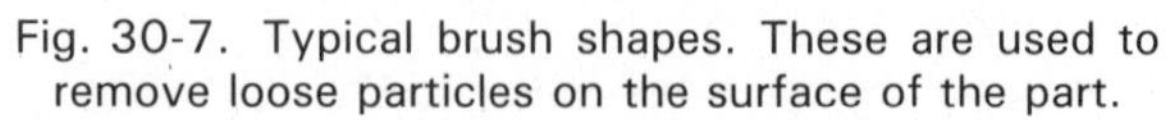

Fig. 30-7. Typical brush shapes. These are used to remove loose particles on the surface of the part.

CHEMICAL PREPARATION PROCESSES

Chemicals may be used to prepare materials for finishing. The chemicals are solvents for the materials to be removed. They will dissolve the unwanted material on the surface. The chemicals, along with the dissolved matter, can then be rinsed from the surface.

The chemicals are usually applied by dipping the parts in a tank or by spraying. The parts and the chemicals are often left to soak. When the materials are cleaned in an acid, as in Fig. 30-8, it is often called *pickling*. Caustic solutions are used to clean metals in the *alkaline cleaning* process. *Ultrasonic cleaning* (cleaning with sound waves) can be used to increase the speed of the cleaning action.

The parts or the chemicals can also be heated to increase the speed of the chemical cleaning action. *Steam cleaning* uses blasts of steam and detergent to remove soil. *Vapor degreasing* uses hot solvent vapors for cleaning.

Flame cleaning uses an oxyacetylene flame for cleaning oxidation and scale from metal surface. The rapid heating causes the impurities to break off the surface.

Fig. 30-8. Metal baking pans, ready for finishing, wait on a rack in front of a pickling (acid) tank. (Mirro)

Electrochemical preparation processes

Electropolishing is the main electrochemical preparation process. It uses an electric current and a chemical to produce the desired result.

The part is placed in a hot alkaline solution called an electrolyte. The part is attached to the positive charged terminal of a direct current source. It becomes the anode. Another metal part is attached to the negative terminal. It becomes the cathode.

Electricity passes between the anode (the part) and the cathode. The current deplates (dissolves) material from the part. The current flows most easily from dry high spots or ridges on the part. These are removed first leaving a smoother surface.

Oxygen bubbles form on the surface of the part and slowly burst. This action develops a scrubbing action which cleans dirt and scale from the part.

Applying finishing materials and coatings

Finishing materials can be applied in many ways.

1. They can be brushed, rolled, sprayed, dipped, or flooded.
2. Electrical current can be used to deposit materials on the surface.
3. Chemicals can be used to convert finishes.

These and other techniques will be discussed in detail in Chapters 31 and 32.

SUMMARY

Finishing practices require that the finishing materials be selected, surfaces be prepared, and finishes applied. The common finishes involve two major categories: organic finishes and inorganic finishes. Organic finishes are the common paints, enamel, or other synthetic coatings. Inorganic finishes are metal and ceramic materials used to coat a material.

Materials must be prepared for finishing. A thorough cleaning is required. Parts may also be smoothed and polished. Mechanical, chemical, and electrochemical practices are used to prepare materials.

The finish may be a layer of the product which is chemically treated. This finish is called a converted coating. A layer of material may also be applied to the product. These coatings can be brushed, sprayed, dipped, plated, or otherwise applied to the base material. The end result is a product which can better withstand the environment. It can resist elements around it.

STUDY QUESTIONS – Chapter 30

1. Define finishing.
2. List and describe the basic steps in finishing.
3. What are converted surface finishes?
4. What is a coating?
5. What are the major inorganic coating materials?
6. What are the common organic coating materials?
7. List the three types of material preparation processes.
8. List and describe the five ways to mechanically prepare material for finishing.
9. What are pickling and alkaline cleaning?
10. Describe electropolishing.

Chapter 31

INORGANIC COATING PROCESSES

Coatings are applied for decorative or nondecorative purposes of both. The decorative functions requires a uniformly smooth and attractive appearance. Nondecorative coatings are applied to resist corrosion, reduce wear, repel moisture, lubricate, and resist impact.

There are a large number of inorganic coatings. They are either metallic or ceramic, Fig. 31-1.

Metallic coatings, as you have already learned, can be a layer of material applied to a part or product or they can be converted finishes (special treatment of the surface of the part to give it different properties). Metallic finishes include chromium plated metal, plastic components, anodized aluminum, oxide-coated steel, and galvanized iron or steel.

Ceramic coatings are layers of inorganic nonmetallic material. They may be applied to ceramic or metal-based materials. Typical ceramic-coated

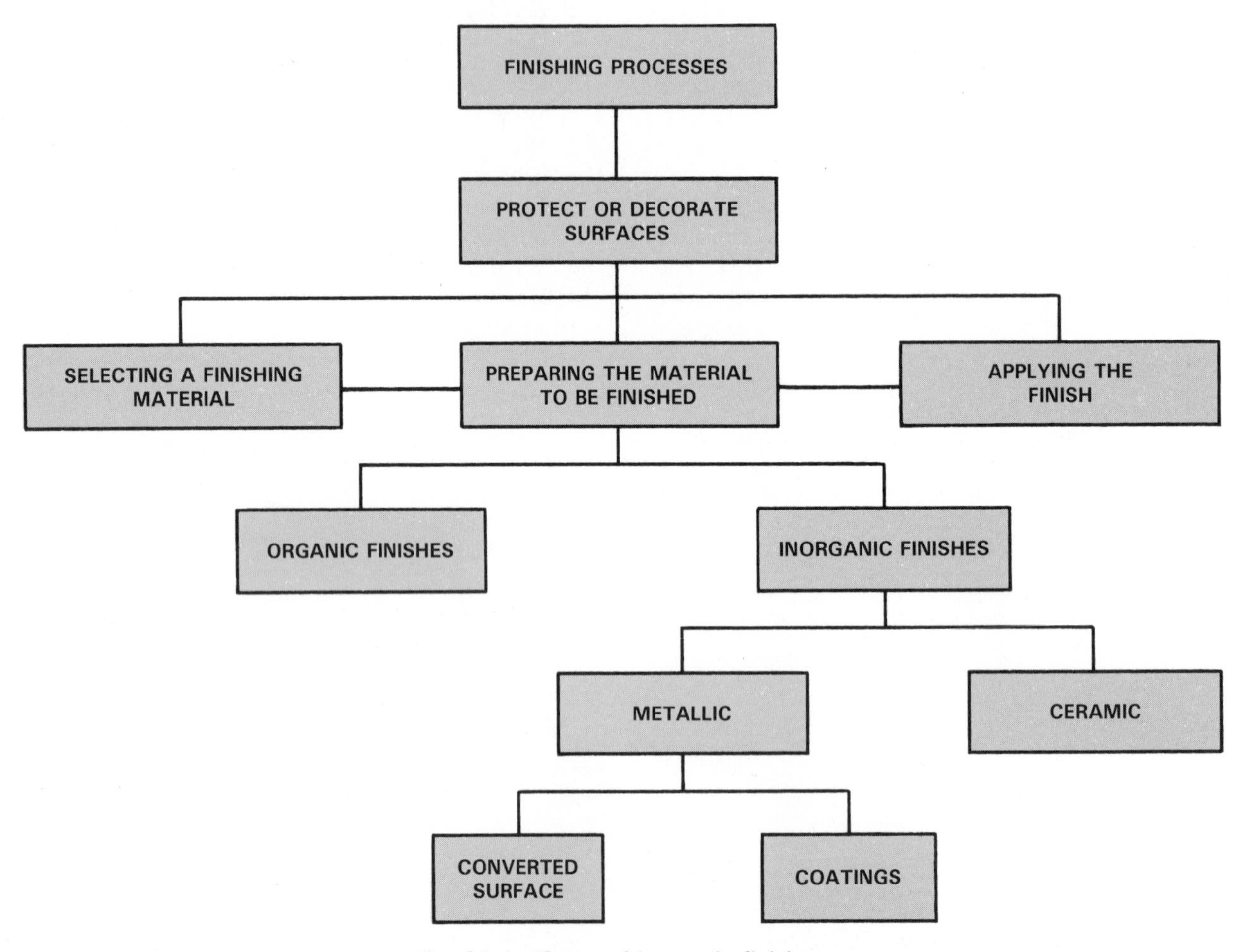

Fig. 31-1. Types of inorganic finishes.

products are dinnerware, glazed floor tile, bathroom fixtures, and spark plug bodies.

There are four basic combinations of inorganic coating on base materials:
1. Metal-on-metal.
2. Metal-on-plastic.
3. Ceramic-on-ceramic.
4. Ceramic-on-metal.

METALLIC CONVERSION FINISHES

When metals are treated chemically to produce a protective skin, no additional metal is added. Only the outside layers are changed to improve their appearance or durability. The skin becomes a solid layer of a chemical compound which has the necessary qualities of a finish. This type of finish is called a *chemical conversion* or conversion coating. The four most common conversion coatings are: phosphate, chromate, oxide, and anodic coatings.

PHOSPHATE CONVERSION COATINGS

Metals can be treated in a bath of a metal phosphate and phosphoric acid. Typical of phosphates used are those of iron, zinc, lead, and manganese. (The phosphate is a salt of the metal for which it is named.) The surface of the part will enter into a chemical reaction with the bath. The result will be a part with a layer of phosphate on its surface, Fig. 31-2.

The phosphate conversion process will work only on clean parts. This requires one or more cleaning and rinsing stages followed by the phosphating step. The phosphated parts are also thoroughly rinsed and dried after receiving the conversion coatings, Fig. 31-3.

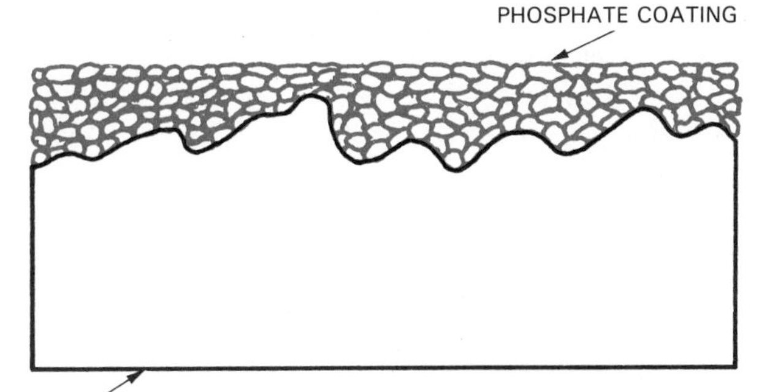

Fig. 31-2. Phosphate coatings protect metal surfaces against corrosion from electrolytic actions.

Phosphate coatings are used for two main purposes:
1. They provide a base for painted, oiled, or waxed surfaces. Zinc phosphates can be applied to iron, steel, zinc plated (galvanized) or cadmium plated materials. The coating improves the adhesion of additional layers of material.
2. Phosphate coatings also reduce friction on the material. This characteristic improves the deep-drawing properties of sheet metal.

CHROMATE COATINGS

Chromate coatings are produced by proprietary methods. The exact solutions used are company secrets. However, all contain two main ingredients: an acid and chromium ions.

Chromate conversion coating processes are most often used on aluminum and zinc and on cadmium

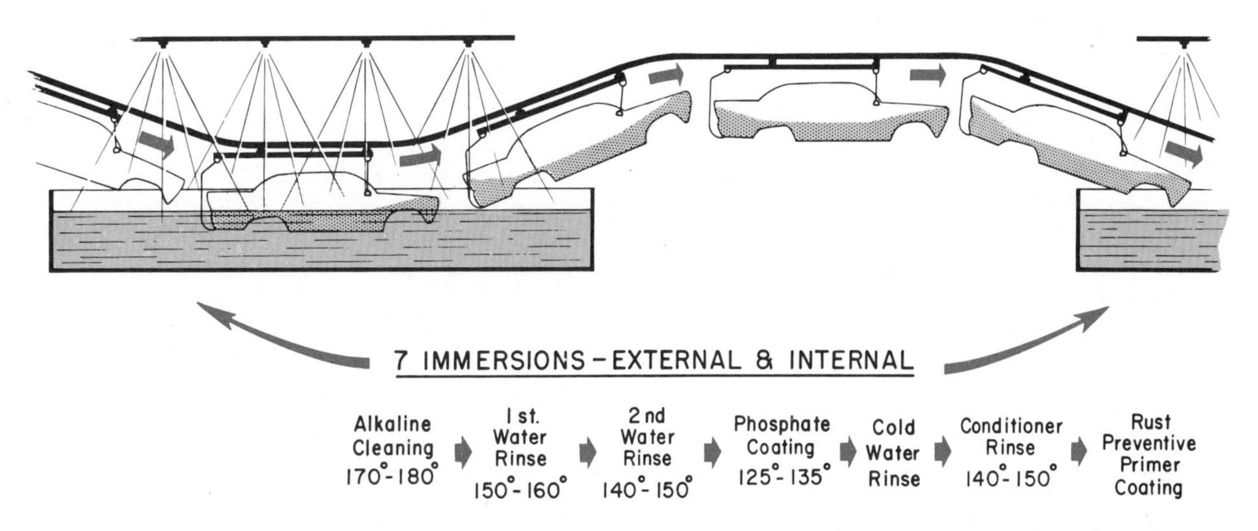

Fig. 31-3. Steps in rustproofing an automobile body. Phosphate coatings are used as a base for paint. (George Koch Sons, Inc.)

coated steel. Magnesium and tin-coated steel may also receive a chromate coating. The process involves covering the metal with the proprietary solution. The parts may be dipped or brushed, sprayed, or roller coated.

The resulting thin (about 0.00002 in.) chromate layer provides a decorative appearance and better paint adherence. The chromating solution may contain a dye to give color to the converted coating.

OXIDE COATINGS

Immersion of parts in a hot caustic soda and nitrite or nitrate solution will produce a layer of oxide on their surfaces. This layer has an attractive appearance and good paint-holding qualities.

The process can be used on steel, aluminum, magnesium, copper, brass, zinc, and zinc alloys. Titanium is given an oxide coating to improve its forming characteristics.

ANODIC CONVERSION COATINGS

A stable film may be produced on a metal by using anodizing, an electrochemical process. Anodizing produces a decorative and protective layer of oxide on the metal component. Aluminum and magnesium are the most commonly anodized metals. The oxide layer may be clear or, by adding dye to the solution, colored.

Anodizing requires the use of a tank containing an acidic solution. The part is placed in the tank while attached to the positive terminal of a power source. The part, thus, becomes the anode. The current flows from cathode to the anode causing a chemical reaction. Oxygen in the electrolyte (anodizing solution) combines with the part. This creates the protective oxide coating, Fig. 31-4. Anodizing is used on a broad range of products in the aircraft, automotive, houseware, sporting goods, furniture, and jewelry industries.

METALLIC COATING FINISHES

Metals can be deposited on plastic or metal parts to provide a decorative or protective finish. The layer separates the base material from the environment. It can also provide an attractive surface. The principal techniques used to apply metallic coatings are:

1. Electroplating.
2. Electroless and immersion plating.
3. Dipping.
4. Metallizing.
5. Diffusion coatings.

ELECTROPLATING

Electroplating uses electrochemical action to deposit metal on a base material. Electroplating is used to coat large quantities of metal and plastic parts, Fig. 31-5. The coating can be used to:

1. Provide an attractive surface appearance on inexpensive base materials.
2. Improve the wear qualities of the material.
3. Increase the thermal conductivity of a material.
4. Improve a material's ability to be soldered.
5. Build up worn surface areas.

The electroplating process is the same for all materials. Some details of the process are adjusted when different materials are plated.

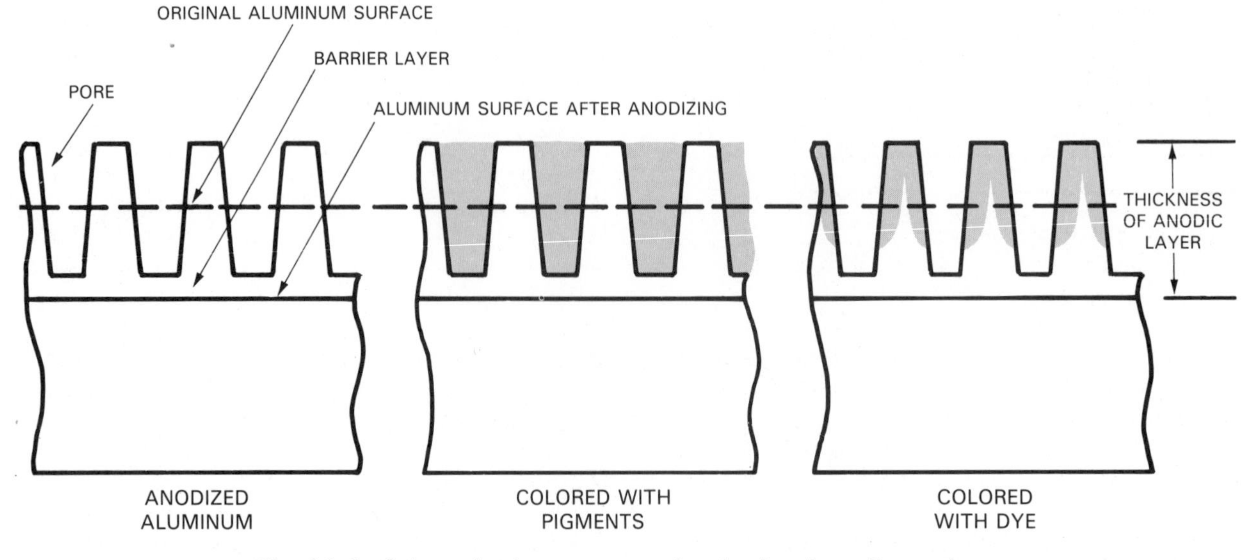

Fig. 31-4. Schematic shows untreated and colored anodic coatings.

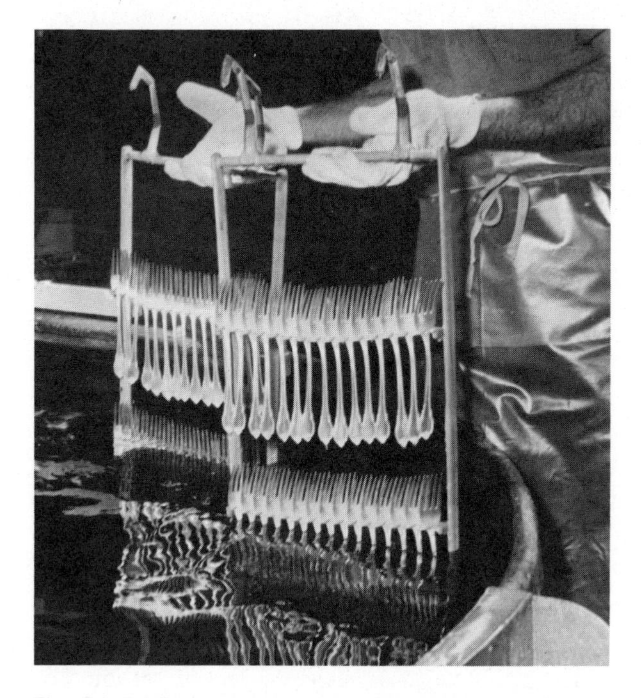

Fig. 31-5. Forks rotate in an electroplating tank containing a cyanide solution and pure silver.

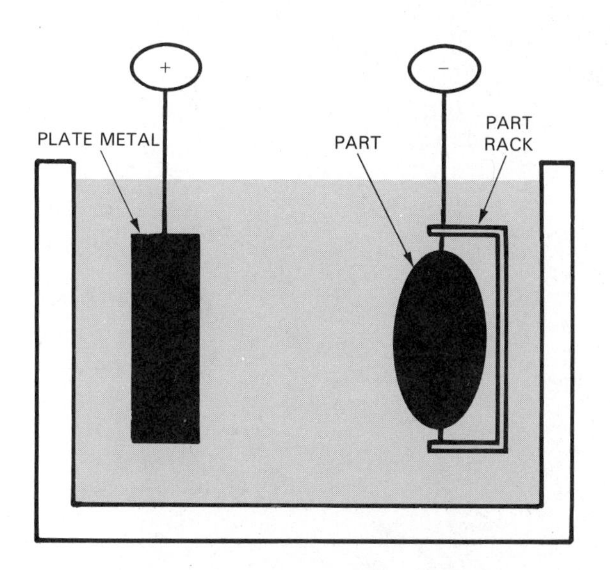

Fig. 31-6. Schematic shows basic setup of the electroplating process.

The basic circuit, shown in Fig. 31-6, has three active components. The work is the cathode (negatively charged). The parts are attached to conductive racks. An electrical charge is then applied. The strip of metal to be deposited on the work is the anode (has a positive charge).

The anode and cathode are suspended in an electrolyte. Metallic ions (positively charged) leave the anode and enter the electrolyte solution. The negatively charged part attracts the positively charged metal ions. They move toward the work and are deposited on the surface.

The rate of metal deposition (plating) is controlled by current density. The stronger the current becomes, the faster the plating action. The thickness of the metal deposit is controlled by the current used and the plating time.

Electroplating is used to deposit metals on metals or plastics. The common plating materials are cadmium, chromium, silver, gold, copper, tin, and zinc.

Plating requires several major steps. The parts are first completely cleaned. Plastic parts are roughed up or deglazed. The parts are then plated.

Often multiple-metal plating is used. Layers of different metals are applied in separate steps. For example, copper is applied to improve adhesion. A surface coat of chromium may then be applied for appearance and durability.

Plastic parts must be treated with chemicals before plating. The treatment causes the surface to conduct electricity and, therefore, promote the plating action.

Electroplating generally produces a metallic layer about 0.0003 to 0.0006 in. thick. Thicker layers of chromium can be applied to build up worn parts. This proccess, called *hard chromium plating,* is often used to restore bearing surfaces and shaft diameters. Layers of chromium up to 0.010 inch thick are applied directly to the base metal for this purpose. The plated areas are then machined back to the original tolerance for the surface.

Electroless and immersion plating

Electroless and immersion are methods for applying very thin coatings of metal without using electricity. Instead, the parts are placed in a bath containing ions (atoms) of the plating metal. Through a chemical reduction process, the metal ions are deposited on the surface of the part.

A very uniform film of plating metal slowly builds up on the part. The coating, however, can be impure. It often contains atoms from the electrolyte. The deposit is, therefore, an alloy rather than the pure metal with which electroplating coats the part.

Electroless and immersion plating can be used to deposit coatings of nickel, gold, tin, copper, cobalt, and zinc. These coatings can be applied to copper, copper based alloys, aluminum, steel, and a number of plastic compounds.

Dipping

Relatively thick coatings of metal may be applied to base material through hot dipping. Mild steel is often hot dip coated. Steel is easily attacked by the elements in the environment. To prevent this destruction of the metal by moisture and other forces, a protective skin is applied. Four principal metals are used in hot dip processes: zinc, tin, lead, and aluminum. Of the four, zinc is most often used.

Hot dipping requires several steps. First the base metal is thoroughly cleaned. Then a dipping in flux insures coverage and bonding. From the flux bath the metal is placed in a bath of molten metal. The hot metal forms an alloy at the point of contact with the base material. The result is a sandwich of base material and pure coating metal. These are separated by an alloy of the two metals. The excess coating metal is allowed to drip from the product as the coating cools.

The hot dip process, using zinc, is called galvanizing. A batch or continuous process, it is used on a wide range of steel products. Products may be individually dip coated and continuous sheet may be fed through cleaning, fluxing, and dipping tanks. See Fig. 31-7.

Metallizing

Decorative and protective metallizing coatings are produced by a process called *metallizing*. The coating material is applied as fine particles on metal, ceramic, or plastic parts.

Three principal types of metallizing techniques are used:

1. Wire metallizing or flame spraying.
2. Plasma-arc spraying.
3. Vacuum metallizing.

Wire metallizing, Fig. 31-8, involves melting the coating material in a special torch or gun. The metal or ceramic material is fed into an oxyacetylene flame. The 5000°F (2760°C) flame melts and atomizes the material. The atomized material, carried to the part by a blast of compressed air, cools as it strikes the part, Fig. 31-9.

Metallizing or combustion flame spraying is used to repair and build up worn parts and to provide a corrosion resistant coating. The coating thickness

Fig. 31-7. A continuous tin plating line. Sheet material is cleaned, fluxed, and dipped as it moves through the line in a continuous web. (Kaiser Steel Corp.)

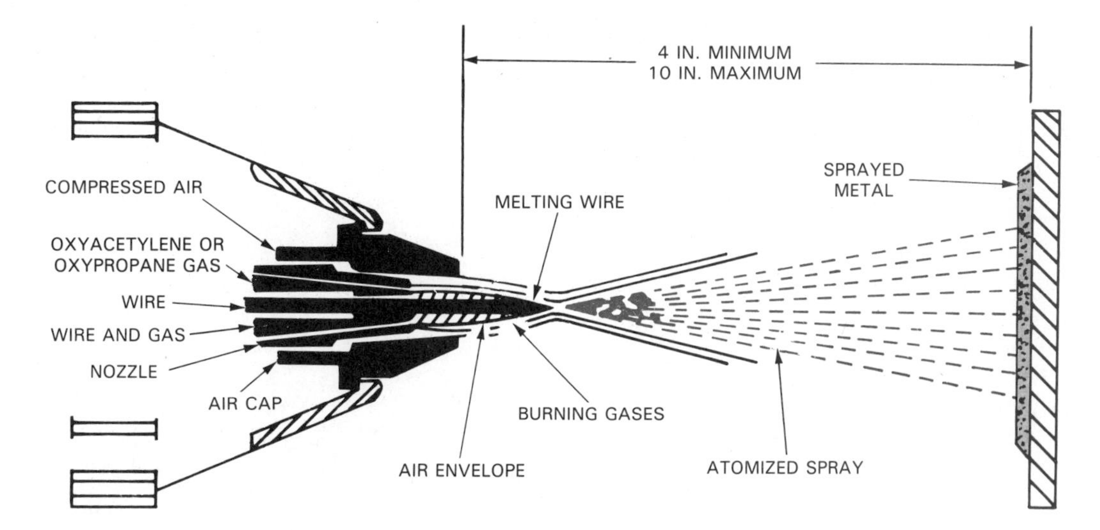

Fig. 31-8. A wire-metallizing flame spray gun uses flammable gas mixture to melt wire. Molten metal is sprayed onto workpiece. (METCO Inc.)

Fig. 31-9. A wire-metallizing gun in operation. Note that wire is fed from back. (METCO Inc.)

generally is less than 1/8 in. However, there is little limit to the maximum deposit thickness.

Plasma-arc spraying uses an arc between two electrodes to heat a gas (usually argon), Fig. 31-10. The heated gas accelerates to supersonic speed. Powdered coating material is fed into the stream of hot gas. The gas melts the material and carries it to the part.

Temperatures up to 30,000°F (16 649°C) are created with plasma-arc spraying equipment. This temperature melts materials which cannot be melted and deposited by other means. Ceramics, tungsten carbide, tungsten, molybdenum, and nickel-chromium alloys are among the materials which can be plasma-arc sprayed. These and other materials can be deposited on a wide variety of metal, plastic, and ceramic products.

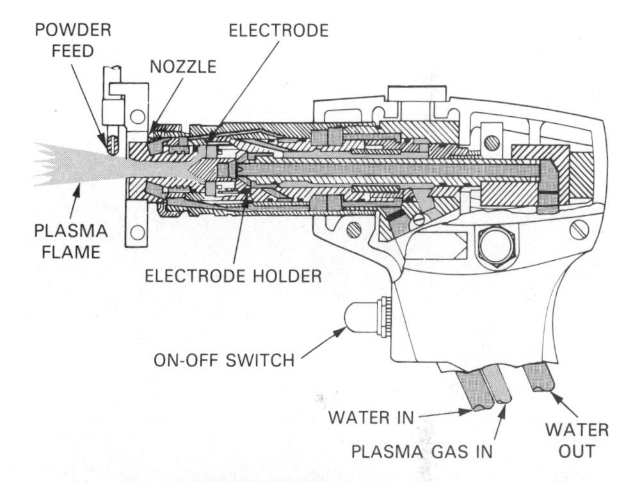

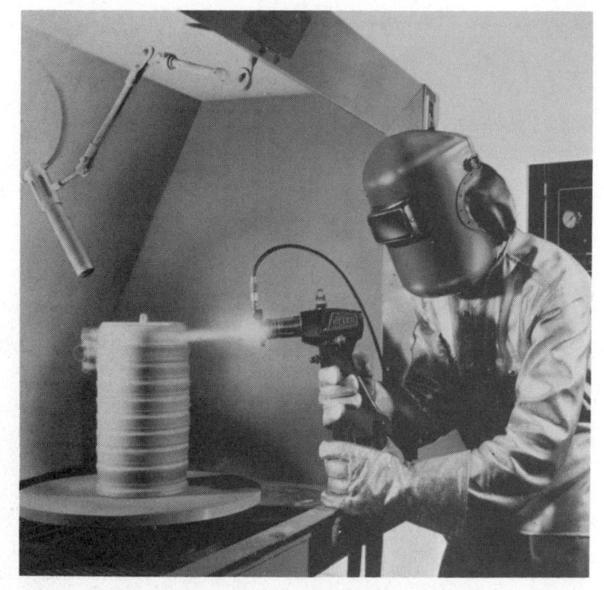

Fig. 31-10. A plasma-arc gun. Top. Schematic. Bottom. Being used to coat a product. (METCO Inc.)

Vacuum metallizing vaporizes the coating material using heated tungsten filaments. The parts and the vaporizing unit are in a high-vacuum chamber. The vaporized metal floats in the vacuum and deposits on the slowly rotating parts. The process produces a thin, highly uniform coating on metal, glass, plastic, textiles, and paper.

Vacuum metallizing, Fig. 31-11, can be used to coat automotive trim, microelectronic circuits, reflectors for flash guns and automotive lights, hardware, and costume jewelry. It deposits a film of high purity in thicknesses from less than one-millionth in. up to 0.002 in.

Ion-sputtered coating and *chemical vapor plating* processes also use metal vapors. The ion-sputtered or arc-wire process uses high voltage electrical current to vaporize the metal. Two wires are fed into the back of the gun, Fig. 31-12. An arc is struck and the ends of the metal wire vaporize. A blast of compressed air hits the molten metal and carries it to the surface to be coated.

Fig. 31-11. Large rotating back has been removed from a vacuum metallizing chamber for loading. (Dow Chemical Co.)

Fig. 31-12. An arc wire metallizing gun applies a finish to a spinning part. (METCO Inc.)

Chemical vapor plating (gas or pyrolytic plating) vaporizes a metal based chemical in a separate chamber. The vapors are then released into a chamber containing a heated part. The vapors enter into a chemical reaction when they touch the hot part. This reaction leaves a coating of metal on the part. Ion-sputtered coating is applied in a weak vacuum. Chemical vapor plating can be done at atmospheric pressure or in a vacuum.

The ion-sputtered process deposits a very pure film of material. It will apply the same coatings as vacuum metallizing. Chemical vapor coatings are used for semiconductor applications and to produce wear and oxidation resistant surfaces.

Diffusion coatings

Diffusion coatings get their name from their being absorbed into the base material of the product or part. They are applied by packing heated parts with a coating metal that may be powdered, solid, liquid, or gaseous. This creates a harder or a more corrosion resistant surface. Diffusion coating processes were discussed in Chapter 25 under "Surface hardening."

CERAMIC COATING FINISHES

Inorganic nonmetallic coatings provide an excellent finish for many products. These ceramic materials are durable and highly resistant to the environment. Ceramic coatings include three major categories:
1. Porcelain enamels.
2. Glazes.
3. Refractory ceramic coatings.

Porcelain enamels are applied to metals or ceramics. Glazes are used only on ceramic products.

PORCELAIN ENAMELS

Porcelain enamel is a glasslike inorganic coating bonded to a metallic or ceramic base material. Porcelain enamel is any of a wide variety of silicate glass.

The enamel is fired (heated at temperatures from about 1000° to 2000°F (538° to 1093°C), depending on the base material and its use. The firing is required to melt and bond the enamel to the metallic or ceramic base. Porcelain enameled sheet metal is fired at about 1500°F (816°C). This material, known as a vitreous coating, is used on many appliances, Fig. 31-13.

Porcelain enamel is very durable and scratch resistant. It can be applied in a number of fade resistant colors. Porcelain enameled surfaces are easy to clean and resist chemicals and heat.

Many metals can be finished with porcelain enamel. The most common ones are steel, aluminized steel,

Fig. 31-13. Washer tops are being pickled before porcelain enamel is applied. (Whirlpool Corp.)

stainless steel, cast iron, aluminum, and copper.

Porcelain enamel is a mixture of four major components:
1. Glass former.
2. Flux.
3. Opacifier.
4. Colorant.

Glass former, a base for the enamel, is a silica which naturally melts and fuses at about 3100°F (1704°C). *Flux* is used to reduce the melting point of the glass former. *Opacifier* is used to make the transparent glass former and flux more opaque. *Colorants* are added to produce a colored porcelain enamel. Metallic oxides and metallic carbonates are used as colorants.

The porcelain enamel mixture is first changed into *frit,* a ground up glass. The manufacture of the frit involves several steps:
1. The dry glass formers, fluxes, opacifiers, and colorants are measured and mixed (dry batched).
2. The mixture is melted in a crucible (frit) furnace.
3. The molten glass is poured into water. This rapid quenching (cooling) breaks the glass into small pieces.
4. The broken glass is ground into a fine powder in a ball mill.
5. The iron impurities are removed by magnetic separation.
6. The frit is separated by particle size and stored ready for use.

The actual application of the porcelain enamel requires two preparation stages. Then the finish is applied and fired.

The base material (workpiece) must be clean and dry. Glass products are enameled shortly after forming and are usually clean. Metals are chemically cleaned and thoroughly rinsed.

In the second preparation stage, steel parts are coated with a solution of nickel sulfate. An electrochemical reaction is used to give the material a flash coat of nickel. The surface coating of nickel insures a good bond between the porcelain enamel and the base metal.

Porcelain enamel can be applied either wet or dry. If applied wet, it must be mixed with water. This produces a mixture called *slip*.

From one to three coats of finish may be applied. Several thin coats may be used to produce a very uniform coating of finish. However, a single coat is sometimes used because it is faster and less expensive.

Wet application uses slip which is applied by spraying or dipping. Slip is used for both metal and ceramic products. It may be silkscreened on glass to produce printed bottles, signs, and other decorative products. The slip is allowed to dry, leaving a powder coat.

In the dry application process the workpiece is heated then coated with powdered frit. The powder sticks to the hot surface. The dry process is used for heavy steel and cast-iron products.

The coated products are then fired in a kiln. The heat causes the enamel to melt, flow, and fuse. As the product cools the enamel solidifies.

Metal products can be quickly fired. Glass products must be slowly preheated, fired, and annealed (cooled).

Porcelain enamels are used by manufacturers of kitchen appliances, heating equipment, farm silos, water heaters, building panels, glass blocks, bottles, sanitary ware, and chemical ware.

GLAZES

Like porcelain enamels, glazes are fired, glassy material. They are exterior coatings applied only to clay materials where they:

1. Produce a hard surface.
2. Allow for easy cleaning.
3. Decorate the surface.
4. Waterproof the product.

Glazes and porcelain enamels are very much alike. Both contain silica glass formers, fluxes, colorants, and opacifiers. Glazes may also contain:

1. Stiffeners to keep the flux from flowing during firing.
2. A matting agent to reduce the gloss of the glaze.

The preparation and application of glaze is like the preparation and wet application of porcelain enamel. A frit is prepared and a glaze slip is produced. The slip is spray, brush, or dip applied. When the surface is dry, the product is preheated, fired, and cooled in a batch or continuous kiln.

A wide range of dinnerware, decorative products, wall and mosaic tile, electrical insulators, sanitary ware, scientific ware, and art pieces are finished with glazes, Fig. 31-14. The glaze will provide a protective and, often, a decorative finish.

Fig. 31-14. Glazes may be applied by dipping as shown here or by spraying liquid glaze or glaze crystals onto hot ware. (Lennox Inc.)

REFRACTORY CERAMIC COATINGS

Refractory ceramic coatings are layers of ceramic materials that provide thermal (heat) protection. For mild temperatures or short periods of time, thin ceramic coatings may be applied by the metallizing processes described earlier, Fig. 31-15. They may also be applied cold.

The materials used are proprietary. Generally they are a ceramic compound with a cement bonding agent added.

The mixture is usually applied with a trowel or special spray equipment. The paste is then heat cured.

Cold applied ceramic coatings are resistant to corrosive environments, oxidation, and abrasion. They are used in engine manifolds, as flame dampers, and in turbine and jet engine cases.

Fig. 31-15. A plasma-arc spray gun is used to apply a wear-resistant ceramic coating on a computer part. Spray is remote controlled. (IBM)

SUMMARY

Inorganic coatings include all metallic and ceramic finishes. They may be metal surfaces which have been chemically treated or converted. They are also layers of additional metals or ceramics applied to the product.

Metal coatings may be applied by electroplating, immersion plating, dipping, metallizing, and diffusion. Ceramic coatings include porcelain enamels, glazes, and refractory coatings. They can be applied to either metals or ceramics.

Inorganic coatings are applied to protect and, in many cases, to beautify the surface. They may be applied to metallic, ceramic, and plastic surfaces.

STUDY QUESTIONS – Chapter 31

1. What are the two reasons for applying a finish?
2. List and describe the two types of metallic coatings.
3. List and describe the major types of metallic conversion finishes.
4. List and describe the major metallic coating processes.
5. List and describe the major ceramic coating processes.

Chapter 32

ORGANIC COATING PROCESSES

The term, organic finishes, covers a wide range of materials which form a film over a base material. This film is designed to adhere to the product.

Most organic finishes are vehicles or binders and pigments held in a solvent, diluent, or thinner. The vehicle or binder is the material which bonds to the surface to provide the desired protection. The pigment provides the color. The diluent carries the binders and pigments to the base material.

TYPES OF ORGANIC FINISHES

Organic finishes can be classified by:
1. Types of coatings.
2. Method of film formation.
3. Types of binders.

TYPES OF COATINGS

Organic coatings, Fig. 32-1, include paint, varnish, enamel, and lacquer. Paint is any polymeric coating. When applied to a surface it provides a continuous film. The coating is made from *binders,* which can be natural or synthetic (manufactured). (Binders or vehicles are the part of a paint which binds or "cements" the particles of pigment together. They are the liquid nonvolatile portion.)

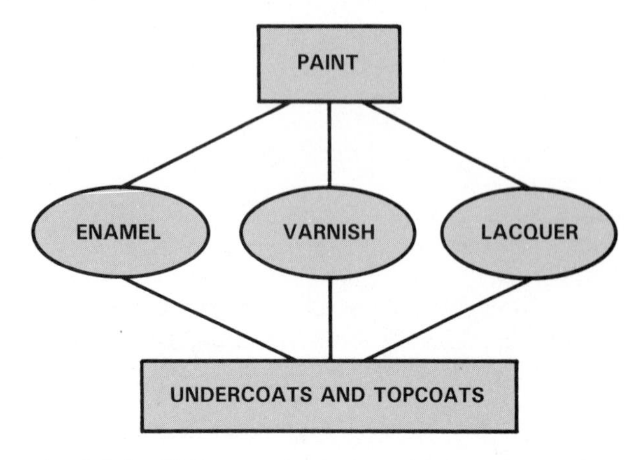

Fig. 32-1. There are four types of organic coating materials.

Vehicles are often mixed with a coloring agent to produce a colored film. The agent is called a *pigment.*

Other materials are added to promote curing (drying) and to improve the finishing characteristics of the paint. The binders, pigments, and additives are mixed with a solvent.

Varnish, a clear finish, is a mixture of oil, resin, solvent, and a drier. There are four types:
1. Oil varnishes – made from natural gums or resins in oil, turpentine or mineral spirits, and a drier.
2. Polyurethane varnishes – made from synthetic resins and polymers that are thinned with mineral spirits.
3. Spirit varnishes – made with resins and gums and a solvent like alcohol or turpentine.
4. Acrylic varnishes – composed of acrylic-resin glycols, mineral spirits, and water.

Enamel is a varnish with pigment added. It may have oil, water, or some other solvent as a base. Both varnishes and enamels dry by polymerization.

Lacquer is a solvent based coating. It dries by solvent evaporation.

Paints are used for both *undercoats* and *top coats.* Undercoats, often called primers, are used under top coats to:
1. Improve the bond between the base material and the top coats.
2. Cover minor surface defects in the base material.
3. Provide a smooth surface for top coats.
4. Reduce corrosion of the base material.

Top coats are the surface films applied to the product. They are chosen to meet a stated need. A wide variety of top coats are available. Each has unique abilities to:
1. Produce an attractive surface.
2. Resist environmental attack.
3. Provide a durable surface.
4. Be applied easily.
5. Dry effectively.

METHODS OF FILM FORMATION

Organic finish films are created through three major actions: solidification of melted films, evaporation of solvents, and polymerization of monomers and polymers, Fig. 32-2. Most organic films are polymers made of repeating organic chains. These chains are made from the binders in the finishing materials.

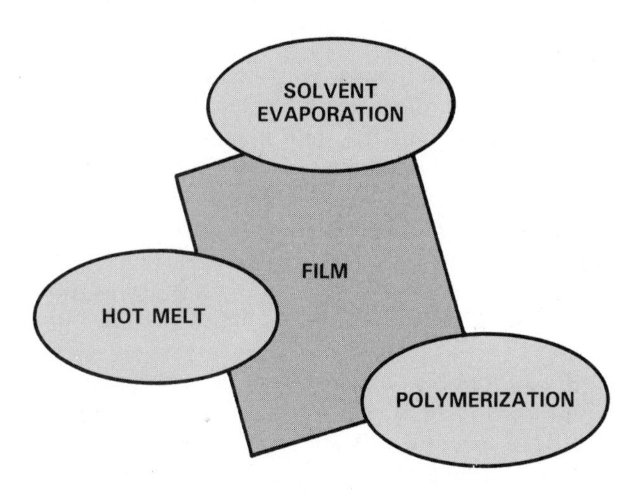

Fig. 32-2. Three methods can be used to produce an organic film.

Some films are already long-chain polymers. They are carried to the surface by solvents or are in sheet form. Solvent based materials dry or cure as the solvent evaporates.

Organic film materials, softened by heat, are brought into contact with the base material. They adhere to the surface as the film cools. Other films are produced as simple chains become more complex. This action, called polymerization, was described in Chapter 3. The growth of the polymer chains can be controlled and increased through the use of chemical catalysts and/or heat.

TYPES OF BINDERS OR VEHICLES

The binders or vehicles used in organic finishing vary considerably. A background in organic chemistry is needed to fully understand the curing or drying of these materials. This level of understanding is beyond the scope of this text. However, a basic introduction to the major types of vehicles is appropriate. These vehicles fall into three basic categories:

1. Polymerization type vehicles.
   a. Oils and oleoresinous materials.
   b. Alkyd resins.
   c. Phenolic resins.
   d. Water thinnable resins.
2. Catalyzed polymerization type vehicles.
   a. Epoxy.
   b. Polyurethane.
3. Solvent evaporation type vehicles.
   a. Vinyl resins.
   b. Acrylics.
   c. Cellulosics.

Oils and oleoresinous vehicles

Natural oils are probably the oldest vehicle for organic finishes. Linseed and tung oils have been used for many years. Known as *drying oils,* these vehicles react with oxygen in the air as they polymerize. They form a relatively hard, tough surface.

Oleoresinous binders, a more recent development, are made from natural drying oils and synthetic resins. These binders form the base for most modern oil paints and varnishes.

Oil paints are a mixture of these vehicles and metallic pigments. The most common pigments are powdered lead, titanium, and aluminum compounds. The pigments provide the color and hide the base material (vehicle).

Oil and oleoresinous coatings dry slowly. They are used primarily on large structures where drying time is not critical.

Alkyd resins

Alkyd resins are polyesters of selected oils, alcohols, or organic acids. They are the most widely used polymerizing vehicles. While producing a very durable surface, alkyd paints retain their gloss, resist moisture, sunlight, and other weathering elements.

Alkyd resin binders are soluble in aliphatic and aromatic solvents such as mineral spirits. Alkyd resin coatings are suitable for interior and exterior applications on both metal and wood products. They are used in: house paints, machine enamels, automotive finishes, metal siding and wall panel paints, farm implement finishes, and outdoor furniture paints.

Phenolic resins

Phenolic resins, produced primarily through a reaction of phenol and formaldehyde, are usually mixed with drying oils to form a phenolic varnish. The varnish is used for floor finishes and to paint ships, bridges, and storage tanks.

Aluminum and various colored pigments are added for paint applications. Iron oxide and zinc chromate pigments are added to produce phenolic primers.

Water thinnable vehicles

A number of vehicles can be thinned by water. The vehicles are either suspended (emulsion) or dissolved in the water. Water emulsions are used mostly to coat paper and for construction applications. Dissolved vehicles or solution-type paints can be used for industrial coatings on wood and metal. These vehicles present no fire hazard and emit no toxic fumes.

The *electrodeposition* process, which will be discussed later, has created much interest in these materials. The resins form a film as the water evaporataes. Then the material polymerizes into a waterproof film. This two-stage action can take place in the air or in drying ovens.

Epoxy resins

Epoxy finishes are coatings which must be polymerized using a catalyst. They are the major two-part coating material. Epoxy resins are thermosetting resins which come in a number of forms. Each form has its own specific application. Most epoxy paints are highly resistant to chemicals and adhere well to most surfaces.

Polyurethanes

Polyurethanes are finishes which are cured by a catalyst. The polymerization action can be activated by moisture, air, or a chemical additive. Many types of polyurethanes are used for finishing coats. Generally, they produce a film highly resistant to weather and chemical attack.

Polyurethanes are produced in both one-part and two-part finishes. Two-package finishes are mixed just before use. Polyurethane finishes are used for both wood and metal products.

Vinyl resins

As a group, polyvinyl resins do not adhere well to metal if allowed to air dry; however, heat curing improves adhesion. Vinyl resin finishes are especially good for underwater applications. This type of finish dries by solvent evaporation.

Acrylic resins

Acrylic resins are thermoplastic materials. They are most often suspended in a solvent and dry by solvent evaporation. Acrylic finishes have excellent resistance to weather and ultraviolet light (sunlight). They are clear compounds that do not discolor under heat.

Acrylic resins are used in automotive and aircraft lacquers and for clear, durable finishes on chromium plated parts.

Thermosetting (set by heat) acrylics are also available. They are baked to produce a very environmentally resistant finish.

Cellulosic polymers

Cellulosic polymers are the base for rapid-drying cellulose lacquers. These finishes are a combination of nitrocellulose, plasticizers, and a rapidly evaporating solvent. Lacquer provides a hard, quick-drying surface. The durability, gloss, flexibility, and adhesion of the film can be modified by adding various resins and plasticizers.

Cellulose nitrate lacquers have been replaced in many applications by alkyds and acrylic lacquers. The material, however, is still widely used in finishing wood cabinets, and wood and metal furniture.

Silicones

Silicone based paints resist temperatures up to 700°F (371°C). This factor makes them useful for coating ovens, mufflers, and industrial equipment. Silicone films also produce a clear finish for baking dishes. The inorganic silicone resins are usually combined with alkyd, phenolic, amine, or other organic resins to increase adhesion to metal surfaces.

APPLYING ORGANIC COATINGS

Organic finishing materials can be applied using any of 10 basic techniques:

1. Brush coating.
2. Roll coating.
3. Dip coating.
4. Flow coating.
5. Curtain coating.
6. Spray coating.
7. Electrocoating.
8. Printing.
9. Powder coating.
10. Heat transfer coating.

BRUSH COATING

Brushing requires considerable skill and is a very slow process. Careful selection of the brush and paint is essential for good results. Brush coating is used mostly on construction projects and for maintenance.

ROLL COATING

Hand roll coating is familiar to most people. It is used to paint interior and exterior walls and ceilings. The roller is usually covered with a synthetic material having a nap. The length of the nap regulates the amount of paint carried by the roller.

Industrial roll coating also uses a cylindrical applicator to apply organic coating to continuous flat surfaces. A number of roll coating machines are available. These, as shown in Fig. 32-3, can coat one or both sides.

The basic roll coating machine, as shown in Fig. 32-4, contains several parts:

Fig. 32-3. A continuous sheet of material is roll coated.

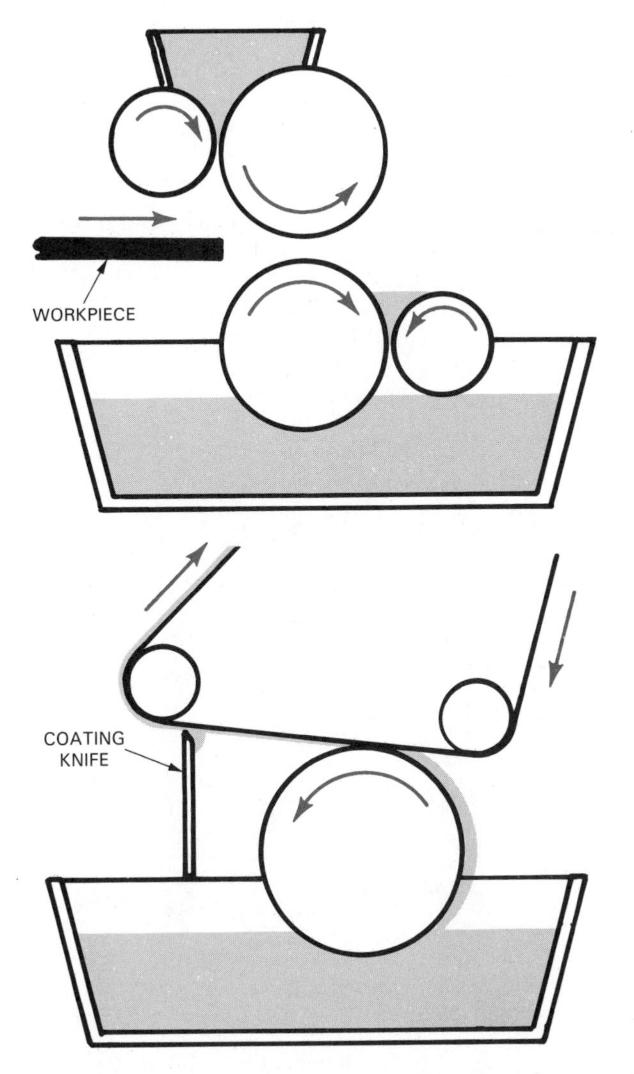

Fig. 32-4. Types of roll coating machines. Top. Coating thickness is controlled by second roller. Bottom. Leveling blade controls thickness.

1. A container for the coating material.
2. A fiber-covered material application roll.
3. A roll or knife to regulate the thickness of the coating material.

The coating material is picked up by the fiber covered roll. The amount of the material on the roll may be regulated by a second roller. The workpiece is passed between the application roll and an idler roll. For two-sided coating, both rolls are coating application rolls. The motion between the fibers on the application roll and the work applies the finishing material. The fibers flex upon contact and work the finish into the surface of the work.

A second type of roll coating machine also uses fiber application rolls. However, the thickness of the finish coating is controlled by a leveling blade. The blade is adjusted to an exact distance above the work. The blade wipes excess finish from the sheet material. Roll coating is used to coat flat and coiled metal, paper, plastic, plywood, and wood fiberboard (wood composites). This process provides a fast, economical method of finishing large quantities of flat uniform material.

DIP COATING

In dip coating, parts or products are submerged in a container of coating material. The items are then lifted from the finishing material and allowed to drip over the container. The work is then air or oven cured. See Fig. 32-5.

Fig. 32-5. A structural steel part is being removed from a dip coating tank. The curing oven is to the left. (George Koch Sons, Inc.)

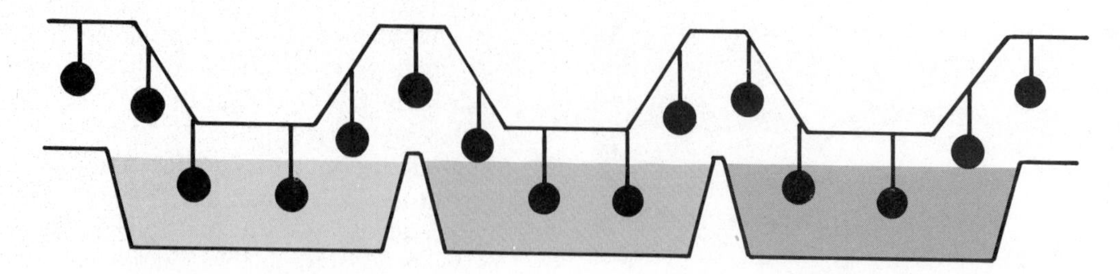

Fig. 32-6. Schematic shows a three-stage cleaning and dip coating process on a conveyor system.

Dip coating may also be continuous. Conveyor fed parts may be dipped, withdrawn, and drained in an unending process, Fig. 32-6.

Dip coating provides a rapid way to coat flat and irregular parts. The coating material must be carefully selected. Control of its viscosity and solvent is necessary for proper coating. Also, almost all dip coats are somewhat uneven. The lower edge of the product will have some buildup of finishing material. This buildup can be reduced by draining the material to a single point rather than an edge.

Dip coating is used most often when colors are rarely changed. Separate tanks are usually used for each color. Dip coating provides an efficient way to prime coat complex parts and products. It also is used for applying glazes to ceramic products and finish coats to many metal products.

FLOW COATING

Flow coating and dip coating are similar. Both processes coat the product, then drain off the excess. They differ in the way the finishing material is applied. Flow coating runs or flows the finish over the workpiece from nozzles or tubes.

Most flow coating is done in an open-ended tank or chamber, Fig. 32-7. The products are mounted on conveyor racks that move through the tank or chamber as parts are coated. As the parts move beyond the coating nozzles, the excess finishing material drains. The conveyor then moves the parts from the finishing chamber to a drying area or into curing ovens.

Like dip coating, flow coating is fast. However, it does not require large tanks of finishing material. Flow coating efficiently coats exterior surfaces but is not effective for enclosed areas. It is useful in applying primer coats to exterior surfaces.

Fig. 32-7. Mail boxes are being flow coated. They are hung on a conveyor for maximum coverage. Note the heavy spray just above the parts.
(George Koch Sons, Inc.)

CURTAIN COATING

Curtain coating is a special flow coating process. It applies a continuous "curtain" of finishing material onto horizontal surfaces. An infeed belt carries the

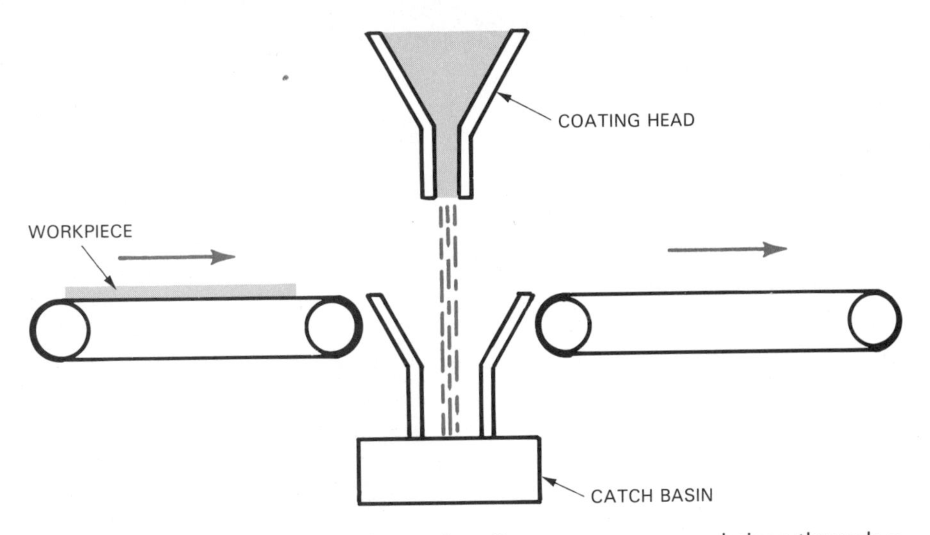

Fig. 32-8. Schematic of curtain coating. Conveyor moves workpiece through a "curtain" of finishing material.

sheet into the coating area, Fig. 32-8. A second belt carries the product away from the finishing machine. The thickness of the coating layer can be controlled by the speed of the conveyor, the volume of material in the curtain, and the composition of the finishing material.

Curtain coating can be used to coat sheet stock and parts. Plywood, fiberboards, and wood cabinet doors are typical of products which lend themselves to the curtain coating process.

SPRAY COATING

Spray coating is the most common method of applying organic finishes. There are several types of spray coating processes. The most important of these methods are:

1. Air spraying.
2. Airless spraying.
3. Electrostatic spraying.

Air spraying

Air spraying, Fig. 32-9, is the most widely used method for spray coating products. The equipment usually includes: a pressurized paint tank, an air supply, and a spray gun. The air supply forces the paint from the pressure tank or pot through hoses to the gun. Compressed air from the air supply also atomizes the paint as it flows through the nozzle of the gun, Fig. 32-10. The particles of paint are propelled

Fig. 32-9. Left. Paint spray booth with paint being applied with a hand-held gun. Right. A spray booth with two automatic guns. (Tonka Toys and Binks Inc.)

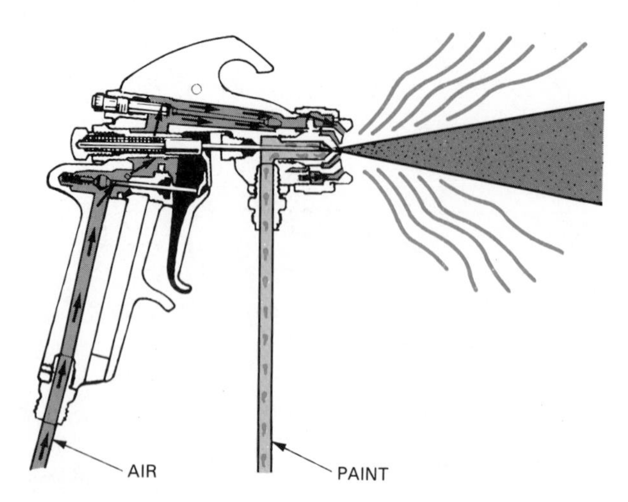

Fig. 32-10. Diagram shows a typical suction feed spray gun. Air pressure to gun creates suction to lift paint to the nozzle and then pressure carries the paint particles to the workpiece. (Dow Chemical Co.)

by the pressure to the surface of the product.

For small applications, a gun with its own paint container may be used. The fluid in the can is either forced by air pressure or pulled by vacuum into the nozzle area.

Air spraying is quick and flexible. Complex shapes can be coated. Major disadvantages include overspray (spray missing the product) and solvent fumes. Both the overspray and the fumes must be collected and removed from the air. Filters and curtains of water are often used for this task, Fig. 32-11. Air spraying can be done with hand-controlled guns or automatic spraying machines.

Airless spraying

Because air spraying traps air in the finishing material, there are often problems with bubbles and very small holes in the organic coating. Airless spraying overcomes this problem. The finishing material is forced through supply lines to the gun under high pressures. Common operating pressures are 1000 to 1500 psi. The fluid is forced through a small opening in the nozzle onto the work. Since expanding air (pressurized air leaving a nozzle) is not used, less overspray is produced. There is little bouncing of air on the surface of the product. Therefore, the finishing material has a less turbulent path to the area being finished.

Electrostatic spraying

Electrostatic spraying eliminates most of the waste and pollution of overspray, Fig. 32-12. There are several electrostatic spraying processes. In all of them the coating material takes on an electrical charge while

Fig. 32-11. Top. Fumes and overspray may be collected by filtering the air. Bottom. Wall of water flowing down the back of the spray booth serves the same purpose. (George Koch Sons, Inc.)

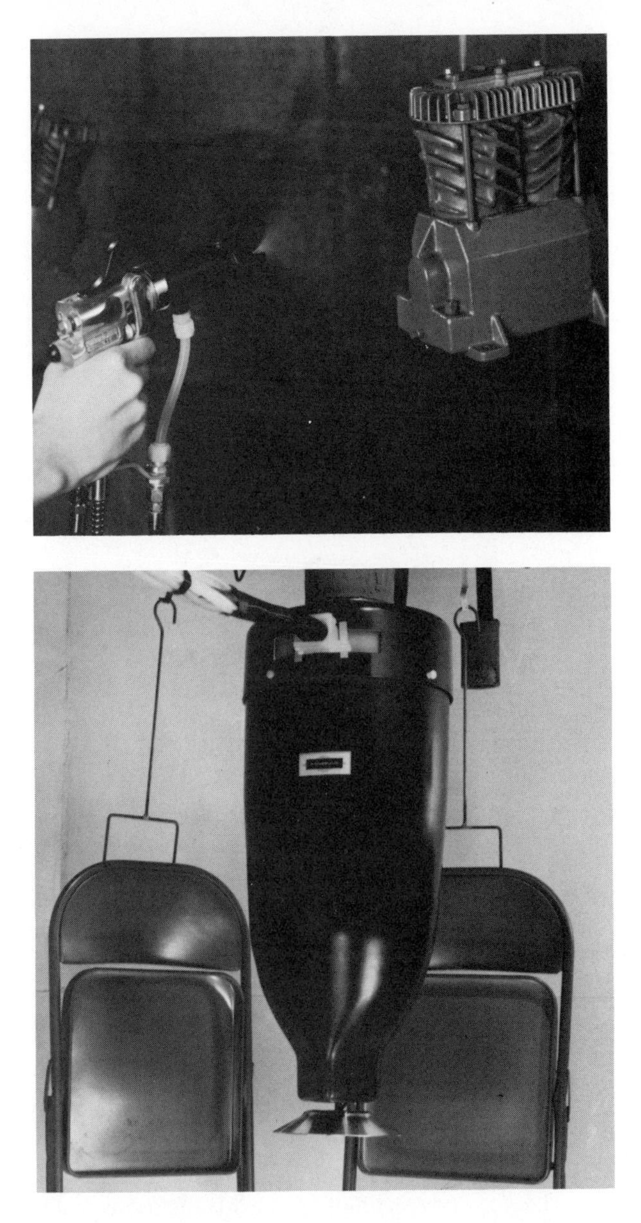

Fig. 32-12. The electrostatic spraying process may use a gun, as at top, or a high-speed turbobell, bottom. (Ransburg Corp.)

the part is given an opposite charge. Since unlike electrical charges attract each other, the finish is drawn to the product, Fig. 32-13. The paint spray will actually wrap around the product, coating all sides though sprayed from only one side.

Metals require no preparation for electrostatic spraying. Wood, however, will not normally take an electrical charge. It must first be coated with a clear base material.

Electrostatic spraying is used for many metal products such as bicycle frames, kitchen appliances, and metal cases. In-place maintenance painting can also be done by this process. Hospital and office furniture can be recoated in their rooms.

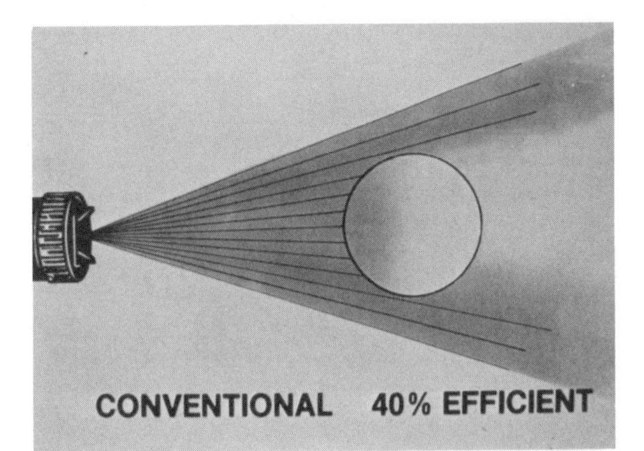

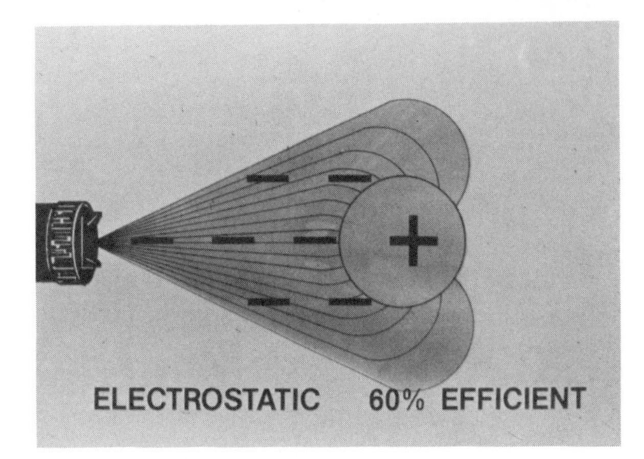

Fig. 32-13. Comparison between spray techniques. Top. Conventional. Bottom. Electrostatic spraying. (Ransburg Corp.)

ELECTROCOATING

Electrocoating is a special dip coating process. The product is placed in a tank of water-thinnable coating material. The part is attached to one side of a dc power source. It becomes one electrode, usually the anode. The tank, or another electrode, is attached to the other dc lead. Current passes between the electrodes in the tank. The electrical current causes several actions in the tank. As a result of these actions, a thin film of finishing material is deposited on the product.

The process, shown in Fig. 34-14, produces a very even finish. The coating is uniform over the product regardless of its shape. Few toxic fumes are produced and little paint is wasted.

The typical electro-coating process involves cleaning the product, applying the finish, rinsing, and drying. Only one coat of finish can be applied by this process. The organic coating is an insulation film and prevents additional layers.

The electrocoating process is used for undercoating a number of products. Automobile bodies are usually primed this way.

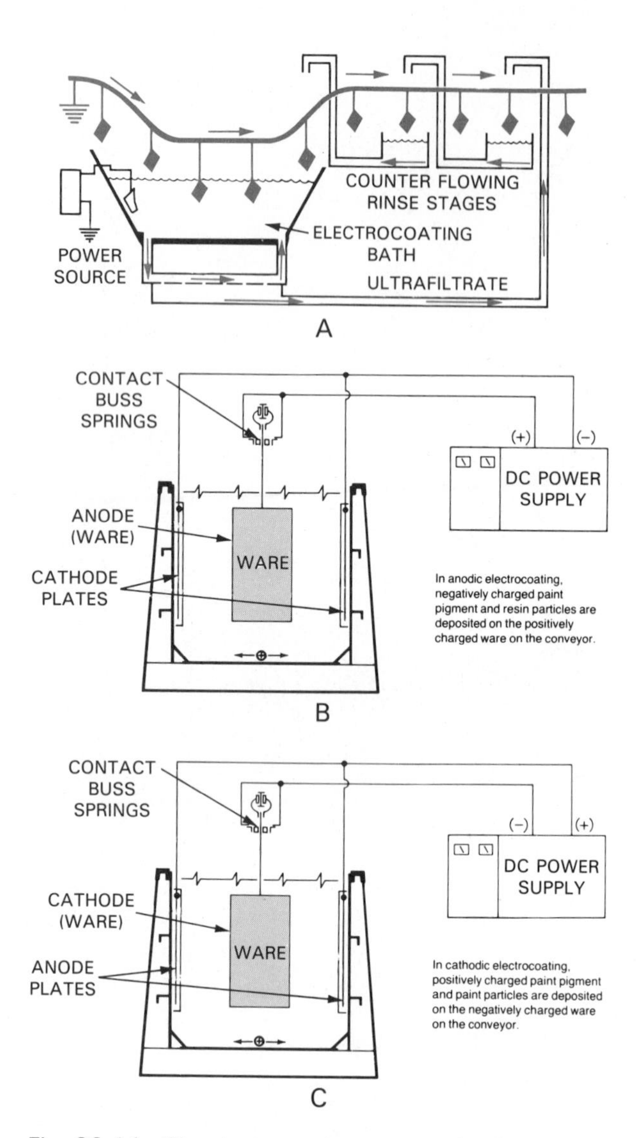

Fig. 32-14. The electrocoating process. A—Schematic shows setup of the line. B—Electrical schematic of system where part is the anode. C—Electrical hookup when part is the cathode.

PRINTING

Organic coatings may also be applied using common printing processes. Wood grain patterns and artistic designs are but two examples of printed coatings. The most commonly used processes include:

1. Offset lithography—printing from a flat surface onto which the design has been placed photochemically. This process is used to imprint thin plastic films and metal foils.
2. Rotogravure printing—printing from a surface into which the pattern has been etched, Fig. 32-15. This process is used to print wood grain on fiberboard paneling.

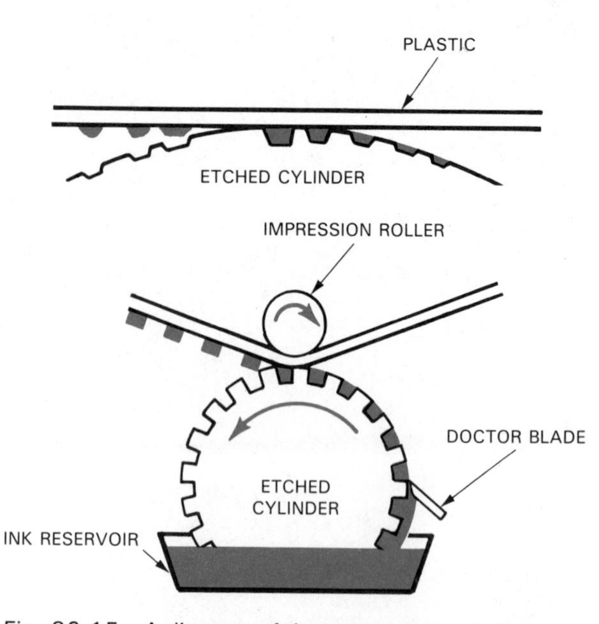

Fig. 32-15. A diagram of the rotogravure printing process. Image areas have been etched below the surface. (Dow Chemical Co.)

3. Screen printing—printing by forcing ink through a design on a screen, Fig. 32-16. Screen printing is used to print signs, fabrics, and other decorative patterns.

POWDER COATING

In powder coating the part is covered with a layer of powdered organic resins. This coating is heat fused and cured.

The coatings may be applied by several methods. The *fluidized bed* technique suspends the resins in air. Compressed air, Fig. 32-17, is pumped into a box. The air acts as a fluid for the resin. The part is heated and suspended in the air/resin fluid. The hot part melts

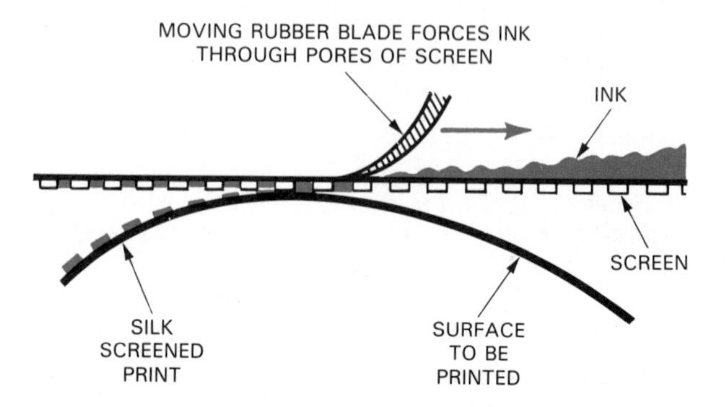

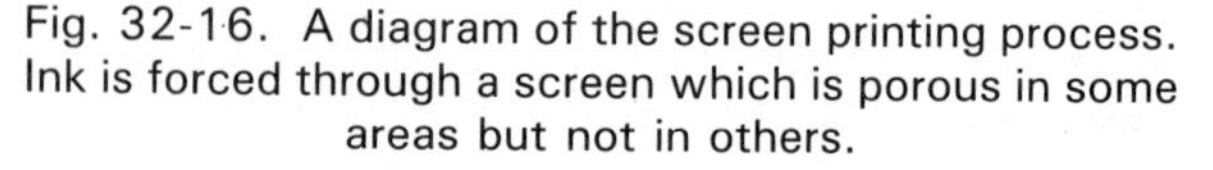

Fig. 32-16. A diagram of the screen printing process. Ink is forced through a screen which is porous in some areas but not in others.

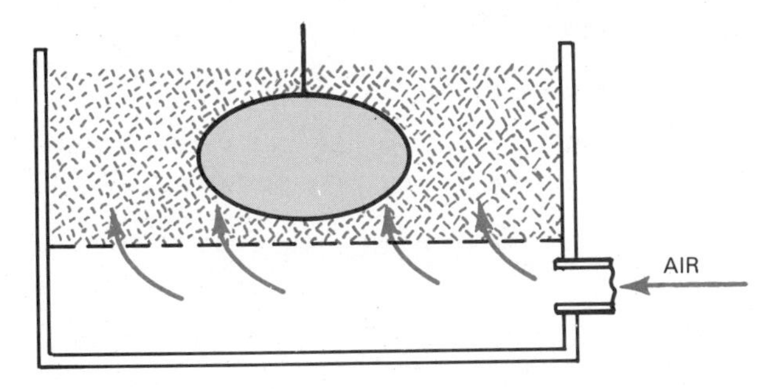

Fig. 32-17. A schematic of the fluidized-bed powder coating process. The heat of the suspended part melts floating resin particles causing them to flow over the part's surface.

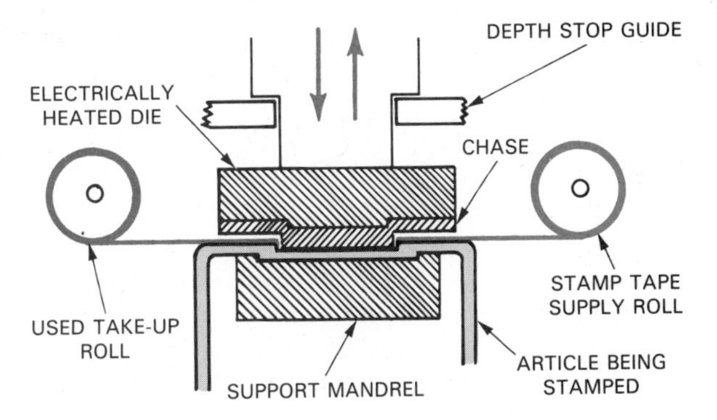

Fig. 32-18. The hot stamping process is often used for decorating plastic. (Dow Chemical Co.)

the particles which come into contact with it. The heat of the part will cause the resin to melt and flow. Upon cooling, the part will have a uniform organic coating.

An electrical charge suspends the resins in the *electrostatic fluidized bed* process. A charged diffuser plate produces a cloud of resin above a fluidized bed. Parts with an opposite charge are moved through the charged cloud on a conveyor. The resin particles are attracted to the part, giving it an even coating. As in the fluidized bed process, parts must be heated to cause the resins to flow and fuse. Thus, parts must be moved through an oven to spread and cure the coating.

Electrostatic powder spraying pumps powdered resins through a gun. These resins are given an electrical charge. They are then sprayed on a surface having an opposite charge. The process is like electrostatic spraying and produces a very even coating. The coating on the product is heat fused either by the hot part or by placing the cold, sprayed part in an oven.

HEAT TRANSFER COATING

Heat transfer coating is a special powder coating process. The powders are placed on a carrier, usually a polyester film. The film is placed coated side down on the product. A hot platen or roll is placed on the film. The resins are fused on the surface of the product. The polyester film is then removed.

Heat transfer or hot stamping is used to decorate a number of products made of wood, plastic, leather, paper, metal, and fiberboard. For example, book titles are gold stamped onto their binding, wood grain finishes are applied to plastic television cabinets, Fig. 32-18, and type is hot stamped on plastic signs.

CURING ORGANIC FINISHES

When natural or air drying is too slow, baking ovens, Fig. 32-19, are used to speed the process. The ovens may be convection type like a home baking oven, and can be heated with electricity, oil, or gas.

Fig. 32-19. Ovens are often used to cure wood and metal products. Top. Wood being cured. Bottom. Painted metal assemblies are moved by conveyor into a convection oven. (George Koch Sons, Inc.)

Infrared electric bulbs can also be used. They provide rapid heating of the product surface without heating the surrounding air.

Another curing process uses radiation. However, the shielding is difficult and very expensive equipment is required.

Open flames can be used on a few finishes and base materials. Both must be able to withstand the effects of the flame and the rapid heating.

SUMMARY

Organic coatings are very popular finishes. They are used for undercoatings and top coats. They form a film through solidification, evaporation, or polymerization. Organic finishing materials are made up of binders or vehicles, solvents, and additives. These films are developed by a number of processes including: brushing, rolling, dipping, flow and curtain coating, spraying, electrocoating, powder coating, and heat transfer. The film is usually cured after application. The heat for curing can be produced in convection, radiation, or infrared ovens. Flames are occasionally used for curing purposes.

STUDY QUESTIONS – Chapter 32

1. What are solvents, vehicles, and pigments?
2. Briefly describe paint, varnish, enamel, and lacquer.
3. What are the three ways films are formed?
4. What are the major organic coating materials?
5. List and describe:
   a. Brush coating.
   b. Roll coating.
   c. Dip coating.
   d. Flow coating.
   e. Curtain coating.
   f. Spray coating.
   g. Electrocoating.
   h. Printing.
   i. Powder coating.
   j. Heat transfer coating.
6. How are organic finishes cured?

These steel doors with intricate louvers would be hard to coat by spraying or brushing. They are coated without holidays or runs by immersing them in a 22,000-gallon paint tank. (Peachtree Door Co.)

SECTION 8
PROCESS DESIGN AND CONTROL

This section deals with the efficient deployment of resources to produce salable products at a profit. As reflected in the chapters following, this requires careful planning. This involves:

- Selection and sequencing of operations, an engineering function.
- Use of automation.
- Quality control.

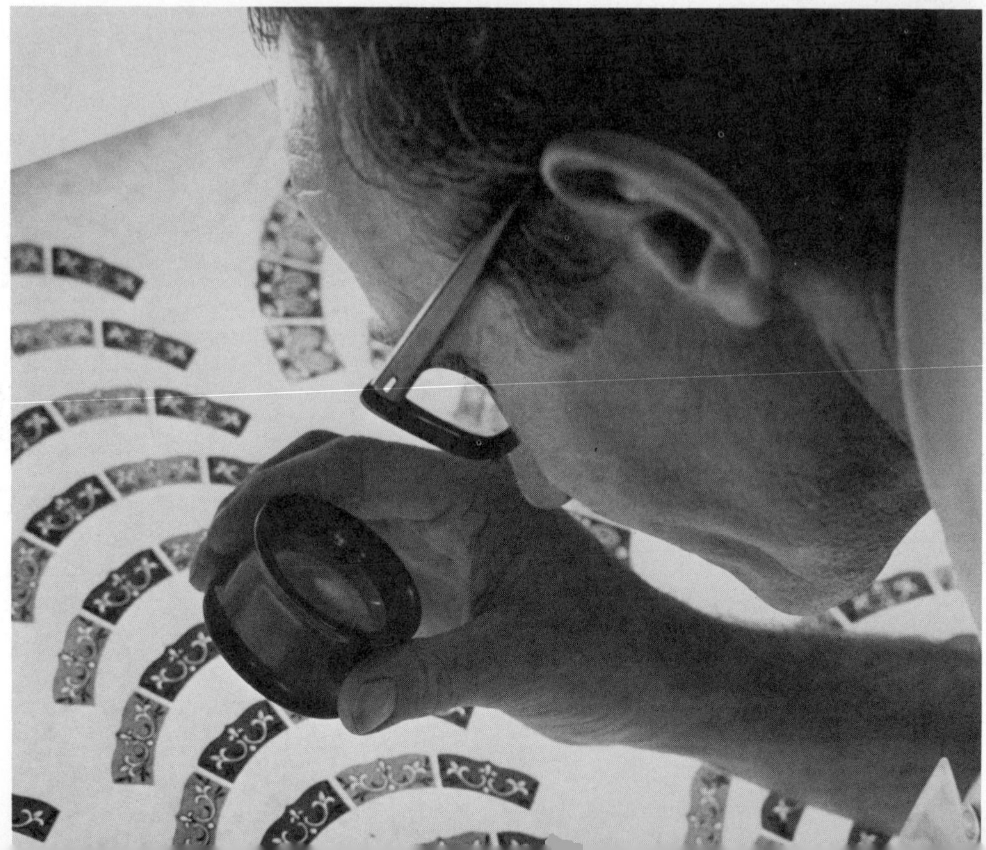

Chapter 33

INTRODUCTION TO PROCESS PLANNING

Manufacturing is a managed, human, productive activity. Its goal is to efficiently change the form of materials.

Materials are processed to add to their worth. Steel is worth more than iron ore, coke, and limestone. Likewise, a metal filing cabinet is worth more than sheet and bar stock.

Form utility is the adding of worth to a product or material. This does not just "happen"; it is planned. Successful manufacturing is carefully managed. Resources are efficiently used. Money is allocated to employ people to operate equipment which is used to process materials, Fig. 33-1. These processed materials are called products or goods. They are sold in the

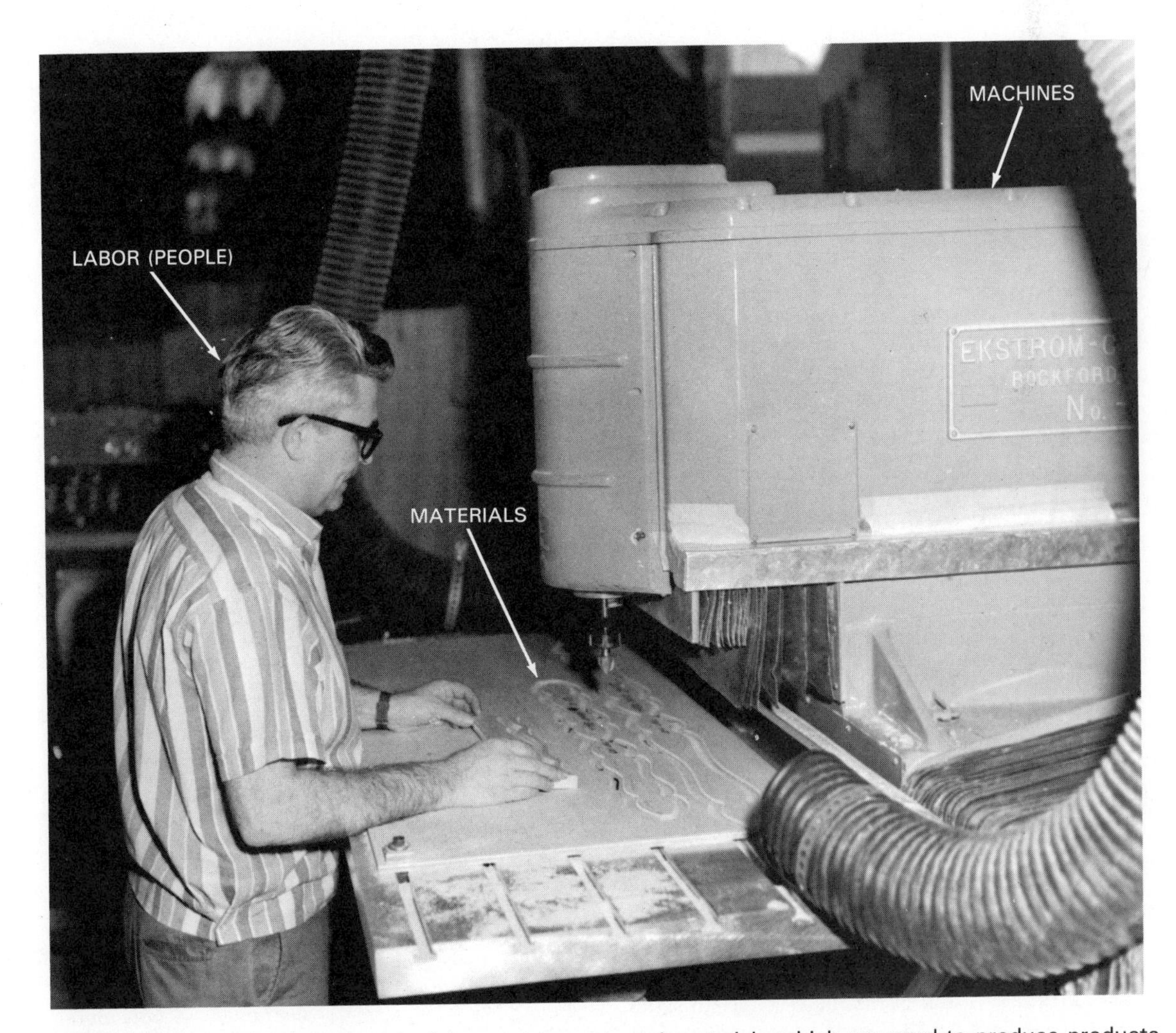

Fig. 33-1. Money is used to buy the labor, equipment, and materials which are used to produce products.

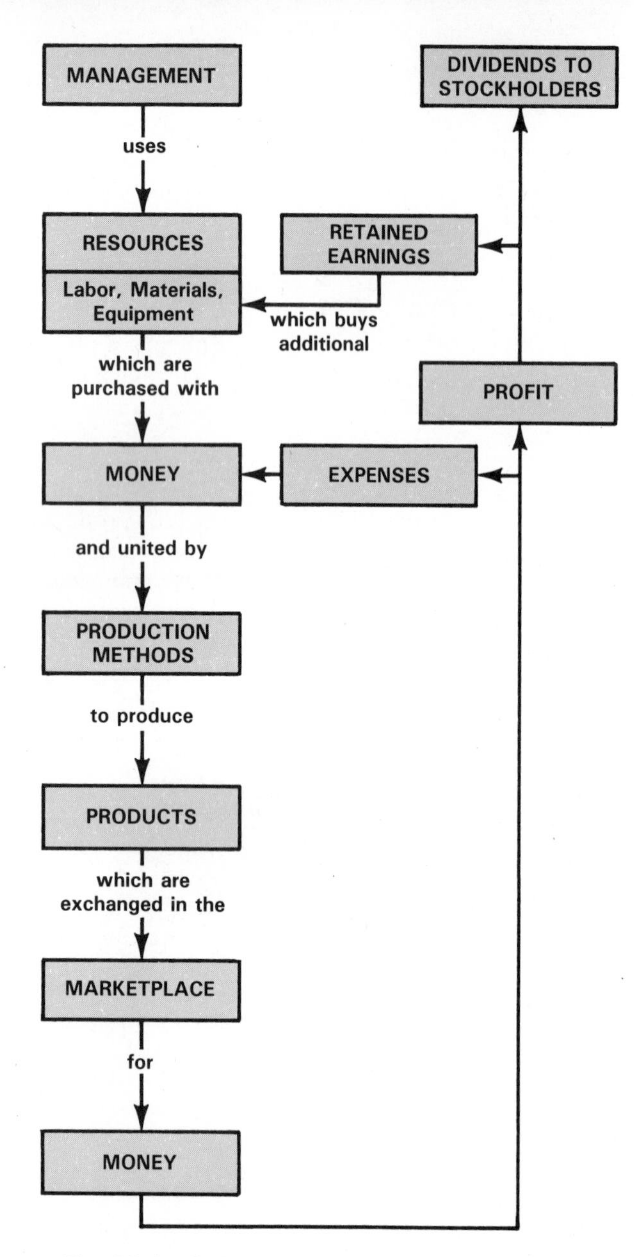

Fig. 33-2. Schematic of the form utility system.

marketplace for money. This money is called sales income. Excess over the cost of production is profit.

The system which uses money to make money is shown in Fig. 33-2. Money pays for resources which are used to make products. These products are sold and, hopefully, make a profit for the company.

As you have seen, processing—the changing of the form of materials—is a key activity in manufacturing. However, knowing how to process materials is not enough. Knowledge of casting, forming, separating, conditioning, assembling, and finishing is an essential base, but performing these processes efficiently is all important. Processes must be selected, ordered, automated, and controlled. They must be planned to the extent that they produce an economical and functional product. The output must, in the eyes of the customer, meet a need and be worth the selling price.

THE DESIGNER IN PROCESS PLANNING

The design of production processes starts with the product designer. The way the designer defines the customer's needs and meets them may directly affect manufacturing decisions. Product designs may dictate the selection of one manufacturing process over another. The design may make a casting process impossible to use or require that the part be machined.

The economics of processing are, therefore, important to design. Designers must know which processes will provide the least expensive and most functional product. For example, if a part can be forged it is generally cheaper to make than one which is machined. Of course, quantity is also a factor. The cost of forging dies must be charged against the parts produced. When only a few parts are needed machining may be more economical.

The economics of process selection, therefore, are complex. The cost of equipment, pay rates of workers, anticipated scrap rate, maintenance costs, environmental control costs, production rates, and many other factors must be considered. It is therefore very important that designers carefully consider manufacture as they design products.

In fact, designers, as shown in Fig. 33-3, consider three major factors as they design a product. They must design for selling, for function, and for manufacture. The final design should be the best blend of these requirements.

DESIGN FOR SELLING

A product must sell if the company is to be profitable. The customer must readily see that the product is appealing, has value, and will work, Fig. 33-4.

The design represents a delicate balance among

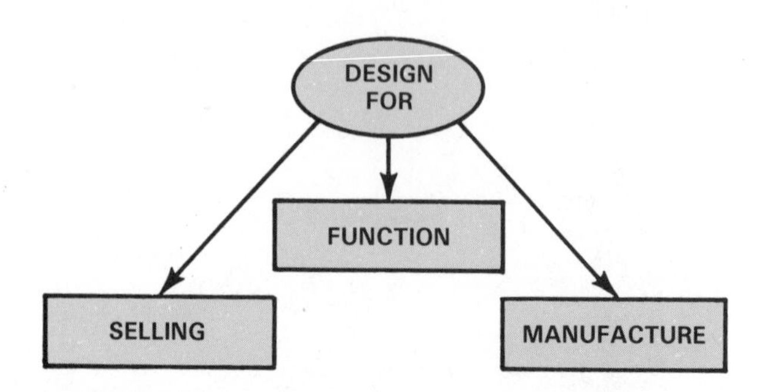

Fig. 33-3. These major elements are considered during the product design activity.

Fig. 33-4. This barbecue represents a product which is appealing, has value, and will work. (Stahl Specialty Co.)

these factors. Some products have high visual appeal (designed for looks). A wall decoration must look good to the customer. The "will work" requirement almost totally relates to its ability to fit into a decor. The opposite is true of a ball bearing. It must work well under a set of operational conditions. Appearance is a secondary factor.

All products must have value. This does not mean they are necessarily inexpensive. Customers must feel that the product is worth the selling price. A product which generates a generally negative response in the marketplace is poorly designed.

DESIGNING FOR FUNCTION

Products must be designed to function (do what they are designed to do). Still, design for function is broader than "will it work?" Surely a product must function. It must open a can, be comfortable to sit in, or meet some other operational requirement. But the product must also be designed for easy operation, convenient maintenance, efficient repair, and safe use. The design should place controls in convenient locations. The operator must be considered. Comfortable levers and knobs which are easy to reach are important. The force required to operate these controls must also be considered. Frequent lubrication requires easy access to grease fittings and oil holes. Parts which must be routinely replaced should be easy to reach and replace. In short, the product should meet its mission with the least amount of maintenance and operator discomfort.

DESIGNING FOR MANUFACTURE

The design factor most closely related to process planning is design for manufacture. The designer must consider many manufacture-related factors during the development of the product. As discussed earlier, the economics of processing must be considered. The equipment, skill of the workers, material and tooling costs, and complexity of the parts are all critical design factors.

Standardization of parts and fasteners is another consideration. Designs should use as many standard parts as possible. A part that can be used in several products is better than noninterchangeable parts. Screws, bolts, and other fasteners should be standardized. The fewer the number of sizes used the better.

The designer must also be concerned about tolerances, the amount a part can deviate from the stated size and still be acceptable. A tolerance should be as large as possible without affecting the part's operation. The part must function in the product, fit in the tooling, and fit in packages. Within these limitations, the looser the acceptable tolerance the better. It would be unwise to hold car doors to the same tolerance as engine cylinder diameters. If we did, the doors on a car built in a cold plant could not be opened after a day in the hot desert sun. The door would expand and wedge in its opening. On the other hand, bearings must be held to close tolerance or they would not fit on shafts.

SUMMARY

Designers are important cogs in the process planning cycle. They are the first step in the evolution of a product. They move product ideas from the mind's eye into reality. Designers sketch and refine product concepts. They then prepare the drawings and specifications which are the "road maps" for production. The drawings are the product represented by line and symbol. Drawings are the designers picture of the final product. The drawings are the conclusion of an effort to design a product which will function, attracts buyers, and can be manufactured.

The drawings are the base for:

1. Selecting and sequencing processes.
2. Automating processes.
3. Controlling product quality. These three major tasks will be discussed in the final chapters of this book.

STUDY QUESTIONS – Chapter 33

1. What is form utility?
2. Name the basic resources used in manufacturing products.
3. What major factors are considered for:
   a. Design for selling?
   b. Design for function?
   c. Design for manufacture?

Chapter 34

SELECTING AND SEQUENCING OPERATIONS

Manufacturing requires a series of operations to change materials into products. These operations must be carefully selected and sequenced for an efficient material processing system. This system is a part of the larger product production system.

Product Research and Development uses a *design development system* to change ideas in a designer's mind into drawings. Product Manufacture uses *material processing systems* to change materials into products as specified by the drawings. Marketing uses a *promotion and selling system* to exchange products for money. The work of research and development and of marketing is of little worth without efficient material processing.

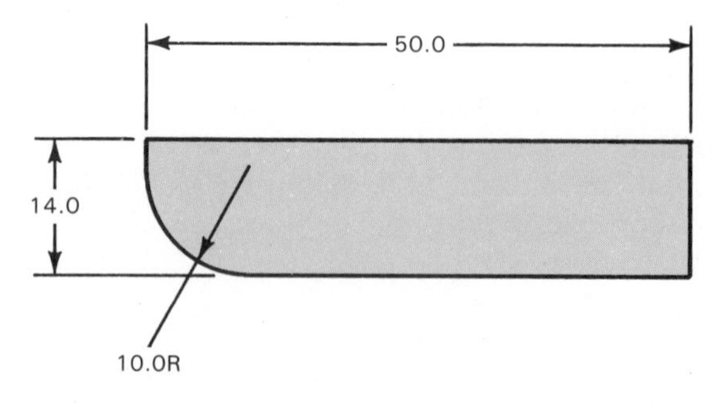

Fig. 34-1. A typical orthographic drawing. It shows overall size of a part and the shape (radius) of the rounded corner. (General Motors Corp.)

ANALYZING PRODUCTS FOR MANUFACTURE

Establishing an efficient material processing system requires planning. The first task is to analyze the drawings, bills of materials, and specification sheets. The planner must know:

1. The size and shape of each part.
2. Surface finish required.
3. Tolerances to be held.
4. Materials to be used.
5. Number of parts and products needed.

SIZE AND SHAPE OF PARTS

Each part has its own geometry. It has a general size, a thickness, width, and length which can be determined from the drawing, Fig. 34-1. Most parts also have a contour. They may be angled, curved, or otherwise formed to provide a shape.

The process planner must determine how to provide this size and shape. Each feature may be produced by one or several different processes. From a careful study and listing of the size and shape requirements, a rough list of possible operations can be developed. Alternative ways of producing each product feature are determined.

SURFACE FINISH

Each manufacturing operation produces its own surface finish or texture. For example, sand casting produces a somewhat rough finish while injection molding produces a smooth part. Superfinishing produces a very fine finish. Special surface finish requirements are listed on the drawing, Fig. 34-2. The analysis of surface finish needs helps the process planner select the most economical process. Some processes will provide the desired finish immediately. Others will need to be followed by additional machining, such as sanding, grinding, and polishing.

TOLERANCES

A third factor which must be considered in selecting processes is dimensional control: how close must a part be held to the dimensions on the drawings. This closeness to dimension is called the *tolerance*. In other

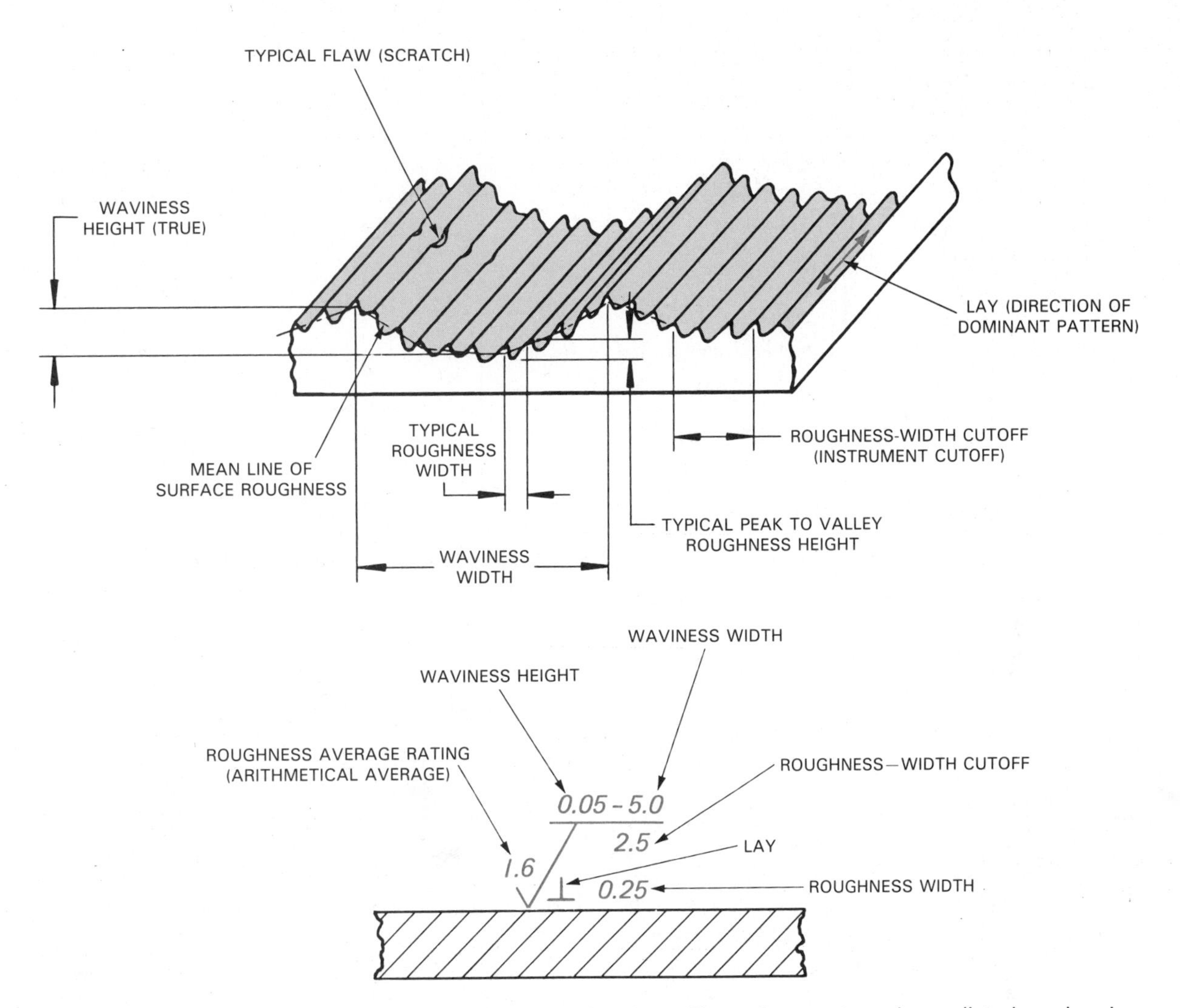

Fig. 34-2. An explanation of the typical surface specifications. These features are always listed on drawings.

words, how far can a part vary from the specified dimension and still be acceptable.

No two parts produced are exactly alike. They all vary in some measure. Some variation does not affect the usefulness of the part.

The product drawings, as shown in Fig. 34-3, will tell the process designer the amount of variation allowed. The task, then, is to select processes which will hold these tolerances.

Some processes are considered precision operations. They can hold close tolerances. Other processes are not very precise. For example, laser cutting is very precise while oxyacetylene flame cutting is not. The process designer will select the process which gives the desired result. It would be foolish to produce parts within .0001 in. when ± 1/4 in. is all that is needed. Added precision is generally more costly. Tooling is usually more expensive, scrap rates higher, and operator skill more demanding.

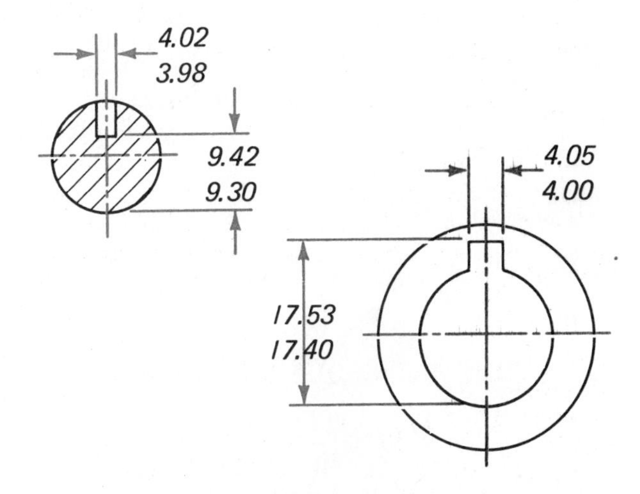

Fig. 34-3. Typical drawing showing tolerancing of a part feature. In this part, any keyseat within the limits stated on the drawing is acceptable. (General Motors Corp.)

PART NO.	NUMBER NEEDED	PART NAME	SIZE T	X W	X L	MATERIALS
G101	1	Base*	3/4	4 1/2	15 3/4	Pine or redwood
G102	1	Small end*	3/4	4 1/2	2	Pine or redwood
G103	1	Large end*	3/4	4 1/2	3	Pine or redwood
G104	2	Rod	1/4		18	Mild steel
G105	1	Ball	1		1 1/4	Welding rod
	4	Screws	No. 8		1 1/4	Reject ball
	2	Rod assembly pins	No. 16			Bearing
						Brads
		*Base and ends cut from 3/4 x 4 1/2 x 21 piece (1 x 10 ripped in half)				

Fig. 34-4. The bill of materials for a part or product must be studied to see what processes are needed to "work" the material.

MATERIALS TO BE USED

The bill of materials, Fig. 34-4, lists the materials to be used. Obviously this material will be a factor in selecting the processes. Wood cannot be cast or flame cut. Some metal alloys are hard to machine while others cannot be easily welded. Each material has its own set of appropriate processes. The processes to be used should be matched with the material to be processed. The planner must know the properties of the material and the capabilities of the processes.

QUANTITY NEEDED

Processes must be selected with production levels in mind. Processes requiring expensive tooling generally are not practical for short runs. Closed die forging are not used to produce one part unless that is the only way to provide the desired size and shape. And if it is used, the entire cost of the dies must be charged against that one part.

The economics of quantity play a significant role in selecting the proper process. The process must produce a part which meets both physical and economic requirements.

SELECTING AND SEQUENCING OPERATION

The analysis of the product provides the base for selecting the proper operations for each part. Process planners use their judgement to make the final process selections. They must consider part size and shape, surface finish needed, tolerances required, materials to be processed, and quantities to be produced. They then must describe and sequence these operations.

OPERATION SHEET

The first step in operation sequencing is to describe how the part is to be manufactured. This description is summarized on an *operation* or *route sheet.* This sheet lists the operations to be used to produce the part. The operations are listed in proper order. Also listed are the equipment to be used and special tooling needed for each operation. Often, setup times and production rates are included. Fig. 34-5 shows plans for a simple product and Fig. 34-6 is an operation sheet for one of its parts. An operation sheet would be developed for each part.

The operation sheet provides valuable information for determining product flow in the plant. The sequence of operations listed assists in developing plant layouts or routing parts through an existing plant layout.

Operation sheets are used to set up production of new parts. They can also be used to study existing production systems. As parts are produced, workers and supervision may, over time, change procedures to meet unique situations. These changes become part of the system. However, not all changes are for the good. A change which is made because of a production problem may not be the best solution.

From time to time, new production techniques, and tools are developed. They may be better than the old ones. Therefore, production systems must be periodically reviewed. The operation sheet may be the base for such a study.

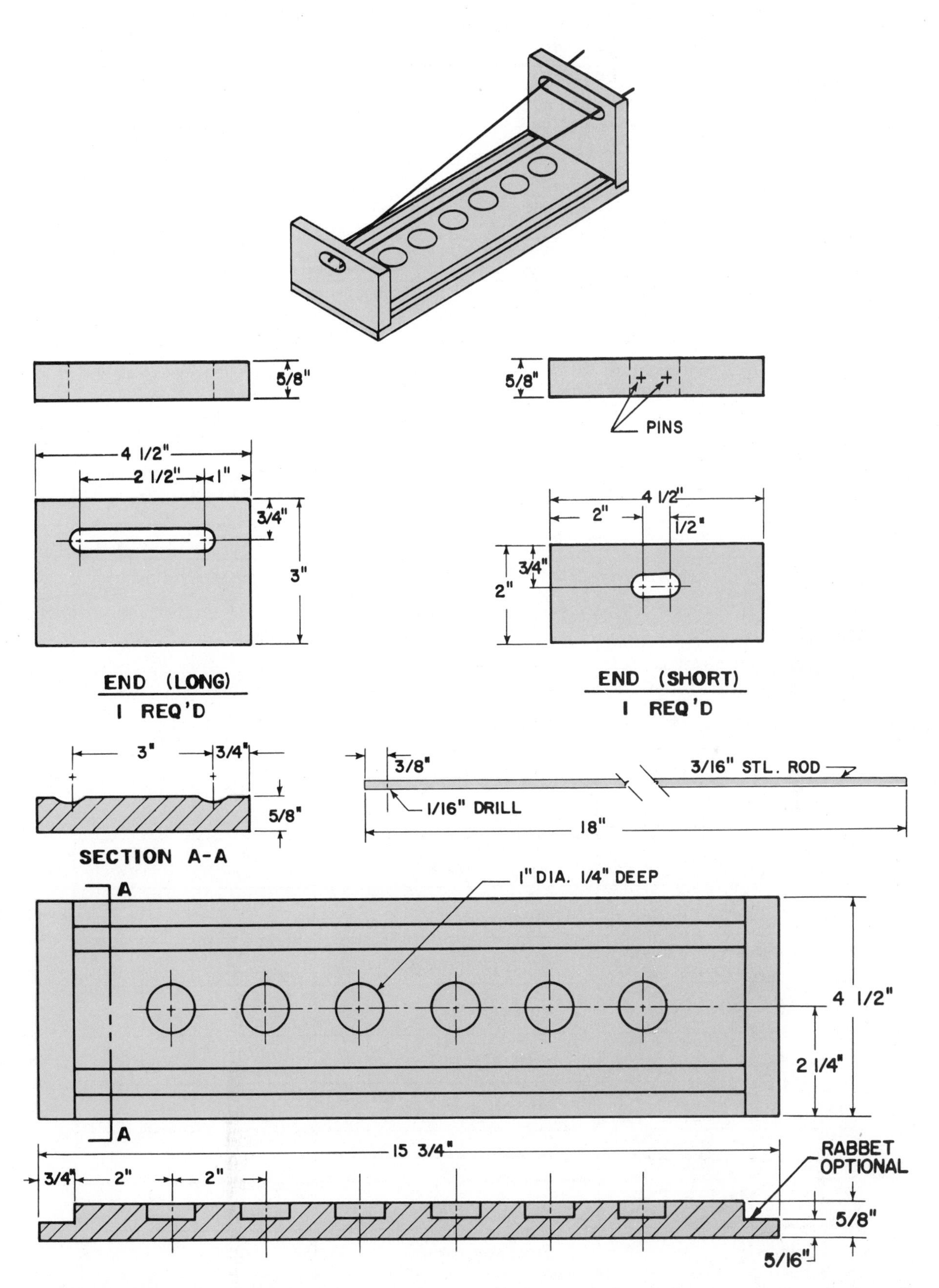

Fig. 34-5. Drawings of even very simple products must include enough descriptions so operation sheet can be developed.

OPERATION SHEET

PART NAME Gravity Game Base **DATE EFFECTIVE** ______

PART NO. GG103 **MATERIAL** Pine **DRAWING NO:** 103

OPERATION NO.	OPERATION DESCRIPTION	MACHINE	TOOLING	SET-UP TIME	PRODUCTION RATE
B-0	Cut to length	Table saw	Fixture B-0	10 min.	180/hr
B-1	Drill score holes	Drill press	Jig B-1 1'' Forstner bit	5 min.	120/hr
B-2	Rout ball gutter	Router	Jig B-2 Cove bit	15 min.	150/hr
B-3	Drill screw holes	Portable drill	Jig B-3 #8 Screwmate	2 min	200/hr
B-4	Sand edges	Orbital sander	100 grit paper	1 min	180/hr

Fig. 34-6. An operation sheet contains detailed information about machines and tooling needed as well as information needed about production and setup time.

PRODUCT NAME: *GRAVITY DEFIER* FLOW BEGINS *BASE-CUT TO LENGTH* FLOW ENDS *FINISHED PART* DATE *4/1/*

PREPARED BY: *R. T. WRIGHT* APPROVED BY: *RTW*

PROCESS SYMBOLS AND NO. USED: OPERATIONS *4* INSPECTIONS *1* TRANSPORTATIONS *6* DELAYS ______ STORAGES *1*

TASK NO.	PROCESS SYMBOLS	DESCRIPTION OF TASK	MACHINE REQUIRED	TOOLING REQUIRED
MB–1		*MOVE TO DRILL PRESS*		
B–1		*DRILL SCORE HOLE*	*DRILL PRESS*	*JIG B–1 1 IN. BIT*
MB–2		*MOVE TO ROUTER*		
B–2		*ROUT BALL GUTTER*	*HAND ROUTER*	*JIG B–2 COVE BIT*
MB–3		*MOVE TO DRILL*		
B–3		*DRILL SCREW HOLES*	*DRILL PRESS OR HAND DRILL*	*JIG B–3 NO. 8 SCREWMATE*
MB–4		*MOVE TO SAND*		
B–4		*SAND FACES AND EDGES*	*ORBITAL SANDER*	*GAGE I–1*
MB–5		*MOVE TO INSPECTION*		
I–1		*INSPECT*		
MB–6		*MOVE TO STORAGE*		
		STORE FOR ASSEMBLY		

Fig. 34-7. A flow process chart includes all operations and handling of parts.

FLOW PROCESS CHARTS

Another tool for studying production systems is the *flow process chart.* It, too, shows the sequence of operations. However, the flow process chart, as shown in Fig. 34-7, includes inspection, delay, storage, and transporting activities. These activities are shown by special symbols, Fig. 34-8, and are placed in proper sequence on the flow process chart.

This chart provides a larger picture of the production process. It shows all the activity involved in changing a piece of material into a part. Each part would have its own flow process chart.

OPERATION PROCESS CHART

A third way to show the order of operations is an operation process chart, Fig. 34-9. Both the operation

Operation — object is changed in its chemical or physical makeup; it is assembled or disassembled.

Transportation — object is moved from one place to another.

Inspection — quality is checked.

Delay — object is held awaiting the next operation.

Storage — object is placed in a protected location.

Fig. 34-8. Flow process symbols are a type of "shorthand" for describing what is being done to the part.

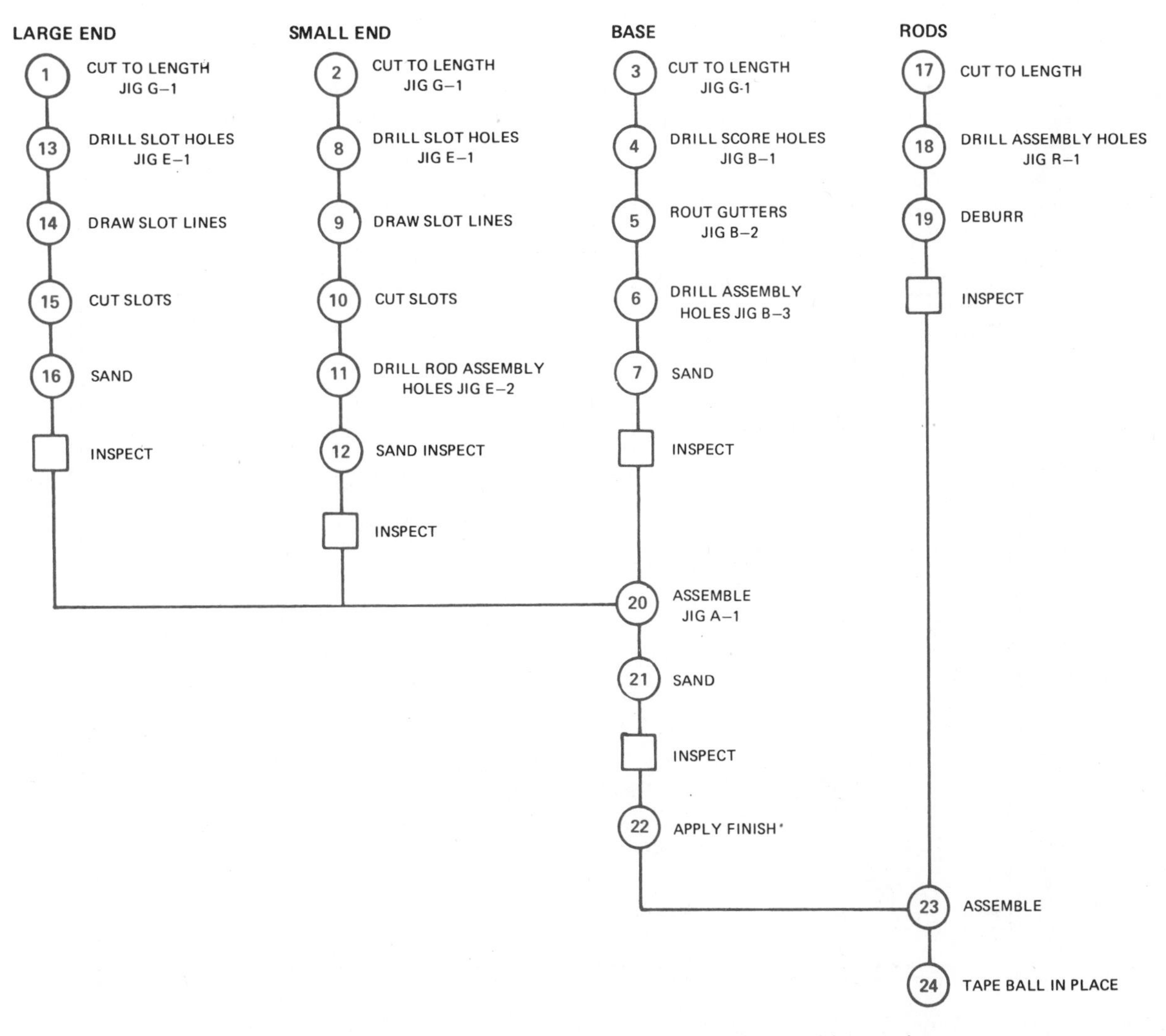

Fig. 34-9. Operation process chart shows operations and inspections.

sheet and flow process chart showed the processes required to produce a single part. The operation process chart shows the operations and inspections needed to produce an entire product. The assembly of parts into subassemblies and final assemblies are shown. The chart pictures the operations needed to produce each part and then the assembling operations used to complete the product.

The operation process chart is particularly valuable in determining when parts must be ready for assembly and the sequence for assembly operations.

FLOW DIAGRAM

The *flow diagram* is the last of the important process planning tools. As shown in Fig. 34-10, it superimposes the information from the flow and operation process charts onto a plant layout. The flow diagram allows the process planner to see if material flow in the plant is proper. Excessively long moves or crossing flow lines can be easily seen. These factors can lead to inefficiency.

SUMMARY

Selecting and sequencing operations is vital to efficient production. Proper analysis of product drawings is the first step. This activity is followed by description of the operations on route or operation sheets for each part. Flow process charts add necessary inspection, delay, storage, and transportation activities to the list of operations. The operation process charts cover all operations and inspections needed to produce a multipart product. Finally, the flow of materials through the plant is shown in a flow diagram.

Each of these activities or tools play a role in developing an efficient system to process materials. They help the process planner mesh many considerations and tasks into a single productive endeavor.

STUDY QUESTIONS – Chapter 34

1. List and describe the factors considered in analyzing a product for selecting manufacturing processes.
2. What is tolerance?
3. What information is contained on an operation sheet?
4. What is a flow process chart?
5. Draw and define the standard flow process symbols.
6. What is an operation process chart?
7. Describe a flow diagram.

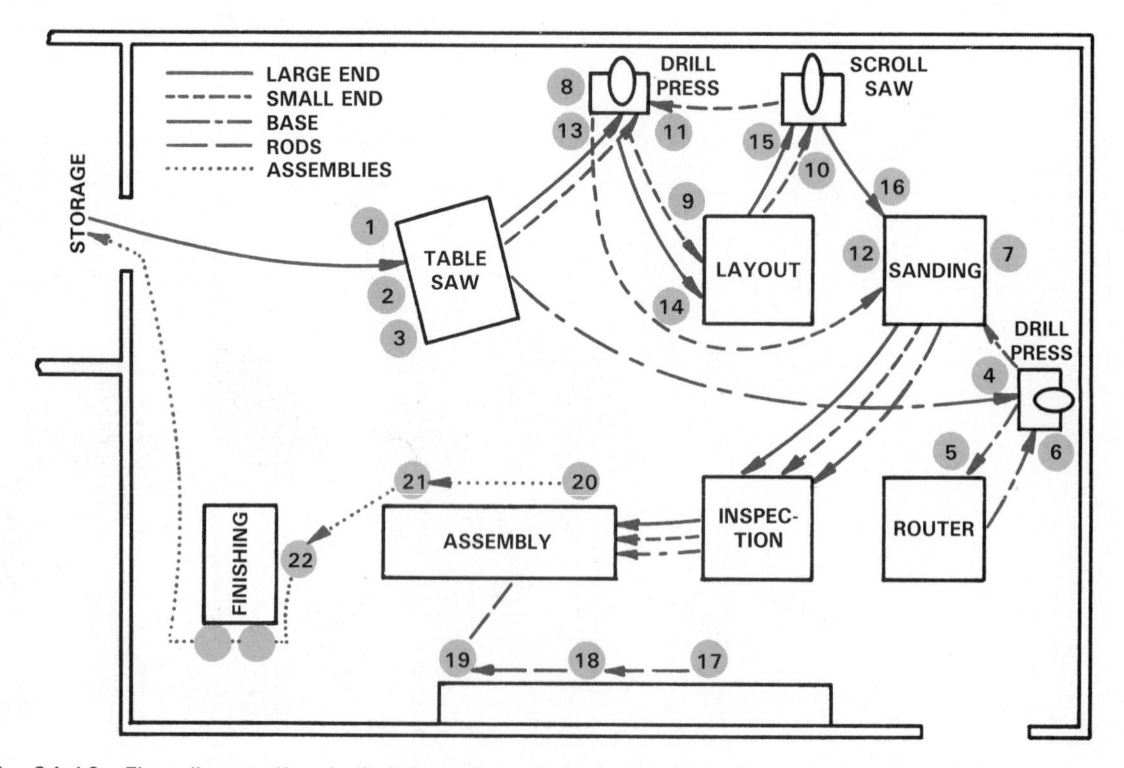

Fig. 34-10. Flow diagram "tracks" all parts through production and allows planners to bring different parts to the assembly operation at the right time.

Chapter 35

AUTOMATING PROCESSES

Inventors and innovators have constantly sought ways to make manufacturing more efficient. Oliver Evans developed continuous processing. Eli Whitney implemented the interchangeable parts concept. Henry Ford introduced the moving assembly line to automobile manufacturing.

More recent developments have replaced human labor with automatic machinery. Direct human control of processing is being replaced with automatic control. This process is known by several names. A common term for automatic control is AUTOMATION. This term was coined by Delmar Harder of Ford Motor Company shortly after World War II. It has been used in so many ways that it has no exact definition. However, it usually means automatically controlled manufacture.

This chapter takes a specific view of automation and is restricted to activities designed to make machines more automatic. Computer-aided design, computer-integrated manufacture, robotics, and other similar technologies deserve to be studied as subjects by themselves. This text is a study of the *methods* used to process industrial materials.

AUTOMATION

Automation in manufacturing is rapidly changing the way parts and products are manufactured. Generally two types of automation are used.

FIXED AUTOMATION

Fixed automation, Fig. 35-1, is a technique which designs specific equipment to complete a sequence of operations. Automotive assembly lines, engine block transfer machining lines, and oil refineries are

Fig. 35-1. These automobile catalytic converters are being assembled on a fixed automation welding line. (AC Spark Plug-GMC)

Fig. 35-2. A scale model of a plant layout with transfer machines. Note the machine units at right angles to the central transfer device.
(Oldsmobile Div., General Motors)

examples of fixed automation. These systems, as shown in Fig. 35-2, are characterized by:
1. High initial cost.
2. High production rates requiring high product demand.
3. Inflexibility.
4. Simple basic operations with complex material-handling (transfer) equipment.
5. Difficulty in coordinating basic operation performed at each workstation.

PROGRAMMABLE AUTOMATION

Newer *programmable automation* systems include numerical control (NC), computer numerical control (CNC), and direct numerical control (DNC). These systems may be supplemented by automatic tool changers, automatic workpiece handlers, robots, and adaptive control systems. All these elements may be interconnected to produce the "factory of the future." This computer-integrated automatic factory is not a futuristic dream but a present reality. Machine tools, videotape records, and other products are now being produced on totally integrated manufacturing systems.

Automatic or automated systems for material processing are built from a number of separate technologies. Not all systems use all these features. However, the most advanced systems will. These technologies are:
1. Automatic machine control.
2. Automatic tool changing.
3. Automatic material handling.

AUTOMATIC MACHINE CONTROL

All machines have a system to produce cutting and feed motions. These systems vary. A table holding the workpiece may be moved horizontally or vertically under a rotating cutting tool. A single-point tool may be fed along a rotating workpiece. A rotating cutting tool may be fed into a stationary workpiece. These examples describe milling, turning, and drilling. They are but three of several hundred different types of operation performed on materials.

In each case, the cutting and feed motions are controlled, Fig. 35-3. The earliest machines depended entirely on human judgment and control. The operator forced the work against the cutting tool. This technique is still used in free-hand grinding and bandsawing of metal and in many of the simpler woodworking operations.

Later machines used tables fed by lead screws to uniformly feed the work. The operator turned handwheels attached to the lead screw. Each turn of the wheel in most metal cutting machines produced table or spindle movement of 0.100 in. The wheel had 100 evenly spaced graduations on its collar. The operators could control the movement to within a few thousandths of an inch, depending upon their skill in reading the marks on the wheels.

Many operations still require human control of the machine. The work is fed by workers who depend upon their judgment and skill for accuracy.

Shortly after World War II, a new system of machine control was introduced into manufacturing. This system became known as *numerical control.* It is capable of operating a machine through a series of instructions which are coded using numbers and other symbols. The codes position the machine tool or the workpiece without human assistance.

Secondary numerical control commands can select a tool in a turret, set machine speeds, cause the spindle to turn in a certain direction, set feed rates, and turn coolants on or off. These machines can be common numerical control (NC), computer numerical control (CNC) or direct numerical control (DNC).

The numerical control unit does not change the basic way the machine operates. An NC, CNC, or DNC milling machine and a manually operated mill will generate chips in the same manner. The part is not machined by numerical control. The machine is

Fig. 35-3. A—The simplest operations require the operator to feed the material into the cutter. B—More complex machines use operator-controlled table feeds. C—The most advanced machine uses NC units to control machine operations. (Delta International Machinery Corp., Cincinnati Milacron, and Coherent)

numerically controlled. The control has been removed from human hands and turned over to a numerical control unit, Fig. 35-4.

The result is more efficient production. The machines cycle at a uniform pace. Operator judgments and possible lapses in concentration do not affect the quality of the parts being produced.

All individuals make mistakes. When these mistakes happen during the operation of machines, scrap and rework results. For these and other reasons, automatically controlled machines, even though they are more expensive, are rapidly replacing manually controlled machines. Today there are numerically controlled (NC, CNC, and/or DNC) lathes, milling machines, flame cutters, grinders, routers, EDM machines, bending machines, laser drilling machines, and the list goes on. Any machine which must be controlled by humans can be controlled by NC units.

THE NC MACHINE

A numerical control machine contains three distinct parts. These, as shown in Fig. 35-5, are:

1. Data input device.
2. Control unit.
3. Basic machine to be controlled.

Fig. 35-4. The metal shaper on the left is controlled by an operator while the one on the right is NC controlled. Both machines produce a chip in exactly the same manner. (Rockwell Machine Tool Co. and Cincinnati Milacron, Inc.)

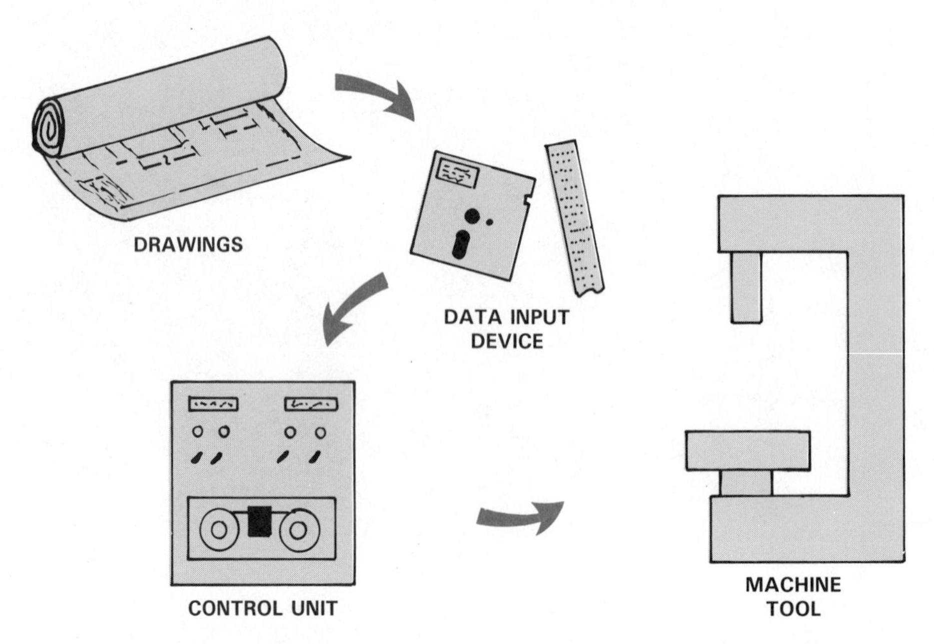

Fig. 35-5. NC systems include a data input device, a control system, and a machine.

DATA INPUT DEVICE

The NC system is designed to produce a part as described on an engineering drawing. But the system must be told what to do. Location, size, and geometric (shape) data on the drawings must be prepared for the system. The machine must be directed to produce the part with all its features.

The data for all numerically controlled machines are developed around a coordinate system, Fig. 35-6. This system uses the Cartesian or rectangular coordinates. It allows a person to describe any point in space using three coordinates; X, Y, and Z. The X and Y coordinates are on the same plane. They are perpendicular (at 90 degrees) to each other. The point at which they intersect is called the *origin*. The Z coordinate is also perpendicular to both the X and Y coordinates and passes through the origin. Each of the coordinates may be either plus (+) or minus (−). They may extend on each side of the origin and therefore form an eight-section box. Any point can be located in one of these sectors. For example, the point X =2, Y = 3, and Z =4, (Fig. 35-7, is found by:

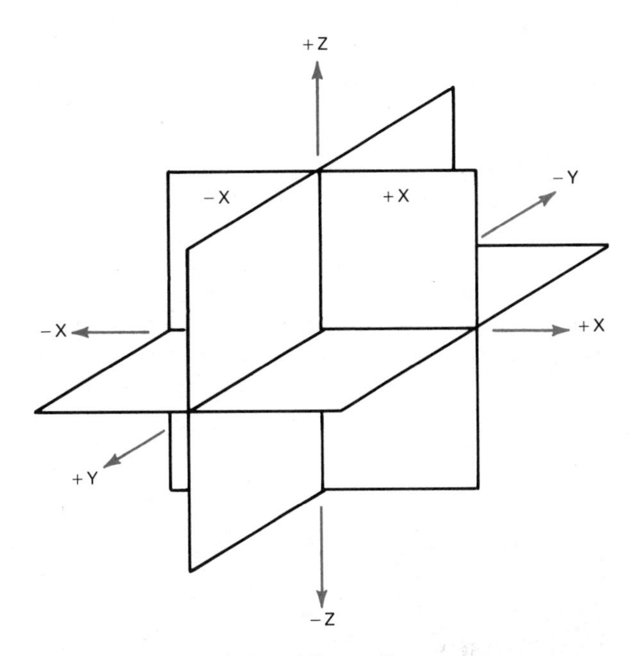

Fig. 35-6. The X, Y, and Z coordinate system can provide machine control to machine a part in three dimensions.

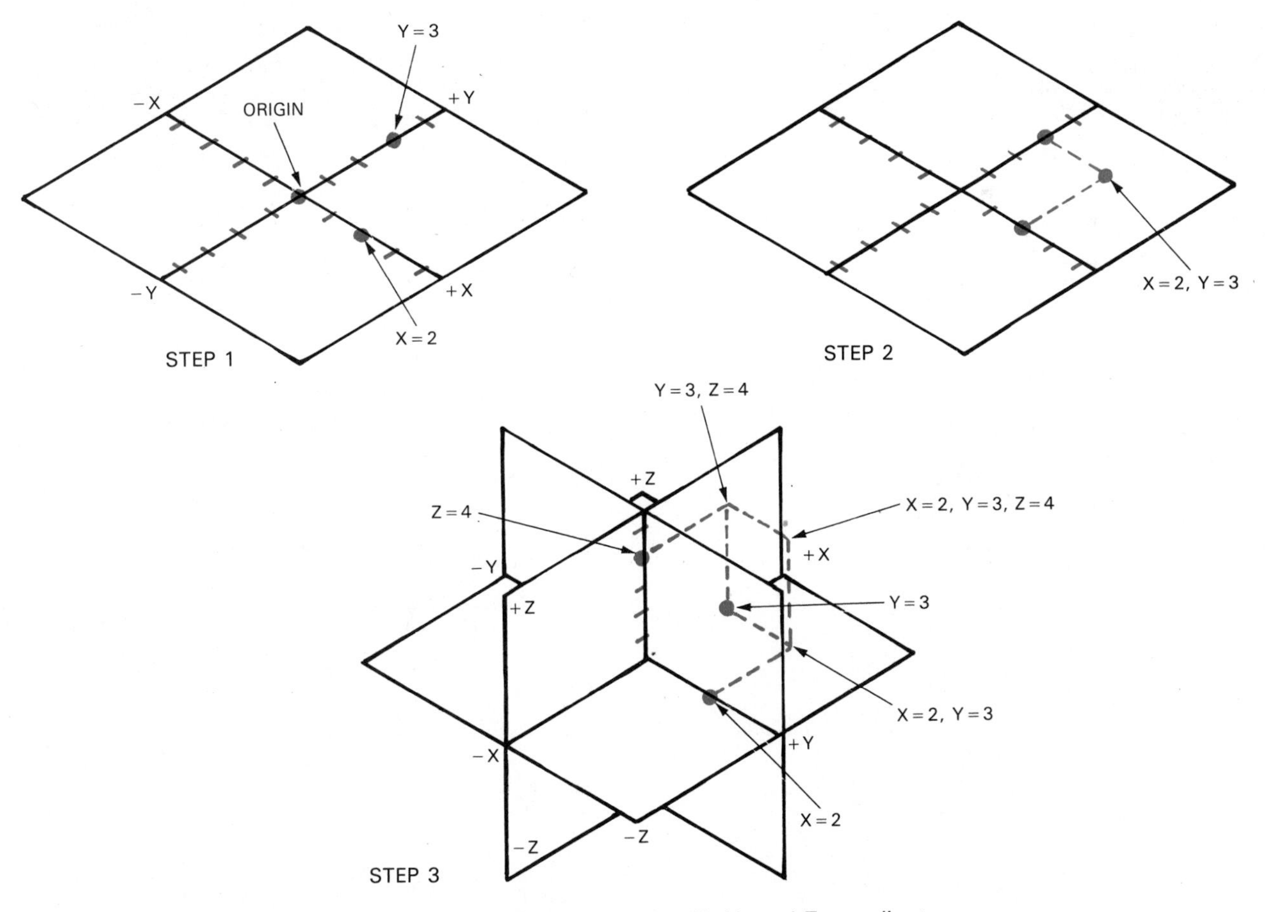

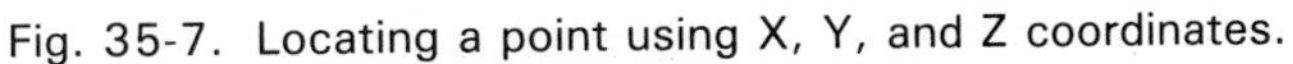
Fig. 35-7. Locating a point using X, Y, and Z coordinates.

1. Locating 2 on the X axis and 3 on the Y axis.
2. Finding the point where X = 2 and Y = 3 cross when the points are extended from the coordinate line.
3. Locating Z = 4, which is four units above the X, Y plane, then extending the Z = 4 point until it is directly above the point where X = 2, Y = 3 intersect.

In numerically controlled machines the X, Y, and Z axes have special meaning. The X axis for the machine is horizontal and parallel to the working-holding device (table, fixture, etc.). The Z axis is parallel to the principal machine spindle or arbor. The Y axis is at right angles (perpendicular) to both the X and Z axes. Fig. 35-8 shows the X, Y, and Z axes of some common machines.

NC machines may have up to five axes controlled. On these machines secondary movement of slides which are parallel to the X, Y, and Z axes are designated U, V, or W. The U axis is parallel to the X axis, V with Y, and W with Z, Fig. 35-9. Rotary motion of the work or toolholder is designated by A, B, or C. For example, the motion of a rotating table is designated as "C" on a vertical profile and contouring milling machine.

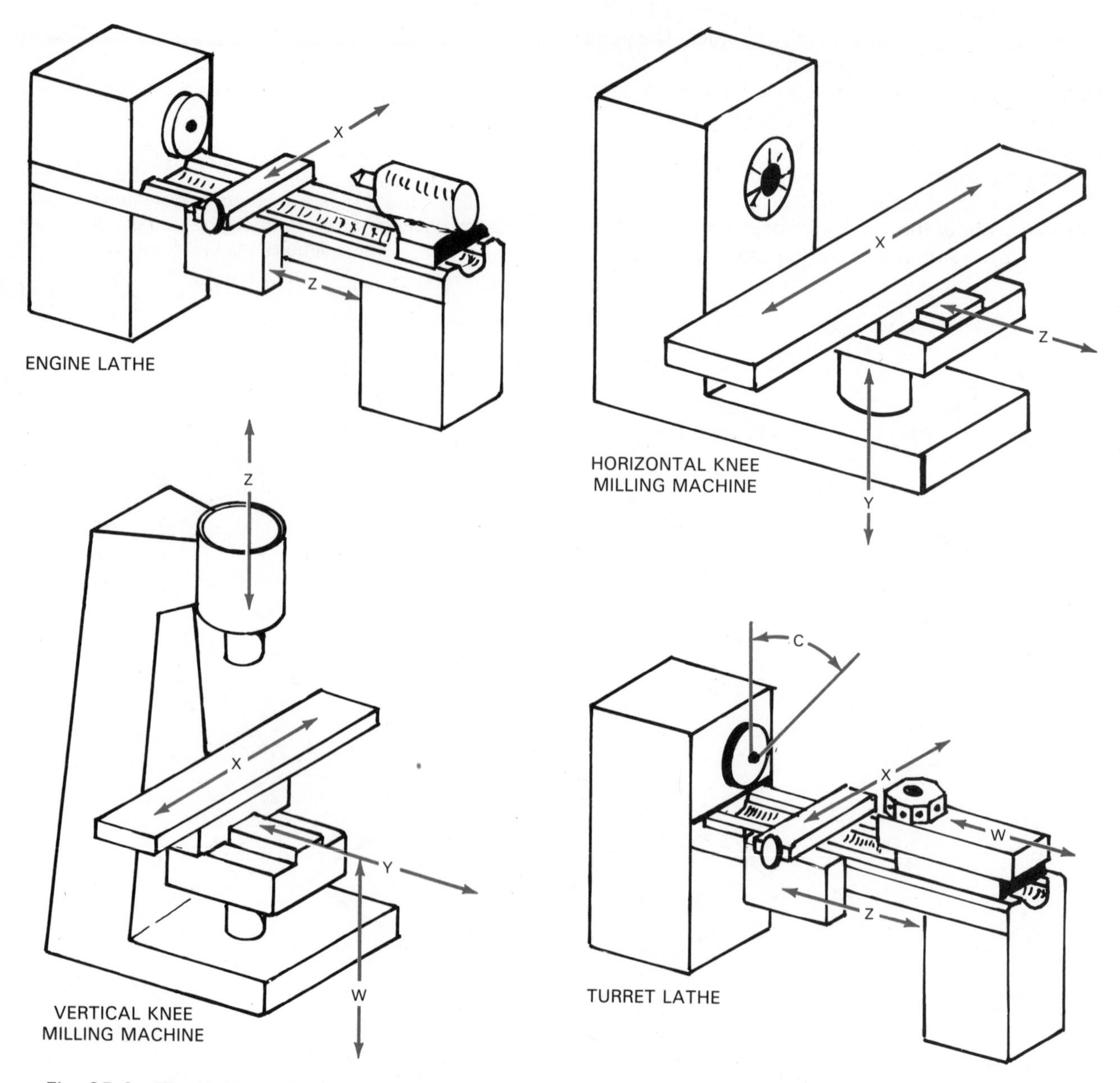

Fig. 35-8. The X, Y, and Z axes of some common machines. The Z axis is always parallel to the principal spindle or arbor of the machine.

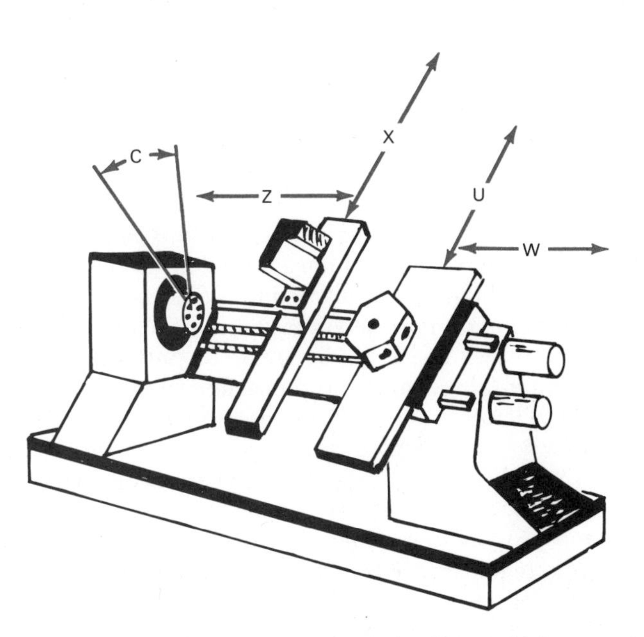

Fig. 35-9. A four-axis slant bed lathe. Note which axes are parallel.

The movement of the tool or work on the axes is independently controlled using data processed by the numerical control unit. This data may be called a program of instructions. These instructions are recorded and stored by several methods for numerically controlled machines. These methods include punched tape, punched cards, magnetic tape, remote computer storage, bubble memory, diskettes (floppy discs) and control unit memory systems. Three of these systems will be discussed:

1. Tape inputs.
2. Direct data input into the control unit memory.
3. Programming using computer languages on floppy discs.

Tape input

A widely used data input device is the punched tape. The tape, usually an inch wide, has eight channels or columns. Holes are punched in these channels, Fig. 35-10, to give data to the machine control unit. The location and number of holes punched will depend on the system used. Presently there are four common standards for NC punched tapes. These are EIA (Electronics Industries Association) RS-273-A, RS-273B, RS-244-A, and the ASCII (American Standard Code for Information Exchange) code which is also called RS-358. The differences among these systems are not important for this general discussion. However, each system divides the tape into blocks along its length. Each block is a complete set of instructions for one task. The block may give commands for:

1. X dimension.

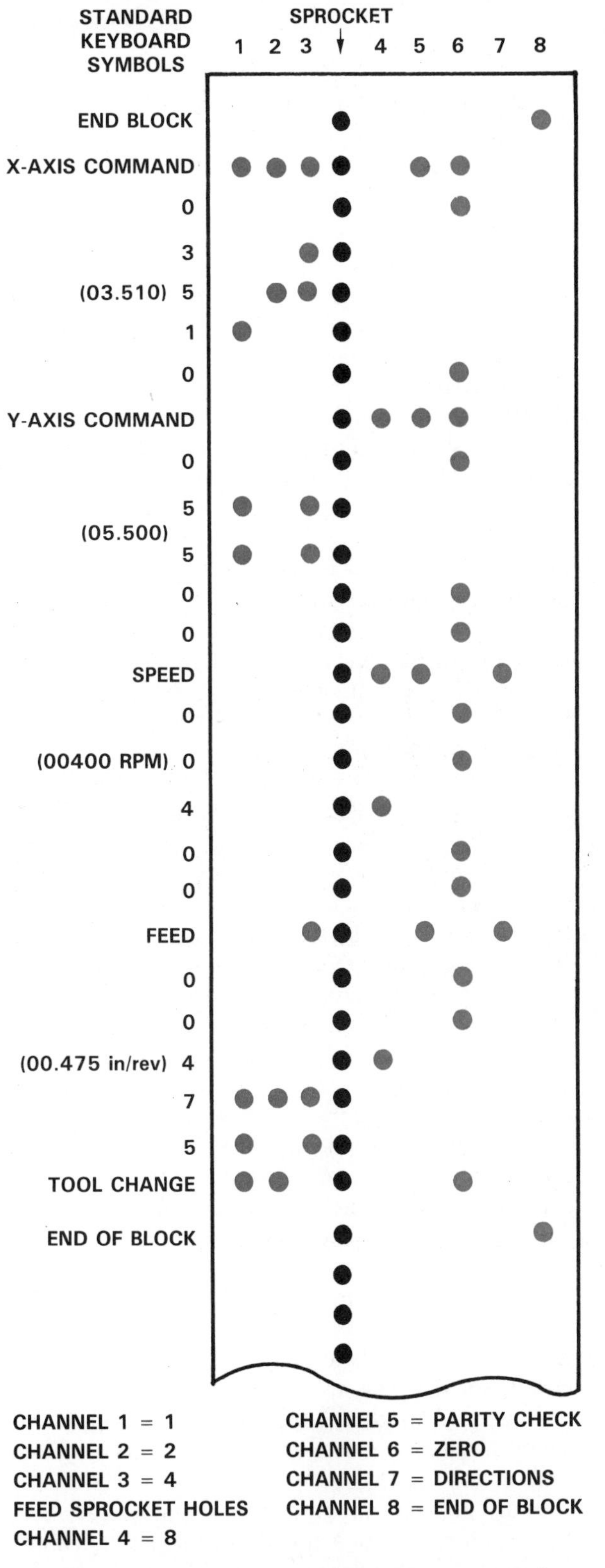

Fig. 35-10. A sample of a block of instructions in a punched NC tape. Note how each channel either gives a number value, an instruction, or is used for some dedicated function. Plus or minus for coordinates is designated by punches placed in channels 5, 6, and 7. A punch in all three means "+" and a punch in channel 7 only indicates a "−."

2. Y dimension.
3. Speed.
4. Feed.
5. Tool change.
6. Coolant start or stop.
7. Machine start or stop.

Even though the data is automatically fed into the machine, the tape is developed by a process planner and a machine programmer who often work together. Drawings are analyzed for the proper X, Y, Z coordinates. Materials for processing are considered. The proper machine and cutting tools are selected; speeds and feeds for each operation are calculated. Coolant requirements are determined; then a program is written and checked. This process is often done with the aid of computers and information from a CAD (computer-aided design) system.

After all the data is collected and analyzed, the tapes are punched and checked. The tape is then ready to provide information to the NC machine.

Manual data input

Simple machining operations can be programmed on the shop floor. This activity is called *manual data input* (MDI) and places machining information directly into the machine's control unit. Standard CNC systems have limited capability for manual data entry. Recently several systems have been developed for full shop floor programming of two-axis lathes, two and three axis milling machines and machining centers.

MDI may use a full alphanumerical (letters and numbers) keyboard, a simpler system asking the operator to enter number values for X, Y, and Z axes, or video screen (cathode ray tube) displays which lead the operator step-by-step through the programming activity. The MDI unit may store the information only within the control unit or may produce a tape as well.

Programming languages

Most numerical control operations are now programmed on a computer, then moved to the machine. Often, these programs are stored and, as needed, moved to the machine control unit on a floppy disc. These are the same discs used by personal computers (microcomputers).

Many different programming languages are used for programming numerically controlled machines. These are a study in themselves and specific descriptions are beyond the scope of this book. However, these languages fall into two main groups:

1. General purposes languages.
2. Special NC languages.

Each of these languages has its own vocabulary. A series of words and symbols communicate meanings. These words and symbols are combined into individual statements making up a specific instruction. These instructions can be processed (understood) by a computer and converted into machine control directions.

The use of computer languages allows easy editing and sharing of information among drafting, engineering, machining, and inspecting activities.

General purpose languages are designed for varied applications. They can be used on a number of different machines and for nonmachining applications. FORTRAN and BASIC are the two most commonly used general purpose languages.

Specialized languages have also been developed for NC processors. They are designed to reduce learning and programming time. Some common specialized NC languages are APT, CONPAC, AUTOSPOT, and SPLIT.

MACHINE CONTROL UNITS (MCU)

The significant difference between NC, CNC, and DNC machines is in their data input and machine control unit which uses the data. The same basic machine tool can be controlled by any of the numerical control systems.

There are two basic types of machine control units:

1. Numerical control (NC) which is hardware controlled.
2. Computer controlled (CNC or DNC) which is software controlled.

The earliest machine control unit was developed for a specific machine and was "hard wired" to it (wired directly to the machine). All elements to control the machine were built into the machine control unit. These units were found on the NC (numerically controlled) machines.

After the mid 1970s, computer control units became popular. Many activities which had to be built into the hardware of the older NC units could now be put in the software (programs). These new CNC units could operate many different machines. The basic control elements of the unit remain unchanged for each machine. Only an "executive" program, which directs these elements, changes. The CNC unit is the most common today.

When each machine has its own independent computer control unit, the machine is called CNC (Computer Numerical Control). However, when several CNC machines are controlled by a central computer directly wired to the machines, the system is called DNC (Direct Numerical Control). Often DNC systems control drafting, machining, inspection, inventory control, and other functions from a single computer.

NC control units

The NC machine control unit provides the bridge which connects the work of the programmer with the machine. It converts the information stored on a punched tape, floppy disc, or other medium, into instructions for the machine.

The most common NC control unit reads a punched paper tape. The unit uses one of three basic methods to convert the coded holes in the tape into directions for the machine control unit. An explanation of these conversion methods follows:

1. Light passes through the holes in the tape as the tape passes through the tape reader unit. The light is converted to electrical energy when it strikes a photoelectric cell. These pulses of electrical energy form the code for the control unit. This method is the most used.
2. Contacts in the tape reading unit protrude through the holes in the tape and complete an electrical circuit. The circuit is broken as the tape moves on, forming a code of electrical pulses.
3. Vacuum tubes under each hole convert the holes into a code. Whenever a hole appears over a tube, air is drawn in which reduces the vacuum in the tube. This change in air pressure is converted into an electrical signal.

Computer control units (CNC)

The computer numerical control unit generally uses a general purpose mini- or microcomputer to process data. The computer stores an application program in its memory. This program makes the computer "think" like a laser cutting machine, a lathe, or a machining center.

The application program gives the control unit its "software" title. The program, not direct wiring, gives the machine tool its operating instructions.

Most CNC units input their data from standard NC punched tape. The data is stored in the computer memory. This has several advantages including:

1. The tape is read once, then the information is used to produce as many parts as required. NC machines must read the tape for each part.
2. The program stored in the computer memory can be corrected or updated by manual (keyboards, dial settings, switches, etc.) entry. This editing feature is not available on standard NC.
3. The computer controlled system can become a part of a totally integrated system more easily than NC systems. Data exchanged from several computer control units can control machine tools, material handling units, and inspection devices so that they work in concert with one another.

The CNC control system has the major components shown in Fig. 35-11:

1. Operator interface.
2. Control unit.
3. Machine interface.

The operator interface unit is any unit which will input information into the control unit. These units place stored, coded data into the computer memory. Paper tapes are the most common method. This technique is often supplemented by an operator station which has video displays (CRT), dials, switches, and control devices, Fig. 35-12. Other interface devices can be magnetic tape or discs, punch cards, or a larger general-purpose host computer.

The control unit makes all the "decisions" during

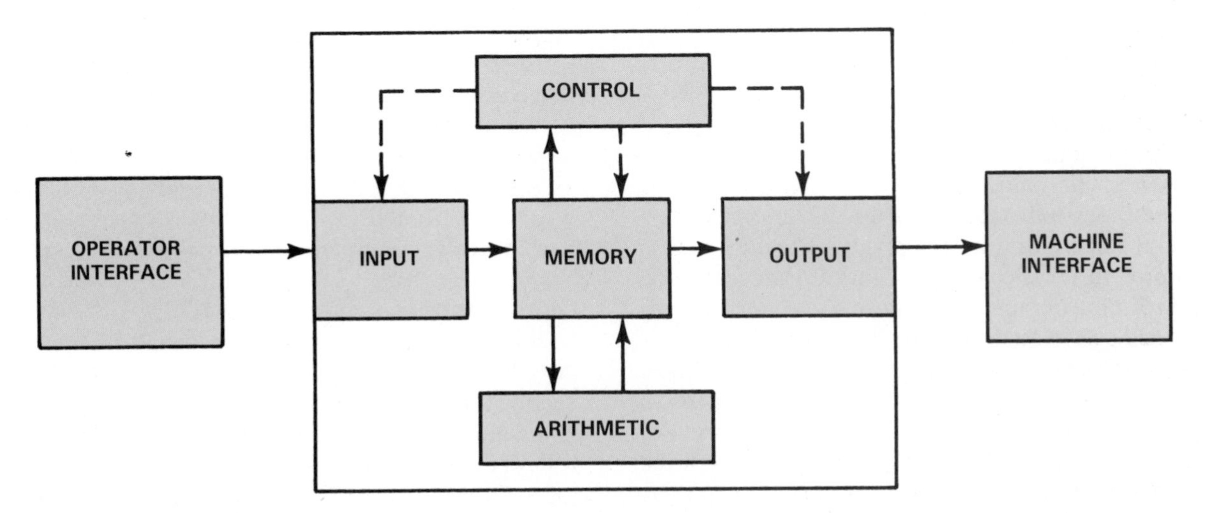

Fig. 35-11. Schematic of a computer control system. The major components are the operator interface which lets a person give information or commands to the control unit (which is the second component). The third component, the machine interface, carries out the commands processed through the control unit.

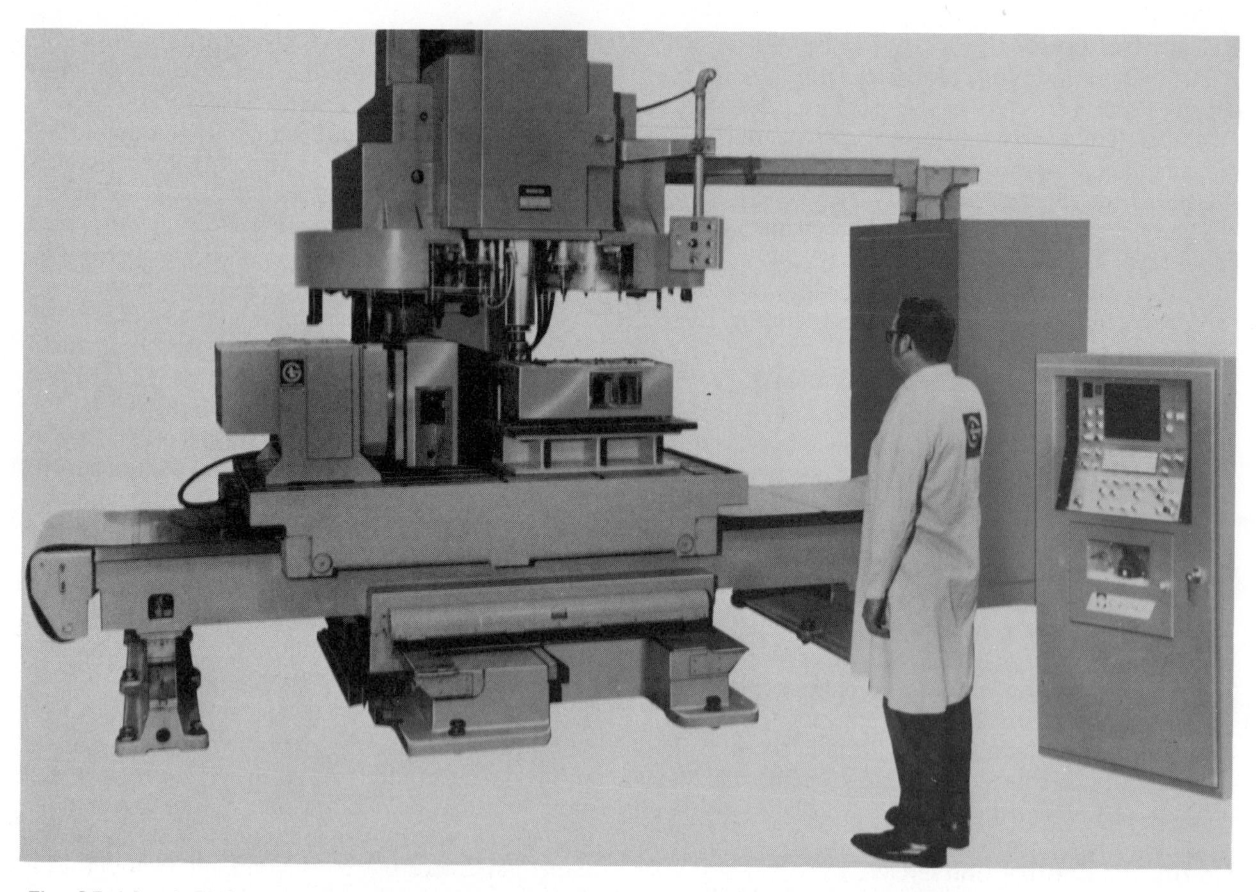

Fig. 35-12. A CNC machining center. Note the control unit behind the operator. The tape reader is in the lower panel and the operator station occupies the upper part of the unit. (Giddings and Lewis)

the operation. It changes speeds and feeds, positions the workpiece, directs tools to be selected and changed, controls the use of coolants, and performs many other tasks. The control unit, itself, has five major parts:

1. Input unit.
2. Memory unit.
3. Output unit.
4. Control unit.
5. Arithmetic unit.

The input unit receives all instructions and data from the operator interface devices. The information is stored in the memory unit until it is needed. The control unit obtains instructions from the memory unit and acts upon them. It sends orders to other units to cause them to follow the programmed instructions. The arithmetic unit completes any mathematical calculations needed to complete the instructions. The output unit transmits machine instructions from the computer control unit to the machine through the machine interface.

The machine interface devices monitor and control the machine's operation. These devices include:

1. Limit switches to determine the location of a machine member.
2. Servomechanisms, Fig. 35-13, to move machine elements such as the table and spindle. Components of servomechanisms include hydraulic or electric motors, gear trains, and transducers (convert energy from one form to another).
3. Temperature sensors to measure conditions throughout the machine.
4. Pressure switches to determine condition of oil, air, or other fluids.
5. Control valves to regulate flow of fluids.

Direct computer control (DNC)

The Electronics Industries Association defines direct numerical control (DNC) as any system connecting a group of numerically controlled machines to a common computer memory. A single computer will send instructions to the machines as they need them. Originally the DNC system was expected to run many machines from a single computer memory. However, CNC machines have their own memories. Therefore, modern DNC systems store all the applica-

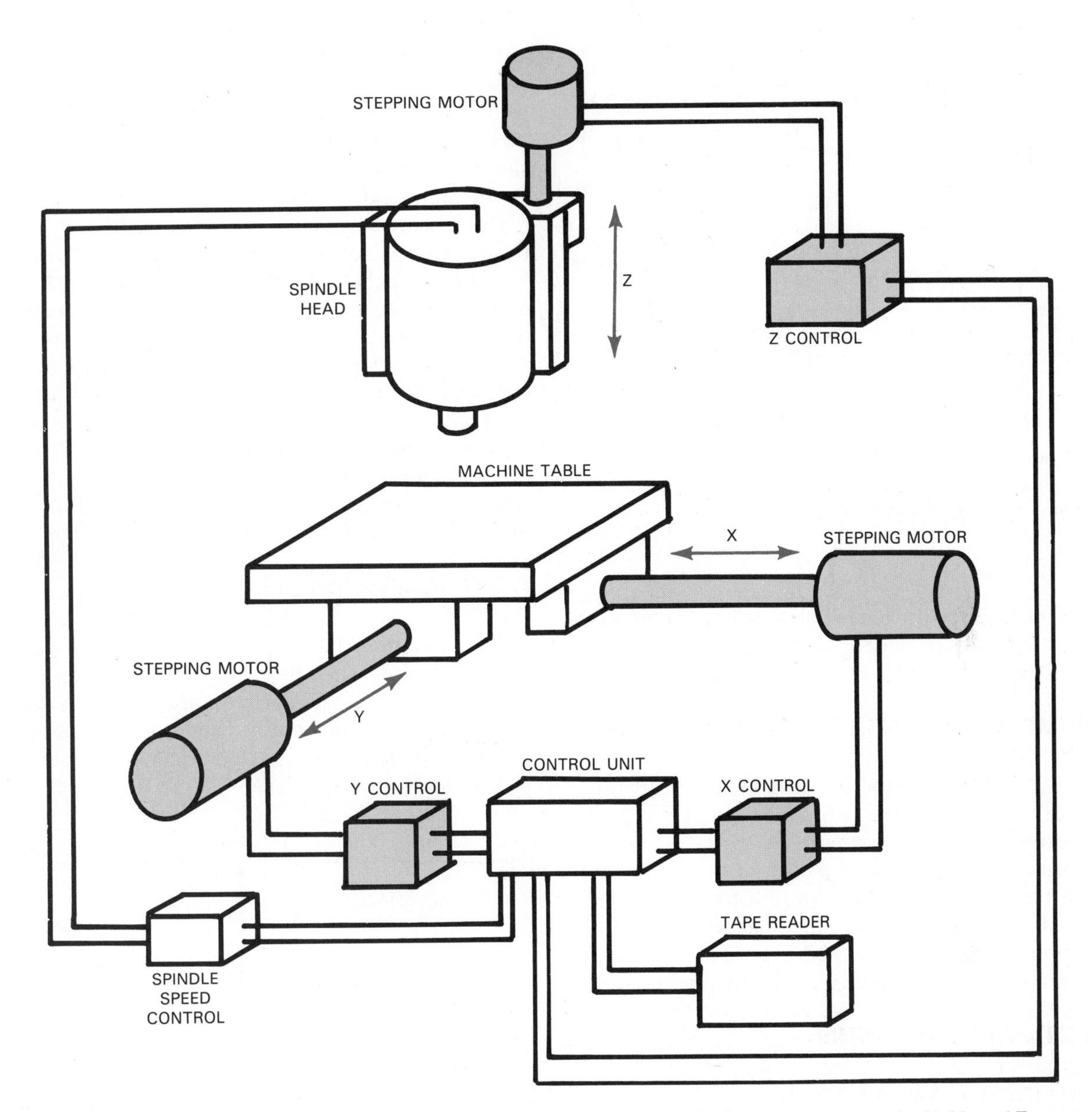

Fig. 35-13. Servomechanisms like these (in color) control table and spindle movements on the X, Y, and Z axes.

tions programs for producing parts and products in a central computer. Specific programs are down loaded (communicated) to individual machine memory units. The machines then produce the parts as programmed.

The central computer serves two major functions. It stores many application programs and distributes them to machines as needed. This has caused some people to change the meaning of DNC from direct numerical control to *distributed numerical control.*

Adaptive control

Another computer control system is called *adaptive control.* It is integrated with CN or CNC machines to enable them to produce products at their most efficient rate.

This system causes the machine to adapt itself to the conditions it encounters. For example, when a hole is drilled manually the operator uses a slow feed rate as the drill first engages the work. After the cut is started, a faster rate can be used. As the drill starts to break through the material, the feed rate is again reduced.

Adaptive control, through force, vibration, electrical current, voltage, or heat sensors will automatically adjust machine feed rates. The system, shown in Fig. 35-14, will keep the machine operating at its optimum level. The cutters will not stall and

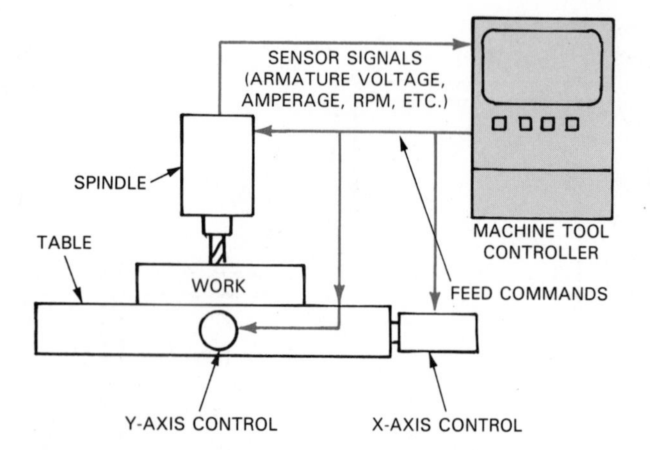

Fig. 35-14. Adaptive control system senses conditions such as vibration or heat buildup caused by the operation and adjusts feed rates to get the optimum cutting operation.

fewer tools will be broken in holes. Adaptive controls extend tool life, reduce tool breakage, and increase productivity.

NUMERICAL CONTROL PROGRAMMING

Numerical control machine programming capabilities can be divided into two major types:

1. Positioning (point-to-point).
2. Contouring (continuous path).

At one time these were two distinct types of machines. An NC machine either had point-to-point or continuous path capabilities. Today many machines are a combination of the two systems.

POSITIONING MACHINES

Positioning machines generally:

1. Perform an operation at a specific point (positioning).
2. Cut along a straight line between two points (point-to-point).

Positioning machines move the cutter over the point of operation while it is clear of the work. Then the operation, such as drilling, reaming, tapping, or boring is performed. Some machines position the work under the tool by use of the table cross slides. Others move the spindle to the proper location over the work. Still others move both the tool and the work.

The movement does not follow a predictable line. For example, a machine which moves the table for position uses two cross slides, each working independently. The NC unit moves each slide at its maximum rate to position the workpiece. The distance each slide must travel will determine the path.

For example, assume the slide on the X axis must move only two units while the Z-axis slide must move four units. The X-axis slide will complete its movement in about one-half the time taken by the Z-axis slide. Fig. 35-15 shows the approximate paths for some sample positioning moves.

Once the workpiece is in its correct relationship with the cutting tool, one or more operations may be performed. A single tool may be fed into the work. Or a turret may be used with several tools being rotated into position then fed into the work. Two-axis machines can position the work but require manual tool feed. Three-axis machines can both position the work and provide tool feed.

Point-to-point machines are special positioning machines which can cut a straight line along either of its two positioning axes or at 45° to the axes. The machine cannot adjust cutter depth during the cut.

CONTOURING MACHINES

The contouring machine can control the movement from three axes while the tool is in contact with the work. Typically these machines are milling or turning machines. They contour and profile the part. Other continuous path machines include flame cutters, welders, and grinders.

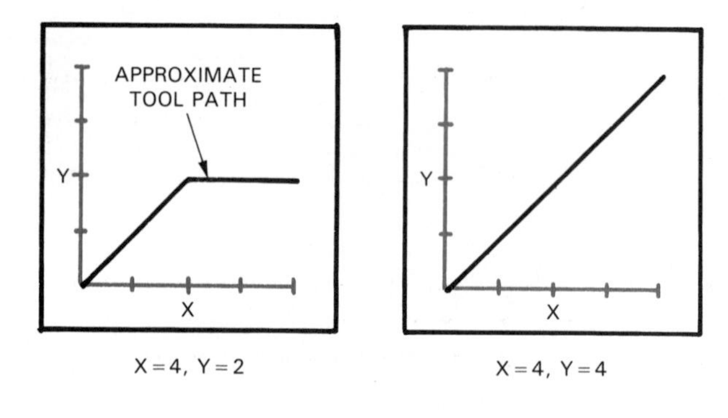

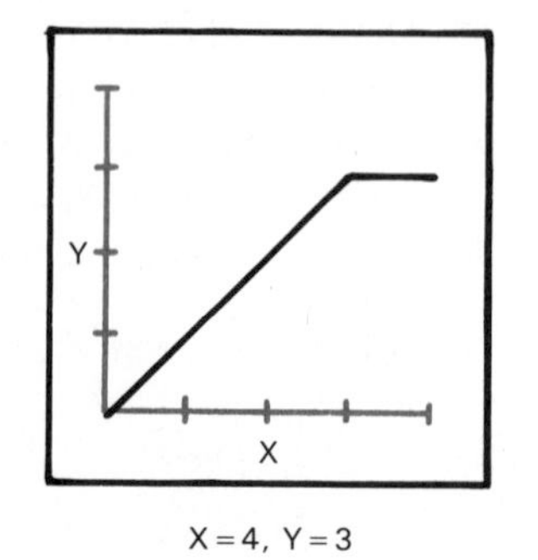

Fig. 35-15. Approximate paths for some movements by a NC positioning machine. Both axes move at their maximum speed.

Each axis of movement must be independently programmed for control. The systems are somewhat complex. It is virtually impossible to plot every point along a curved line which is to be held to within a tolerance of plus or minus 0.001 in. The X, Y, an Z coordinates would have to be plotted for each 0.001 in. of line length. Instead, the machine has a built in *interpolator* (interprets movement) which controls the shape of the line between established points. These interpolators may be linear, circular, or parabolic. For example, the linear interpolator changes the curve to a series of straight lines, Fig. 35-16. A circle interpolator converts the lines to a series of complete or partial circles while a parabolic interpolator uses complete or partial parabola. The use of the interpolator allows the programmer to plot fewer points along the line of cut. The distance between established points and the interpolator used will determine how close an actual cut is to the desired shape.

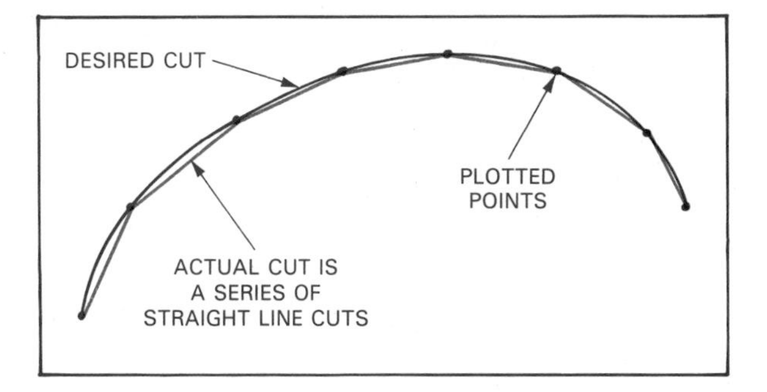

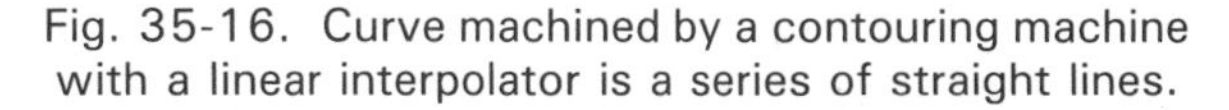
Fig. 35-16. Curve machined by a contouring machine with a linear interpolator is a series of straight lines.

AUTOMATIC TOOL CHANGING

A common feature of modern numerically controlled machines is automatic tool changing. Machines which automatically complete a series of operations will often require several different tools. This is particularly true with the newer machines such as the

Fig. 35-17. A four-axis machining center. Note the part on the auxiliary table in the foreground. (Giddings and Lewis)

machining center. This machine, Fig. 35-17 , is designed to face and end mill, contour, profile, drill, tap, bore, counterbore, and ream.

These machines, along with other NC and CNC machines, are often equipped with an automatic tool changer, Fig. 35-18. The changes vary from manufacturer to manufacturer but will follow a procedure similar to the one shown in Fig. 35-19. This procedure has the changer arm:

1. Select the proper tool in the drum or cassette.
2. Move the tool to machine spindle.
3. Remove old tool from the machine spindle.
4. Rotate new tool into position.
5. Insert new tool into the machine spindle.
6. Move the old tool back to the storage drum.
7. Store old tool in the storage drum.
8. Await next tool change.

Automatic tool changing adds greater flexibility to all numerically controlled machines.

AUTOMATIC MATERIAL HANDLING

Mechanized material handling was one of the earliest efforts in automating the factory. Oliver Evans' automatic flour mill had a series of interconnected machines. The wheat was conveyed to the top floor of the mill. Then chutes and conveyors moved the material through the milling, sifting, cooling, and bagging operations.

Today a wide variety of material-handling devices are in use. They may be used to move parts and material from machine to machine, from storage to processing areas, and from processing areas to storage. Material handling devices also play an integral role in using numerical controlled machines. They are used to load machines, move parts through stations on complex machines, and to unload machines. The typical material-handling devices, as shown in Fig. 35-20, are:

Fig. 35-18. A horizontal machining center. Note tool changer and storage magazine on the left. (Kearney and Trecker)

A. When program calls for new tool, storage drum rotates until the needed tool is next to the changer arm. Meanwhile, the spindle head moves into a position to receive the new tool.

B. Jaw-type tool gripper prepares to take the tool out of the storage drum.

C. Changer arm transports tool along conveyor track which runs between the drum and the machine spindle.

D. Changer arm continues along track until it reaches the tool change position.

E. Changer arm grips the shank of the tool in the spindle with its second set of gripper jaws.

F. Spindle retracts to give clearance for removal of the old tool.

G. After spindle is clear of shank, tool changer arm rotates 180 degrees.

H. Changer arm positions new tool for insertion in the spindle.

I. Spindle advances to receive the new tool.

J. Changer arm jaws release the tool now secured in the spindle. Old tool is shuttled back to the storage drum where it will be stored in its assigned spot. This completes the cycle.

Fig. 35-19. The operation of a typical tool changer. (Lucas Machine Div., Litton Industries)

A

D

B

E

C

F

Fig. 35-20. Common material handling devices include these. A—Gravity conveyors. (Owens-Illinois) B—Belt conveyors. (Lenox China). C—Powered roller conveyors. D—Overhead cranes. (Aluminum Assn.) E—Carousel conveyors. (Aluminum Assn.) F—Robots. (Cincinnati Milacron, Inc.)

1. Machine-to-machine devices.
   a. Gravity conveyors: roller, wheel, chute.
   b. Powered conveyors: roller, belt, chain.
   c. Overhead cranes and monorails.
   d. Circular or carousel conveyors.
   e. Elevators: bucket and tray.
   f. Pipes and augers.
2. Machine-loading devices.
   a. Hopper and magazine.
   b. Transfer devices.
   c. Robots.

The machine-to-machine devices were the first type of material-handling mechanisms developed. They have been used for many years and are familiar to most people. The machine-loading devices are of more recent design.

Hopper feeds are a common machine-loading, material handling-devices. They position and hold parts in an orderly fashion.

One major type of hopper feed uses rotating or vibrating drums. These drums cause the parts to move or tumble until they enter an opening, Fig. 35-21. The parts are then moved into a machine feed channel.

Fig. 35-22. Magazine type hopper on an automatic wood lathe. (Goodspeed Machine Co.)

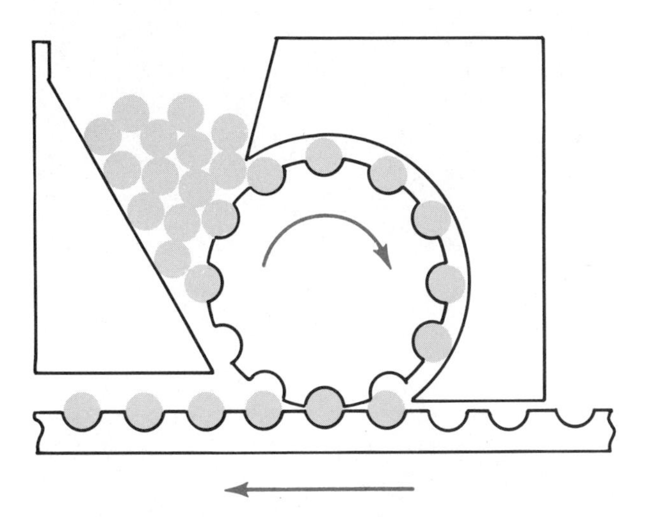

Fig. 35-21. One type of rotary feed hopper. Slots in drum pick up parts and deposit them on the conveyor.

A second hopper feed requires the parts to be manually loaded. This type of feed, often called a magazine, holds the parts in position for later use. The parts are held much like bullets in a pistol magazine, Fig. 35-22.

Transfer devices are used with complex special purpose machines called a transfer machine. The transfer machine is made up of a number of individual machine units. Each unit performs one or more operations on a part. The parts are moved from station to station by an indexing power unit. Refer, again, to Fig. 35-2. It shows a scale model of a plant layout with transfer machines.

Robots are one of the newest of all devices for automating plants. A robot is a programmable part handling or work performing device. Robots are easily programmed to do repetitive tasks.

The simplest robot is a pick-and-place device. It will *pick* up a part from a set location and *place* it in another set location. Pick-and-place systems are used widely for machine loading, Fig. 35-23, and assembling.

Fig. 35-23. A pick-and-place robot is being used to load and unload parts into a lazer welding machine. (Coherent Industrial Systems)

More complex robots can have up to six motions. The arm can move up and down and swing right and left as shown in Fig. 35-24. In addition, the wrist rotates on the X, Y, and Z axes, Fig. 35-25.

Robots can replace many less desirable routine activities. Such unchallenging and/or dangerous jobs as loading machines (Fig. 35-23), spraying paint, Fig. 35-26, handling hot or hazardous materials and welding are easily performed by robots.

SUMMARY

The selection of manufacturing processes will be greatly influenced by the use of automatic equipment. The parts of the truly automatic factory are being developed and refined. The production facility which is totally controlled by computers may soon be commonplace. Computer controlled machines, assembly devices, material handlers, and tool changers are already developed. Flexible manufacturing systems are already a reality. The future is bright for the companies which use modern management and material processing techniques.

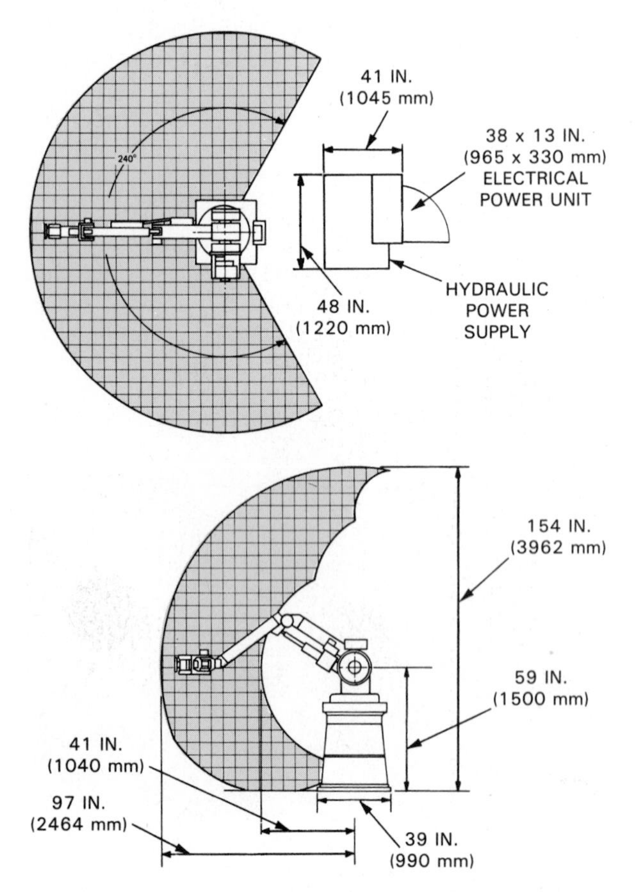

Fig. 35-24. A top and side view of the arm reach flexibility of a robot. (Cincinnati Milacron, Inc.)

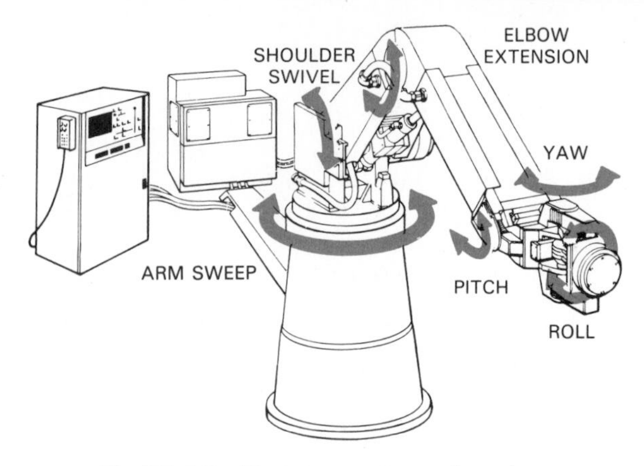

Fig.35-25. The movements of a robot. (Cincinnati Milacron, Inc.)

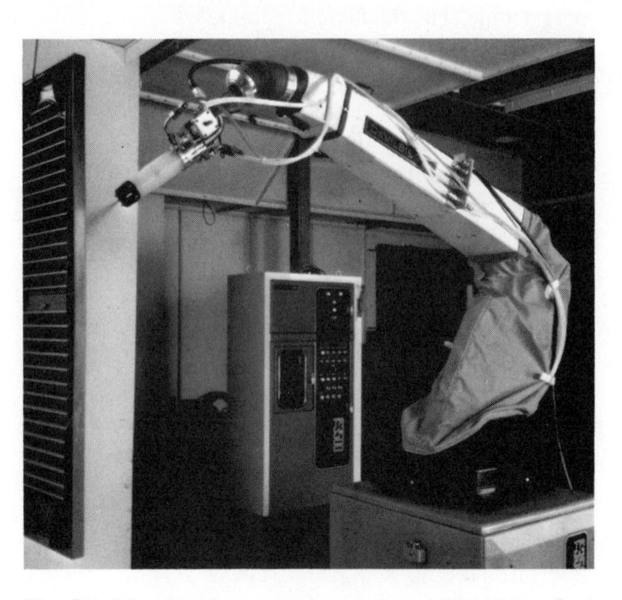

Fig. 35-26. A robot spray system. (DeVilbiss Co.)

STUDY QUESTION — Chapter 35

1. Define the terms, fixed automation and programmable automation.
2. What do the symbols NC, CNC, and DNC mean?
3. List and describe the three distinct parts of an NC machine.
4. What are the three major axes controlled on an NC machine?
5. Describe the three methods an NC machine control unit uses to read a tape.
6. List three advantages of a CNC unit over an NC unit.
7. What is adaptive control?
8. What is the difference between positioning machines and contouring machines?
9. Describe the operation of a typical automatic tool changer.
10. List the typical material-handling devices.

Chapter 36

PLANNING FOR QUALITY

The controlling of quality is directly related to the processing of materials. The producing of industrial materials and finished products is of little value if the output fails to meet a need. It must have a fitness of use. It must have quality.

Quality does not mean precision. It means that the fineness built into the product matches the intended use. A paper cup which will hold liquids for 30 days is overengineered. It does not fit the intended use of holding a liquid for a short period of time. The 30-day cup would probably cost more to produce and therefore fail because suitable cups could be purchased cheaper.

Microchips, nails, toothbrushes, and toasters all have a required quality. No two requirements are the same, but all are matched to the product's intended use, Fig. 36-1.

The developing of a product which meets quality standards requires several activities. These activities, shown in Fig. 36-2, include:

1. Designing.
2. Manufacturing.
3. Using.

DESIGNING FOR QUALITY

The design is a critical force in maintaining quality in a product. Poor design decisions can reduce the quality of the product or increase the reject rate during inspection.

Designers must have quality in mind as products are developed. The physical structure of the product must be such that quality can be maintained. Care must be exercised to ensure that the product will be functional. The design must have the strength, rigidity, or other features required.

The design should also support manufacture. The shape or contour of the surface should allow for accurate processing. The shape should consider draft for castings, flats for drilling, Fig. 36-3, surfaces for clamping in tooling, to name only a few examples.

Fig. 36-1. The quality requirements of the carbide cutting tool on the top are more rigid than for the jars coming off the production line on the bottom. (Kennametal and Glass Packaging Institute)

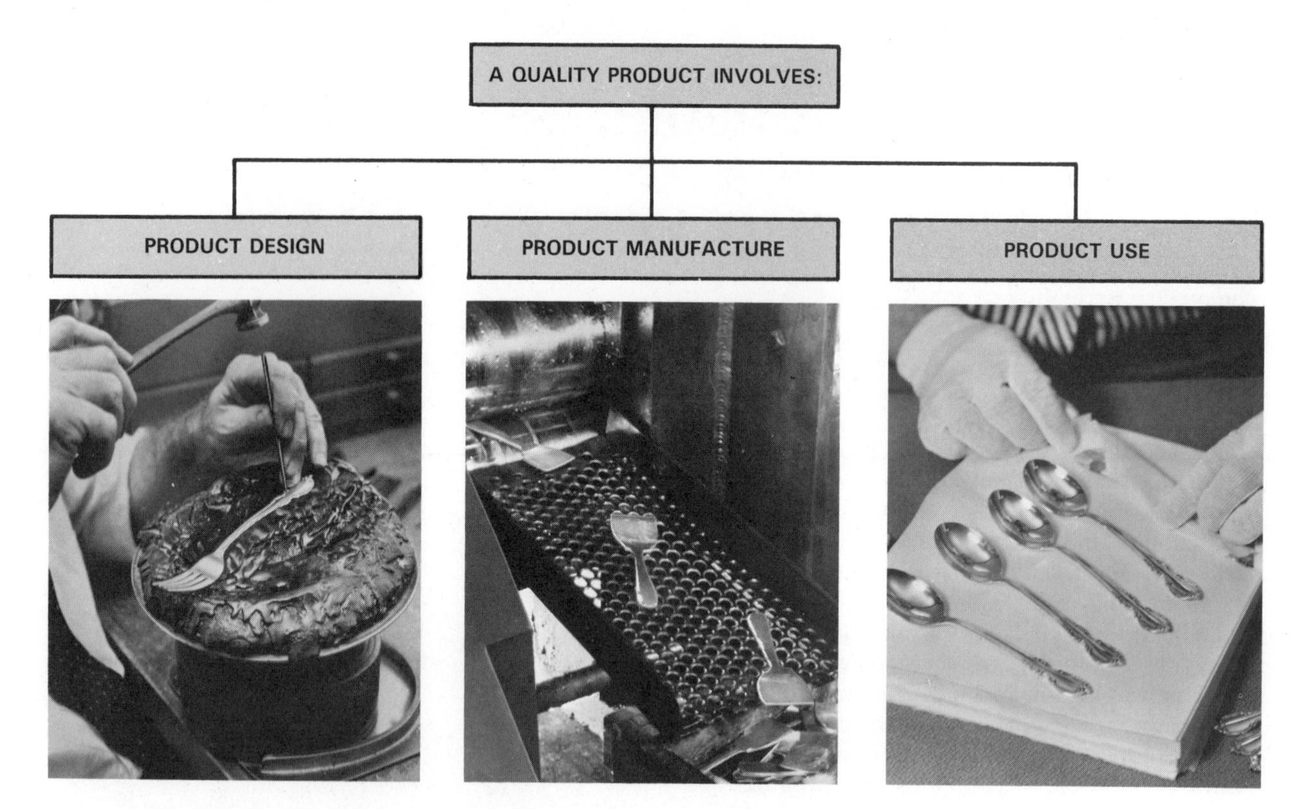

Fig. 36-2. Areas of quality control.

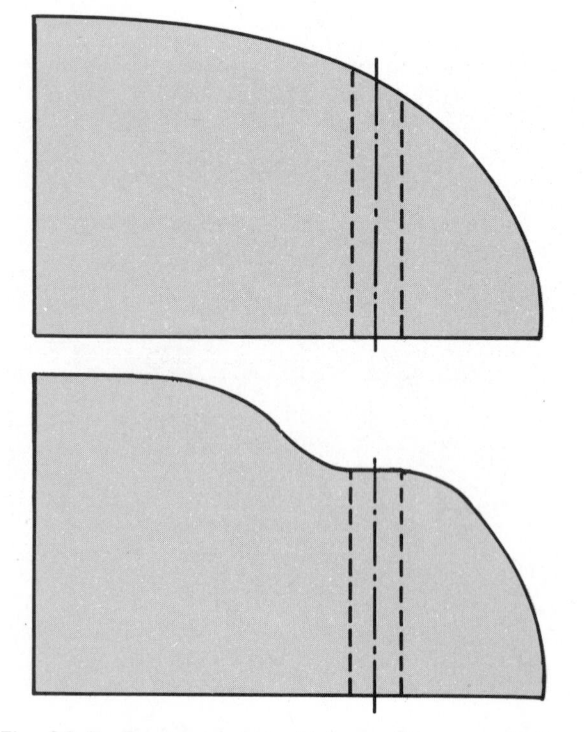

Fig. 36-3. Designs must aid in quality control. The lower design provides a flat area for the drill to start its cut. The upper design would cause the drill to wander off the starting point.

MANUFACTURING TO QUALITY STANDARD

Maintaining quality during manufacture involves two major tasks: motivating workers to produce quality products and controlling the products.

MOTIVATING FOR QUALITY

The concern for quality production has spread from inspection to a more complete program. It has become evident that merely removing defective products from the production line is not enough. "Inspecting out" defects is economically less efficient than "building it right for the first time."

This realization has given birth to quality control motivation programs. Until recently programs have been management centered. Management has established and operated the systems. Badges, slogans, posters, and other materials were furnished by management. They also organized and conducted motivational meetings. The worker received the information and hopefully became motivated. The goal was to convince the worker that quality was important. These programs were operated under a number of names such as "zero defects." However, they were less than totally successful.

A more recent development are "quality circles."

This concept was first developed, but not used, in the United States. The Japanese adopted and refined the process. It was then reintroduced into American industry.

With quality circles, management and workers work together to meet company goals. A quality circle is a group of employees who meet regularly. They are volunteers who want to help the company succeed. The quality circle identifies, investigates, and solves problems.

The use of quality circles has many benefits:

1. Higher quality output.
2. Increased productivity.
3. Lower costs.
4. Reduced scrap and waste.
5. Better communications.
6. Greater job satisfaction.
7. Improved worker attitudes.

CONTROLLING PRODUCTION

Highly motivated workers will produce better products. However the system must still be controlled and monitored (checked). Quality checks must be made.

The quality control of production, as shown in Fig. 36-4, involves three major phases.

The *materials* coming into the system must be subjected to quality control, Fig. 36-5. The purchasing or procurement department of the company must first order the proper materials. Drawings and specification sheets must be checked to determine material

Fig. 36-5. Clay is being checked before entering a chinaware production line. (Syracuse China)

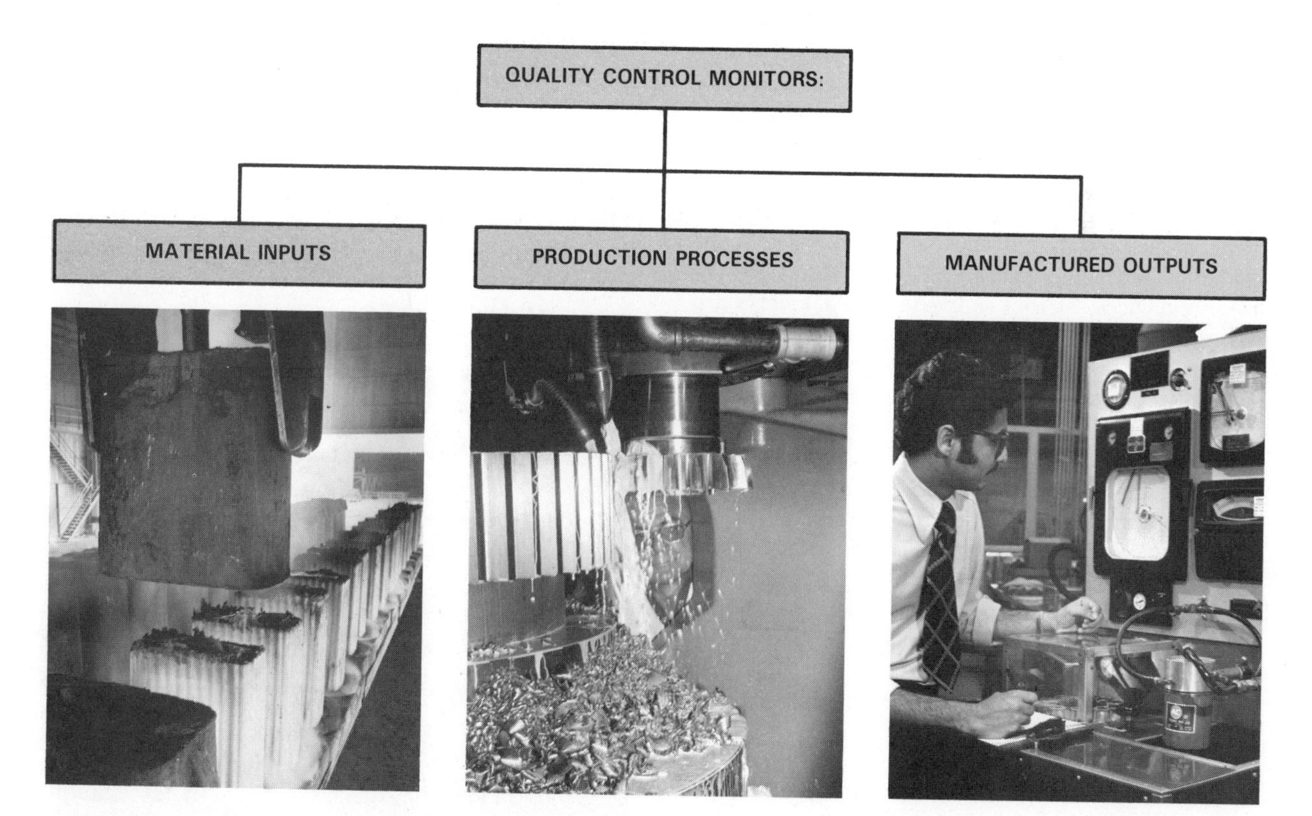

Fig. 36-4. Three areas are involved in maintaining quality standards. Left. Incoming raw materials must be inspected. Defective materials are scrapped or returned. Center. Manufacturing processes must produce products to engineering standards. Right. Finished products, like this oil filter, must be measured and/or tested to see if they meet the standards. (USX, Fellows Corp. and AC Spark Plug)

needs. These materials are then ordered. Upon arrival, the materials must be inspected. Defective materials are returned or scrapped. Only materials which meet specifications should be released for manufacture. Obviously, a quality product cannot be manufactured from substandard materials.

Controlling manufacturing processes involves holding production within the stated standards. These standards are a result of design and engineering activities. Drawings and specification sheets were developed. These communicate sizes, shapes, tolerances, fits, and allowances. These must be produced or maintained during manufacture.

To insure the standards are kept, quality control inspection procedures are used. They compare manufacturing results against established standards. This may involve nondestructive practices such as gauging, photographing through X-rays, and sonic measuring. None of these practices damage the product. In destructive testing, some parts or products may be destroyed. The parts may be subjected to tensile, abrasion, or hardness tests. Products may be operated until they fail. The test results are used to ensure that the product will work.

Testing will find faults. Most companies hope there will be few and that they are simple manufacturing errors rather than complex design problems. The error caused during the processing phase come from two major sources:

1. Direct machine variance.
2. External variance.

Machine errors may be a result of wear, misalignment, or lack of rigidity. The machine may flex under load or it may have play in its feed systems, workholding devices, or tool supports. These factors are basically beyond the control of the operator. Maintenance is usually required to solve the problem.

Other product variations have nothing to do with the machine. They are outside or external to the machine itself. These factors include poor tool design, tool wear, variation between tools of the same size, improper cutting speeds and feeds, variations in the workpieces, and operator error. All those factors require operator decisions. Worn tools need replacing. Speeds and feeds must be correctly set. Poorly designed tools must be returned with suggestions for change. In short, judgments have to be made.

The control of quality during production requires four steps, Fig. 36-6:

1. Interpreting of standards.
2. Inspecting of parts, assemblies, and products.
3. Reporting results from inspections.
4. Correcting manufacturing actions as needed.

Inspections may be made by sensors or by human action. Sensors may check dimensions, measure

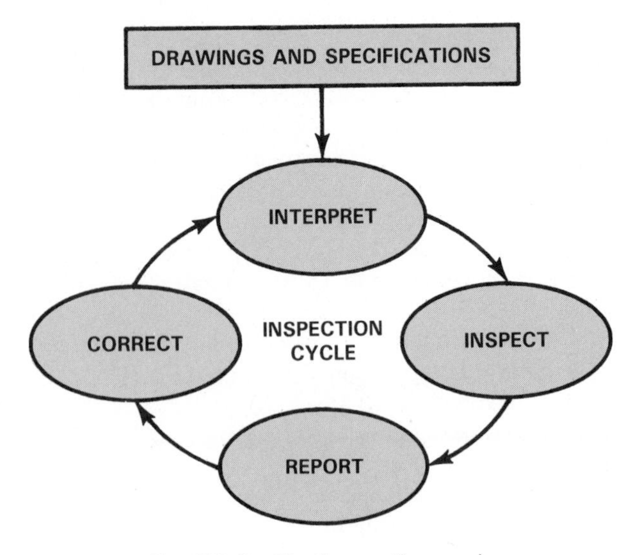

Fig. 36-6. The inspection cycle.

volume, or determine weight. These actions are usually continuous and are reported to a central control room, a data processing center, or the machine control unit. These reports provide the information for needed corrective action. Automatic machines may correct themselves or shut down if product quality is out of standard. Individuals monitoring the reports on dials, gages, or other readouts may make the correction.

The output of manufacturing processes may also be inspected by specially trained employees, Fig. 36-7. The inspectors gage the products and prepare reports. Production managers review the reports and implement necessary corrective actions.

Fig. 36-7. Specially trained employees inspect china tableware before final glazing. (Syracuse China)

Many human inspections are not continuous. Instead, a part of the total output is inspected. The process is called sampling. It is based on mathematical laws of probability. If a sample of the total production run is inspected, an estimate of the total reject or poor part for the run can be made. This total is not actual but an estimate. As long as this estimated reject rate is within predetermined limits, the entire run is accepted. If not, additional sampling may be made. Unless the reject rate is within the limits the whole run or lot is rejected.

Sampling is used in high volume production. The cost of inspecting every nail, screw, or corn flake would be too high. It would cost more to inspect the products than to throw away an occasional defective lot. Complex products such as computers, automobiles, and heat pumps will be more thoroughly inspected.

The level of inspection is a balance between the cost of inspection and the costs and potential hazards caused by products which fail. We are more ready to accept a screw without threads than an aircraft engine which stops running.

PROPERLY USING PRODUCTS

The quality of a product is affected by its use. Products must be shipped, installed, used, and maintained with quality in mind. Products will fail when abused. Engines require oil for lubrication. Lawn mowers are not designed to cut rocks. A household sewing machine will not withstand constant use in an upholstery shop. Every product has maintenance requirements and use limitations. These must be communicated to the customer. Owner's manuals provide this information.

SUMMARY

Quality control is a process that insures all products fit their use. It involves designing a product with proper specifications. These specifications are communicated to production personnel. They design a plant and select processes which should produce products that fit the specifications. The output is then measured. The results of these inspections are reported to management. Necessary corrective action is taken on the basis of the reports.

Quality control ensures the products will function. It involves designing for quality, manufacturing for quality and using products properly.

STUDY QUESTIONS – Chapter 36

1. Name the three activities which are part of meeting quality standards for a product.
2. List the two major tasks involved in maintaining quality during manufacture.
3. What is a quality circle?
4. What are the benefits of a quality circle?
5. List the four steps of quality control during production.

SAFETY IN MANUFACTURING

ACCIDENT PREVENTION

There are two aspects of safety to be considered in preventing accidents and injury while carrying on the processes of manufacturing. One is providing a safer environment in which to work. One part of this is proper housekeeping; another is providing guards and other barriers that keep workers' hands, feet, and other parts of their bodies away from machine parts or operations which could cause injury or death.

Just as important is the second aspect which is training of workers and securing their cooperation in working safely. All the safeguards in the world will not prevent accidents unless the people in such an environment are trained and willing to observe safe working provisions.

BEFORE THE JOB STARTS

No person should begin a job until he or she has been properly trained in safety. The plant manager must make sure that all employees know what hazards exist while working with the materials and equipment in the plant. They should also know how the hazards are controlled or eliminated.

Everyone should know that:

1. No one is expected to perform any job or task until properly instructed on how to do it.
2. No one should undertake a task that appears to be unsafe.
3. Safety devices, both mechanical and electronic, must be kept in place when equipment is operated.
4. Everyone is responsible for reporting unsafe conditions to management.
5. Accidents or illness, however slight, should be reported at once.
6. Observance of some safety rules and wearing of some safety equipment or clothing is a condition of employment.

PLANT CLEANUP

Proper housekeeping in the plant should include:

1. Collection and disposal of rubbish.
2. Providing proper trash containers.
3. Ensuring proper disposal and/or storage of flammable materials.
4. Maintaining unobstructed exits.
5. Providing adequate lighting for the tasks in which workers are engaged.
6. Setting up of neat, well-planned workstations.

METAL FORMING SAFETY

Bone fractures and accidental amputations are common injuries among workers in metal-forming industries. The typical accident occurs as the operator reaches into the tool area while loading, unloading, or holding a part being processed. Accidental finger amputations are most often caused by a failure of the operator to use a brush or an appropriate hand tool. Often, gloves, or long, bulky clothing catch in tooling and cause injury. Burrs and sharp edges of workpieces can cause severe cuts necessitating the wearing of gloves (in stamping operations, for example).

Machines should be designed and selected for the job. The forming operations should be planned to eliminate the need for workers to reach into the area where the tool is operating. If this is not possible, provision should be made to guard the machine so that operators cannot place a hand in the machine area when a slide or ram or other device begins to operate.

Operating controls should be convenient, clearly identifiable, and, except for "stop" controls, should be protected from accidental activation. Controls should be located where it is safe for the operator to activate them.

Inspection of guards and other safety devices

should be included as part of a regular maintenance program. Maintenance of these items should include proper adjustment, replacement of critical parts before failure, and the keeping of maintenance records.

Safety measures are regulated today by the Occupational Safety and Health Act (OSHA). Stamping, forging, and pressure operations may have to be engineered so that no hands are in the die or in the stamping area during operation. Several methods or devices may be used to provide this type of safety measure. Among these are:

1. Straps or cables that pull the operator's hands to safe distances once the process is initiated.
2. Automated feeds.
3. Two-handed controls.
4. Shields.

OPERATOR SAFETY IN FORMING OPERATIONS

Casting

- Additional metal dropped into a molten charge will produce splashing, resulting in horrible burns. Never attempt it.
- Guard against accidental contact between water and molten metal. It will result in a violent explosive reaction.
- Wear protective clothing and goggles when pouring a casting.
- Use care in placement of hot castings. They can cause severe burns and may start fires.
- Keep the foundry area clear and clean.
- Never stand directly in front of a mold during pouring. Steam, generated during pouring, can cause severe burns. A too-moist mold can cause a molten metal to spurt from the mold.
- Molds for large castings should be clamped or weighed down to prevent floating which could result in the molten metal escaping from the mold at the parting line.
- Never pour molten metal into a plaster mold until certain that all traces of moisture have been removed from the mold.

Forming hot and cold metals

- When handling stock, look out for burrs. If possible remove them beforehand.
- Wear gloves to protect hands.
- When using gas-fired heating equipment such as forges, follow the manufacturer's safety precautions for usage. In general:
  1. Make certain the gas valve is closed.
  2. Start the air blower and open the air valve slightly.
  3. While applying the lighter (never use a match), open the gas valve slowly. Stand to one side and do not look into the forge during this operation.
  4. After the forge is lighted, adjust air and gas valves for best heat-up speed and most economical operation.

Soldering and brazing

- Always wear approved eye protection against flux or spattering of hot materials.
- Avoid touching just-soldered or brazed joints.
- Provide good ventilation during these processes.
- When mixing acid-water solutions, always pour acid slowly into water; never water into acid.

Welding

- Protect eyes during all welding operations. Wear an approved mask or goggles.
- Never light a torch with a match; use a spark lighter.
- Torches should be lighted following safety instructions for the type of gas welding system in use.
- Wear welder's gloves, leather apron, and long sleeves. Do not wear canvas shoes.
- Keep flammable materials away from the welding work area.
- Use extreme care in handling just-welded materials.
- Arc welding should not be attempted without using a welding shield. To do so is to risk serious and permanent eye damage.

OPERATOR SAFETY IN MACHINING OPERATIONS

- Avoid wearing loose clothing, neckties, and jewelry while operating machine tools.
- Remove chips with a brush, not the hands.
- Work only with sharp tools.
- Clean up spilled cutting fluids.
- Wear safety glasses.
- Keep hands clear of blades and other cutting tools.
- Do not operate equipment unless all guards and safety features are properly installed.
- Treat all injuries promptly.
- Clamp workpieces securely.
- Pay attention to what you are doing. Never turn your attention away from the machine while it is operating. Do not distract others who are operating equipment.
- Turn off the machine before making adjustments.
- Never reach over or around rotating tools.

TECHNICAL TERMS

A

Abrasive jet machining: Similar to sand blasting processes, this process propels fine abrasive grains at speeds of 500 to 1000 feet per second against the work.

Adhesive bonding: Use of glues or cements to hold two parts together.

Alloy: A pure metal which has additional metal or nonmetal elements added while molten.

Aluminum oxide: Abrasive formed by heating bauxite ore in the presence of small amounts of coke and iron filings at about 3510°F.

Annealing: Softening of meals by heat treatment.

Anodizing: A conditioning process which applies an oxide coating to aluminum. It is done electrolytically in an acid solution with equipment similar to that used for electroplating.

Arc welding: A group of welding processes used to melt and weld metal. They use the heat of an electric arc with or without filler metal.

Atomization: In powdered metallurgy, a primary process that breaks up metal into small, solid particles.

Automatic bar machine: Another name for an automatic screw machine which was first designed for high-volume production of screws and other threaded fasteners.

Automatically programmed tools (APT): Computer program to control machine tools.

Axis: A real or imaginary centerline passing through an object about which the object could rotate.

B

Barrel finishing: Polishing of small parts or products by placing them in a six or eight-sided barrel with an abrasive media.

Bending stress: Stress from both tensile and compressive forces that are not distributed uniformly. Resistance to bending may be called stiffness.

Bessemer Converter: A basic steel-making furnace. The process utilizes a furnace in which molten pig iron is refined by a burning gas.

Billet: A piece of semifinished iron or other metal made by rolling an ingot or bloom.

Binary: A numbering system used in computer language utilizing a base of two.

Blasting: Method of cleaning or surface roughening. Accomplished by projecting a stream of sharp angular abrasives against the material.

Bloom: A semicompleted metal form in which the cross-section is relatively square.

Blow molding: Plastic forming process; softened plastic tube is inflated to fill a cavity.

Bonding: Fastening parts together by either adhesion or cohesion.

Brine: A salt water solution which is occasionally used as a quenching medium in the cooling of steel.

Brinell hardness: Accurate measure of hardness of metal made with a instrument. Measurement is made as a hard steel ball is pressed into the smooth surface at standard conditions.

Brittleness: Quality of a material. It causes the material to develop cracks with little bending (deformation) of the material.

Broaching: An internal or external machining process for flat, round, or contoured shapes. The tool has a series of teeth that gradually increase in height.

C

C-frame press: A style of press in forging representing one of two major designs. Shaped like the letter "C," it allows access to the dies on three sides.

Calendaring: A plastics process which produces thin sheet material or coats a material with plastic. The sheet is formed continuously.

Carbon: An element which, when combined with iron, forms various kinds of steel.

Casting and molding: Family of manufacturing processes in which an industrial material is usually changed to its liquid or plastic state before it is poured into or injected into a mold cavity and allowed to solidify.

Cellulose: One of the two major components making up wood. It is the solid matter.

Centrifugal casting: A casting process which uses spinning force to place the molten metal into the mold.

Ceramic pressing: Process of shaping dinnerware, electrical insulators, and art objects using high pressure (up to 10,000 psi).

Ceramics: Complex crystalline compounds in which unit cells often contain three or more different atoms. Includes important materials such as the clays, the glasses, cement, and concrete.

Chemical milling: A type of chemical machining that can remove large amounts of materials from part or all of a workpiece. The process includes cleaning the part, masking off the area not being machined, etching the unmasked areas, and removing the mask.

China. Translucent clay product containing large quantities of quartz, clay, and feldspar crystals.

Chip removing tools: Tools used in separating processes. These include lathe tools, shaper and planer tools, drills, reamers, taps, broaches, milling cutters, jointer and planer cutting heads, saw blades, router and shaper cutters, grinding wheels, abrasive belts and discs, and abrasive cutoff discs.

Cladding: Somewhat thick layer of material applied on a surface. Application improves resistance to corrosion.

Climb milling: Method of feeding work into the milling cutter in the same direction as tool rotation.

Closed loop system: In computer operation, a system in which the output is fed back for comparison with the input.

Coating: A relatively thin layer of material applied to a part's surface to prevent corrosion, wear, or temperature scaling.

Cofusions of aluminum oxide and zirconium: A manufactured abrasive made by heating process. It is effective in grinding stainless steels.

Cohesion: Sticking together through attraction of like molecules of a material.

Cohesive bonding: Fastening of parts together through welding. Bonding occurs through attraction of the molecules of the materials.

Coining: Closely related to sizing, it gives a material size and a surface impression.

Coke: A purified form of coal used in the manufacture of iron and steel.

Cold welding (CW): Use of high pressure alone to force metal parts to fuse.

Cold work: Metal part given a permanent strain by an outside force while the metal is below its recrystallization temperature.

Cold working: Bending (deforming) force placed on a temperature lower than its recrystallization temperature.

Collet: A holding device for stock which must be held very accurately in a lathe.

Compound: A material composed of two or more elements that are chemically joined.

Compressive strength: Greatest stress developed in material while under compression.

Computer-aided design (CAD): A drafting system which uses a computer to create and/or modify a design.

Computer-aided manufacturing (CAM): A system using computer to control a manufacturing process.

Computer numerical control (CNC): Using a dedicated computer to control some aspects of a numerically controlled machine tool.

Conditioning processes: Operations which improve strength, toughness, or other mechanical properties needed by the product to serve its purpose.

Continuous casting: Method of casting metal in an open-ended mold. Metal is fed into and cools in the mold in a continuous form.

Continuous manufacture: Assembly-line production. Company produces one product over a long period of time using assembly-line method.

Coordinates: The position or relationship of points on planes; usually refers to Cartesian Coordinate System.

Core: A body of sand or other material that is formed to a desired shape and placed in a mold so that it produces a cavity or opening in a casting.

Counterbore: An operation which enlarges a hole to a given depth and diameter.

Converted surface finishes: Thin layers of materials which are actually part of the metal part or product. The material is chemically treated to change the nature of the outside layer.

Copolymerization: The chemical binding of unlike molecules into a larger molecules. Many new plastics are of this type.

Corrosion: The reactions between the chemicals and a material in an enviroment.

Creep: The flow or plastic deformation of metals which occurs after long periods of stress below the yield strength.

Croning process: An old name for the shell mold

process.

Cubic boron nitride (CNB): An abrasive formed synthetically under high pressure and temperature.

Cupola: Blast furnace shaped like a vertical cylinder; used in making gray iron.

Custom production: Products are produced only as orders are received for them.

Cutter path: The path taken by the center of a cutter throughout a program.

Cutting fluid: A liquid designed to cool and lubricate the cutting tool. It improves the quality of the surface finish, as well.

Cyaniding: A process of case-hardening a ferrous alloy by heating it in molten cyanide. This causes the metal to absorb carbon.

D

Dendrite: Crystal development often found in cast metals as they slowly cool through the solidification range.

Deposition rate: Weight of material applied in a unit of time, usually expressed in lbs/hr or kg/hr.

Depth of fusion: Depth to which base metal melts during welding.

Diamond abrasives: Either natural or manufactured, the hardest of all materials. Natural stones unsuited for gens are crushed; synthetics are manufactured specifically for abrasives.

Die casting: A mold casting technique in which molten metal is introduced into a metal mold or die. High presures—up to 100,000 psi—are used.

Die forging: Parts shaped by use of dies in forging operations.

Down milling: Milling operation in which the cutter rotates in the same direction as the feed.

Drag: Bottom half of the flask used in green-sand mold production.

Drain casting: Also called slip casting. A process by which a suspension of clay in water is introduced into a mold and allowed to dry until a proper wall thickness is achieved. Then, remaining slip is poured out of the mold.

Drape forming: A thermoforming technique in which heated plastic sheet is drawn over a male mold. The plastic is clamped and heated, then dropped onto the mold.

Draw bench: A forming machine used for cold reduction of bar stock. Used to form wire and tubing.

Drawing: In general, the operation used to produce cones, cups, boxes and shell-like parts.

Drive fit: A force or pressure fit between two parts. One of the several classes of fits.

Drop forging: Forming hot metal between aligned impression or cavity dies using a drop hammer.

Ductile cast iron: One of the five categories of cast iron having superior properties to gray and white cast iron. It is also called "nodular cast iron."

E

Earthenware: Clay product made of kaolin which is fired at a relatively low temperature. Used for dinnerware and drain tile.

Electric arc furnace: One of the basic steel-making furnaces especially valuable in the production of high alloy steel.

Electric induction furnace: Used in the manufacture of cast iron, utilizes electricity as its sources of power.

Electrical discharge wire cutting (EDWC): Uses a taut, traveling, conductive wire, usually copper or brass, as the cutting tool. A spark between the wire and the workpiece actually does the cutting.

Electrochemical grinding: A modification of electrochemical machining. A metal bonded diamond-grit wheel is the electrode. The metal binder in the abrasive wheel acts as the cathode (negatively charged element). The diamonds "brush" away the freed ions formed in the process.

Electrochemical machining: Use of a hollow electrode to remove metal. "Cutting" is done as an electric current removes electrons from the workpiece.

Electrocoating: A special dip coating process in which the part and the coating material have opposite electrical charges. The coating material is attracted to the part and produces a uniform covering.

Electrohydraulic forming (EHF): Process which converts electrical energy into mechanical energy to form metallic parts. The mechanical energy is transmitted by water to the workpiece.

Electrolytic deposition: An electroplating process used for producing powdered metals used in manufacturing processes.

Electromagnetic forming: Metalforming process using the force of an expanding magnetic field.

Electron beam machining: A thermal machine process that has grown out of electron beam welding. Cutting is done by high-speed beams of electrons.

Emery and corundum: An abrasive made up of aluminum oxide and iron oxide. Emery is 50 to 60 percent aluminum oxide while corundum has about 75 to 90 percent.

Engineering materials: Solid materials such as wood, metals, and plastics which can be formed and cut.

Expansion fit: The reverse of shrink fit. The piece being fitted is placed in liquid nitrogen or dry ice until it shrinks enough to fit into the mating piece.

Explosive forming: Use of rapidly expanding gases from an explosive charge to produce enough force to shape metal over a shaped die cavity.

Extrusion: A metalworking process which forces metal through a hollow, shaped die mounted on a pressure chamber. Metals may be extruded or cold.

F

Faceplate: A holding device used on both metal and wood lathes. On the wood lathe it eliminates use of screw holes in the bottom of the workpiece.

Fatigue limit: A stress limit below which a material can be expected to withstand any number of stress cycles.

Feed rate: Speed at which a material passes through a machining operation. It is expressed as speed of movement over a unit of time (for instance 10 fpm).

Ferrous: A family of metals. Iron is the major ingredient.

Flame hardening: A surface-hardening technique. The hardness is accomplished by heating tahe outside surface with a direct flame and then cooling it by a sudden application of a cold liquid.

Flask: A wooden or metal form consisting of a cope (the top portion) and a drag (the bottom portion). It is used to hold the sand that forms the mold.

Fly cutter: A single point tool fitted in an arbor. While inexpensive to make it is relatively inefficient because only one point does the cutting.

Forging: Metallic shapes made by either hammering or squeezing the original piece of metal.

Forming: Shaping metal to a desired form. The change does not intentionally change the thickness of the metal.

Four-slide press: In metal forging, a press having rams positioned at 90 degrees to each other. Four forming operations can be performed. The slides operate independently of each other.

Full mold casting: Casting process which uses a pattern made of expanded polystyrene. The pattern is destroyed during casting.

Fusion welding: Any type of welding that uses fusion as part of the process.

G

Garnet: Any of several silicate minerals which may be used as gemstones or abrasives.

Gas metal arc welding (GMAW): Arc welding which uses a continuously fed consumable electrode and a shielding gas. Sometimes called MIG.

Gate: the point at which molten metals enter the mold cavity.

Glass: An amorphous (without structure) material made of silica with coloring, firing, and other agents added.

Grain: Any portion of a solid which has external boundaries and a regular internal atomic lattice arrangement.

Gray cast iron: One of the five types of cast iron.

Guerin process: A forming-metal method in which metal sheet is forced to conform to the shape of a male die by the application of force to a confined rubber pad.

H

Hammer forging: Deforming of workpiece by repeated blow from a heavy weight.

Hardening: Heating and quenching certain iron-base alloys to produce a hardness superior to that of the untreated material.

Hardfacing: Filler material placed on a surface to toughen the surface so it can resist abrasion, erosion, wear, corrosion, galling, or impact wear.

High carbon steel: Carbon steel containing approximatley 0.5 percent to 2.0 percent carbon.

High energy rate forming (HERF): A metal forming technique involving the release of a source of high energy such as explosives, electricity, or pneumatic-mechanical materials.

Hobbing: Cutting gear teeth with a hob. The gear blank and hob rotate together during the cutting operation.

Hot forming: Operations performed on metal while it is above the recrystallization temperature of the metal. These operations may include bending, drawing, forging, heading, piercing, and pressing.

Hot shortness: A weakness of metal which occurs in the hot forming range.

Hot working: Shaping of metal at a temperature and rate which will not cause strain hardening.

I

Indexing: A term describing the correct spacing of holes, slots, etc., on the periphery of a cylindrical piece by using a dividing or indexing head.

Inert gas: A gas which does not normally mix with the base metal or filling metal.

Investment casting: A casting process utilizing wax or frozen mercury patterns which are melted out of the mold before casting.

Iron ore: A mined mineral which has a high iron content. It is used as a basic ingredient in the manufacture of all alloys of iron and steel.

J

Job-lot manufacture: Not produced continuously but in limited production runs.

Joint: Point at which two pieces are joined in an assembly.

K

Kilopascal (kPa): SI metric unit of pressure equal to one thousand pascals.

Knee: A support for the saddle and table of column- and knee-type milling machines.

Knurling: A cold forming operation which places a diamond-shaped pattern on round objects such as knobs and locking devices.

L

Lap joint: Joint in which the edges of the two metals to be joined are overlapped.

Laser beam machining (LBM): A machining process that removes material through intense concentrated heat. The process uses a highly focused beam of monochromatic (one color) light to produce the heat.

Laser beam welding (LBW): Process in which single-frequency light beams concentrate a small spot of heat to fuse small, light, metal materials.

Limestone: A mined mineral used as a basic ingredient in the manufacture of iron and steel. Its basic purpose is to remove impurities.

Loose patterns: Exact copies of the part being duplicated with proper allowances added.

Lower transformation temperature: Temperature at which a metal structure starts changing to a different structure.

M

Machine control unit (MCU): Part of a numerical control system that "translates" instructions and controls a machine. Sometimes referred to as the director or controller.

Machine language: In CNC, the set of symbols and characters, and the rules for combining them, which translates the instructions or information to be processed by the machine.

Magnetic chuck: A device using magnetic fields to hold work during machining (grinding).

Malleable castings: Cast forms of metal which have been heat treated to reduce brittleness.

Manufactured abrasives: Synthetic abrasives produced in an electric furnace. The six principal types are: silicon carbide, aluminum oxide, aluminum oxide/zirconium oxide cofusion, sintered bauxite, cubic boron nitride (CNB), and diamonds.

Marform: A metal drawing process that forms sheet metal by using a movable steel punch and a rubber headed ram.

Match plate: A foundry production pattern usually made of metal. It consists of a plate with matching halves of the pattern mounted on each side.

Metallic bonding: Bonds formed by metallic atoms giving up the electrons in their outer shells to a common pool of electrons. The pool is called an "electron cloud." The cloud is evenly distributed throughout the metallic structure.

Milling machines: Machines which can produce flat or curved surfaces by progressively forming and removing chips.

N

NC/CNC lathes: Programmable lathes. NC lathes are controlled by a punch tape which is "read" by a control unit that translates or turns the tape commands into signals for control of the machine. CNC lathes are controlled by computer commands.

Necking: Machining a groove around a cylindrical shaft using a lathe operation.

Nitriding: A case-hardening technique. A ferrous alloy is heated in an atomosphere of ammonia or in contact with a nitrogenous material. Absorption of nitrogen hardens the surface.

Numerical control (NC): A method of controlling the actions of a machine by numbers. Numbers are supplied by paper tape, magnetic tape, computer memory, or by other means.

O

Open hearth furnace: A basic steel-making furnace using a giant hearth exposed to a powerful gas flame to melt the ingredients of the hearth. At one time, it was the primary process in the manufacture of steel. Today, though still used, it is not so popular as the basic oxygen furnace.

Oxyacetylene cutting: Oxyfuel gas cutting process using an oxygen and acetylene flame for heat and a jet of oxygen to oxidize the molten metal and form a cut.

Oxyacetylene welding (OAW): Method of oxyfuel gas welding in which oxygen and acetylene combine and burn to provide heat.

Oxyfuel gas cutting: Cutting metal with an oxygen jet and a preheating flame which combines oxygen and a fuel gas.

Oxyfuel gas welding: Method of welding whose heat comes from combining and burning oxygen and a fuel gas to create the required heat.

Oxygen lance cutting (LOC): An oxyfuel gas cutting process which heats base metal and then removes molten metal with jets of oxygen from an iron pipe.

P

Part program: A computer program that contains instructions for the operation to be performed by a machine tool.

Peening: An operation that involves the mechanical working of metal with hammerlike blows from a round-headed tool.

Permanent mold: A mold used repeatedly for the production of similar castings.

Physical properties: The basic features of a material which can be measured or observed. Includes: size, shape, density, and porosity.

Plasma: Gas that has been heated to extremely high temperatures until partially ionized. This condition enables it to conduct an electrical current.

Plaster mold: A mold made from a gypsum plaster slurry poured around an aluminum pattern. When hardened the mold halves are opened and the pattern removed.

Plastic welding: A low-heat welding process which softens and fuses synthetic plastic materials.

Polymerization: The chemical joining by covalent bonding, or like molecules into larger molecules.

Polymers: A type of materials formed by combining small molecules into larger molecules. Natural polymers include wood, wool, and cotton. Synthetic polymers are the different types of plastics.

Porcelain: Clay product similar to china which is produced without the addition of certain fluxes. It is fired at very high temperatures. Used for chemical ware and surface finishes.

Powder metallurgy (P/M): Forming of parts from powdered metals.

Programming: The act of preparing a detailed sequence of operating instructions for a particular machining operation.

Pure oxide refractories: Ceramic material consisting of oxides of selected metals such as aluminum, magnesium, and zirconium. Used for pyrometer tubes, spark plug insulators, wire drawing dies, valve seats, and crucibles.

Q

Quench hardening: Hardening an iron alloy by austenitizing followed by rapid cooling so that some or all of the austenite is changed to martensite.

Quenching: Rapid cooling of metal after heat treating.

R

Readout: A numerical (number) display of the actual position of a machine slide or a tool.

Ream: Finishing a drilled hole to close tolerance with a reamer.

Refractories: Crystalline body materials noted for their resistance to high temperatures. A type of ceramic material.

Relief angles: Angles which keep cutting tools from rubbing on the surface of the workpiece.

Resistance spot welding (RSW): Resistance welding process which uses the resistance to the flow of electricity to create enough heat for fusion. Small spots are welded between two opposing electrodes.

Resistance welding (RW): Process using the resistance of the metals to electric current as the source of heat.

Rockwell hardness tester: A test device which measures the hardness of materials based on depth of penetration of a standardized force.

Roll forming: Metalworking designed to convert primary stock into other products. Process is mostly confined to tube and pipe making.

Rotary table: A milling attachment that allow a workpiece to be rotated. It consists of a circular worktable turned by a handwheel through a worm gear. The hub of the handwheel is graduated in degrees, permitting precise spacing of holes, slots, grooves, etc., around the piece.

Rotational molding: A plastics molding process designed around a machine which can rotate a mold on two axes which are perpendicular to each other.

Runners and gates: Passages in a mold which conduct the molten materials from the sprue to all parts of the mold cavity.

S

Sand mold casting: A process involving pouring molten metal into a cavity that has been formed in a sand mold.

Scratch hardness: A fast, simple method of measuring hardness without a precision machine. The metal is normally scratched by the edge of a tool or object. Then the hardness of the sample is judged by its scratch resistance.

Shrink fit: A fit in which the outer number is expanded by heating to permit assembly of the inner member. A tight fit results as the cooling outer member shrinks. A very tight fit is permanent.

Silicon: Nonmetallic element used in steel production.

Silicon carbide: A crystal used for abrasives which approaches diamonds in hardness.

Sintering: Bonding process for metal powders that have been compacted by heating them to a certain temperature.

Slab: A semicompleted steel form in which the width is appreciably greater than the thickness.

Single-cut file: A file with evenly spaced, parallel teeth across its width. These are placed at a 65 to 80 degree angle.

Slip: A shear type failure of a ductile material caused when rows of atoms slide past each other during stress. Also, a slurry used in manufacture of ceramic ware.

Spinning: Process in which a disc is formed into a

cone while being rotated on a spinning lathe.

Sprue: Opening in a mold through which the molten material is poured in.

Stainless steel: An alloy of iron containing at lest 11 percent chromium and some nickel.

Stamping: Generally, a term including all press work operations on sheet metal. More narrowly defined, the producing of shallow indentations in sheet metal.

Steel: A material composed mostly of iron, less than 2.0 percent carbon and, normally, small percentages of other elements.

Straight-side press: One of two major designs for forging presses. It is designed to exert pressures of over 1000 tons—well above the capacity of C-frame presses which are limited to forces under 200 ton.

Strain: The measure of change which takes place in size or shape of a body due to force being placed on it.

Strength: The ability of a metal to resist forces or loads. It is normally measured in force per unit area.

Stress: The intensity of force inside a body which will resist a change in its shape. Measurement is in psi or pascals.

Superficial Rockwell hardness test: Surface hardness test used on thin section or small parts, or where a large hardness impression might be harmful.

Surface hardening: A heat-treating process in which the surface of the metal is made very hard but the interior core remains soft and ductile.

Swaging: A forging process which reduces the diameter or tapers round bars, pipe, or tubing through a hammering action. Often called rotary swaging.

T

Tap: The bitlike tool used to cut internal threads.

Tape control: The control of N/C equipment using a 1 in. wide punched tape. Tapes are read by photoelectric cells or electrical contacts.

Tape feed: A device or mechanism that will feed N/C tape to be read or punched.

Tensile strength: The most tensile (pulling) stress that a material is capable of withstanding without breaking when a load is gradually and uniformly applied.

Thermoforming: A plastics-forming operation using heat, pressure, and a shaping device to change flat sheet plastic into three-dimensional objects.

Thermoplastic materials: Materials with polymer chains held together by weak bonds. As a result, an increase of temperature weakens the bonds while colder temperatures strengthen the bonds.

Thermosetting plastics: Plastics having strong bonds between the polymer chains. Such plastics resist breakdown from heat and pressure.

Thread rolling: Applying a thread to a bolt or screw by rolling it between two grooved die plates, one of which is in motion, or between rotating circular rolls.

Tolerance: The deviation permitted from a basic dimension.

Toughness: Work per unit volume needed to fracture a metal. Toughness is equal to the total area under stress-strain curve. In practice, touchness is often considered as resistance to shock or impact—a dynamic property.

Tungsten: A rare metallic element with very high melting point (approximately 3410°C).

Turret lathe: Semiautomatic lathe which has several tools mounted on a rotating fixture called a turret. It can perform several operations by switching from one tool to another.

U

Up milling: Milling operation in which the cutter is rotated in the opposite direction of the workpiece.

V

Vixen-cut file: A file with a series of curved, parallel teeth.

W

Water absorption: Tendency of polymers and ceramic materials to absorb water. This action increases the material's weight and volume.

White cast iron: One of five categories of cast iron which is comparatively hard and brittle.

Wrought iron: A material composed almost entirely of iron. For practical purposes it contains no carbon.

Y

Yield point: In mild or medium-carbon steel, this is the point where a marked increase in deformation occurs without an increase in load. Also called the proportional limit.

Yield strength: The amount of stress at which a material shows a specified permanent plastic yielding.

INDEX

D

E

F

G

H

I

J

K

L

M

N

Q

R

S

T

U

V

ACKNOWLEDGEMENTS

I wish to acknowledge the contributions of the many people and companies that provided help or illustrations. Special thanks to:
My wife, Phyllis, for typing the manuscript.
The following manufacturers who supplied illustrations for the cover:
Bridgeport Machines Division of Textron Inc.
Cincinnati Milacron
Kearney & Trecker Corporation
General Electric Company
Precision Castparts Corporation

R. Thomas Wright